The Wiley Handbook of Evolutionary Neuroscience

The Wiley Handbook of Evolutionary Neuroscience

Edited by

Stephen V. Shepherd

WILEY Blackwell

This edition first published 2017
© 2017 John Wiley & Sons, Ltd

Registered Office
John Wiley & Sons, Ltd, The Atrium, Southern Gate, Chichester, West Sussex, PO19 8SQ, UK

Editorial Offices
350 Main Street, Malden, MA 02148-5020, USA
9600 Garsington Road, Oxford, OX4 2DQ, UK
The Atrium, Southern Gate, Chichester, West Sussex, PO19 8SQ, UK

For details of our global editorial offices, for customer services, and for information about how to apply for permission to reuse the copyright material in this book please see our website at www.wiley.com/wiley-blackwell.

The right of Stephen V. Shepherd to be identified as the author of the editorial material in this work has been asserted in accordance with the UK Copyright, Designs and Patents Act 1988.

All rights reserved. No part of this publication may be reproduced, stored in a retrieval system, or transmitted, in any form or by any means, electronic, mechanical, photocopying, recording or otherwise, except as permitted by the UK Copyright, Designs and Patents Act 1988, without the prior permission of the publisher.

Wiley also publishes its books in a variety of electronic formats. Some content that appears in print may not be available in electronic books.

Designations used by companies to distinguish their products are often claimed as trademarks. All brand names and product names used in this book are trade names, service marks, trademarks or registered trademarks of their respective owners. The publisher is not associated with any product or vendor mentioned in this book.

Limit of Liability/Disclaimer of Warranty: While the publisher and authors have used their best efforts in preparing this book, they make no representations or warranties with respect to the accuracy or completeness of the contents of this book and specifically disclaim any implied warranties of merchantability or fitness for a particular purpose. It is sold on the understanding that the publisher is not engaged in rendering professional services and neither the publisher nor the author shall be liable for damages arising herefrom. If professional advice or other expert assistance is required, the services of a competent professional should be sought.

Library of Congress Cataloging-in-Publication Data

Names: Shepherd, Stephen V., 1978– editor.
Title: The Wiley handbook of evolutionary neuroscience / edited by Stephen V. Shepherd.
Description: Chichester, West Sussex, UK : John Wiley & Sons, 2016. |
    Includes bibliographical references and index.
Identifiers: LCCN 2016019971 | ISBN 9781119994695 (cloth) | ISBN 9781118316573 (epub) |
    ISBN 9781118316610 (Adobe PDF)
Subjects: LCSH: Neurosciences. | Neuropsychology. | Brain–Evolution.
Classification: LCC QP355.2 .W55 2016 | DDC 612.8–dc23
LC record available at https://lccn.loc.gov/2016019971

A catalogue record for this book is available from the British Library.

Cover image: selvanegra/Gettyimages

Set in 10/12pt Galliard by SPi Global, Pondicherry, India

10 9 8 7 6 5 4 3 2 1

Contents

List of Contributors		vii
Preface		ix
Acknowledgments		xiii
1	The Brain Evolved to Guide Action *Michael Anderson and Anthony Chemero*	1
2	The Evolution of Evolutionary Neuroscience *Suzana Herculano-Houzel*	21
3	Approaches to the Study of Brain Evolution *Jon H. Kaas*	38
4	Intraneuronal Computation: Charting the Signaling Pathways of the Neuron *Jorge Navarro, Raquel del Moral, and Pedro C. Marijuán*	49
5	The Evolution of Neurons *Robert W. Meech*	88
6	The First Nervous System *Nadia Riebli and Heinrich Reichert*	125
7	Fundamental Constraints on the Evolution of Neurons *A. Aldo Faisal and Ali Neishabouri*	153
8	The Central Nervous System of Invertebrates *Volker Hartenstein*	173
9	Nervous System Architecture in Vertebrates *Mario F. Wullimann*	236
10	Neurotransmission—Evolving Systems *Michel Anctil*	279
11	Neural Development in Invertebrates *Roger P. Croll*	307
12	Forebrain Development in Vertebrates: The Evolutionary Role of Secondary Organizers *Luis Puelles*	350

13	Brain Evolution and Development: Allometry of the Brain and Arealization of the Cortex *Diarmuid J. Cahalane and Barbara L. Finlay*	388
14	Comparative Aspects of Learning and Memory *Michael Koch*	410
15	Brain Evolution, Development, and Plasticity *Rayna M. Harris, Lauren A. O'Connell, and Hans A. Hofmann*	422
16	Neural Mechanisms of Communication *Julia Sliwa, Daniel Y. Takahashi, and Stephen V. Shepherd*	444
17	Social Coordination: From Ants to Apes *Anne Böckler, Anna Wilkinson, Ludwig Huber, and Natalie Sebanz*	478
18	Social Learning, Intelligence, and Brain Evolution *Sally E. Street and Kevin N. Laland*	495
19	Reading Other Minds *Juliane Kaminski*	514

Index   526

List of Contributors

Michel Anctil
Université de Montréal

Michael Anderson
Franklin & Marshall College

Anne Böckler
Würzburg University

Diarmuid J. Cahalane
Cornell University

Anthony Chemero
University of Cincinnati

Roger P. Croll
Dalhousie University

Raquel del Moral
Instituto Aragonés de Ciencias de la Salud

A. Aldo Faisal
Imperial College London

Barbara L. Finlay
Cornell University

Rayna M. Harris
University of Texas at Austin

Volker Hartenstein
University of California, Los Angeles

Suzana Herculano-Houzel
Vanderbilt University

Hans A. Hofmann
University of Texas at Austin

Ludwig Huber
University of Vienna

Jon H. Kaas
Vanderbilt University

Juliane Kaminski
University of Portsmouth

Michael Koch
University of Bremen

Kevin N. Laland
University of St Andrews

Pedro C. Marijuán
Instituto Aragonés de Ciencias de la Salud

Robert W. Meech
University of Bristol

Jorge Navarro
Instituto Aragonés de Ciencias de la Salud

Ali Neishabouri
Imperial College London

Lauren A. O'Connell
Harvard University

Luis Puelles
University of Murcia

Heinrich Reichert
University of Basel

Nadia Riebli
University of Basel

Natalie Sebanz
Central European University

Stephen V. Shepherd
The Rockefeller University

Julia Sliwa
The Rockefeller University

Sally E. Street
University of Hull

Daniel Y. Takahashi
Princeton University

Anna Wilkinson
University of Lincoln

Mario F. Wullimann
Ludwig-Maximilians-Universität Munich

Preface

A confession:
It is dangerous to anthropomorphize, but I do it. Watching dogs yawn and stretch, birds strut in courtship, fish gasp in air, or even plants leaning their leaves toward the sun—I know this can get ridiculous—I picture my body behaving in the same manner as their own, withdrawing from or yearning toward the very same things, and imagine that my own solitary feelings might have a mirror out there, in another (however alien) mind.

I am not even that contrite. I have no direct access to other minds. I only intimately know my own consciousness. Still, when I see other organisms that look like, behave like, and sometimes even talk like myself, I assume their behaviors, like the same behaviors expressed in my own body, are accompanied by mental experiences. In most animals, behaviors—including our human reports of subjective experience—appear to depend critically on the integrity and features of these animals' brains. I thus gamble that observing the brain is like observing the mind, and that when other animals' brains exhibit similar relationships to the world as my own, they and I may be experiencing similar mental states.

But can we formalize these likenesses? Looking at a human and a monkey brain, we can see they look similar; comparing our lifestyles, we see many similar behaviors. Their and our own patterns of brain activity change in similar ways across behavioral states, perceptual experiences, and motor decisions. Even animals whose brains are wildly different from our own can achieve behaviors we once thought uniquely human: witness a New Caledonian crow making tools, or bees describing through dance where to forage. Are there fundamental features—genetic modules, cellular structures, neural networks, or environmental interactions—which constrain how life mediates complex behavior? Conversely, when confronted with an organism whose sensory environment and affordances differ from our own—say, a naked mole rat, fruit fly, or slime mold—might shared biological features suggest shared mental processes?

Our best hope of understanding other species lies in our shared evolutionary histories. The dog, crow, fly, monkey, bee, rat, and man—and the redwood, the paramecium, the mushroom, the slime mold—all arose from common ancestors through natural selection, through the wandering and winnowing of a braided stream. We interacted with our shared environments and with each other, affected and affecting in kind. Many animals have, in their behavioral toolkits, mechanisms for responding adaptively to other animals. It may yet prove that the human of theory of mind is an elaboration on concepts shared with other animals, rather than a unique and unforeshadowed invention.

Life interacts with the world, maintaining and propagating itself, and does so in a bewildering variety of ways; conceptual boundaries are rarely well-defined. Even the neuron, like the atom, proves entangled and divisible, neither self-determined nor ontologically distinct. To understand how animals coordinate their behavior—and how this behavioral governance can go awry in our own species—we must understand brains. But the brains of different species take astonishingly different forms in the zebra finch and the zebra fish, in nematodes, molluscs and man. In this volume, 33 authors attempt to wrangle order from this chaos, outlining how we can understand the commonalities and complexities of brains across the full span of animal forms.

The Wiley Handbook of Evolutionary Neuroscience is designed to function both as a reference for researchers and as a textbook for the advanced undergraduate or starting graduate student. It is roughly organized into five sections: an introduction to evolutionary neuroscience and comparative methods; a section on biological computation and brain origins; a comprehensive overview of brain structure and development; an exploration of how brains change through evolution and experience; and a discussion of how brains interact with one another.

Introduction and Methods

The first section opens with a philosophical essay by Anderson and Chemero (Chapter 1), grappling with a central issue in evolutionary neuroscience: Why should brains exist at all? The authors argue that even the most complex brains must be understood as relational and action-oriented, rather than objective and computational.

Suzana Herculano-Houzel follows, describing how the very concept of "evolutionary neuroscience" has evolved over time (Chapter 2). In particular, she describes how we have been misled by the Aristotelian *scala naturae*, in which animals are organized along an axis leading from more primitive to more human-like brains. In particular, she argues that understandings of brain scaling have been distorted by anthropocentric measures of brain structure which place humans at the pinnacle.

This first section concludes with a chapter by Jon Kaas outlining methods for researching brain evolution (Chapter 3). Taking mammals as a case study, he describes how fossils and comparative analyses can be integrated to infer historical changes in brain structure and function. Finally, he describes the interaction of phylogenetic, developmental, and biological constraints in shaping brain evolution.

Biological Computation and Brain Origins

In Chapter 4, Navarro and colleagues describe how molecular mechanisms of computations are woven from information-processing pathways which predate the evolution of neurons. In particular, they argue that the cornerstones of eukaryotic information processing are pathways for solute detection, solvent detection, cell-cycle control, and cytoskeletal remodeling.

Robert Meech continues by addressing the origins of neurons (Chapter 5), contrasting the nonneuronal reflexes of sponges with the neurally-mediated behaviors of jellyfish. He argues that neurons should primarily be defined not by their structure but by their function, and lists some of the crucial integrative behaviors mediated by these simplest neural networks.

In the following chapter, Riebli and Reichert address the centralization of these diffuse neural networks into the first brains (Chapter 6). In particular, they point to use of homologous molecular pathways for establishing dorsoventral and antero-posterior brain patterning across organisms as evidence supporting monophyletic brain origins.

Given the extensive conservation of molecular mechanisms of brain formation and function, it seems reasonable to wonder whether all brains must deal with similar computational constraints. Faisal and Neishabouri address this issue head-on (Chapter 7), focusing on the issue of noise: How must brains balance energy versus entropy in processing information?

Brain Structure and Development

The central and most challenging section of the book covers the organization of the two key elements of neural networks: structures and synapses. Volker Hartenstein (Chapter 8) begins by describing the organizing principles of nervous systems, their diverse structural organization across deuterostome and protostome invertebrates, and some examples of how these structures mediate locomotor and perceptual functions. Mario Wullimann (Chapter 9) continues, describing the structural bauplan of more familiar brains—those of vertebrates, including mammals and birds. Comparative anatomy can emphasize the patterns in neural filaments at the expense of the synapse. Michel Anctil fills this (literal) gap by describing how neurotransmitters have evolved across animal clades (Chapter 10).

Concluding the section, Roger Croll (Chapter 11) and Luis Puelles (Chapter 12) describe how these systems change during development in both invertebrates and vertebrates. In describing the development of invertebrate larva, Croll surveys nearly the full range of programs for early brain development. Puelles, by turn, focuses on the vertebrate forebrain, detailing the sequence of core mechanisms which parcel and pattern the telencephalon.

Evolution and Experience

Two chapters address how the brain responds to the environment across and within generations.

First, Cahalane and Finlay (Chapter 13) revisit the theme of allometry, first raised in Chapter 2, providing a contrasting perspective from Herculano-Houzel. Focusing on the developmental mechanisms that pattern mammalian cortex, they show that small, commonly occurring modifications have widespread consequences for individuals and lineages, suggesting they are a major target for evolutionary modification in response to environmental pressures.

In the next chapter (Chapter 14), Michael Koch looks at ways in which individual brains change in response to their environment. He highlights nondeclarative memory, which shares common mechanisms (and, potentially, analogous structures), across invertebrate and vertebrate lineages.

Interacting Brains, Interacting Minds

In the final section, authors examine how these mechanisms and levels of analysis work together to mediate behavioral interactions between organisms. In Chapter 15, Harris and colleagues describe the switches that organize neural systems across species—some deeply conserved—and their role in mediating social decision making. In particular, they describe how social experience can impact brain function at multiple timescales, from the momentary to the developmental to the inter-generational.

Sliwa and colleagues develop this theme in **Chapter 16**, examining how species have evolved to read and respond to signals produced by their own (and other) species. They open with a review of signaling systems, describe the neural mechanisms reported in primates, and conclude by discussing how these systems may generalize across species.

When I first envisioned this book, I hoped to move from the evolution of neurons to that of brains, to that of minds. Since minds are intangible, this necessitates thinking about the biological and cultural evolution of a *theory* of minds. Three short chapters conclude the book, with Bockler and colleagues (Chapter 17) addressing the behavior of coordinated groups, Street and Laland (Chapter 18) addressing social learning and cultural evolution, and Juliane Kaminski (Chapter 19) concluding with the seemingly-unique evolution, in our own species, of a theory of mind.

Acknowledgments

I owe a deep debt to the contributors to this volume, who were often asked to summarize large bodies of knowledge in too few words and with too little time. I have been a sometimes negligent and sometimes overbearing editor: Should you chance upon an awkward turn of phrase, trust it is my fault and not theirs.

I am grateful to Michael Platt for my scientific training; to Asif Ghazanfar for suggesting I edit this book; and to Winrich Freiwald for encouraging from me the time and mental space to complete it. Additionally, several previous authors and editors have guided my thinking on these subjects. Richard Dawkin and Yan Wong's *The Ancestor's Tale* brilliantly resolves the temptation toward *scala naturae* thinking by reversing the narrative: moving from our own species toward our ancient and shared origins. Contributor Jon Kaas edited the monumental *Evolution of Nervous Systems* (and its condensate, *Evolutionary Neuroscience*), which are unrivaled in scope. Larry Swanson's *Brain Architecture* and Georg Streidter's *Principles of Brain Evolution* are excellent attempts to distill the astonishing diversity of nervous systems into a coherent set of principles. Finally, John Allman's *Evolving Brains* remains my favorite introductory neuroscience text: No book better captures the excitement of the field.

Finally, I am personally indebted to my parents, Stan and Kris Shepherd, for their indulgence of their tidepool-swimming, woods-wandering, book-stealing, moody, spoiled son; and to Drew Orr, for bringing light and life to my sometimes dark days.

1

The Brain Evolved to Guide Action

Michael Anderson and Anthony Chemero

1.1 Introduction

In the 19th century, major movements in both psychology and neuroscience were profoundly influenced by Darwin. William James argued for a view of psychology which ultimately came to be known as functionalism; in neuroscience, Herbert Spencer and Santiago Ramon y Cajal argued that we needed to study the mind and brain as adaptations to the environment. In both cases, this evolutionary approach forced a focus on the role of the brain in action guidance. These approaches were revived at the end of the 20th century in the form of embodied cognitive science, which focuses on the importance of action in understanding cognition. Embodied cognitive science calls for an understanding of the brain as having evolved initially for perception and action. It suggests that even complex cognitive abilities such as language and reasoning will use neural resources which initially evolved to guide action. We close by providing evidence that this is, in fact, how the human brain evolved.

1.2 William James and the Functionalist Tradition

In the *Principles of Psychology* (1890), William James described a plan of research for psychology that put front and center both the brain and evolution by natural selection. In the introductory chapter, James contrasted his approach with that of prior (nonscientific) psychologists by pointing out the necessity of the brain for the existence of any experience at all.

> The fact that the brain is the one immediate bodily condition of the mental operations is indeed so universally admitted nowadays that I need spend no more time in illustrating it, but will simply postulate it and pass on. The whole remainder of the book will be more or less of a proof that the postulate was correct. (James, 1890, p. 4)

Modern psychologists of the late 19th century, James wrote, had to be "cerebralists" (p. 5). At the same time, however, James felt that psychology could not be *only* about the brain.

> it will be safe to lay down the general law that *no mental modification ever occurs which is not accompanied or followed by a bodily change*. The ideas and feelings, e.g., which these present printed characters excite in the reader's mind not only occasion movements of his eyes and nascent movements of articulation in him, but will some day make him speak, or take sides in a discussion, or give advice, or choose a book to read, differently from what would have been the case had they never impressed his retina. Our psychology must therefore take account not only of the conditions antecedent to mental states, but of their resultant consequences as well. (James, 1890, p. 5)

Focusing on the brain as the "immediate bodily condition" of the mind, then, required that we understand the brain in light of its (eventual) connections to actions that we engage in.

This last point is a consequence of Darwin's influence on James. Following Herbert Spencer (1855), James thought the purpose of the mind is to adapt to us to the environment. As Spencer put it:

> The fundamental condition of vitality, is, that the internal order shall be continually adjusted to the external order. If the internal order is altogether unrelated to the external order, there can be no adaptation between the actions going on in the organism and those going on in its environment: and life becomes impossible. (Spencer, 1855: §173)

Such adaptation occurs only via action that adjusts the body so that it fits in with the world. Thus, for James, the subject matter of psychology had to be *every* aspect of our mental life, understood in the context of how it adapts us to the environment. It is this feature of Jamesian psychology that led him and his followers to be condescendingly called *functionalists* (Titchener, 1898), because they believed that the way to do psychology was to understand thoughts, habits, emotions, etc. in terms of their adaptive function. Because James was a cerebralist, the same has to be true of the parts of the brain that are these thoughts', habits', and emotions' "immediate bodily conditions". Indeed, Chapter 2 of the *Principles* is called "The Functions of the Brain."

James's combination of functionalism and cerebralism, then, committed him to specific views concerning the evolution of the brain. To understand consciousness, for example, would be to understand how consciousness adapts an animal to its environment. But this adaptation to the environment can only be understood in terms of the other aspects of the animal's life right now, over developmental time, and over phylogenetic time.

> It is very generally admitted, though the point would be hard to prove, that consciousness grows the more complex and intense the higher we rise in the animal kingdom. That of a man must exceed that of an oyster. From this point of view it seems an organ, superadded to the other organs which maintain the animal in the struggle for existence; and the presumption of course is that it helps him in some way

in the struggle, just as they do. But it cannot help him without being in some way efficacious and influencing the course of his bodily history. (James 1890, p. 138)

Brains evolved to guide adaptive action, and human-specific actions must result from evolutionary "superaddition" on to the abilities of human ancestors.

Jamesian functionalism and cerebralism were the dominant views in American psychology for roughly the first half of the 20th century, up until the cognitive revolution. The counterpart view in the neurosciences was not so long-lived.

1.3 Ramon y Cajal's Functionalist Neuroscience

Like James, Spanish neuroanatomist Santiago Ramon y Cajal was influenced by Spencer's evolutionary approach to understanding brain and behavior. Spencer argued that one had to approach the investigation of life and mind taking fundamental principles into account: First was the primacy of adaptation, the continual adjustment of inner to outer conditions. Second was a principle of growth and development, whereby both an organism's repertoire of responses and the biological structures supporting them increase in number, diversity, and complexity. Organisms evolve and develop by becoming at one and the same time more differentiated and more integrated or coordinated in both structure and behavior. It is from these parallel developments (and not from either acting alone) that the increasing complexity of organisms emerges over time.

> In the progress from an eye that appreciates only the difference between light and darkness, to one which appreciates degrees of difference between them, and afterwards to one which appreciates differences of colour and degrees of colour—in the progress from the power of distinguishing a few strongly contrasted smells or tastes, to the power of distinguishing an infinite variety of slightly contrasted smells or tastes ... in all those cases which present merely a greater ability to discriminate between varieties of the same simple phenomenon; there is increase in the speciality of the correspondence without increase in its complexity... But where the stimulus responded to, consists, not of a single sensation but of several; or where the response is not one action but a group of actions; the increase in speciality of correspondence results from an increase in its complexity. (Spencer 1855: §154)

Finally, there was the principle of continuity, which stated that new developments emerge from, build upon, and (partly) preserve what came before. This implied not just that organisms can be arrayed on a biological and psychological continuum, with many differences in degree but few fundamental discontinuities between the mental powers of "higher" and "lower" organisms, but also that, within each organism, the higher mental faculties develop from and rest upon the foundations of the lower. As Robert Wozniak commented:

> The implications of these evolutionary conceptions ... are clear. The brain is the most highly developed physical system we know and the cortex is the most developed level of the brain. As such, it must be heterogeneous, differentiated, and complex. Furthermore, if the cortex is a continuous development from sub-cortical structures,

the sensory-motor principles that govern sub-cortical localization must hold in the cortex as well. Finally, if higher mental processes are the end product of a continuous process of development from the simplest irritation through reflexes and instincts, there is no justification for drawing a sharp distinction between mind and body. The mind/body dichotomy that for two centuries had supported the notion that the cerebrum, functioning as the seat of higher mental processes, must function according to principles radically different from those descriptive of sub-cerebral nervous function, had to be abandoned. (Wozniak, 1992)

Ramon y Cajal took Spencer's principles to heart, and clearly saw them reflected in the neural structures that he was so adept at describing. It is perhaps easiest to start with his summary of three trends in the evolution of neural organization that he observed. The first was a "proliferation of neurons and neuronal processes that … increased the complexity of relationships between various tissues and organs" (Ramon y Cajal, 1904/1995, p. 11). Such proliferation was necessitated by the increase in the number and complexity of *other* cells in the organism that is observed over evolutionary time. As Ramon y Cajal pointed out, an increase in the size and complexity of an organism without an attendant increase in the number of neural cells it possesses would precipitate a decrease in sensory acuity and presumably in agility as well, given the increase in the ratio between body parts and the sensory and motor neurons that would serve them. Ramon y Cajal tied neural development especially closely to the motor system:

> Once it has appeared, the nervous system comes to direct the muscular system through a series of actions and reactions. Indeed, because of the concurrent specializations that occur in animals, both the nervous system and the muscular system not only appear together, but are also functionally interdependent. (Ramon y Cajal 1904/1995, p. 5)

The second evolutionary trend detailed by Ramon y Cajal was "an adaptive differentiation of neuronal morpolology and fine structure." The third was "a progressive unification of the nervous system, a concentration of its elements into neural masses" (Ramon y Cajal 1904/1995, p. 12)—that is, the emergence of central ganglia including the brain and spinal cord. The effect of this centralization is crucial to function:

> Motor neurons that before were peripheral and isolated from one another are now juxtaposed in a single, central nucleus; they are transversely integrated, to use Herbert Spencer's phrase. … the sensory neurons can excite all of the aggregated motor neurons, and only a few additional expansions are necessary for the sensory arborizations to expand their spheres of motor influence (Ramon y Cajal 1904/1995, p. 14)

In what must appear a paradox to those accustomed to understanding the brain in terms of the localization of psychological faculties, the anatomic consolidation that Ramon y Cajal described permits function to be *less* localized, even as the supporting tissues become more central. This arrangement makes perfect sense when one expects the brain to be, rather than a collection of organs with distinct local functions, a structure establishing functional *relationships* between cells to coordinate the organism's interaction with its environment.

Ramon y Cajal argued that coordination, control, and complexity are achieved via the emergence of two new classes of neural cells in addition to sensory and motor neurons: association neurons and psychomotor neurons. Association neurons mediate the link between sensory and motor cells, allowing the emergence of complex responses to sensory stimuli.

> With the association neuron, multicellular organisms become true animals. Sensory stimuli, even if localized to one point on the integument, are no longer isolated … The association pathways that interrelate various muscle fields and the areas of the integument with which they are connected are by no means randomly distributed. Evolution and adaptation have determined their organization, and the precision of their distribution is such that each stimulus received by a sensory cell causes the animal to respond with what Exner has called a *combination of movements*, that is to say, with a complex movement that is appropriately coordinated for the animal's self-preservation and procurement of nutritional requirements. (Ramon y Cajal 1904/1995, pp. 5, 7)

Psychomotor neurons were understood by Ramon y Cajal to be exceptionally powerful and centralized association neurons, able to exert their influence over an extraordinarily broad range of circumstances and behaviors. Psychomotor neurons are able to modulate behavior based not just on external stimuli, but also on internal conditions, and not just on current stimulation but also past experience.

> In the evolution of the nervous system, this element, which underlies the still largely unexplored world of psychological (psychic) phenomena, is a more recent addition than the association neuron. It too is interpolated between sensory and motor neurons, but at a distance, and is generally located in one particular ganglion: the cerebral ganglion of invertebrates and the cerebral cortex of vertebrates.... The empire of the psychomotor neuron, together with the various ganglia distributed throughout the body, constitute the organism's newest and most useful weapons in the struggle for survival. (Ramon y Cajal 1904/1995, p. 8)

Ramon y Cajal's choice of metaphor is striking, for, in his view, the psychomotor neuron truly does rule over vast swaths of behavior. Two things especially are important to note: the first is that the regulatory capacity of the psychomotor neuron is made possible only because the centralization of neural structures permits single cells to quickly and specifically affect a wide range of inputs and outputs; the second is that the power of psychomotor neurons does not lie in their *intrinsic* properties but rather in their defining functional *relationships*.

> Wherein lies the superiority—the supremacy—of the cephalic ganglion? In our view it derives from the inherent superiority of the functional relationships established between the external world and this ganglion. Let us explain. The abdominal ganglia are linked to the sensory nerve cells that relay simple, rather poorly defined and crude tactile and thermal sensations from the integument. The cephalic ganglion, in

contrast, is connected to the very specialized neurons that subserve vision, hearing and smell, and this receives preorganized patterns (including more complex temporal and spatial information) that provide the most accurate representations of the external world. This difference in type of connections is mostly responsible for the preeminence of the cerebral ganglion. And the eye and the ear are the major artisans of this preeminence. In essence these organs are *computational devices*, to use Max Nordau's pleasing expression, that select in a very specific way from the middle range of the immensely broad energy spectrum those wavelengths for which they are adapted. (Ramon y Cajal 1904/1995, p. 8)

Interestingly, for Ramon y Cajal the precision and accuracy with which the sense organs represent the external world obviates the need for central structures to do so.

The cerebral cortex of vertebrates, and the cerebral ganglion of invertebrates, do not need to create images; complete images are formed by the sense organs and supplied instead to the cerebral cortex or cerebral ganglion in highly refined ways that actually reflect the intensity and all the subtle nuances inherent in the excitatory stimuli. In the final analysis, the marvelous structural organization of the eye and ear is the primary reason for the dominant position of the cerebral cortex. (Ramon y Cajal 1904/1995, pp. 8–9)

There is much that is striking in Ramon y Cajal's perspective. First is his focus not on intrinsic function or localized faculties in the regions of the brain but rather on the establishment of functional relationships. Indeed, Ramon y Cajal took this perspective so seriously that he was led to predict the outcome of experiments—different in detail but identical in intent—first performed over 80 years after the time of his writing (e.g., Sur, Garraghty, & Roe, 1988):

Insights provided by the evolution of central neural centers have now so convinced us of the preeminent role played by the nature of their relationship to the external world that we are tempted to propose the following: If by some capricious and seemingly impossible developmental anomaly the optic nerve should end in the spinal cord, visual sensations would be elaborated in the region occupied by motor neurons! (Ramon y Cajal 1904/1995, p. 9)

It is worth emphasizing an important consequence of this focus on neural relationships: Differences in neural morphology should not be taken to indicate differences in intrinsic function but rather to indicate different abilities to establish sorts of functional connections or coordination. It is only for this reason that anatomic, morphological differentiation can effect increasing functional—which is to say behavioral—complexity.

Second, we see a recognition here of the importance of peripheral structures to cognition, not just as input channels but as organs of cognition in their own right. Indeed, it would not be inappropriate to see, in Ramon y Cajal's insight that sense organs play a role in selecting and structuring stimuli, a precursor to current recognition of the importance of bodily activity and morphology in cognitive processes (Anderson, 2003; Barrett, 2011; Chemero, 2009), including such recently emerging notions as morphological computation (Paul, Lungarella, & Iida 2006). Cephalopods, for instance, take advantage of various limb properties to make the inverse kinematics problem they must solve to compute limb movements much simpler than it would otherwise be in their extremely flexible extremities (Hochner, 2012).

Third, and finally, there is the fundamental orientation toward action:

> What utilitarian goal has nature (which never seems to act in vain) pursued in forcing nervous system differentiation to these lengths? ... [T]he refinement and enhancement of reflex activity, which protects the life of both the individual and the species. ... Such reflexes constitute the fundamental repository of neural adaptations that provide an animal with the necessities of life ... To the hierarchy of increasingly more complex reflexes—irritability in protozoa, simple reflexes in lower vertebrates, and more complex reflexes in higher invertebrates and vertebrates—one must add the all-powerful psychic reflex of vertebrates, and especially the higher vertebrates. In the latter ... neural and nonneural structures are not simply under the influence of external stimuli; they are also subject to internal stimuli arising from control centers within the organism itself. (Ramon y Cajal 1904/1995, p. 16)

For Ramon y Cajal, the *telos* of cognition is action, and for this reason even "complex and deferred responses ... are true reflexes" (Ramon y Cajal 1904/1995, p. 17). Since in our time we tend to reserve the term "reflex" for those simple, stereotyped (and generally spinally mediated) motor responses to strong, simple stimuli, it would be easy to dismiss Ramon y Cajal's view here as not reflecting the true complexity of the brain's function. In point of fact, given his insistence on a "hierarchy" of reflexes, and the more general point that evolution tends to preserve, adapt and enhance existing structure and function, what he appears to have in mind is a functional arrangement not unlike the subsumption architecture proposed by Rodney Brooks (1991), whereby simpler, specific reflex responses are modulated or suppressed by higher "reflexes" that reflect more general sensory-motor coordination. It is, in any case, clear that Ramon y Cajal imagined overall brain function was achieved via the establishment of a hierarchical continuum of sensorimotor control processes, all aimed at "conferring advantage in the struggle for survival" (Ramon y Cajal 1904/1995, p. 17).

1.4 Embodied Cognition

As noted above, functionalism was something like the orthodoxy in American psychology until the 1950s, when the "cognitive revolution" happened. The cognitive revolution replaced the functionalist ideas with ideas drawn from the Cartesian, rationalist tradition. The idea that thinking is computation occurring within the brain runs counter to the functionalist focus on the place of thinking in action and in evolutionary context; it is also, at best, neutral with respect to cerebralism. The idea that thinking is computation encourages a lack of interest in the brain. This is the case because computational processes are multiply realizable, which is to say that the same computational processes can occur in many different media. The web browser Firefox, for example, can run in the Mac, Windows, and Linux operating systems and on very different computer hardware. Despite differences in implementation, it is, in an important sense, the same software. Similarly, the purported computational processes that implement, for example, face recognition can be implemented differently in different brains, and could even be implemented on a computer, while still being the same software. This encouraged cognitive scientists to ignore details about the brain when they proposed computational mechanisms for cognition (e.g., Fodor, 1975). In doing so, they rejected the cerebralism of the functionalist tradition. At the same time, computational cognitive science abstracted away from the details of action and bodily control. If cognition is a computational process, it is natural to treat the body as a mere peripheral device, like keyboard or modem, that provides information about the environment to the central processor that does the real cognitive work. Completing the rejection of the functionalist tradition is the antipathy that many of the founders of cognitive science have shown to evolution by natural selection. Infamously, Chomsky argued that the human language facility could not have evolved by natural selection (Chomsky, 1988). More recently, Jerry Fodor has gone from arguing that evolution by natural selection cannot explain how thoughts have meaning (1990) to arguing that evolution by natural selection cannot explain the nature of cognition (2000) to arguing that evolution by natural selection is simply ill-conceived (2007).

In the 1980s, psychology began to reclaim its functionalist foundations. As Bechtel, Abrahamsen, and Graham (1999, p. 75) put it, cognitive science moved "outwards into the environment and downwards into the brain." The movement downwards into the brain was sparked by the introduction of drastically improved neural imaging techniques—including the introduction of positron emission tomography (PET) in the late 1970s (Sokoloff et al., 1977) and functional magnetic resonance imaging (fMRI) in the early 1990s (Ogawa, Lee, Kay, & Tank 1990)—and with the renaissance of artificial neural network modeling (Rumelhart, McClelland & PDP Research Group, 1986). These innovations allowed cognitive scientists and psychologists to focus in on the details of neural activity, at the same time strongly suggesting that these details really do matter. With the rise to prominence of cognitive neuroscience in the 1980s, Jamesian cerebralism was back. The simultaneous move outwards into the environment was initiated by the publication of Gibson's posthumous *The Ecological Approach to Visual Perception* (1979), especially its more widely available second edition (1986). Gibson argued that the primary function of perception is the guidance of action, and because of that, the primary perceivables are affordances, or opportunities for action. From this strongly evolutionary perspective, the object of psychological inquiry was not the brain as computer, but rather perceptual systems—which include the brain, sensory surfaces, and moving body of an animal—surrounded by their information-rich environments. Gibson's view was an explicit reclamation of the functionalist focus on evolution and the role of perception and cognition in controlling action.

In many ways, however, the real beginning of the movement known as "embodied cognition" came a few years later in the form of Rodney Brooks's "Intelligence without representation" (1991). In that paper, Brooks used an explicitly evolutionary argument to shift the focus of cognitive science from abstract thinking to the control of action. Brooks presented a timeline of evolutionary highlights, from the appearance of single-celled organisms approximately 3.5 billion years ago, to the appearance of vertebrates approximately 500 million years ago, to the advent of written language about 5,000 years ago. Though the intervening decades have revised some of these dates, Brooks's point stands. As he put it, the majority of evolutionary "research and development" was spent on getting from single-celled organisms to vertebrates, which is to say getting from living things to creatures with sophisticated control of their actions. From this, Brooks concludes that the bulk of intelligence is perception and action, with language and other human-specific abilities mere icing on the cake.

The movement that followed was a restoration of Jamesian functionalism and rejection of the abstraction away from the brain and the body which came with the cognitive revolution. The details about the way the brain works *are* important to understanding cognition; cognition and the brain must be understood in their evolutionary context. In this evolutionary context, it is clear that for most of the history of life on earth, the primary function of nervous systems has been the control of action (e.g., Anderson, 2003; Barrett, 2011; Chemero, 2009; Clark, 1997). The current work in embodied cognitive science that arose from these sources (among many others) is broad-based, incorporating work in robotics, simulated evolution, developmental psychology, perception, motor control, cognitive artifacts, phenomenology, and, of course, theoretical manifestos. Given this variety of subject matter, there is also variety in theoretical approach. The following tenets, though, are more or less universally held among embodied cognitive scientists.

1.4.1 Interactive Explanation and Dynamical Systems

Explaining cognitive systems that include aspects of the body and environment requires an explanatory tool that can span the agent–environment border. Many embodied cognitive scientists use dynamical systems theory. That is, many (though not all) proponents of embodied cognitive science take cognitive systems to be dynamical systems, best explained using the tools of dynamical systems theory. A dynamical system is a set of quantitative variables changing continually, concurrently, and interdependently in accordance with dynamical laws that can, in principle, be described using equations. To say that cognition is best described using dynamical systems theory is to say that cognitive scientists ought to try to understand cognition as intelligent *behavior* and to model intelligent behavior using a particular sort of mathematics, most often sets of differential equations. Dynamical systems theory is especially appropriate for explaining cognition as interaction with the environment because single dynamical systems can have parameters on each side of the skin. That is, we might explain the behavior of the agent in its environment over time as coupled dynamical systems, using something like the following equations from Beer (1995):

$$\dot{X}_A = A(X_A; S(X_E))$$
$$\dot{X}_E = E(X_E; M(X_A))$$

where A and E are continuous-time dynamical systems, modeling the animal and its environment, respectively, and $S(x_E)$ and $M(x_A)$ are coupling functions from environmental variables to animate parameters and from animate variables to environmental parameters, respectively. It is only for convenience (and from habit) that we think of the organism and environment as separate; in fact, they are best thought of as forming just one nondecomposable system, U. Rather than describing the way external (and internal) factors cause changes in the organism's behavior, such a model would explain the way U, the system as a whole, unfolds over time.

1.4.2 Changing the Role of Representations

Although embodied cognitive science's main modeling tool, dynamical systems theory, is neutral about mental representations, with few exceptions embodied cognitive scientists are representationalists. The representations they call on are indexical-functional (Agre and Chapman, 1987), pushmi-pullyu (Millikan, 1995), action-oriented (Clark, 1997), emulator (Churchland, 2002; Grush, 1997, 2004), or guidance representations (Anderson & Rosenberg, 2008). Action-oriented representations differ from representations in earlier computationalist theories in that they necessarily represent things in a nonneutral way, as geared to an animal's actions, as affordances. Action-oriented representations are more primitive than other representations, in that they can lead to effective behavior without requiring separate representations of the state of the world and the cognitive system's goals. That is, the perceptual systems of animals need not build an objective representation of the world, which can then be used by the action-producing parts of the animal to guide behavior; instead, the animal produces representations that are geared from the beginning toward the adaptive actions it aims to perform.

1.4.3 Intelligent Bodies, Scaffolded Environments, Fuzzy Borders

Given this minimized role of mental representation, it is a challenge to explain complex, intelligent behavior. In embodied cognitive science, some intelligence is "off-loaded" from the brain to the body and environment. As Ramon y Cajal noticed more than a century ago, our bodies are well-designed tools, making the jobs of our brains much easier. For example, our kneecaps limit the degrees of motion possible in our legs, easing balance and locomotion. It is only a small exaggeration to say that learning to walk is easy for humans because our legs already know how (see Thelen & Smith, 1994 and Thelen, 1995). This off-loading goes beyond the boundaries of our skin: The natural environment is already rich with affordances and information that can guide behavior. In interacting with and altering this environment, as beavers do when they build dams, animals enhance these affordances. Kirsh and Maglio (1994, see also Kirsh, 1995) show that manipulating the environment often aids problem solving. Their example is of Tetris players rotating zoids on-screen, saving themselves the work of mental rotation. Hutchins (1995) shows that social structures and well-designed tools allow humans to easily accomplish tasks that would otherwise be too complex. Many of us therefore believe that cognitive systems are not confined to the brain or body, but include aspects of the environment (Anderson, Richardson, & Chemero, 2012; Clark, 1997; Hutto, 2005; Hutto & Myin, 2012; Menary, 2007; Rowlands, 2006). Clark even argues (2003) that external tools including phones, computers, language, and so on are so crucial to human life that we are literally cyborgs, partly constituted by technologies. Echoing Ramon y Cajal's view of the sense organs, embodied cognitive scientists argue that the functional relationships that enable intelligent action are not among merely neural components, but instead integrate neurons, bodies, and environment.

These tenets of embodied cognition, of course, have consequences for the functional organization of the brain. We consider these in the next section.

1.5 Embodied Cognition and the Brain

In light of the discussion above, we would like to suggest three principles that together define a functionalist neuroscience. A functionalist believes that (1) the functional architecture of the brain has been established by natural selection and, more particularly, via a process marked by both functional differentiation and continuity; (2) our complex and diverse behavioral repertoire is supported primarily by the brain's ability to dynamically establish multiple different functional coalitions, coordinating both neural partnerships and external resources; and (3) the brain is fundamentally action-oriented, with its primary purpose to coordinate the organism's ongoing adjustments to external circumstances.

Despite the growing interest in, and information about, the details of neural processing, much of the cognitive neuroscience of the last two decades has still been guided by cognitivist principles and the computer metaphor for the brain (see Miller, 2003 and Posner, Petersen, Fox, & Raichle, 1988 for discussion). Thus, a truly functionalist cognitive neuroscience has yet to emerge. There is, however, some suggestive work that points the way.

1.5.1 Brains Evolve through Elaboration

Consider the first principle, that the functional architecture of the brain should reflect an evolutionary history marked by both functional differentiation and also the incorporation and reuse of existing structures for new purposes. If the last century of neuroscience has established anything, it is that the various regions of the brain are functionally differentiated. For most of that time, this fact has been taken to indicate that each region of the brain is highly functionally specialized, implementing a single cognitive operation (e.g., Posner et al., 1988 and Kanwisher, 2010). Recent work, however, has characterized brain regions in a multi-dimensional manner that highlights functional differences while recognizing that, in point of fact, each region of the brain appears to be active in multiple diverse circumstances (Anderson, Kinnison, & Pessoa, 2013; Hanson & Schmidt, 2011; Poldrack, Halchenko, & Hanson, 2009).

Indeed, it is at this point well established that individual regions of the brain support many different tasks across multiple task categories, as would be predicted by the principle of continuity. For instance, although Broca's area has been strongly associated with language processing, it turns out to also be involved in many different action- and imagery-related tasks, including movement preparation (Thoenissen et al., 2002), action sequencing (Nishitani, Schürmann, Amunts, & Hari, 2005), action recognition (Decety et al., 1997; Hamzei et al., 2003; Nishitani et al., 2005), imagery of human motion (Binkofski et al., 2000), and action imitation (Nishitani et al., 2005; for reviews, see Hagoort, 2005; Tettamanti & Weniger, 2006). Similarly, visual and motor areas—long presumed to be among the most highly specialized in the brain—have been shown to be active in various sorts of language processing and other "higher" cognitive tasks (Damasio & Tranel, 1993; Damasio, Grabowski, Tranel, Hichwa, & Damasio, 1996; Glenberg & Kaschak, 2002; Hanakawa et al., 2002; Martin, Haxby, Lalonde, Wiggs, & Ungerleider, 1995; Martin, Wiggs, Ungerleider, & Haxby, 1996; Martin, Ungerleider, & Haxby, 2000; Pulvermüller, 2005; see Schiller, 1996 for a related discussion). Excitement over the discovery of the fusiform face area (Kanwisher, McDermott, & Chun, 1997) was quickly tempered when it was discovered that the area also responded to cars, birds, and other stimuli (Gauthier, Skudlarski, Gore, & Anderson, 2000; Grill-Spector, Sayres, & Ress, 2006; Rhodes, Byatt, Michie, & Puce, 2004; Hanson & Schmidt, 2011).

Recent meta-analyses of neuro-imaging results have tended to support this emerging picture of a functionally differentiated, but not functionally specialized, brain. For example, Russell Poldrack (2006) estimated the selectivity of Broca's area by performing a Bayesian analysis of 3,222 imaging studies from the BrainMap database (Laird, Lancaster, & Fox, 2005). He concluded that current evidence for the notion that Broca's area is a "language" region is fairly weak, in part because it was more frequently activated by nonlanguage tasks than by language-related ones. Similarly, several whole-brain statistical analyses of large collections of experiments from BrainMap (Laird et al., 2005), Neurosynth (Yarkoni, Poldrack, Nichols, Van Essen, & Wager, 2011), and other sources demonstrate that most regions of the brain—even fairly small regions—appear to be activated by multiple tasks across diverse task categories (Anderson, 2010; Anderson et al., 2013; Anderson & Penner-Wilger, 2013).

That this apparent functional diversity is a reflection of the evolutionary history of the brain is supported by some interesting features of the *pattern* of use and reuse of individual regions of the brain across multiple circumstances. For instance, it appears that, *ceteris paribus*, older regions of the brain tend to be used in more tasks—presumably because they've been around for longer, and have thus had more opportunity to be incorporated into multiple functional coalitions (Anderson 2007). In addition, more recently emerged cognitive functions, such as language, appear to be supported by more and more widely scattered brain regions than do evolutionarily older functions such as vision and attention (Anderson, 2010; Anderson & Penner-Wilger, 2013). Again, this makes sense in light of both differentiation and continuity, for the later a given cognitive process or behavioral competence emerges, the greater the number and diversity of neural structures that will be available to support the new competence, and there is little reason to expect structures with the necessary functional properties will always be near one another in the brain.

1.5.2 Cognition Does Not Respect Boundaries

This brings us to the second principle, that achieving behavioral competence is a matter of establishing the right functional coalitions to support the tasks in question. There is little work that specifically investigates the neural supports for the incorporation of external resources into cognitive processing. Just as we create physical tools like hammers, knives, and levers to augment our physical capacities, so too we have invented cognitive artifacts to augment our mental ones, perhaps none more important than the written symbols and other tokens we manipulate in mathematical processing. And we *do* manipulate and interact with them as tools: we write them, move them, strike them out, gesture at and over them. How we marshal the internal resources of memory and perception along with the external resources of pencil, paper, and space to solve mathematical problems is a sensorimotor skill that is very poorly understood behaviorally, and not at all neuroscientifically (Clark, 1997; Landy & Goldstone, 2009). This is a lacuna that the field should begin to address.

However, there is certainly evidence that, *within* the brain, cognitive function is a matter of flexibly assembling the right coalition of neural partners. Some suggestive evidence for this possibility comes from a meta-analysis of more than 1100 neuroimaging studies across 10 task domains: It was demonstrated that, although many of the same regions of the brain were used and reused in multiple tasks across the domains, the regions cooperated with one another in different patterns in each task domain (Anderson & Penner-Wilger, 2013).

Experimental work investigating temporal coherence in the brain also points in the direction of large-scale modulation of neural partnerships in support of cognitive function. For instance, there is evidence relating changes in the oscillatory coherence between brain regions (local and long-distance) to sensory binding, modulation of attention, and other cognitive functions (Varela, Lachaux Rodriguez, & Martinerie, 2001; Steinmetz et al., 2000). Two early findings illustrate the basic notion well: Friston (1997) demonstrated that the level of activity in posterior parietal cortex determined whether a given region of inferotemporal cortex was face-selective, that is, its functional properties were modulated by distributed neural responses. Likewise, McIntosh et al. (1994) investigated a region of inferotemporal cortex and a region of prefrontal cortex that both support face identification and spatial attention. McIntosh et al. showed that during the face-processing task the inferotemporal region cooperated strongly with a region of superior parietal cortex; while during the *attention* task, that same region of parietal cortex cooperated more strongly with the prefrontal area. Similar patterns of changing functional connectivity are observed over developmental time, which suggests that acquiring new skills involves changes to both local and long-distance functional partnerships (Fair et al., 2009; Superkar, Musen, & Menon 2009). It seems reasonable to predict that such results will continue to emerge, given the increasing interest in network-oriented approaches to the brain (Sporns, 2011).

1.5.3 Brains Function to Guide Adaptive Action

Finally we consider the third principle, that the brain should be understood as an action-oriented system. It is here, perhaps, that there is the most work left to be done to establish a functionalist neuroscience: The computer metaphor for the brain still dominates the cognitive neurosciences, leading researchers to interpret the neural activity observed during experiments as reflecting information processing, rather than action coordination (but see Anderson, 2015). It is nevertheless worth highlighting one recent line of work in the neural bases of decision making, to illustrate what a more action-oriented approach to the brain might look like.

Prevailing models of decision making have generally inherited from the cognitivist approach to the mind the notion that decision making involves first building an objectively specified world-model, then generating possible action plans, deciding between them (action selection), and finally determining how the motor system will enact them (action specification). But, as we saw above, from the functionalist embodied perspective, perception naturally assesses the adaptive values of current organism–environment relationships and detects opportunities for changing those values through action. If we perceive the world in terms of the opportunities for action that it affords, then deciding what to do might be largely a matter of choosing which perception–action path—which affordance—has the highest predicted return.

> The proposal made here is that *the process of action selection and specification occur simultaneously* and continue even during overt performance of movements. That is, sensory information arriving from the world is continuously used to specify currently available potential actions... From this perspective, behaviour is viewed as a constant competition between internal representations of the potential actions which Gibson (1979) termed 'affordances'. Hence, the framework presented here is called the "affordance competition hypothesis." (Cisek, 2007, p. 1586)

Although it is imprecise to talk of "representing" affordances—an affordance is the perceivable relationship between an organism's abilities and features of the environment (Chemero, 2003; 2009)—clearly it is the notion of internal (neural) competition between possible courses of action that is the center of Cisek's account. There are two central tenets to his hypothesis, both of which fit in nicely with the functionalist approach to the brain being described here:

1 **The process of action selection and specification occur continuously and in parallel.** Because the organism's brain evolved to support interactive behavior, and perception is to be understood in terms of the detection of opportunities for action, it stands to reason that the process of selecting and specifying actions is a continuous, ongoing part of simply perceiving and acting in the world. Elaborating this notion, Cisek and Kalaska (2010) write:

   > Schmolesky et al. (1998) showed that neural responses to simple visual tasks appear quickly throughout the dorsal visual system and engage putatively motor-related areas such as FEF [frontal eye fields] in as little as 50ms. This is significantly earlier than some visual areas such as V2 and V4. ... In a reaching task, population activity in PMd responds to a learned visual cue within 50ms of its appearance (Cisek & Kalaska 2005). Such fast responses are not purely visual because they reflect the context within which the stimulus was presented. For example, they reflect whether the monkey expects to see one or two stimuli (Cisek & Kalaska 2005), reflect anticipatory biases or priors (Coe et al. 2002; Takikawa et al. 2002), and can be entirely absent if the monkey already knows what action to take and can ignore the stimulus altogether (Crammond & Kalaska 2000). In short, these phenomena are compatible with the notion of a fast dorsal specification system that quickly uses novel visual information to specify the potential actions most consistently associated with a given stimulus (Gibson 1979; Milner & Goodale 1995). (Cisek & Kalaska, 2010, p. 285)

2 **For any given behavior, both processes occur in the same regions of the brain.** Regional differentiation in the brain reflects different capacities for managing certain classes of sensorimotor transformation, rather than (for instance) specializations for perceptual discrimination, decision making, and action execution. Cisek and Kalaska (2010) write:

   > Studies on the neural mechanisms of decision making have repeatedly shown that correlates of decision processes are distributed throughout the brain, notably

including cortical and subcortical regions that are strongly implicated in the sensorimotor control of movement. Neural correlates of decision variables (such as payoff) appear to be expressed by the same neurons that encode the attributes (such as direction) of the potential motor responses used to report the decision, which reside within sensorimotor circuits that guide the online execution of movements. These data and their implications for the computational mechanisms of decision making have been the subject of several recent reviews (Glimcher 2003; Gold & Shadlen 2007; Schall 2004). (Cisek & Kalaska, 2010, p. 270)

This overlap has behavioral consequences that are hard to explain within the classical framework. For instance, one body of evidence shows that the trajectory of reaching movements depends on the amount of separation of the targets in space: Subjects move directly to the chosen target when the options are far apart, but initially reach between the targets when they are close together, and only later veer to their selection (Ghez et al., 1997). Such behavior is naturally explained as a side-effect of the similarity of the response options in the case of the close targets; the nearly identical neural patterns generated by the similar affordances will tend to reinforce one another, perhaps even merging for a time, and it is only when the differences in the reach trajectory between the options become more pronounced as the hand approaches the targets that the two possibilities become distinguishable and competitive (Cisek, 2006; Erlhagen & Schöner, 2002). Similarly, when subjects in two alternative forced choice tasks are asked to indicate their decision by moving a hand or a cursor to the chosen location, the subject's confidence in their choice predicts aspects of their movement of including endpoint and peak velocity (McKinstry et al., 2008). Cisek and Kalaska (2010) conclude:

These findings are difficult to reconcile with the idea that cognition is separate from sensorimotor control (Fodor 1983) but make good sense if the continuous nature of the representations that underlie the selection of actions has been retained as selection systems evolved to implement increasingly abstract decisions. (Cisek & Kalaska, 2010, p. 283)

This perspective illustrates not just the notion of cognitive continuity, but also the potential power of action-oriented representations in interpreting neuroscientific data. If brains are action-oriented systems, then whatever representations it trucks in should reflect this functional inheritance. Cisek believes that the decision-making literature points in precisely this direction:

The affordance competition hypothesis ... differs in several important ways from the cognitive neuroscience frameworks within which models of decision making are usually developed. Importantly, it lacks the traditional emphasis on explicit representations which capture knowledge about the world. For example, the activity in the dorsal stream and the fronto-parietal system is not proposed to encode a representation of objects in space, or a representation of motor plans, or cognitive variables such as expected value. Instead, it implements a particular, functionally motivated mixture of all these variables. From a traditional perspective, such activity appears surprising because it does not have any of the expected properties of a sensory, cognitive or motor representation. It does not capture knowledge about the world in the explicit descriptive sense expected from cognitive theories and has proven difficult to interpret from that perspective... However, from the perspective of affordance competition, mixtures of sensory information with motor plans and cognitive biases make perfect sense. Their functional role is not to describe the world, but to mediate adaptive interaction with the world. (Cisek, 2007, p. 1594)

For a functionalist neuroscience to fully emerge, this action-oriented perspective must be taken up into every corner of the field. Given the apparent challenges facing the current framework, and the importance of integrating the study of the mind and brain more fully with the ecological and evolutionary biology, this is a possibility we should both welcome and encourage (Anderson, 2010; 2015; Anderson, Richardson, & Chemero 2012).

1.6 Conclusion

Our point in this chapter has been that a focus on evolution in psychology and neuroscience has to come with a focus on action. Our hominid predecessors were able to walk, run, forage, and avoid predators, but were presumably not able to do logic or read. The brain evolved to guide action, not write sonnets. This fact has consequences for how one does both psychology and neuroscience. We have pointed to 20th-century functionalist traditions in psychology and neuroscience where action has, indeed, been the focus. The rejuvenation of functionalism in psychology, via the embodied cognition movement, is beginning to be carried over into the neurosciences. Future neuroscience has to take our evolutionary history into account, and, to do so, it must become more action-oriented.

Acknowledgments

Michael Anderson was a fellow at the Center for Advanced Study in the Behavioral Sciences at Stanford University, and supported by a sabbatical leave from Franklin & Marshall College during the writing of this chapter. He gratefully acknowledges the support.

References

Agre, P., & Chapman, D. (1987). Pengi: An implementation of a theory of activity. *Proceedings of the Sixth National Conference on Artificial Intelligence* (pp. 268–272). Menlo Park, CA: AAAI Press.

Anderson, M. L. (2003). Embodied cognition: A field guide. *Artificial Intelligence*, 149, 91–130.

Anderson M. L. (2007). Evolution of cognitive function via redeployment of brain areas. *The Neuroscientist*, 13(1), 13–21.

Anderson M. L. (2010). Neural reuse: A fundamental organizational principle of the brain. *Behavioral and Brain Sciences*, 33(4), 245–266.

Anderson M. L. (2015). *After phrenology: Neural reuse and the interactive brain*. Cambridge, MA: MIT Press.

Anderson, M. L., Kinnison, J., & Pessoa, L. (2013). Describing functional diversity of brain regions and brain networks. *Neuroimage*, 73, 50–58.

Anderson M. L., & Penner-Wilger, M. (2013). Neural reuse in the evolution and development of the brain: Evidence for developmental homology? *Developmental Psychobiology*, 55(1), 42–51.

Anderson, M. L., Richardson, M. J., & Chemero, A. (2012). Eroding the boundaries of cognition: Implications of embodiment. *Topics in Cognitive Science*, 4(4), 717–730.

Anderson, M. L., & Rosenberg, G. (2008). Content and action: The guidance theory of representation. *Journal of Mind and Behavior*, 29, 55–86.

Barrett, L. (2011). *Beyond the brain: How body and environment shape animal and human minds*. Princeton, NJ: Princeton University Press.

Bechtel, W., Abrahamsen, A., & Graham, G. (1998). The life of cognitive science. In W. Bechtel & G. Graham (Eds.), *A companion to cognitive science* (pp. 1–104). Oxford: Basil Blackwell.

Beer, R. (1995). A dynamical systems perspective on agent–environment interactions. *Artificial Intelligence*, 72, 173–215.

Binkofski, F., Amunts, K., Stephan, K. M., Posse, S., Schormann, T., Freund, H. J., ... Seitz, R. J. (2000). Broca's region subserves imagery of motion: A combined cytoarchitectonic and fMRI study. *Human Brain Mapping*, 11(4), 273–285.

Brooks, R. (1991). Intelligence without representation. *Artificial Intelligence*, 47, 139–159.

Chemero, A. (2009). *Radical embodied cognitive science*. Cambridge, MA: MIT Press.

Chomsky, N. (1988). *Language and the problems of knowledge. The Managua lectures*. Cambridge, MA: MIT Press.

Churchland, P. S. (2002). *Brain-Wise*. Cambridge, MA: MIT Press.

Cisek, P. (2006). Integrated neural processes for defining potential actions and deciding between them: A computational model. *Journal of Neuroscience*, 26(38), 9761–9770.

Cisek, P. (2007). Cortical mechanisms of action selection: The affordance competition hypothesis. *Philosophical Transactions of the Royal Society B, Biological Sciences*, 362, 1585–1599.

Cisek, P., & Kalaska J. F. (2005). Neural correlates of reaching decisions in dorsal premotor cortex: Specification of multiple direction choices and final selection of action. *Neuron*, 45(5), 801–814.

Cisek, P., & Kalaska J. F. (2010). Neural mechanisms for interacting with a world full of action choices. *Annual Review of Neuroscience*, 33, 269–298.

Clark, A. (1997). *Being there*. Cambridge, MA: MIT Press.

Clark, A. (2003). *Natural born cyborgs*. New York, NY: Oxford University Press.

Coe, B., Tomihara, K., Matsuzawa, M., & Hikosaka, O. (2002). Visual and anticipatory bias in three cortical eye fields of the monkey during an adaptive decision-making task. *Journal of Neuroscience*, 22(12), 5081–5090.

Crammond, D. J., & Kalaska, J. F. (2000). Prior information in motor and premotor cortex: Activity during the delay period and effect on premovement activity. *Journal of Neurophysiology*, 84(2), 986–1005.

Damasio, A., & Tranel, D. (1993). Nouns and verbs are retrieved with differently distributed neural systems. *Proceedings of the National Academy of Sciences of the USA*, 90, 4957–60.

Damasio, H., Grabowski, T. J., Tranel, D., Hichwa, R. D., & Damasio, A. R. (1996). A neural basis for lexical retrieval. *Nature*, 380, 499–505

Decety, J., Grezes, J., Costes, N., Perani, D., Jeannerod, M., Procyk, E., ... Fazio, F. (1997). Brain activity during observation of actions. Influence of action content and subject's strategy. *Brain*, 120, 1763–1777.

Erlhagen, W., & Schöner, G. (2002). Dynamic field theory of movement preparation. *Psychological Review*, 109(3), 545–572.

Fair, D. A., Dosenbach, N. U. F., Church, J. A., Cohen, A. L., Brahmbhatt, S., Miezin, F., ... Schlaggar, B. L. (2007). Development of distinct control networks through segregation and integration. *Proceedings of the National Academy of Sciences of the USA*, 104, 13507–13512.

Fodor, J. (1975). *The language of thought*. London: Thomas Crowell.

Fodor J. (1983). *The modularity of mind: An essay on faculty psychology*. Cambridge, MA: MIT Press.

Fodor, J. (1990). *A theory of content and other essays*. Cambridge, MA: MIT Press.

Fodor, J. (2000). *The mind doesn't work that way*. Cambridge, MA: MIT Press.

Fodor, J. (2007). Why pigs don't have wings. *London Review of Books*, 29(20), 19–22.

Friston K. J. (1997). Imaging cognitive anatomy. *Trends in Cognitive Science*, 1, 21–27.

Gauthier, I., Skudlarski, P., Gore, J. C., & Anderson, A. W. (2000). Expertise for cars and birds recruits brain areas involved in face recognition. *Nature Neuroscience*. 3(2), 191–197.

Ghez, C., Favilla, M., Ghilardi, M. F., Gordon, J., Bermejo, R., & Pullman, S. (1997). Discrete and continuous planning of hand movements and isometric force trajectories. *Experimental Brain Research*, 15(2), 217–233.

Gibson, J. J. (1979). *The ecological approach to visual perception*. Hillsdale, NJ: Erlbaum.

Glenberg, A. M., & Kaschak, M. P. (2002). Grounding language in action. *Psychonomic Bulletin and Review*, 9, 558–565.

Glimcher, P.W. (2003). The neurobiology of visual-saccadic decision making. *Annual Review of Neuroscience*, 26, 133–179.

Gold, J. I., & Shadlen, M. N. (2007). The neural basis of decision making. *Annual Review of Neuroscience*, 30, 535–574.

Grill-Spector, K., Sayres, R., & Ress, D. (2006). High-resolution imaging reveals highly selective nonface clusters in the fusiform face area. *Nature Neuroscience*, 9(9), 1177–1185.

Grush, R. (1997). The architecture of representation. *Philosophical Psychology*, 10, 5–24.

Grush, R. (2004).The emulation theory of representation: Motor control, imagery, and perception. *Behavioral and Brain Sciences*, 27, 377–442.

Hagoort, P. (2005). On Broca, brain and binding. *Trends in Cognitive Sciences*, 9(9), 416–423.

Hamzei, F., Rijntjes, M., Dettmers, C., Glauche, V., Weiller, C., & Büchel, C. (2003). The human action recognition system and its relationship to Broca's area: An fMRI study. *NeuroImage*, 19, 637–644.

Hanakawa, T., Honda, M., Sawamoto, N., Okada, T., Yonekura, Y., Fukuyama, H., & Shibasaki, H. (2002). The role of rostral Brodmann area 6 in mental-operation tasks: An integrative neuroimaging approach. *Cerebral Cortex*, 12, 1157–1170.

Hanson, S. J., & Schmidt, A. (2011). High-resolution imaging of the fusiform face area (FFA) using multivariate non-linear classifiers shows diagnosticity for non-face categories. *NeuroImage*, 54, 1715–1734.

Hochner, B. (2012). An embodied view of octopus neurobiology. *Current Biology*, 22, R887–892.

Hutchins, E. (1995). *Cognition in the wild*. Cambridge, MA: MIT Press.

Hutto, D. (2005). Knowing what? Radical versus conservative enactivism. *Phenomenology and the Cognitive Sciences*, 4, 389–405.

Hutto, D., & Myin, E. (2012). *Radicalizing enactivism*. Cambridge, MA: MIT Press.

James, W. (1890). *The principles of psychology*. New York, NY: Henry Holt and Company.

Kanwisher, N. (2010). Functional specificity in the human brain: A window into the functional architecture of the mind. *Proceedings of the National Academy of Sciences of the USA*, 107(25), 11163–11170. doi:10.1073/pnas.1005062107

Kanwisher, N., McDermott, J., & Chun, M. M. (1997). The fusiform face area: A module in human extrastriate cortex specialized for face perception. *Journal of Neuroscience*, 17(11), 4302–4311.

Kirsh, D. (1995). The intelligent use of space. *Artificial Intelligence*, 72, 31–68.

Kirsh, D., & Maglio, P. (1994). On distinguishing epistemic from pragmatic action. *Cognitive Science*, 18, 513–549.

Laird A. R., Lancaster, J. L., & Fox P. T. (2005). BrainMap: The social evolution of a functional neuroimaging database. *Neuroinformatics*, 3, 65–78.

Landy, D. H., & Goldstone, R. L. (2009). How much of symbolic manipulation is just symbol pushing. In *Proceedings of the 31th Annual Conference of the Cognitive Science Society* (pp. 1318–1323). Amsterdam: Cognitive Science Society.

Martin, A., Haxby, J. V., Lalonde, F. M., Wiggs, C. L., & Ungerleider, L. G. (1995). Discrete cortical regions associated with knowledge of color and knowledge of action. *Science*, 270, 102–105.

Martin, A., Ungerleider, L. G., & Haxby, J. V. (2000). Category-specificity and the brain: The sensorymotor model of semantic representations of objects. In M. S. Gazzaniga (Ed.), *The new cognitive neurosciences* (2nd ed., pp. 1023–1036). Cambridge, MA: MIT Press.

Martin, A., Wiggs, C. L., Ungerleider, L. G., & Haxby, J. V. (1996). Neural correlates of category-specific knowledge. *Nature*, 379, 649–652.

McIntosh, A. R., Grady, C. L., Ungerleider, L. G., Haxby, J. V., Rapoport, S. I., & Horwitz, B. (1994). Network analysis of cortical visual pathways mapped with PET. *Journal of Neuroscience*, 14, 655–666.

McKinstry, C., Dale, R., & Spivey, M. J. (2008). Action dynamics reveal parallel competition in decision making. *Psychological Science*, 19(1), 22–24.

Menary, R. (2007). *Cognitive integration*. New York, NY: Palgrave.

Miller, G. A. (2003). The cognitive revolution: A historical perspective. *Trends in Cognitive Sciences*, 7(3), 141–144.

Millikan, R. (1995). Pushmi-pullyu representations. In J. Tomberlin (Ed.) *Philosophical perspectives, 9* (pp. 185–200). Atascadero, CA: Ridgeview.

Milner, A. D., & Goodale, M. A. (1995). *The visual brain in action*. Oxford: Oxford University Press

Nishitani, N., Schürmann, M., Amunts, K., & Hari, R. (2005). Broca's region: From action to language. *Physiology*, 20, 60–69.

Ogawa, S., Lee, T. M., Kay, A. R., & Tank, D. W. (1990). Brain magnetic resonance imaging with contrast dependent on blood oxygenation. *Proceedings of the National Academy of Sciences of the USA*, 87, 9868–9872.

Paul, C., Lungarella, M., & Iida, F. (2006). Morphology, control and passive dynamics (Editorial of special issue on morphology, control and passive dynamics). *Robotics and Autonomous Systems*, 54(8), 617–618.

Poldrack, R. A. (2006). Can cognitive processes be inferred from neuroimaging data? *Trends in Cognitive Sciences*, 10, 59–63.

Poldrack, R. A., Halchenko, Y., & Hanson, S. J., (2009). Decoding the large-scale structure of brain function by classifying mental states across individuals. *Psychological Science*, 20, 1364–1372.

Posner, M. I., Petersen, S. E., Fox, P. T., & Raichle, M. E. (1988). Localization of cognitive operations in the human brain. *Science*, 240 (4859), 1627–1631.

Pulvermüller, F. (2005). Brain mechanisms linking language and action. *Nature Reviews Neuroscience*, 6, 576–582.

Ramón y Cajal, S. (1904/1995). *Histology of the nervous system* (Vol. 1, N. Swanson & L. W. Swanson, Trans.). New York, NY: Oxford University Press.

Rhodes, G., Byatt, G., Michie, P. T., & Puce, A. (2004). Is the fusiform face area specialized for faces, individuation, or expert individuation? *Journal of Cognitive Neuroscience*, 16(2), 189–203.

Rowlands, M. (2006). *Body language*. Cambridge, MA: MIT Press.
Rumelhart, D. E., McClelland, J. L., & PDP Research Group. (1986). Parallel distributed processing: Explorations in the microstructure of cognition, Vols. 1–2. Cambridge, MA: MIT Press.
Schall, J. D. (2004). On building a bridge between brain and behavior. *Annual Review of Psychology*, 55, 23–50.
Schiller, P. H. (1996). On the specificity of neurons and visual areas. *Behavioral Brain Research*, 76, 21–35.
Schmolesky, M. T., Wang, Y., Hanes, D. P., Thompson, K. G., Leutgeb, S., Schall, D. J., & Leventhal, A. G. (1998). Signal timing across the macaque visual system. *Journal of Neurophysiology*, 79(6), 3272–3278.
Sokoloff, L., Reivich, M., Kennedy, C., Des Rosiers, M. H., Patlak, C. S., Pettigrew, K. D., ... Shinohara, M. (1977). The [14C]deoxyglucose method for the measurement of local cerebral glucose utilization: Theory, procedure, and normal values in the conscious and anesthetized albino rat. *Journal of Neurochemistry*, 28(5), 897–916.
Spencer, H. (1855). *Principles of psychology*. London: Williams and Norgate.
Sporns, O. (2011). *Networks in the brain*. Cambridge, MA: MIT Press.
Steinmetz, P. N., Roy, A., Fitzgerald, P. J., Hsiao, S. S., Johnson, K. O., & Niebur, E. (2000). Attention modulates synchronized neuronal firing in primate somatosensory cortex. *Nature*, 404, 187–190.
Supekar, K. S., Musen, M. A., & Menon, V. (2009). Development of large-scale functional brain networks in children. *PLOS Biology*. Retrieved from http://journals.plos.org/plosbiology/article?id=10.1371/journal.pbio.1000157
Sur, M., Garraghty, P. E., & Roe, A. W. (1988). Experimentally induced visual projections into auditory thalamus and cortex. *Science*, 242, 1437–1441.
Takikawa, Y., Kawagoe, R., & Hikosaka, O. (2002). Reward-dependent spatial selectivity of anticipatory activity in monkey caudate neurons. *Journal of Neurophysiology*, 87(1), 508–515.
Tettamanti, M., & Weniger, D. (2006). Broca's area: A supramodal hierarchical processor? *Cortex* 42, 491–494.
Thelen, E. (1995). Time-scale dynamics and the embodiment of an embodied cognition. In R. Port & T. van Gelder (Eds.), *Mind as motion* (pp. 69–100). Cambridge, MA: MIT Press.
Thelen, E., & Smith L. B. (1994). *A dynamic systems approach to the development of cognition and action*. Cambridge, MA: MIT Press.
Thoenissen, D., Zilles, K., & Toni, I. (2002). Differential involvement of parietal and precentral regions in movement preparation and motor intention. *Journal of Neuroscience*, 22, 9024–9034.
Titchener, E. B. (1898). The postulates of a structural psychology. *Philosophical Review*, 7, 449–465.
Varela, F., Lachaux J. P., Rodriguez E., & Martinerie J. (2001). The brainweb: Phase synchronization and large-scale integration. *Nature Reviews Neuroscience*, 2(4), 229–239.
Wozniak, R. H. (1992). *Mind and Body: René Descartes to William James*. Bethesda, MD: National Library of Medicine; Washington, DC: American Psychological Association. Retrieved from https://serendip.brynmawr.edu/Mind/Adaptation.html
Yarkoni, T., Poldrack, R. A., Nichols, T. E., Van Essen, D. C., & Wager, T. D. (2011). Large-scale automated synthesis of human functional neuroimaging data. *Nature Methods*, 8(8), 665–670.

2

The Evolution of Evolutionary Neuroscience

Suzana Herculano-Houzel

2.1 The Evolution of "Evolution"

The history of evolution is as long as the history of the Earth—and yet, "evolution" hasn't always been there. Before "evolution," naturalists framed their thoughts on the assumption of a fixed *scala naturae* as conceived by Aristotle: a strict hierarchical structure of all that is, descending from God down to minerals ("the great chain of being"; Lovejoy, 1964), with animals arranged in between "according to the degree of perfection of their souls" (Bunnin & Yu, 2004).

Once it appeared, the concept of evolution itself evolved—that is, changed over time—and along with it have evolved the questions and interpretations posed by neuroscientists. In the 19th century, the uncovering of growing numbers of particular fossils in different geological strata led to the concept of the mutability over time of the panoply of beings that had lived, and evolution came to be conceptualized by Charles Darwin (Darwin, 1859).

In the light of evolution, the *scala naturae* became a phylogenetic scale that organisms supposedly ascended as they evolved, over time, from simple to complex. Thus reasoned Ludwig Edinger, by many considered the father of comparative neuroanatomy, when he formulated at the end of the 19th century a unified theory of brain evolution that combined Charles Darwin's 1859 concept of evolution with the then current version of Aristotle's *scala naturae*. Edinger viewed evolution as progressive and linear: from fish to amphibians, reptiles, birds, then mammals—culminating with humans, naturally, in an ascent from "lower" to "higher" intelligence. In the process, the brains of extant vertebrates supposedly retained ancestral structures; for that reason, and in the face of progressive evolution, the comparison of the brain anatomy of extant species would reveal the origin of more recent structures. This supposed evidence of "past lives" in modern brain structures resonated with the Law of Recapitulation proposed by Étienne Serres, supported by Étienne Geoffroy Saint-Hilaire (1830), and formulated by Ernst Haeckel in the aphorism "ontogeny recapitulates phylogeny" (1866). Haeckel claimed that the development of more recent ("advanced") species passes through successive stages represented by adult forms of older (more "primitive") species.

The Wiley Handbook of Evolutionary Neuroscience, First Edition. Edited by Stephen V. Shepherd.
© 2017 John Wiley & Sons, Ltd. Published 2017 by John Wiley & Sons, Ltd.

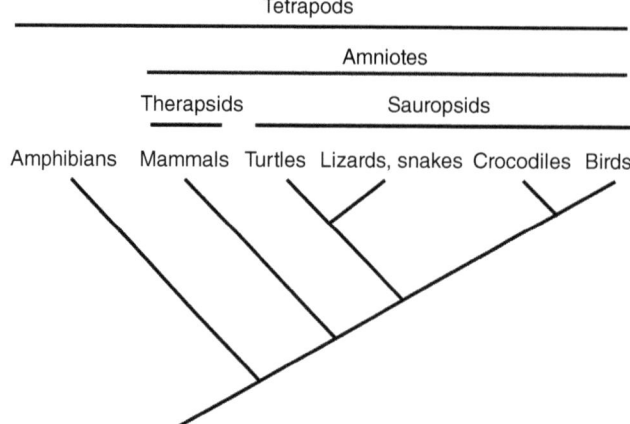

Figure 2.1 Phylogeny of Tetrapods.
Phylogeny of tetrapods places modern reptiles (including birds), the living descendants of sauropsids, as a sister group to modern mammals, the living descendants of therapsids.

Recapitulation was refuted both at the level of embryogenesis and of brain evolution in the 20th century (reviewed in Gould, 1977). Darwin himself acknowledged that early embryonic stages may be similar across related species, but are not similar to the adult forms of those species—a view that is shared by modern evolutionary developmental biology (see below). In 1922, Walter Garstang advanced the idea that differences among adult animal species arise because of evolutionary modifications in their development program—that is, that phylogeny occurs through changes in ontogeny. This amounted to the exact opposite of what Haeckel had advocated in 1866. More recent evidence against recapitulation was the recognition that mammals and birds/reptiles are sister groups—that is, that mammals don't derive from reptiles as they are today, just as humans don't derive from modern monkeys, and also that the last common ancestral form to all mammals was not a reptile (Carroll, 1988). Rather, mammals arose from ancestral therapsids, while reptiles (including the later birds) arose from ancestral sauropsids, and both therapsids and sauropsids were sister branches of the ancestral stem amniote (see Figure 2.1; see also Carroll, 1988; Evans, 2000). The notion of an ancestral "reptilian" brain has been hard to shake off, however (see below).

Modern evolutionary biologists also realize that phylogenetic trees are actually not trees, much less ladders, but rather erratic bushes with branches growing here and there in divergent directions (Gould, 1989), only some of which last through the ages. Edinger's phylogenetic scale also fails in the face of secondary simplification: the fact that species do not always "progress" into more complex beings in evolution (Jenner, 2004). Regardless, however, *scala naturae* thinking persisted in the neurosciences (Hodos and Campbell, 1969), reflecting a lack of proper training in evolutionary biology (Striedter, 2009).

Other inversions in the evolutionary tree have been driven by discoveries made by molecular phylogenetics: the use of differences and similarities in the coding and noncoding sequences of particular genes to establish likely evolutionary relationships amongst them. Hence, the categorization of modern animals went from division into acoelomates, pseudocoelomates, and coelomates to, instead, division of Bilateria into protostomes (Lophotrochozoa and Ecdysozoa) and deuterostomes (echinoderms, hemichordates, urochordates, cephalochordates, and chordates) (Halanych et al., 1995; Hervé, Lartillot, & Brinkmann, 2005). The modern grouping is based on phylogenetic relationships unsuspected from simple morphological studies, upturning several popular theories on the evolution of the nervous system, and leads to the proposition of the Urbilateria as an ancestral group (De Robertis and Sasai, 1996).

Now disabused of the idea that ontogeny recapitulates phylogeny, modern evo-devo (evolutionary developmental biology) views animal evolution as the result of changes in the developmental process, as envisioned by Garstang (1922). The study of comparative neurobiology thus still fuels the modern search for the origins of nervous system diversity—no longer through the search for successive, progressive steps in development, but, instead, by looking for those evolved modifications in development that gave rise to different adult life forms.

2.2 Evolution of "the Nervous System"

The nervous system is a characteristic of animals—although not of all animals, as Placozoa and sponges lack any semblance of a nervous system. Cnidaria and Ctenophora have a distributed network of nerve cells throughout their bodies. It is only in bilaterians that the nervous system assumes a cord-like structure, although one that is arranged differently in protostomes and deuterostomes: It is situated ventrally in the former, and dorsally in the latter, raising the issue of how the nervous system was arranged in the common ancestor of bilaterians. One popular view was that this ancestor had a ventral nerve cord which became inverted in deuterostomes, as proposed by Geoffroy Saint-Hilaire and by Anton Dorhn (Gerhart, 2000)—a view in line with the mistaken but popular concept of progressive phylogenesis, in which the "simpler" protostomes arose before deuterostomes instead of simultaneously. The remaining—and then apparently less likely—possibility was that a nerve cord was formed twice independently: once (ventrally) in protostomes and again (dorsally) in deuterostomes (see Figure 2.2).

Comparing the expression of genes in *Drosophila* and mouse for extracellular signals that provide positional information in embryonic development, Eric de Robertis and Yoshiki Sasai showed in 1996 that the ventral cord of the insect and the dorsal cord of the vertebrate express similar genes, and are thus putatively homologous on molecular bases. Dorsal-ventral patterning in insects and vertebrates therefore appears to be controlled by homologous morphogens with mutually antagonistic actions. As a consequence, they argued that the position of the mouth (which is also homologous in protostomes and deuterostomes) changed between lineages, causing "dorsal" and "ventral" surfaces to become inverted (that is, "the animal lies on its back"). De Robertis and Sasai (1996) coined the term Urbilateria for the earliest bilaterally symmetrical animals. They proposed Urbilateria had a ventral nerve cord, but "turned over" by changing the location of the mouth in deuterostomes. This was based on the (incorrect) assumption that the protostome form would have been ancestral, in line

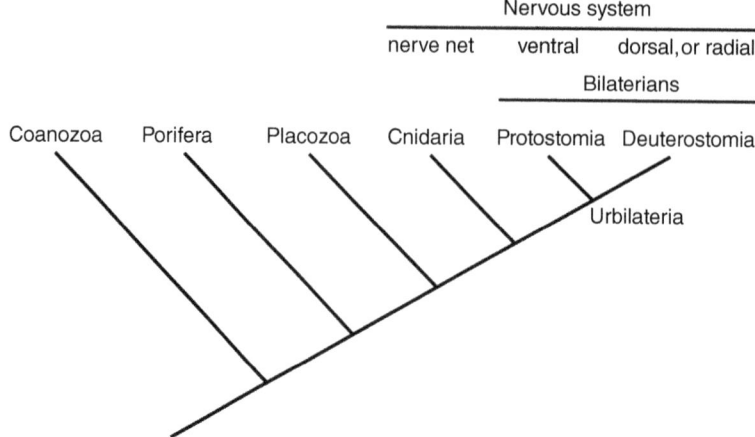

Figure 2.2 Organization of Nervous Systems.
Amongst animals, only Cnidaria, Protostomia, and Deuterostomia have a nervous system, which is organized in cords only in the latter two groups (the bilaterians).

with Geoffroy Saint-Hilaire's idea of an inversion of the dorsal-ventral axis in deuterostomes.

The origin of the bilateral nervous system thus goes back at least to the Urbilateria. Based on the simple comparison between mammals and *Drosophila*, Hirth and Reichert proposed a single, monophyletic brain origin, with a tripartite brain (comprising the hindbrain, forebrain/midbrain, and an intervening boundary region) and extended central nervous system already evident in the last common bilaterian ancestor (Hirth et al., 2003; Hirth & Reichert, 2007). But how to reconcile this monophyletic origin of all bilaterally symmetric animals with the many different shapes and types of nervous systems represented by extant Bilateria, which include insects, cephalopods, tunicates, mammals, and even the radially organized adult nervous systems of echinoderms?

The tripartite brain of bilaterians is supposed to have arisen after the diversion of the cnidarian and protostome/deuterostome lineages (Hirth, 2010), circa 630 million years ago (Erwin, 2009; Peterson et al., 2004). Alternately, cladistic analysis suggests that neurons, centralized nervous systems, and brains arose independently as many as seven times amongst Bilateria (Moroz, 2009). This analysis indicates that Urbilateria did not possess a tripartite brain, and probably no brain at all, but rather an uncentralized nerve net, similar to that maintained by Cnidarians today. In this scenario, modular mechanisms of development are invoked to explain why homologous genes might organize brains that are not, themselves, homologous (Moroz, 2009). Importantly, positing that the ancestral Urbilaterian nervous system was a neural net turns the otherwise improbable dual independent origins of ventral (protostome) and dorsal (deuterostome) nervous systems into the most parsimonious scenario, in which homologous molecular modules acted independently to form the two nonhomologous nervous systems (Northcutt, 2010).

Along the same lines, the similar patterns of Hox gene expression in the vertebral spinal cord and ventral nerve cord of insects, and of Pax-related genes in forebrain, eye, and hindbrain (Halder, Callaerts, & Gehring, 1995; Harris, 1997) in vertebrates and invertebrates do not necessarily make these structures homologous across the two clades. Rather, these are other instances in which the same genes may be used independently in the evolution of brain structures (Northcutt, 2010). In this case, the tripartite urbilaterian brain—considered by Heinrich Reichert, Frank Hirth, and Antonio Simeone to have had paired eyes, forebrain, a midbrain–hindbrain boundary (equivalent to the deuto–tritocerebrum boundary of invertebrates), and a hindbrain, with the telencephalon and most of the modern midbrain absent—would not have existed, and tripartite brains would have arisen independently, but from the same genetic modules, in vertebrates and invertebrates at the Ur/Bilateria branching (Northcutt, 2010).

A similar conceptual reorganization has taken place in our understanding of the evolution of the vertebrate brain. Amongst deuterostomes, the vertebrate brain was previously considered to be an invention exclusive of extant cephalochordates, given that urochordates (tunicates) lack a well-organized brain. Recently, however, Pani et al. (2012) found that genes involved in patterning the neurectoderm in vertebrates are also expressed in a hemichordate. This finding suggests that the genetic programs that were eventually modified into patterning the vertebrate brain already existed in an ancestral creature that lived over 600 million years ago and survived into Cambrian times, but then degenerated in amphioxus and tunicates, remaining (but patterning divergent structures) in hemichordates and vertebrates.

Besides having a well-organized brain, vertebrates differ from other chordates in that only the former are active, mobile predators, directed by an array of specialized sense organs, most of which are concentrated in the head. These sense organs are formed in development from neurogenic epidermal placodes, or thickenings of the ectoderm. The neural crest also contributes to the formation of vertebrate sense organs and other structures, and it was on this basis that Carl Gans and Glenn Northcutt proposed in 1983 that the elaboration of neural crest and neurogenic placodes was the seminal event in the origin of vertebrates, producing an enlarged brain and paired eyes and so leading to the formation of the new vertebrate head (Gans & Northcutt, 1983). Obviously, however, the head is not a vertebrate invention, as many protostome invertebrates also exhibit a well-defined head. It will be interesting to watch as continued research unveils the similarities and differences that go into building protostome and deuterostome heads.

Finally, proteomic studies of the molecular components of mammalian synapses, and their homologous proteins in other species, point to ancestral molecular machinery in unicellular organisms that existed well before the evolution of metazoans and neurons (the protosynapse; Ryan & Grant, 2009). For example, cadherins and ephrin receptors are also found in choanoflagellates, GABA and metabotropic glutamate receptors are found in Poriferans, and several synaptic proteins are found in Fungi (Ryan and Grant, 2009). Thus, regardless of when tripartite brains, a CNS, or even a neuron first appeared in evolution, the ursynapse most likely predated it, appearing sometime after the branching of Poriferans but before the branching of Cnidarians. Tomás Ryan and Seth Grant (2009) propose that, in this "synapses first" scenario, it may have been the evolution of the synapse that led to the evolution of the neuron.

2.3 New Understandings of Brain Structure

The main divisions of the human central nervous system—spinal cord, medulla, pons, cerebellum, diencephalon, mesencephalon, and telencephalon—are recognizable in all vertebrates. Amongst these structures, however, the telencephalon differs the most across species. Ludwig Edinger proposed in 1908 that the preeminence of the telencephalon in mammals, and particularly in humans, was a sign of the human evolutionary status as "highest" amongst animals. At that time, evolutionary relationships among vertebrates placed mammals as the most recently evolved group. That, however, was to change later in the 20th century, as mammals (the only remaining Therapsids) came to be recognized as a sister group to reptiles/birds (Sauropsids), rather than as their descendants (see Figure 2.1; Carroll, 1988).

But at the beginning of the 20th century, and in line with the idea of progressive evolution through gradual increases in complexity and size from fish to amphibians, to reptiles, to birds, to mammals—culminating with humans, of course—Edinger suggested that each new vertebrate group in evolution acquired a more advanced cerebral subdivision, much as the earth's geological strata formed over time. He thus proposed that an ancestral brain (the palaeoencephalon, or "striatum") controlled instinctive behavior, and had been followed by the addition of a newer brain (the neoencephalon, or pallium, or "cortex"), which controlled learned and intelligent behavior (Edinger, 1908). The ensuing view was to become dominant in neuroscience, codified in an important comparative neuroanatomy text (Kappers, Huber, & Crosby, 1936): that the primordial telencephalon of fishes had a small pallium (a palaeocortex) and a larger subpallium (the palaeostriatum), to which an archistriatum and archicortex were added in reptiles. Birds would have evolved a hypertrophied striatum, but not any further pallial regions; in contrast, mammals were thought to have evolved the latest and greatest achievement, on top of the primitive palaeo- and archicortices: the "neocortex." Thus, the striatal structures in fish, reptiles, and avians would correspond to the mammalian striatum, and their limited pallium would be a mostly olfactory version of what would become the mammalian neocortex (reviewed in Jarvis et al., 2005).

The (mistaken) idea that the neocortex was a recent mammalian invention gained popularity when neuroanatomist Paul MacLean evoked them in his view of a "triune brain" (MacLean, 1964, 1990), consisting of a reptilian complex (from the medulla to the basal ganglia) which evolved first, to which was added a "paleomammalian" complex (the limbic system), and finally a neomammalian complex (the neocortex). The intuitive (but incorrect) equation of evolution with progress, along with the alluring notion of a primitive reptilian brain, supposedly incapable of anything as complex as what a mammalian neocortex can achieve, attracted much attention from the popular media once it was used in Carl Sagan's popular book, *The Dragons of Eden* (Sagan, 1977). Building on the "evolutionary" version of the *scala naturae*, Edinger thus established the basis of a nomenclature that was used for an entire century to define the cerebral subdivisions of all vertebrates—and one that, through the words of MacLean, to this day influences popular concepts of brain evolution.

In parallel to changing views on vertebrate evolution, neuroanatomy slowly accumulated evidence against Edinger's progressive school of telencephalic advancement. Path-tracing and behavioral studies in the mid-1960s found that, like the mammalian neocortex, the neostriatum and hyperstriatum of the avian dorsal ventricular ridge

(DVR) receive sensory input from the thalamus, and carry out the same type of information processing as is performed by cortical layers. The archistriatum and hyperstriatum give rise to descending projections to premotor and motor neurons, like the mammalian cortico-bulbar and cortico-spinal pathways. Moreover, also like the mammalian neocortex, the avian DVR is crucial in motor control and sensori-motor learning (reviewed in Jarvis et al., 2005).

These functional similarities were captured in the nuclear-to-layered hypothesis formulated by Harvey Karten in 1969 (Karten, 1969, 1991) that proposed that the striatum of birds and the neocortex of mammals are homologous. According to this hypothesis, the common ancestor of birds, reptiles, and mammals had cells organized into a globular structure, the DVR (the ensemble of hyperstriatum, neostriatum, and archistriatum) and this structure was reorganized into a laminar pallium early in the mammalian lineage, while maintaining the functional connectivity that determines distinct functional areas and relationships.

With the advent of studies of patterns of the expression of Hox genes and related early transcription factors, the homology between the avian DVR and the mammalian cortex gained increasing support. In a series of comparative studies, Luis Puelles, John Rubinstein, and their colleagues found similar expression patterns in developing mammalian and avian brains (Puelles et al., 2000; Puelles & Rubenstein, 2003). For instance, EMX1 and PAX6 are expressed both in the avian DVR and in the mammalian dorsal claustrum and basolateral amygdala. The bird "striatum," or DVR, is therefore equivalent in gene expression to the mammalian pallium—although not to the dorsal pallium, as originally proposed by Harvey Karten. While the expression of similar genes in morphologically different structures of bird and mammalian brains is considered by many as evidence of the homology of these structures, it must again be kept in mind that these may be genetic modules co-opted in evolutionarily independent ways in birds and mammals, as in protostomes and deuterostomes.

More recently, the nuclear-to-laminar hypothesis was tested directly and supported by the finding that genetic markers expressed in cells in mammalian cortical layers 4 and 5 are indeed expressed in thalamic input and brainstem output nuclei of the avian DVR (Dugas-Ford et al., 2012). Thus, the neuronal circuitry of the avian DVR (now "pallium"; see below) does feature cell types with the connectional and molecular properties of neocortical input and output neurons.

In 2004, in the face of this new wealth of neurochemical, anatomical, and functional data, and in an attempt to expunge neuroscience of *scala naturae* thinking, a consortium of comparative neurobiologists revised the neuroanatomical nomenclature in avians to replace the terms neostriatum, archistriatum, and paleostriatum (which suggested that brains evolved by the sequential addition of new brain regions) with neutral terms (Jarvis et al., 2005; Reiner et al., 2004a, 2004b). The 2004 Avian Brain Nomenclature Consortium concluded that "the avian telencephalon is organized into three main, developmentally distinct domains that are homologous in fish, amphibians, reptiles, birds and mammals: pallial, striatal and pallidal domains" (Jarvis et al., 2005, p. 155). Subdivisions were then named within each domain with terms based on homologies, topology, and other recognizable roots, eliminating all phylogeny-alluding prefixes (palaeo-, archi-, and neo-) that erroneously implied the relative age, or evolutionary order, of each subdivision. Thus, the avian hyperstriatum, neostriatum, and archistriatum were renamed hyperpallium and mesopallium; nidopallium; and arcopallium, with a neighboring amygdaloid complex. The similarly corrected terms for mammalian pallium (isocortex and allocortex) have, however, not been universally accepted. The Consortium notes that "neocortex" is still an appropriate term only if used to refer to the uniqueness of the six-layered pallium in mammals, and not to imply that it evolved from an older cortex, nor that it is the newest-evolved pallial organization.

Indeed, now that birds are recognized as a branch off sauropsids, which arose much later than the branching between therapsids and sauropsids that separated future reptiles from future mammals, the avian hyperpallium (a nuclear structure) is considered to have evolved more recently than the mammalian six-layered cortex (Evans, 2000). Because the six-layered cortex is shared by all living descendants of the therapsids (the mammals), it was presumably inherited from the common therapsid ancestor over 200 million years ago. In parallel, the nuclear pallium of birds and reptiles was presumably present in the common sauropsid ancestor equally long ago. Thus, the reptilian structure formerly known as the DVR is just as derived a structure as the mammalian layered pallium. Notably, the presence of a recognizable DVR in fish strongly suggests a nuclear organization of the ancestral pallium in the stem amniotes that gave rise to sauropsids (and then reptiles, and then birds) and therapsids (and then mammals), implying a nuclear-to-layered mechanism is indeed necessary to account for the evolution of the mammalian cortex. In turn, the true basal ganglia of birds (with which the DVR has often been confused) has been found to be organized in a similar, conserved way comparable to the basal ganglia of mammals and other vertebrates (Reiner, Medina, & Veenman, 1998).

2.4 New Understandings of Brain Size

Larger animals, be they vertebrate or invertebrate, tend to have larger brains. This relationship was recognized and formulated as early as in 1762, when Albrecht von Haller proposed what became known as Haller's rule: that larger animal species have larger brains, which are, however, *relatively* smaller in proportion to body size (von Haller, 1762). This rule was later confirmed by Georges Cuvier (1801). In 1867, Johann Friedrich von Brandt linked this reduction in the ratio between brain mass and body mass to the changing ratio between body volume and surface area in larger animals (Rensch, 1960). The relationship gained mathematical treatment in 1937, when von Bonin used linear regression of log values of brain mass and body mass to describe that the former varies as a power law of the latter, with an allometric exponent of $2/3$, consistent with von Brandt's suggestion. Von Bonin was building on the concept of allometry, newly coined by Julian Huxley, which acknowledged that the mass of body parts scales as a power function of body mass (that it, it scales with body mass raised to an exponent, called the allometric exponent), a relationship that can be turned linear by plotting log values of each variable. Von Bonin (1937) has since often been credited with the introduction of objective mathematical and statistical methods to studies of brain evolution.

Harry Jerison also became interested in allometry from the point of view of Huxley's "Theory of Growth" (Huxley, 1932), through the calculation of allometric exponents that relate the scaling of body parts such as the brain to overall body mass. In the framework of the Theory of Growth, the finding of such an allometric exponent indicates that the growth pattern during the development of an individual species predicts not only relationships between body parts in that species, but also among adults of different species. Working with a sample of mammalian species, Jerison reported an allometric exponent of 0.73 (Jerison, 1955), closer to the value of 0.75 that later became more accepted for mammals as a whole (Martin, 1990), although more recent work has pointed out that this allometric exponent is particular to each mammalian order (Martin & Harvey, 1985). Moreover, it is now recognized that the growth exponents that apply within a species do not necessarily apply across species, for instance depending on whether significant body growth happens after brain growth is over (Riska & Atchley, 1985).

Regardless of the flawed assumption of similar growth patterns within and across species, the recognition of an allometric exponent that described brain mass allowed a new concept to emerge: that of brain enlargement *relative to the size expected for a given body mass*. This is a measurement of how large a species' brain is compared to how large it was predicted to be from the species' body size, or, put in mathematical terms, the residual value of the brain x body mass allometric function. One of these new calculations was based on a wealth of new data collected and published between the 1960s and 1980s by Heinz Stephan's group in Ludwig Edinger's laboratory in Frankfurt, Germany (reviewed in Stephan et al., 1981a). In 1969, by assuming (in good Edingerian tradition) that primates and most if not all modern placental mammals had their phylogenetic origin in insectivore-like animals, Stephan and Andy calculated what they called "progression indices": that is, a measure of how much modern species would have distanced themselves from the "primitive" state. True to their advisor's "progressive" spirit, Stephan and Andy consider that "a fairly reliable and characteristic feature of directed progressive evolution is the concentration, enlargement and differentiation of the nervous system …. This development culminated in mammals, and especially in primates" (Stephan & Andy, 1969, p. 372). Their "progression indices" use the allometric relationship between brain volume (or the volume of each particular structure) and body mass pertaining to "basal insectivores" to then measure "how many times larger a given brain structure of a certain species is than the corresponding structure in a typical basal insectivore of the same body weight" (Stephan & Andy, 1969, p. 373).

As they expected, they found that the neocortex shows the strongest "progression," in an "ascending primate scale," while the olfactory bulb was the only structure found to regress. Progression was particularly high, and actually highest, in the human neocortex. In exemplary circular logic, they conclude that the progressive index of the neocortex "represents the best cerebral criterion presently available for the classification of a given species in a scale of increasing evolutionary stages" (p. 375). Since the progression indices of all other structures are very low compared to the neocortex, they further conclude that "the uncommonly large neocortex of the human represents indubitably the morphological substrate for the very high and complex functional capacity of his central nervous system" (Stephan & Andy, 1969, p. 376).

The same concept of progression indices was formulated independently as the encephalization quotient (EQ) by Harry Jerison (1973), though with the explicit purpose of serving as an indicator of intelligence both in human evolution and across primate and nonprimate species. Over time, the EQ became widely accepted as a standard for comparing species, with the expectation that it served as a better proxy for cognitive capacity than absolute brain size (for instance, Marino, 1998; Sol, Duncan, Blackburn, Cassey, & Lefebvre, 2005), as this excess brain mass, in Jerison's view, should be available for functions other than those related to bodily demands. This expectation, however, was not founded on correlation with actual measures of behavioral capacity, but rather on the fact that for four decades, it was only in EQ, and not in absolute or relative brain size, that the human species stood out in comparison to all others (Herculano-Houzel, 2011; Marino, 1998). A recent analysis based on measures of behavioral capacity indicated that, among primates, simple brain size is a much better correlate of general cognitive abilities than EQ (Deaner, Isler, Burkart, & van Schaik, 2007).

The mid-1900s saw many other attempts at comparative studies of brain volumes (e.g. Count, 1947; Finlay & Darlington, 1995; Hofman, 1985; Haug, 1987; Rockel, Hiorns, & Powell, 1980; Tower, 1954; Zhang & Sejnowski, 2000). However, those were based on motley crews of species as different as ferret, cow, opossum, elephant, insectivore, and human, all put together in one package, with no respect for phylogenetic relationships, and actually with the unstated but clear assumption that all mammalian brains scaled in the same way (Count, 1947; Haug, 1987). In this respect, the comparative analysis of brain volumes initiated by Stephan's group had the enormous advantage of being systematic and clade-specific. The group organized their wealth of data on volumes of brain structures of 76 mammalian species (28 insectivores, 21 prosimians, and 27 simians) into tables made available to all scientists for comparison and analysis (Stephan et al., 1981a). The large dataset fulfilled its purpose time and again, as more and more independent groups used their data on brain volumes in insectivores, primates, and also chiropterans (Stephan et al., 1981a,1981b) for analysis. One of the most influential external analyses of their dataset was published by Barbara Finlay and Richard Darlington (1995) and suggested a regular pattern of linked changes in volume across brain structures in evolution, which they attributed to a highly conserved order of neurogenesis across mammals, in correlation with the relative enlargement of structures as brain size increases (see Chapter 13). However, later analyses of the same dataset by different groups also revealed strong evidence of mosaic evolution, in which structures with major anatomical and functional links evolve together independently of evolutionary changes in other structures (Barton & Harvey, 2000).

2.5 Comparative Brain Mapping: Wally Welker's School of Cortical Cartography

In its 19th-century origins, neuroscience had known comparative studies of various mammalian brains by the hands of anatomists such as Franz Gall, François Leuret, and Louis Pierre Gratiolet, who mapped the folds and fissures of the cerebral cortex, and physiologists like David Ferrier, who used electrical stimulation to map the motor cortex in primates and dogs.

In the 20th century, however, neurobiology and the study of its relationship to behavior were largely confined to a single, domesticated species—the rat—while the occasionally studied locusts and pigeons were dismissed as not "real animals" in the case of the former, or an evolutionary dead-end in the case of the latter (Zeigler, 2011, p. 1). However, comparative neuroanatomy and neurophysiology was to gain a new impulse in the 1960s and 1970s, when Wally Welker put the newly available microelectrodes to use to map the cerebral cortex and other brain structures of different mammalian species.

By systematically studying the somatosensory system of as many mammalian species as he could, Welker showed that behavioral specializations across species correlate with differentially enlarged sensory representations of the behaviorally important appendages and other body parts throughout the sensory pathway leading to the apparently distorted "raccunculus," "hyraxunculus," "llamunculus," or "simiunculus" representation of the body in the somatosensory cortex (Welker & Campos, 1963; Welker & Carlson, 1976; Welker et al., 1976).

Besides the concept of a direct relationship between functional neuroanatomy and behavior, Welker's legacy also includes the brains of the dozens of species he collected and processed and which have since been the subject of investigations by many independent researchers. These brains are still available for study in the Comparative Mammalian Brain Collection (www.brainmuseum.org).

Wally Welker also made history by influencing several other neuroscientists interested in the evolution of the nervous system. One of these was Jon Kaas, who overlapped with Welker in Clinton Woolsey's lab at the University of Wisconsin. Moving further the combined neuroanatomical and physiological approach, Kaas developed the flattened preparation of the cerebral cortex (lovingly nicknamed the "roadkill preparation") which allowed the visualization of the cortex as a whole, including functional areas otherwise buried within sulci and the analysis of their characteristics, such as borders, distribution of neurochemical markers, and patterns of connectivity as well as their comparison across species (Gould & Kaas, 1981). A similar flattened preparation was later developed for the analysis of images obtained with PET and MRI (reviewed in Fischl, Sereno, & Dale, 1999). Kaas's former students Kenneth Catania and Leah Krubitzer continue to perpetuate the field of comparative functional neuroanatomy through their studies of specialized sensory systems such as the tactile appendages of the star-nosed mole (Catania, 1995) and of the evolution of cortical areas (Krubitzer, 2000).

2.6 The Human's Place in Nature: All Brains Are Not Made the Same

In regard to the human brain, much of comparative and evolutionary neuroscience has been based on two contradictory assumptions. On the one hand, it has been assumed that all brains, including the human brain, are built with the same basic cellular constitution, with a similar relationship between brain size, number of neurons, and neuronal density (the inverse of average neuronal size) that would not have changed over the course of mammalian brain evolution. Along the same lines, influential models of mammalian brain evolution have been built on assumptions of cortical uniformity across regions and species (e.g., Rockel et al., 1980), with cortical expansion occuring through the lateral addition of modules sharing identical numbers of cells (e.g., Rakic, 1988) and with a constant fraction of cortical neurons connected through the white matter (e.g., Zhang & Sejnowski, 2000).

On the other hand, it has been tacitly agreed that evolutionary rules, while applying to every other species, might not apply to humans. Hence the frequency with which the human brain has been considered "special," an outlier: in having the largest brain size relative to what it "should" have for the human body mass (Jerison, 1973; Marino, 1998), in having an unusually large prefrontal cortex (Smaers et al., 2011), or in having a particular type of spindle-shaped neuron in its cerebral cortex (which many others species are now known to share; Nimchinsky et al., 1999).

Harry Jerison's concept of encephalization, which put humans on top—at last—created a trend in the field of adjusting brain size (and densities, and glia/neuron ratios, and cognitive measurements, and everything else) for body mass. Absolute values became more and more disregarded, as if body mass were a determinant variable on which even the most basic aspects of brain morphology and function depended, and to which they should be normalized. Similarly, the relative size of the cerebral cortex within the brain became a standard for comparisons across species, as if it provided a proxy for the relative functional importance of the structure. This emphasis creates paradoxes, such as expecting very large and small cerebral cortices with similar relative sizes to have the same "relative importance" across species. Recently, a meta-analysis showed that relative sizes are not meaningful, at least amongst primates: the best correlate of cognitive ability across non-human primates is absolute brain size, not encephalization or relative size of the cerebral cortex (Deaner et al., 2007).

Our recent finding that not all brains are made the same, with different relationships between brain size and number of neurons across mammalian orders, has provided a new conceptual basis for comparative neuroanatomy: One that regards brain size as a consequence of developmental programs that generate brains according to different scaling rules across clades (Herculano-Houzel, 2011; Herculano-Houzel, Manger, & Kaas, 2014). Our systematic analysis of the numbers of neurons and other cells that compose the brain of dozens of mammalian species has shown, for instance, that primate brains hold many more neurons than rodent brains of a similar size—and the human brain is no outlier, having the number of neurons that is expected for a generic primate of its brain size (Herculano-Houzel, 2012). We thus view the human brain as remarkable, yet not extraordinary, with notable cognitive abilities that can be attributed to the enormous number of neurons in its cerebral cortex, regardless of its relative size, its degree of encephalization, or the relative volume of the prefrontal area (Barton & Venditti, 2013; Semendeferi et al., 2001)—a number of neurons that is shared by no other animal, primate or otherwise, possibly due to a change in diet that allowed our direct ancestors to circumvent the metabolic limitation imposed by a raw foods diet (Fonseca-Azevedo & Herculano-Houzel, 2012). This new framework paves the way to studies of the genetic and nongenetic mechanisms that relate numbers of neurons to average neuronal cell size, and to the changes in those mechanisms that generate diversity in evolution, all the while constrained by a set of scaling rules (Herculano-Houzel et al., 2014).

2.7 Conclusions and Perspectives

The evolution of evolutionary neuroscience is a story as much of scientific achievements as it is of conceptual revolutions: from *scala naturae* to branched evolution, from homologies as evidence of conserved development to evidence of independent use of the same genes in different evolutionary branches, from veiled or explicit anthropocentrism to the analysis of humans as just another species. While some extrapolations have been informative, neuroscientists have learned that some are not.

Vertebrate and invertebrate brains may seem equally tripartite, but may not be homologous, just like bird and mammalian brain structures may not be homologous, despite expressing similar genes; body size may correlate with brain size, but probably does not determine it; and the human brain is not an enlarged mouse brain, although it is, in many senses, an enlarged primate brain. While there is still a lot to be learned from comparing turtle, frog, chicken, opossum, mouse, monkey, and human brains, modern comparative neuroanatomy has yet to shed the *scala naturae* bias of progression and embrace evolution more systematically: not as the means to the human brain, but as the way to diversity.

Acknowledgments

The author is grateful to Jon Kaas for ten years of informal guidance in the field of neuroanatomy. Supported by CNPq, FAPERJ, MCT/INCT and the James S. McDonnell Foundation.

References

Barton R. A., & Harvey, P. H. (2000). Mosaic evolution of brain structure in mammals. *Nature*, 405, 1055–1058.

Barton, R. A., & Venditti, C. (2013). Human frontal lobes are not relatively large. *Proceedings of the National Academy of Science of the USA*, 110, 9001–9006.

Bunnin, N. & Yu, J. (Eds.) (2004). *Blackwell Dictionary of Western Philosophy*. Oxford: Blackwell.

Carroll, R. L. (1988). *Vertebrate paleontology and evolution*. New York, NY: Walther Freeman & Co.

Catania, K. C. (1995). Magnified cortex in star-nosed moles. *Nature*, 375, 453–454.

Count, E. W. (1947). Brain and body weight in man: Their antecedents in growth and evolution. *Annals of the New York Academy of Sciences*, 46, 993–1122.

Cuvier, G. (1801). *Leçons d'anatomie comparée* (Ed. C. Dumeril). Retrieved from https://archive.org/details/leonsdanatomiec00dumgoog

Darwin, C. (1859). *On the origin of species by means of natural selection*. John Murray, London.

Deaner, R. O., Isler, K., Burkart, J., & van Schaik, C. (2007). Overall brain size, and not encephalization quotient, best predicts cognitive ability across non-human primates. *Brain Behavior and Evolution*, 70, 115–124.

de Robertis, E. M., & Sasai, Y. (1996). A common plan for dorsoventral patterning in Bilateria. *Nature*, 380, 37–40.

Dugas-Ford, J., Rowell, J. J., & Ragsdale, C. W. (2012). Cell-type homologies and the origins of the neocortex. *Proceedings of the National Academy of Sciences of the USA*, 109, 16974–16979.

Edinger, L. (1908). The relations of comparative anatomy to comparative psychology. *Journal of Comparative Neurology and Psychology*, 18, 437–457.

Erwin, D. H. (2009). Early origin of the bilaterian developmental toolkit. *Philosophical Transactions of the Royal Society B Biological Sciences*, 364, 2253–2261.

Evans, S. E. (2000). Contribution to "General Discussion." In G. R. Bock & G. Cardew (Eds.), *Evolutionary Developmental Biology of the Cerebral Cortex* (pp. 109–113). John Wiley & Sons, Chichester.

Finlay, B. L., & Darlington, R. B. (1995). Linked regularities in the development and evolution of mammalian brains. *Science* 268, 1578–1584.

Fischl, B., Sereno, M. I., & Dale, A. M. (1999). Cortical surface-based analysis. *NeuroImage*, 9, 195–207.

Fonseca-Azevedo, K., & Herculano-Houzel, S. (2012). Metabolic constraint imposes tradeoff between body size and number of brain neurons in human evolution. *Proceedings of the National Academy of Sciences of the USA*, 109, 18571–18576.

Frahm, H. D., Stephan, H., & Stephan, M. (1982). Comparison of brain structure volumes in Insectivora and Primates. I. Neocortex. *Journal für Hirnforschung*, 23, 375–389.

Gans, C., & Northcutt, R. G. (1983). Neural crest and the origins of vertebrates: A new head. *Science*, 220, 268–273.

Garstang, W. (1922). The theory of recapitulation: A critical restatement of the biogenic law. *Journal of the Linnaean Society of London, Zoology*, 35, 81–101.

Geoffroy Saint-Hilaire, E. (1830). *Principes de philosophie zoologique, discutée en mars 1830, au sein de l'Académie Royale des Sciences* [Principles of zoological philosophy, discussed in March 1830 at the Royal Academy of Sciences]. Paris: Pichon et Didier.

Gerhart, J. (2000). Inversion of the chordate body axis: Are there alternatives? *Proceedings of the National Academy of Science of the USA*, 97, 4445–4448.

Gould, S. J. (1977). *Ontogeny and phylogeny*. Cambridge, MA: Harvard University Press.

Gould, S. J. (1989). *Wonderful life. The Burgess shale and the nature of history*. New York, NY: W. W. Norton & Co.

Gould, H. J., & Kaas, J. H. (1981). The distribution of commissural terminations in somatosensory areas I and II of the grey squirrel. *Journal of Comparative Neurology*, 196, 489–504.

Haeckel, E. (1866). *Generelle Morphologie der Organismen* [General Morphology of Organisms]. Berlin: Georg Reimer.

Halanych, K. M., Bacheller, J., Liva, S., Aguinaldo, A. A., Hillis, D. M., & Lake, J. A. (1995). 18S rDNA evidence that the lophophorates are protostome animals. *Science*, 267, 1641–1643.

Halder, G., Callaerts, P., & Gehring, W. J. (1995). Induction of ectopic eyes by targeted expression of the eyeless gene in Drosophila. *Science*, 267, 1788–1792.

Haug, H. (1987). Brain sizes, surfaces, and neuronal sizes of the cortex cerebri: A stereological investigation of man and his variability and a comparison with some mammals (primates, whales, marsupials, insectivores, and one elephant). *American Journal of Anatomy*, 180, 126–142.

Harris, W. A. (1997). Pax-6: Where to be conserved is not conservative. *Proceedings of the National Academy of Sciences of the USA*, 94, 2098–2100.

Herculano-Houzel, S. (2011). Brains matter, bodies maybe not: the case for examining neuron numbers irrespective of body size. *Annals of the New York Academy of Sciences*, 1225, 191–199.

Herculano-Houzel, S. (2012). The remarkable, yet not extraordinary, human brain as a scaled-up primate brain and its associated cost. *Proceedings of the National Academy of Sciences of the USA*, 109, 10661–10668.

Herculano-Houzel, S., Manger, P. R., & Kaas, J. H. (2014). Brain scaling in mammalian brain evolution as a consequence of concerted and mosaic changes in numbers of neurons and average neuronal cell size. *Frontiers in Neuroanatomy*, 8, 77.

Hervé, P., Lartillot, N., & Brinkmann, H. (2005). Multigene analyses of bilaterian animals corroborate the monophyly of ecdysozoa, lophotrochozoa, and protostomia. *Molecular Biology and Evolution*, 22, 1246–1253.

Hirth, F. (2010). On the origin and evolution of the tripartite brain. *Brain Behavior and Evolution*, 76, 3–10.

Hirth, F., Kammermeier, L., Frei, E., Waldorf, U., Noll, M., & Reichert, H. (2003). An urbilaterian origin of the tripartite brain: Developmental insights from Drosophila. *Development*, 130, 2365–2373.

Hirth, F., & Reichert, H. (2007). Basic nervous system types: One or many? In J. H. Kaas (Ed.), *Evolution of nervous systems* (Vol. 1, pp. 55–72). Amsterdam: Academic Press.

Hodos, W., & Campbell, C. B. G. (1969). Scala naturae: Why there is no theory in comparative psychology. *Psychological Review*, 76, 337–350.

Hofman, M. A. (1985). Size and shape of the cerebral cortex in mammals. 1. The cortical surface. *Brain Behavior and Evolution*, 27, 28–40.

Huxley, J. S. (1932). *Problems of relative growth*. London: Allen & Unwin.

Jarvis, E. D., Güntürkün, O., Bruce, L. L., Csillag, A., Karten, H., Kuenzel, W., ... Butler, A. (2005). Avian brains and a new understanding of vertebrate brain evolution. *Nature Reviews Neuroscience*, 6, 151–159.

Jenner, R. A. (2004). When molecules and morphology clash: Reconciling conflicting phylogenies of the Metazoa by considering secondary character loss. *Evolution and Development*, 6, 372–378.

Jerison, H. J. (1955). Brain to body ratios and the evolution of intelligence. *Science*, 121, 447–449.

Jerison, H. J. (1973). *Evolution of the brain and intelligence*. New York, NY: Academic Press.

Kappers, C. A., Huber, C. G., & Crosby, E. C. (1936). *Comparative anatomy of the nervous system of vertebrates, including man*. New York, NY: Hafner.

Karten, H. J. (1969). The organization of the avian telencephalon and some speculations on the phylogeny of the amniot telencephalon. In J. Petras & C. Noback (Eds.), *Comparative and evolutionary aspects of the vertebrate nervous system* (pp. 164–179). New York, NY: New York Academy of Sciences.

Karten, H. J. (1991). Homology and evolutionary origins of the "neocortex." *Brain Behavior and Evolution*, 38, 264–272.

Krubitzer, L. A. (2000). How does evolution build a complex brain? *Novartis Foundation Symposium*, 228, 206–200.

Lovejoy, A. O. (1964). *The great chain of being: A study of the history of an idea*. Cambridge, MA: Harvard University Press.

MacLean, P. D. (1964). Man and his animal brains. *Modern Medicine*, 2, 95–106.

MacLean, P. D. (1990). *The triune brain in evolution: Role in paleocerebral functions*. New York, NY: Plenum Press.

Marino, L. (1998). A comparison of encephalization between odontocete cetaceans and anthropoid primates. *Brain, Behavior and Evolution*, 51, 230–238.

Martin, R. P. (1990). *Primate origins and evolution: A phylogenetic reconstruction*. Princeton, NJ: Princeton University Press.

Martin, R. D., & Harvey, P. (1985). Brain size allometry: Ontogeny and phylogeny. In W. Jungers (Ed.), *Size and scaling in primate biology*. New York, NY: Plenum Press.

Moroz L. L. (2009). On the independent origins of complex brains and neurons. *Brain Behavior and Evolution*, 74, 177–190.

Nimchinsky, E. A., Gilissen, E., Allman, J. M., Perl, D. P., Erwin, J. M., & Hof, P. R. (1999). A neuronal morphologic type unique to humans and great apes. *Proceedings of the National Academy of Sciences of the USA*, 96, 5268–5273.

Northcutt, R. G. (2010). Cladistic analysis reveals brainless Urbilateria. *Brain Behavior and Evolution*, 76, 1–2.

Pani, A. M., Mullarkey, E. E., Aronomicz, J., Assimacopoulos, S., Grove, E. A., & Lowe, C.J. (2012). Ancient deuterostome origins of vertebrate brain signalling centres. *Nature*, 483, 289–294.

Peterson, K. J., Lyons, J. B., Nowak, K. S., Takacs, C. M., Wargo, M. J., & McPeek, M. A. (2004). Estimating metazoan divergence times with a molecular clock. *Proceedings of the National Academy of Sciences of the USA*, 101, 6536–6541.

Puelles, L., & Rubinstein, J. L. R. (2003). Forebrain gene expression domains and the evolving prosomeric model. *Trends in Neuroscience*, 26, 469–476.

Puelles, L., Kuwana, E., Puelles, E., Bulfone, A., Shimamura, K., Keleher, J., ... Rubenstein, J. L. (2000). Pallial and subpallial derivatives in the embryonic chick and mouse telencephalon, traced by the expression of the genes Dlx-2, Emx-1, Nkx-2.1, Pax-6 and Tbr-1. *Journal of Comparative Neurology*, 424, 409–438.

Rakic, P. (1988). Specification of cerebral cortical areas. *Science*, 241, 170–176.

Reiner, A., Medina, L., & Veenman, C. L. (1998). Structural and functional evolution of the basal ganglia in vertebrates. *Brain Research Reviews*, 28, 235–285.

Reiner, A., Perkel, D. J., Bruce, L. L., Butler, A. B., Csillag, A., Kuenzel, W., ... Jarvis, E. D. (2004a). The avian brain nomenclature forum: Terminology for a new century in comparative neuroanatomy. *Journal of Comparative Neurology*, 473, E1–E6.

Reiner, A., Perkel, D. J., Bruce, L. L., Butler, A. B., Csillag, A., Kuenzel, W., ... Jarvis, E. D. (2004b). Revised nomenclature for avian telenchephalon and some related brainstem nuclei. *Journal of Comparative Neurology*, 473, 377–414.

Rensch, B. (1960). *Evolution above the species level*. New York, NY: Columbia University Press.

Riska, B., & Atchley, W. R. (1985). Genetics of growth predicts patterns of brain-size evolution. *Science*, 229, 668–671.

Rockel, A. J., Hiorns, R. W., & Powell, T. P. (1980). The basic uniformity in structure of the neocortex. *Brain*, 103, 221–244.

Ryan, T. J., & Grant, S. G. N (2009). The origin and evolution of synapses. *Nature Reviews Neuroscience*, 10, 701–712.

Sagan, C. (1977). *The dragons of Eden*. New York, NY: Random House.

Semendeferi, K., Lu, A., Schenker, N., & Damasio, H. (2002). Humans and great apes share a large frontal cortex. *Nature Neuroscience*, 5, 272–276.

Smaers, J. B., Steele, J., Case, C. R., Cowper, A., Amunts, K., & Zilles, K. (2011). Primate prefrontal cortex evolution: Human brains are the extreme of a lateralized ape trend. *Brain Behavior and Evolution*, 77, 67–78.

Sol, D., Duncan, R. P., Blackburn, T. M., Cassey, P., & Lefebvre, L. (2005). Big brains, enhanced cognition, and response of birds to novel environments. *Proceedings of the National Academy of Sciences of the USA*, 102, 5460–5465.

Stephan, H., & Andy, O. J. (1969). Quantitative comparative anatomy of primates: An attempt at a phylogenetic interpretation. *Annals of the New York Academy of Sciences*, 167, 370–386.

Stephan, H., Frahm, H., & Baron, G. (1981a). New and revised data on volumes of brain structures in insectivores and primates. *Folia Primatologica*, 35, 1–29.

Stephan, H., Nelson, J. E., & Frahm, H. D. (1981b). Brain size comparison in Chiroptera. *Journal of Zoological Systematics and Evolutionary Research*, 19, 195–222.

Striedter, G. F. (2009). History of ideas on brain evolution. In J. H. Kaas (Ed.), *Evolutionary neuroscience*, (Vol. 1). New York, NY: Associated Press.

Tower, D. B. (1954). Structural and functional organization of mammalian cerebral cortex: The correlation of neurone density with brain size. Cortical neurone density in the fin whale (Balaenoptera physalus L.). with a note on the cortical neurone density in the Indian elephant. *Journal of Comparative Neurology*, 101, 19–51.

von Bonin, G. (1937). Brain weight and body weight in mammals. *Journal of General Psychology*, 16, 379–389.

von Haller, A. (1762). *Elementa physiologiae corporis humani* [Elements of the physiology of the human body] (Vol. 4). Lausanne: Grasset.

Welker, W. I., & Campos, G. B. (1963). Physiological significance of sulci in somatic sensory cerebral cortex in mammals of the family procyonidae. *Journal of Comparative Neurology*, 120, 19–36.

Welker, W. I., & Carlson, M. (1976). Somatic sensory cortex of hyrax (Procavia). *Brain Behavior and Evolution*, 13, 294–301.

Welker, W. I., Adrian, H. O., Lifshitz, W., Kaulen, R., Caviedes, E., & Gutman, W. (1976). Somatic sensory cortex of llama (Lama glama). *Brain Behavior and Evolution*, 13, 284–293.

Zeigler, H. P. (2011). Wally Welker and neurobehavioral evolution: An appreciation and bibliography. *Annals of the New York Academy of Sciences*, 1225, 1–13.

Zhang, K., & Sejnowski, T. J. (2000). A universal scaling law between gray matter and white matter of cerebral cortex. *Proceedings of the National Academy of Sciences of the USA*, 97, 5621–5626.

3

Approaches to the Study of Brain Evolution

Jon H. Kaas

3.1 Introduction

The variability in structure and function across the brains of present-day vertebrates ranges from very simple nervous systems to an astonishing variety of complex forms that allow great differences in behavioral specializations and range (see Kaas, 2007). Understanding how these different nervous systems evolved from those of ancient ancestors both addresses a natural curiosity and helps us understand the relationship of brains to behavior. Such understanding may change the way we think about our place in the biological world and how we plan our future. This review considers some of the strategies we use to study brain evolution. Some of us are especially concerned about the evolution of the large and very impressive human brain, and the question of human origins has intrigued humans across cultures from at least the emergence of modern humans. While this question can be a focus for some, it is important to have a broader understanding of nervous systems, as comparisons reveal general principles, and provide another way of evaluating theories of brain organization and function for any extant species, including our own. In reality, much of what we know about human brains, for example, comes from comparative studies of other primates and, more broadly, other animals, together with assumptions about the significance of shared features and how human brains evolved. As we readily make such assumptions to support the validity of various lines of research—the current emphasis on mouse brains, for example—a broad understanding of brain evolution can be very helpful in guiding interpretations of results. For practical reasons, this review focuses on ways of studying the evolution of brains in mammals. There is no obvious starting point, but a reasonable start is with the brains of the first mammals and their fossil kin. The methods of study discussed here can be applied to other lines of evolution.

3.2 The Structure of the Mammalian Radiation

Studies of brain evolution largely depend on comparisons of traits observed in the brains of extant animals and on inferences made about brains from skull endocasts from the fossil record. For such studies, it is very important to have an understanding of the phylogeny of the group of interest, mammals in this case, that is reliable.

Approaches to the Study of Brain Evolution 39

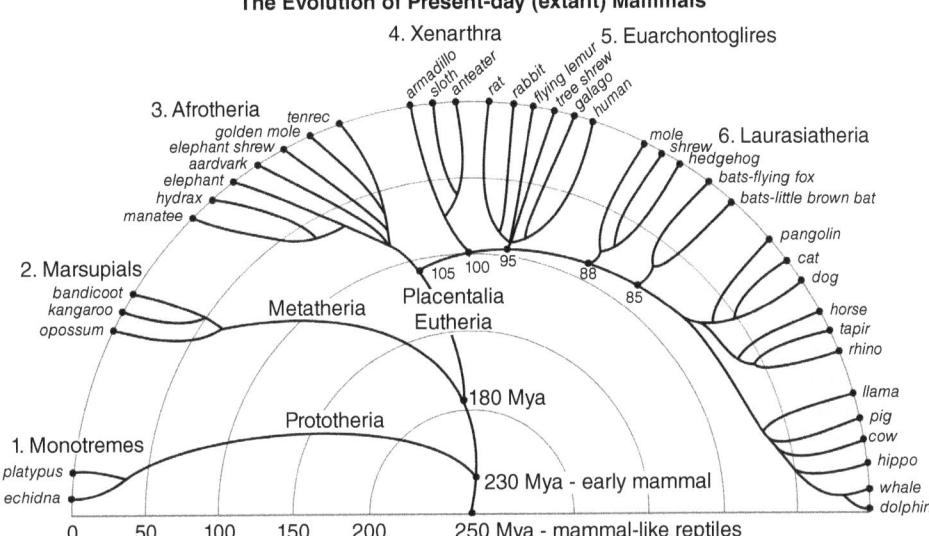

Figure 3.1 The Phyletic Radiation of Extant Mammals.
The evidence supports the division of current mammals into six major clades or superorders. The times of divergence are estimates based on the fossil record and molecular differences. Based on Murphy et al., 2004.

Here we are fortunate in that the use of comparative molecular data has produced powerful evidence and considerable agreement on phylogenetic relationships, especially for mammals (e.g., Bininda-Emonds et al., 2007; Murphy, Pevzner, & O'Brien, 2004). In addition, the fossil record has continued to improve, and has added critical time points to the understood molecular evidence for diversification and radiation.

All present-day mammals appear to have evolved from a single source, with subsequent offspring differing from parents in various ways, gradually accumulating differences and ultimately producing the great mammalian radiation with six major branches and many sub-branches (see Figure 3.1), leading to over 250 extant species of primates alone, as well as the thousands of species of the especially successful bat and rodent radiations. Traditional concepts of how mammals are related to each other via common ancestors at various times in the past were inferred from similarities and differences in body anatomy, often from bones as these were usually all that was preserved in extinct species. But there are long-recognized problems with this approach as members of different lines of evolution may come to resemble each other as they adapt to similar environments (convergent evolution), and separate lines of evolution may continue to resemble each other as they retain ancestral features or specialize in similar ways (parallel evolution). For example, some investigators incorrectly concluded that members of one of the two major branches of the bat radiation (megabats or fruit bats) were really primates because of proposed similarities and not at all closely related to other bats, the microbats or echolocating bats, even though this meant that microbats and megabats would have evolved wings separately (see Kaas & Preuss, 1993 for discussion). As another example of confusion arising from existence of similar traits, it has been extremely difficult to determine the evolutionary relationships of a number of small, insect-eating mammals because they closely resemble each other.

As early mammals were small and insect-eating, many of these shared morphological traits were retained or reacquired in these separate groups of mammals that we now know are not closely related, although they had been grouped as "insectivores."

Although studies of anatomical features in extant species and fossils have provided much valuable information about evolutionary trees, modern phylogenetic trees are largely based on genetic similarities and differences stemming from studies of genetic material and gene products. As for the use of anatomical features, the more gene products considered in a study the better, for genes change (mutate) at different rates and are under variable selection pressure. In general, the phylogenetic trees based on genetic relationships are considered to be much more accurate than those based on a more limited range of anatomical features, while the times of branch points of different lines of evolution can be determined by referring to fossil evidence that can be dated and then estimating rates of genetic change. Recently, the oldest known skeleton of a eutherian mammal has been dated at about 125 million years old (Ji et al., 2002). Obviously, the origin of any species would predate the oldest discovered fossil.

The present consensus is that present-day mammals all stem from an early mammalian ancestor some 250–280 million years ago (mya). Over time, mammals evolved to form six major surviving branches, the rare monotremes (platypus and echidna) that retained the primitive trait of egg-laying, the marsupials (roughly 6–7 percent of extant species) with primitive, short gestation periods, and four branches of eutherian mammals with placentas and prolonged gestation. Within each of these major branches or clades, there have been many subsequent branches. According to modern revisions, microbats and megabats remain together, but in Laurasiatheria rather than in Euarchontoglires with primates. Shrews, hedgehogs, and moles remain as members of the original order Insectivora, now renamed Eulipotyphia in the Laurasiatherian Superorder, while tenrecs and golden moles are in Afrosoricida of a different Superorder, Afrotheria. Tree shrews (Scandentia) and elephant shrews (Macroscelidea), once thought to be closely related to insectivores, are now in separate superorders (Euarchontoglires and Afrotheria).

This revised phylogeny with times of origin is essential to accurately reconstructing the evolution of mammalian brains using a comparative, cladistic approach. A clade is any group of organisms that have all originated from a common ancestor. The common ancestor defines the clade. Thus, a clade is one branch of the evolutionary tree of any size. All mammals form a clade, and all primates form a smaller clade. The Insectivores of former phylogenetic trees do not form a clade. Likewise, fish or reptiles are useful terms, but they do not denote a clade, as all members based on a common ancestor are not included. Fish do not include all the vertebrates that evolved from fish, and reptiles do not include the birds that evolved from reptiles. Correctly identifying members of a clade is important because comparisons across members of clades are used to infer those traits that the common ancestor passed on to the extant members of the clade. For example, a variable feature of brains, the corpus callosum, is found in all eutherian (placental) mammals, but not in any noneutherian mammals (monotremes and marsupials). Thus, we infer that the corpus callosum emerged as a feature or character of the common ancestor of all eutherian mammals, but not before the branching of the marsupials from the eutherians. A misclassification of some marsupials, such as the marsupial mole, as eutherian would confound the process of identifying ancestral traits, as did the recent claim that megabats are primates, a claim that has now been rebutted by molecular (genetic) evidence.

3.3 What We Learn from the Fossil Record

Fossils are useful as they provide an opportunity to look directly into the past. What we learn about the evolution of mammalian brains from the fossils of extinct species is limited by the fact that the soft tissues of the body, especially the brain, seldom fossilize. Fortunately, brains of mammals fill the skull rather tightly, so that the skull cavity reflects the size and shape of the brain and, for larger brains, even the locations of some of the major fissures of the brain may be revealed, and such fissures may mark functional divisions of the brain (Kaas, 2009; Radinsky, 1976; Ronan et al., 2014; Welker, 1990; Zilles, Palomeros, Gallagher, & Amunts, 2013).

From the endocasts and reconstruction of the shapes of brains from the inner surface of brain cavities of fossilized skulls, we have a good overall concept of how brains changed in size and shape from early mammals over 150 million years ago to modern humans and other extant mammals (de Sousa & Wood, 2007; Jerison, 1973, 2007; Kielan-Jaworowska et al., 2004). Early mammals were typically small and they had small brains, even for their small body size. Brains, like other body organs, increase in size with body size, but not at the same rate (Jerison, 1973; 2007). Thus, they tend to occupy proportionately less of the total body mass as bodies get bigger. Some mammals have larger brains than expected for body size, and this feature is called encephalization. The empirical quotient of the deviation from expected brain size based on body size is the encephalization quotient, with higher values suggesting greater brain capacity. However, at least in primates, body size is not a good predictor of brain size, as some primates have evolved smaller body sizes without a corresponding decrease in brain size, while other species have much larger males than females without a corresponding difference in brain size (Herculano-Houzel & Kaas, 2011). The forebrain of early mammals was dominated by proportionately large olfactory structures, the olfactory bulb and the olfactory, piriform cortex. A small cap of neocortex was separated from piriform cortex by a shallow rhinal sulcus (dimple), which sometimes could be seen in the skull endocasts (Kielan-Jaworoska, Cifelli, & Luo). There was little room in this small cap for much visual, auditory, and somatosensory cortex, so olfaction was obviously the most important sense. Orbit morphology indicated that eyes were small and vision was less important in these early, apparently nocturnal mammals. Teeth were suitable for grasping small, mainly insectivorous prey, and herbivory was probably limited. The middle ear bones were separate from the jaw mandible in early mammals, allowing high-frequency hearing, and thereby social communication in ranges beyond the hearing of reptilian predators (Allman, 1999). Brain sizes did not increase much until the extinction of dinosaurs about 65 mya (Smith et al., 2010), although the major branches of the mammalian radiation were established before that time (Ji et al., 2002; Murphy, Pevzner, & O'Brien, 2004). After the extinction of dinosaurs and many other animals, many lines of evolution led to bigger mammals with larger brains that were more varied in shape and had more fissures. In early primates, the occipital and temporal visual regions of cortex were enlarged, denoting an increased role for vision. Eye sockets were large, suggesting an adaptation for nocturnal vision. Later, anthropoid monkeys emerged with reduced eye sockets, indicating that they were diurnal, the temporal lobe expanded even further, and the snout was reduced to allow for better frontal vision. Apes became distinct from monkeys at least 30 million years ago, modern apes have larger brains with some asymmetries in the shape and length of the lateral fissure, suggesting that there was some degree of specialization of each cerebral hemisphere. Brains increased rapidly in size over the last two million years in the hominin line leading to modern humans (de Sousa & Wood, 2007), which have enlarged parietal, temporal, and prefrontal regions of cortex. Further, asymmetries in fissure patterns of the two hemispheres suggesting modifications in cortical organization in the hominin line that predated those reflecting the acquisition of language.

Other skeletal features can tell us indirectly about brain functions and behavior in our close, extinct hominin relatives. Dental micro-wear as well as modifications in teeth provided much information about diet (Scott et al., 2005; Ungar & Sponheimer, 2011), and carbon isotope data can indicate the type of plants that were eaten. In tool-using Neanderthals, knife-cut angles on teeth made while separating bites of meat demonstrated that most were right-handed (Frayer et al., 2012). The boney canal for the hypoglossal nerve that supplies the muscles of the tongue are large in early modern humans and in Neanderthals, suggesting that both had vocal capabilities compatible with language (Kay, Cartmill, & Balow, 1998). These are just some of the inferences about the brain and neural mechanisms for behavior that are possible from the fossil record.

We can learn even more from the fossil record as we come to understand better the meaning of brain fissures that can be found in endocasts. It has been long known that in the somatosensory cortex of some carnivores, including extant species and in the skull endocasts of extinct species (Radinsky, 1976; 1977), that fissures mark the boundaries between the cortical representation of some body parts, such as those of the forepaw and face (Welker, 1990). We also know that, at least in some mammals, there are relatively few intrinsic connections between forelimb and face representations (Kaas, 2009). Information on cortical connections led Van Essen (2007) to develop a theory that postulates that the connections within and between cortical areas create tension within the expanding cortex during development that causes cortex to buckle between highly interconnected regions, producing gyri, while fissures emerge between poorly interconnected regions. Thus, the poorly interconnected face and hand region of primary somatosensory cortex is sometimes marked by a fold. The important conclusion, however, is that fissure patterns can tell us something about the functional organization of the cortex of the brains of extinct mammals (Radinsky, 1976).

3.4 Deducing Brain Evolution from the Comparative Studies of the Brains of Extant Mammals

Modern mammals are not necessarily completely modern. As the mammalian radiation progressed from a few early mammals to the great number of species and individuals today, some parts or features of some of their brains were likely to have been retained from various ancestors, while others were likely to have been modified or were relatively new. The method for distinguishing features of brains that are new from old, and reconstructing the probable features of brains of ever more distant relatives of any extant species is a comparative approach that depends on a cladistic analysis of the characters (features) of the brains of a suitable range of extant species. The general approach was outlined by Hennig (1966) in which he sought to better reveal phylogenetic relationships and better define clades by considering across compared species the presence or absence of as many characters (traits or features) as possible. A character can be any observable feature. For the brain, this could be the corpus callosum or the lamination pattern of the dorsal lateral geniculate nucleus. However, not every character accurately reveals evolutionary relationships. For example, electroreception systems have evolved independently several times. The more characters that support a specific conclusion, the better. But, as discussed previously in this review, the many comparisons that are possible at the molecular and DNA levels have led to greatly revised and presumably highly accurate phylogenetic trees, and an analysis of brain characters is unlikely to add to present frameworks. Instead, we reverse the goals of the comparative process, and assume that phylogenetic relationships are known, and use this information to interpret the meanings of similarities and differences in brain organization. By assuming that the brains of any two extant species will be more alike the more closely they are related, and that similarities likely reflect the retention of a character from a common ancestor, the features of the brains of recent to ever more distant common ancestors can be reconstructed by determining which features are distributed across members of a small clade with a recent common ancestor to ever larger clades with ever more distant ancestors. We could, for example, consider those features of brains that are shared by all hominins as evidence that these features were present in the first hominins as they diverged from apes. (This effort would be seriously limited by the fact that modern humans are the only extant hominin.) Or we could more productively consider all anthropoid primates, all primates, all eutherian mammals, or all mammals as progressively larger clades with earlier and earlier origins. Rules for best reconstructing ancestral traits have emerged (e.g., Brooks & McLennan, 1991; Cunningham, Omland, & Oakley, 1998) based on maximal parsimony criteria and optimization procedures. But, in practice, the cladistic comparative approach can be applied productively in a less formal way.

One problem with implementing a full cladistic approach is that many brain features can be difficult to identify without extensive investigation, and researchers can still disagree after years of investigation. For example, it is only recently that most investigators have come to the agreement that all primates have a third visual area known as V3 (Lyon & Kaas, 2001). A related problem is that it can be quite costly in terms of manpower, equipment, and supplies when experiments that include microelectrode recording and histological processing are needed to identify the presence or absence of a brain character. For example, it is both difficult and costly to determine if the retinotopic representation of the contralateral eye in the superior colliculus is of the complete retina, as in most mammals, or of the nasal hemi-retina, as in primates (Lane, Allman, Kaas, & Miezin, 1973). In addition, it may be difficult or impossible to examine some or many numbers of a clade, as they may be extinct, endangered and rare, protected, or unavailable in some other way. For example, tarsiers form the sister taxon of anthropoid primates, and it is therefore important to know as much as possible about tarsier brains, but tarsiers are an endangered species and unavailable for experimental study. Thus, we are limited to postmortem histological studies of brain organization (Wong, Collins, & Kaas, 2010). For such reasons, we need to be practical and choose carefully from the species available for study those that will provide the most valuable information (Kaas, 2002). While, in principle, detailed studies of the brains of any of the members of the six major clades of the mammalian radiation could provide useful information about the organization of the brains of early mammals, studies of some extant mammals are more likely to provide the most useful information. As the brains of the first mammals were small with little neocortex, it seems reasonable to focus on members of the clades with small brains and little neocortex, such as opossums, armadillos, tenrecs, hedgehogs, and rats (Kaas, 2011), as the brains of elephants, humans, and whales have obviously changed greatly from the ancestral form. In the same manner, the brains of prosimian galagos are more likely to resemble those of early primates than the brains of chimpanzees or humans. In some ways, this truncated approach seems similar to the early efforts of Eliot Smith (Smith, 1910) and Le Gros Clark (Clark, 1959) as they used a sequence of extant primates and other mammals at progressively "higher levels" of brain differentiation to represent the evolutionary history of the human brain (Preuss, 2000). Thus a potential weakness of selecting favorable species for studies of brain evolution is that derived characters can be mistaken for primitive features when they are in a generally more primitive brain. But such incorrect inferences can still be avoided by extending the scope of the comparison to additional species in order to obtain further evidence.

3.5 Understandings of Brain Evolution Based on Developmental Patterns and Biological Constraints

Another approach to understanding the ways in which brains evolve is to determine the constraints that occur during brain development on the formation of new phenotypes. As one example, it has been possible to demonstrate that the brains of rodents (Herculano-Houzel, Mota, & Lent, 2006) and primates (Herculano-Houzel, Collins, Wong, & Kaas, 2007) gain neurons in proportion to other cells (mainly glia) in different ways as species with increasingly larger brains are considered. Neuron numbers increase at a lower rate than brain size in rodents, while neurons get bigger and other cells remain the same size. Thus, neuronal densities decrease and glia-to-neuron ratios increase with brain size in the rodent clade. In contrast, neuron numbers increase almost isometrically with brain size in primates, as neurons do not increase in average size and the neuron-to-glia cell ratio remains relatively stable. The apparent validity of this scaling rule for primates (or for rodents) allows one to make inferences or predictions about the numbers of neurons in any primate brain, once the size of that brain is known. Thus, it was possible to state with some confidence the numbers of neurons in the brains of extinct hominins (Herculano-Houzel & Kaas, 2011). However, it is yet unclear why one scaling rule applies to rodents, and probably most mammals, and another to primates.

As another example of rules that apply to brain development and evolution, Finlay and Darlington (1995) were able to present evidence that the proportions of brains of most mammals change in orderly and predictable ways as they range from small to large. In general, larger brains have proportionately more neocortex and less brain stem. This observation suggested to Finlay and Darlington that the parts of the brain that mature later in development, such as neocortex, become proportionately larger because they continue to grow over longer developmental times (late makes great).

Finally, there are changing design problems and solutions as brains range in size (Kaas, 2000). Mammals evolved from a small common ancestor, and the overall tendency has been to evolve larger bodies with larger brains (Baker, Meade, Pagel, & Venditti, 2015). Neurons communicate with each other over axonal connections and synapses. Larger brains have more neurons and longer distances between different parts of the brain than smaller brains. As species with larger brains evolve, neurons in those larger brains would have more neurons with which to make connections, and connections would be over longer distances, unless the ways neurons interconnect are modified. Long connections are especially costly if the speed of computations is to be maintained, as distance is time in the nervous system. To speed up conduction times in longer axons and dendrites, they need to be thicker (Bekkers & Stevens, 1970). Thicker and longer axons and dendrites take up more space. In addition, if neurons are to maintain connections with a fixed proportion of the total neural population, axons would need more branches, and this would also take more space. Thus, maintaining a fixed pattern of connectional and areal organization would not work very well as brains became bigger. Brains would devote more and more of their mass to connections, and they would gain fewer and fewer neurons and less and less computational power with each increase in brain size. To some variable extent, larger brains may be less efficient than smaller brains. But also, there would be selection pressure for solutions to the larger brain problem. The number of long connections can be minimized by keeping most connections local. This would mean that larger brains should have more subdivisions, the cortical areas and subcortical nuclei in which neurons can interact over short distances. Large areas with longer intrinsic connection distances would be few, and small areas would dominate. In addition, very large areas would be subdivided into modules where neurons interconnect with each other, but not with other classes of modules within the area. Functionally related areas that need to be densely interconnected would be adjacent or close. Long, fast connections between the cerebral hemispheres via thick axons would be few. Thus, functionally related networks would emerge in each cerebral hemisphere, and communication between the two hemispheres would be reduced (Ringo, Doty, Demeter, & Simard, 1994). Long connections between distant structures would still be important, but they would be few. The large human brain seems to have been modified in ways to reduce the connection problem. Systems for language and other functions are concentrated in a single hemisphere. The corpus callosum in humans has proportionally fewer axons than expected from a fixed proportion relative to cortical size, and few of its axons are thick and fast conducting. In addition, humans have many cortical areas, perhaps as many as 200 or more, and their large primary areas such as V1 have not increased in size over the course of hominin evolution, although the brain size has increased threefold (Kaas & Preuss, 2014).

In contrast, some mammals appear to be smaller than their ancestors and have smaller brains, which presents a different problem: do you reduce cortical areas and subcortical nuclei in proportion to brain size, giving up resolution, or lose some structures to allow others to maintain a functionally significant size? The tiny brain of one of the smallest of mammals, the masked shrew, near the lower size limit for mammals, appears to have fewer cortical areas than any other well-studied mammal, while preserving primary sensory areas (Catania, Lyon, Mock, & Kaas, 1999). Thus, mammals that revert to a smaller body and brain size may adapt by losing some cortical areas in order to maintain others at a large size.

3.6 Studies of Brain Development

Comparative studies of how brains develop can reveal mechanisms of phylogenetic change. Phylogenetic differences are generated by changes in aspects of brain development (Molnár et al., 2014). Such studies are indicating the ways in which the six-layered neocortex of mammals may have been generated, and how larger brains with more neocortex get more neurons and spread over a larger surface area, while suggesting ways that more cortical areas and subcortical nuclei may have emerged. Such studies also reveal important differences in genes and gene expression that lead to phenotypic specializations. They can also lend support for some theories of brain evolution over others. Finlay and Darlington (1995) provide an example of how large sets of empirical observations on how the proportions of brain parts relative to brain size can be explained in part by an overall model of how brain development relates to brain size. Thus, there is great value in comparing sequences of brain development as well as mature brains within and across clades.

References

Allman, J. (1999). *Evolving brains.* New York, NY: W.H. Freeman & Co.
Baker, J., Meade, A., Pagel, M., & Venditti C. (2015). Adaptive evolution toward larger size in mammals. *Proceedings of the National Academy of Sciences of the USA*, 112, 5093–5098.
Bekkers, J. M., & Stevens, C. F. (1970). Two different ways evolution makes neurons larger. *Progress in Brain Research*, 83, 37–45.
Bininda-Edmonds, O. R. P., Cardillo, M., Jones, K. E., MacPhee, R. D. E., Beck, R. M. D., Grenyer, R., ... Purvis, A. (2007). The delayed rise of present-day mammals. *Nature*, 446, 507–512.
Brooks, D. R., & McLennan, D. A. (1991). *Phylogeny, ecology, and behavior.* Chicago, IL: University of Chicago Press.
Catania, K. C., Lyon, D. C., Mock, O. B., & Kaas, J. H. (1999). Cortical organization in shrews: Evidence from five species. *Journal of Comparative Neurology*, 410, 55–72.
Clark, W. E. L. G. (1959). *The antecedents of man.* Edinburgh: Edinburgh University Press.
Cunningham, C. W., Omland, K. E., & Oakley, T. H. (1998). Reconstructing ancestral character states: A critical reappraisal. *Trends in Ecology & Evolution*, 13, 361–366.
de Sousa, A., & Wood, B. (2007). The hominin fossil record and the emergence of the modern human central nervous system. In T. M. Preuss & J. H. Kaas (Eds.), *The evolution of primate nervous systems* (pp. 291–336). London: Elsevier.
Finlay B. L., & Darlington, R. B. (1995). Linked regularities in the development and evolution of mammalian brains. *Science*, 2681, 1578–1584.
Frayer, D. W., Lozano, M., Castro, B., Carbonell. E., Arsuaga, J. L., Radovcic, J., ... Bondioli, L. (2012). More than 500,000 years of right-handedness in Europe. *Laterality* 17, 51–69.
Hennig, W. (1966). *Phylogenetic systematics.* Urbana, IL: University of Illinois Press.
Herculano-Houzel, S., Collins, C. E., Wong, P., & Kaas, J. H. (2007). Cellular scaling rules for primate brains. *Proceedings of the National Academy of Sciences of the USA*, 104, 3562–3567.
Herculano-Houzel, S., Kaas, J. H. (2011). Gorilla and orangutan brains conform to the primate cellular scaling rules: Implications for human evolution. *Brain Behavior and Evolution*, 77, 33–44.
Herculano-Houzel, S., Mota, B., & Lent, R. (2006). Cellular scaling rules for rodent brains. *Proceedings of the National Academy of Sciences of the USA*, 103, 12138–12143.
Jerison, H. J. (1973). *Evolution of the brain and intelligence.* New York: Academic Press.

Jerison, H. J. (2007). What fossils tell us about the evolution of the neocortex. In J. H. Kaas & L. A. Krubitzer (Eds.), *Evolution of nervous systems* (Vol. 3). *Mammals* (pp. 1–12). London: Elsevier.

Ji, Q., Luo, Z. X., Yuan, C. X., Wible, J. R., Zhang, J. P., & Georgi, J. A. (2002). The earliest known eutherian mammal. *Nature* 416(6883), 816–822.

Kaas, J. H. (2000). Why is brain size so important: Design problems and solutions as neocortex gets bigger or smaller. *Brain and Mind* 1, 7–23.

Kaas, J. H. (2002). Convergences in the modular and areal organization of the forebrain of mammals: implications for the reconstruction of forebrain evolution. *Brain Behavior and Evolution*, 59, 262–272.

Kaas, J. H. (ed.) (2007). *Evolution of nervous systems: A comprehensive reference*. London: Elsevier.

Kaas, J. H. (2009. Cerebral fissure patterns. In L. R. Squire (Ed.), *Encyclopedia of neuroscience* (3rd ed, pp. 793–800). Oxford: Elsevier.

Kaas, J. H. (2011). Reconstructing the areal organization of the neocortex of the first mammmals. *Brain Behavior and Evolution*, 78, 7–21.

Kaas, J. H., & Preuss, T. M. (1993). Archontan affinities as reflected in the visual system. In F. Szalay, M. Novacek, M. McKenna (Eds.), *Mammal phylogeny* (pp. 115–128). New York, NY: Springer-Verlag.

Kaas, J. H., & Preuss, T. M. (2014). Human brain evolution. In L. R. Squire (Ed.), *Encyclopedia of neuroscience* (3rd ed, pp. 901–918). Oxford: Elsevier.

Kay, R. F., Cartmill, M., & Balow, M. (1998). The hypoglossal canal and the origin of human vocal behavior. *Proceedings of the National Academy of Sciences of the USA*, 95, 5417–5419.

Kielan-Jaworowska, Z., Cifelli, R. L., & Luo, Z.-X. (2004). *Mammals from the age of dinosaurs*. New York, NY: Columbia University Press.

Lane, R. H., Allman, J. M., Kaas, J. H., & Miezin, F. M. (1973). The visuotopic organization of the superior colliculus of the owl monkey (Aotus trivirgatus) and the bush baby (Galago senegalensis). *Brain Research*, 60(2), 335–349.

Lyon, D. C., & Kaas, J. H. (2001). Connectional and architectonic evidence for dorsal and ventral V3, and dorsomedial area in marmoset monkeys. *Journal of Neuroscience*, 21(1), 249–261.

Molnár, Z., Kaas, J. H., deCarlos, J. A., Hevner, R. F., Lein, E., Němec, P. (2014). Evolution and development of the mammalian cerebral cortex. *Brain Behavior and Evolution*, 83(2), 126–139.

Murphy, W. J., Pevzner, P. A., & O'Brien, J. O. (2004). Mammalian phylogenomics comes of age. *Trends in Genetics*, 20(12), 631–639.

Preuss, T. M. (2000). Taking the measure of diversity: comparative alternatives to the model-animal paradigm in cortical neuroscience. *Brain Behavior and Evolution*, 55, 287–299.

Radinsky, L. (1976). Cerebral clues. *Natural History*, 85, 54–59.

Radinsky, L. (1977). Brains of early carnivores. *Paleobiology*, 3, 333–349.

Ringo, J. L., Doty, R. W., Demeter, S., & Simard, P. Y. (1994). Time is of the essence: A conjecture that hemispheric specialization arises from interhemispheric conduction delay. *Cerebral Cortex*, 4, 331–343.

Ronan, L., Voets, N., Rue, C., Alexander-Bloch, A., Hough, M., Mackey, C., … Fletcher, P. C. (2014). Differential tangential expansion as a mechanism for cortical gyrification. *Cerebral Cortex*, 24, 2219–2228.

Scott, R. S., Ungar, P. S., Bergstrom, T. S., Brown, C. A., Grime, F. E., Teaford, M. F., & Walker, A. (2005). Dental microwear texture analysis shows within-species diet variability in fossil hominins. *Nature*, 436, 693–695.

Smith, F. A., Boyer, A. G., Brown, J. H., Costa, D. P., Dayan, T., Ernest, S. K. M., … Uhen, M. D. (2010). The evolution of maximum body size of terrestrial mammals. *Science*, 330, 12216–12219.

Smith, G. E. (1910). Some problems relating to the evolution of the brain. Lecture II. *The Lancet*, 88, 147–153.

Ungar, P. S., & Sponheimer, M. (2011). The diets of early hominins. *Science*, 334, 190–193.

Van Essen, D. C. (2007). Cerebral cortical folding patterns in primates: Why they vary and what they signify. In J. H. Kaas & T. M. Preuss (Eds.), *Evolution of nervous systems* (Vol. 4). *Primates* (pp. 267–276). London: Elsevier.

Welker, W. I. (1990). Why does cerebral cortex fissure and fold? In E. G. Jones & A. Peters (Eds.), *Cerebral cortex*. New York: Plenum.

Wong, P., Collins, C. E., & Kaas, J. H. (2010). Overview of sensory systems of Tarsius. *International Journal of Primatology*, 31, 1002–1031.

Zilles, K., Palomeros-Gallagher, N., & Amunts, K. (2013). Development of cortical folding during evolution and ontogeny. *Trends in Neuroscience*, 36, 275–284.

4

Intraneuronal Computation

Charting the Signaling Pathways of the Neuron

Jorge Navarro, Raquel del Moral, and Pedro C. Marijuán

4.1 Introduction: The Centrality of Intracellular Signaling

A bewildering variety of signaling components and pathways co-occur in neurons—more than in any other living tissue. During the 1970s and 1980s, perhaps the "golden age" of molecular neurobiology, the sheer variety of receptors, channels, transducers, protein kinases, and many other components discovered in neurons was seen as an "impenetrable jungle" (Crick, 1979). Neural tissue appeared special in many ways (Allman, 1999; John & Miklos, 1988): It consumes up to ten times as much metabolic energy as average, keeps the largest number of genes in an active state of expression, synthesizes the highest total amount of protein, and contains the highest number of specialized cell types—more than 100 in humans, over half our total, despite taking up merely 2% our body mass. What was so special about the evolutionary function of the neuron? Why were the molecular mechanisms involved so complex and so expensive? The pioneering discussions in that "golden age" about the function of the neuron and the origins of nervous systems may count among the most elegant debates in the modern history of neurobiology (Bullock & Horridge, 1965; Horridge, 1968; Mackie, 1990; Pavans de Ceccatty, 1974).

While it can be asked of any tissue how its specialized function is necessary to the multicellular organism, this question is particularly cogent in the extreme case of neuronal complexity. Roughly speaking, the tissular function of neurons deals with receiving signals via neurotransmitters and other sources, maintaining a transmembrane electrical potential, transporting the electrical disturbances of the membrane potential along dendrites and axonal prolongations, and maintaining a vast population of actualized synaptic contacts. In engineering terms, this implies a complex molecular management of both an extended electrical-computational network and a distributed memory system. However, the molecular mechanisms and subsystems implementing these functions have been, and continue to be, highly elusive. When the concept of a cellular *signaling system* was established in early 1990s (Cohen, 1992; Egan & Weinberg, 1993), it became clear that the enormous variety of membrane and cytoplasmic neuronal mechanisms discovered during the two previous decades were caught

---

The Wiley Handbook of Evolutionary Neuroscience, First Edition. Edited by Stephen V. Shepherd.
© 2017 John Wiley & Sons, Ltd. Published 2017 by John Wiley & Sons, Ltd.

up in that category. Moreover, *signaling* is the appropriate label to characterize the special function and complexity of neuronal tissues, as they express more (and more diverse) signaling elements than any other eukaryotic tissue or cellular specialization.

The complexity of signaling pathways expressed in neurons did not arise from scratch. A good portion of neurons' signaling system was directly inherited from prokaryotes, but new additions were incorporated through *bricolage*, cobbled together through complex controlling apparatuses foreign to prokaryotes. Functionally speaking, however, the relative simplicity attributed to prokaryotic cells is only apparent, at least as far as their signaling capabilities are concerned. As will be discussed later, prokaryotic cells have only three main classes of "component-system" arrangements for signaling purposes, but they are instantiated in about *one or two hundred* different pathways for each cell, acting as mostly independent channels for the introduction and processing of external information. In contrast, the cells of complex eukaryotes, such as vertebrates and mammals, are endowed with several dozen major classes of component-system arrangements (main signaling pathways), but these comprise *thousands* of specific molecular implementations in different tissues, particularly within the nervous system. Amidst all that complexity, however, there is a deep evolutionary coherence in the way signaling pathways operate in eukaryote neurons. Making evolutionary sense of this diversity will be the goal of this chapter.

Genomic studies of transition from prokaryotes to eukaryotes and the origins of synaptic contacts are providing important new insights into the origins of nervous systems themselves (Miller, 2009) and into the evolution of advanced signaling control, with a good portion traceable to humble bacterial ancestors (Aravind, Anantharaman, & Iyer, 2003; Aravind, Iyer, & Koonin, 2006). In what follows, we will not depart from the basic idea that there is an *informational continuum* of signaling tools and strategies, and that any attempt to explain neuronal complexity must connect with the evolutionary trajectory of signaling systems and with the origins of nervous systems as one of the obligate, earliest inventions of eukaryotic multicellularity. A synthetic path must be established among a forest of intricate discoveries in systems biology, signaling science, synthetic biology, genomics, proteomics, molecular biology, evolutionary neurobiology, neurophysiology, and so on.

Two distributed *memory systems* were invented by metazoa: an immune system recognizing "molecular configurations" and a nervous system recognizing "sensorimotor configurations." Both adapt the organism to an open-ended and fast-changing environment, and they maintain a continuous dialog between them (Arnsten, 2009; Steinman, 2004). It is intriguing that a good portion of their informational organization and molecular signaling cascades is still unknown. This brief chapter is a small step in bringing them out of the shadows: It is meant to serve as a synopsis of the basic eukaryotic pathways deployed in neurons and as an (admittedly speculative!) rationale for their evolution.

4.2 How Signaling Resources Evolved in the Transition from Prokaryotic to Eukaryotic

An information revolution took place in cellular systems around 1200 million years ago. It was preceded and made possible by an energy revolution derived from the symbiotic capture of mitochondria, as Lynn Margulis so forcefully argued in her theory of endosymbiosis (Margulis, 1970). The data are staggering: An average protozoan has nearly 5000 times more metabolic power than a single bacterium, supporting a genome several thousand times larger, with over two orders of magnitude more energy devoted to the expression and translation of each gene (Lane & Martin, 2010). Whereas prokaryotes had already made a start towards eukaryotic-style cellular complexity, they could not exhibit more than one complex trait at a time, given the implicit energy costs. Novel protein foldings, protein interactions, and regulatory cascades were required to integrate the segregated traits already explored in bacteria—innovations including a separate nucleus, dynamic cytoskeleton, endocytosis, linear chromosomes, introns and exons, massive intracellular and intercellular signaling, etc. The increase in protein repertoire by the "last eukaryotic common ancestor" (LECA) was dramatic: It represented some 3,000 novel gene families—the most intense phase of gene invention since the origin of life: "If evolution works like a tinkerer, evolution with mitochondria works like a corps of engineers" (Lane & Martin, 2010, p. 933).

A heavy investment in signaling resources was necessary in order to produce a new kind of life cycle amenable to controlled dissociation or *modularization* amidst newly complex internal and external happenstances. Most cellular functions had to change from a temporal context to a spatial one, tightly controlled by signals: While some functions were delayed or directly suppressed, others became augmented and specialized (Nedelcu & Michod, 2004). The decoupling of cell division from genuine cell reproduction, organizing successive levels of differentiation potency during development, was one of the central achievements. The ability to govern cellular reproduction permissively or suppressively depending on signaling contexts made possible the advent of true multicellularity (Davidson, 2006; 2010).

Four main resources were used in the expansion of eukaryotic signaling systems, four "roots" that supported the fast elaboration of new eukaryote capabilities. Two were directly taken from existing prokaryotic stock:

- Prokaryotic signaling pathways devoted to *detection of solutes*, comprising receptors, protein kinases, phosphatases, and regulated transcription factors.
- Prokaryotic apparatuses for *solvent sensing* (which maintain osmotic homeostasis by counteracting the Donnan effect) comprising stretch-gated ion channels, voltage-gated ion channels, ligand-gated channels, water transporters, and pumps.

Another two signaling avenues were related to new cellular subsystems supporting the enlarged eukaryotic complexity. Co-opted for signaling purposes, these control apparatuses modularized the deployment of cellular functions and became facultative components of the emergent complex network of signaling pathways:

- The cell-cycle control system, comprising hierarchies of protein kinases, checkpoints, cyclins, and protein degradation systems.
- The cytoskeleton and endocytic matrix, providing mechanical support, adhesion, and force-field detection on the one side; compartments, inner transportation, and vesicle formation on the other.

Lynchpins of prokaryotic metabolism provided key substances previously involved in detecting the energetic state of the cell (cAMP, cGMP) and in the synthesis and integrity of membrane systems (IP3, DAG, arachidonic acid, ceramide acid), as well as the key ionic effector, Ca^{++}. They were reused inside the eukaryotic signaling pathways as *second messengers* or "symbolic molecules" to amplify the information flow, conveying integrative messages by diffusing through localized regions of a far bigger cell, in connection to diverse membrane systems, compartments, and inner transportation mechanisms.

Before entering into analysis of these different signaling resources, either segregated or nonexistent in prokaryotic physiology, we must address the fundamental question: What genetic strategy made possible the assembly of eukaryotic signaling systems?

4.2.1 Recombination Is a Central Theme of Signaling System Evolution

The accelerated evolution of the new signaling systems was fundamentally based on protein-domain recombination. The signaling innovations of eukaryotic cells, crucial to nervous systems, were not due to any of these previous "roots" in isolation, but rather to their integration into larger systems of cross-linked pathways. Osmotic tools (i.e., ion channels) were liberally interwoven with protein receptors for solute detection, with hierarchical chains of protein kinases and with endosomal pathways for protein recycling culminating in ubiquitylation and degradation. The postsynaptic processing of the neurotransmitter glutamate is one of the best examples of how these heterogeneous signaling resources were recombined (as we will analyze later). The multidomain structure of most involved enzymes, proteins, channels, and receptors, in addition to the flexibility of their binding properties, made possible the evolution of interconnected pathways (Pawson & Nash, 2003). In particular, the arrangement of signaling components into integrated *scaffolds*, themselves subject to domain recombination, was a high-level way to exploit both specificity and convergence in pathways.

Scaffold proteins avoid the entropic cost of diverse signaling molecules finding specific partners in a solution; they also afford easy regulation by external signals which modify the association of proteins along the scaffold, and so offer a simple, flexible strategy for regulating the selectivity of individual pathways, shaping signaling regimes and achieving new responses from preexisting components (Good, Zalatan, & Lim, 2011). Being themselves modular, composed of multiple interaction domains and tandem repeat proteins assembled through recombination, scaffolds provided an elegant evolutionary solution to the prokaryotic conflict between signaling specificity and the efficient coordination of information flow through intracellular networks. From an evolutionary perspective, scaffolds were a crucial functional element supporting signal-cascade recombination and the integration of eukaryotic signaling systems.

However, basic aspects of early eukaryotic evolution remain poorly understood. Phylogenomic reconstructions show that the characteristic complexity of eukaryotes, both in structure and signaling pathways, arose without any apparent intermediate grades of complexity between the widely separated levels of organization (Koonin, 2010). Multicellularity in general, and nervous systems specifically, evolved quickly by redeploying and recombining the communication tools of unicellular eukaryotes such as yeast. Thus, most components regulating structural plasticity in the synaptic pathways of mammals have analogous roles in unicellular responses to environmental cues (ions, nutrients, repellents) and in pheromonal communication among unicellular eukaryotes (Emes et al., 2008). In point of fact, one of the major evolutionary challenges would be explaining the signaling system of the LECA. How did this system evolve, and how much of modern signaling systems are conserved from this common ancestor?

As new experimental studies and bioinformatic analyses of protein domain architecture have suggested, a number of bacterial and eukaryotic signaling proteins share similar mechanisms (Aravind et al., 2003, 2006; Koonin, 2010). In fact, the evidence suggests that laterally transferred domains of prokaryotic provenance (bacterial and archaeal) have contributed to the evolution of important sensory pathways related to stretch, light, nitric oxide, and redox signaling, as well as to central developmental pathways such as Notch, cytokine, and cytokinin signaling. The phagotrophic lifestyle of the LECA and of primitive eukaryotes could have served as a conduit for such lateral transfers, as well as for the endosymbiotic origins of mitochondria and chloroplasts themselves (Aravind et al., 2003, 2006). In terms of mechanisms (protein domains and architectures) the parallels between prokaryotic and eukaryotic signaling systems are startling, but it is in the flexible arrangement of functional modules and the intercombination of previously segregated systems where the major expansions of eukaryotic signaling complexity have occurred.

Of the four evolutionary forces acting on genes (mutation, recombination, selection, and drift), *domain recombination* seems to have been the prevailing creative force in eukaryotic signaling (Alm, Huang, & Arkin, 2006). Additional factors, such as intron–exon gene organization, differential splicing, and whole genome duplication (at least two instances) provided eukaryotes with far greater resources than prokaryotes to explore and leverage domain recombination. Taking into account that more than 45% of the eukaryotic genome is an accretion of residual transposons and retrotransposons, both ancient and recent, eukaryotes have explored quite a number of potential domain-recombination events as well as multiple opportunities to fine-tune their genomic directionality and control (Lander et al., 2001). The remarkable expansion of behavioral and cerebral complexity achieved by vertebrates is inseparable from the increasing sophistication achieved by their signaling systems through these evolutionary processes, augmented by the conservation and expansion of non-coding *cis* DNA sequences devoted to fine-tuning of gene control (Carroll, 2005; Davidson, 2010; Lynch et al., 2011). In the case of mammalian brains, recombination strategies seem to have been incorporated even deeper—in ontogenetic development. Recombination via retrotransposons and mobile elements that are active during cerebral development has become a new way to generate additional molecular complexity and *mosaicism*, particularly in the signaling capabilities of cortical and hippocampus´ neurons (Baillie et al., 2011; Kang et al., 2011; Lander et al., 2001).

We now examine in more detail the main resources and avenues—the "four evolutionary roots"—that eukaryotic signaling systems have put together, either through direct prokaryotic heredity or through functional co-option and modularization.

4.3 Four Evolutionary Roots of Eukaryotic Signaling Systems

4.3.1 The Prokaryotic "Detection of Solutes"

The detection of external substances to be processed as signals (rather than as metabolic substrates) occurs in almost all prokaryotic cells. Indeed, it may well be considered the first evolutionary resource in the assemblage of eukaryotic signaling systems. In prokaryotes, a variety of molecular systems are involved in the detection of solutes, ranging from simple transcription-sensory regulators (a single protein comprising two domains), such as the well-known *embR*, *alkA* or *furB*, to systems of several components and to interconnected pathways that regulate key stages of the cell cycle, such as latency, pathogenesis, replication, and dispersion. We have proposed a basic taxonomy of bacterial signaling systems centering on "the 1-2-3 scheme" (Marijuán, Navarro, & del Moral, 2010) (see Figure 4.1.):

- The first level of signaling complexity corresponds to simple regulators, the "one-component systems" (1CS). Actually, most cellular proteins involved in cellular adaptation to changing environments, in a general sense, could be included as participants in this primary category (Galperin, 2005). Around one hundred different 1CS elements may be present in a moderately complex prokaryotic cell.
- Increasing the scale of complexity, the "two-component systems" (2CS) appear, which include a *histidine kinase* protein receptor and an independent *response regulator*. Conventionally, they are considered the central paradigm for prokaryotic signaling systems, and a number of intercellular communication processes between different prokaryotic and eukaryotic species are carried out by these specialized systems. A few dozen 2CS pathways may coexist in a prokaryote.
- To maintain conceptual coherence, an additional category, the "three-component system" (3CS) should apply to two-component systems that employ an additional non-kinase receptor to activate a protein kinase. Very few prokaryotic pathways

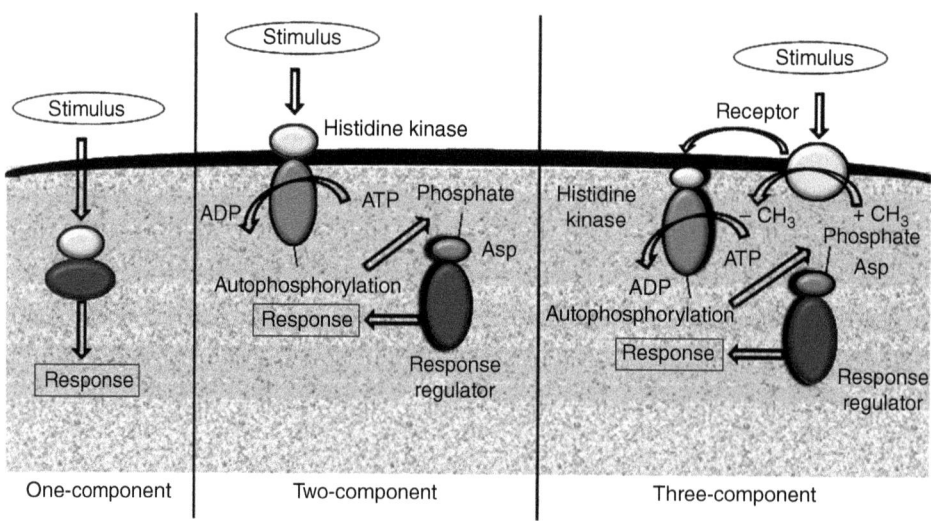

Figure 4.1 The Three Characteristic Signaling Pathways Developed by Prokaryotes.
The external stimulus is perceived either by an internal receptor–transducer (left part), or by a transmembrane histidine kinase that connects with a response regulator (middle), or by an independent receptor associated to the histidine kinase (right part). Adapted from Marijuán 2010. Reproduced with permission of Elsevier.

show a 3CS arrangement, but they are very important ones (e.g. chemotaxis) and are usually subject to further regulations, such as the variable methylation of receptors.

The signaling costs and benefits of 1CS, 2CS, and 3CS systems determine their functional deployment. The relative disadvantage of 1CSs stems from the fact that they detect their stimuli almost exclusively in the cytosol (including environmental cues such as light, gases, and other small molecules); afterwards they act on DNA-binding in more than 80% of cases (Grigoroudis, Panagiotidis, Lioliou, Vlassi, & Kyriakidis, 2007; Ulrich, Koonin, & Zhulin, 2005). The evolutionary strategy to overcome the limitations of these single signaling elements has consisted in dividing the individual protein in two, putting one half on the membrane while the other remains a soluble cytosolic regulator, and keeping both linked via a phosphotransfer relay (Ulrich et al., 2005). Two-component systems (2CSs) thus arise as a good evolutionary solution enlarging signaling performance (sensitivity, amplification, adaptability) beyond the capabilities of single cytosolic receptor–transducers (1CS).

However, it must be recognized that the multitude of single proteins acting as 1CSs comprise the best-represented signaling strategy within many bacteria, both pathogenic and free-living. Moreover, 1CSs are the most primitive components endowed with signaling functions, their signaling dynamics are easily circulated among the multiple metabolic and transcriptional circuits, and they may straightforwardly interact with more complex signaling pathways. Thus, 1CS become a fundamental factor in the organization of "bacterial intelligence." In point of fact, the relative predominance of 1CS cytosolic detection versus 2CS extracellular detection has been used to discriminate "introverted" from "extroverted" prokaryotes: those that are focused on controlling the inner metabolic complexity versus those that predominantly monitor the rapidly changing external environment. That ratio is also a way to gauge the extra metabolic complexities incorporated in the life cycle of the prokaryote (Galperin, 2005).

In the transition to eukaryotic signaling, most simple 1CSs were relegated to strictly metabolic functions; however, cytosolic and nuclear detection by single proteins remains the primary intracellular mechanism for steroid and hormone signaling, as well as other cases where specificity and security in molecular recognition become essential. While few 2CSs remain active in eukaryotes (mostly plants and yeasts), they have been massively replaced in animals by serine/threonine and autophosphorylated tyrosine receptors, which are very rare in prokaryotes. The reason is that phosphorylation by histidine protein kinases involves recognition of a 3-dimensional folded surface, and is thus less amenable to recombination than serine/threonine kinases, which additionally recognize unstructured linear motifs (Kiel, Yus, & Serrano, 2010); similar evolutionary flexibility favors tyrosine kinases over 2CSs. Unconventional 2CSs *do* exist in bacteria: ECF proteins (related to "sigma factors") and some eukaryotic-like serine/threonine protein kinases and phosphatases that are used mainly to control cell shape and to manipulate the signaling pathways of eukaryotic host organisms. Interestingly, there seem to be a significant number of 3CSs among the eukaryotic-like serine/threonine protein kinases present in many pathogenic bacteria.

As we will see in the pathways of Table 4.1. (§4.4.2), the machinery of *solute detection*, as adopted by eukaryotes, builds directly off 2CS and 3CS. In particular, ligand binding by transmembrane receptors and ion channels is of outmost importance for information processing in nervous systems.

4.3.2 Counteracting the Donnan Effect: "Sensing the Solvent"

The Donnan (or Donnan-Gibbs) effect refers to the osmotic disequilibrium that a living cell experiences due to the fact that it contains predominantly negative charges (i.e., nucleic and amino acids moieties) separated from the liquid environment by a semipermeable membrane. As a result of the opposing osmotic and ionic influences generated, a series of ionic and solvent exchanges follow with the net result that the membrane swells and finally bursts. To prevent this catastrophe, one of the earliest inventions of living cells was a series of molecular mechanisms actively counteracting the Donnan effect: stretch-activated channels, voltage-gated channels, ionic pumps (Na/K), aquaporins, etc. These mechanisms manipulate ionic/osmotic equilibria to restore the appropriate levels of mechanical stress, electrical potential, and ionic concentration gradients across the cell membrane.

Neurons are the great specialists in this ancestral toolkit, as usual with ample recombination of elements, using it to assemble an information processing system based on the generation and circulation of electrical perturbations along a network of membrane potentials. Ion channels associated with ligand receptor domains serve as fundamental portals for most neurotransmitters, particularly the fast-acting ones (glutamate, acetylcholine, glycine, GABA). In addition, voltage-gated channels of K^+, Na^+, Ca^{++}, and Cl^- classes integrate and propagate electrical perturbations of the membrane potential across neurites and cell bodies. The kinetic response properties of each channel and its state transitions—as well as those of any transient "inactive" states—are carefully fine-tuned by means of differential splicing, so that dozens and dozens of slightly different types are deployed as needed for the different functions. For instance, there are over 80 mammalian genes that encode potassium channel subunits (Alberts et al., 2002), each subject to post-transcriptional and functional modifications, to provide channel types specific to each intraneuronal microdomain. This exquisite control over electrical dynamics along the neuron makes possible sophisticated computations, but is extremely energetically expensive. The well-known Na/K pump, part of the prokaryotic inheritance, is a key element within the osmotic/ionic toolkit maintaining the membrane potential of neurons—and represents the highest metabolic expense of the nervous system, up to 2/3 of the total energy budget of the brain (Alberts et al., 2002).

The fast changes in membrane potential governed by these pathways support cellular excitability in neuronal, muscular, renal, cardiac and epithelial tissues. However, it is not only cellular excitability that emerges from the osmotic toolkit—a new form of sensory detection has also been derived from the osmotic sensors of prokaryotes. The *mechanosensitive* (MS) ion channels initially acted as "emergency valves" in primitive prokaryotic cells, but evolved in multicellulars (animal, fungal, and plant) into an extraordinary range of mechanosensors and transient receptor proteins (TRPs) detecting gravity, growth, blood pressure, muscle tension, sound, thirst, heat, and so on (Kung, Saimi, & Martinac, 1990; Kung, 2005)

In point of fact, the receptors of sensory neurons can be broadly classified as either derived from toolkits for *solute detection* (vision, smell, taste, hormones, pheromones, nutrients, neurotransmitters, neuropeptides) or *solvent detection* (sound, touch, pressure, texture, proprioception, osmolarity, heat, volume, vibration) (Kung et al., 1990). In the first case, receptors associated with G protein pathways are the most frequent sensory detection mechanisms, while in the second case, mechanosensitive channels and TRPs become the central tools. Mechanical channel-based senses differ from other senses in the close association of their receptors with the *membrane lipids*, explaining their higher susceptibility to anesthetics, and with the *cytoskeleton*, explaining their ability to transduce forces acting on cellular structures into electrical currents.

Therefore, in the same way that a number of eukaryotic receptors and pathways derive from the solute detection systems of prokaryotes (the "1-2-3" scheme), many eukaryotic sensory systems, notably the MS channels and TRP subfamilies (TRPV, TRPC, TRPA, TRPP, etc.), derive from homologous MS channels of bacteria and archaea used in solvent detection (Kung, 2005). The impact of these ancient mechanisms is impressive: Most of the electrical signaling toolkit used in neural computations is directly derived from the osmotic toolkit of prokaryotes.

4.3.3 The Interface between Signaling and Cell-Cycle Control

Discussing the role of the prokaryotic cell cycle in signaling demands a change of perspective. Unlike the two previously discussed roots of eukaryotic signaling systems, the cell-cycle machinery cannot be considered as a mere subcomponent or toolkit for neural computation. Rather, the cell cycle acts as the main "consumer" of the intracellular signaling system and as the cellular "chief executive"—it is central in the eukaryotic signaling network, linking sensory signaling pathways to powerful effector systems. The cell-cycle machinery inherited from prokaryotes seems to have projected itself or erupted towards the cell surface, accessing relevant guidance cues from the external environment and, in so doing, to have subordinated many previously independent pathways for solute- and solvent-detection.

The most powerful set of protein kinases in the signaling network is directly associated with mitotic control: the MAPK cascade (MAPKKK, MAPKK, MAPK). Depending on the cellular context, this cascade can be divided into three branches: MAPK/ERK, SAPK/JNK, p38/MAPK. Whatever receptors and transducers happen to be associated with the successive kinase hierarchies of these cascades come to occupy a highly-privileged position in the control of the cell cycle and the life and death decisions of a neuron–and these associations are quite flexible, involving a variety of inputs (see Figure 4.2.). For example, one of the functions attributed to MAPK cascades is to integrate information by crosslinking signaling pathways at appropriate levels of amplification within the kinase hierarchy.

Control of the different phases of the cell cycle (G1, S, G2, M) and their respective transitions is a crucial piece of the modular organization of multicellular organisms. Such population control is ultimately based on a cloud of internal and external signals, usually of opposed signs (activators vs. inhibitors), which carefully regulate the reproduction and differentiation of cells and tissues. These interactions require the cell cycle control system to "erupt" toward the cellular surface in metazoa. Most of the signaling pathways listed in Table 4.1 (§4.4.2) share information with cyclins, MAPK cascades, phosphatases, checkpoints, and other protein complexes involved in the cell-cycle control. Throughout these pathways, multiple *growth factors* and *apoptotic factors* are secreted locally from neighboring cells, from more distant

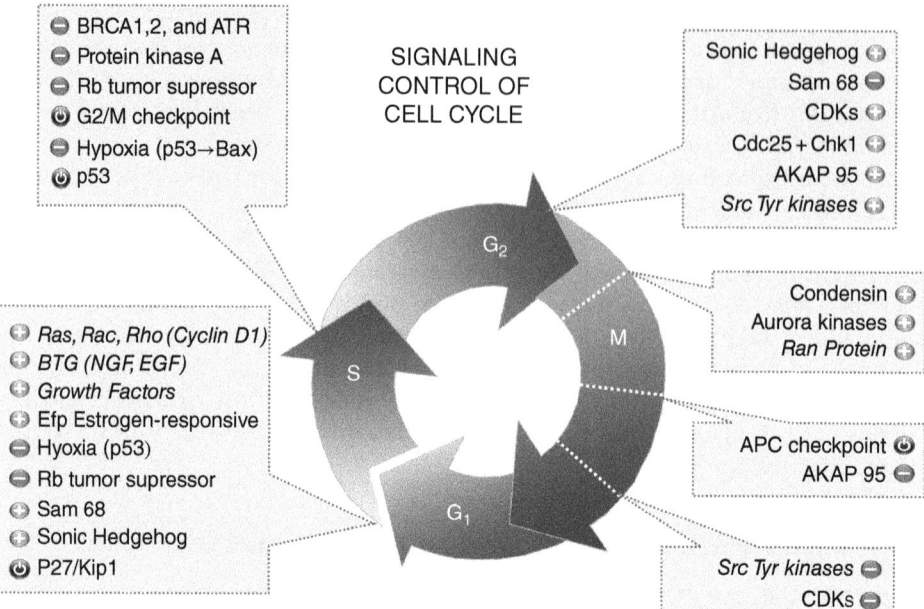

Figure 4.2 Cell-cycle Control.
Transitions between the different phases of the cell cycle (*G1*, gap; *S*, synthesis; *G2*, interphase gap; *M*, mitosis) are tightly controlled. The modular organization of the multicellular organism allows space-time separation between cell cycle phases, mediated by a number of controlling signaling pathways. The signaling control is ultimately based on a cloud of internal and external signals, usually of opposed signs (activators vs. inhibitors), that carefully regulate the reproductive and specialization trajectories of cells and tissues. In the figure, activating signaling pathways promoting cell-cycle progress bear the ⊕ sign, while the inhibiting ones bear the ⊖ sign. The sign ⊙ denotes cellular checkpoints, which can result in progress or arrest depending on incoming factors. The signaling pathways associated with MAP kinases appear in *italics*. Marijuán 2013. Reproduced with permission of Elsevier.

paracrine sites, and through interactions with the extracellular matrix. The balance between growth factors and apoptotic factors determines the developmental patterning of the multicellular organism by propelling cellular growth, eliminating transformed, senescent, or redundant cells, and generally keeping organs and tissues within functional bounds—processes which are especially crucial (and complex) in neural synapses, which behave much like autonomous cellular subunits. Thus, the elimination of decaying synapses, as well their maintenance, growth, and motion along the dendritic arbor, is strongly related to the balance between signaling pathways, including cell-cycle control mechanisms, at the postsynaptic site. These mechanisms play a crucial role in synaptic plasticity and thus in learning, usually by activating or inhibiting synaptic development based on the relative pattern of electrical activity between the dendrite and axon of the postsynaptic neuron.

As a reminder of the symbiotic origins of eukaryotes, the cell-cycle control network mediating irreversible apoptosis is localized to mitochondria—specifically, to the *Bcl-2* protein linked with the integrity of the respiratory chain. This makes a great deal of evolutionary sense, involving mitochondrial compatibility with the host cell as a form of selection for cells and organism with functional respiration (Blackstone & Green, 1999; Lane, 2011). The metabolic centrality of mitochondria makes them an important target for a number of signaling pathways, a crossroads where metabolic state, cell-cycle state, and external signals convene into fundamental "checkpoints" deciding cellular fate. Notwithstanding their metabolic and cellular importance, mitochondria do not directly signal to other cellular subsystems. The surrender of almost all mitochondrial genes to the cell nucleus genome has deprived mitochondria of their own independent downstream signaling—even in the apoptotic case just mentioned, key proteins belong to the host cell.

4.3.4 The Cytoskeleton and Endocytic Matrix: Signaling Incorporation of Mechanical and Membrane-Remodeling Systems

A dynamic cytoskeleton and a dynamic membrane are primordial traits of eukaryotes. Probably they coevolved, as they are functionally interdependent—and presumably both derived from the phagocytic life style of the LECA and early eukaryotes. With their increasing complexity, these two cellular systems have represented important mechanisms for both upstream and downstream extension of the primordial signaling pathways discussed above.

4.3.4.1 Endocytosis and the vesicular trafficking of receptors and messengers. Endocytosis is one of the cellular processes most tightly associated with signaling dynamics; together they might be conceptualized as a single process central to the cellular "master plan" of the eukaryote ("two sides of the same coin", according to Scita & Di Fiore, 2010). The integration of signaling molecules into the vesicle dynamics of membranes is crucial. In a number of pathways, recycling of receptors to and from the plasma membrane by means of endocytic/exocytic cycles serves to regulate receptor density and thus signal sensitivity; AMPA and NMDA glutamate receptors are among the best known instances (§4.5.2). Besides the recycling of receptors, the differentiation of endocytic routes serves as an additional way of regulating signaling activity. For instance, clathrin-mediated endocytosis targets receptors for recycling and signaling continuance, while non-clathrin-mediated endocytosis targets them for degradation (Mills, 2007; Scita & Di Fiore, 2010).

Neurons intensively utilize the properties derived from the endocytic matrix. Through membrane recycling, "signaling endosomes" play a variety of intraneuronal functions. They replenish the cell surface with unbound ligand receptors, resensitize G-protein-coupled receptors, provide scaffolding for ligands and signaling complexes, generate unique signals barred at the plasma membrane, control the trafficking of integrins during cell migration, and convey molecular signals over long distances via microtubule-mediated rapid transport (Scita & Di Fiore, 2010). Additionally, endocytosis and exocytosis mediate intercellular communication through the release of exosomes—microvesicles emitted extracellularly, loaded with hundreds of mRNA and microRNA classes, which can reprogram a cell when taken up (Valadi et al., 2007). It is plausible to speculate that the cerebrospinal fluid serves as a conduit for exosome-mediated extracellular communication.

4.3.4.2 Cytoskeleton: Sensing tissular force fields. Without an active cytoskeleton, endocytosis does not make evolutionary sense, and vice versa. Functionally, they need each other. The eukaryotic cytoskeleton, more than a mere physical scaffold, is a dynamic and adaptive structure whose components and regulatory proteins are in a constant flux, governed by a number of intra- and intercellular signals (Fletcher & Mullins, 2010). The cytoskeleton spatially organizes the contents of the cell, connects the internal milieu with the forces of the external environment, and generates coordinated forces allowing movement and shape change. In much the same way that elaboration of prokaryotic osmotic mechanisms allowed eukaryotes to harness electrical fields and forces, the elaboration of the cytoskeletal mechanisms allowed the harnessing of mechanical forces for functional purposes. The eukaryotic cytoskeleton is a locus for the integration of biochemical signals, genetic programs, and force fields acting on the cell. In multicellulars, cell–cell adhesion and contractility emerge as critical determinants of cell fate, regulating both morphology and motility. Relationships with neighboring cells (through c-adherins) and extracellular matrix (through integrins) are mediated by dedicated signaling pathways which implement a combinatorial strategy to guide complex development using relatively few modular components (Montell, 2008). The convergence of signaling modules and cytoskeletal states allows coherent interactions between intracellular governance and the tissular environment. The mechanical property of *tensegrity* (the dynamic, self-organizing equilibrium based on opposed forces of compression and tension in the cytoskeleton) is another key principle that has been incorporated into both dynamic neuronal function and central nervous system development (Ingber, 1998).

In spite of all this apparent complexity, the incorporation of mechanical factors within signaling is a simplifying, integrative force: The nearly-instantaneous propagation of mechanical force through the cytoskeleton nicely complements the slower diffusive dynamics of chemicals. Thus, diverse neuronal phenomena such as migration, axonal growth, axonal transportation, synaptic generation and displacement, as well as synaptic elimination, are tightly regulated by the cytoskeletal signaling paths and their modular combinations. In a middle ground between osmotic and mechanical functions, *gap junctions* and *tight junctions* provide continuity between the electrical, ionic, and osmotic states of neighboring cells—for example, directly coupling the electrical potential (and physical positioning) of excitatory and inhibitory neurons. Among the cytoskeleton components, the system of microtubules (MT) has been highlighted as one of the fundamental partners in the learning and memory processes, providing far more than mere physical support to the regulation of dendritic spines. In developing neurons, actin and MT systems act together to support growth and differentiation of neurites, and both are of key importance in mature synapses, too, supporting the regulation of spines during learning-associated circuit remodeling (Jaworski et al., 2009). Bold hypotheses suggest further computational roles for MTs, based on their quantum mechanical properties, in processes ranging from unicellular decision-making (Clark, 2010a, 2010b) to the emergence of consciousness itself (Hameroff, 2010; Hameroff & Penrose, 1996).

4.3.4.3 Adaptability of signaling configurations. The networks governing the cytoskeleton and the endocytic matrix highlight the difficulty of establishing rigorous functional distinctions between signaling and non-signaling processes. What is the frontier between signaling and the cytoskeleton, or between signaling and endocytosis? The flexible modularization of gene networks and protein networks suggests frontiers depend on cellular context (Siso-Nadal, Fox, Laporte, Hébert, & Swain, 2009). We have drawn a line, for convenience of discussion, but a permissive view of intracellular signaling networks could be extended further to include proteolysis, protein degradation, and autophagy. The autophagy network is one of the biggest and most complex functional systems of the cell (Behrends, Sowa, Gygi, & Harper, 2010): Ubiquitylation—the addition of ubiquitin by specialized chains of ubiquitin ligases— is recognized as a major system for covalent modification of signaling proteins, not only targeting components for degradation in proteasomes, but in the process shaping signaling, transportation, and cellular remodeling networks. Many other subsystems contribute to signaling: phosphorylation, acetylation, methylation, SUMOylation, glycosylation, lipidation, and so on. While phosphorylation and ubiquitylation occur in the whole cell, acetylation and methylation are common modifications of nuclear proteins (histones), and lipidation and glycosylation are often associated with membrane signaling elements. Regulation of protein folding also provides important constraints on intracellular signaling: The unfolded protein response (UPR) senses aberrant protein forms, mostly in the lumen of the endoplasmic reticulum, and activates signal pathways which activate restorative gene-expression programs or induce apoptosis if the stress cannot be mitigated (Walter & Ron, 2011).

We want to emphasize that a number of *intracellular* signaling elements impinge upon the canonical (and non-canonical) *extracellular* signaling pathways described above, and comprise an essential aspect of the signaling system's plasticity, as mediated by differential splicing, scaffolding, histone coding, gene coactivation, etc. Molecular recognition, as performed by protein domains, is the fundamental phenomenon supporting combinatorial integration of signaling pathways, though it passes unnoticed in most functional analysis (Marijuán, 2001a). It depends on the tissues and the cellular circumstances of a signaling pathway, e.g. that a particular scaffold will have one molecular configuration or another, leading to total or partial incorporation, downstream, of elements from another signaling subsystem. That inherent flexibility, supported by specific molecular recognition within appropriate kinetic regimes, will be a central idea to keep in mind for §4.4, as we list and categorize the most important, "canonical," signaling pathways. Intracellular signaling, like the cell itself, is fluid and adaptive, advancing with the life cycle of the cell and with the development of local tissues in lockstep with the life cycle of the organism as a whole (Marijuán, 2002).

4.4 Fundamental Signaling Pathways in Neuronal Development and Physiology

Building on these four roots, there are more than 20 major classes of "canonical" signaling pathways in eukaryotes. As specialists in information processing, neurons keep most, if not all, of these pathways in action. While some pathways work exclusively in early development, others show up during organogenesis and others take charge of specialized physiological functions of neurons and partnered tissues. Specialized pathways survey the integrity of tissues and their "healing" responses including inflammation, apoptosis, and necrosis. We speculate that the overall organization of these signaling pathways proceeds from a signaling "master plan", as previously discussed [§4.3.4.1]. Evolutionary commonalities seen at the macroscopic level—like structural *bauplans*—also exist at the level of signaling pathway organization, and particularly in their functional rearrangement by evolutionary changes in the genetic addresses of transcriptional enhancers (Carroll, 2005; Loehlin & Werren, 2012). Though such a master plan is far from unraveled (though cf. *cellular bauplans* in Mojica et al., 2009), most embryonic development involves just a handful of signaling pathways.

4.4.1 The Prototypical Signaling Pathway

What does a prototypical eukaryotic, neuronal signaling pathway look like? Figure 4.3 captures the most general traits (cf. the conventional prokaryotic paths in Figure 4.1). Functionally, three main features can be distinguished: reception events at the membrane, processing by cytoplasmic mechanisms, and transcription of nuclear targets.

- *Membrane receptors.* External signals, or first messengers, may enter the cell through multiple gates: direct membrane crossing; G-protein-coupled receptors; ionic channels, activated by means of voltage, external ligands, internal ligands, or stretch; tyrosine kinases; serine/threonine kinases; phospatases; multidomain peptides; and other enzyme-associated receptors. Overall, on the order of 2000 receptors are coded in the human genome, and their variety is amplified further through alternative splicing and post-translational modification. In some cases (e.g. steroids, neuromodulators, and opiates) an incoming signal may activate different receptors at the same time, crossing to the cytoplasm through multiple pathways producing different effects. In other cases, a branching or "multiplexing" of pathways occurs, and several non-canonical pathways respond alongside the canonical one after activation of a common receptor; depending on the cellular context, only a subset of these alternative branches will be functionally relevant (Hyduke & Palsson, 2010; Liu, Slotine, & Varabais, 2011; Scott & Pawson, 2009). Needless to say, the distinction between canonical and non-canonical paths is not always very clear, and is often based simply on precedence of discovery.
- *Cytoplasmic processors.* As in artificial circuits, signals registered by receptors have to be filtered, processed, and amplified so that they can be conveyed to downstream effectors with appropriate timing and amplitude. The classical Weber–Fechner law—that sensed variations are logarithmically transformed by the internal processing mechanisms—seems to be one of the most common themes. The structure of enzyme kinetics (substrate, product, and effector relationships) easily provides for logarithmic transformation, but the variety of mechanisms involved is endless: scaffolds, enzyme complexes and networks, kinase cascades, ion channels, second messengers, endocytosis, proteasomes, cytoskeletal MTs, microfilaments, and so forth, not to mention the retinue of kinases, phosphatases, phospholipases, adenyl and guanyl cyclases, phosphodiesterases, cyclins, proteases, caspases, and other enzyme classes involved! Consequently, the dynamic regimes emerging at the end of each signaling pathway (even without considering cross-talk between them) do not necessarily maintain any formal relationship with the class and intensity of the received signal; the effects of the original perturbation become systemic and nonlinear (Davidson 2010; Liu et al., 2011; Scott & Pawson, 2009). In other cases, the recruitment of processing components into scaffolds and protein

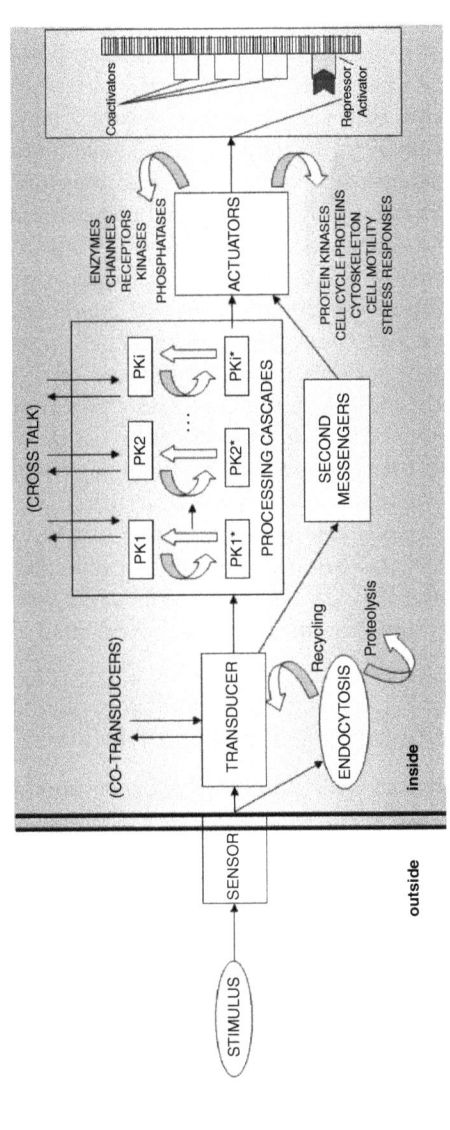

Figure 4.3 Prototypical Signaling Pathways of Multicellularity.
From left to right, a stimulus in the intercellular space binds to a transmembrane receptor (sensor) on its extracellular domain. Upon binding, the receptor undergoes a transient modification of its cytoplasmic domain; this effect triggers a transient modification of a series of proteins in the cell, each one acting as an intermediate in the signal transduction pathway (signal processing), with characteristic hierarchies of protein kinases and second messengers. The last components are actuators or effectors that activate or inhibit proteins and channels that control several cellular functions, notably gene expression by means of transcriptional switches that may interact with several coactivation partners. The whole of the biochemical changes produced in the cell represent the response to the received signal—its *molecular meaning*. Marijuán 2013. Reproduced with permission of Elsevier.

complexes may prevent cross signaling between pathways or may commit highly multifunctional molecules to very particular functions (Good et al., 2011).
- *Nuclear targets.* One of the essential consequences of signaling is the specific activation of gene targets by downstream effectors working as transcription factors: This is the stage where overlap among pathways is most likely to occur. Specific signaling pathway response elements (SPREs) act as *transcriptional switches*, whereby target genes become activated in the presence of signaling but repressed in its absence. *Default repression, activator insufficiency*, and *cooperative activation* constitute the central methods of transcriptional control shared among the major pathways (Barolo & Posakony, 2002; Pires-daSilva & Sommer, 2003). Default repression refers to the transcriptional repression of target genes in the absence of signaling; in general, the SPRE-binding transcription factors are converted from default repressors to activators upon arrival of the signal. Figure 4.3 shows how the arrival of the final response element of the pathway transforms the standing transcription factor and activates it. The presence of numerous co-activators, working together as logical machines to enact combinatorial regulation of transcription, ensures that expression activity is restricted to specific groups of cells or functional contexts (Davidson, 2006, 2010). In addition, there has to be a permissive arrangement of the euchromatin–heterochromatin expression state, also regulated via signaling pathways acting on the histone code, so that the cooperative activation of the transcriptional switch will lead to appropriate gene expression. Epigenetic mechanisms are integrated here, acting as permanent markers on DNA itself (C methylation), accompanied by transient modifications in the amino acids of histones, for example by Lys acetylation, Lys and Arg methylation, Ser phosphorylation, and Lys and Arg ubiquitylation. This capability to selectively prepare the expression state of chromatin regions becomes one of the essential aspects of the signaling system's guidance.

4.4.2 The Catalog of Eukaryotic Signaling Pathways

In Table 4.1, and the appendix at the end of this chapter, we have built upon Gerhart's scheme (1999), slightly modifying it to include new categories including stress and criticality, electro-molecular osmotic mechanisms, Hippo, the complement cascade, and apoptotic mechanisms. Needless to say, any attempt at cataloguing eukaryotic signaling pathways is a highly risky exercise: variation across cell types, exuberant pathway diversity, facultative mixing of control mechanisms, and the mechanistic complexity of the paths all conspire to preclude general classification. Table 4.1 lists major pathways and highlights their core mechanisms and functionalities. In the appendix, an abbreviated description and some comments (with emphasis on neuronal implications) have been included. The following main groupings have been distinguished:

- *Early development:* Wnt, Hedgehog, Notch, TGF-β, Neurotrophins.
- *Mid development and organogenesis:* Integrins, Cadherins, Nuclear Hormones, Reelin.
- *Tissue physiology:* Guanylate Cyclase, G-Protein Coupled Receptors, Electro-molecular Transmission.
- *Stress and criticality:* NFκB, Cytokines, Autophagy, Apoptosis, Hippo, Complement Cascade.

Table 4.1 The Major Eukaryotic Signaling Pathways (21 of them), Highlighting the Main Mechanisms and Functionalities.

Signaling system pathways	Components (to DNA-binding)	Main functions	References
Early Development and Morphogenesis			
Wnt pathway (Glycoproteins)	Frizzled and LDL co-receptors → β-catenin → TCF/LEF Non-canonical: Wnt/Ca^{2+}, Wnt/PCP	Cell proliferation, migration, polarity, neural differentiation, axonal growth. Ciliogenesis, craniofacial development and regeneration.	Dale, Sisson, & Topczewski, 2009 Eisenmann, 2005 Reya & Clevers, 2005
Hedgehog pathway (Hh proteins)	Ptc/Ptch1 → 7TM co-receptor Smo→Ci/Gli	Cell growth, tissue homeostasis, neural tube differentiation, loss of neural stem cells, embryonic formation, tumor initiation and growth.	Bermann et al., 2002 Hausmann, von Mering, & Basler, 2009 Lum & Beachy, 2004
Notch pathway (Notch/Delta proteins)	DSL TM proteins→ NICD → CBF1/CSL/Rbp-j → MAML1	Cell-fate specification, differentiation, self-renewal, proliferation, and apoptosis, stem cells maintenance, vertebrate segmentation.	Aulehla & Herrman, 2004 Ersvaer, Hatfield, Reikvam, & Bruserud, 2011 Schwanbeck, Martini, Bernoth, & Just, 2010
TGFβ pathway (TGFβ superfamily proteins)	Ser/Treo kinase (type I y II) →SMAD cascade Non-canonical: Ras/MAPKs (ERKs & SAPKs)	Cell division, differentiation, migration and adhesion; neuronal development and remodeling, synapse formation and growth, programmed cell death, suppressor of carcinogenesis.	Derynck & Zhang, 2003 Massagué & Gomis, 2006 Mulder, 2000 Roberts & Mishra, 2005
Trk kinase pathway (neurotrophins)	$p75^{NTR}$→Small G proteins →Ras/MAPK Non-canonical: PLC, PI3K	Cell survival, proliferation, axon and dendrite growth and patterning, cytoskeleton assembly & remodeling, synaptic strength and plasticity.	Huang & Reichardt, 2003 Reichardt, 2006 Segal, 2003

(*Continued*)

Table 4.1 (Continued)

Signaling system pathways	Components (to DNA-binding)	Main functions	References
Mid-development and Organogenesis			
Integrin pathway (Glycoprotein)	ECM proteins → FAK → Ras/Raf → MEK → ERK Non-canonical: actin, JNK, AKT/PKB, RLC	Differentiation, proliferation, cell shape and migration, cytoeskeletal organization, maintenance and cell survival.	Martin et al., 2002 Moser, Legate, Zent, & Fässler, 2009 Stevens & George, 2005 Yee, Weaver, & Hammer, 2008
Cadherin pathway (Glycoprotein)	PDZ-domain proteins → protein phosphatase → protein kinase → actin Non-canonical: Rho family GTPases, Wnt, RTK	Embryonic development, tissue morphogenesis and homeostasis. Synaptogenesis, synaptic plasticity and synapse morfogenesis.	Arikkath & Reichardt, 2008 Hulpiau & van Roy, 2009 Wheelock & Johnson, 2003 Yagi & Takeichi, 2011
Nuclear hormone receptor pathway	NHR (monomer/homodimer/RXRheterodimer) → HREs	Cell growth and cycle progression, apoptosis, hypothalamic–pituitary–adrenal axis, neuro-endocrine stress response, autonomic nervous system, development, metabolic homeostasis.	Aranda & Pascual, 2001 Beildeck, Gelmann, & Byers, 2010
Reelin pathway (Glycoprotein)	ApoER2/VLDLR → DAB1 → SFK → NMDAR Non-canonical: Cdk5, BLBP/Notch1	Neuronal migration and positioning in the developing brain, modulation of synaptic plasticity, induction and maintenance of LTP, stimulation of dendrites and dendrites spine, migration of neuroblast and neurogenesis.	Akopians et al., 2008 Chameau et al., 2009 Durakoglugil, Chen, White, Kavalali, & Herz, 2009 Niu, Yabut, & D'Arcangelo, 2008
Tissue Physiology			
Guanylatecyclase pathway (hormones, toxins, free radicals, calcium…)	GCs+NO→GMPc→ PK/PDE cascades GCp→GMPc→ PK/PDE cascades	Neuronal signal transduction, vascular smooth muscle relaxation, inhibition of platelet aggregation, electrolytic homeostasis.	Lucas et al., 2000 Hofmann, Feil, Kleppisch, & Schlossmann, 2006

G-protein coupled receptor (large G proteins) pathways	Adenyl cyclase→ cAMP → EPACs/PKA/CNG channels GPCRs+ Src-related Kinase + PI3K+ Shc → Ras →MAPK	Cellular responses to hormones and neurotransmitters, immune responses, cardiac and smooth muscle contraction and blood pressure regulation, proliferation, tissue remodeling and repair, inflammation, angiogenesis, normal cell growth and cancer.	Dorsam & Gutkind, 2007 Gavi, Shumay, Wang, & Malbon, 2006 Rosenbaum et al., 2009
Electrical Transmission			
Gap junctions	Hemichannels or connexones permanently opened allowing unspecific ionic flow.	Direct electrical transmission between neurons (and also glial cells).	Kandel, Schwartz, & Jessell, 2000 Kelsell, Dunlop, & Hodgins, 2001 Willecke et al., 2002
Stretch-activated channels	Mechanotransducer channels sensing membrane stress and allowing non-specific ionic flow.	Vibration sensing, pressure, stretch, touch, heat sensation, hearing, osmotic and blood pressure, and propioceptive sensation.	Kung, 2005 Purves et al., 2008 Yin and Kuebler, 2010
Voltage-activated channels	Voltage sensor channels allowing specific ionic flow (sodium, potassium, calcium, chloride, proton).	Generation and transmission of electrical signals in central and peripheral neurons, glia, skeletal muscle, heart, kidney, vascular tone, hormonal control, etc.	Alberts et al., 2002 Catterall, 2000 Kandel et al., 2000
Ligand-gated channels	Transmembrane ion channels opened or closed in response to the binding of a ligand or neurotransmitter, and allowing selective ionic flow (sodium, potassium, calcium, chloride, proton).	The main neurotransmitters (glutamate, acetylcholine, GABA, glycine, serotonin, etc.) represent the basic controls of brain function and the most general means of communication between neurons.	Barry & Lynch, 2005 Kandel et al., 2000 Swijsen, Hoogland, & Rigo, 2009

(*Continued*)

Table 4.1 (Continued)

Signaling system pathways	Components (to DNA-binding)	Main functions	References
Stress and Criticality			
Toll-like receptors pathway	TLR(1-2, 4-13) *MyD88dependent*+ IRAK kinases → TRAF6 → TAK1 →IKKs →NFκB TLR (3-4) *TRIF dependent*+ TBK/RIP1→ IRF3/TAK1-NFκB	Apoptosis, cell mediated immunity, bacterial death, autoimmune diseases (inflammatory process).	Kumar, Kawai, & Akira, 2009 O'Neill, Fitzgerald, & Bowie, 2003 O'Neill, 2008
Cytokine receptors (cytoplasmic tyrosine kinases) pathway	TypeI: JAK→STATs→IRF→ISRE TypeII: JAK→STAT1→GAS *Non-canonical*: JAK→ MAPK/PI3K/AKT	Inflammatory and immunological response, cell proliferation and haematopoiesis; regulated survival of injured neurons, neurite elongation, and re-establishment of neuronal connections.	Miyajima, Kitamura, Nobuyuki, Takashi, & Kenichi 1992 O'Sullivan, Liongue, Lewis, Stephenson, & Ward, 2007
Autophagy pathway	mTOR→ ULK complex (ULK1, ULK2, mAtg13)→ FIP200→ Class III PI3K complex (hVps34, Beclin 1, p150, Atg14L)	Stress survival and longevity, elimination and recycling of damaged cellular components generated in response to induced oxidative stress or during normal aging, promoting constant cellular renewal, cell growth, development, and homeostasis.	Valentino & Pierre, 2006 Levine & Kroemer, 2008 Mizushima, Levine, Cuervo, & Klionsky, 2008 Yang & Klionsky, 2010
Apoptosis pathway	Bcl-2 family (Bax, bak, Noxa, PUMA)→ cit. C, Smac/Diablo, Omi/HtrA2→ Caspase 9 & Efector Caspases *Non-canonical*: TNF receptor Superfamily (CD95/Fas/Apo, TNF-R1), Sphingomyelin-Ceramide	Embryogenesis and the maintenance of tissue homeostasis during the stage adult, pathological conditions or in healthy tissue (for example, its participation in the development of nervous system and immune, or neurodegenerative dementias).	Almeida, 2012Brunelle & Letai, 2009 Gamen, Anel, Piñeiro, & Naval, 1998 Kolesnick & Golde, 1994

The Hippo pathway	DCHS½→FAT4→FRMD6/Mer/ KIBRA→Mst½ →Mob1→Lats ½ →YAP/TAZ	Stem cell and progenitor self-renewal, cell proliferation, antiapoptosis Organ Size control	Huang, Wu, Barrera, Matthews, & Pan, 2005 Zhao, Li, & Guan, 2010 Pan, 2010
Complement Cascade	*Classical*: IgG/IgM → C1q → C4/ C2 → C3 →C5 *Non-canonical*: Lectin: MBL → MASPs→ C4/C2 → C3 *Alternative*: C3-H2O → Factor B/ Factor D/Properdin → C5	Immunoregulatory functions (enhancing humoral immunity, modifying T cell immunity, shaping the development of the natural Ab repertoire, regulating tolerance to nuclear self Ags such as DNA and chromatin). Pathogenic role (initiation and regulation of inflammatory response, opsonization and phagocytosis, systemic organ ischemia/reperfusion) and autoimmune diseases.	Alexander, Anderson, Barnum, Stevens, & Tenner., 2008Stevens et al., 2007 Thurman & Holers, 2006

Marijuán 2013. Reproduced with permission of Elsevier.

None of those pathways acts in isolation. Cells are rarely exposed to agonists individually—they receive constant extracellular stimulation—and through cross-talk, multiple inputs interact to produce different results in different contexts. For instance, a developing neuron exposed to two agonists fostering mutually-exclusive responses (e.g. growth vs. apoptosis) must decide which signal to follow (§4.3.3), with selection of the winner also suppressing signal efficacy of the loser.

This balancing (and symmetry breaking) between opposed pathways is used to navigate the process of development, defining neuronal morphology and physiology. Frequent players are Wnt and Hedgehog, Hippo and Wnt, Notch and Hedgehog, Hippo and TGF-β, etc. Particularly in neurons, interconnections between pathways are highly nonlinear, enmeshed in networks and circuits of fiendish complexity. One of the most important developmental steps in the embryo refers to what is called the epithelial-mesenchymal transition (EMT), crucial for the formation of many tissues and organs as well as for physiological processes such as wound healing (not to mention the initiation of cancer metastasis). As a result of this transformation, well-stacked epithelial cells lose their polarity and adhesion-junctions and gain invasive and migratory capabilities, becoming mesenchymal cells. Processes such as gastrulation, neural tube formation, heart formation, as well as different types of cancer, depend on these signaling pathways, basically a series of inputs and outputs around Par complex signaling which involve Wnt, Notch, TGF-β, Tyrosine kinases, and the cytoskeleton (McCaffrey & Macara, 2011). In other cases, mesenchymal cells experience the reverse process (MET) in order to participate in the formation of epithelial mesodermal organs. The flexible conjugation of both EMT and MET events is an essential feature of metazoan developmental.

In multicellular and vertebrate development, signaling is everywhere: from the earliest steps of axis specification, to the diverse kinds of morphogenesis, organogenesis, and growth in the embryo; from sexual maturation and regular tissue renovation to the ongoing physiology in the adult (Gerhart, 1999). Actually, each phylum is characterized by a body plan, *bauplan*, which is a unique topological configuration of secreted signals, active signaling pathways, and expressed genes, all of them dynamically self-organized along the development and life cycle of the individual. Probably, as we have already mentioned, a signaling "master plan" could be envisioned too; but it could hardly be described by any formal expression.

As stated in §4.1, neurons contain a bounty of signaling elements surpassing that found in any other eukaryotic tissue or cellular specialization: Among all eukaryotic tissues, they are the most specialized in intracellular computation. Table 4.1 and the appendix make clear that ALL signaling pathways are potentially involved in one or another aspect of nervous system development, structure, or function. Having to deal with that most subtle biological stuff—*information*—and organize a macroscopic system for memory probably represents an evolutionary challenge of the highest order, and has involved the most complex and sophisticated signaling solutions.

4.5 Intracellular Signaling at the Synapse

Although the concept of the synapse as a communication nexus between neurons is more than 100 years old, and although it has been one of the most intensely studied neuronal components, its origins, evolution, and dynamic control of structural elements remain poorly understood. Proteomic studies have provided key steps toward identifying the many hundred proteins which are involved, and their functional interactions. Two aspects of these new studies are important here: on one side, how a variety of synaptic classes have developed from a "protosynapse" which predated metazoa; and, on the other, how neuronal and immune synapses share structural and functional elements.

It nowadays appears that "protosynapses" originating in the LECA evolved further structural and functional complexity as their information processing mechanisms were modularized, diversified and recombined by multicellulars. From temporary portals where unicellular partners exchanged information about metabolic and reproductive states, eukaryotic synapses have evolved towards stable adhesive junctions between cells across which information is relayed by directed secretions. The cells of both the nervous and immune systems rely on synapses as a central communicative tool (Dustin & Colman, 2002; Steinmann, 2004). In each of these systems, synapses are built around a microdomain structure including central active zones of endocytosis and exocytosis surrounded by adhesion domains. Both systems have produced an ample variety of synaptic types, based on specific molecular recognition events, intercellular adhesion mechanisms, positional stability, and directed secretion for communication. In the neuron, the most complex, characteristic, and best-studied neuronal synapses are the excitatory ones.

4.5.1 Structural Components of Excitatory (Post)Synaptic Sites

In the dendrites and soma of pyramidal neurons, thousands of excitatory and inhibitory synapses are made by axons from almost as many neurons. Most synapses are excitatory, and release the neurotransmitter glutamate to postsynaptic contacts, often on dendritic spines. Each presynaptic terminal forms a junction with one (or at most two) postsynaptic spines. Spines are characterized by an expanded head connected to a dendrite shaft by a narrow neck. Within their tiny dimensions—1-3 μm long by 1 μm diameter—spines enclose astonishing complexity: The postsynaptic proteome comprises more than 1000 protein components in mammals, while the presynaptic proteome "only" comprises a few hundred (Ryan & Grant, 2009). The spine's size and variety of components are on the order of a complete prokaryotic cell, including local protein synthesis, dynamic cytoskeletal reorganization, and highly active processes of endocytosis and protein replacement/degradation.

On the presynaptic side, a secretory apparatus is activated by appropriate electro-molecular signaling events, typically incoming action potentials. On the postsynaptic side, therefore, a receptor apparatus transduces the presence of neurotransmitters in the synaptic cleft into relevant intracellular signals. The gap or cleft between the two synaptic sides is not structural but functional. Rather than being an empty space, it is filled with electron-dense material consisting of adhesion molecules, cadherins (or integrins at immune synapses), neuroligins, extracellular parts of receptors, and other filamentous material spanning the synaptic cleft. There is evidence that cadherins and neuroligins (among other partners) act as synaptic specifiers, in the molecularrecognition phase of synaptogenesis, and then as adhesive, later, to glue together the pre- and postsynaptic sides of the cleft (Cohen-Cory, 2002).

Most of the postsynaptic processing machinery is contained in a highly organized ensemble attached to the membrane, the *postsynaptic density* (PSD). This superstructure contains scaffolds and protein complexes associated with both glutamate receptors and actin microfilaments; it also incorporates relevant enzymes, proteins, and ion channels belonging to several major pathways (Baron et al., 2006; Okabe, 2007). Microtubules enter into spines only in a dynamic, EB3-associated form, acting to regulate spine morphology and maturation by controlling the levels of F-actin within spines. They are essential for the maintenance of spine morphology and maturation (Jaworski et al., 2009).

The ultimate goal of these synaptic structures is to regulate the plastic responses that mediate memory for the behavioral consequences of processed signals. Memories can be reinforced or erased by changing synaptic strength, either through presynaptic mechanisms, which alter neurotransmitter release in response to neural activity and are frequently short-term, or through the machinery of the PSD. The PSD orchestrates most of the synaptic changes necessary for plasticity in information processing and memory formation, usually in the form of long-term potentiation (LTP) and depression (LTD) (Murakoshi & Yasuda, 2012; Newpher & Ehlers, 2009). A simplified summary of the main signaling events includes the stages below, but see also Chapter 14 for more on comparative mechanisms of memory.

The most intensively studied instances of LTP and LTD dynamics correspond to the mammalian hippocampus CA3-CA1 glutamate excitatory synapses. Four main classes of glutamate receptors have been identified here: AMPA (α-amino-3-hydroxy-5-methyl-4-isoxazolepropionic acid, or "quisqualate"), plus the associated KA (kainate) and NMDA (N-Methyl-D-aspartate) receptors, as well as the mGluRs (metabotropic glutamate receptors).

The highly complex subcellular distribution of glutamate receptors is tightly controlled, and is central to regulation of synaptic efficacy. Many molecular agents participate in PSD regulatory complexes, including PI3K, Rac, Rap, non-receptor tyrosine kinases, etc. Of outmost importance are the *neurotrophins*, a family of neuronal growth factors that include NGF, BDNF, NT-3, and NT-4/5, among others, which are recognized by receptor tyrosine kinases (see Table 4.1 and the appendix). Neurotrophins were initially characterized as growth factors promoting neuronal survival and differentiation, but also participate in important aspects of synaptic development and function, including the regulation of activation-induced plasticity. In addition, they deeply influence axonal and dendritic morphology through their effect on cytoskeletal components and related signaling pathways. The three most relevant pathways converge on MAP kinases: the AMPA-NMDA system, adhesive neurexins/neuroligins, and actin/MT cytoskeletal components. Protocadherins mediating dendritic self-avoidance, as well as many other signaling pathways related to neuromodulation, pheromonal signaling, interleukins, and death factors also participate in spine regulation (Blitzer, Iyengar, & Landau, 2005; Lefebvre, Kostadinov, Chen, Maniatis, & Sanes, 2012; Manabe, 2002; Murakoshi and Yasuda, 2012; Newpher and Ehlers, 2009; Silver & Kanichay, 2008). As if all this signaling complexity was not sufficient, it is accompanied by an unexpected phenomenon: local protein synthesis within the spine itself.

4.5.2 Local Protein Synthesis in Spines: Molecular Markers of Plasticity

Among the many puzzles in synaptic plasticity, one of the most relevant is how the activation of, and ensuing calcium influx through, NMDA receptors can mediate both LTP and LTD—either strengthening or weakening (or even eliminating) the host synapse. Just changing the relative timing of pre- and postsynaptic activation by a few tens of milliseconds can reverse the direction of synaptic modification ("spike-timing dependent plasticity"). Synaptic remodeling thus comprises one of most complex electro-molecular phenomena. In addition to all the signaling pathways already discussed, it demands also receptor-complex regulators such as calcineurin and specialized phosphatases, local involvement of ubiquitin and proteasomes, and a variety of other molecular components (Bingol & Shuman, 2006; Haas, Miller, Friedman, & Broadie, 2007; Hughes, 2012; Paolicelli et al., 2011). Numerous switching mechanisms seem to cooperate at different levels, signaling, structural, degradation, transcriptional, and translational, with this last involving not only central protein synthesis but also local protein synthesis in spines.

Within spines, a variety of tagging molecules indicate synaptic fates to otherwise "blind" crews of cellular construction and repair mechanisms. Several candidates for tagging have been proposed: second messengers, protein kinases, adhesion molecules, polymerized actin, and dynamic MTs. Essentially, these synaptic tags guide local translation by organizing the release of mRNAs from specific granules in dendritic accumulation sites and their capture by highly stimulated synapses (with some effects spilling over to local neighbors). Thereafter, *in situ* protein synthesis takes place. This local protein synthesis ensures the appropriate molecules are directed to their proper postsynaptic destinations. Different patterns of activation lead to the release of different kinds of mRNAs, leading to correspondingly specific patterns of protein synthesis, which, with accompanying proteasome function, leads either to consolidation or weakening of specific synaptic structures (Blitzer et al., 2005). Since some of the synthesized proteins impact further protein synthesis, positive feedback loops are created which maintain the newly set synaptic strength. In this way, tagging molecules mediate an elaborate space-time regulatory framework for distributed computation and mnemonic representation (Bingol & Shuman, 2006; Ho, Lee, & Martin, 2011; Murakoshi & Yasuda, 2012; Silva, Zhou, Rogerson, Shobe, & Balaji, 2009).

As a matter of fact, synaptic spines possess the fundamental equipment required for protein synthesis, from ribosomes and mRNA transport to articulate membranous systems, as well as numerous molecular components belonging to the protein translational machinery. All of these have been localized in spines. Synaptic protein synthesis thus seems to us to be not only quite plausible, but the most parsimonious and coherent way to reconcile disparate explanations for synaptic remodeling. It also makes sense in the light of behavioral and ecological evidence regarding the different time scales and rhythms involved in learning consolidation, in the efficient allocation of synaptic memory resources, and in the organization of behavioral cycles (activity, fatigue, rest, sleep). The results from these relatively recent studies are likely to have a considerable impact on computational neuroscience and on the neurological understanding of memory disorders (Blitzer et al., 2005; Destexhe & Marder, 2004; Feldman & Brecht, 2005; Ho et al., 2011; Silva et al., 2009). Nonetheless, the occurrence of protein synthesis in postsynaptic structures was quite an unexpected finding. Symbolically, one of the most basic and "primitive" characteristic of life, protein synthesis, makes itself present in one of the most sophisticated evolutionary achievements, the synapse, where it plays an essential role in support of informational process—under the functional guidance of a motley crew of signaling pathways.

4.6 Concluding Comments: Molecular Tools for the Evolution of "Social Brains"

This chapter has just sketched the evolutionary history of signaling. Starting from the 1-2-3 scheme of prokaryotes, we described how four roots grew into the 20-odd groups of major eukaryotic signaling pathways. We further observed that synapses form a bridge between cells' intracellular signaling pathways: first describing the protosynapses which predated metazoa, the neuronal and immune synapses which share evolutionary origins as well as structural and functional elements, the postsynaptic density in excitatory synapses, and the occurrence of local protein synthesis in these synapses.

Most likely, the original function of nervous systems' components was not computational *per se*, but osmotic, then trophic, developmental, sensorimotor, and only much latter cognitive. As nervous systems abstract their function toward pure information processing, they entangle more and more signaling components and pathways. The evolution of the postsynaptic proteome is a case in point, from unicellular eukaryotes to complex invertebrates and vertebrates. Ancestral protosynapses would have contained only a small fraction of the mechanisms and pathways we described above (§4.5.1) for modern excitatory neuronal synapses. All organisms have to respond to their environment, but the ability to alter these responses, to sense and manipulate the environment itself, involves really marked differences in the structural complexity of nervous systems and even more in synaptic proteomes. Why should more complex behavioral repertoires demand all that extra signaling complexity in synapses? Why should synapses evolve along those intricate paths?

On the one side, we tend to forget the essential cognitive mission of the nervous system: to guide the realization of a life cycle within a particular ecological niche. This obviously implies sensitivity to numerous physical and chemical variables as well as to locomotion in an open-ended environment. The computational problems involved, even just in self-motion, are orders of magnitude beyond a reasonable description (Brooks, 2001). Implicit in these sensorimotor computations should be an evaluation of the fitness consequences for the organism: All inner and outer sensations, motor behaviors, rhythms, and time scales should be played with by the nervous system to capture the ever-changing landscape of adaptive fitness—and to resolve that variability appropriately (Marijuán, 2001b). The panoply of heterogeneous signaling pathways at the synapse becomes the essential element to achieving, by electro-molecular means, behavioral responses that coordinate sensory, locomotor, metabolic, digestive, respiratory, immune, circulatory, reproductive, etc. mechanisms. Synapses "compute"; but they do so multifunctionally and multidimensionally, with capability to adjust their computation via signaling pathways to the changing reference of a life cycle in progress.

On the other side, complex brains contain a number of specialized circuits, columns, maps areas, and structures. The computational functions of all those specialized localizations are different, approximately in the same way than their synaptic proteomes are. In particular, thinking in the most complex collection of such computing areas, the human cortex, the way to grant an adequate supply of raw molecular complexity for the numerous cortical specializations is far from trivial (Kang et al., 2011; Rakic, 2009; Ryan & Grant, 2009). Quite intriguingly, parental conflicts on maternal versus paternal epigenetic influences are directly involved in the establishment of such molecular complexity (Badcock & Bernard, 2008; Crespi & Badcock 2008; Gregg, Zhang, Butler, Haig, & Dulac, 2010). Some of the neurobiological problems recurrent in the development of the social brain (autism, schizophrenia, depression, fibromyalgia) have been linked to aberrant outcomes in the epigenetic imprint of important signaling/synaptic components (Alter et al., 2009; Kaati, Bygren, Pembrey, & Sjostrom, 2007; Wilkinson, Davies, & Isles, 2007). Further, the mobilization of retrotransposons L1 and Alu during early embryo and later brain development (see §4.2.1) is another example of how somatic genome mosaicism that may reshape the neuronal circuitry underlying both normal and abnormal neurobiological processes. Certainly, there has been no resource spared in the development and functioning of our complex brain.

Weaving together these strands, we may conclude that the information-processing complexity derives from the need to guide realization of a complex life cycle, using the resources afforded by synaptic proteomes. It reaches a climax when the main environment surrounding the individual is an extended tightly-knitted society—where individuals are specifically relating to each other through synaptic clouds or *engrams* of mutual memories–bonds. Thereafter, charting the dynamics and evolution of "social decision-making networks" in vertebrates (O'Connell & Hoffmann, 2012), and particularly in anthropoids and humans, becomes one of the highest multidisciplinary challenges: Molecularly, computationally, and behaviorally. In other words, the great challenge regarding our social brain continues to be bridging the gap from molecules and physiology to human behavior and human consciousness. Signaling, we have argued, becomes an essential bridging avenue.

Herein we have covered the theme of signaling, the informational engine that drives both social and neural complexity, only lightly. There is an extraordinary parsimony—and elegance—in the way evolution shaped and recombined the earliest osmotic tools of prokaryotes, together with solute-detection, cytoskeletal and endocytosis systems, to subserve electro-molecular processing of information. A system ultimately capable of mediating such subtle things as the "memorable" stuff of human experience.

Appendix: Catalog of Eukaryotic Signaling Pathways

4.A.1   Pathways of Early Development

WNT PATHWAY: Wnt (messenger) proteins are a large family of Cys-rich secreted glycoproteins that bind to members of the Frizzled (FZD) family of 7-transmembrane receptors. The binding of Wnt to its receptors leads to activation of at least three distinct pathways: 1) the canonical β-catenin pathway, 2) the planar cell polarity pathway, and 3) the calcium pathway. Through the canonical pathway, Wnt controls cell fate determination, and through the non-canonical pathways controls cell movement and tissue polarity. In vertebrates, dorso-ventral patterning of the developing neural tube is achieved by the counteracting activities of morphogenetic signaling gradients set up by the canonical Wnt/β-catenin pathway acting in the roof plate and by Sonic Hedgehog in the ventral floor plate and notochord (Gli3 transcription factor). Besides, the pathway participates in numerous neuronal processes such as migration, ciliogenesis, axonal growth, craniofacial development and regeneration. It involves more than 120 molecules and close to 50 transcriptionally regulated genes.

HEDGEHOG PATHWAY: The Hedgehog family of messenger proteins is the central regulator of a number of developmental, morphogenetic and physiological processes. Vertebrates are known to have three Hedgehog genes that perform specialized functions and show different spatio-temporal expression patterns: Desert hedgehog, Indian hedgehog, and Sonic hedgehog. Binding of Hedgehog ligands to their receptors, Patched 1 and 2, prevents inhibition of a 7-transmembrane receptor called Smoothened (Smo), leading to activation of Gli family of transcription factors (Gli 1-3). Signaling through this pathway is essential for the development of most tissues and organs; its aberrant activation has been associated with a number of human malignancies including carcinoma of lung, esophagus, pancreas and prostate. The pathway plays a fundamental role in neural tube differentiation, embryonic formation, and loss of neural stem cells. It involves three genes transcriptionally regulated and around different 40 molecules.

NOTCH PATHWAY: Highly conserved along evolution, the Notch signaling pathway is essential in cell-fate determination, tissue patterning, cell differentiation, proliferation, and cell death. Proteins of the Notch families are single-pass transmembrane proteins that function both as cell surface receptors and as nuclear transcriptional regulators. There are four Notch receptors in humans (Notch 1-4) that bind to a family of five ligands (Jagged 1-2 and Delta-like 1-3). Signaling activation occurs upon ligand receptor-binding on two adjacent cells. Signaling through the Notch receptors induces cleavage of the extracellular domain by an ADAM family metalloprotease followed by a cleavage within the transmembrane domain by gamma secretase complex, translocation of the cytosolic domain into the nucleus, and gene expression. Notch proteins are important in lineage specification and stem cell maintenance, as well as vertebrate segmentation. Aberrant Notch signaling has been linked to a number of malignancies including leukemias, lymphomas, and carcinomas of the breast, skin, lung, cervix and kidneys. More than 80 different molecules are involved in this pathway, with 23 genes transcriptionally regulated.

TGF-β PATHWAY: TGF-β receptors belong to a subfamily of membrane-bound serine/threonine kinases which are designated as Type I or II based on their structural and functional properties. The corresponding TGF-β ligands belong to a large superfamily of cytokines (TGF-β Superfamily) that includes bone morphogenic proteins, activins, inhibin, growth/differentiation factors, Nodal, and several other structurally-related polypeptides. Mammals express three TGF-β ligand isoforms (TGF-β1, TGF-β2, and TGF-β3) that can homodimerize or heterodimerize before binding to the receptors. In addition to activating the canonical Smad2/3-dependent signaling, there is an ever expanding list of non-canonical signaling molecules stimulated by TGF-β: PI3K, AKT, mTOR, integrins and focal adhesion kinase, members of the MAP kinase (ERK1/2, JNK), and p38 MAPK small GTP-binding proteins (Ras, Rho, Rac1). They play essential roles in regulating virtually all aspects of mammalian development and differentiation, and in maintaining mammalian tissue homeostasis. Some of the neural processes where they participate are: neural development, neural remodeling, synapse formation and growth. The TGF-β pathway includes more than 200 molecules and regulates transcriptionally more than 600 genes.

NEUROTROPHIN RECEPTORS PATHWAY: Neurotrophins are a family of closely related proteins that control many aspects of survival, development, and function of neurons in both the peripheral and the central nervous systems. Each of the four mammalian neurotrophins activates one or more of the three members of the tropomyosin-related kinase family of receptors (TrkA, TrkB and TrkC). Through Trk receptors, neurotrophins activate Ras, PI3-kinase, phospholipase C-γ1 and signaling pathways controlled through these proteins, such as MAP kinases. In addition, each neurotrophin activates p75 neurotrophin receptor (p75NTR) of the tumour necrosis receptor superfamily, which results in activation of the nuclear factor-κB (NF-κB) and Jun kinase as well as other signaling pathways. Continued presence of the neurotrophins is required in the adult nervous system, where they control synaptic function and plasticity, and sustain neuronal survival, morphology and differentiation.

4.A.2   Mid Development and Organogenesis

INTEGRIN PATHWAY: Integrins are cell-surface transmembrane heterodimeric (α/β) glycoproteins that mediate external cell-cell and cell-matrix (substratum) interactions and provide internal linkage to the cytoskeleton and to a variety of signaling pathways. The family includes at least 24 heterodimers assembled from 18 alpha and 8 beta subunits. These heterodimers contain binding sites for divalent magnesium and calcium that facilitate the binding of ligands. Integrins mediate signaling from the extracellular space into the cell via adaptor molecules such as focal adhesion kinase (FAK), integrin-linked kinase (ILK), cysteine-histidine rich protein (PINCH), and non-catalytic tyrosine kinase adaptor protein 2 (Nck2). Activated integrins participate in a variety of processes including survival/apoptosis, cell-cycle progression, proliferation, cell shape, polarity, adhesion, migration and differentiation. Plastic expression of different integrin subunits also controls the different stages of neural development, whereas in the adult integrins regulate synaptic stability. Integrins along with the immunoglobulin superfamily, selectins, cadherins, and mucins comprise the five major groups of cell adhesion molecules.

CADHERIN PATHWAY: Cadherins are a superfamily of transmembrane (receptor) proteins grouped by the presence of one or more cadherin repeats in their extracellular domains. Arrays of these domains form the intermolecular surfaces responsible for the formation of cadherin-mediated cell-cell interactions. The cadherin intracellular domain is a site for the assembly of a macromolecular complex that links the adhesion interface to the actin cytoskeleton. In epithelial cells, classic cadherins together with three catenins form a core functional unit, the cadherin-catenin complex (CCC), which is a major component of the apical junctions formed between these cells. This pathway is also involved in important neuronal processes such as embryonic development, synaptogenesis, synapse morphogenesis, self avoidance of synaptic contacts, and synaptic plasticity.

NUCLEAR HORMONE PATHWAY: Messengers of this pathway are small lipophilic molecules, such as steroid and thyroid hormones or the active forms of vitamin A (retinoids) and vitamin D, that are playing an important role in metabolism, cell growth, cell cycle progression, differentiation, apoptosis, development, reproduction, and immunity. Their binding to nuclear hormone receptors (NHR), which are transcription factors, mediates changes in gene expression via interaction with nuclear proteins that act as co-activators and co-repressors. The human genome contains close to 50 members of this superfamily. They have been classified into seven subfamilies; namely, NR1, NR2, NR3, NR4, NR5, NR6 and NR0. Nuclear receptors are characterized by a central DNA-binding domain (DBD), composed of two highly conserved zinc fingers that target the receptor to specific DNA sequences known as hormone response elements (HRE). The C-terminal half of the receptor encompasses the ligand-binding domain (LBD), which is responsible for hormone recognition and ensures both specificity and selectivity of the physiologic response. Nuclear hormone pathways are in charge of the communication between the nervous system and the immune system, including the hypothalamic–pituitary–adrenal axis, hormones of the neuroendocrine stress response, and the autonomic nervous system.

REELIN PATHWAY: Reelin is a large secreted extracellular matrix protein that helps regulate processes of neuronal migration and positioning in the developing brain by controlling cell–cell interactions. One class of Reelin receptor includes the VLDL receptor and the ApoER2; another class is the cadherin-related neuronal receptors (CNR receptors), which activates the Fyn tyrosine kinases. Besides its important role in early development, reelin continues to work in the adult brain, where it modulates synaptic plasticity by enhancing the induction and maintenance of long-term potentiation. It also stimulates dendrite and dendritic spine development, and regulates the continuing migration of neuroblasts generated in adult neurogenesis sites like subventricular and subgranular zones. It is found not only in the brain, but also in the spinal cord, blood, and other body organs and tissues.

4.A.3 Tissue Physiology

G-PROTEIN COUPLED RECEPTORS: The majority of transmembrane signal transduction in response to hormones and neurotransmitters is mediated by G protein-coupled receptors (GPCRs). They are among the largest and most diverse protein families in mammalian genomes (more than 800 GPCRs are listed in the human genome). Upon receptor activation, the G protein exchanges GDP for GTP, causing the dissociation of the GTP-bound α and β/γ subunits, and triggering diverse signaling cascades. Receptors coupled to different heterotrimeric G protein subtypes can utilize different scaffolds to activate the small G protein/MAPK cascade, employing at least three different classes of Tyr kinases. Src family kinases are recruited following activation of PI3Kγ by β/γ subunits; they are also recruited by receptor internalization, crossactivation of receptor Tyr kinases, or by signaling through an integrin scaffold involving Pyk2 and/or FAK. GPCRs can also employ PLCβ to mediate activation of PKC and CaMKII, which can have either stimulatory or inhibitory consequences for the downstream MAPK pathway. Among the physiological processes regulated by this pathway are: normal cell growth, proliferation, tissue remodeling and repair, inflammation, angiogenesis, immune responses, and cancer. Moreover, GPCRs are the principal signal transducers for the senses of sight and smell.

GUANYLATE CYCLASE PATHWAY: This is an atypical signal transduction which is also regulated by a gas messenger (NO). Processes controlled by the NO/cGMP system include: smooth muscle relaxation and blood pressure regulation; platelet aggregation and disaggregation; and neurotransmission both peripherally, in non-adrenergic, non-cholinergic (NANC) nerves, and centrally, in the processes of long term potentiation and depression. All signal transduction through sGC (and pGC as well) takes place through an increased concentration of cGMP, involving cGMP-dependent protein kinase, cGMP-regulated phosphodiesterase, and cGMP-gated ion channels. The latter channels are found in the retina and in the olfactory epithelium where they are involved, respectively, in visual phototransduction and in olfaction.

ELECTRICAL TRANSMISSION PATHWAYS: Under this heading, a series of pathways and molecular apparatuses related to counteracting the Donnan effect should be included (§4.3.2). Though they participate in a myriad of cellular and organismic processes, most of their action takes place within physiological settings, carrying on nearly all of the functions related to electrical excitability. By definition, the whole neuroelectrical processing is based on pathways controlled by these "Donnan apparatuses". The most important classes are:

- Gap junction pathways
- Stretch-activated channel pathways
- Voltage-activated channel pathways
- Ligand-gated channel pathways.

4.A.4 Stress and Criticality (Apoptosis and Necrosis)

NFκB (TOLL LIKE RECEPTOR) PATHWAY: The family of Toll-like receptors (TLRs) detects a wide variety of microbial components and elicits innate immune responses. All TLR signaling pathways culminate in activation of the transcription nuclear factor KappaB (NFκB). NFκB is a heterodimeric protein composed of different combinations of the Rel family of transcription factors, whose members are involved mainly in stress-induced, immune, and inflammatory responses. NFkB is an important regulator in cell fate decisions, such as programmed cell death and proliferation control, and is critical in tumorigenesis. NF-κB can be activated by exposure of cells to lipopolysaccharides, inflammatory cytokines such as TNF (Tumor Necrosis Factor) or IL-1 (Interleukin-1), lymphokines, oxidant-free radicals, inhaled particles, viral infection, UV irradiation, and B or T-cell activation. NFκB family members have also been implicated in neoplastic progression and in the formation of neuronal synapses.

CYTOKINE RECEPTOR (CYTOPLASMIC TYROSINE KINASES) PATHWAY: The cytokine receptor superfamily has been divided into several families based on their structure and activities, including type I cytokine receptors, type II cytokine receptors, TNF receptor family, chemokine receptors, TGF-beta receptors, and members of the immunoglobulin superfamily. These receptors share extracellular motifs but have limited similarity in their cytoplasmic domains. Although lacking catalytic domains, signaling by cytokine receptors depends upon their association with the Janus kinases (JAKs), which couple ligand binding to tyrosine phosphorylation of signaling proteins recruited to the cytokine receptor complex. At the end of these signaling proteins there is a unique family of transcription factors named the signal transducers and activators of transcription (STATs). Cytokines are essential communication instruments of the immune system and the inflammatory response, but they also play an important role in the survival of injured neurons and in neurite elongation or for the re-establishment of connection. Injured neurons prepare the involved signaling machinery at an early phase of the regenerative process, accelerating for the neuron to respond to cytokines that may regulate survival and/or neurite elongation.

AUTOPHAGY PATHWAY: The kinase mTOR is a critical regulator of autophagy induction, with positive regulation of mTOR (Akt and MAPK signaling) suppressing autophagy, and negative regulation of mTOR (AMPK and p53 signaling) promoting it. Several pro-apoptotic signals, such as TNF, TRAIL, and FADD, also induce autophagy. Additionally, Bcl-2 inhibit Beclin-1-dependent autophagy, thereby functioning both as a pro-survival and as an anti-autophagic regulator. Autophagy is generally activated by conditions of nutrient deprivation but has also been associated with physiological as well as pathological processes such as development, differentiation, neurodegenerative diseases, stress, infection, and cancer. In the central nervous system, the activation of autophagy has been shown to be protective in certain chronic neurodegenerative diseases but deleterious in acute neural disorders such as stroke and hypoxic/ischemic injury.

APOPTOSIS PATHWAY: Several pathways lead to caspase activation and apoptosis: 1) TNF/Fas-family cytokine receptors, 2) mitochondrial release of cytochrome c, and, 3) granzyme B-mediated cleavage of caspases in the context of cytolytic T cell responses. The endoplasmic reticulum (ER) participates in the initiation of apoptosis induced by the unfolded protein response (UPR) and by aberrant Ca^{2+} signaling. Damage to the ER can trigger caspase activation, and certain members of the caspase family are associated with the ER (e.g., caspase-12) or the Golgi (e.g., caspase-2). Bax inhibitor-1 (BI-1) is an antiapoptotic protein that contains several transmembrane domains, localizes to ER membranes, and is conserved in both animal and plant species. This pathway has a fundamental participation in the development of the nervous system and the immune system, and is also involved in the appearance of neurodegenerative dementias.

HIPPO PATHWAY: The Hippo pathway, evolutionarily highly conserved, controls cell proliferation, apoptosis, and organ size in response to changing cell density levels. At relative low cell density, transcription co-activators YAP and TAZ bind transcription factors to induce expression of genes that favor cell growth and proliferation. As cell density increases, interaction between membrane-bound upstream hippo regulators activates cytoplasmic kinases Mst1/2 and LATS1/2. Activated Mst kinase (the eponymous Hippo in Drosophila) associates with the adaptor WW45 and stimulates the downstream LATS kinase, which phosphorylates YAP and TAZ, that can no longer help promote transcription of genes that favor increased cell growth and proliferation. Other Hippo protein interactions at the cell surface involve Dachsous and Fat cadherins, as well as Mst activation by upstream regulators Merlin and FRMD6. Concerning its action in the nervous system, the Hippo pathway is involved in establishing and maintaining dendritic fields of sensory neurons, in tiling of neuroepithelial cells, and in glial proliferation.

COMPLEMENT CASCADES PATHWAY: The complement system is an important proteolytic cascade in blood plasma that acts as a mediator of innate immunity and as a nonspecific defense mechanism against pathogens. It consists of three different ways of simultaneously activating the membrane attack complex: the classical pathway, which forms part of adaptive immunity; and the alternative and the lectin pathways, which form part of innate immunity. The classical pathway is stimulated by antigen-antibody complexes; the alternative pathway spontaneously activates on contact with pathogenic cell surfaces; and the mannose-binding lectin pathway recognizes specific mannose sugars that are present on bacterial cell surfaces. All of these pathways generate a crucial enzymatic activity that, in turn, generates the effector molecules of complement. The main consequences of complement activation are opsonization of pathogens, recruitment of inflammatory and immunocompetent cells, and the direct killing of pathogens. In the nervous system, this pathway is associated with neuroinflammation and with the elimination of inappropriate synapses during development.

Acknowledgments

We would like to acknowledge the Journal *BioSystems* for giving us permission to reproduce some texts, two figures, and one table from the following paper: Marijuán, del Moral, & Navarro, J. (2013) On eukaryotic intelligence: Signaling system's guidance in the evolution of multicellular organization. *BioSystems* 114: 8–24.

References

Akopians, A. L., Babayan, A. H., Beffert, U., Herz, J., Basbaum, A. I., & Phelps, P. E. (2008). Contribution of the Reelin signaling pathways to nociceptive processing. *European Journal of Neuroscience*, 27, 523–537.

Alberts, B., Johnson, A., Lewis, J., Raff, M., Roberts, K., & Walter, P. (2002). *Molecular biology of the cell* (4th ed). New York, NY: Garland Science.

Alexander, J. J., Anderson, A. J., Barnum, S. R., Stevens, B., & Tenner, A. J. (2008). The complement cascade: Yin–Yang in neuroinflammation—neuro-protection and -degeneration. *Journal of Neurochemistry*, 107(5), 1169–1187.

Allman, J. (2000). *Evolving brains*. New York, NY: Scientific American Library, W H Freeman.

Alm, E., Huang, K., & Arkin, A. (2006). The evolution of two-component systems in bacteria reveals different strategies for niche adaptation. *PLoS Computational Biology* 2, 1329–1342.

Almeida, A. S., Queiroga, C. S., Sousa, M. F., Alves, P. M., & Vieira, H. L. (2012). Carbon monoxide modulates apoptosis by reinforcing oxidative metabolism in astrocytes: Role of Bcl-2. *Journal of Biological Chemistry*, 287(14), 10761–10770.

Alter, M. D., Gilani, A. I., Champagne, F. A., Curley, J. P., Turner, J. B., & Hen, R. (2009). Paternal transmission of complex phenotypes in inbred mice. *Biological Psychiatry*, 66, 1061–1066.

Aranda, A., & Pascual, A. (2001). Nuclear hormone receptors and gene expression. *Physiological Reviews*, 81(3), 1269–1304.

Aravind, L., Anantharaman, V., & Iyer, L. M. (2003). Evolutionary connections between bacterial and eukaryotic signaling systems: A genomic perspective. *Current Opinion in Microbiology*, 6(5), 490–497.

Aravind, L., Iyer, L. M., & Koonin, E. V. (2006). Comparative genomics and structural biology of the molecular innovations of eukaryotes. *Current Opinion in Structural Biology*, 16(3), 409–419.

Arikkath, J., & Reichardt, L. F. (2008). Cadherins and catenins at synapses: Roles in synaptogenesis and synaptic plasticity. *Trends in Neuroscience*, 31(9), 487–494.

Arnsten, A. F. T. (2009). Stress signaling pathways that impair prefrontal cortex structure and function. *Nature*, 10, 410–422.

Aulehla, A., & Herrmann, B. G. (2004). Segmentation in vertebrates: Clock and gradient finally joined. *Genes & Development*, 18, 2060–2067.

Badcock, C., & Bernard, C. (2008). Battle of the sexes may set the brain. *Nature*, 454(7208), 1054–1055.

Baillie, J. K., Barnett, M. W., Upton, K. R., Gerhardt, D. J., Richmond, T. A., De Sapio, F., ... Faulkner, G. J. (2011). Somatic retrotransposition alters the genetic landscape of the human brain. *Nature*, 479(7374), 534–537.

Barolo, S., & Posakony, J. W. (2002). Three habits of highly effective signaling pathways: Principles of transcriptional control by developmental cell signaling. *Genes and Development*, 16(10), 1167–1181.

Baron, M. K., Boeckers, T. M., Vaida, B., Faham, S. J., Gingery, M., Sawaya, M. R., ... Bowie, J. U. (2006). An architectural framework that may lie at the core of the postsynaptic density. *Science*, 311, 531–535.

Barry, P. H., & Lynch, J. W. (2005). Ligand-gated channels. *IEEE Transactions on NanoBioscience*, 4(1), 70–80.

Behrends, C., Sowa, M. E., Gygi, S. P., & Harper, J. W. (2010). Network organization of the human autophagy system. *Nature*, 466 (7302), 68–76.

Beildeck, M. E., Gelmann, E. P., & Byers, S. W. (2010). Cross-regulation of signaling pathways: An example of nuclear hormone receptors and the canonical Wnt pathway. *Experimental Cell Research*, 316(11), 1763–1772.

Berman, D. M., Karhadkar, S. S., Hallahan, A. R., Pritchard, J. I., Eberhart, C. G., Watkins, D. N., ... Beachy, P. A. (2002). Medulloblastoma growth inhibition by hedgehog pathway blockade. *Science*, 297(5586), 1559–1561.

Bingol, B., & Schuman, E. M. (2006). Activity-dependent dynamics and sequestration of the proteasome in dendritic spines. *Nature*, 441, 1144–1148.

Blackstone, N. W., & Green, D. R. (1999). The evolution of a mechanism of cell suicide. *Bioessays*, 21(1), 84–88.

Blitzer, R. D., Iyengar, R., & Landau, E. M. (2005). Postsynaptic signaling networks: cellular cogwheels underlying long-term plasticity. *Biological Psychiatry*, 57, 113–119.

Brooks, R. A. (2001). The relationship between matter and life. *Nature*, 409, 409–411.

Brunelle, J. K., & Letai, A. (2009). Control of mitochondrial apoptosis by the Bcl-2 family. *Journal of Cell Science*, 122, 437–441.

Bullock TH & Horridge GA (1965). *Structure and function in the nervous systems of invertebrates* (Vol.1). New York, NY: W. H. Freeman.

Carroll, S. B. (2005). *Endless forms most beautiful: The new science of evo devo and the making of the animal kingdom*. New York, NY: W. W. Norton & Company.

Catterall, W. A. (2000). Structure and regulation of voltage-gated Ca2+ channels. *Annual Review of Cell and Developmental Biology*, 16, 521–555.

Chameau, P., Inta, D., Vitalis, T., Monyer, H., Wadman, W. J., & van Hooft, J. A. (2009). The N-terminal region of reelin regulates postnatal dendritic maturation of cortical pyramidal neurons. *Proceedings of the National Academy of Sciences of the USA*, 106(17), 7227–7232.

Clark, K. B. (2010a). Bose-Einstein condensates form in heuristics learned by ciliates deciding to signal "social" commitments. *BioSystems*, 99(3), 167–178.

Clark, K. B. (2010b). On classical and quantum error-correction in ciliate mate selection. *Communicative & Integrative Biology*, 3(4), 374–378.

Cohen, P. (1992). Signal integration at the level of protein kinases, protein phosphatases and their substrates. *Trends in Biochemical Sciences*, 17, 408–413.

Cohen-Cory, S. (2002). The developing synapse: Construction and modulations of synaptic structures and circuits. *Science*, 298, 770–775.

Crespi, B., & Badcock, C. (2008). Psychosis and autism as diametrical disorders of the social brain. *Journal of Behavioral and Brain Science*, 31(3), 241–261.

Crick, F. H. (1979). Thinking about the brain. *Scientific American*, 241, 219–232.

Dale, R. M., Sisson, B. E., & Topczewski, J. (2009). The emerging role of Wnt/PCP signaling in organ formation. *Zebrafish*, 6(1), 9–14.

Davidson, E. H. (2006). *The regulatory genome: Gene regulatory networks in development and evolution*. Burlington, MA: Academic Press.

Davidson, E. H. (2010). Emerging properties of animal gene regulatory networks. *Nature*, 468(7326), 911–920.

Derynck, R., & Zhang, Y. E. (2003). Smad-dependent and Smad-independent pathways in TGF-beta family signaling. *Nature*, 425(6958), 577–584.

Destexhe, A., & Marder, E. (2004). Plasticity in single neuron and circuit computations. *Nature*, 431, 789–795.

Dorsam, R. T., & Gutkind, J. S. (2007). G-protein-coupled receptors and cancer. *Nature Reviews Cancer*, 7(2), 79–94.

Durakoglugila, M. S., Chena, Y., Whiteb, C. L., Kavalalic, E. T., Herz, J. (2009). Reelin signaling antagonizes β-amyloid at the synapse. *Proceedings of the National Academy of Sciences of the USA*, 106(37), 15938–15943.

Dustin, M. L., & Colman, D. R. (2002). Neural and immunological synaptic relations. *Science*, 298(5594), 785–789.

Egan, S. E., & Weinberg, R. A. (1993). The pathway to signal achievement. *Nature*, 365, 781–783.

Eisenmann, D. M. (2005). Wnt signaling. *WormBook: The Online Review of C. elegans*. Bookshelf ID: NBK19669. Retrieved from http://wormbook.org/chapters/www_wntsignaling/wntsignaling.html

Emes, R. D., Pocklington, A. J., Anderson, C. N., Bayes, A., Collins, M. O., Vickers, C. A., ... Grant, S. G. (2008). Evolutionary expansion and anatomical specialization of synapse proteome complexity. *Nature Neuroscience*, 11(7), 799–806.

Ersvaer, E., Hatfield, K. J., Reikvam, H., & Bruserud, Ø. (2011). Future perspectives: Therapeutic targeting of notch signaling may become a strategy in patients receiving stem cell transplantation for hematologic malignancies. *Bone Marrow Research*, Article ID 570796. doi: 10.1155/2011/570796

Feldman, D. E., & Brecht, M. (2005). Map plasticity in somatosensory cortex. *Science*, 310(5749), 810–815.

Fletcher, D. A., & Mullins, R. D. (2010). Cell mechanics and the cytoskeleton. *Nature*, 463, 485–492.

Galperin, M. Y. (2005). A census of membrane-bound and intracellular signal transduction proteins in bacteria: Bacterial IQ, extroverts and introverts. *BMC Microbiology*, 5, 1–19.

Gamen, S., Anel, A., Piñeiro, A., & Naval, J. (1998). Caspases are the main executioners of Fas-mediated apoptosis, irrespective of the ceramide signaling pathway. *Cell Death and Differentiation*, 5, 241–249. Retrieved from http://www.jneurosci.org/external-ref?access_num=10.1016/0092-8674%2894%2990147-3&link_type=DOI

Gavi, S., Shumay, E., Wang, H., & Malbon, C. C. (2006). G-protein coupled receptors and tyrosine kinases: Crossroads in cell signaling and regulation. *Trends in Endocrinology & Metabolism*, 17, 46–52.

Gerhart, J. (1999). 1998 Warkany lecture: Signaling pathways in development. *Teratology*, 60, 226–239.

Good, M., Zalatan, J. G., Lim, W. A. (2011). Scaffold proteins: Hubs for controlling the flow of cellular information. *Science*, 332(6030), 680–686.

Gregg, C., Zhang, J., Butler, J. E., Haig, D., & Dulac, C. (2010). Sex-specific parent-of-origin allelic expression in the mouse brain. *Science*, 329(5992), 682–685.

Grigoroudis, A. I., Panagiotidis, C. A., Lioliou, E. E., Vlassi, M., & Kyriakidis, D. A. (2007). Molecular modeling and functional analysis of the AtoS-AtoC two-component signal transduction system of Escherichia coli. *Biochimica et Biophysica Acta*, 1770(8), 1248–1258.

Haas, K. F., Miller, S. L. H., Friedman, D. B., & Broadie, K. (2007). The ubiquitin-proteasome system postsynaptically regulates glutamatergic synaptic function. *Molecular and Cellular Neuroscience*, 35, 64–75.

Hameroff, S. (2010). The "conscious pilot"—dendritic synchrony moves through the brain to mediate consciousness. *Journal of Biological Physics*, 36(1), 71–93.

Hameroff, S. R., & Penrose, R. (1996). Orchestrated reduction of quantum coherence in brain microtubules: A model for consciousness. In S. R. Hameroff, A. Kaszniak, & A. C. Scott (Eds.), *Toward a Science of Consciousness* (pp. 507–540). Cambridge, MA: MIT Press.

Hausmann, G., von Mering, C., & Basler, K. (2009). The hedgehog signaling pathway: Where did it come from? *PLoS Biology*, 7(6), e1000146. doi:10.1371/journal.pbio.1000146.

Ho, V. M., Lee, J. A., & Martin, K. C. (2011). The cell biology of sinaptic plasticity. *Science*, 334, 623–628.

Hofmann, F., Feil, R., Kleppisch, T., & Schlossmann, J. (2006). Function of cGMPdependent protein kinases as revealed by gene deletion. *Physiological Reviews*, 86(1), 1–23.

Horridge, G. A. (1968). The origins of the nervous system. In G. H. Bourne (Ed.), *The structure and function of nervous tissue* (pp. 1–31). New York, NY: Academic Press.

Huang, E. J., & Reichardt, L. F. (2003). TRK receptors: Roles in neuronal signal transduction. *Annual Review of Biochemistry*, 72, 609–642.

Huang, J., Wu, S., Barrera, J., Matthews, K., & Pan, D. (2005). The Hippo signaling pathway coordinately regulates cell proliferation and apoptosis by inactivating Yorkie, the Drosophila homolog of YAP. *Cell*, 122, 421–434.

Hughes, V. (2012). The constant gardeners. *Nature*, 485, 570–572.

Hulpiau, P., & van Roy, F. (2009). Molecular evolution of the cadherin superfamily. *The International Journal of Biochemistry & Cell Biology*, 41(2), 349–369.

Hyduke, D. R., & Palsson, B. Ø. (2010). Towards genome-scale signaling network reconstructions. *Nature Reviews Genetics*, 11(4), 297–307.

Ingber, D. E. (1998). The architecture of life. *Scientific American*, 278(1), 48–57.

Jaworski, J., Kapitein, L. C., Gouveia, S. M., Dortland, B. R., Wulf, P. S., Grigoriev, I., ... Hoogenrad, C. C. (2009). Dynamic microtubules regulate dendritic spine morphology and synaptic plasticity. *Neuron*, 61, 85–100.

John, B., & Miklos, G. L. G. (1998). *The eukaryotic genome in development and evolution*. London: Allen and Unwin.

Kaati, G., Bygren, L. O., Pembrey, M., & Sjostrom, M. (2007). Transgenerational response to nutrition, early life circumstances and longevity. *European Journal of Human Genetics*, 15, 784–790.

Kandel, E. R., Schwartz, J. H., & Jessell, T. M. (2000). *Principles of neural science* (4th ed.). New York, NY: McGraw-Hill.

Kang, H. J., Kawasawa, Y. I., Cheng, F., Zhu, Y., Xu, X., Li, M., ... Sestan, N. (2011). Spatio-temporal transcriptome of the human brain. *Nature*, 478(7370), 483–489.

Kelsell, D. P., Dunlop, J., & Hodgins, M. B. (2001). Human diseases: Clues to cracking the connexin code? *Trends in Cell Biology*, 11(1), 2–6.

Kiel, C., Yus, E., & Serrano, L. (2010). Engineering signal transduction pathways. *Cell*, 140, 33–47.

Kolesnick, R., & Golde, D. W. (1994). The sphingomyelin pathway in tumor necrosis factor and interleukin-1 signaling. *Cell*, 77, 325–328.

Koonin, E. V. (2010). The origin and early evolution of eukaryotes in the light of phylogenomics. *Genome Biology*, 11(5), 209. doi:10.1186/gb-2010-11-5-209

Kumar, H., Kawai, T., & Akira, S. (2009). Toll-like receptors and innate immunity. *Biochemical and Biophysical Research Communications*, 388(4), 621–625.

Kung, C. (2005). A possible unifying principle for mechanosensation. *Nature*, 436(4), 647–654.

Kung, C., Saimi, Y., & Martinac, B. (1990). Mechano-sensitive ion channels in microbes and the early evolutionary origin of solvent sensing. *Current Topics in Membranes & Transport*, 36, 145–153.

Lander, E. S., Linton, L. M., Birren, B., Nusbaum, C. Zody, N. C., Baldwin, J., ... International Human Genome Sequencing Consortium (2001). Initial sequencing and analysis of the human genome. *Nature*, 409(6822), 860–921.

Lane, N. (2011). The cost of breathing. *Science*, 334, 184–185.

Lane, N., & Martin, W. (2010). The energetics of genome complexity. *Nature*, 467, 929–934.

Lefebvre, J. L., Kostadinov, D., Chen, W. V., Maniatis, T., & Sanes, J. R. (2012). Protocadherins mediate dendritic self-avoidance in the mammalian nervous system. *Nature*, 488(7412), 517–521.

Levine, B., & Kroemer, G. (2008). Autophagy in the pathogenesis of disease. *Cell* 132(1), 27–42.

Liu, Y. Y., Slotine, J. J., & Varabais, A. L. (2011). Controllability of complex networks. *Nature*, 473, 167–173.

Loehlin, D. W., & Werren, J. H. (2012). Evolution of shape by multiple regulatory changes to a growth gene. *Science*, 335, 943–947.

Lucas, K. A., Pitari, G. M., Kazerounian, S., Ruiz-Stewart, I., Park, J., Schulz, S., … Waldman, S. A. (2000). Guanylyl cyclases and signaling by cyclic GMP. *Pharmacological Reviews*, 52(3), 375–413.

Lum, L., & Beachy, P. A. (2004). The hedgehog response network: Sensors, switches, and routers. *Science*, 304(5678), 1755–1759.

Lynch, V. J., May, G., & Wagner, G. P. (2011). Regulatory evolution through divergence of a phosphoswitch in the transcription factor CEBPB. *Nature*, 480(7377), 383–386.

Mackie, G. O. (1990). The elementary nervous system revisited. *American Zoologist*, 30, 907–920.

Manabe, T. (2002). Does BDNF have pre– or postsynaptic targets? *Science*, 295, 1651–1653.

Margulis, L. (1970). *Origin of eukaryotic cells*. New Haven, CT: Yale University Press.

Marijuán, P. C. (2001a). Molecular recognition and the organization of the living cell. *Symmetry: Culture and Science*, 12(3–4), 407–423.

Marijuán, P. C. (2001b). Cajal and consciousness: An introduction. *Annals of the New York Academy of Sciences*, 929, 1–10.

Marijuán, P. C. (2002). Bioinformation: Untangling the networks of life. *BioSystems*, 64(1–3), 111–118.

Marijuán, P. C., del Moral, R., & Navarro, J. (2013). On eukaryotic intelligence: Signaling system's guidance in the evolution of multicellular organization. *BioSystems*, 114, 8–24.

Marijuán, P. C., Navarro, J., & del Moral, R. (2010). On prokaryotic intelligence: Strategies for sensing the environment. *BioSystems*, 99, 94–103.

Martin, K. H., Slack, J. K., Boerner, S. A., Martin, C. C., & Parsons, J. T. (2002). Integrin connections map: To infinity and beyond. *Science*, 296(5573), 1652–1653.

Massagué, J., & Gomis, R. R. (2006). The logic of TGF-beta signaling. *FEBS Letters*, 580(12), 2811–2820.

McCaffrey, L., & Macara, I. G. (2011). Epithelial organization, cell polarity and tumorigenesis. *Trends in Cell Biology*, 21(12), 727–735.

Mills, I. G. (2007). The interplay between clathrin-coated vesicles and cell signaling. *Seminars in Cell and Developmental Biology*, 18(4), 459–470.

Miyajima, A., Kitamura, T., Nobuyuki, H., Takashi, Y., & Kenichi, A. (1992). Cytokine receptors and signal transduction. *Annual Review of Immunology*, 10, 295–331.

Mizushima, N., Levine, B., Cuervo, A. M., & Klionsky, D. J. (2008). Autophagy fights disease through cellular self-digestion. *Nature*, 451(7182), 1069–1075.

Mojica, N. S., Navarro, J., Marijuán, P. C., & Lahoz-Beltra, R. (2009). Cellular "bauplans": Evolving unicellular forms by means of Julia sets and Pickover biomorphs. *BioSystems*, 98, 19–30

Montell, D. J. (2008). Morphogenetic cell movements: Diversity from modular mechanical properties. *Science*, 322(5907),1502–1505.

Moser, M., Legate, K. R., Zent, R., & Fässler, R. (2009). The tail of integrins, talin, and kindlins. *Science*, 324(5929), 895–899.

Mulder, K. M. (2000). Role of Ras and Mapks in TGFβ signaling. *Cytokine & Growth Factor Reviews*, 11(1–2), 23–35.

Murakoshi, H., & Yasuda, R. (2012). Postsynaptic signaling during plasticity of dendritic spines. *Trends in Neuroscience*, 35(2), 135–143.

Nedelcu, A. M., & Michod, R. E. (2004). Evolvability, modularity, and individuality during the transition to multicellularity in volvocalean green algae. In G. Schlosser & G. Wagner (Eds.), *Modularity in development and evolution* (pp. 468–489). Chicago, IL: University of Chicago Press.

Newpher, T. M., & Ehlers, M. D. (2009). Spine microdomains for postsynaptic signaling and plasticity. *Trends in Cell Biology*, 19(5), 218–227.

Niu, S., Yabut, O., & D'Arcangelo, G. (2008). The reelin signaling pathway promotes dendritic spine development in hippocampal neurons. *The Journal of Neuroscience*, 8, 28(41), 10339–10348.

O'Connell, L. A., & Hofmann, H. A. (2012). Evolution of a vertebrate social decision-making network. *Science*, 336(6085), 1154–1157.

O'Neill, L. A. (2008). The interleukin-1 receptor/Toll-like receptor superfamily: 10 years of progress. *Immunological Reviews*, 226, 10–8.

O'Neill, L. A., Fitzgerald, K. A., & Bowie, A. G. (2003). The Toll–IL-1 receptor adaptor family grows to five members. *Trends in Immunology*, 24(6), 286–289.

Okabe, S. (2007). Molecular anatomy of the postsynaptic density. *Molecular and Cellular Neuroscience*, 34, 503–518.

O'Sullivan, L. A., Liongue, C., Lewis, R. S., Stephenson, S. E., & Ward, A. C. (2007). Cytokine receptor signaling through the Jak-Stat-Socs pathway in disease. *Molecular Immunology*, 44(10), 2497–506.

Pan, D. (2010). The hippo signaling pathway in development and cancer. *Developmental Cell*, 19(4), 491–505.

Paolicelli, R. C., Bolasco, G., Pagani, F., Maggi, L., Scianni, M., Panzanelli, P., ... Gross, C. T. (2011). Synaptic pruning by microglia is necessary for normal brain development. *Science*, 333(6048), 1456–1458.

Pavans de Ceccatty, M. (1974). The origin of the integrative system: A change in view derived from research on coelenterates and sponges. *Perspectives in Biology and Medicine*, 17, 379–390.

Pawson, T., & Nash, P., (2003). Assembly of cell regulatory systems through protein interaction domains. *Science*, 300(5618), 445–52.

Pires-daSilva, A., & Sommer, R. J. (2003). The evolution of signalling pathways in animal development. *Nature Reviews Genetics*, 4(1), 39–49.

Purves, D., Augustine, G. J., Fitzpatrick, D., Hall, W. C., LaMantia, A. S., McNamara, J. O., & White, L. E. (2008). *Neuroscience* (4th ed.). Sunderland, MA: Sinauer Associates.

Rakic, P. (2009). Evolution of the neocortex: A perspective from developmental biology. *Nature Reviews Neuroscience*, 10, 724–735.

Reichardt, L. F. (2006). Neurotrophin-regulated signaling pathways. *Philosophical Transactions of the Royal Society B*, 361(1473), 1545–1564.

Reya, T., & Clevers, H. (2005). Wnt signaling in stem cells and cancer. *Nature*, 434(7035), 843–850.

Roberts, A., & Mishra, L. (2005). Role of TGFbeta in stem cells and cancer. *Oncogene*, 24, 5667.

Rosenbaum, D. M., Rasmussen, S. G. F., & Kobilka, B. K. (2009). The structure and function of G-protein-coupled receptors. *Nature*, 459, 356–363.

Ryan, T. J., & Grant, S. G. (2009). The origin and evolution of synapses. *Nature Reviews Neuroscience*, 10(10), 701–712.

Schwanbeck, R., Martini, S., Bernoth, K., & Just, U. (2011). The Notch signaling pathway: Molecular basis of cell context dependency. *European Journal of Cell Biology*, 90(6–7), 572–581.

Scita, G., & Di Fiore, P. P. (2010). The endocytic matrix. *Nature*, 463(7280), 464–473.

Scott, J. D., & Pawson, T. (2009). Cell signaling in space and time: Where proteins come together and when they're apart. *Science*, 326, 1220–1224.

Segal, R. A. (2003). Selectivity in neurotrophin signaling: Theme and variations. *Annual Review of Neuroscience*, 26, 299–330.

Silva, A. J., Zhou, Y., Rogerson, T., Shobe, J., & Balaji, J. (2009). Molecular and cellular approaches to memory allocation in neural circuits. *Science*, 326, 391–395.

Silver, R. A., & Kanichay, R. T. (2008). Neuroscience. Refreshing connections. *Science*, 320(5873), 183–184.

Siso-Nadal, F., Fox, J. J., Laporte, S. A., Hébert, T. E., & Swain, P. S. (2009). Cross-talk between signaling pathways can generate robust oscillations in calcium and cAMP. *PLoS One*, 4(10), 7189.

Steinman, L. (2004). Elaborate interactions between the immune and nervous systems. *Nature Immunology*, 5, 575–581.

Stevens, B., Allen, N. J., Vazquez, L. E., Howell, G. R., Christopherson, K. S., Nouri, N., … Barres, B. A. (2007). The classical complement cascade mediates CNS synapse elimination. *Cell*, 131(6), 1164–1178.

Stevens, M. M., & George, J. H. (2005). Exploring and engineering the cell surface interface. *Science*, 310(5751), 1135–1138.

Swijsen, A., Hoogland, G., Rigo, J. M. (2009). Potential role for ligand-gated ion channels after seizure-induced neurogenesis. *Biochemical Society Transactions* 37(6), 1419–1422.

Thurman, J. M., & Holers V. M. (2006). The central role of the alternative complement pathway in human disease. *The Journal of Immunology*, 176(3), 1305–1310.

Ulrich, L. E., Koonin, E. V., & Zhulin, I. B. (2005). One-component systems dominate signal transduction in prokaryotes. *Trends in Microbiology*, 13, 52–56.

Valadi, H., Ekström, K., Bossios, A., Sjöstrand, M., Lee, J. J., & Lötvall, J. O. (2007). Exosome-mediated transfer of mRNAs and microRNAs is a novel mechanism of genetic exchange between cells. *Nature Cell Biology*, 9(6), 654–659.

Valentino, L., & Pierre, J. (2006). JAK/STAT signal transduction: Regulators and implication in hematological malignancies. *Biochemical Pharmacology*, 71(6), 713–721.

Walter, P., & Ron, D. (2011). The unfolded protein response: From estress pathway to homeostatic regulation. *Science*, 334(6059), 1081–1086.

Wheelock, M. J., & Johnson, K. R. (2003). Cadherin-mediated cellular signaling. *Current Opinion in Cell Biology*, 15(5), 509–514.

Wilkinson, L. S., Davies, W., & Isles, A. R. (2007). Genomic imprinting effects on brain development and function. *Nature Reviews Neuroscience*, 8(11), 832–843.

Willecke, K., Eiberger, J., Degen, J., Eckardt, D., Romualdi, A., Güldenagel, M., … Söhl, G. (2002). Structural and functional diversity of connexin genes in the mouse and human genome. *Biological Chemistry*, 383(5), 725–737.

Yagi, T., & Takeichi, M. (2000). Cadherin superfamily genes, Functions, genomic organization, and neurologic diversity. *Genes & Development*, 14, 1169–1180.

Yang, Z., & Klionsky, D. J. (2010). Mammalian autophagy: Core molecular machinery and signaling regulation. *Current Opinion in Cell Biology*, 22(2), 124–131.

Yee, K. L., Weaver, V. M., & Hammer, D. A. (2008). Integrin-mediated signaling through the MAP-kinase pathway. *IET Systems Biology*, 2(1), 8–15.

Zhao, B., Li, L., & Guan, K. L. (2010). Hippo signaling at a glance. *Journal of Cell Science*, 123, 4001–4006.

5

The Evolution of Neurons
Robert W. Meech

5.1 Introduction

In a letter to Charles Darwin dated July 20, 1875 George Romanes speculated about the evolutionary origins of the nervous system. He wrote:

> My dear Mr. Darwin, …. physiologists have been so long accustomed to associate the phenomena of reflex action with some such distinguishable system, I was afraid that they might think me rather audacious in propounding the doctrine, that there is such a thing as reflex action without well-defined structural channels for it to occur in. But if you have found something of the same sort in plants, of course I shall be very glad to have your authority to quote.

Reflexes might well be the basis of all behavior (Sechenov, 1863 [1965]) but Romanes worried that medusae appeared to exhibit them without the mediation of nerves.[1]

> And I think it follows deductively from the general theory of evolution, that reflex action ought to be present before the lines in which it flows are sufficiently differentiated to become distinguishable as nerves.

Believing that he and Darwin had each discovered a proto-nervous system, Romanes concluded that reflex function was not dependent on the prior evolution of neurons. Instead, reflex function appeared early in the course of evolution and then the neuron, the structure that enabled such functions to be coordinated efficiently, followed on.

Some 35 years later G. H. Parker (1911) proposed that nervous systems came into being in three successive stages. At the outset there were independent effectors such as the myocytes of the sponge osculum or the nematocytes of the cnidaria.[2] Then receptor cells evolved from undifferentiated epithelia. Finally "central nervous organs would begin to differentiate in the region between receptors and effectors." (Parker, 1911, p. 224) Initially they would be "organs of transmission whereby the whole musculature could be brought into coordinated action from a single point on its surface (Parker, 1911 pp. 224–225)." Later they would become "the storehouse for the nervous experience of the individual." (Parker, 1911 p. 225) With this theory Parker (1910) consciously displaced an older idea,

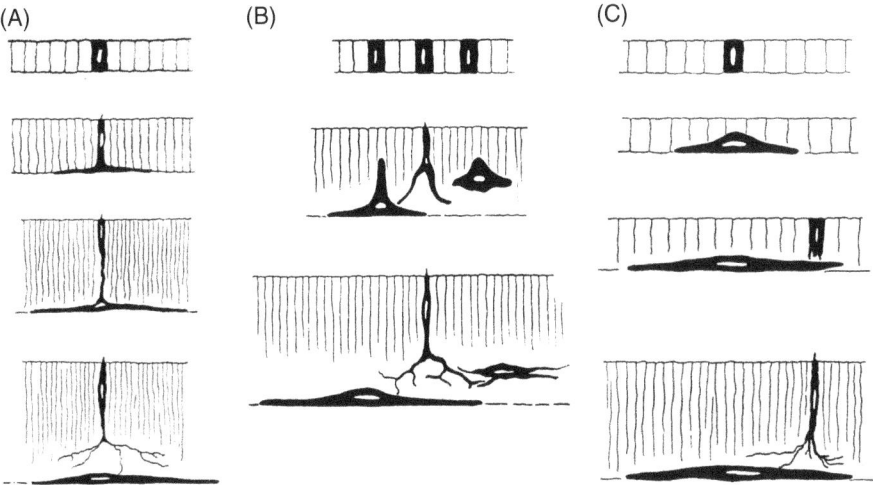

Figure 5.1 Diagram to Illustrate the Way in which Undifferentiated Epithelial Cells Give Rise to the Components of a Reflex Arc according to (A) Kleinenberg (1872); (B) Hertwig and Hertwig (1878); and (C) Parker (1910).
(A) a single cell in an epithelial layer (top) differentiates to form a neuromuscular cell which then separates to give a neuron and a muscle cell. (B) three separate epithelial cells differentiate to form a muscle cell, a nerve cell, and a ganglion cell. (C) one epithelial cell differentiates to form an independent effector; subsequently a second epithelial cell differentiates to form a nerve cell. Parker (1910) finds no evidence for Kleinenberg's neuromuscular cells and opposes the Hertwigs' hypothesis of simultaneous differentiation of nerve and muscle, because of evidence in sponges of contractile elements without nerves. Whether the sponge's immediate ancestor might have already evolved nerves is a matter of current debate (see Ryan & Chiodin, 2015). Adapted from Parker, 1911.

proposed by Oscar and Richard Hertwig (1878) that nerve and muscle cells had evolved simultaneously (see Figure 5.1).

Carl Pantin (1956, p. 173) favored the Hertwig's earlier idea on the grounds that "The Metazoan behavior machine did not evolve cell by cell and reflex by reflex. From its origin it must have involved the structure of the whole animal." Pantin (1956) and Passano (1965) followed Sherrington in believing that integration was an essential property of any nervous system, the first appearance of which would be as a network of neurons because a simple three-component reflex would be incapable of producing an integrated output.

In the 1960s it became generally recognized that animals did more than simply respond to sensory input. Even the simplest nervous systems could generate activity internally. One proposal was that assemblages of electrically coupled "protomyocytes" might develop unstable "pacemaker" regions, becoming specialized for activity initiation with "The specialization of the nerve cell for the conduction … seen as a secondary development" (Passano, 1965, p. 308). Based on his studies on the cnidaria, Passano suggested that this secondary development might arise from regions of the neuronal nerve net that had become specialized for "through-conduction," rather in the way a bypass might be installed once traffic through the city center becomes too congested.

Another important development in the 1960s was the discovery of epithelial conduction (Mackie, 1964, 1965; Mackie & Passano, 1968; Roberts, 1969) and the possibility that neuronal conduction evolved from excitable epithelia as a means to give precision to a system that would otherwise generate only generalized responses to excitation (Horridge, 1968; Mackie, 1970). Other proposals at about that time included the idea that impulse conduction was secondary to the delivery of chemical agents to distant sites (Lentz, 1968). Such chemicals might perhaps include trophic agents delivered by what we now call axonal transport.

Carl Pantin, while emphasizing the complexity of "the neuro-muscular machine" that executes behavior in higher animals, recommended studying its origin and evolution and encouraged the reexamination of "the capabilities of simple systems in the light of contemporary physiological knowledge" (Pantin, 1965, pp. 171, 178). To begin the process we need to define exactly what a neuron is, but this turns out to be a surprisingly difficult task. The ideas of Romanes, the Hertwigs, Pantin, and Passano, as summarized above, suggest that such a definition should encompass:

- the early stages of evolution when the functions of the nervous system might be performed by non-neuronal structures (Romanes, 1875).
- the connection of independent effectors with receptor cells (Parker, 1919).
- the difficulty of making a simple three component reflex produce an integrated output (Pantin, 1956).
- the ability to initiate repetitive activity (Passano, 1965).

A definition that fully satisfies all criteria may well be impossible but as a start: "A neuron is a constituent of the nervous system; its function is to communicate with other cells by way of electrical and/or chemical signals; it is a form of communication that usually takes place via synapses." The word "communication" bears examination: It creates the impression that neurons simply relay what Sherrington (1906) called "states of excitement." However, as living systems became more complex, and their sense of the environment expanded, it became essential to integrate the different streams of information, or there would be no guarantee of an appropriate response. The word "communication" thus carries a lot of baggage—it needs to convey a capacity for carrying highly processed information.

The aim of this chapter is to consider the evolution of both aspects of neuronal function: signal transmission and integration of information. It begins with an account of the *Porifera* and the question of whether the sponges contain the elements of a proto-nervous system. This is followed by a discussion of whether neurons may have evolved more than once. The publication of the genomes of the ctenophores *Mnemiopsis leidyi* (Ryan et al., 2013) and *Pleurobrachia bachei* (Moroz et al., 2014) lends some support to this possibility as it appears that neurons may have evolved independently in the Ctenophora and the Cnidaria. Either this, or "much of the genetic machinery necessary for a nervous system was present in the ancestor of all extant animals" (Ryan et al, 2013, p. 1342)[3] being secondarily lost in the lineage that lead to the sponges. Much of the rest of the review is taken up with a summary of the properties of the cnidarian nervous system, because it is here that we have the most clues to nervous system origins.

5.2 Non-neuronal Reflexes in Porifera

According to Robert Grant (1936, p. 108) sponges lack "perceptible nervous or muscular filaments or organs of sense." The key word here is "perceptible": Only after many years of careful histological study was it finally established that sponges really do not possess nerve-like elements (Jones, 1962; Pavans De Ceccatty, 1974), which is *not* to say that they do not exhibit reflexes or conduct electrical impulses and show evidence of contractile apparatus (see Leys & Meech, 2006 for references).

Almost all sponges are filter-feeders, the water being pumped by beating flagella through a system of delicate canals. The danger is that these canals are susceptible to damage by sediment in the incoming water, a problem that different classes of sponge counter in different ways. The cellular sponges (*Calcarea* and *Demospongiae*) regulate the flow by compressing the flagellated chambers or by contracting the canal system (see Leys & Meech, 2006), while the glass sponge *Rhabdocalyptus dawsoni* (*Hexactinellida*) provides an entirely different model based on an electrical mechanism.

Hexactinellids are not the easiest animals to record from, but they do recognize, and accept, grafted cells. It is from these grafts that all-or-nothing Ca^{2+}-based action potentials (APs) may be recorded in response to electrical stimulation (Leys & Mackie, 1997; Leys, Mackie, & Meech 1999). These APs, which last at least 5 seconds and propagate at a speed of 0.17–0.3 cm/s, are associated with arrested water pumping (see Figure 5.2). Much of the soft tissue in *Rhabdocalyptus* consists of a single syncytium, which means that impulses can propagate freely. Other sorts of cells are attached to the reticulum by "plugged junctions." As these junctions are little more than protein-filled holes, impulses can pass through them too and arrest the beating of flagellated cells. Presumably the entry of Ca^{2+} during the impulse affects the ciliary beating in much the same way that it does in *Paramecium* (Naitoh & Eckert, 1969) and that is what brings water pumping to a halt.

It is evident that complex defensive behavior, precisely coordinated if somewhat slow, can arise without the participation of neurons. The freshwater sponge *Ephydatia muelleri* provides another example. It protects itself from damaging sediment by a non-neural reflexive "sneeze" (Elliott & Leys, 2007). The sponge first inflates; there is a period of hiatus, and then an explosive contractile spasm causes accumulated indigestible matter to be expelled through the osculum. Observation of this stereotyped inflation–contraction behavior was made possible by the transparency of the preparation; the authors were able to watch cells crawling through the sponge interior and note that they arrested their movement for as much as 10 minutes while the wave of contraction passed by, as if the spasm was coordinated by a diffusible chemical messenger (Elliott & Leys, 2007). The propagated contractions can be mimicked using low levels of glutamate and prevented with the metabotropic glutamate receptor antagonist, L-2-amino-3-phosphonopropionate (L-AP3) or the competitive inhibitor kynurenic acid (Elliott & Leys, 2010).

Sponge larvae exhibit another form of behavior coordinated without nerves. According to Nielsen (2008, p. 254), humans, as well as all other eumetazoans, are "descendents of a derived sponge larva." Most sponge larvae are covered by a ciliated columnar epithelium (Maldonado & Bergquist, 2002) whose beat provides the force for forward movement. The larvae of *Reniera* are negatively phototactic in the first

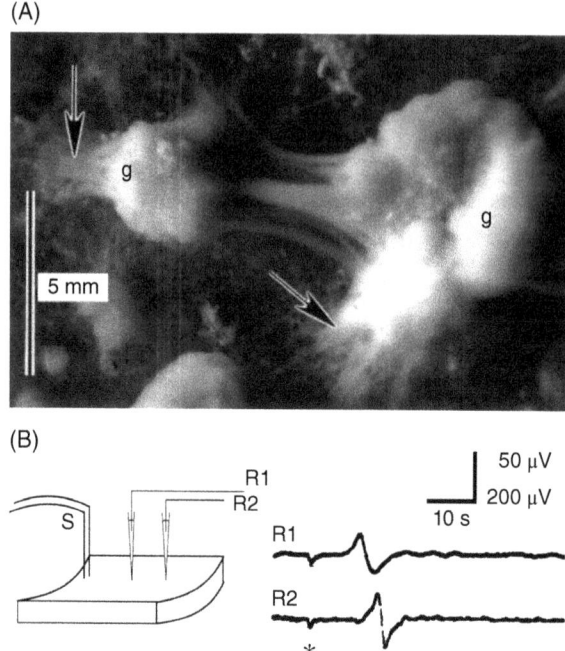

Figure 5.2 Impulse Generation in the Glass Sponge *Rhabdocalyptus dawsoni*.
(A) Dispersed and reaggregated sponge cells (g) 24 hours after having been placed on the surface of a piece of the "donor" sponge. The two parts of the graft are connected by "toffee"-like strands (arrows) that establish syncytial continuity between graft and sponge. (B) Conduction time: an impulse initiated at (S) propagates past recording electrodes (R1, R2). Stimulating and recording electrodes arranged in a line as shown; the calculated conduction velocity was about 0.3 cm s^{-1}. An asterisk marks the stimulus artifact. From Leys, Mackie, & Meech, 1999.

few hours of life (Leys & Degnan, 2001). In addition to the short cilia already mentioned, larvae have a "tuft" of pigmented epithelial cells, each of which gives rise to a single long cilium (Maldonado, 2006). These long cilia, which are assumed to act as a kind of rudder, take up "relaxed," "contracted" or "expanded" arrangements depending upon whether the light intensity is increased or decreased. How light brings about the transformation is not known but the end result is that a sensory input is integrated with directed movement, enabling *Reniera* larvae to settle in shaded regions of a reef.

Whatever their position on the phylogenetic tree, sponges make a good, if somewhat static, living in the absence of a nervous system. Even if the sponge's immediate ancestor had the genetic machinery to build a nervous system, the evolution of a rigid skeleton in the presence of a relatively stable environment would make rapid response times unnecessary; diffusion-based chemical coordination would be quite sufficient. Consequently, the slow electrical conduction seen in *Rhabdocalyptus* should not be thought of as a degraded version of neural conduction but as an example of how easy it is for natural selection to assemble the molecular architecture needed for propagating signals.

5.3 The Ctenophore Enigma

Ctenophores are voracious predators whose locomotion depends on the beating of (usually) eight bands of cilia, an arrangement that makes them highly maneuverable (Tamm, 2014). They can hunt their prey by smell ("chemokinetically") (Swanberg, 1974) and can escape from predators (Kreps, Purcell, & Heidelberg, 1997). Feeding responses are triggered when prey contacts either the lips (*Beroe*) or the tentacles (*Pleurobrachia*). Their orientation is influenced by information from a statocyst in the aboral organ.

It is unfortunate that we know so little about the ctenophore nervous system and its electrophysiology. What little we do know has had to be inferred from anatomical and behavioral studies. The ctenophore nervous system takes the form of two distinctly separate nerve nets (Jager et al., 2011) with a high density of nerve elements at the aboral sensory pole and along the ciliary bands; bipolar and tripolar neurons are generally distributed under and between the ectodermal epithelial cells (Hernandez-Nicaise, 1973). In *Beroe* the nerve net is denser around the lips and in *Pleurobrachia* and *Hormiphora* two thick strands of fibers and neurons connect the aboral organ and the tentacles. There are frequent and highly differentiated synaptic contacts suggestive of chemical transmission and electrical events resembling excitatory synaptic potentials have been recorded from isolated muscle fibers (see Meech, 2015). To judge from the density of nerve elements any integration of the animal's responsive movements must take place either at the ciliary bands, which are subject to a variety of sensory inputs relating to prey capture and balance, or at individual muscle fibers in the body wall.

Unlike the cnidaria (see later), which have a myoepithelium, ctenophores have "true" muscle cells each with its own neuronal input. In *Beroe ovata*, muscle cells are of three main types, longitudinal, radial, and circular (Hernandez-Nicaise & Amsellem, 1980), each fiber running freely within the transparent extracellular matrix, the mesogloea. Longitudinal fibers may be many centimeters long in large specimens because some travel the whole length of the body; radial fibers are often as much as 40 µm in diameter. These multinucleate cells resemble vertebrate smooth muscles in that they lack striations or a transverse system of T-tubules (Hernandez-Nicaise & Nicaise, 1986) and they contract upon stretching. Their sarcoplasmic reticulum makes up less than 1% of the myofilament volume (Cario, Malaval, & Hernandez-Nicaise, 1995), contraction depending on influx of Ca^{2+} from the bathing medium rather than intracellular Ca^{2+} release. The characteristics of the ion channels in the membrane of each of the different classes of muscle establish the ion currents that mold the form of the muscle AP. This sets the amount of calcium entering the cell and therefore contributes significantly to its contractile properties (Hernandez-Nicaise, Mackie, & Meech, 1980; Bilbaut, Meech, & Hernandez-Nicaise, 1988; Bilbaut, Hernandez-Nicaise, Leech, & Meech, 1988).

Radial fibers are capable of generating a train of APs in response to a maintained depolarization, whereas excitation of longitudinal muscle elicits only a single plateau-shaped AP and a short-lived contractile response such as might open the mouth during swallowing. It is the maintained tension generated by the radial fibers as they contract against the viscoelastic mesogloea that keeps the animal stiff, and this suggests that the nerve net must provide continuous excitation. Whether ctenophore neurons do any more than simply distribute excitation remains a key question.

5.4 The Cnidarian Nervous System

"I think everyone will feel it to be obviously true that if ganglionic action is ever to receive any considerable elucidation, the medusae are by far the most promising structure to yield it." (Romanes, 1880)

Textbooks have, for far too long, given the impression that nervous systems evolved in a simple progression from nerves to nerve nets, from nerve nets to ganglia, and from simple ganglia to the complexities of the human brain, with each stage awaiting dramatic changes in the host species' genome. In fact, a more likely scenario is that early metazoans contained collections of genes that afforded nervous system construction—not in the early stages as a necessity, but certainly as a possibility—with the form of the first nervous systems depending as much on the properties of their host's life cycle, development, and skeleton as on limitations imposed by their neuronal components.

Two examples will illustrate the links between body plan, life cycle, and neuronal function. According to Passano & Pantin (1955), the form of the sea anemone *Calliactis parasitica* resembles a giant sense organ. The whole body is structured like an integrative surface with a thin tentacular disc sitting ear-drum-like on a stout cylindrical column. Sensory information is gathered with maximum sensitivity from all over the anemone's surface. Most sea anemones have a sedentary existence and, as we shall see, neurons in the form of a diffuse nerve net are well-adapted to connect the sensory input, in the form of vibrations, to the necessary motor output.

In contrast, Cnidaria that go through a medusoid stage have evolved the more condensed nervous systems that permit the rapid, complex behavior necessary for survival as a free-living entity. Like other ambush foragers (Colin, Costello, & Klos, 2003; Dabiri, Colin, Katija, & Costello, 2010), the hydrozoan *Aglantha digitale* (Figure 5.3A) lives by "sink-fishing" spending part of its time sinking with its tentacles fully extended trapping prey, and part of its time regaining height in the water column by performing a series of slow swims. During the swimming phase of the cycle it rights itself with a series of asymmetrical contractions, and then swims upward with its tentacles contracted (Mackie, 1980). As shown in Figure 5.3C *Aglantha* can also generate fast swims to escape from predators (Donaldson, Mackie, & Roberts, 1980). This varied behavior emanates from a highly organized nervous system. *Aglantha* has nerve tracts that run from the tentacle-fringed margin to control its lips during feeding; has pacemaker neurons responsible for rhythmic swimming; has inhibitory synapses to coordinate swimming and feeding; and has giant motor axons that support its rapid escape (Mackie & Meech, 1985; Mackie, Marx, & Meech, 2003; Meech & Mackie, 1995; Roberts & Mackie, 1980). Furthermore, its electrically coupled myoepithelium, which is organized into "fields" by a grid of lateral neurons (Figures 5.3B, 5.4C; Kerfoot, Mackie, Meech, Roberts, & Singla, 1985), is capable of graded and regionalized contractions.

The way the nervous system is organized in each of the two main cnidarian body patterns broadly defines their behavioral capabilities. In the usually sedentary anthozoan polyp, the nervous system takes the form of a diffusively conducting nerve net, although some condensation is seen in some sea anemones and corals that exhibit quite complex, apparently coordinated, movements. Condensation is more obvious in

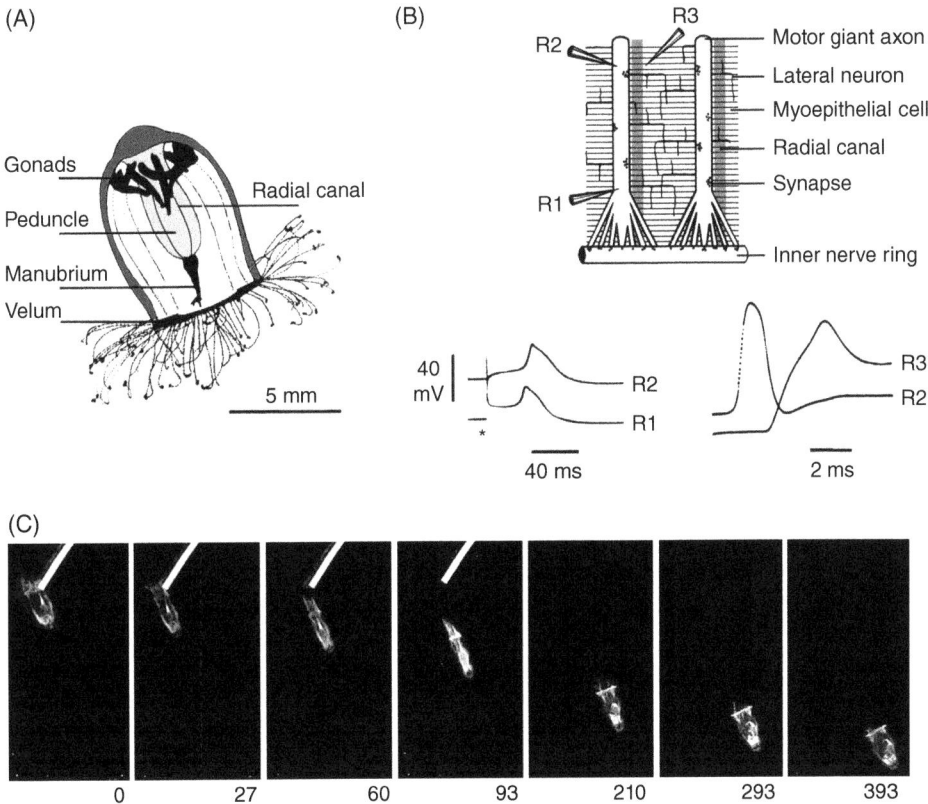

Figure 5.3 Swimming in *Aglantha digitale*.
(A) Transverse section through *Aglantha digitale* (line drawing from a photograph by Claudia Mills). The animal is resting at the end of a swimming cycle. Many fine tentacles extend from the margin at the base of the bell. The margin also contains the neurons responsible for generating slow swimming. (B) Diagram showing two motor giant axons in synaptic contact with the myoepithelium and electrically coupled to an array of lateral motor neurons. The radial canal (shaded area) runs in the endoderm below. Intracellular recording sites R1, R2, and R3 are shown. From Kerfoot et al., 1985. Inset (below right): overshooting action potential recorded at site R2 with the corresponding electrical event in the nearby myoepithelium, R3. Axon resting potential, −62 mV; myoepithelium resting potential, −77 mV. Inset (below left): low-amplitude slow swim spike recorded at sites R1 and R2 in response to a brief depolarizing current pulse (*); a small increase in the intensity of the injected current was sufficient to elicit an overshooting action potential in the same axon; axon resting potential, −72 mV. Mackie and Meech, 1985. (C) Video frames of an escape swim elicited by a glass probe used to disturb vibration receptors at the base of the tentacles; time (ms) after stimulus shown at the bottom of each frame. Frames were captured at 1/300 second. During an escape swim the jellyfish bell contracts uniformly along its length but during recovery the base recovers before the mid-bell. Meech, 2015.

the medusa of the Scyphozoa, Cubozoa and Hydrozoa, and these more accessible modular nervous systems, have revealed more of their integrative mechanisms. In this review I shall focus, almost exclusively, on examples from the Hydrozoa with only a short summary to show some differences in the Scypho- and Cubozoa.

5.4.1 Complex Behavior and the Anthozoan Nerve Net

When Carl Pantin started his classic sea anemone experiments it was believed that a nerve net would conduct excitation in all directions according to the strength of the stimulus; in other words, the greater the strength of the stimulus, the farther its effect propagated. The implication was that the electrical signal died away as it traveled, rather than being the all-or-nothing phenomenon that Adrian (1912) had demonstrated in frog nerves.

A visit to Naples marine station gave Pantin (1935a, 1935b, 1935c) the opportunity to correct this idea. Following a series of electrophysiological experiments on crab nerves he decided to compare them with the closure reflex of the sea anemone *Calliactus parasitica*. He soon realized that the size and nature of the reflex was completely independent of the strength of individual stimuli. In fact, it was controlled simply by the frequency and number of shocks, just like the "all-or-nothing" responses of vertebrate axons. The secret of Pantin's success was that he used physiological stimuli; previous workers had used either individual shocks or high-frequency shocks that ignored the refractory period. A striking feature of the sea anemone nervous system was its very low background activity; in a normal resting state an impulse was recorded less than once in every 10 seconds. Even when artificially stimulated, a rate of one AP/second evoked a maximal response. "The whole scheme is thus 100 to 1000 times slower than that which characterizes the nervous organization of the vertebrates." (Pantin, 1935b, p. 154)

Calliactus has a very strong defensive withdrawal response which can be elicited using either mechanical or electrical stimuli. The main muscles involved are the ectodermal longitudinals in the tentacles and the endodermal marginal sphincter, which is at the top of the column and closes it off like a draw-string bag (Hall & Pantin, 1937; Pantin, 1935a). Pantin found that although a single stimulus produced no response at all, multiple stimuli produced a graded response, the amplitude and extent of which depended on frequency. Having already realized that cnidarian nerves were all-or-nothing systems Pantin concluded that although a single impulse might not cause contraction by itself, it could facilitate transmission. "The nerve net consists of units that behave like true nerve. It is characteristic that it tends to conduct stimuli in all directions; but its most striking feature is the extreme degree to which facilitation is developed both within the net and between the net and the muscles." (Pantin, 1935a, p. 136) Facilitation between the net and the muscles would explain why two or more stimuli were needed to initiate contraction, while facilitation within the net would mean that the contraction would spread out with increasing numbers of stimuli.

There is a striking absence of information about the mechanism of this facilitation in either the nerve net or at the neuromuscular junction. It is possible that it is a presynaptic phenomenon but this is far from certain. A form of neuromuscular facilitation studied in the dyphid siphonophore *Chelophyes* arises from the progressive inactivation of a repolarising K^+ current during a series of muscle APs (Inoue, Tsutsui, & Bone, 2005), and a similar inactivating current has been recorded from isolated *Calliactis* muscle cells (Holman & Anderson, 1991). Muscle APs recorded by Holman and Anderson (1991) get progressively longer during a train as expected from a decline in the availability of repolarizing K^+ current. Since the strength of contraction depends upon the amount of Ca^{2+} influx the longer the AP the stronger the contraction will be. If this mechanism provides the basis for facilitation in *Calliactis*, it should fade as the muscle K⁺ channels recover from inactivation, but no data is available for this characteristic at present.

Although superficially similar in form, different species of sea anemone have evolved markedly different ways of carrying out defensive withdrawal. In *Metridium senile* (plumose anemone) the major muscles involved are the longitudinal retractors of the mesenteries. These muscles contract and withdraw the oral disk and it is only after this withdrawal is completed that the sphincter contracts and provides a cover. In another species, *Anemonia viridis* (snakelocks anemone), Pantin could find no reaction that was obviously protective; the sphincter is barely developed in this species and the disk does not close (Pantin, 1935b). Stimuli that might produce withdrawal in *Calliactis* or *Metridium* appear to evoke a feeding response in *Anemonia*. There is clearly scope for exploring the links between the nature of the environmental stresses, the mechanism of defense and the molecular basis for facilitation in each of these different species.

5.4.2 Mechanisms of Neuronal Integration in Medusae

Most jellyfish are radially symmetrical with either a bell or umbrella shape (see Figures 5.3A and 5.5A for bell-shaped hydrozoan jellyfish). The body wall consists of two layers of epithelium separated by an elastic jelly-like layer, the mesogloea (Figure 5.4A). The outer (exumbrella) epithelium is a layer of simple epithelial cells. The inner (subumbrella) epithelium is made up of a sheet of myoepithelial cells each of which consists of an elongated muscle with a cell body connected to it by a short neck (Figure 5.4B). The bell, or umbrella, is fringed with tentacles and hanging down within the subumbrella cavity is the peduncle with the mouth, or manubrium, at its end. The mouth leads to the radial canals which travel up the peduncle towards the bell apex and then on down the subumbrella to the rim or margin. The radial canals are lined with beating cilia, which distribute the products of digestion around the body. They divide up the subumbrellar myoepithelium into sections and are continuous with the ring-shaped lateral canal, which travels all around the margin of the bell. Also at the margin, on either side of the mesogloea, are the inner and outer nerve rings.

From a simple bell-shaped structure the long course of natural selection has crafted a wide range of morphological variants. Each body form specializes in a particular foraging strategy and associated with these strikingly different patterns of behavior are variations in the neuronal ground plan. Some jellyfish, *Aequorea victoria* for example, are more umbrella-shaped (oblate), the polar axis being shorter than the diameter. Others, such as *Polyorchis* and *Aglantha* are taller than they are wide. *Aequorea* forages/swims for more than 80% of the time (Colin & Costello, 2002; Colin et al., 2003) using a rowing motion to set up a steady feeding current that draws prey into the entrapping fringe of tentacles; *Polyorchis* and *Aglantha* engage in "sink-fishing," a type of foraging that uses intermittent jet-propelled swimming for which their more streamlined prolate form is well suited.

The advantage to the swimming exhibited by oblate jellyfish is that it can be maintained for long periods. The disadvantage is that the trapped prey is so far removed from the mouth that, for transfer from the tentacles to take place, swimming must cease and the rim of the bell be drawn inwards. In *Aequorea* this is achieved by the contraction of discrete bands of radial muscles that run down the subumbrella from the top of the bell to the margin (Satterlie, 1985). Naturally, "sink-fishing" has its

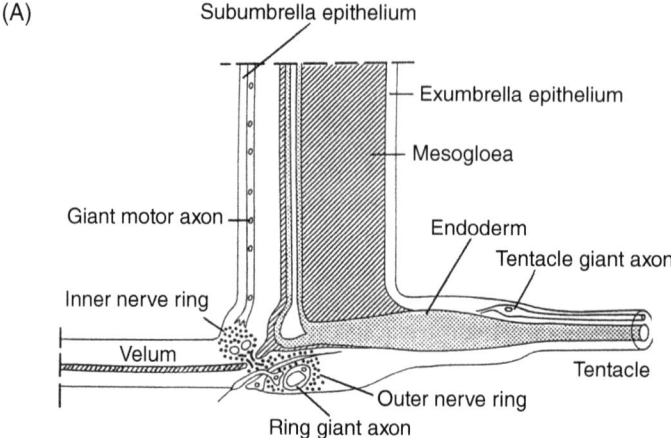
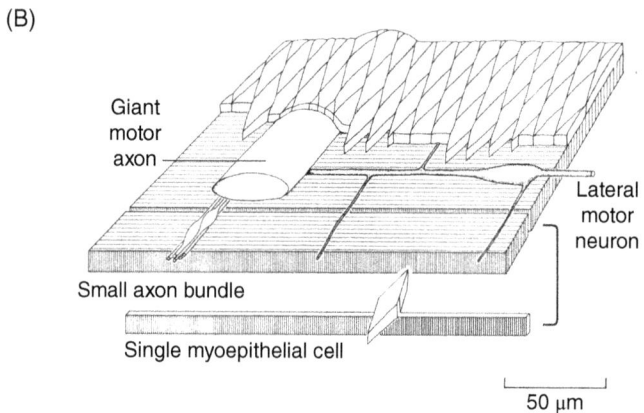
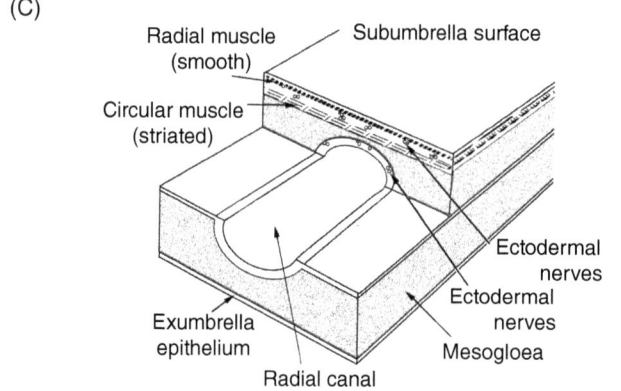

Figure 5.4 Hydrozoan Anatomy.
(A) Per radial section to show the marginal nerves. Running in the nerve rings at the base of the velum are about 800 neurones; most are less than 1 μm in diameter, but the motor giant, ring giant, and tentacle giant axons are significantly larger. The nerves that cross the mesogloea at the base of the velum connect the nerve rings. Hair cells make contact with the ring giant axon directly. Connections passing around the sides of the tentacle connecting the tentacle giant axon to the outer nerve ring have been omitted. Also omitted is the small nerve bundle that runs beside the motor giant. Mackie & Meech, 1995a.

advantages too. One is that during the non-swimming phase jellyfish such as *Aglantha* deploy an extended field of tentacles to trap their prey. The tentacles are lined with beating cilia that set up water currents to draw the prey inwards (Mackie, Nielsen, & Singla, 1989). A disadvantage is that without swimming *Aglantha* will sink continuously in the water column. To swim upward again, it must shorten its widespread field of tentacles so as to reduce the effects of drag (Mackie, 1980).

Before examining the molecular basis of some of these behaviors in more detail, some account is necessary of the subumbrellar musculature and its neuronal innervation. Despite a relatively simple ground plan (or perhaps because of it) there are a large number of variations. In *Neoturris breviconis* (Figure 5.4C) the entire subumbrellar surface is covered by both circular striated muscle and radial smooth muscle while in other species, such as *Stomatoca atra*, the radial smooth muscle is confined to a thickened band above each of the radial canals. In *Aglantha*, radial muscle is completely absent from the subumbrella although it remains in the manubrium and the velum.

The radial muscles that play such an important role in *Aequorea* in bringing trapped prey to the manubrial lips also have an important defensive role. Many hydromedusae respond to agitation by exhibiting a behavior described by Hyman (1940) as "crumpling." The response, which can be elicited by mechanical or electrical stimulation of the exumbrella, involves the contraction of the radial muscles of the bell and the longitudinal muscles of the tentacles so that the margin and tentacles are drawn up into the subumbrellar cavity. Crumpling is absent in *Aglantha*, which does not have radial muscle in its bell and which exhibits another form of defense: escape swimming (see Figure 5.3C; Donaldson et al., 1980).

In *Polyorchis*, the organisation of the swim motor neuron network, which extends in an arch at the apex of each subumbrellar quadrant (Lin, Gallin, & Spencer, 2001), suggests that excitation of the electrically coupled muscle sheet spreads from all four sides. The synchronous firing of the swim motor neurons around the nerve ring ensures that all quadrants contract together, producing symmetrical swimming movements. In *Aglantha* the motor axons travel up the bell next to each of the (in this case) eight radial canals. Symmetrical contraction during escape swimming is accomplished by an unusual ring-shaped neuron, the ring giant, located in the outer nerve ring. This provides an almost coincident input into the eight giant motor axons. Excitation then spreads into the myoepithelium from neuromuscular junctions distributed along each axon. Electrical coupling between myoepithelial cells ensures that the depolarizing current spreads around the bell, but the spread of excitation is also promoted by lateral neurons (see Figure 5.3B) that are electrically coupled to the motor axons and innervate the myoepithelium in discrete, mostly non-overlapping,

Figure 5.4 (*Continued*) (B) Diagram to show the position of a giant motor axon within the myoepithelium between the muscles and their cell bodies. An isolated myoepithelial cell is shown at the bottom of the figure to clarify its structure. The muscles run in a circular direction. Lateral motor neurons, which also run below the cell bodies, make electrical contact with the giant axon and innervate the myoepithelium in distinct fields. The small axon bundle runs parallel to the giant axon. The radial canal which runs within the endoderm, parallel to the giant axon is not shown. Adapted from Kerfoot 1985. (C) Schematic layout of nerves and muscles in the subumbrella of *Neoturris breviconis* to show their relationship to the radial canal. Note the presence of circular smooth muscle. Adapted from Mackie & Meech, 2008.

fields (Kerfoot et al., 1985; Weber, Singla, & Kerfoot, 1982). As a consequence depolarizing junction potentials may be recorded at all sites in the myoepithelium.

In *Aequorea*, swimming movements are somewhat less symmetrical than in *Aglantha*. The myoepithelium is divided by 50 or more radial canals, and excitation is largely confined to individual stimulated segments. Swims are initiated by a burst of APs in the equivalent of the swim system in the inner nerve ring. In the subumbrellar myoepithelium the AP burst is translated into a flurry of junction potentials giving rise to a single long-lasting AP (Satterlie, 2008). As with *Aglantha* there are junction potentials at all sites, and the widespread presence of neurites suggests that the spread of excitation within the myoepithelium depends on a subumbrellar nerve net. Local inhibition of this nerve net can prevent swimming contractions in restricted regions of the bell (Satterlie, 2008).

Although we might question speculations about neuronal evolution based on present-day morphologies, it seems not unreasonable to conclude that the first medusa was broadly bell-shaped and used its myoepithelium to swim and its tentacles to trap food. That being so, the rich variety summarized in this section gives us the means to explore the relationship between body form, behavior and neuromuscular function, so as to identify recurring mechanisms of integration that may have arisen in the earliest nervous systems. The main areas of interest are: (1) swimming, (2) control of tentacle length, (3) defense, (4) swimming inhibition and mouth manipulation during feeding.

5.4.2.1 Swimming. Among the Hydrozoa one of the best worked out neural systems is in the anthomedusa *Polyorchis penicillatus* (Figure 5.5A) where the swim pacemakers are sufficiently large for intracellular recording (Spencer, 1981) and voltage-clamp (Przysiezniak & Spencer, 1989). The trachymedusa *Aglantha digitale*, with its 40 μm diameter motor axons, has also been extensively studied under voltage and patch-clamp (Meech & Mackie, 1993a, 1993b). A major difference between the two animals is that in *Aglantha* the pacemaker neurons and the swim motor neurons are separate entities whereas in *Polyorchis* the pacemaker neurons directly innervate the myoepithelium.

5.4.2.1.1 Control by the nerve ring of hydrozoan jellyfish. The pacemaker activity that provides for rhythmic swimming in hydrozoan jellyfish arises from nerve rings at the base of the bell. In *Polyorchis* (Figure 5.5A) the pacemaker is affected by ambient light levels and includes as many as three interconnected modules, each consisting of an electrically coupled ring of neurons (Spencer & Arkett, 1984). As indicated in Figure 5.5B, activity in the 'swim system" is modified by inputs from the "B system" and the "O system." In other jellyfish, such as *Aglantha*, the pacemaker system is unaffected by light.

5.4.2.1.2 The "swim system." When *Polyorchis* swim, motor neurons show slow, apparently endogenous, baseline oscillations even when high levels of Mg^{2+} are used to block synaptic and muscular activity (Satterlie & Spencer, 1983). They are electrically coupled via gap junctions (Anderson & Mackie, 1977; Spencer, 1975) and, as in other examples of electrically coupled neurons, would be expected to respond only to inputs that affect many cells in the network simultaneously (Willows & Hoyle, 1969). In fact the inputs need not be exactly coincident because their EPSPs take about

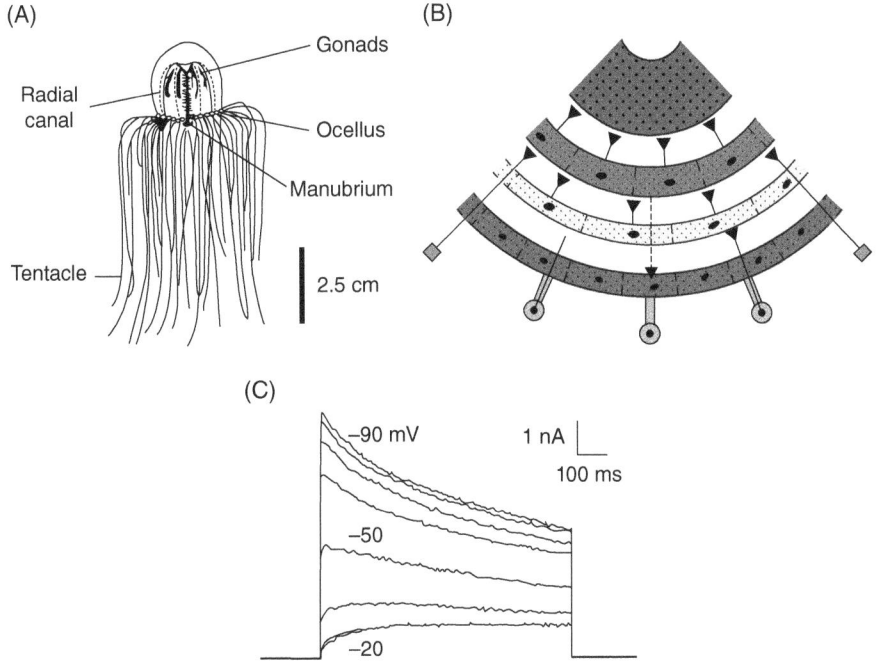

Figure 5.5 Modular Nature of *Polyorchis penicillatus* Nervous System.
(A) Line drawing of a *Polyorchis* specimen at rest. (B) Partial representation of the ring-shaped neuronal networks showing known synaptic contacts. The muscle epithelium (blue) is directly excited by the swim motor neurons (red) in the inner nerve ring. They in turn receive an excitatory synaptic input from neurons of the "B" system (yellow) in the outer nerve ring and from unicellular receptors (orange). Excitatory inputs to the "B" system are shown arising in the ocelli (purple). Also shown is an excitatory pathway from the swim motor neurons to the "O" system (green) in the outer nerve ring; other connections may be present. Excitatory synapses are indicated by filled triangles (▼). All three systems consist of a ring of electrically coupled nerve cells. Adapted from Spencer & Arkett 1984. (C) Inactivating K⁺ currents recorded under voltage clamp in response to test commands to +50 mV. Each test command preceded by a conditioning command lasting 1 s. The superimposed current traces show the effect of the conditioning level (range −90 to −20 mV; 10 mV steps) on the availability of the inactivating current. At −20 mV the inactivating component (IK_{fast}) is absent, leaving IK_{slow}. Adapted from Przysiezniak & Spencer, 1994. (*See insert for color representation of the figure*).

750 ms to reach a maximum and have an overall duration of 1.5 second (Spencer & Arkett, 1984; Mackie, Meech, & Spencer, 2012). This slow time course is an inevitable consequence of the electrical coupling and the low-pass filtering arising from the series resistance and capacitance to ground of the postsynaptic cell membrane (Bennett, 1966; Bennett & Zukin, 2004; Spencer, 1981).

Na⁺, Ca²⁺ and K⁺ channels have been studied in dissociated swim system neurons under whole cell voltage-clamp (Przysiazniak & Spencer, 1994). The outward K⁺ current (IK_{fast}), which flows only transiently because it undergoes rapid inactivation, resembles a current thought to dominate the latter stages of the inter-spike interval in repetitively firing neurons (Connor & Stevens, 1971). In *Polyorchis*, IK_{fast} is half-inactivated with the membrane set to −52 mV (Przysiazniak & Spencer, 1994) although the inward Ca²⁺ and Na⁺ currents show little inactivation at the same voltage (Grigoriev et al., 1996; Przysiazniak & Spencer, 1992). Thus the long-lasting EPSPs can markedly affect the availability of pacemaker current while having little effect on channels necessary for spiking (see Meech, 2015).

5.4.2.1.3 The "B system."
Although the primary role of the "B system" is the regulation of tentacle length (see §5.4.2.2), swimming is frequently preceded by a twitch contraction of the tentacles. B system neurons provide an input to the swim system from all around the nerve ring (see Figure 5.5B). They are electrically coupled and, as a consequence, AP spiking is synchronous throughout (Anderson & Mackie, 1977; Spencer & Arkett, 1984). As in the swim system, B system neurons appear to be endogenously active (Satterlie & Spencer, 1983).

Although a single B system impulse rarely induces activity in the swim system, a study by Passano (1973) on another well-studied hydrozoan species *Sarsia tubulosa* found that double pulses almost always do. A pair of B system impulses initiated a swim system impulse after a characteristic delay that depended on the inter-pulse interval. The delay was surprisingly long; about 750 ms if the B system pulses were 1 s apart. Presumably swim initiation arises from the combined effects of EPSP summation plus inactivation of *Sarsia*'s version of IK_{fast}. This explanation would account for the presence of two processes each following a different time course (Passano, 1973).

Leonard (1982) finds that swim bouts in *Sarsia* are separated by highly irregular intervals even in an aquarium. She suggests that *Sarsia*'s "B system" equivalent consists of a number of independent pacemakers, each responsive to local conditions. Although ineffective when acting alone, once synchronized they could activate the swim pacemaker by advancing its cycle. Longer intervals might arise if the initiation of a swim bout depends on two or more rapidly oscillating pacemakers becoming active at the same time.

5.4.2.1.4 The "O system."
Swimming in *Polyorchis* is markedly affected by the level of ambient light. An important, but by no means exclusive, action of light is on a second system of coupled neurons in the outer nerve ring, the "O system" (Spencer & Arkett, 1984). Photosensitivity in the O system remains even when the animal's ocelli are removed. Thus, either the O system is itself light-sensitive or it receives an input from extra-ocular photoreceptors such as those described by Satterlie (1985). The swim and O systems appear to have common synaptic inputs and are inhibited by activity in the ectodermal epithelium.

5.4.2.1.5 Swimming in *Aglantha digitale*.
The trachymedusa *Aglantha digitale* is apparently unique in that it is capable of two forms of jet-propelled swimming: escape and slow swimming (Figures 5.3 & 5.6). The two forms depend on the same set of motor nerves, with the different strengths of contraction arising from differences in the propagating signal: a fast overshooting Na⁺-dependent AP and a slower, low-amplitude Ca²⁺-dependent "spike" (Figure 5.3B; Mackie & Meech, 1985; Meech & Mackie, 1995).

During an escape swim rapidly depolarizing synaptic potentials take the axon membrane beyond the threshold for the Na⁺-dependent AP. In mature specimens these impulses conduct at a velocity of about 4 m s⁻¹ which means that they travel the 2 cm from base to the apex of the bell in 5 ms. They excite fast-rising, large-amplitude junction potentials in the nearby myoepithelial cells with a synaptic delay of less than 1 ms at 10 °C (Figure 5.3B; Kerfoot et al., 1985). The electrical coupling between myoepithelial cells ensures that the current spreads evenly across the myoepithelium but the process is aided by excitation in the lateral neurons. Because of the speed of conduction up the bell, contraction is uniform from top to bottom and round the circumference of the animal (Figure 5.3B). The base of the animal therefore forms itself into a "nozzle," which increases the thrust generated during the swim, albeit with a significant increase in energy consumption.

Slow swimming is part of the feeding cycle. During slow swimming, pacemaker cells in the inner nerve ring set off slowly rising, low-amplitude synaptic potentials in the giant motor axons which activate low threshold "T"-type calcium channels and give rise to low amplitude Ca^{2+}-dependent impulses (Mackie & Meech, 2000; Meech & Mackie, 1993a, 1995). These low-amplitude spikes (Figure 5.3B) propagate along the motor axon with their peak just below the threshold of the Na^+-dependent AP. In larger specimens the contraction is confined to the upper half of the bell (Figure 5.6) and is strongest near the motor axons (Meech, 2015). The strength of contraction depends directly on the amplitude of the potential change at the neuromuscular junction, which in turn reflects the rate of rise of the presynaptic signal. Maintaining a wide opening for expelled water during slow swimming ensures that the velocity is low and provides for more efficient energy use.

The regional differences in contraction strength seen during a slow swim require an explanation, as they are unexpected in an electrically coupled myoepithelium. Intracellular recordings in the giant motor axon show that the fully formed Ca^{2+} spike appears some 4 mm away from the marginal synapses in fully mature specimens (bell height approx. 20 mm). At closer sites the high conductance generated by the pacemaker synaptic potential partially "short circuits" the voltage-gated Ca^{2+} spike. Patch-clamp studies show that these low amplitude potentials activate only low threshold Ca^{2+} channels in the muscle membrane with the partially short-circuited spikes producing the weakest responses. It appears that the myoepithelium, just like the motor giant axon, has two thresholds, and this helps to explain the increased strength of contraction observed during escape swims. When the overshooting AP generates a junction potential in the myoepithelium it activates the high threshold voltage-gated channels and the associated Ca^{2+} influx is large enough to generate a strong contraction (Meech & Mackie, 2006).

5.4.2.1.6 Swimming in siphonophores. There is one further morphological complication to consider; not all the Hydrozoa are single-belled swimmers. In colonial variants (the siphonophores) there are multiple nectophores (swimming bells). These are medusoid in form but lack many of the structures present in hydrozoan medusae. The functions normally carried out by tentacles, gonads, manubrium are instead performed by other highly specialized members of the colony. Species such as *Chelophyes appendiculata* (Figure 5.7A) and *Muggiaea atlantica* have two swimming bells, whereas *Nanomia bijuga* has a variable number of back-to-back nectophores (Figure 5.8A).

5.4.2.2 Control of tentacle length. The control of tentacle length has been explored in a variety of different hydrozoan jellyfish including *Eperetmus typus* (Mackie & Mackie, 1963), *Leuckartiara octona* (Russell, 1953), and *Bougainvillia superciliaris*

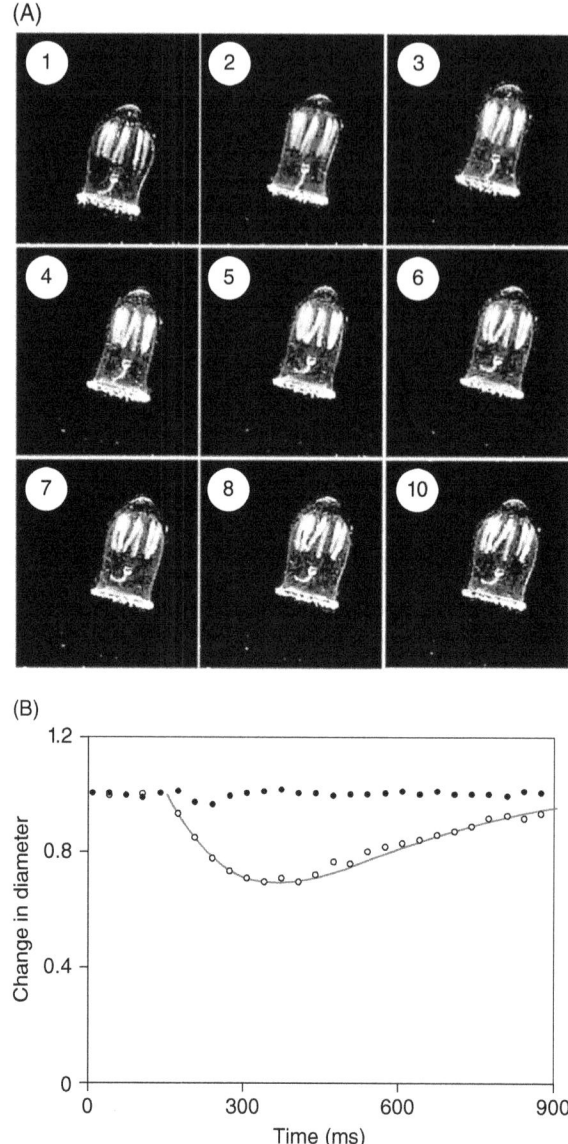

Figure 5.6 Slow Swim in *Aglantha digitale*.
(A) Video frames at 0.1 s intervals of a single slow swim. Frames were captured at 1/300 s. Mature specimen about 2 cm long. (B) Change in diameter at the base of the bell (closed circles) is significantly less than that at the mid point (open circles). Data measured from individual video frames including those in A. Line through the data drawn according to an equation for the performance of a damped oscillator: $x = A_0 \exp^{-bt} (\sin 2\pi t/\tau)$, where x is the change in bell diameter, A_0 is the maximum diameter change in an un-damped system, b is the damping coefficient, t is time and τ is the period of the swim. At the base of the bell, the values during the slow swim are 0.18 (A_0), 0.0038 (b), 1250 (τ). Temperature, 10 °C. Meech, 2015.

(Agassiz, 1849). In each case it appears that tentacle length is under the control of what in *Polyorchis* is called the "B system" (see §5.4.2.1.2). As noted above, the onset of swimming is usually preceded by tentacle shortening. The delay between tentacle shortening and swimming ranges from 1 second in *Proboscidactyla* (Spencer, 1975) to 3 seconds in *Muggiaea* (Meech & Mackie, unpublished).

Species that employ "sink-fishing" have different strategies for managing their elaborate tentacle net. APs in the myoepithelium of *Chelophyes* during a swim sequence show a gradual increase in duration and as a consequence the swims show a gradual increase in strength (Figure 5.7B; Inoue et al., 2005). The efficiency of this strategy arises from the relationship between energy expended and the product of drag and velocity (Vogel, 1994). The weaker swims early in the sequence reconfigure the stem and tentacles so that when the animal has taken up a more streamlined configuration, less energy is dissipated by the stronger swims. In *Muggiaea* swims have equal strength

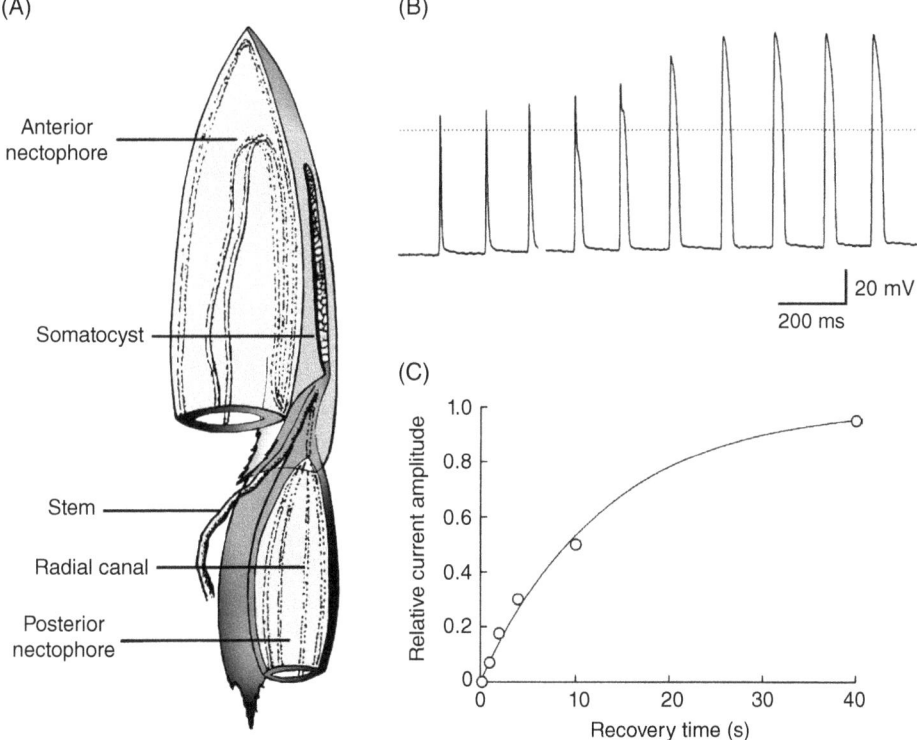

Figure 5.7 Swimming in *Chelophyes appendiculata*.
(A) Drawing showing the anterior swimming bell (anterior nectophore), the posterior swimming bell (posterior nectophore), and the origin of the trailing stem and tentacles. Adapted from Fewkes, 1880. (B) continuous intracellular record from myoepithelial cell during a series of 10 spontaneous swims; the dotted line shows 0 mV. The preparation was bathed in an artificial seawater consisting of (in mmol·l^{-1}) 450 NaCl, 9 KCl, 10 CaCl$_2$, 50 MgCl$_2$ and 15 Na-HEPES buffer (pH 7.8). (C) The recovery of K$^+$ current amplitude with time. The line drawn through the points is an exponential curve with a time constant of 13.2 s. The preparation was bathed in nominally Ca^{2+}-free artificial sea water containing 4·µmol·l^{-1} TTX, and the pipette solution was K-aspartate. (B, C) Inoue, Tsutsui & Bone, 2005.

but the swim sequence is delayed until the tentacle net is fully contracted. Recent observations (Meech & Mackie, unpublished) suggest that the alternative strategies employed by *Chelophyes* and *Muggiaea* are linked to differences in the ion channel make-up of their myoepithelial cells (see §5.5.2).

The complexity of the dual system of swimming exhibited by *Aglantha digitale* is reflected in the neuronal control over its tentacles. *Aglantha* has a cloud of tentacles all of which are capable of both fast (twitch) and slow (postural) contractions. Activation of the slow system produces a graded response in the longitudinal muscle of the tentacle, which can range from a slight curling of the tip to the formation of tight coils. The faster system (the tentacle giant axon) which conducts at 60–90 cm s^{-1} has a high threshold and fires only once or twice after stimulation. The slow system conducts at 15–20 cm s^{-1} and fires in bursts (Bickell-Page & Mackie, 1991). Although the tentacle net is not as asymmetric as in the siphonophores, the escape reflex would nevertheless be significantly compromised if the drag from the trailing tentacles was not minimized. During an escape swim APs in the ring giant axon excite not only the motor axons but also the fast tentacle system (Donaldson et al., 1980; Roberts & Mackie, 1980) causing the tentacles to rapidly shorten. Roberts and Mackie (1980) have suggested that tentacle and ring giants are in electrical continuity allowing the tentacles to contract shortly before the swim (Meech & Mackie, 1995).

Slow swimming, performed by *Aglantha* when "sink-fishing," takes place in bursts of activity of variable length. There is no tentacle shortening before swimming gets started (Mackie, 1980). Instead *Aglantha* uses the *Chelophyes* strategy in that its early slow swims are somewhat weaker than later ones. During later swims the tentacles "shorten and curl inwards" (Mackie, 1980, p. 1551). The slow tentacle system responsible for these postural changes is under the control of the relay system, a neuronal pathway in the inner nerve ring (Mackie & Meech, 1995a, 1995b). The relay system is in turn excited by the pacemaker system so that during a sequence of slow swims, the tentacles go through a series of graded tonic contractions. The pacemaker system not only initiates the impulses in the relay system, but also accelerates them (to about 24 cm s^{-1}) by an unknown mechanism (cf. the "piggyback" mechanism of *Nanomia*; §5.4.3.1) so that tentacles all around the margin respond within about 70 ms, to match the initial contraction of the bell.

Propagation in the relay system is accelerated even further by an impulse in the ring giant axon. Although the accelerated velocity (~41 cm s^{-1}) does not compare to that of the ring giant axon itself (~200 cm s^{-1}) the coupling does ensure that during escape swims fast twitch and slow postural contractions of the tentacles are coordinated all around the bell margin. The relay system also excites the carrier system in the outer nerve ring. When the pacemaker, relay, and carrier systems fire together, they may elicit summating synaptic potentials in the ring giant axon and cause it to spike. If it does so, it fires the tentacle giant so that a stronger tentacle contraction occurs within the sequence of slow swims.

5.4.2.3 Swimming inhibition and mouth manipulation during feeding. When prey is introduced into the mouth of *Aglantha digitale*, swimming is inhibited, probably because high-velocity water flow would detach captured prey (Mackie et al., 2003). Prey, which includes a variety of small planktonic organisms (Arai, McFarlane, Saunders, & Mapstone, 1993), is trapped by tentacles at the base of the bell. Once trapped, the manubrium bends toward the food and engulfs it with its lips prior to ingestion. In flatter species, such as *Aequorea*, a local contraction of the margin is necessary to bring the food close enough to the mouth for transfer to take place (Satterlie, 1985). In *Aglantha* there are no radial muscles on the underside of the bell, but they persist in the manubrium, where they bring about "pointing," described below. Here, they are innervated by bundles of small axons that run in the subumbrella alongside each motor giant axon as seen in Figure 5.4B (Mackie, Singla, & Stell, 1985).

When *Aglantha* preparations are stimulated at the point where a small axon bundle joins the margin, the manubrium accurately "points" to the stimulation site. Pointing is just as accurate for stimuli at intermediate positions as if the small axon bundles are linked by pathways within the nerve ring. We suppose that food captured near the exit of the axon bundle travels directly to the manubrium and its associated muscle band but as impulses spread around the nerve ring this excitation fails to be conveyed to other manubrial muscle bands. It seems that pointing requires the small axon bundle to fire impulses repeatedly producing summated responses in the muscle. However, the later impulses travel more slowly than the first. Hence the further they travel the more separated they will be and the weaker any summated responses in the manubrium (Mackie et al., 2003). See §5.4.3.1 and Figure 5.8B for a similar mechanism in *Nanomia*.

5.4.2.4 Defense. When given an electrical or noxious stimulus, the exumbrella of *Sarsia* produces a series of propagating impulses that are conducted from cell to cell, via electrical junctions (Mackie & Passano, 1968). This capacity for epithelial conduction turns the entire surface of the jellyfish bell into an early warning sensor, prompting a defensive "crumpling" maneuver (Hyman, 1940; Spencer, 1971, 1975). During crumpling, swims are inhibited and the radial muscles that overlie the radial canals contract so that the margin is drawn up into the subumbrella cavity. In *Polyorchis* epithelial APs are associated with long-lasting inhibitory post-synaptic potentials (IPSPs) in the swim motor neurons.

The trachymedusan *Aglantha digitale* does not have radial muscle in its bell. Instead it responds to a noxious stimulus by performing an escape swim (Donaldson et al., 1980). Input from vibration receptors at the base of the tentacles excites the ring giant axon producing a large overshooting AP which excites all eight motor giant axons.

Noxious stimuli also provoke escape in the siphonophore *Nanomia cara* (Figure 5.8A). In this case escape depends on contractions in the double row of back-to-back nectophores that act together as the colony's swimming unit. A neural input to the subumbrellar myoepithelium causes each nectophore to generate a jet of water producing forward swimming. However if the front of the colony makes contact with an obstruction, such as the air–water interface, the radial muscles on either side of the velum contract, the water flow is deflected forwards and swimming is reversed (Mackie, 1964). Whether an animal undergoes forward or reverse swimming will depend on whether excitation travels via an exclusively nerve pathway or via the exumbrella epithelium (Mackie, 1978). How the epithelial impulse excites the subumbrella and the different muscles of the velum is unclear.

Another siphonophore, *Chelophyes appendiculata* responds to a disturbance in the water with a series of escape swims attaining an instantaneous velocity of as much as 30 cm s^{-1}. Chelophyes has two bell-shaped nectophores (Figure 5.7A). Both nectophores contribute to these escape swims, but the smaller posterior nectophore plays only a minor role. Normally its activity is confined to providing occasional contractions to maintain the position of the colony in the water column.

5.4.3 Interactions between Epithelia and Nerves

It is tempting to regard "crumpling" as an example of a proto-nervous system in action. We might suppose that early in the evolution of the Cnidaria, undifferentiated epithelia gained contractile elements (becoming one of G.H. Parker's "independent effectors," see §5.1) while at the same time acting as a sensory surface (Mackie, 1970). The ability to conduct APs through the epithelium would mean that information could be conveyed directly from sensory to contractile regions. The difficulty with this scenario is that King and Spencer (1981) have shown that during "crumpling" the final link between epithelium and muscle is a neuronal one. So although "crumpling" might reflect a time when excitable epithelia completed the reflex arc, the evolution of muscle cells may have been responsible for relegating epithelial conduction to its present subsidiary role.

Evidence for a carry-over from a stage in evolution, when independent effectors were coordinated by "neuroid" conduction comes from George Mackie's work on endodermal luminescence in *Euphysa*, and exumbrellar luminescence in *Hippopodius* (Mackie, 1976a; 1991). These examples, together with his discovery of the "Piggyback Effect" in *Nanomia* (Mackie, 1976b), might represent "Parkerian" preneural evolution but for the fact that, among the jellyfish, conducting epithelia are found only in the Hydrozoa (Mackie, Anderson, & Singla, 1984). Furthermore, as we saw in §5.1, natural selection appears to have engineered electrogenesis in so many cell types that it may have evolved in epithelia simply as a nervous system adjunct. I include examples here because excitable epithelia provide fascinating insights into integration in action.

5.4.3.1 The piggyback effect. The piggyback effect describes the observation that a rapidly propagating impulse in one system appears to carry what is normally a slowly propagating impulse in another system along with it. Piggybacking serves to coordinate fast and slow responses and was first observed by Mackie (1976b) in his studies on the stem of *Nanomia*. The stem is a muscular thread of tissue that connects the different individuals that make up the siphonophore colony (Figure 5.8A). It contains two rapidly conducting neural pathways, together with a double myoepithelium consisting of an excitable endoderm and a contractile ectoderm. The two large (30 μm) nerve axons that make up the neuronal pathways are electrically coupled to a distributed nerve plexus whose role is to spread excitation over the muscle sheet (Mackie, 1984).

Impulses in the neuronal system elicit from the ectodermal muscle both graded postural changes and fast contractions. If the two axons fire APs close together there is a summated response in the myoepithelium near to the stimulus site. However the two axons conduct at different rates, so that as the APs travel along the stem they become separated in time and the summated response in the ectoderm gradually dissolves into two weaker contractions (Figure 5.8B).

In addition to the fast conduction in the giant axons, there is a slower system of conduction through the endodermal epithelium. Here a series of spontaneous impulses produce slow postural changes in the stem. If the fast and slow systems are stimulated to fire together, propagation in the epithelial system (normally about $30 \, \text{cm} \, \text{s}^{-1}$) is greatly speeded up.

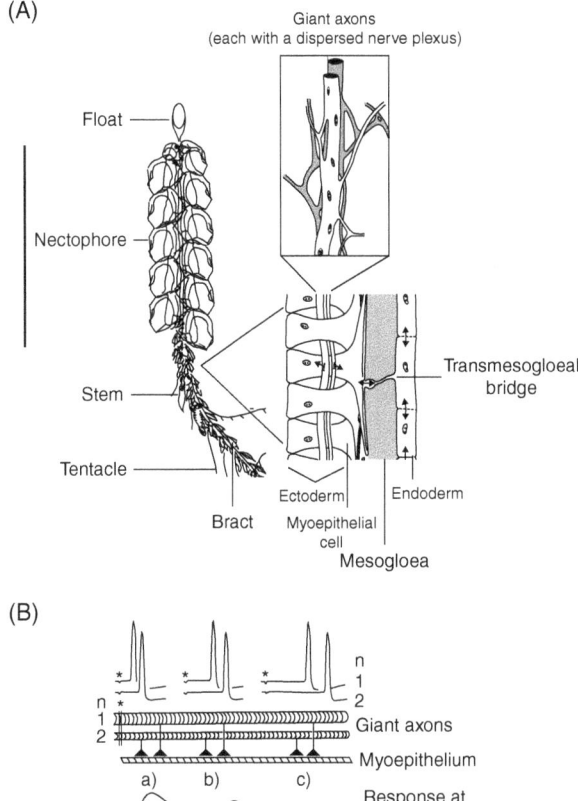

Figure 5.8 Impulse Conduction in the Stem of *Nanomia cara*. (A) Colony at rest showing its upper section containing eleven nectophores and a short length of stem. The scale bar is 2 cm. At higher magnification are insets showing a diagrammatic representation of the organization of the stem and the dispersed nerve plexuses associated with each giant axon. Adapted from Mackie, 1964; 1984. (B) Mechanism for gradation of muscle response intensity with conduction from the stimulus (position and time indicated by an asterisk). Each pair of traces represents internally recorded neuronal impulses (top) and graded responses near a neuromuscular junction in the myoepithelium (bottom). The neuronal trace consists of action potential from n1 (the largest giant axon and most rapidly propagating) followed by that in n2. These overshooting action potentials elicit chemically mediated excitatory junction potentials in the ectodermal myoepithelium. Junction potentials close to the stimulation site sum to give twitch contractions. Overshooting epithelial impulses in the endoderm also contribute to the graded depolarization in the myoepithelium via electrical connections through the mesogloea. Adapted from Mackie 1984.

George Mackie suggests that current passes through transmesogloeal bridges (seen in sections prepared for EM and shown in Figure 5.8A) between the endoderm and the ectodermal muscle. When it flows from the ectoderm it can initiate an epithelial impulse in the endoderm. However, the endoderm will only reach threshold in the vicinity of the stimulus, because it is only here that the currents are large enough. Further away from the stimulus, the muscle responses are weaker and will no longer carry the endodermal impulse. Consequently it will "fall off the piggyback" and its conduction velocity will revert to 30 cm s^{-1}.

The behavioral significance of piggybacking in *Nanomia* is that postural changes and fast contractions are tied together in regions of the stem where the fast contractions are strong, that is, near the point of stimulation. But, as the excitation spreads and the twitch fades, the postural movements revert to their normal rates of propagation. Piggybacking is, therefore, a means of coordinating fast and slow movements.

5.4.4 Schyphozoa and Cubozoa Compared with the Hydrozoa

As we have seen, swimming in jellyfish is driven by pacemaker neurons located at the bell margin. However in the hydromedusae they are electrically coupled and distributed around the inner nerve ring, while in the scypho- and cubomedusae they are localized in, or near, discrete ganglion-like nerve centers called rhopalia. Pacemaking arises at multiple sites (Passano, 1982). In cubomedusae there are four rhopalia, while scyphomedusae have eight or more. Experimental analysis shows that individual pacemakers have a quite irregular rhythm, while modeling studies suggest that an interconnected system, in which a leading pacemaker resets the remainder, is far more stable (see Horridge, 1959; Passano, 1982). The swim frequency of a coupled system is somewhat faster than the natural rhythm of individual pacemakers because each cycle is driven by the first pacemaker to reach threshold. Such a system has the advantage that a synaptic input can lead to rapid upward or downward adjustments in frequency.

5.4.4.1 Control by rhopalia in scyphomedusae. The scyphomedusae, sometimes called "true" jellyfish, rarely swim in short bursts, but instead exhibit longer periods of continuous activity, albeit with a highly variable inter-swim interval (Horridge, 1959). In contrast to the almost synchronized contraction seen in hydrozoan jellyfish, swimming in scyphomedusae consists of a slow wave of contraction originating from one of the ganglion-like nerve centers, or rhopalia, at the bell margin (Romanes, 1877). There are at least eight of these nerve centers, each of which shows pacemaker activity. The rhythm of a single center is less regular than that of the whole animal (Horstmann, 1934), a finding Horridge (1959) attributed to the fact that a group of coupled pacemakers will be driven by the fastest member, thus shortening the time between swims.

Rhopalia, which also contain photoreceptive cells, olfactory pits, and a statocyst (Hyman, 1940), are connected by a motor nerve net. This is a two-dimensional network of large neurons which covers the subumbrellar surface and transmits excitation in any direction from pacemaker neurons to the swimming musculature and from one rhopalium to another (Anderson & Schwab, 1981, 1983). Communication between network neurons in *Cyanea capillata* depends upon bidirectional synapses, in which either side of the synaptic junction is able to release a chemical transmitter (Anderson, 1985).

Activity in a second nerve net, called the "diffuse nerve net" (Horridge, 1956), sets off tentacle contraction (Romanes, 1885), triggers slow manubrial reflexes (Passano, 1973) and modulates activity in the motor nerve net (Horridge, 1956). In many ways the scyphozoan diffuse nerve net corresponds to the hydrozoan B system, in that it can initiate swim impulses (Passano, 1965). When studied in *Cassiopea xamachana*, there was, a significant delay between excitation of the diffuse nerve net and the appearance of the swim pulse (Passano, 1965), just as with the B system in *Sarsia*. Unlike the situation in *Sarsia*, however, a single impulse in the diffuse nerve net produces a swim impulse 40% or more of the time. This has made it possible to study the interaction in more detail. Passano (1965) found that the delay depended on the interval between the impulse in the diffuse nerve net and the preceding pacemaker event. Perhaps low-threshold bidirectional chemical synapses in Scyphomedusae perform the functions proposed for electrical junctions in pacemaker resetting in the Hydrozoa (see §5.4.2.1.2; see also Meech, 2015).

5.4.4.2 Control by rhopalia in Cubomedusae. Cubomedusae are capable of a wide range of complex behavior, including obstacle avoidance, courtship and mating (Lewis & Long, 2005). *Carybdea rastonii* swims continuously unless all four of its rhopalia are removed (Conant, in Berger, 1900; Satterlie, 1979; Satterlie & Spencer, 1979). Each rhopalium is connected to the nerve ring by a muscular stalk (Passano, 1982) and pacemaker activity appears to arise from the outer surface near the origin of the stalk (Satterlie, 1979). Impulses spread to the subumbrellar muscle and also act on other rhopalia to suppress their activity (Satterlie, 1979). In addition to being the site of pacemaker activity, each rhopalium has two large eyes equipped with lenses and two pairs of ocelli. When appropriately excited the eyes are able to uniquely modify the pacemaker output (Garm & Mori, 2009; Petie, Garm, & Nilsson, 2013) making steering possible.

Muscle potentials are graded events; the larger the muscle potential, the stronger the force of contraction. Facilitation of the muscle response, which is at a maximum with interpulse intervals of 0.5 second, disappears when the interval is increased to 1.5 seconds (Satterlie, 1979). From what we know of the ionic basis of electrical potentials in other cnidarian muscle cells (Inoue et al., 2005), it is likely that repolarization depends on a rapidly inactivating K^+ current. With shorter intervals between stimuli, recovery from inactivation would be incomplete, muscle events would be enhanced and contractions facilitated.

Satterlie and Nolen (2001) suggest that the great navigational powers of the cubomedusae arise from the fact that their semi-independent pacemakers can respond appropriately to asymmetrical stimuli. An absolute linkage between pacemakers would prevent asymmetric swimming such as may be called for when the animal is tracking prey. Furthermore turning, which is achieved through enhanced local contractions (Gladfelter, 1973) and variations in the shape of the valarium (Petie et al., 2013), appears to depend in part on facilitated muscle responses. So although having more pacemakers would increase swim regularity, the natural swim frequency would be also raised enforcing a matching rise in the rate of decay of facilitation, otherwise the musculature would be permanently facilitated. According to Satterlie and Nolen, four pacemakers provide the best compromise for cubomedusae.

5.4.5 Summary

Neural coordination in the Hydrozoa is so reliant on electrical junctions that their absence in the Scyphozoa and Cubozoa raises important evolutionary questions. The contribution of low-threshold bidirectional chemical synapses may provide a partial answer, but in the following summary of hydromedusan behavior, neuronal modules are seen to be coordinated by a variety of electrical field events.

1. In swimming jellyfish, excitation spreads rapidly from each neuromuscular junction because electrical junctions allow currents to flow from cell to cell across the myoepithelium.
2. In *Polyorchis*, the low-pass filtering arising from the electrical coupling between pacemaker neurons extends synaptic potentials sufficiently for them to affect the activation/inactivation state of the pacemaker mechanism.
3. In some species the low-pass property of the pacemaker system is responsible for the delayed onset of swimming necessary for the management of the tentacle net.
4. In species that show defensive crumpling, electrical junctions transform the exumbrella into a sensory surface.
5. In species that escape swim, electrical pathways between neuronal modules help to coordinate the movements of the tentacles via piggybacking. Piggybacking brings slow changes in posture together with faster twitch contractions.

5.5 The Evolution of Neural Integration

5.5.1 Requirements for an Integrative Nervous System

It is my contention that neuron-like structures cannot be considered truly neuronal unless they participate in an integration network responsible for generating coordinated behavior. Examination of the Cnidaria suggests that nervous systems evolved in a modular fashion with components being responsible for specific functions such as swimming, feeding, and defense. The aim of the present section is to review how these modular neuronal systems came to perform in an integrated fashion.

We might speculate that the evolution of an integrating nervous system started with the ability to send *all-or-nothing* sensory messages rapidly over long distances. Although initially protective, such simple binary behavior would be likely to generate frequent "false alarms" so that significant energy savings would become possible with a system that could convey small changes in conditions so as to produce appropriately *graded responses*. Regular changes in the availability of food would favor the evolution of rhythmically active neurons (*rhythmic activity*) as a way of efficiently programming foraging activity. To forage, an exploring organism must select from incompatible modes of behavior (*mode selection*), such as feeding and escape. To do this incoming information must be integrated and preceding activity remembered and evaluated (*memory of preceding activity states*). The following section considers these stages in more detail.

5.5.1.1 All-or-nothing regenerative activity. In a simple reflex arc, an effective stimulus elicits an all-or-nothing impulse leading to an all-or-nothing response. On the face of it, escape swimming in *Aglantha* is a fine example of this fundamental form of activity. Yet *Aglantha*'s escape swim is more than a simple reflex; it is a highly efficient form of behavior in which the myoepithelium is coordinated to maximize thrust. Several morphological innovations facilitate this response: The ring giant axon minimizes the conduction time around the base of the bell; the motor nerves, fused into eight giant axons, conduct impulses to the apex of the bell in milliseconds; electrical junctions promote the spread of excitation through the epithelium.

In combination, these adaptations cause the entire muscle sheet to contract and force the contents of the bell through the "nozzle" which forms at the bell margin, all within 100 ms. As this is happening, the tentacular net withdraws to help reduce drag. The key ingredient is speed of response.

Aglantha's escape swim is a good illustration of the multiplicity of the variables that determine neuronal function, but how did such a system evolve? Did it come together step-by-step or appear fully-formed, like Athena from the head of Zeus? Clearly *Aglantha*'s ancestors bequeathed it the potential for constructing electrical junctions, fused axons, and so on, but the protective power would not have emerged until the components came together in a single "package."

5.5.1.2 Graded responsiveness. Once cells evolved the means to register fine variations in stimulus strength, motor responses could be more than simple "all-or-nothing" reflexes. In sea anemones contractile responses are graded depending on the frequency of impulses in the motor nerve net. Yet even in this simplest of reactive systems, there is the means of grading not only the strength but also the area of muscle contraction. This is achieved by having a facilitating system that requires multiple impulses for response initiation. As in *Chelophyes* some forms of facilitation can arise from the properties of the muscles themselves rather than from the neurons that innervate them. In these early forms, integration of response can be as much a product of the musculature as of the nervous system. Hence, coordinated activity may depend on the specific properties of different effectors: Not only does facilitation develop at different rates in different muscles, but the muscle's contractile properties also vary. This allows different species to evolve markedly different responses to their environment. Stimuli that produce a strong defensive withdrawal in *Calliactus* (an animal that is exposed to buffeting by the elements) will instead initiate a feeding response in the snakelocks anemone.

Where graded responsiveness is the product of neuronal control it arises from a range of different mechanisms. In *Chelophyes* it comes about not only through facilitating muscle responses but also by coordinating multiple nectophores. Swimming that maintains the colony's position in the water column is normally confined to the smaller, posterior nectophore (see Figure 5.7); contributions from the anterior nectophore are normally restricted to escape swims. As with *Nanomia* (see Figure 5.8), the swim velocity depends on whether the nectophores are excited together. In jellyfish with a single bell, swims are more uniform in strength. The exception is *Aglantha*: Here the swimming velocity depends on the type of propagating impulse in the motor axons (Figure 5.3B; Mackie & Meech, 1985). Regional differences in the strength of contraction in *Aglantha* arise in part because the myoepithelium is divided into fields by lateral nerves, in part because the ion channels in the muscle cells interact to provide for different thresholds (Meech, 2015), and in part because the spread of current is not isotropic: Current spreads further in the circular direction (Kerfoot et al., 1985).

In general terms, graded responsiveness depends on an interaction between the active and passive properties of participating neurons. Active properties depend on the various voltage-gated ion channels that contribute to electrogenesis; passive properties depend on the anatomical structures that influence the spread of current. In most invertebrates the structure and electrical properties of the effectors themselves play an important integrative role.

5.5.1.3 Rhythmic activity. The nature of the solar system is such that our environment changes in a daily and seasonal fashion—often making us fret ineffectually in rush hour traffic jams. Nevertheless, the regularity of the changes in the availability of food would have given animals with the ability to generate rhythmic behavior a competitive "edge." Rhythmically active neurons would also make it possible for swimming jellyfish to avoid energy losses arising through water turbulence. Their coordinated rhythmic swimming means that much of the energy stored by the mesogloea during a swim contraction is regained as the bell refills with water.

At the same time, any rhythmically active animal will be at a disadvantage unless it can respond rapidly to local events by speeding up its pacemaker or slowing it down. In hydromedusae the pacemaker is a single ring of electrically coupled neurons with synchronously firing APs that drive almost symmetrical contractions of the swimming bell. Because of electrical coupling, the synaptic events are slow enough to influence the inactivation state of the ion channels that determine the AP interval. This provides the system with a finely tuned frequency control mechanism, although the scope for steering is somewhat limited.

Unlike the hydromedusae, swim pacemakers in the scypho- and cubo-medusae are localized in discrete nerve centers (rhopalia). As with the hydromedusae, the pacemaker frequency depends on the state of the voltage-gated ion channels that determine the AP interval, but because the pacemaker currents develop so slowly, a prolonged barrage of summated "conventional" synaptic events would be required to bring about any changes in channel state. The overall control of swim frequency therefore appears to be less "centralized" than in the hydrozoa in as much as it depends upon interactions between rhopalia and the facilitated state of the swim musculature.

5.5.1.4 Mode selection. Interactions between systems of neurons can be described as "integrated" if the outcome combines incoming sensory data with whatever internal state the system occupies at the time. In one of the earliest demonstrations of inhibition in the Cnidaria, Horridge (1956) showed that in many hydrozoans, stimulation of the muscles concerned with feeding prevented spontaneous swimming. In *Aglantha* where inhibition occurs when food touches the manubrial lips (Mackie et al., 2003), the inhibitory pathway consists of a nerve plexus that runs within the walls of the radial canals ending up in the inner nerve ring at the base of the bell (Mackie & Meech 2008). In *Polyorchis* pacemaker neurons in the swim system exhibit long lasting IPSPs when the nerve plexus is stimulated (Mackie et al., 2012). We suspect that this allows the mouth to get a firm grip on the food so that when swimming is resumed, there is less risk of it being swept out of the subumbrellar cavity.

Other examples of IPSP activity in Cnidaria include those in the swim systems of *Aequorea*, associated with contractions of the radial muscle during feeding (Satterlie, 1985, 2008), and those in *Polyorchis*, associated with the effects of light and crumpling. It is notable that in general IPSPs modify ongoing pacemaker activity although in *Aglantha* an EPSP-based selection mechanism operates. Here, pacemaker neurons elicit low-amplitude excitatory post-synaptic potentials (EPSPs) in the motor axon that are just sufficient to exceed the slow swim spike threshold while a somewhat larger EPSP arising from vibration receptors at the margin of the bell (Arkett et al., 1987) sends the axon membrane to the more depolarized of its two thresholds and elicits an escape swim.

5.5.1.5 "Memory" of preceding activity states. Many voltage-gated channels pass into an inactivated state upon depolarization. This property turns out to be of critical importance in the process of registering prior activity. Inward current inactivation ensures that, following an impulse, neurons become temporarily refractory, enforcing the direction the impulses travel through the nervous system and also affecting their rate of propagation. Propagation rates depend on the amplitude of the inward current (among other things), and so APs elicited during an axon's relative refractory period will have a lower conduction velocity.

Facilitation of propagation during repeated stimulation, sometimes called superexcitability, is widespread in both vertebrate and invertebrate neurons (see, for example, Bullock, 1951; Grundfest & Glasser, 1938). Facilitation of conduction was also reported by Pickens (1974) in a sea anemone (*Calliactus*). Among the many explanations for superexcitability is the proposal that it arises from an increased external K^+ concentration, but AP propagation could well become faster if K^+ channels recover more slowly from inactivation than Na^+ or Ca^{2+} channels. This would produce a short term increase in net inward current.

5.5.2 Integration in a Modular Nervous System

For Bullock and Horridge (1965), integration is a set of processes resulting in an output that is some function of the input without being identical to it. They compiled a list of integrative properties of neurons that were open to natural selection:

1 Regional differences in membrane composition (number, class and time constants of ion channels).
2 Proximity of each synaptic input to the AP initiating zone.
3 Nature of synaptic inputs (excitatory or inhibitory).
4 Scale of the region that summates graded potentials (larger regions with longer time constants sum over longer periods).
5 After-effects of activity (facilitation, temporal summation, or post-tetanic potentiation).

The Cnidaria provide examples of most of these five categories, although little is known about "regional differences in membrane composition." Data on agonist-activated channels is absent and there is a marked shortage of data on the distribution of voltage-gated ion channels. In a patch-clamp study of *Aglantha* giant motor axons, five classes of voltage-gated channel were found to be separately clustered in the membrane (Meech & Mackie, 1993a, 1993b, 1995). When fine-tipped micropipettes were used to sample the channel distribution, K^+ channels were routinely recorded whereas Na^+ and Ca^{2+} channels were rarely found. Na^+ and Ca^{2+} channels could only be seen by sampling with much larger (10 μm) tipped pipettes as if they were separated by some distance. The K^+ channels also seemed to be clustered into three separate classes. Each class had clearly distinct kinetic properties and voltage dependence, and could be assigned a specific role in ensuring the independent propagation of the slow and fast swim impulses. What was most remarkable was that all three classes of K^+ channels appeared to have the same conductance. Perhaps their properties arise from different β subunits as reported for some other K⁺ channels (see Brenner, 2014).

Although not specified by Bullock and Horridge (1965), it is sometimes valuable to compare the ion channel makeup of equivalent cell types across jellyfish species. Although there is little information from neurons, the integrative contribution of the motor effectors themselves is becoming clearer. For example, a comparison of the ion channel composition of myoepithelial membranes from *Chelophyes* and *Muggiaea* shows significant differences. During a sequence of swims in *Chelophyes*, APs increase in duration because repolarization depends on a single rapidly inactivating K⁺ current; this increase in AP duration yields increased swim strength, part of the animal's strategy for dealing with its extended tentacle net (see §5.4.2.2). *Muggiaea*, by contrast, contracts its tentacle net before swimming; its swims are of uniform strength and accordingly its myoepithelium is equipped with more than one class of K⁺ channel (Meech & Mackie, unpublished).

In category two of Bullock and Horridge's treatment (1965), the AP initiating zone is fixed, with any variation being due to its proximity to the synaptic input. In fact, in *Aglantha* the reverse can be true. The escape and slow swim spikes are initiated at different points along the motor giant axon while both synaptic inputs are located at the bell margin. The differences in the site of spike initiation come about because the currents associated with the slow swim AP are of such low amplitude that they can be short-circuited by a high conductance at the synapse.

The best data concerning the interaction between excitatory and inhibitory inputs (category three) is from the pacemaker system of *Polyorchis*. Excitation from the B system interacts with an inhibitory input that originates in the manubrium or from the exumbrella epithelium. The region that summates these potentials (category four) is a ring of coupled neurons that behaves like a low-pass filter. This means that the time course of the synaptic potentials is greatly slowed, so that they not only move the membrane potential relative to the AP threshold but also have extended effects on the availability of the different voltage-gated channels.

In the cnidaria there are a number of examples of the way in which electrical activity can produce long-term aftereffects (category five), the prolonged APs in *Chelophyes* myoepithelium being just one of them. Another is the observation that repeated impulses are often found to propagate at different rates. In the E system of *Aglantha* the second of a pair of impulses (1 second apart) propagates more slowly than the first; for impulses five seconds apart, propagation is accelerated. On this basis we might add a further variable to Bullock and Horridge's list: the distance between impulse initiation and its site of action. A primary example is the role played by the giant axons in the stem of *Nanomia*. Here APs in the two axons propagate at different rates but generate summed effects in the myoepithelium (see Figure 5.8). The further they travel the less effective the summed input and so the twitch response is focused around the spike initiation point. The pointing response in *Aglantha* is another example: Here the mouth is accurately directed towards food at the periphery because radial muscles in the mouth contract most strongly in response to impulses that have travelled the shortest distance from the initiation site, which is where the prey is located. A similar effect in some corals (subclass Hexacorallia) accounts for the pattern of polyp retraction (Shelton, 1975).

5.6 The First Neurons

Any examination of neuron evolution comes up against the question of how to recognize a neuron. My working definition (in the Introduction) focused on what neurons "do" as opposed to what neurons "are." If I avoided including any restrictions based on structure, it was precisely because neuronal structure is so variable that the definition would, on the one hand, be so broad as to be meaningless or, on the other hand, so narrow as to be far too exclusive. Having reached the end of the review, however, some mention of neuronal structure is inescapable. Structure intrudes into every example, influenced as it is by species life cycle, development, and body form. Perhaps I can revise my working definition of a neuron in a way that recognizes the role of structure but retains the focus on function. As redefined it becomes: "A constituent of the nervous system whose function is to communicate with other cells by way of electrical and/or chemical signals, a form of communication that usually takes place via synapses and depends critically on electrical fields generated by a precise anatomical structure."

Acknowledgments

This manuscript is the outcome of the many stimulating discussions I have enjoyed during my collaboration with George Mackie. I thank him for them, for the use of his drawings in the Figures, and for all the support he has freely given me. I thank the Director and staff of the Friday Harbor Laboratories, Washington, USA for their skilled help and warm hospitality. In particular I thank Claudia Mills for her unfailing interest and help. Research funding was provided by the Natural Sciences and Engineering Research Council of Canada (Grant No. OGP0001427 01). Earlier work was supported by grants from The Royal Society and The Wellcome Trust.

Notes

1. Quite why Romanes was unable to see nerves in jellyfish is not clear. In fact Louis Agassiz had "almost certainly" described nerves in medusae in 1850 (see Mackie, 2004) but the Darwinists lead by Thomas Henry Huxley clearly mistrusted this work, perhaps because Agassiz had argued so strongly against Darwin's view of evolution. Huxley had stated categorically: "I have not observed any indubitable trace of a nervous system in the Medusae" (Huxley, 1849, p. 424) and it was not until 1865 that his friend Ernst Haeckel persuaded him otherwise. At the time of his letter to Darwin, Romanes was unaware of Haeckel's work or of Huxley's change of heart and he continued to present what he thought was the Huxley line until Huxley drew the Haeckel work to his attention (see Romanes, 1876).
2. More recently Parker's belief that nematocytes act as "independent effectors" (Parker & van Alstyne, 1932, p. 342) has been overtaken by electron microscopy showing both efferent and afferent synaptic contacts (Holtmann & Thurm, 2001; Westfall, 1996) and the fact that mechanical stimulation of one nematocyte induces electrical responses in it as well as in neighbouring nematocytes (Thurm et al., 2004).
3. Pisani et al. (2015) discuss the assumption, often made during phylogenetic tree construction, that amino acids have an equal probability of substituting at all sites in a protein. This assumption ignores known constraints on amino acids such as their hydrophobicity and size.

In reality, membrane-bound amino acids may be less liable to change than hydrophilic amino acids that project into the cytoplasm. Ignoring such factors would "overestimate the number of amino acids a site can accept, and therefore underestimate the probability of convergent evolution toward identical amino acids in unrelated species." When Pisani et al. (2015) re-evaluated datasets from Ryan et al. (2013) and Moroz et al. (2014) they found little support for the hypothesis that ctenophores are the sister group to all other animals

References

Adrian, E. D. (1912). On the conduction of subnormal disturbances in normal nerve. *Journal of Physiology (London)*, 45, 389–412.

Agassiz, L. (1849). Contributions to the natural history of the Acalephae of North America Part 1. On the naked-eyed medusae of the shores of Massachusetts in their perfect state of development. *Memoirs of the American Academy of Arts and Sciences*, 4, 221–312.

Anderson, P. A. V. (1985). Physiology of a bidirectional, excitatory, chemical synapse. *Journal of Neurophysiology*, 53, 821–835.

Anderson, P. A., & Mackie, G. O. (1977). Electrically coupled, photosensitive neurons control swimming in a jellyfish. *Science*, 197, 186–188.

Anderson, P. A.V., & Schwab, W. E. (1981). The organization and structure of nerve and muscle in the jellyfish *Cyanea capillata* (Coelenterata; Scyphozoa). *Journal of Morphology*, 170, 383–399.

Anderson, P. A. V., & Schwab, W. E. (1983). Action potential in neurons of motor nerve net of *Cyanea* (Coelenterata). *Journal of Neurophysiology* 50, 671–683.

Arai, M. N., McFarlane, G. A., Saunders, M. W., & Mapstone, G. M. (1993). Spring abundance of medusae, ctenophores and siphonophores off southwest Vancouver Island: Possible competition or predation on sablefish larvae. *Canadian Technical Report of Fisheries and Aquatic Science*, 1939, 1–37.

Arkett, S. A., Mackie, G. O., &., Meech, R. W. (1987). Hair cell mechanoreception in the jellyfish *Aglantha digitale*. *Journal of Experimental Biology*, 173, 188–204.

Bennett, M. V. L. (1966). Physiology of electrotonic junctions. *Annals of the New York Academy of Sciences*, 137, 509–539.

Bennett, M. V. L., & Zukin, R. S. (2004). Electrical coupling and neuronal synchronization in the mammalian brain. *Neuron*, 41, 495–511.

Berger, E. W. (1900). *Physiology and histology of the cubomedusae including Dr. F. S. Conant's notes on the physiology*. Baltimore, MD: The Johns Hopkins University Press.

Bickell-Page, L. R., & Mackie, G. O. (1991). Tentacle autotomy in the hydromedusa *Aglantha digitale*: An ultrastructural and neurophysiological investigation. *Philosophical Transactions of the Royal Society of London B*. 331, 155–170.

Bilbaut, A., Meech, R. W., & Hernandez-Nicaise, M.-L. (1988a). Isolated giant smooth muscle fibres in *Beroe ovata*. *Journal of Experimental Biology*, 135, 343–362.

Bilbaut, A., Hernandez-Nicaise, M.-L., Leech, C., & Meech, R. W. (1988b). Membrane currents that govern smooth muscle contraction during body movement in a ctenophore. *Nature*, 331, 533–535.

Brenner, R. (2014). Knockout of the BK β2 subunit reveals the importance of accessorizing your channel. *Journal of General Physiology*, 144, 351–356.

Bullock, T. H. (1951). Facilitation of conduction rate in nerve fibres. *Journal of Physiology*, 114, 89–97.

Bullock, T. H., & Horridge, G. A. (1965). *Structure and function in the nervous system of invertebrates* (Vol. I). San Francisco: W. H. Freeman.

Cario, C., Malaval, L., & Hernandez-Nicaise M.-L. (1995). Two distinct distribution patterns of sarcoplasmic reticulum in two functionally different giant smooth muscle cell of *Beroe ovata*. *Cell & Tissue Research*, 282, 435–443.

Colin, S. P., & Costello, J. H. (2002). Morphology, swimming performance and propulsive mode of six co-occurring hydromedusae. *Journal of Experimental Biology*, 205, 427–437.

Colin, S. P., Costello, J. H., & Klos, E. (2003). In situ swimming and feeding behavior of eight co-occurring hydromedusae. *Marine Ecology Progress Series*, 253, 305–309.

Connor, J. A., & Stevens, C. F. (1971). Prediction of repetitive firing behaviour from voltage clamp data on an isolated neurone soma. *Journal of Physiology*, 213, 31–53.

Dabiri, J. O., Colin, S. P., Katija, K., & Costello, J. H. (2010). A wake-based correlate of swimming performance and foraging behavior in seven co-occurring jellyfish species. *Journal of Experimental Biology*, 213, 1217–1225.

Donaldson, S., Mackie, G. O., & Roberts, A. (1980). Preliminary observations on escape swimming and giant neurons in *Aglantha digitale* (Hydromedusae: Trachylina). *Canadian Journal of Zoology*, 58, 549–552.

Elliott, G. R. D., & Leys, S. P. (2007). Coordinated contractions effectively expel water from the aquiferous system of a fresh water sponge. *Journal of Experimental Biology*, 210, 3736–3748.

Elliott, G. R. D., & Leys, S. P. (2010). Evidence for glutamate, GABA and NO in coordinating behaviour in the sponge, *Ephydatia muelleri* (Demospongiae, Spongillidae). *Journal of Experimental Biology*, 213, 2310–2321

Fewkes, J. W. (1880). Contributions to a knowledge of the tubular jellyfishes. *Bulletin of the Museum of Comparative Zoology Harvard*, 6, 127–146.

Garm, A., & Mori, S. (2009). Multiple photoreceptor systems control the swim pacemaker activity in box jellyfish. *Journal of Experimental Biology*, 212, 3951–3960.

Gladfelter, W. G. (1973). A comparative analysis of the locomotory systems of medusoid Cnidaria. *Helgoländer wissenschaftliche Meeresuntersuchungen*, 25, 228–272.

Grant, R. E. (1936). Animal kingdom. In R. B. Todd (Ed.) *The cyclopaedia of anatomy and physiology* (pp. 107–118). London: Sherwood, Gilbert and Piper.

Grigoriev, N. G., Spafford, J. D., Przysiezniak, J., & Spencer, A. N. (1996). A cardiac-like sodium current in motor neurons of a jellyfish. *Journal of Neurophysiology*, 76, 2240–2249.

Grundfest, H., & Gasser, H. S. (1938). Properties of mammalian nerve fibers of slowest conduction. *American Journal of Physiology*, 123, 307–318.

Hall, D. M., & Pantin, C. F. A. (1937). The nerve net of the actinozoa. V. Temperature and facilitation in *Metridium senile*. *Journal of Experimental Biology*, 14, 71–78.

Hernandez-Nicaise, M.-L. (1973). The nervous system of Ctenophora. III. Ultrastructure of synapses. *Journal of Neurocytology*, 2, 249–263.

Hernandez-Nicaise, M-L., & Amsellem, J. (1980). Infrastructure of the giant smooth muscle fiber of the ctenophore *Beroe ovata*. *Journal of Ultrastructure Research*, 72, 151–168.

Hernandez-Nicaise, M.-L., Mackie, G. O., & Meech, R. W. (1980). Giant smooth muscle cells of *Beroe*. Ultrastructure, innervation, and electrical properties. *Journal of General Physiology*, 75, 79–105.

Hernandez-Nicaise, M.-L., & Nicaise, G. (1986). Structural evidence for contractile units in the giant smooth muscle cell of *Beröe*. *Cell Motility and the Cytoskeleton*, 6, 153–158.

Hertwig, O., & Hertwig, R. (1878). *Das Nervensystem und die Sinnesorgane der Medusen*. Leipzig: F. C. W. Vogel.

Holman, M. A., & Anderson, P. A. V. (1991). Voltage-activated ionic currents in myoepithelial cells isolated from the sea anemone *Calliactis tricolor*. *Journal of Experimental Biology* 161, 333–346.

Holtmann, M., & Thurm, U. (2001). Mono- and oligo-vesicular synapses and their connectivity in a cnidarian sensory epithelium (*Coryne tubulosa*). *Journal of Comparative Neurology*, 432, 537–549.

Horridge, G. A. (1956). The nerves and muscles of medusae V. Double innervation in Scyphozoa. *Journal of Experimental Biology*, 33, 366–383.

Horridge, G. A. (1959). The nerves and muscles of medusae. VI. The rhythm. *Journal of Experimental Biology*, 36, 72–91.

Horridge, G. A. (1968). *Interneurons*. San Francisco: W. H. Freeman.

Horstmann, E. (1934). Untersuchungen zur Physiologie der Schwimmbewegungen der Scyphomedusen. *Pflügers Archiv für die gesamte Physiologie des Menschen und der Tiere*, 334, 406–420

Huxley, T. H. (1849). On the anatomy and the affinities of the family of the medusae. *Philosophical Transactions of the Royal Society of London*, 139, 413–434.

Hyman, L. H. (1940). Observations and experiments on the physiology of medusae. *Biological Bulletin*, 79, 282–296.

Inoue, I., Tsutsui, I., & Bone, Q. (2005). Long-lasting potassium channel inactivation in myoepithelial fibres is related to characteristics of swimming in dyphid siphonophores. *Journal of Experimental Biology*, 208, 4577–4584.

Jager, M., Chiori, R., Alié, A., Dayraud, C., Quéinnec, E., & Manuel, M. (2011). New insights on ctenophore neural anatomy: Immunofluorescence study in *Pleurobrachia pileus* (Müller, 1776). *Journal of Experimental Zoology (Molecular and Developmental Evolution)* 316, 171–187.

Jones, W. C. (1962). Is there a nervous system in sponges? *Biological Reviews of the Cambridge Philosophical Society*, 37, 1–50.

Kerfoot, P. A. H., Mackie, G. O., Meech, R. W., Roberts, A., & Singla, C. L. (1985). Neuromuscular transmission in the jellyfish *Aglantha digitale*. *Journal of Experimental Biology*, 116, 1–25.

King, M. G., & Spencer A. N. (1981). The involvement of nerves in the epithelial control of crumpling behaviour in a hydrozoan jellyfish. *Journal of Experimental Biology*, 94, 203–218.

Kleinenberg N. (1872). *Hydra – Eine anatomisch-entwicklungsgeschichtliche Untersuchung*. Leipzig: Wilhelm Engelmann.

Kreps, T. A., Purcell, J. E., & Heidelberg, K. B. (1997). Escape of the ctenophore *Mnemiopsis leidyi* from the scyphomedusa predator *Chrysaora quinquecirrha*. *Marine Biology*, 128, 441–446.

Lentz, T. L. (1968). *Primitive nervous systems*. New Haven, CT: Yale University Press.

Leonard, J. L. (1982). Transient rhythms in the swimming activity of *Sarsia tubulosa* (Hydrozoa). *Journal of Experimental Biology*, 96, 181–193.

Lewis, C., & Long, T. A. F. (2005). Courtship and reproduction in *Carybdea sivickisi* (Cnidaria: Cubozoa). *Marine Biology*, 147, 477–483.

Leys, S. P., & Degnan, B. M. (2001). Cytological basis of photoresponsive behavior in a sponge larva. *Biological Bulletin*, 201, 323–338.

Leys, S. P., & Mackie, G. O. (1997). Electrical recording from a glass sponge. *Nature* 387, 29–30.

Leys, S. P., Mackie, G. O., & Meech, R. W. (1999). Impulse conduction in a sponge. *Journal of Experimental Biology*, 202, 1139–1150.

Leys, S. P., & Meech, R. W. (2006). Physiology of coordination in sponges. *Canadian Journal of Zoology*, 84, 288–306

Lin, Y.-C. J., Gallin, W. J., & Spencer, A. N. (2001). The anatomy of the nervous system of the hydrozoan jellyfish, *Polyorchis penicillatus*, as revealed by a monoclonal antibody. *Invertebrate Neuroscience*, 4, 65–75.

Mackie, G. O. (1964). Analysis of locomotion in a siphonophore colony. *Proceedings of the Royal Society of London. Series B*, 159, 366–391.

Mackie, G. O. (1965). Conduction in the nerve-free epithelia of Siphonophores. *American Zoologist*, 40, 439–353.

Mackie, G. O. (1970). Neuroid conduction and the evolution of conducting tissues. *The Quarterly Review of Biology*, 45, 319–332.

Mackie, G. O. (1976a). Propagated spikes and secretion in a coelenterate glandular epithelium. *Journal of General Physiology*, 68, 313–325.
Mackie, G. O. (1976b). The control of fast and slow muscle contractions in the siphonophore stem. In G. O. Mackie (Ed.), *Coelenterate Ecology and Behavior* (pp. 647–659). New York, NY: Plenum.
Mackie, G. O. (1978). Coordination in physonectid Siphonophores. *Marine Behaviour and Physiology*, 5, 325–346.
Mackie, G. O. (1980). Slow swimming and cyclical "fishing" behavior in *Aglantha digitale* (Hydromedusae, Trachylina). *Canadian Journal of Fisheries and Aquatic Sciences*, 37, 1550–1556.
Mackie, G. O. (1984). Fast pathways and escape behavior in Cnidaria. In R. C. Eaton (Ed.), *Neural basis of startle behavior*. New York, NY: Plenum.
Mackie, G. O. (1991). Propagation of bioluminescence in *Euphysa japonica* hydromedusae, (Tubulariidae). *Hydrobiologia*, 216/217, 581–588.
Mackie, G. O. (2004). The first description of nerves in a cnidarian: Louis Agassiz's account of 1850. *Hydrobiologia*, 530/531, 27–32.
Mackie, G. O., Anderson, P. A. V., & Singla, C. L. 1984. Apparent absence of gap junctions in two classes of Cnidaria. *Biological Bulletin*, 167: 120–123.
Mackie, G. O., & Mackie, G. V. (1963). Systematic and biological notes on living hydromedusae from Puget Sound. *National Museum of Canada Bulletin*, 199, 63–84.
Mackie, G. O., Marx, R. M., & Meech, R. W. (2003). Central circuitry in the jellyfish *Aglantha digitale*. IV. Pathways coordinating feeding behaviour. *Journal of Exprimental Biology*, 206, 2487–2505.
Mackie, G. O., & Meech, R. W. (1985). Separate sodium and calcium spikes in the same axon. *Nature*, 313, 791–793.
Mackie, G. O., & Meech, R. W. (1995a). Central circuitry in the jellyfish *Aglantha digitale*. I. The relay system. *Journal of Experimental Biology*, 198, 2261–2270.
Mackie, G. O., & Meech, R. W. (1995b). Central circuitry in the jellyfish *Aglantha digitale*. II. The ring giant and carrier systems. *Journal of Experimental Biology*, 198, 2271–2278.
Mackie, G. O., & Meech, R. W. (2000). Central circuitry in the jellyfish *Aglantha digitale*. III. The rootlet and pacemaker systems. *Journal of Experimental Biology*, 203, 1797–1807.
Mackie, G. O., & Meech, R. W. (2008). Nerves in the endodermal canals of hydromedusae and their role in swimming inhibition. *Invertebrate Neuroscience*, 8, 199–209.
Mackie, G. O., Meech, R. W., & Spencer, A. N. (2012). A new inhibitory pathway in the jellyfish *Polyorchis penicillatus*. *Canadian Journal of Zoology*, 90, 172–181.
Mackie, G. O., Nielsen, C., & Singla C. 1989. The tentacle cilia of *Aglantha digitale* (Hydrozoa-Trachylina) and their control. *Acta Zoologica* 70: 133–141.
Mackie, G. O., & Passano, L. M. (1968). Epithelial conduction in hydromedusae. *Journal of General Physiology*, 52, 600–621.
Mackie, G. O., Singla, C. L., & Stell, W. K. (1985). Distribution of nerve elements showing FMRFamide-like immunoreactivity in hydromedusae. *Acta Zoologica*, 166, 199–210.
Maldonado, M. (2006). The ecology of the sponge larva. *Canadian Journal of Zoology*, 84, 175–194
Maldonado, M., & Bergquist, P. R. (2002). Phylum Porifera. In C. M. Young, M. A. Sewell, & M.E. Rice (Eds.) *Atlas of marine invertebrate larvae* (pp. 21–50). San Diego, CA: Academic Press.
Meech, R. W. (2015). Electrogenesis in the lower Metazoa and implications for neuronal integration. *Journal of Experimental Biology*, 218, 1–14.
Meech, R. W., & Mackie, G. O. (1993a). Ionic currents in giant motor axons of the jellyfish, *Aglantha digitale*. *Journal of Neurophysiology*, 69, 884–893.
Meech, R. W., & Mackie, G. O. (1993b). Potassium channel family in giant motor axons of *Aglantha digitale*. *Journal of Neurophysiology*, 69, 894–901.

Meech, R. W., & Mackie, G. O. (1995). Synaptic potentials and threshold currents underlying spike production in motor giant axons of *Aglantha digitale*. *Journal of Neurophysiology*, 74, 1662–1670.

Meech, R. W., & Mackie, G. O. (2006). Ionic currents in the myoepithelium of Aglantha digitale. *FASEB Journal*, 20, A826.

Meech, R. W., & Mackie, G. O. (unpublished). Electrophysiology of swimming in *Muggiaea*.

Moroz, L. (2015). Convergent evolution of neural systems in ctenophores. *Journal of Experimental Biology*, 218, 598–611.

Moroz, L. L., Kocot, K. M., Citarella, M. R., Dosung, S., Norekian, T. P., Povolotskaya, I. S., ... Bruders, R. (2014). The ctenophore genome and the evolutionary origins of neural systems. *Nature*, 510, 109–114.

Naitoh, Y., & Eckert, R. (1969). Ionic mechanisms controlling behavioral responses of *Paramecium* to mechanical stimulation. *Science*, 164, 963–965.

Nielsen, C. (2008). Six major steps in animal evolution: Are we derived from sponge larvae? *Evolution and Development*, 10, 241–257.

Pantin, C. F. A. (1935a). The nerve net of the Actinozoa. I. Facilitation. *Journal of Experimental Biology*, 12, 119–138.

Pantin, C. F. A. (1935b). The nerve net of the Actinozoa. II. Plan of the nerve net. *Journal of Experimental Biology*, 12, 139–155.

Pantin, C. F. A. (1935c). The nerve net of the Actinozoa. III. Polarity and after-discharge. *Journal of Experimental Biology*, 12, 156–164.

Pantin, C. F. A. (1956). The origin of the nervous system. *Pubblicazioni della Stazione Zoologica di Napoli*, 28, 171–181.

Parker, G. H. (1910). The reactions of sponges, with a consideration of the origin of the nervous system. *Journal of Experimental Zoology*, 8, 765–805.

Parker, G. H. (1911). The origin and significance of the primitive nervous system. *Proceedings of the American Philosophical Society*, 50, 217–225.

Parker, G. H. (1919). *The elementary nervous system*. Philadelphia, PA: J. B. Lippincott Company.

Parker G. H., & van Alstyne, M. A. (1932). The control and discharge of nematocysts, especially in *Metridium* and *Physalia*. *Journal of Experimental Zoology*. 63, 329–344.

Passano, L. M. (1965). Pacemakers and activity patterns in medusae: Homage to Romanes. *American Zoologist*, 5, 465–481.

Passano, L. M. (1973). Behavioural control systems in medusae: A comparison between hydro- and schyphomedusae. *Publications of the Seto Marine Biological Laboratory*, 20, 615–645

Passano, L. M. (1982). Scyphozoa and cubozoa. In G. A. B. Shelton, (Ed.), *Electrical conduction and behaviour in "simple" invertebrates*. Oxford: Clarendon Press.

Passano, L. M., & Pantin, C. F. A. (1955). Mechanical stimulation in the sea-anemone *Calliactis parasitica*. *Proceedings of the Royal Society B*, 143, 226–238.

Pavans de Ceccatty, M. (1974). Coordination in sponges. The foundations of integration. *American Zoologist*, 14, 895–903.

Petie, R., Garm, A., & Nilsson, DE. (2013). Valarium control and visual steering in box jellyfish. *Journal of Comparative Physiology A*. 199, 315–324.

Pickens, P. E. (1974). Changes in conduction velocity within a nerve net. *Journal of Neurobiology* 5, 413–420.

Pisani, D., Pett, W., Dohrmann, M., Feuda, R., Rota-Stabelli, O., Philippe, H., ... Wörheide, G. (2015). Genomic data do not support comb jellies as the sister group to all other animals. *Proceedings of the National Academy of Sciences of the USA*, 112, 15402–15407.

Przysiezniak, J., & Spencer, A. N. (1989). Primary culture of identified neurons from a cnidarian. *Journal of Experimental Biology*, 142, 97–113.

Przysiezniak, J., & Spencer, A. N. (1992). Voltage-activated calcium currents in identified neurons from a hydrozoan jellyfish, *Polyorchis penicillatus*. *Journal of Neuroscience*, 12, 2065–2076.

Przysiezniak, J., & Spencer, A. N. (1994). Voltage-activated potassium currents in isolated motor neurons from the jellyfish *Polyorchis penicillatus*. *Journal of Neurophysiology*, 72, 1010–1019.

Roberts, A. (1969). Conducted impulses in the skin of young tadpoles. *Nature*, 222, 1265–1266.

Roberts, A., & Mackie, G. O. (1980). The giant axon escape system of a hydrozoan medusa, *Aglantha digitale*. *Journal of Experimental Biology*, 84, 303–318.

Romanes, G. J. (1875). Letter to Charles Darwin, July 20, 1875. In E. D. Romanes (Ed.) *Life and letters of George John Romanes*. Cambridge: Cambridge University Press.

Romanes, G. J. (1876). The Croonian Lecture: Preliminary observations on the locomotor system of medusae. *Philosophical Transactions of the Royal Society of London*, 166, 269–313.

Romanes, G. J. (1877). Further observations on the locomotor system of medusae. *Philosophical Transactions of the Royal Society of London*, 167, 659–752.

Romanes, G. J. (1880). Concluding observations on the locomotor activity of medusae. *Philosophical Transactions of the Royal Society of London*, 171, 161–202.

Romanes, G. J. (1885). *Jelly-fish, star-fish and sea-urchins*. London: K. Paul, Trench and Company.

Russell, F. S. (1953). *The medusae of the British Isles* (Vol. I). Cambridge: Cambridge University Press

Ryan, J. F, & Chiodin, M. (2015). Where is my mind? How sponges and placozoans may have lost neural cell types. *Philosophical Transactions of the Royal Society B*, 370, 20150059. doi: 10.1098/rstb.2015.0089

Ryan, J. F., Pang, K., Schnitzler, C. E., Nguyen, A.-D., Moreland, R. T., Simmons, D. K., ... Baxevanis, A. D. (2013). The genome of the ctenophore *Mnemiopsis leidyi* and its implications for cell type evolution. *Science*, 342, 1242592. doi: 10.1126/science. 1242592

Satterlie, R. A. (1979). Central control of swimming in the cubomedusan jellyfish *Carybdea rastonii*. *Journal of Comparative Physiology*, 133, 357–367.

Satterlie, R. A. (1985). Central generation of swimming activity in the hydrozoan jellyfish *Aequorea aequorea*. *Journal of Neurobiology*, 16, 41–55.

Satterlie, R. A. (2008). Control of swimming in the hydrozoan jellyfish *Aequorea victoria*: Subumbrellar organization and local inhibition. *Journal of Experimental Biology*, 211, 3467–3477.

Satterlie, R. A., & Nolen, T. G. (2001). Why do cubomedusae have only four swim pacemakers? *Journal of Experimental Biology*, 204, 1413–1419.

Satterlie, R. A., & Spencer, A. N. (1979). Swimming control in cubomedusan jellyfish. *Nature*, 281, 141–142.

Satterlie, R. A., & Spencer, A. N. (1983). Neuronal control of locomotion in hydrozoan medusae. *Journal of Comparative Physiology*, 150, 195–206.

Sechenov, I. (1863 [1965]). *Reflexes of the brain* (S. Belsky, Trans., G. Gibbons, Ed.). Cambridge, MA: MIT Press.

Shelton G. A. B. (1975). Colonial behaviour and electrical activity in the hexacorallia. *Proceedings of the Royal Society, London. B*. 190, 239–256.

Sherrington, C. S. (1906). *The integrative action of the nervous system*. New York, NY: Charles Scribner's Sons.

Spencer, A. N. (1971). Behaviour and electrical activity in Proboscidactyla flavicirrata (Hydrozoa) (Ph.D. thesis). University of Victoria, Canada. (Microfilm No. 8346, Public Archives, Canadian National Library, Ottawa).

Spencer, A. N. (1975). Behavior and electrical activity in the hydrozoan *Proboscidactyla flavicirrata* (Brandt). II. The medusa. *Biological Bulletin*, 149, 236–250.

Spencer, A. N. (1981). The parameters and properties of a group of electrically coupled neurons in the central nervous system of a hydrozoan jellyfish. *Journal of Experimental Biology*, 93, 33–50.

Spencer, A. N., & Arkett, S. A. (1984). Radial symmetry and the organization of central neurones in a hydrozoan jellyfish. *Journal of Experimental Biology*, 110, 69–90.

Swanberg, N. (1974). The feeding behaviour of *Beroe ovata*. *Marine Biology*, 24, 69–76.
Tamm, S. L. (2014). Cilia and the life of ctenophores. *Invertebrate Biology*, 133, 1–46.
Thurm, U., Brinkmann, M., Golz, R., Holtmann, M., Oliver, D., & Sieger, T. (2004). Mechanoreception and synaptic transmission of hydrozoan nematocytes. *Hydrobiologia*, 530/531, 97–105.
Vogel, S. (1994). *Life in moving fluids: The physical biology of flow* (2nd ed.). Princeton, NJ: Princeton University Press.
Weber, C., Singla, C. L., & Kerfoot, P. A. H. (1982). Microanatomy of the subumbrellar motor innervation in *Aglantha digitale* (Hydromedusae, Trachylina). *Cell Tissue Research*, 223, 305–312.
Westfall, J.A. (1996). Ultrastructure of synapses in the first-evolved nervous systems. *Journal of Neurocytology*, 25, 735–746.
Willows, A. O. D., & Hoyle, G. (1969). Neuronal network triggering a fixed action pattern. *Science*, 166, 1549–1551.

6

The First Nervous System

Nadia Riebli and Heinrich Reichert

6.1 Introduction

The origin of the first nervous system is an intriguing enigma. Stated in its simplest form, a nervous system can be defined as a set of interconnected neural cells that process information via electrical and/or chemical signals. In consequence, by definition, the first nervous system evolved after the evolutionary transition from unicellular to multicellular life forms. Since nervous systems allow integration of sensory input and coordination of motor output in a behaviorally relevant manner, there are obviously significant selective advantages in evolving more sophisticated and complex nervous systems. In animal evolution this has led to the emergence of centralized nervous systems which comprise distinct agglomerations of functionally specialized neurons, that may be subdivided into separate parts (ganglia), are interconnected by axon tracts (neuropil) and connect to the periphery via nerves (Arendt, Denes, Jékely, & Tessmar-Raible, 2008). Moreover, in most extant bilaterian animals, nervous system centralization combined with cephalization has resulted in the appearance of brains, which are prominent anterior ganglia that receive major input from sense organs located on the head and send descending motor output to the somatic effector apparatus in the remaining body via nerve cords. In this review, we focus on the evolution of complex nervous systems from simple neural origins and consider evidence from comparative, developmental, and molecular genetic studies that shed light on this fascinating evolutionary process.

6.2 The Ambiguity of Nervous System Origins

A phylogenetic assessment of the origin of nervous systems based on currently available paleontological data is both enlightening and disappointing. On one hand, there is clear fossil evidence for the existence of complex nervous systems, including brains, in bilaterian animals that date back to at least 530–540 MYA (million years ago). Thus, the fossil records for ancestral arthropods and agnathan-like vertebrates indicates that both groups already had brains and central nervous systems with features typical of extant arthropods, which are members of the protostome supergroup, and of extant vertebrates, which are members of the deuterostome supergroup (Fortey, 2000; Holland & Chen, 2001; Ma, Hou, Edgecombe, & Strausfeld, 2012; Northcutt, 2012;

Tanaka, Hou, Ma, Edgecombe, & Strausfeld, 2013). This implies that centralized nervous systems with brains evolved before the protostome–deuterostome split in the urbilaterian ancestor of both major bilaterian supergroups. Centralization of nervous systems must have occurred earlier, probably after the split between bilaterians and radiate animals such as cnidarians which is dated at 600–630 MYA (Peterson et al., 2004). However, fossil evidence for nervous systems from this Precambrian period is scarce and difficult to interpret. Hence, although evidence for the existence of central nervous systems in the early Cambrian is solid, we are left with little information on the origin of the first nervous system from paleontology.

A phylogenetic evaluation of the origin of nervous systems based on comparative neuroanatomical analyses of extant animals is also ambiguous, albeit for different reasons.

First, the nervous systems of all extant animals are by definition modern in that they have had the same amount of time to evolve (hundreds of millions of years). Furthermore, this evolutionary process can lead to both increase and reduction of nervous system complexity. Thus, even when a nervous system appears to be rather simple and "primitive" in neuroanatomical respects, this simple morphology can be due to a secondary loss of more complex structures due to the environmental features that the animal has adjusted to and due to the requirements of its ecological niche. Hence, it is a priori unclear which, if any, of the living animals have nervous systems that reflect the original, primitive nervous system in the Precambrian ancestor of bilaterian and radiate animals.

Second, our understanding of animal phylogeny (see Figure 6.1) is currently in flux—and a source of considerable controversy—mainly due to the interpretation of new data from genetic and genomic analysis. As a result, it is often unclear which group of extant animals is basal and, thus, most likely to have a "primitive" type of nervous system. This is exemplified by the recent dramatic changes in "flatworm" phylogeny and their implications for brain evolution in bilaterians. Flatworms are classically considered to represent the simplest organizational form of all living bilaterians with a true central nervous system and, based on their simple body plans, have been traditionally grouped together in a single phylum at the base of the bilaterians (Bullock & Horridge, 1965; Hyman, 1940). However, subsequent molecular phylogenetic analyses have removed the flatworms from this basal position and placed the entire flatworm phylum within the Lophotrochozoa, one of the two protostome superclades (Adoutte, Balavoine, Lartillot, & De Rosa, 1999; Adoutte et al., 2000). From this molecular phylogenetic viewpoint, there is no reason to assume that the flatworm central nervous system is any more basal than that of the other lophotrochozoan animals. Current molecular phylogenomic studies have now actually split the flatworms into two widely separated clades, the platyhelminth flatworms, which remain embedded among the lophotrochozoan phyla, and the acoelomorph flatworms, which are placed either at the base of the bilaterians or associated with the deuterostomes either as basal deuterostomes or as the sister group of hemichordates and echinoderms (Hejnol et al., 2009; Mwinyi et al., 2010; Philippe, Brinkmann, Martinez, Riutort, & Baguñá, 2007; Philippe et al., 2011). Thus, depending on their precise phylogenetic position, the acoel (but not the platyhelminth) flatworms and their supposedly "primitive" nervous systems are either basal to all bilaterians or basal only to the deuterostomes or highly derived and related to hemichordate nervous systems.

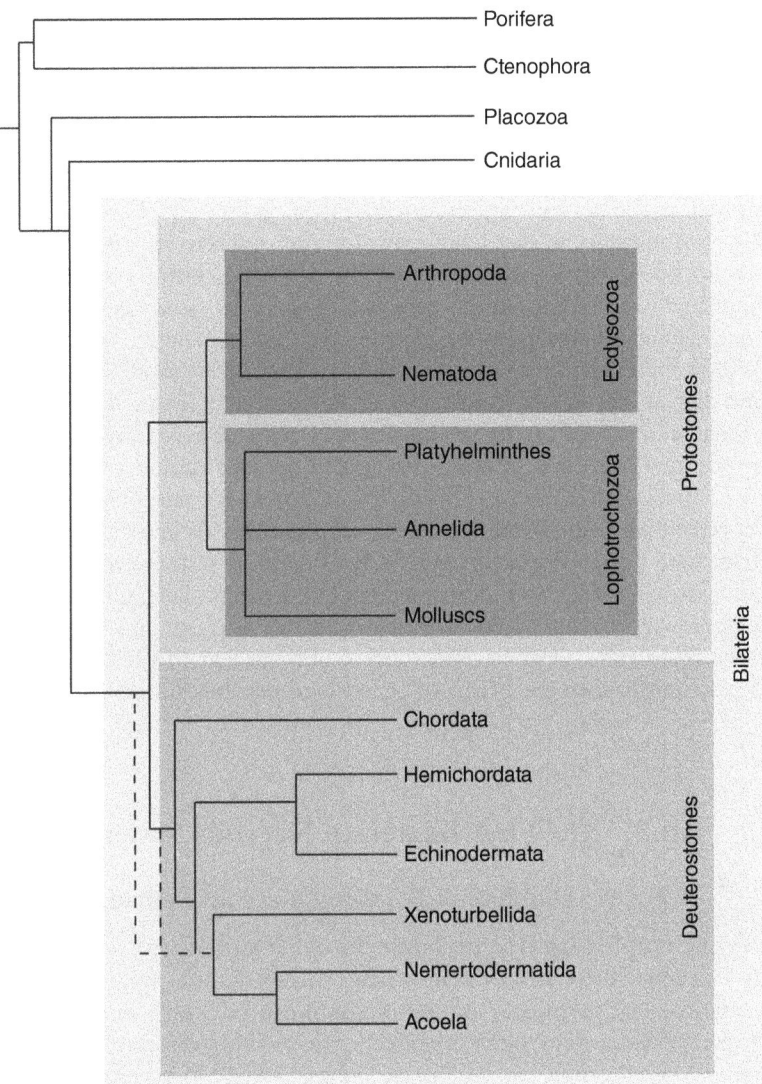

Figure 6.1 Summary Scheme of the Metazoan Phylogeny.
It is widely agreed that the cnidarians are the sister clade to the bilaterian animals. Note that the former flatworm group has been split into the "Acoela" and the "Platyhelmithes" (Philippe et al., 2007). Whereas the Platyhelminthes remained embedded within the Lophotrochozoans, the phylogenetic position of the Acoela is still a matter of debate. The newest studies either place the Xenoturbellida, the Nemertodermatida, and the Acoela at the base of the Bilateria as a sister group to all other bilaterian animals (Hejnol et al., 2009; Mallat, Craig, & Yoder, 2010), at the base of the Deuterostomes, or within the Deuterostomes (Philippe et al., 2011).

Third, it has been found to be very difficult to consider the nervous systems of extant animal groups as primitive, even when they are located on a very basal position within the tree of life. It is widely, but not universally, agreed that Porifera, Cnidaria, Ctenophora, and Placozoa are basal animal groups. Since sponges and placozoans have neither nervous systems nor neurons, they are of limited help in defining the first nervous system. Neurons and nervous systems are present in Cnidaria and Ctenophora, as well as in all other eumetazoan animals, and therefore it has been hypothesized that the first nervous system evolved after the evolutionary separation of the Porifera from the Radiata (reviewed in Lichtneckert & Reichert, 2007). Since this implies that the Cnidaria might be the most basally branching phylum of the Eumetazoa manifesting a nervous system, the nervous organization of these animals has been studied in some detail. These studies show that cnidarian nervous systems are remarkably diverse, ranging from diffuse nerve nets, in which there is little central integration and the sensory input and motor output are processed locally, to clearly centralized nervous systems with ganglion-like nervous centers that are associated with sophisticated sensory organs such as lens eyes (Satterlie, 2011). It is largely arbitrary to consider any one of the diverse nervous system types to be basal and hence "primitive" in this phylum, since even the diffuse nerve-net-like nervous system type might represent the secondary loss of a previously present centralized nervous system.

In view of the problems in elucidating the origin of the nervous system based on classical comparative neuroanatomical analysis, a number of investigations in the last two decades have explored a novel approach to nervous system evolution that combines comparative studies with developmental and molecular genetic analysis. This new integrated approach has revealed considerable insight into the evolutionary origin of the brain and central nervous system of bilaterian animals. Moreover, it has provided new insight into the origin of centralized nervous systems that may also be relevant for understanding the origin of the first metazoan nervous system.

6.3 The First Bilaterian Nervous System

6.3.1 Diversity of Bilaterian Nervous Systems

Many different morphological types and shapes of nervous systems are found in extant bilaterians. All deuterostomes investigated have central nervous systems and peripheral nervous systems. The peripheral nervous systems can be highly variable in structure ranging from the nerve-net type of organization seen in the vertebrate enteric nervous system to the ordered ganglionic organization exemplified by the vertebrate autonomic nervous system. The central nervous system of deuterostome chordates is, in general, less variable in structure. In the chordates, which include the vertebrates, the central nervous system comprises an anterior brain, subdivided into multiple compartment-like substructures, that is associated with sensory organs and is connected to a dorsally located nerve cord which links the brain to the peripheral body parts. In the urochordate tunicates, this type of central nervous system organization is only present in the larva and is radically reduced after metamorphosis in the sedentary adult form. In cephalochordates, the subdivisions in the brain and nerve cord are cryptic but can be revealed with molecular markers (Nieuwenhuys, 2002).

The central nervous systems of the remaining deuterostome phyla are more diverse. Hemichordates, which have been thought to posses only a net-like peripheral nervous system, are now known to have a fully formed central nervous system comprising dorsal as well as ventral nerve cords (Lowe et al., 2003; Nomaksteinsky et al., 2009; reviewed in Benito-Gutierrez & Arendt, 2009; Holland et al., 2013). Echinoderms also possess central nervous systems, which are, however, clearly divergent from those of other deuterostomes due to the secondary acquisition of radial symmetry in these animals (Nieuwenhuys, 2002). Acoel flatworms (whether they are bona-fide deuterostomes or not) and probably the flatworm-like xenoturbellids have a central nervous system comprising an anterior ganglion and multiple nerve cords (Achatz & Martinez, 2012; Bullock & Horridge, 1965; Semmler, Chiodin, Bailly, Martinez, & Wanninger, 2010).

Most protostomes also have both central nervous systems and peripheral nervous systems. Prominent among the central nervous systems are the complex multiganglionic brains and nerve cords of most free-living arthropods, annelids, and molluscs culminating in the remarkably complex brain of cephalopods. These complex nervous systems can be markedly reduced or absent in sedentary or parasitic forms within these and other protostome phyla. Central nervous systems with a somewhat more simple organization, consisting of an anterior ganglion and associated nerve cords, are seen in free-living members of phyla as diverse as platyhelminth flatworms, ribbon worms, tardigrades, chaetognathes, sipunculids, rotifers, ectoprocts, and nematodes (Bullock & Horridge, 1965, Kotikova & Raiikova, 2008). As in deuterostomes, the peripheral nervous systems of protostomes can be highly variable in structure ranging from a diffuse nerve-net type, seen in molluscs, to the ganglionic organization of some of the components of the arthropod peripheral nervous system.

Taken together, these data support the notion implied by current paleontological findings that centralized nervous systems were present in ancestral bilaterians before the protostome–deuterostome split. Moreover, based on comparative neuroanatomical data, these ancestral bilaterian central nervous systems likely consisted of an anterior brain-like ganglion ("protobrain") connected to descending nerve cord-like structures ("protocords") which may or may not have had ganglionic features (Ghysen, 2003). What is not at all obvious from these data is whether the diverse central nervous systems of extant bilaterians evolved separately or if they all had a common urbilaterian origin.

In its ontogeny, the bilaterian central nervous system is a complex three-dimensional structure that develops from a two-dimensional embryonic neuroepithelium. Since this neuroepithelium is located dorsally in most deuterostomes and ventrally in most protostomes, an independent evolutionary origin of the central nervous system in these two animal groups has been postulated (the gastroneuralia–notoneuralia concept; e.g., Brusca & Brusca, 1990). However, more recently (and supporting earlier ideas) the notion that the central nervous systems of protostomes and deuterostomes are homologous and derive from a common ancestral (urbilaterian) brain has been put forward (Arendt & Nübler Jung, 1994; Ghysen, 1992; Reichert & Simeone, 2001). This notion has received considerable support from the astounding conservation of developmental mechanisms that pattern the anteroposterior and dorsoventral axes of the central nervous system in several vertebrate and invertebrate model systems, as well as from the remarkable similarities in developmental origins of neuronal cell types and complex circuitry in bilaterian central nervous systems. Taken together, these recent comparative developmental genetic data indicate that similar mechanisms operate in many major stages of central nervous system formation in vertebrates and invertebrates, implying a monophyletic origin of the centralized bilaterian nervous system (Arendt et al., 2008; Hirth, 2010; Reichert, 2009).

6.3.2 Conserved Mechanisms for Anteroposterior Patterning of the Bilaterian Central Nervous System

During early development, the two-dimensional neuroepithelium that gives rise to the neurons of the bilaterian central nervous system is subdivided into compartment-like domains along both its axes by the regionalized expression of patterning genes. These genes and their respective patterns of expression are comparable in vertebrates and invertebrates. Patterning along the anteroposterior axis involves the cephalic gap genes, which are expressed in the anterior brain, the (homeotic) Hox genes, which are expressed in the posterior brain and nerve cord, and a set of other genes which delimit specific compartment interfaces in the central nervous system (see Figure 6.2).

Cephalic gap genes such as *orthodenticle* (*otd*)/*Otx* and *empty spiracles* (*ems*)/*Emx* encode transcription factors that were originally identified in the embryogenesis of the fruit fly, *Drosophila*, as key patterning elements for anterior cephalic domains (Cohen & Jurgens, 1990; Dalton, Chadwick, & McGinnis, 1989; Finkelstein & Perrimon, 1990). In addition to their role in head development, these genes are expressed in the anterior neuroectoderm of vertebrates and invertebrates and play key, evolutionarily conserved roles in central nervous system patterning (reviewed in Lichtneckert & Reichert, 2008). The most prominent of these is exemplified by *otd*/*Otx* which is expressed in the anterior brain or protobrain of bilaterians as diverse as planarians, nematodes, annelids, molluscs, arthropods, urochordates, cephalochordates, and vertebrates, including mammals (Acampora et al., 2001; Arendt, Technau, & Wittbrodt, 2001; Bruce & Shankland, 1998; Finkelstein, Smouse, Capaci, Spradling, & Perrimon, 1990; Hirth & Reichert, 1999; Lanjuin, VanHoven, Bargmann, Thompson, & Sengupta, 2003; Nederbragt, te Welscher, van der Driesche, van Loon, & Dictus, 2002; Schilling & Knight, 2001; Tomsa & Langeland, 1999; Umesono, Watanabe, & Agata, 1999; Wada, Saiga, Satoh, & Holland, 1998).

Functional studies carried out in fruit fly and mouse show that *otd*/*Otx* genes are required for formation and regionalization of the anterior neuroectoderm in both animals. Mutation of *otd* in *Drosophila* results in defective anterior neuroectoderm specification and failure in formation of stem-cell like neuroblasts in this region (Hirth et al., 1995; Younossi-Hartenstein et al., 1997). Mutation of *Otx2*, one of two *otd* homologs in mouse, results in lack of anterior brain structures due to an impairment in the specification of the anterior neuroectoderm (Acampora et al., 1995). The evolutionary conservation of expression and function of *otd*/*Otx* genes in anterior brain specification is underscored by cross-phylum transgenic experiments in which the mammalian *Otx* genes were expressed in fly *otd* mutants and, inversely, in which *Drosophila otd* was expressed in mouse *Otx* mutants (Acampora et al., 1998; Acampora, Boyl et al., 2001; Acampora, Gulisano, Broccoli, & Simeone, 2001; Leuzinger et al., 1998). In both cases the transgene was able to effect a cross-phylum rescue of brain development. Comparable cross-phylum rescue experiments carried out for the *ems*/*Emx* genes, which are also regionally expressed in anterior brain regions of vertebrate and invertebrate bilaterians, showed that murine *Emx1* can rescue brain defects in fly *ems* mutants (Hartmann, Hirth, Waldorf, & Reichert, 2000; Hirth et al., 1995). Interestingly, an *Emx* transgene from a non-bilaterian cnidarian (*Acropora*) was not able to rescue the *ems* mutant brain defects of *Drosophila*, although it did rescue head patterning defects in the fly mutant (Hartmann et al., 2010).

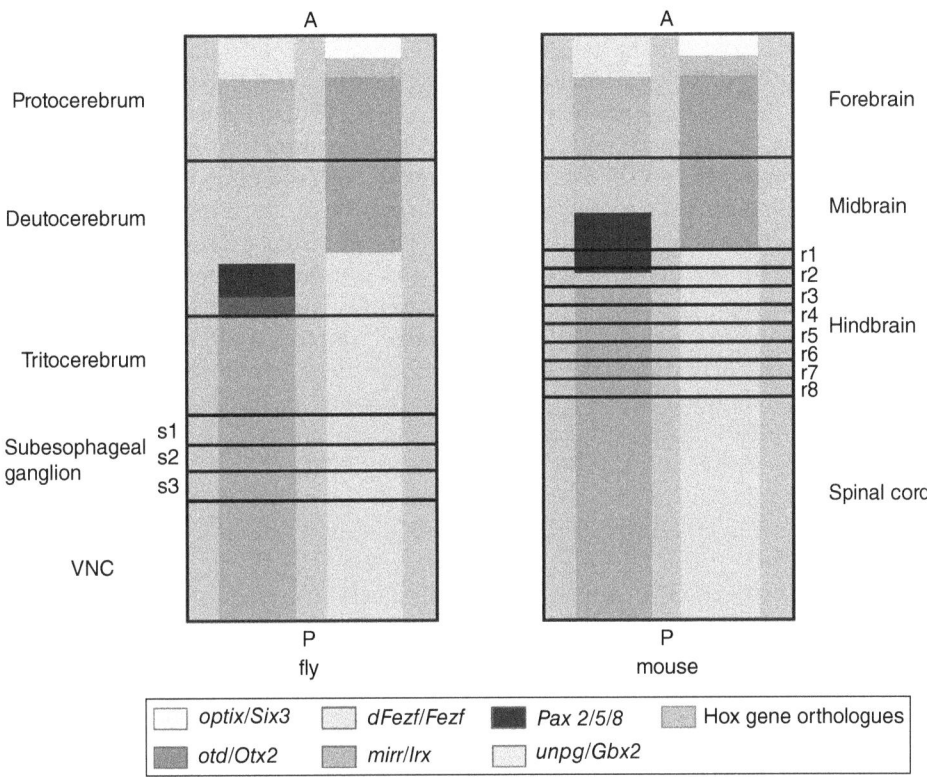

Figure 6.2 Simplified Summary Scheme of the Anteroposterior Order of Conserved Gene Expression in Embryonic CNS Development of Bilaterians.
Dorsoventral patterning is not indicated. Schematic diagram shows the expression of the patterning genes *optix/Six3*, *otd/Otx2*, *dFezf/Fezf*, *mirr/Irx*, *Pax 2/5/8*, *unpg/Gbx2* and Hox gene orthologues in the developing CNS of *Drosophila* and mouse. Expression domains are color-coded. (left) Gene expression in *Drosophila* CNS of embryonic stage 14. Borders of the protocerebral, deutocerebral, tritocerebral, mandibular (s1), maxillary (s2), labial (s3), and ventral nerve cord (VNC) neuromeres are indicated by horizontal lines. (right) Gene expression in mouse CNS of embryonic day 9.5–12.5. Borders of the forebrain, midbrain and the hindbrain and its rhombomeres (r1-r8) are indicated by horizontal lines. In both fly and mouse, an *optix/Six3* expression domain patterns the most anterior CNS region and overlaps with the *otd/Otx2* expression pattern (Steinmetz et al., 2010) which is anterior to the abutting *unpg/Gbx2* expression (Bouillet, Chazaud, Oulad-Abdelghani, Dollé, & Chambon, 1995; Urbach, 2007; Wassarman et al., 1997). In both animals, a *Pax2/5/8-* expression domain is positioned close to the interface between the anterior *otd/Otx2* and the posteriorly abutting *unpg/Gbx2* expression domains (Asano & Gruss, 1992; Hirth et al., 2003; Rowitch & McMahon, 1995). Hox genes orthologues expression follows posteriorly to the *Pax2/5/8* expression domain in both animals (Davenne et al., 1999; Hirth et al., 1998; Lichtneckert & Reichert, 2007). Furthermore, the interface of the relative expression of *dFezf/Fezf* and *mirr/Irx* was reported to be conserved between fly and mouse (Irima et al., 2010; Oliver et al., 1995). Adapted from Lichtneckert and Reichert, 2007. (*See insert for color representation of the figure*).

Hox genes encode a set of evolutionarily conserved homeodomain transcription factors that are involved in the specification of regionalized identity during development (Carroll, 1995); their role in anteroposterior regionalization is thought to have evolved early in metazoan history (Finnerty, 2003). They are generally expressed along the developing anteroposterior body axis in the same order as their arrangement on chromosomes ("co-linearity"). Hox gene expression is especially prominent in the developing central nervous system, which may be the ancestral site of Hox gene action in bilaterians (Hirth & Reichert, 2007). Hox genes are expressed in an ordered set of domains in the developing central nervous system of bilaterians as diverse as acoels, nematodes, annelids, molluscs, arthropods, urochordates, cephalochordates and vertebrates including zebra fish, chicken, mouse, and human (Carpenter, 2002; Hejnol & Martindale, 2009; Hirth & Reichert, 1999; Hughes & Kaufman, 2002; Hunt et al., 1991; Ikuta, Yoshida, Satoh, & Saiga, 2004; Irvine & Martindale, 2000; Kenyon et al., 1997; Kourakis et al., 1997; Lee, Callaerts, de Couet, & Martindale, 2003; Lumsden & Krumlauf, 1996; Moens & Prince, 2002; Steinmetz, Kostyuchenko, Fischer, & Arendt 2011; Vieille-Grosjean, Hunt, Gulisano, Boncinelli, & Thorogood, 1997; Wada, Garcia-Fernandez, & Holland, 1999; Wilkinson, Bhatt, Cook, Boncinelli, & Krumlauf, 1989).

Mutant analyses of Hox gene action in central nervous system development of fly and mouse reveal a comparable function in specification of regional identity. In *Drosophila*, Hox genes are required for the specification of regionalized neuronal identity in the posterior brain (Hirth, Hartmann, & Reichert, 1998). In mouse, Hox genes are involved in specifying the rhombomeres of the developing hindbrain (Gavalas et al., 1998; Studer, Lumsden, Ariza-McNaughton, Bradley, & Krumlauf, 1996; Studer et al., 1998). This evolutionary conservation of Hox gene action in central nervous system development is emphasized by the fact that cis-regulatory regions driving the specific spatiotemporal expression of Hox genes are interchangeable between insects and mammals (Malicki, Cianetti, Peschle, & McGinnis, 1992; Popperl et al., 1995). Together, these data imply that expression, function, and regulation of Hox gene action in central nervous system development are conserved features of this developmental control gene family.

While Hox genes are expressed in the posterior brain and nerve cord of bilaterians, they are excluded from the region of *otd/Otx2* and *ems/Emx* gene expression in the anterior brain. In vertebrates, a marked boundary region in the developing brain called the midbrain–hindbrain boundary (MHB) is located anterior to the expression domain of the Hox genes, and this region has an essential organizer function in patterning the midbrain and anterior hindbrain (Liu & Joyner, 2001; Rhinn & Brand, 2001; Wurst & Bally-Cuif, 2001). In vertebrates too, the developing MHB is delimited by the interface of the posterior *Otx2* expression domain and an abutting *Gbx2* expression domain, and it is also characterized by the expression of *Pax2/5/8* encoding genes. In *Drosophila*, a comparable boundary region is found in the developing brain anterior to the Hox expression domain; this region is also delimited by the interface of the posterior *otd/Otx2* domain and the abutting *unplugged (unpg)/Gbx2* expression domain, and is similarly characterized by the expression of *Pax2/5/8* (Hirth et al., 2003; Urbach, 2007). Comparable expression patterns of homologs of these genes are found anterior to the Hox expression domains in the developing brains of several other deuterostome and protostome taxa (Holland, 2009; Irimia et al., 2010; Steinmetz et al., 2011; Wada et al., 1998; Wada & Satoh, 2001).

Hence, a defined boundary region between the anterior (*otd*/*Otx2*-expressing) and the posterior (Hox-expressing) parts of the brain, which together have been considered to be representative of a tripartite organization of the ancestral chordate brain, appears to be evolutionarily conserved in bilaterians.

In vertebrate brains, a second region with organizer function is found at the *zona limitans intrathalamica* (ZLI) which develops within the diencephalon at the boundary between the expression domains of the *Fezf* and the *Irx* genes (Irimia et al., 2010). Comparable patterns of abutting gene expression define a ZLI-like boundary zone in the anterior brain of the basal chordate *Amphioxus*, implying that a ZLI-like structure is a conserved feature of chordate brains. Remarkably, a boundary of expression of the homologous insect genes is found in the anterior brain of *Drosophila*, where expression of *dFezf*, restricted to the anterior part of the brain, and expression of *mirr*, the earliest expressed fly *Irx* gene, adjoin to form a gene expression boundary (Irimia et al., 2010). The conserved nature of this ZLI-like interface of gene expression domains provides further support for the conserved nature of brain development in bilaterians. Additional support for this notion is provided by the conserved expression of *optix*/*Six3* genes in a comparable domain at the most anterior tip of the central nervous system neuroectoderm in animals as diverse as vertebrates, insects, and annelids (Oliver et al., 1995; Steinmetz et al., 2010). Thus the *Six3-Otx2* brain patterning system, like the *Fezf-Irx* and *Otx2-Gbx* patterning systems, may also be universal to central nervous system development in bilaterians.

It is noteworthy that comparable anteroposterior patterns of expression in a set of homologous genes are found in the net-like peripheral nervous system of the hemichordate *Saccoglossus* (Lowe et al., 2003). Whether this is also true for the developing central nervous system of this hemichordate is not currently known (Nomaksteinsky et al., 2009).

6.3.3 Conserved Mechanisms for Dorsoventral Pattering of the Bilaterian Central Nervous System

In addition to its anteroposterior axis, the neuroectoderm also has a second axis, which can be considered either mediolateral or dorsoventral, since the plate-like neuroectoderm can extend in a dorsal direction or can give rise to a neural tube through invagination. A second set of patterning genes subdivides the neuroectoderm along this dorsoventral axis in a manner that is conserved in vertebrates and invertebrates. Key among these is a set of homeobox genes, referred to as the columnar genes, which control the formation of longitudinal domains in the neuroectoderm. Moreover, the induction of the neuroectoderm that gives rise to the central nervous system appears to rely on conserved dorsoventral patterning mechanisms that determine the dorsoventral body axis itself as well as the location of the neuroectoderm along that axis (see Figure 6.3).

A set of conserved interacting signaling molecules play key roles in the establishment of dorsoventral polarity during embryogenesis. Central among these are morphogen-like signaling molecules of the Transforming Growth Factor β (TGFβ) family, represented by BMP4 in vertebrates and its homolog Dpp in *Drosophila* (De Robertis, 2008; De Robertis & Sasai, 1996). This BMP signalling pathway appears to be conserved in dorsoventral polarity formation in bilaterian animals such as insects, spiders, vertebrates, amphioxus and annelids, with the exceptions of nematodes and tunicates,

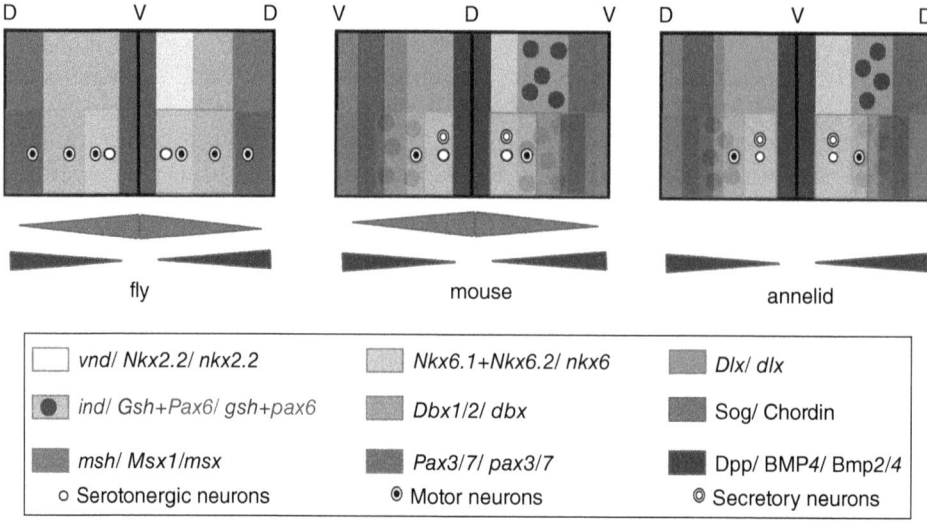

Figure 6.3 Schematic Representation of Examples of Conserved Dorsoventral Genetic Expression Boundaries in a Segmental Part of the Neuroectoderm in Arthropods (Left), Vertebrates (Middle) and Annelids (Right).

The vertebrate neuroectoderm is shown before folding. Anteroposterior patterning is not indicated. The neurogenic region is patterned in a dorsoventral fashion by a set of conserved patterning genes in all three animals, here indicated by color code. Note that the neuroectoderm of each animal is subdivided in two parts at its midline by a black vertical line enabling to show normally overlapping gene expression domains more clearly. At the bottom of the bars the overlap is shown for better comprehension. Within this overlay conserved neuron cell types emerging from this particular region are indicated by different circles (Arendt et al., 2008; Denes et al., 2007; Mizutani & Bier, 2008). The homologous proteins Dpp/ BMP4/ Bmp2/4 (violet) form a dorsoventrally inverted gradient in vertebrates with respect to *Drosophila melanogaster* and *Platynereis dumerilii*. In *Drosophila* and vertebrates, another homologous protein pair, namely Sog/ Chordin (brown) forms an opposing gradient with respect to the Dpp/ BMP4 pattern, where it inhibits Dpp/ BMP4 and therefore enables induction of neurogenesis and with different gradients gives identity to different subdomains of the neuroectoderm (Lichtneckert & Reichert, 2005). The dorsoventral columnar patterning genes are highly conserved between the bilaterian animals (see comparable relative expression domains of *vnd/ Nkx2.2/ nkx2.2* (yellow), *ind/ Gsh/ gsh* (orange), *msh/ Msx1/msx* (red), *Nkx6.1+Nkx6.2/ nkx6* (light green) in *Drosophila*, mouse and *Platynereis*) (Lichtneckert & Reichert 2007; Seibert, Volland, & Urbach, 2009). In the annelid and the mouse neuroectoderm even more similarities compared to Drosophila are apparent, such as the additional *Dbx1/2/ dbx* and *Dlx/ dlx* expression domains, the columnar medial *Pax6* expression (red dots) domain (Mizutani & Bier, 2008), as well as the *Pax3/7* expression which in *Drosophila* is expressed in a strictly segmented fashion (dark green) (Denes et al., 2007). (*See insert for color representation of the figure*).

which both have a modified type of development (Akiyama-Oda & Oda, 2006; Denes et al., 2007; Levine & Brivanlou, 2007; Little & Mullins, 2006; Lowe et al., 2006; Miya, Morita, Ueno, & Satoh, 1996; Mizutani et al., 2005, Mizutani, Meyer, Roelink, & Bier 2006; Sasai, Lu, Steinbeisser, & De Robertis, 1995; Suzuki et al., 1999; Yu et al., 2007). The polarizing action of BMP4/Dpp is antagonized in a spatially restricted manner by a second group of conserved extracellular signaling molecules which include Chordin in vertebrates and its homolog Sog in *Drosophila*

(Holley et al., 1995). The interacting Chordin/Sog and BMP4/Dpp signaling molecules act from opposing dorsoventral poles, and these poles are inverted in vertebrates versus invertebrates such as arthropods and annelids. (This provides strong support for the "dorsoventral inversion" hypothesis brought forward by Geoffroy Saint-Hilaire (1822) which states that the dorsoventral axis of vertebrates and invertebrates are equivalent but inverted; see Arendt & Nübler-Jung, 1994; De Robertis & Sasai, 1996). In addition to its polarizing function, the BMP4/Dpp morphogen suppresses development of the neuroectoderm and this suppressive function is inhibited by Chordin/Sog acting along the induced dorsoventral axis. Hence, in both vertebrate and invertebrate bilaterians, the region of the embryo that forms the neuroectoderm (dorsal in vertebrates, ventral in invertebrates) is the one in which Chordin/Sog is expressed and inhibits invading BMP4/Dpp. Indeed, whenever a central nervous system develops in vertebrates, insects, annelids and cephalochordates it derives from a neuroectoderm on the non-BMP body side (Denes et al., 2007; Levine & Brivanlou, 2007; Mizutani et al., 2005, 2006; Sasai et al., 1995; Yu et al., 2007). This suggests that the functional conservation of the Chordin/Sog and the BMP4/Dpp morphogens in CNS neuroectoderm induction represents a conserved dorsoventral patterning mechanism that was already present in the urbilaterian ancestor of vertebrates and invertebrates.

Following early neuroectoderm induction, a conserved set of homeodomain proteins, encoded by the *vnd/Nkx2.2*, *ind/Gsh* and *msh/Msx1* genes, act in further dorsoventral regionalization of the developing CNS (Chan & Jan, 1999; Cornell & Von Ohlen, 2000). All three genes are expressed in specific, nonoverlapping longitudinal columnar domains along the dorsoventral (or mediolateral) axis of the central nervous system. In *Drosophila*, *vnd* is expressed in a ventral column, *ind* in an intermediate column, and *msh* in a dorsal column of the ventral neuroectoderm; in the mouse, *Nkx2.2* is expressed in a ventral, *Gsh* in an intermediate, and *Msx1* in a dorsal column of the neural tube (Briscoe et al., 1999; Chu, Parras, White, & Jiménez 1998; Hsieh-Li et al., 1995; Isshiki, Takeichi, & Nose, 1997; McDonald, Holbrook, Isshiki, Weiss, Doe, Mellerick, 1998; Pabst, Herbrand, & Arnold, 1998; Pera & Kessel, 1998; Qui, Shimamura, Sussel, Chen, & Rubenstein 1998; Shimamura, Hartigan, Martinez, Puelles, & Rubenstein, 1995; Sussel, Marin, Kimura, & Rubenstein, 1999; Valerius et al., 1995; Wang, Chen, Xu, & Lufkin, 1996; Weiss et al., 1998). In both animals these so-called columnar genes control the formation of corresponding columnar dorsoventral identity domains, and act in neurogenesis at their site of action. These findings suggest that the role of the columnar genes in dorsoventral patterning of the central nervous system might be conserved throughout bilaterians (reviewed in Arendt & Nübler-Jung, 1999; Lichtneckert & Reichert, 2007; Reichert & Simeone, 2001; Urbach & Technau, 2008). In support of this idea, comparable longitudinal domains of expression of homologous columnar genes are observed in the neuroectoderm of the lophotrochozoan annelid *Platynereis*. Even more extensive similarities in putative dorsoventral patterning genes are seen in the annelid versus vertebrate neuroectoderm, in that a columnar *Pax6* expression domain as well as a columnar lateral *Pax3/7* expression domain is apparent in both animals (Briscoe, Pierani, Jessell, & Ericson, 2000; Denes et al., 2007; Ericson et al., 1997; Kriks, Lanuza, Mizuquchi, Nakafuka, & Goulding, 2005). (*Pax3/7* is also expressed in the developing central nervous system of *Drosophila*, albeit in a strictly segmented fashion; Davis et al., 2005; Kammermeier & Reichert, 2001). In all three bilaterian superphyla (Deuterostomes, Ecdysozoa, and Lophotrochozoa), the expression of these patterning genes is sensitive to BMP4, which specifically regulates their expression in a threshold-dependent manner (Denes et al., 2007; Mizutani et al., 2006). Interestingly, BMP4 may play an additional, conserved role in promoting sensory over motor neuron fate at later developmental stages (Denes et al., 2007; Lowe et al., 2006; Mizutani et al., 2006; Rusten, Kantera, Kafatos, & Barrio, 2002; Schlosser & Ahrens 2004; reviewed in Arendt et al., 2008; Mizutani & Bier, 2008).

Given the remarkable degree of conserved mechanisms for patterning the neuroectoderm, it is conceivable that some of the neural cell types that derive from the compartment-like domains of the neuroectoderm might also be conserved in vertebrate and invertebrate bilaterians. Evidence for a conservation of neuron types comes from recent comparative studies of annelid versus vertebrate central nervous system development. Thus, serotonergic projection neurons in the vertebrate hindbrain and those in the *Platynereis* both emerge from the *nkx2.2/nkx6* column, and cholinergic motor neurons with a comparable transcription factor signature emerge from a similar columnar *nkx6/pax6* domain in both vertebrates and annelids (Arendt et al., 2008; Arendt & Nübler-Jung 1999; Briscoe et al., 1999; Denes et al., 2007; Ericson et al., 1997; Pattyn et al., 2003). Similarly, early differentiating neurosecretory cells that produce the conserved neuropeptide arg-vasotocin/neurophysin develop in the anterior *nk2.2* domain of the central nervous system in *Platynereis* and mouse (Arendt, Tessmar-Raible, Snyman, Dorresteijn, & Wittbrodt, 2004). If these observations are indications of a more general conservation of neuronal cell types in bilaterians, then an explanation of these striking similarities based on evolutionary convergence (Moroz, 2009) becomes more and more unlikely, and we are left with the notion of a common, monophyletic origin of the bilaterian central nervous system (see Lichtneckert & Reichert, 2005; Mizutani & Bier, 2008; Reichert & Simeone, 2001). Indeed, there is increasing evidence that even rather complex central neural circuitries might have a common urbilaterian origin.

6.3.4 Common Patterning Mechanisms for Complex Brain Circuitry?

There are obvious differences in the olfactory sense organs of vertebrates and insects; the vertebrate olfactory epithelium is in the nasal cavity while the insect olfactory sensilla are on the antenna. Furthermore, the olfactory receptor molecules are evolutionarily distinct (Benton, Sachse, Michnick, & Vosshall, 2006; Wistrand, Käll, & Sonnhammer, 2006) and also differ somewhat in terms of expression control and activation mechanism between the two clades (Imai, Sakano, & Vosshall, 2010). Nevertheless, the circuit organization of the olfactory system in insects and vertebrates (see Figure 6.4) is remarkably similar in several respects (Hildebrand & Shepherd, 1997; Kay & Stopfer, 2006). First, a given olfactory sensory neuron—also called olfactory receptor neuron—in both flies and vertebrates expresses only a single olfactory receptor out of a large repertoire of olfactory receptor genes. Second, the axons of the olfactory sensory neurons that express a given receptor converge onto the same glomerulus in the primary olfactory center of the brain (vertebrate olfactory bulb, insect antennal lobe). Third, in the glomeruli the sensory neuron axons make synaptic connections with second-order olfactory neurons, the interneurons (vertebrate mitral/tufted cells, insect projection neurons). Moreover, the development of the olfactory circuitry is similar in several respects. For example, in both animal

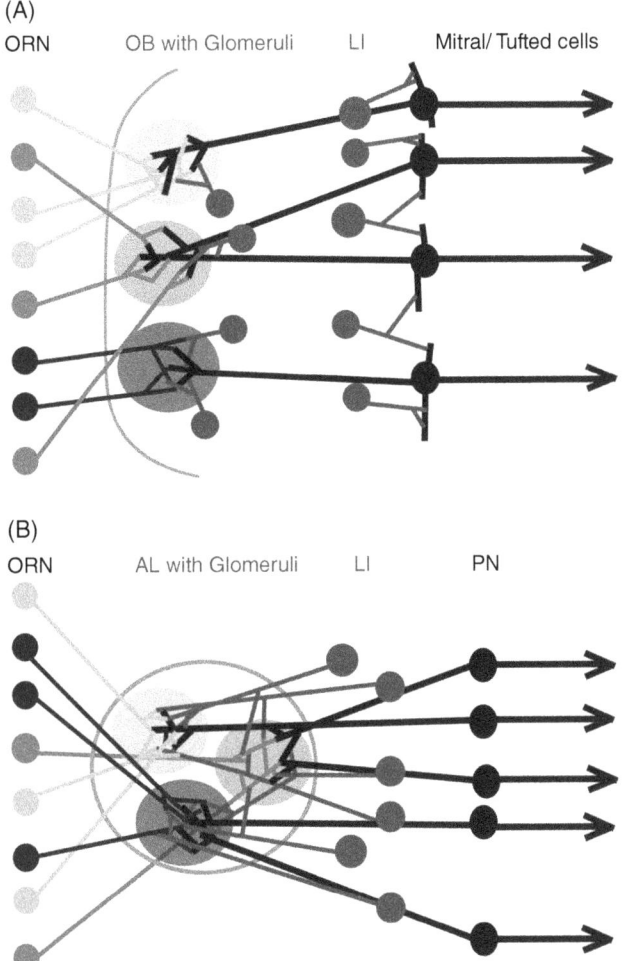

Figure 6.4 General Similarities of Olfactory Circuit Organization in Mammals (A) and Insects (B). ORN expressing the same olfactory receptor project to the same glomerulus in both animals (expressed olfactory receptor type in neurons is indicated by differently colored neurons). In the glomeruli the ORN connect to the dendrites of the mitral/tufted cells in the mammals (A) or PN in insects (B). In both animals, the sensory information is then transmitted by the mitral/tufted cells or the PN into higher brain centres. Different LI interconnect the information from the various glomeruli and process this olfactory information in fly and mouse. AL, antennal lobe; ORN, olfactory receptor neurons; OB, olfactory bulb; LI, local interneurons; PN, projection neurons. Adapted from Kay & Stopfer, 2006. (*See insert for color representation of the figure*).

groups, the *ems/Emx* genes are required for olfactory system development (Bishop, Garel, Nakagawa, Rubenstein, & O'Leary, 2003; Das et al., 2008; Lichtneckert, Nobs, & Reichert, 2008; Mallamaci et al., 1998; Sen, Hartmann, Reichert, & Rodrigues, 2010; Simeone et al., 1992). Furthermore, although the molecules involved are often different, gradients of axon guidance molecules and axon–axon interactions are important in the formation of topographic maps of olfactory receptor neuron (ORN) projections to the olfactory bulb in mouse and in the antennal lobe in flies (Hattori et al., 2007; Imai et al., 2009; Komiyama, Sweeney, Schuldiner, Gracia, & Luo, 2007; Lattemann et al., 2007; Luo & Flanagan, 2007; Sweeney et al., 2007; Zhu et al., 2006). While it is possible that these strikingly similar organizational and developmental features are all the result of convergent evolution, it is equally possible that they are evolutionarily conserved features which reflect the existence of 'primitive' olfactory circuitry in the brain of the urbilaterian ancestors of insects and vertebrates.

As in olfaction, there are also obvious differences in the sense organs for vision in vertebrates and insects; vertebrates posses single-lens eyes that contain ciliary-type photoreceptors and insects have compound eyes that comprise rhabdomeric-type photoreceptors. Despite these differences, there are surprising similarities in the structural and functional organization of the two visual systems (Sanes & Zipursky, 2010). At the circuit level, both fly and vertebrate visual systems comprise a few basic neural cell types that diversify into a high number of subtypes. Moreover, synaptic interconnections among these cells take place in sequentially arranged parallel laminar layers which are linked by orthogonal pathways that originate in the photoreceptors and terminate in higher visual centers of the brain. Indeed, these similarities prompted Cajal and Sanchez (1915) to conclude that the essential plan was maintained with small variations and re-touches of adaptation in the two apparently different types of visual systems. This notion is supported by more recent studies, which indicate that comparable control genes operate in insect and mammalian visual system development; it is also exemplified by the comparable role of the *otd/Otx* cephalic gap genes in the development of the peripheral and central visual systems of flies and mice (Acampora et al, 1998, 1999; Finckelstein et al., 1990; Hirth et al., 1995; Sen, Reichert, & Raghavan 2013; Vandendries, Johnson, & Reinke, 1996). As in the case of the olfactory system, the shared organizational and developmental features might be due to convergent evolution. However, they might also be due to evolutionary conservation of the 'essential plan' of an ancestral visual system that was already present in the urbilaterian brain.

Remarkably, recent evidence suggests that evolutionarily related higher brain centers might also have been present in the urbilaterian ancestor of vertebrates and invertebrates. In higher invertebrates such as annelids and arthropods, the mushroom body of the protocerebrum represents a high-order associative brain center involved in learning and memory. In vertebrates, comparable associative learning and memory functions are carried out by the cerebral cortex and hippocampus, which are developmental derivatives of the pallium. When the expression of a suite of conserved developmental control genes is compared in the developing mushroom body of the annelid *Platynereis* and in the developing pallium of the mouse, similar spatial patterns of expression are observed (Tomer, Denes, Tessmar-Raible, & Arendt, 2010). Based on these results, and in support of earlier findings, it has been proposed that the two higher brain centers are in fact homologous and that the urbilaterian ancestor might already have possessed a "high-order" associative brain center from which the extant mushroom body and pallium evolved (Strausfeld, Hanson, Li, Gomez, & Ito, 1998; Sweeney & Luo, 2010; Tomer et al., 2010). Most recently, a deep homology was suggested between the vertebrate basal ganglion and the arthropod central complex. These brain structures not only share comparable organizational features of neural circuitry but also have comparable functional roles in sensorimotor integration and affective behaviour. Furthermore, during development, both the basal ganglion and the central complex share genetic programs that involve homologous genes with comparable expression patterns and function (Strausfeld & Hirth, 2013).

In each of these examples of complex neural systems (olfaction, vision, learning/memory, and sensorimotor integration), similarities in organization and development between vertebrates and invertebrates might have evolved independently through convergent evolution. Alternatively, they could be due to the evolution of extant systems from a common ancestral system present in the urbilaterian brain. While further comparative studies are needed to resolve this issue, the increasing evidence for an urbilaterian animal that possessed a centralized brain with surprisingly complex sensory and associative brain centers, similar to those of higher vertebrates and invertebrates, provides further support for a monophyletic origin of the bilaterian central nervous system.

6.4 The First Metazoan Nervous System: Insights from Cnidarians

Since comparative and developmental genetic data imply that the last bilaterian common ancestor already possessed a complex centralized nervous system, when in animal evolution did centralization of nervous systems take place? As mentioned above, centralized nervous systems are found in cnidarians, which are one of the most basally branching animal phyla that have a nervous system. (While nervous systems are also present in ctenophorans, there is relatively little data on their organization, function and development, and they will not be considered further in this review.) Hence, the question arises of whether the radial nervous systems of Cnidaria and the bilateral nervous system of bilaterians are evolutionarily related (monophyletic origin of metazoan nervous systems) or not (polyphyletic origin of cnidarian versus bilaterian nervous systems).

Cnidarian neurons are found in the ectodermal and the endodermal cell layers, and in terms of cell types, correspond to sensory cells, motor neurons and ganglionic interneurons (Galliot et al., 2009; Watanabe, Fujisawa, & Holstein, 2009). In terms of neuroanatomy, cnidarian nervous systems show a great deal of variability among species and between their life cycle stages as sessile polyps and swimming medusae (Bullock & Horridge, 1965; Mackie, 2004). Polyps of cnidarians generally have diffuse epithelial nerve nets, but they also display regionalized concentrations of morphologically and neurochemically distinct neuronal subsets (Grimmelikhuijzen & Graff, 1985; Grimmelikhuijzen, Leviev, & Carstensen, 1996; Koizumi et al., 1992; Marlow, Srivastava, Matus, Rokhsar, & Martindale, 2009; Piraino et al., 2011). In medusoid cnidarians such as the swimming jellyfish, an even more complex organization of the radial nervous system is apparent. In addition to the peripheral nerve net, these nervous systems have single or double nerve rings, containing multiple neural cell types in circuitry for coordinated movement control and for processing of sensory information from organs such as statocysts, ocelli, and lens eyes (see Figure 6.5) (Garm, Ekström, Boudes, & Nilsson, 2006; Garm, Poussart, Parkefelt, Ekström, & Nilsson, 2007; Koizumi, 2007; Mackie 2004; Parkefelt & Ekström, 2009; Piraino et al., 2011). Moreover, the rhopalia of some cubomedusan cnidarians represent a ganglionic-like

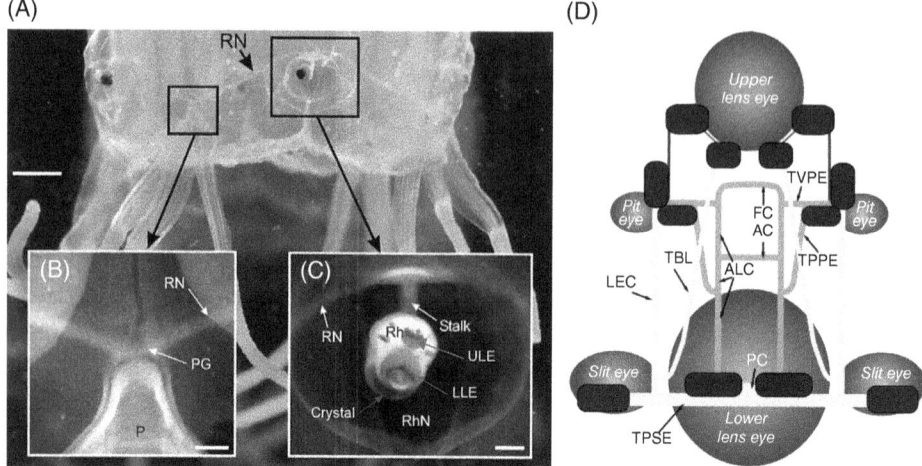

Figure 6.5 Complexity of the CNS of the Cubozoan Jellyfish *Tripedalia cystophora*. (A) The ring nerve RN connects the pedalial ganglion PG (B) with the rhopalia Rh (C) in the central nervous system. (C) the Rh constitute the main sensory structures of cubomedusae. Rh hang within the RhN on a stalk and carry six eyes (only the two lens eyes ULE and LLE are indicated). (A–C) Modified from Garm 2006. Reproduced with permission of Springer Publishing. (D) Schematic overview of commissural connections (light green and violet) between the different neuronal cell groups (dark blue) of the 6 distinct eyes (grey circles) in the rhopalium, indicating the remarkable complexity of this visual and integrating structure. Rh, rhopalium; RhN, rhopalial niche; LLE, large lens eye; ULE, upper lens eye; RN, ring nerve; PG, pedalial ganglion; P, pedalium; AC, anterior commissure; ALC, apical lateral connective; FC, frontal commissure; LEC, lateral exe connective; PC, posterior commissure; TBL, basal lateral tract; TPPE, posterior pit eye tract; TPSE, posterior slit eye tract; TVPE, vertical pit eye tract. Bars indicate 1 mm (A), 100 μm (B, C). Adapted from Parkefelt 2005. Reproduced with permission of John Wiley & Sons, Inc. (*See insert for color representation of the figure*).

centralization of multiple sensory cells together with a premotor pacemaker neuropil (Satterlie, 2010). Remarkably, even the larval forms of some cnidarians can have surprisingly complex regionalized nervous systems. The "crawling" planula larva of the hydrozoan *Clava multicornis* manifests a concentration of different neural cell types at the anterior pole, forming a neural plexus in association with the large number of sensory cells arranged at the anterior tip of the animal (Piraino et al., 2011).

Thus, both in terms of morphology and in terms of function, the nervous systems of cnidarians presents a remarkable degree of centralization supporting the notion that the cnidarian nervous systems might be representative of the first integrated concentrations of nervous tissue in metazoan evolution (Arendt et al., 2008; Bullock & Horridge, 1965; Satterlie, 2011). However, due to their intrinsic radial organization, the nervous systems of cnidarians manifest a different morphological bauplan from that of bilaterian nervous systems. This makes it difficult to assess the degree of evolutionary relationship between cnidarian and bilaterian nervous systems based on anatomy alone. Might a comparative developmental genetic analysis provide additional insight?

Many of the developmental control genes involved in axial patterning in insects and vertebrates are known to be conserved in cnidarians, implying that that a substantial number of these genes were already present in the last common ancestor of bilaterians and cnidarians (Ball, Hayward, Saint, & Miller, 2004; Boero, Scierwater, & Piraino, 2007; Finnerty, 2003; Finnerty, Pang, Burton, Paulson, & Martindale 2004; Galliot, 2000; Jacobs et al., 2007; Kusserow et al., 2005; Technau et al., 2005). Furthermore, there is now increasing evidence for a conservation of developmental patterning mechanisms between cnidarians and bilaterians. For example, the BMP4/Dpp and Chordin/Sog morphogen system appears to be present in cnidarians (Finnerty et al., 2004; Rentzsch, Guder, Vocke, Hobmayer, & Holstein, 2007; Saina, Genikhovitch, Renfer, & Technau, 2009; Technau & Steele, 2011). However, in contrast to the situation in bilaterians, the expression of both signaling components is found on the same side of the secondary ("directive") body axis rather than forming opposing gradients, and no morphological regionalization of the nervous system along the secondary axis of polyps has been observed. Regionalized expression in the developing cnidarian nervous system has been reported for the homologs of the *Emx* genes, *Msx* genes, and *Gsx* genes, and for the latter a functional role in nerve net development has been established (de Jong et al., 2006; Galliot et al., 2009; Miljkovic-Licina, Chera, Ghila, & Galliot, 2007; Miljkovic-Licina, Gauchat, & Galliot, 2004). *Otx* and Hox genes have also been identified in cnidarians. However, their expression patterns vary greatly among different species, and these genes do not seem to be involved in regionalized neuronal versus non-neuronal determination or (in the case of *Otx*) in head development (Chiori et al., 2009; Finnerty et al., 2004; Müller, Yanze, Schmid, & Spring, 1999; Quiquand et al., 2009; Ryan et al., 2007; Smith, Gee, Blitz, & Bode, 1999; Technau & Steele, 2011; Yanze, Spring, Schmidli, & Schmid, 2001).

Currently, it is difficult to interpret this gene expression data in cnidarians in terms of mechanisms for nervous system development. It is even more difficult to draw conclusions as to the possible conservation of these largely uncharted mechanisms in cnidarian nervous system development as compared to those known to operate in bilaterians. Such considerations must wait until further experimental insight is obtained into the developmental genetic origin of the cnidarian nervous system, and until meaningful molecular and mechanistic comparisons with bilaterian nervous system development can be carried out. However, the currently established findings do, at least, suggest that the genetic toolkit which is used to generate the nervous systems of bilaterians is also largely present in these radially symmetric animals. Hence, this genetic toolkit was probably already present in the last ancestor of cnidarians and bilaterians. Whether or not this toolkit was used in the common eumetazoan ancestor to build the first nervous system remains an intriguing enigma.

Acknowledgements

Supported by the SNF.

References

Acampora, D., Avantaggiato, V., Turoto, F., Barone, P. Reichert, H., Finkelstein, R., & Simeone, A. (1998). Murine Otx1 and Drosophila otd genes share conserved genetic functions required in invertebrate and vertebrate brain development. *Development*,125(9), 1691–1702.

Acampora, D., Boyl, P. P., Signore, M., Martinez-Barbera, J. P., Ilengo, C., Puelles, E., ... Simeone, A. (2001). OTD/OTX2 functional equivalence depends on 5 and 3 UTR–mediated control of Otx2 mRNA for nucleo-cytoplasmic export and epiblast-restricted translation. *Development*, 128(23), 4801–4813.

Acampora, D., Gulisano, M., Broccoli, V., & Simeone, A. (2001). Otx genes in brain morphogenesis. *Progress in Neurobiology*, 64(1), 69–95.

Acampora, D., Gulisano, M., & Simeone, A. (1999). Otx genes and the genetic control of brain morphogenesis. *Molecular & Cellular Neuroscience*, 13(1), 1–8.

Acampora, D., Mazan, S., Lallemand, Y., Avantaggiato, V., Maury, A. Simeone, A, & Brulet, P. (1995). Forebrain and midbrain regions are deleted in Otx2–/– mutants due to a defective anterior neuroectoderm specification during gastrulation. *Development*, 121(10), 3279–3290.

Achatz, J. G., & Martinez, P. (2012). The nervous system of Isodiametra pulchra (Acoela) with a discussion on the neuroanatomy of the Xenacoelomorpha and its evolutionary implications. *Frontiers in Zoology*, 9(1), 27.

Adoutte, A., Balavoine, G., Lartillot, N., & De Rosa, R. (1999). Animal evolution. The end of the intermediate taxa? *Trends in Genetics* 15(3), 104–108.

Adoutte, A., Balavoine, G., Lartillot, N., Lespinet, O., Prud'homme, B., & De Rosa, R. (2000). The new animal phylogeny: Reliability and implications. *Proceedings of the National Academy of Sciences of the USA*, 97(9), 4453–4456.

Akiyama-Oda, Y., & Oda H. (2006). Axis specification in the spider embryo: dpp is required for radial-to-axial symmetry transformation and sog for ventral patterning. *Development*, 133(12), 2347–2357.

Arendt, D., Denes, A. S., Jékely, G., & Tessmar-Raible, K. (2008). The evolution of nervous system centralization. *Philosophical Transactions of the Royal Society B, Biological Sciences*, 363(1496), 1523–1528.

Arendt, D., & Nübler-Jung, K. (1994). Inversion of dorsoventral axis? *Nature*, 371(6492), 26.

Arendt, D., & Nübler-Jung, K. (1999). Comparison of early nerve cord Development, in insects and vertebrates. *Development*, 126(11), 2309–2325.

Arendt, D., Technau, U., & Wittbrodt, J. (2001). Evolution of the bilaterian larval foregut. *Nature*, 409(6816), 81–85.

Arendt, D., Tessmar-Raible, K., Snyman, H., Dorresteijyn, A. W., & Wittbrodt, J. (2004). Ciliary photoreceptors with a vertebrate-type opsin in an invertebrate brain. *Science* 306(5697), 869–871.

Asano, M., & Gruss, P. (1992). Pax-5 is expressed at the midbrain–hindbrain boundary during mouse development. *Mechanisms of Development*, 39(1–2), 29–39.

Ball, E. E., Hayward, D. C., Saint, R., & Miller, D. J. (2004). A simple plan—Cnidarians and the origins of developmental mechanisms. *Nature Reviews Genetics*, 5(8), 567–577.

Benito-Gutierrez, E., & Arendt, D. (2009). CNS evolution: New insight from the mud. *Current Biology*, 19(15), R640–642.

Benton, R., Sachse, S., Michnick, S. W., & Vosshall, L. B. (2006). Atypical membrane topology and heteromeric function of Drosophila odorant receptors in vivo. *PLoS Biology*, 4(2), e20. doi 10.1371/journal.pbio 0040020

Bishop, K. M., Garel, S., Nakagawa, Y., Rubenstein, J. L., & O' Leary, D. D. (2003). Emx1 and Emx2 cooperate to regulate cortical size, lamination, neuronal differentiation, Development,of cortical efferents, & thalamocortical pathfinding. *Journal of Comparative Neurology*, 457(4), 345–360.

Boero, F., Schierwater, B., & Piraino, S. (2007). Cnidarian milestones in metazoan evolution. *Integratice & Comparative Biology*, 47(5), 693–700.

Bouillet, P., Chazaud, C., Oulad-Abdelghani, M., Dollé, P., & Chambon, P. (1995). Sequence and expression pattern of the Stra7 (Gbx-2) homeobox–containing gene induced by retinoic acid in P19 embryonal carcinoma cells. *Developmental Dynamics*, 204(4), 372–382.

Briscoe, J., Pierani, A., Jessell, T. M., Ericson, J. (2000). A homeodomain protein code specifies progenitor cell identity and neuronal fate in the ventral neural tube. *Cell*, 101(4), 435–445.

Briscoe, J., Sussel, L., Serup, P., Hartigan-O'Connor, D., Jessell, T. M., Rubenstein, J. L. R., & Ericson, J. (1999). Homeobox gene Nkx2.2 and specification of neuronal identity by graded Sonic hedgehog signalling. *Nature*, 398(6728), 622–627.

Bruce, A. E., & Shankland, M. (1998). Expression of the head gene Lox22–Otx in the leech Helobdella and the origin of the bilaterian body plan. *Developmental Biology*, 201(1), 101–112.

Brusca, R. C., & Brusca, G. J. (1990). *Invertebrates*. Sunderland, MA: Sinauer Associates.

Bullock, T. H., & Horridge, G. A. (1965). *Structure and function in the nervous systems of invertebrates* (Vol. 1). New York, NY: W. H. Freeman.

Cajal, S., & Sánchez, D (1915). Contribución al conocimiento de los centros nerviosos de los insectos. *Trabajos del Laboratorio de Investigaciones biológicos de la Universidad de Madrid*, 13, 1–68.

Carpenter, E. M. (2002). Hox genes and spinal cord development. *Developmental Neuroscience*, 24(1), 24–34.

Carroll, S. B. (1995). Homeotic genes and the evolution of arthropods and chordates. *Nature*, 376(6540), 479–485.

Chan, Y. M., & Jan, Y. N. (1999). Conservation of neurogenic genes and mechanisms. *Current Opinion in Neurobiology*, 9(5), 582–588.

Chiori, R., Jager, M., Denker, E., Wincker, P., Da Silva, C., Le Guyadur, H., & Manuel, M. (2009). Are Hox genes ancestrally involved in axial patterning? Evidence from the hydrozoan Clytia hemisphaerica (Cnidaria). *PLoS One*, 4(1), e4231. doi: 10.1371/journal.pone.0004231

Chu, H., Parras, C., White, K., & Jiménez, F. (1998). Formation and specification of ventral neuroblasts is controlled by vnd in Drosophila neurogenesis. *Genes & Development*, 12(22), 3613–3624.

Cohen, S. M., & Jurgens, G. (1990). Mediation of Drosophila head development by gap-like segmentation genes. *Nature*, 346(6283), 482–485.

Cornell, R. A., & Ohlen, T. V. (2000). Vnd/nkx, ind/gsh, & msh/msx: Conserved regulators of dorsoventral neural patterning? *Current Opinion in Neurobiology*, 10(1), 63–71.

Dalton, D., Chadwick, R., & McGinnis, R. (1989). Expression and embryonic function of empty spiracles: A Drosophila homeo box gene with two patterning functions on the anterior–posterior axis of the embryo. *Genes & Development*, 3(12A), 1940–1956.

Das, A., Sen, S., Lichtneckert, R., Okado, R., Ito, K., Rodrigues, V., & Reichert, H. (2008). Drosophila olfactory local interneurons and projection neurons derive from a common neuroblast lineage specified by the empty spiracles gene. *Neural Development*, 3: 33. doi: 10.1186/1749-8104-3-33

Davenne, M., Maconochie, M. K., Neun, R., Pattyn, A., Chambon, P., Krumlauf, R., & Rijli, F. M. (1999). Hoxa2 and Hoxb2 control dorsoventral patterns of neuronal development in the rostral hindbrain. *Neuron*, 22(4), 677–691.

Davis, G. K., D'Alessio, J. A., & Patel, N. H. (2005). Pax3/7 genes reveal conservation and divergence in the arthropod segmentation hierarchy. *Developmental Biology*, 285(1), 169–184.

de Jong, D. M., Hislop, N. R., Hayward, D. C., Reece-Hoyes, J. S., Pontynen, P. C., Ball, E. E., & Miller, D. J. (2006). Components of both major axial patterning systems of the Bilateria are differentially expressed along the primary axis of a radiate animal, the anthozoan cnidarian Acropora millepora. *Developmental Biology*, 298(2), 632–643.

De Robertis, E. M. (2008). Evo-devo: Variations on ancestral themes. *Cell*, 132(2), 185–195.

De Robertis, E. M., & Sasai, Y. (1996). A common plan for dorsoventral patterning in Bilateria. *Nature*, 380(6569), 37–40.

Denes, A. S., G. Jékely, G., Steinmetz, P. R., Raible, F., Snyman, H., Prud'homme, B., ... Arendt, D. (2007). Molecular architecture of annelid nerve cord supports common origin of nervous system centralization in bilateria. *Cell*, 129(2), 277–288.

Ericson, J., Rashbass, P., Shedl, A., Brenner-Morton, S., Kawakami, A., Van Heyningen, V., ... Briscoe, J. (1997). Pax6 controls progenitor cell identity and neuronal fate in response to graded Shh signaling. *Cell*, 90(1), 169–180.

Finkelstein, R., & Perrimon, N. (1990). The orthodenticle gene is regulated by bicoid and torso and specifies Drosophila head development. *Nature*, 346(6283), 485–488.

Finkelstein, R., Smouse, D., Capaci, T. M., Spradling, A. C., & Perrimon, N. (1990). The orthodenticle gene encodes a novel homeo domain protein involved in the Development, of the Drosophila nervous system and ocellar visual structures. *Genes & Development*, 4(9), 1516–1527.

Finnerty, J. R. (2003). The origins of axial patterning in the metazoa: how old is bilateral symmetry? *International Journal of Developmental Biology*, 47(7–8), 523–529.

Finnerty, J. R., Pang, K., Burton, P., Paulson, D., & Martindale, M. Q. (2004). Origins of bilateral symmetry: Hox and dpp expression in a sea anemone. *Science*, 304(5675), 1335–1337.

Fortey R. A. (2000). *Trilobite: Eyewitness to evolution*. New York, NY: Vintage Books.

Galliot, B. (2000). Conserved and divergent genes in apex and axis development of cnidarians. *Current Opinion in Genetics & Development*, 10(6), 629–637.

Galliot, B., Quiquand, M., Ghila, L., de Rosa, R., Miljkovic-Licina, M., & Chera, S. (2009). Origins of neurogenesis, a cnidarian view. *Developmental Biology*, 332(1), 2–24.

Garm, A., Ekström, P., Boudes, M., & Nilsson, D.-E., (2006). Rhopalia are integrated parts of the central nervous system in box jellyfish. *Cell Tissue Research*, 325(2), 333–343.

Garm, A., Poussart, Y., Parkefelt, L., Ekström, P., & Nilsson, D.-E. (2007). The ring nerve of the box jellyfish Tripedalia cystophora. *Cell Tissue Research*, 329(1), 147–157.

Gavalas, A., Studer, M. Lumsden, A., Rijli, F. M., Krumlauf, R., & Chambon, P. (1998). Hoxa1 and Hoxb1 synergize in patterning the hindbrain, cranial nerves and second pharyngeal arch. *Development*, 125(6), 1123–1136.

Geoffroy Saint–Hilaire, E. (1822). Considérations générales sur la vertèbre. *Mémoires du Muséum d'Histoire Naturelle de Paris* 9, 89–119.

Ghysen, A. (1992). The developmental biology of neural connectivity. *International Journal of Developmental Biology*, 36(1), 47–58.

Ghysen, A. (2003). The origin and evolution of the nervous system. *International Journal of Developmental Biology*, 47(7–8), 555–562.

Grimmelikhuijzen, C. J., & Graff, D. (1985). Arg-Phe-amide-like peptides in the primitive nervous systems of coelenterates. *Peptides*, 6, Suppl. 3, 477–483.

Grimmelikhuijzen, C. J., Leviev, I., & Carstensen, K. (1996). Peptides in the nervous systems of cnidarians: structure, function, & biosynthesis. *International Review of Cytology*, 167, 37–89.

Hartmann, B., Hirth, F., Walldorf, U., & Reichert, H. (2000). Expression, regulation and function of the homeobox gene empty spiracles in brain and ventral nerve cord development of Drosophila. *Mechanisms of Development*, 90(2), 143–153.

Hartmann, B., Muller, M., Hislop, N. R., Roth, B., Tomjenovic, L., Miller, D., & Reichert, H. (2010). Coral emx–Am can substitute for Drosophila empty spiracles function in head, but not brain development. *Developmental Biology*, 340(1), 125–133.

Hattori, D., Demir, E., Kim, H. W., Viragh, E., Zipursky, S. L., Dickson, B. J. (2007). Dscam diversity is essential for neuronal wiring and self–recognition. *Nature*, 449(7159), 223–227.

Hejnol, A., & Martindale, M. Q. (2009). Coordinated spatial and temporal expression of Hox genes during embryogenesis in the acoel Convolutriloba longifissura. *BMC Biology*, 7: 65. doi: 10.1186/1741-7007-7-65

Hejnol, A., Obst, M., Stamatakis, A., Ott, M., Rouse, G. W., Edgecombe, G. D., ... Dunn, C. W. (2009). Assessing the root of bilaterian animals with scalable phylogenomic methods. *Proceedings of the Royal Society, B. Biological Sciences*, 276(1677), 4261–4270.

Hildebrand, J. G., & Shepherd, G. M. (1997). Mechanisms of olfactory discrimination: converging evidence for common principles across phyla. *Annual Review of Neuroscience*, 20, 595–631.

Hirth, F. (2010). On the origin and evolution of the tripartite brain. *Brain Behavior and Evolution*, 76(1), 3–10.

Hirth, F., Hartmann, B., & Reichert, H. (1998). Homeotic gene action in embryonic brain development of Drosophila. *Development*, 125(9), 1579–1589.

Hirth, F., L. Kammermeier, et al. (2003). An urbilaterian origin of the tripartite brain: Developmental genetic insights from Drosophila. *Development*, 130(11), 2365–2373.

Hirth, F., & Reichert, H. (1999). Conserved genetic programs in insect and mammalian brain development. *Bioessays*, 21(8), 677–684.

Hirth, F., & Reichert, H. (2007). Basic nervous system types: One or many? In G. Strieder & J. Rubenstein (Eds.), *Evolution of the nervous system* (Vol.1). *History of ideas, basic concepts and developmental mechanisms* (pp. 56–72). Amsterdam: Elsevier.

Hirth, F., Therianos, S., Loop, T., Gehring, W. J., Reichert, H., & Furukubo-Tokunaga, K. (1995). Developmental defects in brain segmentation caused by mutations of the homeobox genes orthodenticle and empty spiracles in Drosophila. *Neuron*, 15(4), 769–778.

Holland, L. Z. (2009). Chordate roots of the vertebrate nervous system: Expanding the molecular toolkit. *Nature Reviews Neuroscience*, 10(10), 736–746.

Holland, N. D., & Chen, J. (2001). Origin and early evolution of the vertebrates: New insights from advances in molecular biology, anatomy, & palaeontology. *Bioessays*, 23(2), 142–151.

Holland, L. Z., Carvalho, J. E., Escriva, H., Laudet, V., Schubert, M., Shimeld, S. M., & Yu, J.-K. (2013). Evolution of bilaterian central nervous systems: A single origin? *EvoDevo*, 4(1), 27.

Holley, S. A., Jackson, P. D., Sasai, Y, Lu, B., De Robertis, E. M., Hoffmann, F. M., & Freguson, E. L. (1995). A conserved system for dorsal–ventral patterning in insects and vertebrates involving sog and chordin. *Nature*, 376(6537), 249–253.

Hsieh-Li, H. M., Witte, D. P., Szucsik, J. M., Weinstein, M., Li, H., & Potter, S. S. (1995). Gsh-2, a murine homeobox gene expressed in the developing brain. *Mechansims of Development*, 50(2–3), 177–186.

Hughes, C. L., & Kaufman, T. C. (2002). Exploring the myriapod body plan: Expression patterns of the ten Hox genes in a centipede. *Development*, 129(5), 1225–1238.

Hunt, P., Whiting, J., Nonchev, S., Sham, M.-H., Marshall, H., Graham, A., ... Krumlauf, R. (1991). The branchial Hox code and its implications for gene regulation, patterning of the nervous system and head evolution. *Development*, Suppl. 2: 63–77.

Hyman, L. H. (1940). *The invertebrates: Protozoa through ctenophora*. New York, NY: McGraw-Hill.

Ikuta, T., Yoshida, N., Satoh, N., & Saiga, H. (2004). Ciona intestinalis Hox gene cluster: Its dispersed structure and residual colinear expression in development. *Proceedings of the National Academy of Sciences of the USA*, 101(42), 15118–15123.

Imai, T., Sakano, H., & Vosshall, L. B. (2010. Topographic mapping—the olfactory system. *Cold Spring Harbor Perspectives in Biology*, 2(8), a001776. doi: 10.1101/chsperspect. a001776

Imai, T., Yamazaki, T., Kobayakawa, R., Kobayakawa, K., Takaya, A., Suzuki, M., & Sakano, H. (2009). Pre-target axon sorting establishes the neural map topography. *Science*, 325(5940), 585–590.

Irimia, M., Piñeiro, C., Maeso, I., Gómez-Skarmeta, J. L., Casares, F., & Garcia-Fernàndez, J. (2010). Conserved developmental expression of Fezf in chordates and Drosophila and the origin of the Zona Limitans Intrathalamica (ZLI) brain organizer. *Evodevo*, 1(1), 7.

Irvine, S. Q., & Martindale, M. Q. (2000). Expression patterns of anterior Hox genes in the polychaete Chaetopterus: correlation with morphological boundaries. *Developmental Biology*, 217(2), 333–351.

Isshiki, T., Takeichi, M., & Nose, A. (1997). The role of the msh homeobox gene during Drosophila neurogenesis: implication for the dorsoventral specification of the neuroectoderm. *Development*, 124(16), 3099–3109.

Jacobs, D. K., Nakanishi, N., Yuan, D., Camara, A., Nichols, S. A., & Hartenstein, V. (2007). Evolution of sensory structures in basal metazoa. *Integrative & Comparative Biology*, 47(5), 712–723.

Kammermeier, L., & Reichert, H. (2001). Common developmental genetic mechanisms for patterning invertebrate and vertebrate brains. *Brain Research Bulletin*, 55(6), 675–682.

Kay, L. M., & Stopfer, M. (2006). Information processing in the olfactory systems of insects and vertebrates. *Seminars in Cell & Developmental Biology*, 17(4), 433–442.

Kenyon, C. J., Austin, J., Costa, M., Cowing, D.W., Harris, J. M., Honigberg, L., ... Wrischnik, L. A. (1997). The dance of the Hox genes: Patterning the anteroposterior body axis of Caenorhabditis elegans. *Cold Spring Harbor Symposia on Quantitative Biology*, 62, 293–305.

Koizumi, O. (2007). Nerve ring of the hypostome in hydra: is it an origin of the central nervous system of bilaterian animals? *Brain Behavior & Evolution*, 69(2), 151–159.

Koizumi, O., Itazawa, M., Mizumoto, H., Misobe, S., Javois, L. C., Grimmelikhuijzen, C. J., & Bode, R. (1992). Nerve ring of the hypostome in hydra. I. Its structure, development, & maintenance. *Journal of Comparative Neurology*, 326(1), 7–21.

Komiyama, T., Sweeney, L. B., Schuldiner, O, Garcia, K. C., & Luo, L. (2007). Graded expression of semaphorin-1a cell-autonomously directs dendritic targeting of olfactory projection neurons. *Cell*, 128(2), 399–410.

Kotikova, E. A., & Raiikova, O. I. (2008). Architectonics of the central nervous system in Acoela, Plathelminthes, & Rotifera. *Zhurnal Evoliutsionnoi Biokhimii I Fiziologii*, 44(1), 83–93.

Kourakis, M. J., Master, V. A., Lokhorst, D. K., Nardelli-Haefliger, D., Weedon, C. J., Martindale, M. Q., & Shankland, M. (1997). Conserved anterior boundaries of Hox gene expression in the central nervous system of the leech Helobdella. *Developmental Biology*, 190(2), 284–300.

Kriks, S., Lanuza, G. M., Mizuquchi, R., Nakafuka, M., & Goulding, M. (2005). Gsh2 is required for the repression of Ngn1 and specification of dorsal interneuron fate in the spinal cord. *Development*, 132(13), 2991–3002.

Kusserow, A., Pang, K., Sturm, C., Hrouda, M., Lenfer, J., Schmidt, H. A., ... Holstein, T. W. (2005). Unexpected complexity of the Wnt gene family in a sea anemone. *Nature*, 433(7022), 156–160.

Lanjuin, A., VanHoven, M. K., Bargmann, C. I., Thompson, J. K., & Sengupta, P. (2003). Otx/otd homeobox genes specify distinct sensory neuron identities in C. elegans. *Developmental Cell*, 5(4), 621–633.

Lattemann, M., Zierau, A., Schulte, C., Seidl, S., Kuhlmann, B., & Hummel, T. (2007). Semaphorin–1a controls receptor neuron-specific axonal convergence in the primary olfactory center of Drosophila. *Neuron*, 53(2), 169–184.

Lee, P. N., Callaerts, P., de Couet, H. D., & Martindale, M. Q. (2003). Cephalopod Hox genes and the origin of morphological novelties. *Nature*, 424(6952), 1061–1065.

Leuzinger, S., Hirth, F., Gehrlich, D., Acampora, D., Simeone, A., Gehring, W. J., ... Reichert, H. (1998). Equivalence of the fly orthodenticle gene and the human OTX genes in embryonic brain development of Drosophila. *Development*, 125(9), 1703–1710.

Levine, A. J., & Brivanlou, A. H. (2007). Proposal of a model of mammalian neural induction. *Developmental Biology*, 308(2), 247–256.

Lichtneckert, R., Nobs, L., & Reichert, H. (2008). Empty spiracles is required for the development of olfactory projection neuron circuitry in Drosophila. *Development*, 135(14), 2415–2424.

Lichtneckert, R., & Reichert, H. (2005). Insights into the urbilaterian brain: Conserved genetic patterning mechanisms in insect and vertebrate brain development. *Heredity*, 94(5), 465–477.

Lichtneckert, R., & Reichert, H. (2007). Origin and evolution of the first nervous system. In J. H. Kaas (Ed.) *Evolution of nervous systems* (Vol. 1). *Theories, Development, Invertebrates* (pp. 291–315). Amsterdam: Elsevier.

Lichtneckert, R., & Reichert, H. (2008). Anteroposterior regionalization of the brain: Genetic and comparative aspects. *Advances in Experimental Medicine and Biology*, 628: 32–41.

Little, S. C., & Mullins, M. C. (2006). Extracellular modulation of BMP activity in patterning the dorsoventral axis. *Birth Defects Research C, Embryo Today Reviews*, 78(3), 224–242.

Liu, A., & Joyner, A. L. (2001). Early anterior/posterior patterning of the midbrain and cerebellum. *Annu Rev Neurosci* 24: 869–896.

Lowe, C. J., Terasaki, M., Wu, M., Freeman, R. M., Runft, L., Kwan, K., ... Gerhart, J. (2006). Dorsoventral patterning in hemichordates: insights into early chordate evolution. *PLoS Biology*, 4(9), e291. Retrieved from http://journals.plos.org/plosbiology/article?id=10.1371/journal.pbio.0040291

Lowe, C. J., Wu, M., Salic, A., Evans, L., Lander, E., Stange-Thomann, N., ... Kirschner, M. (2003). Anteroposterior patterning in hemichordates and the origins of the chordate nervous system. *Cell*, 113(7), 853–865.

Lumsden, A., & Krumlauf, R. (1996). Patterning the vertebrate neuraxis. *Science*, 274(5290), 1109–1115.

Luo, L., & Flanagan, J. G. (2007). Development, of continuous and discrete neural maps. *Neuron*, 56(2), 284–300.

Ma, X., Hou, X., Edgecombe, G. D., & Strausfeld, N. J. (2012). Complex brain and optic lobes in an early Cambrian arthropod. *Nature*, 490(7419), 258–261.

Mackie, G. O. (2004). Central neural circuitry in the jellyfish Aglantha: A model simple nervous system. *Neurosignals* 13(1–2), 5–19.

Malicki, J., Cianetti, L. C., Peschle, C., & McGinnis, W. (1992). A human HOX4B regulatory element provides head-specific expression in Drosophila embryos. *Nature*, 358(6384), 345–347.

Mallamaci, A., Iannone, R., Briata, P., Pintonello, L., Mercurio, S., Boncinelli, E., & Corte, G. (1998). EMX2 protein in the developing mouse brain and olfactory area. *Mechanisms of Development*, 77(2), 165–172.

Mallatt, J., Craig, C. W., & Yoder, M. J. (2010). Nearly complete rRNA genes assembled from across the metazoan animals: effects of more taxa, a structure-based alignment, & paired-sites evolutionary models on phylogeny reconstruction. *Molecular Phylogenetics and Evolution*, 55: 1–17.

Marlow, H. Q., Srivastava, M., Matus, D. Q., Rokhsar, D., & Martindale, M. Q., (2009). Anatomy and Development, of the nervous system of Nematostella vectensis, an anthozoan cnidarian. *Developmental Neurobiology*, 69(4), 235–254.

McDonald, J. A., Holbrook, S., Isshiki, T., Weiss, J., Doe, C. Q., & Mellerick D. M. (1998). Dorsoventral patterning in the Drosophila central nervous system: The vnd homeobox gene specifies ventral column identity. *Genes & Development*, 12(22), 3603–3612.

Miljkovic-Licina, M., Chera, S., Ghila, L., & Galliot, B.(2007). Head regeneration in wild-type hydra requires de novo neurogenesis. *Development*,134(6), 1191–1201.

Miljkovic-Licina, M., Gauchat, D., & Galliot, B. (2004). Neuronal evolution: analysis of regulatory genes in a first–evolved nervous system, the hydra nervous system. *Biosystems*, 76(1–3), 75–87.

Miya, T., Morita, K., Ueno, N., & Satoh, N. (1996). An ascidian homologue of vertebrate BMPs-5-8 is expressed in the midline of the anterior neuroectoderm and in the midline of the ventral epidermis of the embryo. *Mechanisms of Development*, 57(2), 181–190.

Mizutani, C. M., & Bier, E. (2008). EvoD/Vo: the origins of BMP signalling in the neuroectoderm. *Nature Reviews Genetics*, 9(9), 663–677.

Mizutani, C. M., Meyer, N., Roelink, H., & Bier, E. (2006). Threshold-dependent BMP-mediated repression: A model for a conserved mechanism that patterns the neuroectoderm. *PLoS Biology*, 4(10), e313. doi: 10.1371/journal.pbio.0040313

Mizutani, C. M., Nie, Q., Wan, F. Y., Zhang, Y. T., Vilmos, P., Sousa-Never, R., & Lander, A. D. (2005). Formation of the BMP activity gradient in the Drosophila embryo. *Developmental Cell*, 8(6), 915–924.

Moens, C. B., & Prince, V. E. (2002). Constructing the hindbrain: Insights from the zebrafish. *Developmental Dynamics*, 224(1), 1–17.

Moroz, L. L. (2009). On the independent origins of complex brains and neurons. *Brain Behavior & Evolution*, 74(3), 177–190.

Muller, P., Yanze, N., Yanze, Schmid, V., & Spring J. (1999). The homeobox gene Otx of the jellyfish Podocoryne carnea: role of a head gene in striated muscle and evolution. *Developmental Biology*, 216(2), 582–594.

Mwinyi, A., Bailly, X., Bourlat, S. J., Joondelius, U., Timothy, D., Littlewood, J., Posiadlowski, L. (2010) The phylogenetic position of Acoela as revealed by the complete mitochondrial genome of Symsagittifera roscoffensis. *BMC Evolutionary Biology*, 10: 309.

Nederbragt, A. J., te Welscher, P., van der Driesche, S., van Loon, A. E., & Dictus, W. J. (2002). Novel and conserved roles for orthodenticle/otx and orthopedia/otp orthologs in the gastropod mollusc Patella vulgata. *Development Genes and Evolution*, 212(7), 330–337.

Nieuwenhuys, R. (2002). Deuterostome brains: Synopsis and commentary. *Brain Research Bulletin*, 57(3–4), 257–270.

Nomaksteinsky, M., Rottinger, E., Dufour, H. D., Chettouh, Z., Lowe, C. J., Martindale, M. Q., & Brunet, J.-F. (2009). Centralization of the deuterostome nervous system predates chordates. *Current Biology*, 19(15), 1264–1269.

Northcutt, R. G. (2012). Evolution of centralized nervous systems: two schools of evolutionary thought. *Proceedings of the National Academy of Sciences of the USA*, 109 Suppl. 1, 10626–10633.

Oliver, G., Mailhos, A., Wehr, R., Copeland, N. G., Jenkins, N. A., & Gruss, P. (1995). Six3, a murine homologue of the sine oculis gene, demarcates the most anterior border of the developing neural plate and is expressed during eye development. *Development*, 121(12), 4045–4055.

Pabst, O., Herbrand, H., & Arnold, H. H. (1998). Nkx2-9 is a novel homeobox transcription factor which demarcates ventral domains in the developing mouse CNS. *Mechanisms of Development*, 73(1), 85–93.

Parkefelt, L., & Ekström, P. (2009). Prominent system of RFamide immunoreactive neurons in the rhopalia of box jellyfish (Cnidaria: Cubozoa). *Journal of Comparative Neurology*, 516(3), 157–165.

Parkefelt, L., Skogh, C., Nilsson, D.-E., & Ekström, P. (2005). Bilateral symmetric organization of neural elements in the visual system of a coelenterate, Tripedalia cystophora (Cubozoa). *Journal of Comparative Neurology*, 492(3), 251–262.

Pattyn, A., Vallstedt, A., Dias, J. M., Samad, O. A., Krumlauf, R., Rijli, F. M., … Ericson, J. (2003). Coordinated temporal and spatial control of motor neuron and serotonergic neuron generation from a common pool of CNS progenitors. *Genes & Development*, 17(6), 729–737.

Pera, E. M., & Kessel, M. (1998). Demarcation of ventral territories by the homeobox gene NKX2.1 during early chick development. *Development Genes and Evolution*, 208(3), 168–171.

Peterson, K. J., Lyons, J. B., Nowak, K. S., Takacs, C. M., Wargo, M. J., & McPeek, M. A. (2004). Estimating metazoan divergence times with a molecular clock. *Proceedings of the National Academy of Sciences of the USA*, 101(17), 6536–6541.

Philippe, H., Brinkmann, H., Martinez, P., Riutort, M., & Baguñá, J. (2007). Acoel flatworms are not platyhelminthes: Evidence from phylogenomics. *PLoS One*, 2(1), e717. doi: 10.1371/journal.pone.0000717

Philippe, H., Brinkmann, H., Copley, R. R., Moroz, L. L., Nakano, H., ... Telford, J. (2011). Acoelomorph flatworms are deuterostomes related to Xenoturbella. *Nature*, 470(7333), 255–258.

Piraino, S., Zega, G., di Benedetto, C., Leone, A., Dell'Anna, A., & Pennati, R. (2011). Complex neural architecture in the diploblastic larva of Clava multicornis (Hydrozoa, Cnidaria). *Journal of Comparative Neurology*, 519(10), 1931–1951.

Popperl, H., Bienz, M., Studer, M., Chan, S. K., Aparicio, S., Brenner, S., ... Krumpauf, R. (1995). Segmental expression of Hoxb-1 is controlled by a highly conserved autoregulatory loop dependent upon exd/pbx. *Cell*, 81(7), 1031–1042.

Qui, M., Shimamura, K., Sussel, L., Chen, S., & Rubenstein, J. L. R. (1998). Control of anteroposterior and dorsoventral domains of Nkx–6.1 gene expression relative to other Nkx genes during vertebrate CNS development. *Mechanisms of Development*, 72(1–2), 77–88.

Quiquand, M., Yanze, N., Schmich, J., Schmid, V., Galliot, B., & Piraino, S. (2009). More constraint on ParaHox than Hox gene families in early metazoan evolution. *Developmental Biology*, 328(2), 173–187.

Reichert, H. (2009). Evolutionary conservation of mechanisms for neural regionalization, proliferation and interconnection in brain development. *Biology Letters*, 5(1), 112–116.

Reichert, H., & A. Simeone (2001). Developmental genetic evidence for a monophyletic origin of the bilaterian brain. *Philosophical Transactions of the Royal Society, B. Biological Sciences*, 356(1414), 1533–1544.

Rentzsch, F., Guder, C., Vocke, D., Hobmayer, B., & Holstein, T. W. (2007). An ancient chordin-like gene in organizer formation of Hydra. *Proceedings of the National Academy of Sciences of the USA*, 104(9), 3249–3254.

Rhinn, M., & Brand, M. (2001). The midbrain–hindbrain boundary organizer. *Current Opinions in Neurobiology*, 11(1), 34–42.

Rowitch, D. H., & McMahon, A. P. (1995). Pax-2 expression in the murine neural plate precedes and encompasses the expression domains of Wnt-1 and En-1. *Mechanisms of Development*, 52(1), 3–8.

Rusten, T. E., Cantera, R., Kafatos, F. C., & Barrio R. (2002). The role of TGF beta signaling in the formation of the dorsal nervous system is conserved between Drosophila and chordates. *Development*, 129(15), 3575–3584.

Ryan, J. F., Mazza, M. E., Pang, K., Matus, D. Q., Baxevanis, A. D., Martindale, F. Q., & Finnerty, J. R. (2007). Pre-bilaterian origins of the Hox cluster and the Hox code: Evidence from the sea anemone, Nematostella vectensis. *PLoS One*, 2(1), e153.

Saina, M., Genikhovich, G., Renfer, E., & Technau, U. (2009). BMPs and chordin regulate patterning of the directive axis in a sea anemone. *Proceedings of the National Academy of Sciences of the USA*, 106(44), 18592–18597.

Sanes, J. R., & Zipursky, S. L. (2010). Design principles of insect and vertebrate visual systems. *Neuron*, 66(1), 15–36.

Sasai, Y., Lu, B., Steinbeisser, H., & De Robertis, E. M. (1995). Regulation of neural induction by the Chd and Bmp-4 antagonistic patterning signals in Xenopus. *Nature*, 377(6551), 757.

Satterlie, R. A. (2011) Do jellyfish have central nervous systems? *Journal of Experimental Biology*, 214(Pt 8), 1215–1223.

Schilling, T. F., & Knight, R. D. (2001). Origins of anteroposterior patterning and Hox gene regulation during chordate evolution. *Philosophical Transactions of the Royal Society of London, B. Biological Sciences*, 356(1414), 1599–1613.

Schlosser, G., & Ahrens, K. (2004). Molecular anatomy of placode development in Xenopus laevis. *Developmental Biology*, 271(2), 439–466.

Seibert, J., Volland, D., & Urbach, R. (2009). Ems and Nkx6 are central regulators in dorsoventral patterning of the Drosophila brain. *Development*, 136(23), 3937–3947.

Semmler, H., Chiodin, M., Bailly, X., Martinez, P., & Wanninger, A. (2010). Steps towards a centralized nervous system in basal bilaterians: insights from neurogenesis of the acoel Symsagittifera roscoffensis. *Development, Growth & Differentiation*, 52(8), 701–713.

Sen, S., Hartmann, B., reichert, H., & Rodrigues, V. (2010). Expression and function of the empty spiracles gene in olfactory sense organ development of Drosophila melanogaster. *Development*, 137(21), 3687–3695.

Sen, S., Reichert, H., & Raghavan, V. (2013). Conserved roles of ems/Emx and otd/Otx genes in olfactory and visual system development in Drosophila and mouse. *Open Biology*, 3(5), 120177. doi: 10.1098/rsob.120177

Shimamura, K., Hartigan, D. J., Martinez, S., Puelles, L., & Rubenstein, J. L. (1995). Longitudinal organization of the anterior neural plate and neural tube. *Development*, 121(12), 3923–3933.

Simeone, A., Gulisano, M., Acamporo, D., Stornaiuolo, A., Ramboldi, M., & Boncinelli, E. (1992). Two vertebrate homeobox genes related to the Drosophila empty spiracles gene are expressed in the embryonic cerebral cortex. *The Embo Journal*, 11(7), 2541–2550.

Smith, K. M., Gee, L., Blitz, I. L., & Bode, R. (1999). CnOtx, a member of the Otx gene family, has a role in cell movement in hydra. *Developmental Biology*, 212(2), 392–404.

Steinmetz, P. R., Kostyuchenko, R. P., Fischer, A., Arendt, D. et al. (2011) The segmental pattern of otx, gbx, & Hox genes in the annelid Platynereis dumerilii. *Evolution & Development*, 13(1), 72–79.

Steinmetz, P. R., Urbach, R., Posmen, N., Eriksson, J., Kostyuchenko, R. P., Brena, C. (2010). Six3 demarcates the anterior-most developing brain region in bilaterian animals. *Evodevo*, 1(1), 14.

Strausfeld, N. J., Hansen, L., Li, Y., Gomez, R. S., & Ito, K. (1998). Evolution, discovery, & interpretations of arthropod mushroom bodies. *Learning & Memory*, 5(1–2), 11–37.

Strausfeld, N. J., & Hirth, F. (2013). Deep homology of arthropod central complex and vertebrate basal ganglia. *Science*, 340(6129), 157–161.

Studer, M., Gavalas, A., Marshall, H., Ariza-McNaughton, L., Fijli, F. M., Chambon, P., & Krumlauf, R. (1998). Genetic interactions between Hoxa1 and Hoxb1 reveal new roles in regulation of early hindbrain patterning. *Development*, 125(6), 1025–1036.

Studer, M., Lumsden, A., Ariza-McNaughton, L., Bradley, C., & Krumlauf, R. (1996). Altered segmental identity and abnormal migration of motor neurons in mice lacking Hoxb–1. *Nature*, 384(6610), 630–634.

Sussel, L., Marin, O., Kimura, S., & Rubenstein, J. L. (1999). Loss of Nkx2.1 homeobox gene function results in a ventral to dorsal molecular respecification within the basal telencephalon: evidence for a transformation of the pallidum into the striatum. *Development*, 126(15), 3359–3370.

Suzuki, Y., Yandell, M. D., Roy, P. J., Krishna, S., Savage, C., Ross, R. M., ... Wood, W. B. (1999). A BMP homolog acts as a dose-dependent regulator of body size and male tail patterning in Caenorhabditis elegans. *Development*, 126(2), 241–250.

Sweeney, L. B., Couto, A., Chou, Y. H., Berdnik, D., Dickson, B. J., Luo, L., Ko,iyama, T. (2007). Temporal target restriction of olfactory receptor neurons by Semaphorin–1a/PlexinA–mediated axon–axon interactions. *Neuron*, 53(2), 185–200.

Sweeney, L. B., & Luo, L. (2010) Fore brain: A hint of the ancestral cortex. *Cell*, 142(5), 679–681.

Tanaka, G., Hou, X., Ma, X., Edgecombe, G. D., & Strausfeld, N. J. (2013). Chelicerate neural ground pattern in a Cambrian great appendage arthropod. *Nature*, 502(7471), 364–367.

Technau, U., Rudd, S., Maxwell, P., Gordon, P. M. K., Saina, M., Grasso, L. C., ... Miller, D. J. (2005). Maintenance of ancestral complexity and non–metazoan genes in two basal cnidarians. *Trends in Genetics*, 21(12), 633–639.

Technau, U., & Steele, R. E. Evolutionary crossroads in developmental biology: Cnidaria. *Development*, 138(8), 1447–1458.

Tomer, R., Denes, A. S., Tessmar-Raible, K., & Arendt, D. (1998) Profiling by image registration reveals common origin of annelid mushroom bodies and vertebrate pallium. *Cell*, 142(5), 800–809.

Tomsa, J. M., & Langeland, J. A. (1999). Otx expression during lamprey embryogenesis provides insights into the evolution of the vertebrate head and jaw. *Developmental Biology*, 207(1), 26–37.

Umesono, Y., Watanabe, K., & Agata, K. (1999). Distinct structural domains in the planarian brain defined by the expression of evolutionarily conserved homeobox genes. *Development, Genes and Evolution*, 209(1), 31–39.

Urbach, R. (2007). A procephalic territory in Drosophila exhibiting similarities and dissimilarities compared to the vertebrate midbrain/hindbrain boundary region. *Neural Development*, 2, 23. doi: 10.1186/1749-8104-2-23

Urbach, R., & Technau, G. M. (2008). Dorsoventral patterning of the brain: A comparative approach. *Advances in Experimental Medicine and Biology*, 628, 42–56.

Valerius, M. T., Li, H., Stock, J. L., Weinstein, M. Kaur, S., Singh, G., & Potter, S. S. (1995). Gsh–1: a novel murine homeobox gene expressed in the central nervous system. *Developmental Dynamics*, 203(3), 337–351.

Vandendries, E. R., Johnson, D., & Reinke, B (1996). Orthodenticle is required for photoreceptor cell development in the Drosophila eye. *Developmental Biology*, 173(1), 243–255.

Vieille-Grosjean, I., Hunt, P., Gulisano, M., Boncinelli, E., & Thorogood, P. (1997). Branchial HOX gene expression and human craniofacial development. *Developmental Biology*, 183(1), 49–60.

Wada, H., Garcia-Fernandez, J., & Holland, P. W. (1999). Colinear and segmental expression of amphioxus Hox genes. *Developmental Biology*, 213(1), 131–141.

Wada, H., Saiga, H., Satoh, N., & Holland, P. W. H. (1998). Tripartite organization of the ancestral chordate brain and the antiquity of placodes: insights from ascidian Pax–2/5/8, Hox and Otx genes. *Development*, 125(6), 1113–1122.

Wada, H., & Satoh, N. (2001). Patterning the protochordate neural tube. *Current Opinions in Neurobiology*, 11(1), 16–21.

Wang, W., Chen, X., Xu, H., & Lufkin, T. (1996). Msx3: a novel murine homologue of the Drosophila msh homeobox gene restricted to the dorsal embryonic central nervous system. *Mechanisms of Development*, 58(1–2), 203–215.

Wassarman, K. M., Lewandoski, M., Campbell, K., Joynrt, J. L., Rubenstein, J. L. R., & Martinez, S. (1997). Specification of the anterior hindbrain and establishment of a normal mid/hindbrain organizer is dependent on Gbx2 gene function. *Development*, 124(15), 2923–2934.

Watanabe, H., Fujisawa, T., & Holstein, T. W. (2009). Cnidarians and the evolutionary origin of the nervous system. *Development, Growth, & Differentiation*, 51(3), 167–183.

Weiss, J. B., Von Ohlen, T., Mellerick, D. M., Dressler, G., Doe, C. Q., & Scott, M. P. (1998). Dorsoventral patterning in the Drosophila central nervous system: the intermediate neuroblasts defective homeobox gene specifies intermediate column identity. *Genes & Development*, 12(22), 3591–3602.

Wilkinson, D. G., Bhatt, S., Cook, M., Boncinelli, E., & Krumlauf, R. et al. (1989). Segmental expression of Hox-2 homoeobox–containing genes in the developing mouse hindbrain. *Nature*, 341(6241), 405–409.

Wistrand, M., Käll, L, & Sonnhammer, E. L. (2006). A general model of G protein–coupled receptor sequences and its application to detect remote homologs. *Protein Science*, 15(3), 509–521.

Wurst, W., & Bally-Cuif, L. (2001). Neural plate patterning: Upstream and downstream of the isthmic organizer. *Nature Reviews Neuroscience*, 2(2), 99–108.

Yanze, N., Spring, J., Schmidli, C., & Schmid, V. (2001). Conservation of Hox/ParaHox–related genes in the early Development, of a cnidarian. *Developmental Biology*, 236(1), 89–98.

Younossi-Hartenstein, A., Green, P., Liaw, G.-J., Rudolph, K., Lengyel, J., & Hartenstein, V. (1997). Control of early neurogenesis of the Drosophila brain by the head gap genes tll, otd, ems, and btd. *Developmental Biology*, 182(2), 270–283.

Yu, J. K., Satou, Y., Holland, N. D., Shin-I, T., Kohara, Y., Satoh, N., ... Holland, L. Z. (2007). Axial patterning in cephalochordates and the evolution of the organizer. *Nature*, 445(7128), 613–617.

Zhu, H., Hummel, T., Clemens, J. C., Berdnik, D., Zipursky, S. L., & Luo, L. (2006). Dendritic patterning by Dscam and synaptic partner matching in the Drosophila antennal lobe. *Nature Neuroscience*, 9(3), 349–355.

7

Fundamental Constraints on the Evolution of Neurons

A. Aldo Faisal and Ali Neishabouri

7.1 Introduction

Nervous systems are responsible for perceiving, integrating, and responding to complex and diverse stimuli, such as reading and typing this text. On an abstract level, our brain and a computer have to solve similar computational tasks and, thus, show similarities in their design. However, the brain's basic building blocks are fundamentally different from those of conventional electronics—it uses neurons and synapses as computational components, which are made up of proteins instead of transistors, fat (bilipid membrane) as insulators, and salty water instead of gold or copper as a conducting core.

Processing and transmission of information in neurons is accomplished by altering the membrane potential through movement of ions. The cell membrane is largely impermeable to ions and acts as a capacitance with a finite response speed, determined by the membrane time constant. The finite response range of neurons—signals range over 100 mV in amplitude and less than 1 kHz in action potential frequency—imposes limits on the total information throughput (Stemmler & Koch, 1999). Rates of synthesis, release diffusion, and uptake of chemical transmitters also limit the performance of neural fibres.

Random fluctuations are present at all levels of nervous systems (reviewed in Faisal, Selen, & Wolpert, 2008). Ion channels are subject to thermodynamic noise that may cause their spontaneous opening or closing (Faisal, White, & Laughlin, 2005; White, Klink, Alonso, & Kay, 1998), which is called channel noise. Synaptic vesicle release, diffusion, and molecular interactions (Laughlin, 1989) are all stochastic processes. The existence of these sources of variability undermines reliable processing and transmission of information in neurons. Creation and maintenance of a nervous system is very metabolically expensive: this includes the cost of producing and maintaining neurons, their connections, and support cells (astrocytes, oligodendrocytes, and Schwann cells), to which we must add the cost of generating and propagating neural signals (Attwell & Laughlin, 2001; Harris & Attwell, 2012; Laughlin, de Ruyter van Steveninck, & Anderson, 1998). The metabolic cost of APs in the human brain alone accounts for 22% of the resting metabolic consumption (Alle, Roth, & Geiger, 2009; Laughlin, 2001; Sengupta, Stemmler, Laughlin, & Niven, 2010).

Finally, in the case of very dense circuits such as the brain, or in very small organisms, neural fibres are constrained by volume (see Niven & Farris, 2012, for a review of miniaturization of nervous systems). There is evidence that the wiring of the brain optimizes the volume occupied by axons to reduce metabolic cost and conduction delays (Wang et al. 2008). The size of axons directly interacts with all 4 physical constraints: bigger axons increase the overall volume of nervous system and have a higher associated metabolic cost, while smaller axons conduct APs slowly. Moreover, noise imposes a lower limit on the diameter of axons (Faisal et al., 2005).

How do these differences affect brain function and design? How have they channelled the evolution of the nervous systems? The brain's building blocks are several orders of magnitude less reliable than those of computers, yet computers with the computational capability and reliability of the brain would require a small power plant while the brain completes all its function with less power than a light bulb needs to illuminate this text. The likely reason for the brain's efficiency lies in its design: It uses circuits arranged in large, massively parallel networks and molecular components operating on the nanometer scale. With the advent of synthetic biology and current efforts to engineer living machines *ab initio*, it has become important not only to have a mechanistic understanding of functional and structural drivers of nervous system's evolution but also to uncover the essential the design principles of biological "devices."

We will focus on two fundamental constraints that apply to any form of information processing system, be it a cell, a brain or a computer: 1. Noise (random variability) and 2. Energy (metabolic demand). We will show how these two constraints are fundamentally limited by the basic biophysical properties of the brain's building blocks (protein, fats, and salty water) and link nervous system structure to function. The understanding of the interdependence of information (and its "nemesis" noise) and energy has profoundly influenced the development of efficient telecommunication systems and computers. However, in biology and neuroscience this fundamental relationship between information and energy is little investigated, although it bears important implications for understanding evolution (see Figure 7.1).

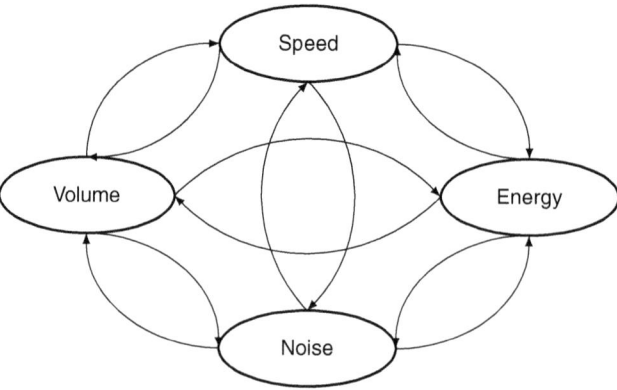

Figure 7.1 Basic Constraints on the Design of Neural Circuits.
Energy, noise, speed, and volume are linked to each other by basic biophysical principles. For example, reducing the diameter of an axon will decrease its volume and its metabolic cost. However, smaller axons are noisier, and conduct action potentials more slowly.

7.2 Noise as a Fundamental Limit on Axon Diameter

When Adrian began to record from neurons in the 1920s, he observed that neural responses were highly variable across identical stimulation trials and only the average response could be related to the stimulus (Adrian, 1928; Adrian & Matthews, 1927). Biology viewed this variable nature of neuronal signaling as "variability", engineers called it "noise." The two terms are closely related, but as we shall see imply two very different approaches to think about the brain—one operating at the systems level, the other operating at the molecular level. On the one side, the healthy brain functions efficiently and reliably, as we routinely experience ourselves. On the other side, variability is a reflection of the complexity of the nervous system.

In the classical view of neurobiology it is implicitly assumed that averaging large numbers of such small stochastic elements effectively wipes out the randomness of individual elements at the level of neurons and neural circuits. This assumption however requires careful consideration for two reasons:

1. Neurons perform highly nonlinear operations involving high-gain amplification and positive feedback. Therefore, small biochemical and electrochemical fluctuations of a random nature can significantly change whole cell responses.
2. Many neuronal structures are very small. This implies that they are sensitive to (and require only) a relatively small number of discrete signaling molecules to affect the whole. These molecules, such as voltage-gated ion channels or neurotransmitters, are invariably subject to thermodynamic fluctuations and hence their behavior will have a stochastic component that may affect whole cell behavior.

All forms of signaling in the brain are in the end controlled by proteins embedded in the cell membrane or within the cell. These proteins operate with an element of randomness due to thermodynamic fluctuations, which can have important consequences in terms of how the nervous system is designed and functions. This suggests that unpredictable random variability (noise) produced by thermodynamic mechanisms (e.g., diffusion of signaling molecules) or quantum mechanisms (e.g., photon absorption in vision) at the molecular level can have a deep and lasting influence on variability present at the system level. We, on the other hand, have come to expect a near deterministic experience of our nervous system. We do not expect to see or hear anything unless there is something to be seen or heard. We generally seem to see the same thing if we look twice. This deterministic experience implies that the design principles of the brain must mitigate or even exploit the constraints set by noise and other biophysical factors.

This prowess can only be fully appreciated when we realize that noise cannot be removed from a signal once it has been added to it. Since signals can easily be lost, and noise easily added, this sets a one-sided limit on how well information can be represented. Noise diminishes the capacity to receive, process, and direct information, the key tasks of the brain. Investing in the brain's design can reduce the effects of noise, but this investment often increases energetic requirements, which is likely to be evolutionary unfavourable.

7.3 Molecular Noise as a Fundamental Limit on Wiring Density

In our brain the action potential (AP) is used as the basic signal for communication in neural networks. The AP is carried by the spread of membrane depolarization along the membrane and is mediated by voltage-gated ion channels: The depolarization is (re)generated by nonlinear voltage-gated sodium conductances acting as positive feedback amplifiers, and is terminated by leak conductances and voltage-gated potassium channels that repolarize the membrane (Hille, 2001; Weiss, 1997).

How small can neurons or axons be made before channel noise effects disrupt action potential signaling? This is clearly a neuronal design question that had no systematic answer as recently as a few years ago—anatomists had previously shown that axons as fine as $0.08\,\mu m$ to $0.1\,\mu m$ are commonly found in the central nervous system. Hille (1970) suggested that in very fine axons the opening of a few sodium channels could generate an AP. Detailed theoretical and simulations (Faisal et al., 2005) showed that spontaneous opening of sodium channels can, in theory, trigger random action potentials below a critical axon diameter of $0.15\,\mu m$ to $0.2\,\mu m$ diameter.

This is because at these diameters the input resistance of a sodium channel is comparable to the input resistance of the axon. The single, persistent opening of a single sodium channel can therefore depolarize the axon membrane to threshold. Below this diameter, the rate at which randomly generated APs appear increases exponentially as diameter decreases (see Figure 7.2A). This will disrupt signaling in axons below a limiting diameter of about $0.1\,\mu m$, as random action potentials cannot be distinguished from signal-carrying action potentials. This limit is robust with respect to parameter variation around two contrasting axon models, mammalian cortical axon collaterals and the invertebrate squid axon. This robustness shows that the limit is mainly set by the order of magnitude of the properties of ubiquitous cellular components, conserved across neurons of different species. The occurrence of random action potentials (RAP) and the exponential increase in RAP rate as diameter decreases is an inescapable consequence of the AP mechanism. The stochasticity of the system becomes critical when its inherent randomness makes it operationally infeasible, that is, when random APs become as common as evoked APs.

7.4 Higher Body Temperature, Lower Neuronal Noise: Why Warmer Brains Are More Reliable

Temperature is not only a key factor in determining the speed of biochemical reactions such as ion channel gating; it also controls the amount of ion channel variability (Faisal & Matheson, 2000; Faisal et al., 2005). While commonly overlooked, temperature—and via its effects on ion channel kinetics, channel noise—can vary greatly across the nervous system: Cold-blooded insects can warm up their body to over $40\,°C$ prior to taking flight, while human extremities and the sensory and motor neurons therein can be exposed to temperature differences of up to $10\,°C$ or more between their dendrites, cell bodies, and axon terminals as they span from cold extremities to the warmer spinal cord.

The rate of RAPs triggered by channels noise counterintuitively *decreases* as temperature increases—just the opposite of what one would expect from electrical Johnston noise. Stochastic simulations (Faisal et al., 2005) showed that RAP rate is inversely temperature dependent in the cortical pyramidal cell and the squid axon which operated at 36 °C and 6.3 °C, respectively. Increasing temperature has a well-known accelerating effect on ion channel kinetics. Higher temperatures speed up the movement of charged gating particles, which, in turn, decreases the time between changes of conformation, i.e., opening and closing of channels. This means that a spontaneously opened channel will spend less time in the "open" state as the temperature increases. This reduction of the duration of spontaneous depolarizing currents means that the membrane is less likely to reach AP threshold (this effect prevails over the increased rate of spontaneous channel openings). In other words, increasing temperature shifts channel noise to higher frequencies where it is attenuated by the low-pass characteristics of the axon (Faisal et al., 2005; Steinmetz, Manwani, Koch, London, & Segev, 2000). This may suggest that increasing temperature allowed homeothermic animals, such as mammals, to develop more reliable, smaller, more densely connected and thus faster neural circuits.

7.5 Channel Noise and Channelopathies

Channelopathies are disorders due to abnormalities in ion channels. Generalized epilepsy with febrile seizures, a condition without clear trigger, is, for instance, associated with a mutation of the b4 subunit in sodium channels, while benign familial neonatal epilepsy is associated with a reduced expression of slow, KCNQ-type K channels. Intriguingly, in a simulation study of action potential initiation under different channelopathies, we found that these altered channel kinetics did not always result in a change of the neuron's average behavior, yet produced clinical symptoms.

We advance the following hypothesis that could provide a general framework of explanation and highlight a novel aspect of "stochastic diseases": Channelopathies may cause ion channels to display the same average behavior, but greatly change the afflicted channel's trial-to-trial variability around this average behavior. Wild-type ion channel fluctuations can cause random, spontaneous APs even in absence of synaptic input (while deterministic models of the same channels do not produce any spontaneous activity). Should channelopathies alter channel kinetics in such a way so as to make ion channels more unreliable (as suggested by our preliminary stochastic simulations of an NaV 1.2 sodium channel mutant), then we would expect to see a greatly increased rate of spontaneous neuronal activity in much larger nerves and neurons.

The altered probabilistic behavior of ion channels—whether in absence of or in addition to changes to the average behavior of the channel—has previously neglected implications for understanding epilepsy and neuropathic as the result of increased unwanted neuronal activity. One way to test this hypothesis is to screen the rapidly growing literature on channelopathies and the relevant ion channel kinetics. These data are typically described as Hodgkin–Huxley-type deterministic kinetics, but can be easily converted into a stochastic Markov-model-type kinetics.

Such studies should investigate the diversity (including channelopathies) of voltage-gated Na⁺ and K⁺ channels from a functional perspective and ultimately in their evolutionary context. While the phylogeny of ion channels has been extensively studied in genetic and molecular terms, a systematic analysis of the functional implications of ion channel variations is so far missing. Taking the view that the primary role of an axon is to transmit information, one can assess the role of an axon's component from a functional perspective. Thus, an ion channel can be thought of as representing a particular choice in trade-offs between the 4 basic constraints on information processing.

7.6 Are There Other Biophysical Limits to Axon Size?

How small can a functioning axon be constructed, given the finite size of its individual components? Faisal et al. (2005) showed, using a volume exclusion argument, that it is possible to construct axons much finer than 0.1 µm diameter (see Figure 7.2). Neural membrane (5 nm thickness) can be bent to form axons of 30 nm diameter because it also forms spherical synaptic vesicles of that diameter. A few essential molecular components are required to fit inside the axon; these include an actin felt work (7 nm thick) to support membrane shape, the supporting cytoskeleton (a microtubule of 23 nm diameter), the intracellular domains of ion channels and pumps (intruding 5 nm to 7 nm), and kinesin motor proteins (10 nm length) that transport vesicles (30 nm diameter) and essential materials (<30 nm diameter). Adding up the cross-sectional areas shows that it is possible to pack these components into axons as fine as 0.06 µm (60 nm). Indeed, the finest known neurites, those of amacrine cells in Drosophila laminae, are about 0.05 µm in diameter, contain microtubules, and connect to extensive dendritic arbours but do not transmit APs. The fact that the smallest known AP-conducting axons are about twice as large as the steric limit to axon diameter (0.1 µm cf. 0.06 µm, see Figure 7.2B), whereas electrically passive axons reach the physical limit, supports our argument that channel noise limits the diameter of AP-conducting axons to about 0.1 µm.

Furthermore, other molecular limits to axon diameter are well below the noise-limited diameter of 0.1 µm, thus AP-conducting axons finer than 0.1 µm could, in theory, exist. Yet anatomical data across many species, invertebrate and vertebrate, including extremely small insects and large mammals, shows an identical lower limit of diameter for AP-conducting axons of 0.1 µm. This suggests that channel noise limits axon diameter, and thus the wiring density of the central nervous system, and therefore ultimately the size of the cortex. Curiously, the anatomical literature (see Figure 7.2C) demonstrated a common lower value for the diameter of axons for over 30 years, yet this was not noticed till a systems biology view on the study on stochastic limits to cell size prompted a search for the smallest known axon diameters (Faisal et al., 2005).

Figure 7.2 Noise Limits the Miniaturization of Unmyelinated Axons.
(A) SAP rate versus axon diameter for a pyramidal cell axon collateral (open triangles, 23 °C; closed triangles, 37 °C) and a squid axon (circle) of 1 mm length. Spontaneous AP rate increases sharply below a critical diameter of 0.15 µm to 0.2 µm. (Inset) Semilogarithmic plot of the data shows the exponential character of the dependence of spontaneous AP rate on diameter below the critical diameter. The arrow highlights how little changing the signal AP rate from 4 to 20 Hz affects the limiting diameter (the diameter at which SAP rate equals half the signal AP rate). (B) Scale drawing illustrating how essential components can be packed into the cross-section of an axon of 50 nm diameter (see text for details). The unfilled circle illustrates the finest known AP-conducting axons, whose diameter, 100 nm, corresponds to the channel-noise limit derived in this study. (C) Diameters of fine AP-conducting axons in a wide range of species and tissues (Berthold & Rydmark, 1978; Braitenberg & Schüz, 1998; Easton, 1971; Guillery, Feig, & van Lieshout, 2001; Heck & Sultan, 2002; Hsu, Tsukamoto, Smith, & Sterling, 1998; Keynes & Ritchie, 1965; Olivares, Montiel, & Aboitiz, 2001; Shepherd & Harris 1998; Small and Pfenninger, 1984; Sugimoto, Fukuda, & Wakakuwa, 1984; Williams & Chalupa, 1983; Wozniak & O'Rahilly, 1981). The finest AP-conducting axons reach the limiting diameter of 0:1 µm (dotted line); the few exceptions are developing fibers of 0:08 µm diameter (arrowhead). Adapted from Faisal et al. 2005. Reproduced with permission of Elsevier publishing.

Fundamental Constraints on the Evolution of Neurons

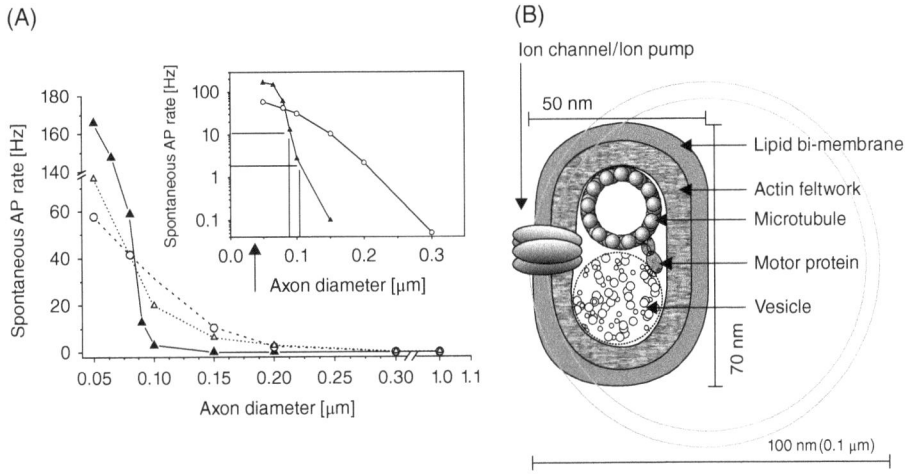

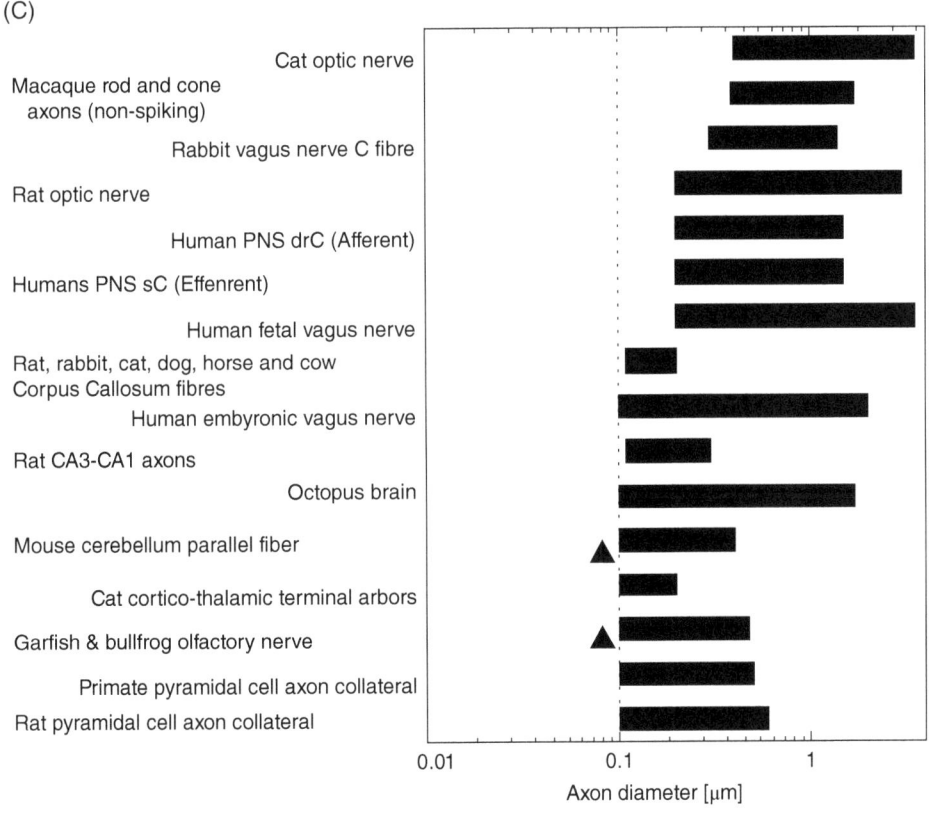

Figure 7.2 (*Continued*)

7.7 Is Behavioral Variability the Cause of Molecular Noise?

Neurons are variable, in that we observe both irregular spontaneous activity (activity that is not related in any obvious way to external stimulation) and trial-to-trial variations in neuronal responses to repeated identical stimuli—and both are often considered signs of "noise" (Shadlen & Newsome, 1995; Softky & Koch, 1993). Whether this neuronal trial-to-trial variability is indeed just noise (defined in the following as individually unpredictable, random events that corrupt signals), a result of the brain being too complex to control the conditions across trials (e.g., the organisms may become increasingly hungry or tired across trials), or rather the reflection of a highly efficient way of coding information, cannot easily be answered. In fact, being able to decide whether we are measuring the neuronal activity that is underlying the logical reasoning and not just meaningless noise is a fundamental problem in neuroscience (Rieke, Warland, de Ruyter van Steveninck, & Bialek, 1997). There are multiple sources contributing to neuronal trial-to-trial variability: deterministic ones, such as changes of internal states of neurons and networks, as well as stochastic ones, noise inside and across neurons (Faisal et al., 2008; White, Rubinstein, & Kay, 2000). To what extent each of these sources makes up the total observed trial-to-trial variability remains unclear.

What has become clear is that to solve this question it is not sufficient to study neuronal behavior experimentally (as this measures only the total variability of the system). Rather, answering this question requires taking a system biology view of neuronal information processing (Faisal et al., 2008). This is because noise is ultimately due to the thermodynamic and quantum nature of sensory signals, neuronal and muscular processes operating at the molecular level. Given that the molecular biology and biophysics of neurones is so well known, it allows us to use stochastic modelling of these molecular components to control and assess the impact of each source of (random) variability at the level of neurons, circuits and the whole organism.

While a neuron will contain many ion channels, typically these ion channels do not interact instantaneously (neurons are not iso-potential) and thus their fluctuations do not average out, as much smaller (and thus noisier) subsets of ion channels are responsible for driving activity locally in the neuron. However, little was known on how the stochasticity of ion channels influences spikes as they travel along the axon to the synapse and how much information arrives there. Experimentally, axonal spike time jitter has previously been only measured in vitro at myelinated cat and frog axons of several micrometers diameter and was in the order of 0.01 ms (Abeles & Lass, 1975; Lass & Abeles, 1975). Biologically accurate stochastic simulations of axons showed (Faisal & Laughlin 2007) that the variability of action potential propagation (measured as spike time jitter) in unmyelinated axons 0.1-0.5 μm in diameter was on the order of 0.1 ms to 1 ms standard deviation over distances of millimetres (see Figure 7.3). Thus, axonal variability can grow several orders of magnitude larger than previously expected and have considerable impact on neural coding.

Channel noise acts in two ways that are implicit to the AP mechanism itself and thus unavoidable. First, only a few sodium channels are involved in driving the AP, when the membrane is between resting potential and AP threshold, and these small Na$^+$ currents are thus subject to large fluctuations. Second, the resting membrane ahead of the AP is far from being at rest, but fluctuates considerably.

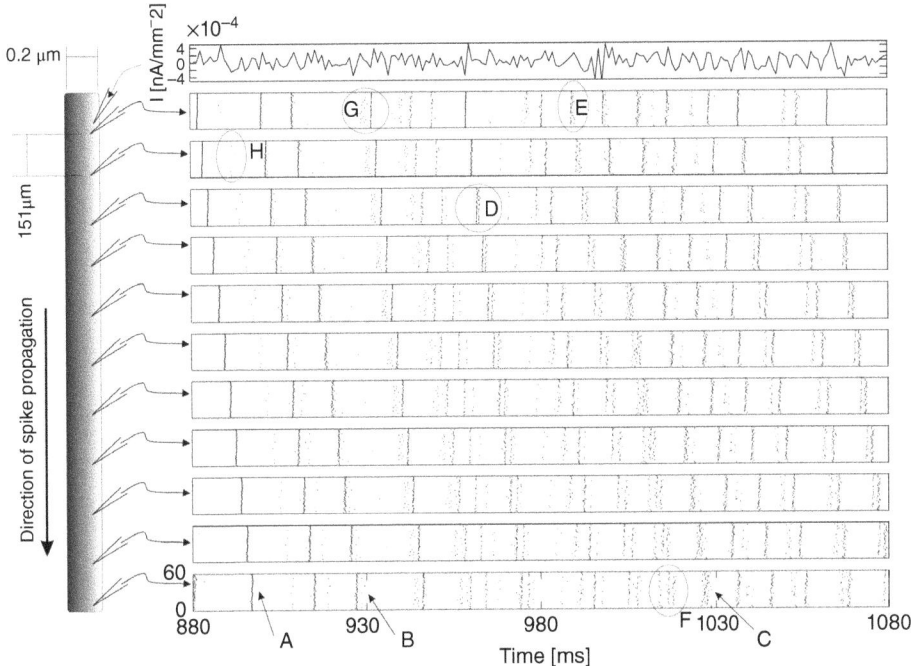

Figure 7.3 Effects of Ion Channels Noise on Propagating APs.
The topmost row shows the stimulus current. Below, each row contains a spike raster plot recorded at equally spaced axonal positions (from the proximal stimulus site at the top to the distal part the axon at the bottom). In each spike raster plot, the precise timing of a spike is marked by a dot on an invisible time line. These time lines are stacked over each other for N = 60 repeated trials. The linear shift visible in the overall spike pattern across rows reflects the APs traveling along the axon. Data based on 10-s trials, squid-type axon of 0.2 μm diameter (average diameter of cerebellar parallel fibers). The timing APs within each set of AP are either unimodal (sets marked by arrows A, B, and C), or are markedly multimodal distributed forming visually distinct groups (set D). In general the timing difference between each group increases as APs travel along the axon (AP sets E and F). APs in a set may be triggered markedly earlier due to differences in the distributions of ion channel states across trials (set G). APs are spontaneously and randomly added (circle H) and in a few cases are even deleted. Adapted from Faisal & Laughlin 2007.

7.8 Channel Noise Impacts Crucial AP Properties

Axons are often thought of as faithful transmission channels for electrical impulses. But advances in experimental methods have allowed to reconsider these assumptions (Debanne, Guerineau, Gahwiler, & Thompson, 1997; Kole, Letzkus, J. J., and Stuart, 2007; Sasaki, Matsuki, & Ikegaya, 2011), and sparked interest about the potential role of the axon as a computational unit in its own right (reviewed in Debanne, 2004; Sasaki, 2013; see also Chapter 5, this volume). This is closely related to the question of how APs are translated at the synapse. Do synapses consider incoming APs as unitary events, or do they use information contained in the waveform to modulate the release of neurotransmitters?

Although the nervous system exhibits variability (noise) at all levels (reviewed in Faisal et al. 2008), it is generally assumed that little variability affects the AP waveform itself as it travels from the soma along the axon to the synapse, which would enable it to transmit information encoded, for example, in the width (Aldworth, Bender, & Miller, 2012; Shu, Hasenstaub, Duque, Yu, & McCormick, 2006) of the AP. However, many unmyelinated axons are very thin (0.1 μm to 0.3 μm diameter, Wang et al., 2008). Examples include cerebellar parallel fibres (average diameter 0.2 μm, Sultan, 2000), C-fibres implicated in sensory and pain transmission (diameter range 0.1 μm to 0.2 μm, Berthold & Rydmark, 1978) and cortical pyramidal cell axon collaterals (average diameter 0.3 μm, Braitenberg & Schutz, 1998). These thin unmyelinated

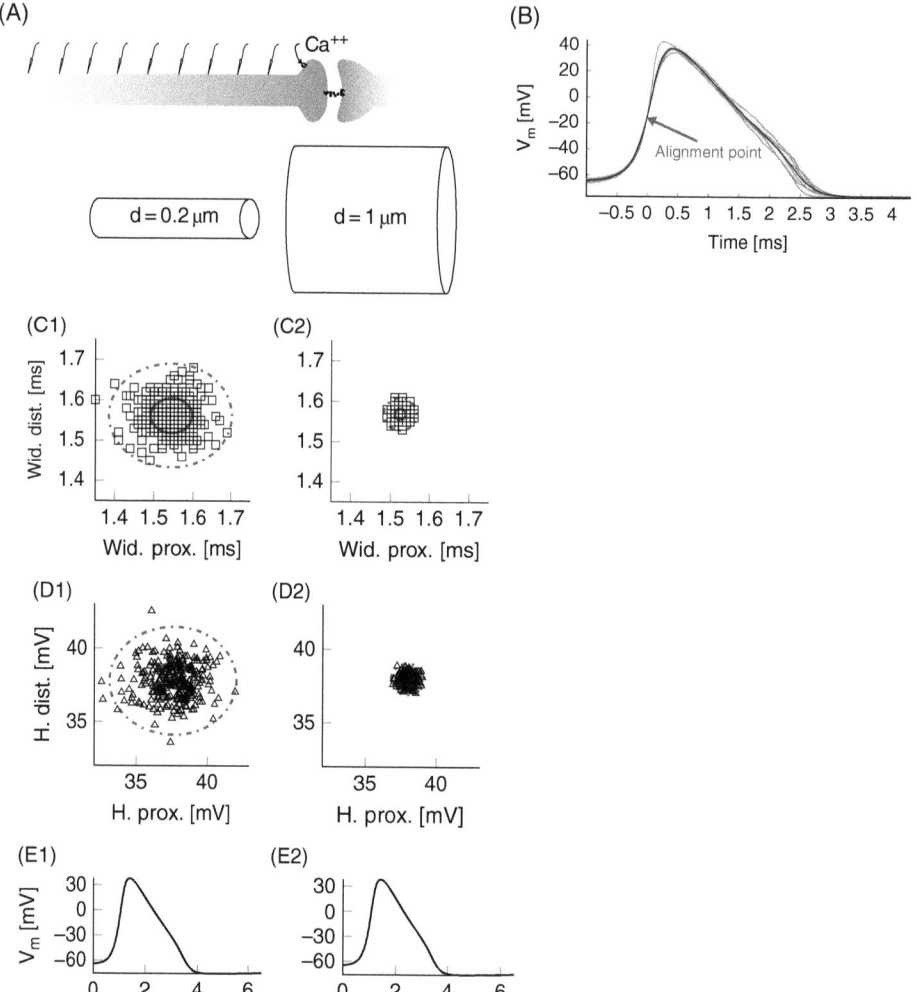

Figure 7.4 Noise Modifies the AP Waveform.
(A) Membrane potential and ionic currents are recorded at regularly placed points along the axon. The subfigures display data for N>200 single APs triggered by identical stimuli and initial conditions in thin squid giant axon model. (B) Due to channel noise, APs triggered in an identical fashion will have different shapes across trials. Here APs in a 0:2 μm diameter axon from 5 trials out of 250 are superimposed. The point at which the membrane potential crossed the half-height line is used to align APs.

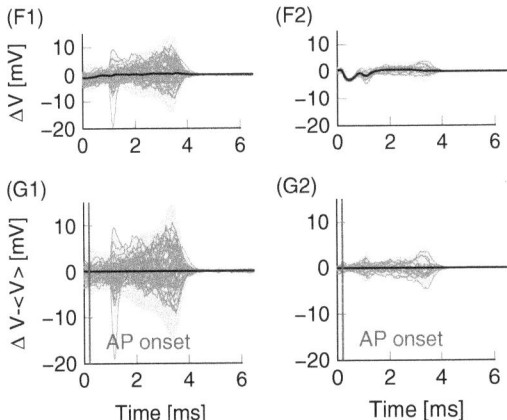

Figure 7.4 (*Continued*) (C) Distribution of AP width and (D) AP height (red circle, 1SD; dotted circle 3SD). (E) Mean waveform of the AP at the proximal site. (F) Pairwise difference between an AP's shape at the proximal and the distal location. The average difference is plotted in thick black, while the light grey shaded area represents the 3SD range. Grey lines represent sample traces plotted individually. (G) Fluctuations around the mean pairwise difference. The average difference is plotted in thick black (0 by definition), while the light grey shaded area represents the 3SD range. Grey lines represent sample traces plotted individually. Adapted from Neishabouri & Faisal 2014. (*See insert for color representation of the figure*).

axons make up most of the local cortical connectivity (Braitenberg & Schutz, 1998) but the variability of the AP waveform in them is unknown. Basic biophysical considerations suggest that axonal noise sources are bound to introduce fluctuations in the shape of the travelling AP waveform in thin axons with immediate consequences for synaptic transmission (Faisal et al., 2005; Sasaki et al., 2011).

Stochastic simulations (Neishabouri & Faisal, 2014) show that in thin unmyelinated axons below 1 μm diameter, commonly found in the CNS and PNS, the travelling waveform of an AP undergoes considerable random variability. This random variability is caused by axonal Na⁺ and K⁺ channel noise which continuously acts during propagation and thus accumulates with distance. The variability of AP width and height, key parameters linked to synaptic efficacy, dramatically increased (the coefficient of variation increasing by a factor of approximately 4, see Figure 7.4) as diameter decreased from 1 μm to 0.1 μm. AP height and width variabilities increase with a power-law as diameter decreases.

7.9 Effects of Channel Noise Spread to Other Neurons

Once the AP arrives at the synapse, the characteristics of its waveform are fundamental in determining the strength and reliability of information transmission (Augustine, 1990, 2001; Borst & Sakmann 1999; Coates & Bulloch, 1985; Delaney, Tank, D. W., and Zucker, 1991; Gainer, Wolfe, Obaid, & Salzberg, 1986; Klein & Kandel, 1980; Llinas, Steinberg, & Walton, 1981; Llinas, Sugimori, & Silver, 1982; Llinas, Sugimori, & Simon, 1992; Niven et al. 2003; Sabatini & Regehr 1997; Spencer, Przysiezniak, Acosta-Urquidi, & Basarsky, 1989; Wheeler, Randall, A., and Tsien, 1996). Because of this influence, the impact of channel noise in thin axons is not limited to the axon itself, but can propagate through the whole network.

This variability in the AP waveform causes some variability in the Ca^{++} influx at the synapse, and through the variability in the instantaneous Ca^{++} concentration, in the synaptic vesicle release rate and total number of vesicles released. Data from the cerebellar granule-to-Purkinje-cell synapse (Sabatini & Regehr, 1997) and simulations using a model of the Calyx of Helb both show that axonal variability may have considerable impact on synaptic response variability. However, it can be easily confounded with synaptic variability in many theoretical spike-timing-dependent plasticity models, or in experimental frameworks investigating synaptic transmission through paired-cell recordings or extracellular stimulation of the presynaptic neuron and intracellular postsynaptic recordings.

This random variability sets limits on synaptic adaptation, and ultimately on learning and memory. One mechanism of synaptic adaptation, for instance, is the subthreshold inactivation of some axonal K^+ channels. This results in wider APs and therefore stronger EPSPs (Kim, Wei, & Hoffman, 2005; Korngreen, Kaiser, & Zilberter, 2005). Random variability in the waveform of APs in very thin axons reduces the accuracy of this mechanism (and any other mechanism intervening on the width of APs). One way to interpret this in the evolutionary context is that there is a trade-off to be made between the thinness of an axon, and the efficiency of mechanisms relying on the preservation of AP waveforms.

7.10 The Brain Must Balance Noise vs. Metabolic Cost

Ion channels could, in principle, be made less sensitive to thermodynamic noise. For example, hyperpolarizing the membrane's resting potential would decrease the rate of RAP because sodium ion channels are less likely to open at lower membrane potentials. Modifying the channels' kinetics—for example, by shifting the activation function's dependency on membrane potential—would fulfil the same role. So why does the brain not simply get rid of channel noise?

The answer to this question comes from a consideration of another crucial constraint on nervous systems, namely energy. The brain is very metabolically expensive. At rest, a human brain accounts for a fifth of our energy consumption (Attwell & Laughlin 2001; Sengupta et al. 2010). The dissipation of heat generated by cerebral activity does not appears to limit the activity (Karbowski, 2009), but the need to supply the nervous system with enough energy for its functions seems to have had a large impact during the evolution of humans (Skoyles, 2014).

Attempts at reducing the impact of channel noise are bound to increase the metabolic demands of the nervous system. Maintaining the resting potential of neurons, for instance, accounts for 11% of the brain's metabolic needs, since ionic pumps that maintain the concentrations of ions at the desired levels require energy to function. Hyperpolarizing the resting potential would require these pumps to activate more, resulting in even higher metabolic cost. In addition, more current would need to cross the membrane to depolarize it enough for an AP to be fired. Therefore, each actual (wanted) AP would be more metabolically costly. Shifting the kinetics of sodium ion channels would have a comparable effect by requiring a larger depolarizing current for each AP the cell "willingly" fires.

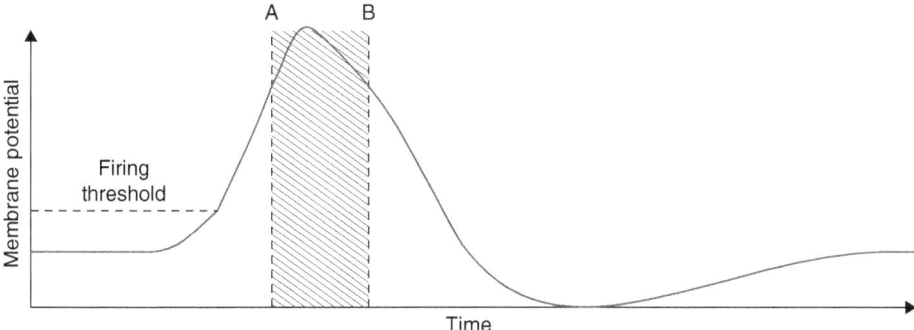

Figure 7.5 Current Overlaps in APs.
A schematic view of the action potential and overlap between currents. (A) Activation of K⁺ channels (B) Inactivation of Na⁺ channels. The hashed area represents the period during which Na⁺ and K⁺ channels are both open, which leads to "wasted" membrane current and therefore inefficient action potentials.

Firing APs accounts for 22% of the brains metabolic requirements (Sengupta et al., 2010), and fast inactivation of sodium channels is crucial from a metabolic cost point of view (Alle et al., 2009; Sengupta et al., 2010). Fast inactivation reduces the overlap between sodium and potassium currents, and reduces the total amount of charges crossing the membrane for the same level of depolarization (Figure 7.5; Alle et al., 2009; Sengupta et al., 2010). But, since sodium channels inactivate rapidly, they also need to activate fast enough so that a significant portion of them are actually open before inactivation kicks in.

Thus, using only energetic constraints, we can make functional predictions (kinetics) about the building blocks of the nervous system. The trade-off between energy and noise is now rather obvious. Sodium channels need to be sufficiently sensitive so that they supply enough current for firing and propagating APs, but this sensitivity also inevitably introduces noise in the nervous system. We can quantify this in a very simplified manner as follows. Consider V_{th}, The threshold at which Na⁺ channels open, and V_{rest}, the resting membrane potential. The amount of energy necessary to bring such a membrane to fire an action potential is proportional to $C_m(V_{th} - V_{rest})$, which is the amount of current required to depolarize the membrane to the V_{th}. This determines, for instance, the amount of synaptic input required to trigger an action potential in the axon initial segment. Hyperpolarizing V_{th} would, in principle, allow for fewer synaptic currents, or fewer Na⁺ channels to ensure reliable transmission. But, on the other hand, the probability of Na⁺ channels opening spontaneously at rest is, to a first approximation, given by the Poisson-Boltzmann equation: $e^{(-(V_{th}-V_{rest})/(k_B T))}$. Therefore, the more depolarized the threshold is, the more unlikely it is for the neuron to fire spontaneously.

These simple observations can form the basis of a prediction about the distribution of different channels on a single neuron. In large segments, like the soma, where a large number of channels are present, it may be beneficial to use very "sensitive" Na⁺ channels in order to ensure reliable AP initiation. On the other hand, in thin axon collaterals, where noise can be more of a constraint, we might expect to see different Na⁺ channels. This would not necessarily mean that we should expect to see, in the soma for instance, channels that "open" at membrane potentials near the resting membrane potential. In fact, as discussed in the case of temperature, faster kinetics reduce the noisiness of channels, and Na⁺ channels in axons are often found to have faster kinetics (Hallermann, de Kock, Christiaan, Stuart, & Kole, 2012), resulting in thinner APs, although this is usually thought of as a way of reducing the overall metabolic cost.

7.11 Homeostatic Limits on Neurite Anatomy

Firing an action potential requires the expenditure of large amount of energy in a very short period, on the order of a few milliseconds. It is impossible for neurons to "supply" this amount of energy on the fly. Instead, the energy necessary for propagating APs is stored in the form of ionic concentration gradients across the membrane. At rest, the concentration of Na⁺ outside mammalian neurons is in the order of 120 mM, whereas the concentration inside is about only 6 mM. This difference creates an osmotic pressure that forces Na⁺ ions through ion channels into the cell. Similarly, the difference of concentration of K⁺ between the inside and outside of neurons causes the displacement of K⁺.

Each AP causes the transfer of a certain amount of ions between the inside and outside of the membrane. It is commonly assumed that the number of ions crossing the membrane is very small compared to the total number of ions involved, as this is the case in classically studied axons, for example, the squid giant axon, with diameters of hundreds of micrometers. However, the mammalian nervous system contains much thinner axons, for example, C-fibres or cortical axon collaterals with diameters of 0.1–0.3 μm.

This is important, because the number of ions crossing the membrane during an AP is proportional to the membrane surface area, and hence the diameter of the axon. The impact of this charge on the ionic concentrations, however, is inversely proportional to the volume of the axon, and hence the square of the diameter. Therefore, as the axons get thinner, the impact of the charges transferred during each AP grows (Qian & Sejnowski 1989).

Since the current generated by Na⁺-K⁺ pumps is much smaller than that due to ion channels during an AP, firing rates of thin axons may be limited by rapid depletion of energy. Each AP fired lowers the difference of ionic concentrations between the inside and the outside of the cell. In particular, the K⁺ reversal potential is progressively depolarized. This results in a depolarization of the resting membrane potential. When the resting potential reaches the firing threshold, a burst of APs is generated until the concentration gradients are completely depleted, likely leading to cell death (see Figure 7.6A).

In the long run—for example, for Purkinje cells which permanently exhibit high firing rates—the influence of pumps becomes crucial for sustaining high firing rates (see Figure 7.6B). The firing rate an axon can sustain indefinitely is independent of diameter (energy store). Instead, it only depends on the cost of individual action potentials (energy expenditure) and the charge transferred by ionic pumps (energy production). However, in the short term—for example, for bursting neurons—homeostatic constraints impose a limit on the firing rate based on the diameter of axons (Neishabouri & Faisal, 2014). Sustaining high firing rates requires a minimum diameter for the axons, so that the energy consumed by the spike train does not deplete the potential energy stored in the ionic concentration gradients, since ionic

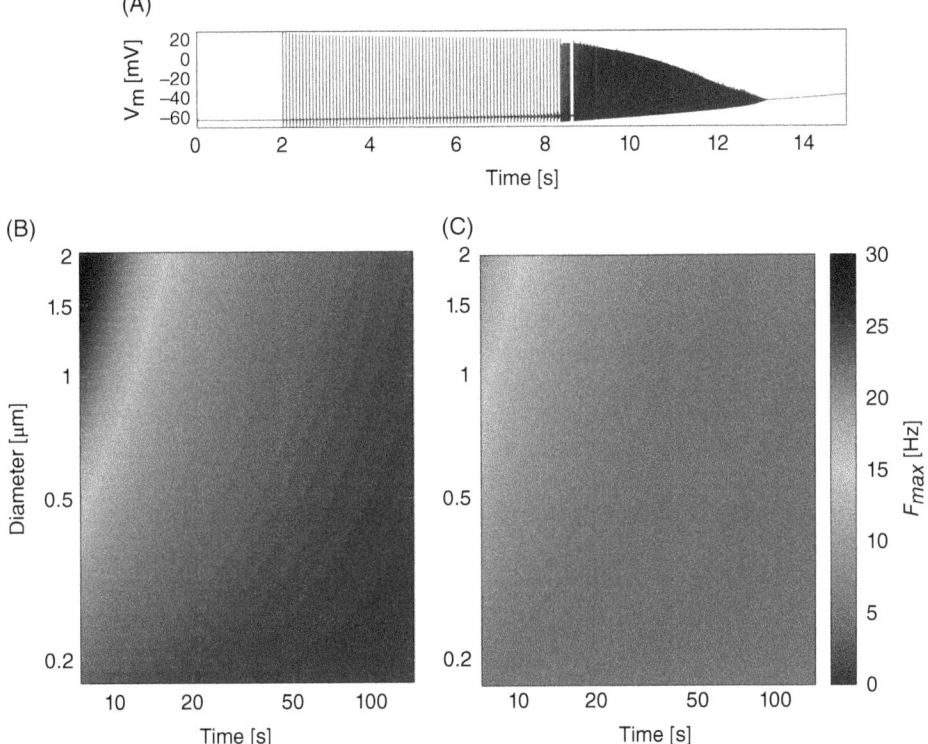

Figure 7.6 Maximum Sustainable Firing Rates.
(A) Firing at too high a firing rate for too long can drive the neuron into a burst which will eventually deplete ionic concentration gradients and lead to cell death. (B) Color map of maximum sustainable firing rates given by fitting a simplified model using parameters from the squid giant axon. Over long time periods, the sustainable firing rate is independent of diameter. (C) For shorter time periods, larger axons can fire at higher rates. The refractory period prevents the axon from firing at more than approx. 40 Hz. (*See insert for color representation of the figure*).

pumps cannot generate a large enough current to have any significant impact over short time scales. This constraint links the function and structure of axons.

We can thus derive the maximum amount of sustainable neuronal activity as a function of axonal diameter. Neurons may need to manage their energy budget more carefully in compact circuits of the cortex. As with noise, energetic considerations pose constraints on the anatomy of axons and limit the miniaturization of neural circuits, the effects of which could be directly observed.

7.12 Conclusion

Understanding the design and architecture of nervous systems, or reverse-engineering the brain, requires two fundamental understandings. The first, perhaps the more immediate one, is how nervous systems receive, process, and transmit information. From the discovery of the action potential (Du Bois-Reymond, 2006 [1848]) to that of neurons (Ramon y Cajal, 1897, Nobel prize in physiology or medicine, 1906), the existence of functional regions in the brain (Broca, 1878), neural codes (Adrian &

Zotterman, 1926, Nobel prize in physiology or medicine, 1932), mechanisms of action potential propagation (Hodgkin & Huxley, 1952, Nobel prize in physiology or medicine, 1963), ion channels (Neher & Sakmann, 1976, Nobel prize in physiology or medicine, 1991) and their structure (MacKinnon, Cohen, Kuo, Lee, & Chait, 1998, Nobel prize in chemistry, 2003), and even of neural network properties, etc., the primary approach has been about understanding how information is perceived, how decisions are made, and how they are transmitted to appropriate destinations.

The work of Hodgkin and Huxley (1952), however, made the investigation of the other, perhaps deeper, question possible. Taking an evolutionary perspective, the function of the nervous system is determined by its ability to generate appropriate behavior. Thus, ultimately, selective pressures on fitness will act on structure-vs.-function relationships of the nervous system, as determined by the basic biophysical properties of brain's components. To come up with the functional phylogeny of nervous systems, we need to answer basic questions: Why do nervous systems function the way they do? In other words, what are the constraints applied by selective pressure, and what trade-offs have those constraints forced during the evolution of nervous systems (see Figure 7.7)? Having a perfect answer to this question would require, in

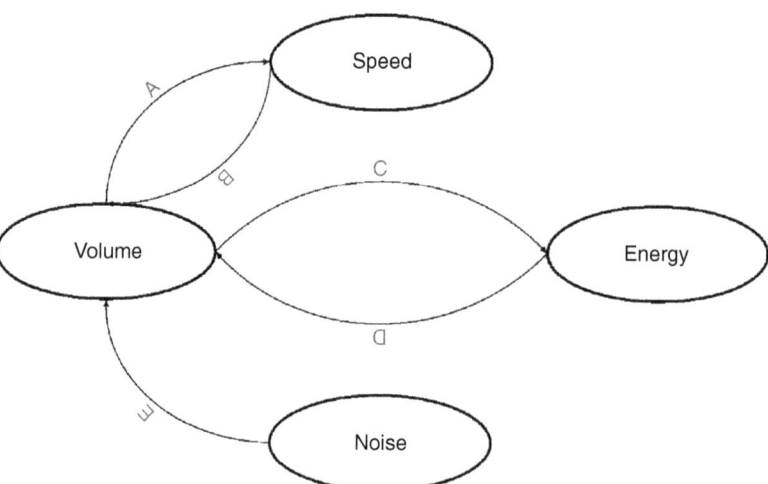

Figure 7.7 Summary of Trade-offs in the Nervous System.
(A) Larger axons conduct action potentials faster (Rushton 1951), but this comes at the expense of volume, which can push neurons further apart from each other. Similarly, myelinated axons conduct action potentials significantly faster than unmyelinated axons (Rushton, 1951; Waxman & Bennett, 1972). (B) The volume of axons limits their maximum sustainable firing rate, and hence their information throughput. (C) The volume of axons directly impacts the metabolic cost of signaling (Alle et al., 2009; Attwell & Laughlin, 2001; Sengupta et al., 2010). The relationship between diameter and cost is linear in unmyelinated axons, but sublinear in myelinated axons. The volume also determines the amount of energy stored in concentration gradients across the membrane. (D) The brain is responsible for 20% of the metabolic needs of the body at rest (Attwell & Laughlin, 2001; Harris & Attwell, 2012; Sengupta et al. 2010). This huge metabolic requirement limits the size of the brain. (E) Noise imposes a lower limit on the diameter of axons (Faisal & Laughlin, 2005), and consequently the miniaturization of nervous systems (Niven & Farris, 2012).

addition to knowing the *how* of nervous system organization and function, also being able to provide a *rationale* behind every choice made during nervous system evolution.

The answer to these questions lies in a comparative study of nervous systems. Different species seem to solve identical problems in different ways, with perhaps the best known example being the squid giant axon providing fast transmission without the use of myelin. It is not clear if this is simply an evolutionary "accident" (the squid's nervous system simply not having access to myelination) or if a deeper reason is involved. After all, myelination has evolved independently in at least 3 lineages, including two invertebrates. This ambiguity might be resolved if we instead turn our attention to a comparative study of various parts of the nervous system in the same organism. Various parts of nervous system are subject to different constraints: for example, the central nervous system (and particularly the brain) has likely been constrained by volume but less by transmission delays, because of the short distances involved. In the peripheral system, on the other hand, transmission delays are crucial for the function of neural fibers. Thanks to the different sets of constraints, and the different trade-offs "made" by nervous systems between these constraints, we may hope to achieve an understanding of the engineering design principles underlying the structure of nervous systems, and how these trade-offs matched the functional constraints these systems have to satisfy.

References

Abeles, M., & Lass, Y. (1975). Transmission of information by the axon: II. The channel capacity. *Biological Cybernetics*, 125, 121–125.

Adrian, E. (1928). *Basis of sensation*. New York, NY: Norton.

Adrian, E., & Matthews, R. (1927). The action of light on the eye. Part I. The discharge of impulses in the optic never and its relation to the electric changes in the retina. *Journal of Physiologyogy*, 63, 378–414.

Adrian, E., & Zotterman, Y. (1926) The impulses produced by sensory nerve endings Part III. *Journal of Physiology*, 1, 465–483.

Aldworth, Z. N., Bender, J. A., & Miller, J. P. (2012). Information transmission in cercal giant interneurons is unaffected by axonal conduction noise. *PLoS One* 7(1), e30115. doi 10.1371/journal.pone.0030115

Alle, H., Roth, A., & Geiger, J. R. P. (2009). Energy-efficient action potentials in hippocampal mossy fibers. *Science*, 325(5946), 1405–1408.

Attwell, D., & Laughlin, S. B. (2001). An energy budget for signaling in the grey matter of the brain. *Journal of Cerebral Blood Flow & Metabolism*, 21(10), 1133–1145).

Augustine, G. J. (1990). Regulation of transmitter release at the squid giant synapse by presynaptic delayed rectifier potassium current. *Journal of Physiology*, 431 (December), 343–364.

Augustine, G. J. (2001). How does calcium trigger neurotransmitter release? *Current Opinion in Neurobiology*, 11(3), 320–326).

Berthold, C. H., & Rydmark, M. (1978). Morphology of normal peripheral axons. In S. G. Waxman, J. D. Kocsis, & P. K. Stys (Eds.), *The axon: Structure, function, and pathophysiology* (pp. 13–48). Oxford: Oxford University Press.

Borst, J. G. G., & Sakmann, B. (1999). Effect of changes in action potential shape on calcium currents & transmitter release in a calyx-type synapse of the rat auditory brainstem. *Philosophical Transactions of the Royal Society B., Biological Sciences*, 354(1381), 347–355.

Braitenberg, V., & Schüz, A. (1998). *Cortex: Statistics and geometry of neuronal connectivity* (2nd ed.). Berlin: Springer.

Broca, P. (1878). Anatomie comparée des circonvolutions céréebrales: Le grand lobe limbique et la scissure limbique dans la série des mammifères. *Revue d'Antropologie*, 1, 384–498.

Coates, C. J., & Bulloch, A. G. M. (1985). Synaptic plasticity in the molluscan peripheral nervous system: Physiology and role for peptides. *Journal of Neuroscience*, 5(10), 2677–2684.

Debanne, D. (2004). Information processing in the axon. *National Review of Neuroscience*, 5(4), 304–316.

Debanne, D., Guerineau, N. C., Gahwiler, B. H., & Thompson, S. M. (1997). Action-potential propagation gated by an axonal I(A)-like K+ conductance in hippocampus. *Nature*, 389(6648), 286–289.

Delaney, K., Tank, D. W., & Zucker, R. S. (1991). Presynaptic calcium and serotonin-mediated enhancement of transmitter release at crayfish neuromuscular junction. *Journal of Neuroscience*, 11(9), 2631–2643.

Du Bois-Reymond, E. (2006 [1848]). Untersuchungen über thierische Elektricität. *Annalen der Physik*, 151(11), 463–464.

Easton, D. (1971). Garfish olfactory nerve: Eeasily accessible source of numerous long, homogeneous nonmyelinated axons. *Science*, 172 (986), 952–955).

Faisal, A. A., & Laughlin, S. B. (2007). Stochastic simulations on the reliability of action potential propagation in thin axons. *PLoS Computational Biology*, 3(5), e79. doi: 10.1371/journal.pcbi.0030079

Faisal, A. A., & Matheson, T. (2000). Coordinated righting behaviour in locusts. *Journal of Experimental Biology*, 204, 637–648.

Faisal, A. A., Selen, L., & Wolpert, D. M. (2008). Noise in the nervous system. *Nature Reviews Neuroscience*, 9(4), 292–303.

Faisal, A. A., White, J. A., & Laughlin, S. B. (2005). Ion-channel noise places limits on the miniaturization of the brain's wiring. *Current Biology*, 15(12), 1143–1149.

Gainer, H., Wolfe, S. A., Obaid, A. L., & Salzberg, B. M. 1986). Action potentials and frequency dependent secretion in the mouse neurohypophysis. *Neuroendocrinology*, 43(5), 557–563.

Guillery, R. W., Feig, S. L., & van Lieshout, D. P. (2001). Connections of higher order visual relays in the thalamus: A study of corticothalamic pathways in cats. *Journal of Computational Neurology*, 438 (1), 66–85).

Hallermann, S., de Kock, C. P. J., Stuart, G. J., & Kole, M. H. P. (2012). State and location dependence of action potential metabolic cost in cortical pyramidal neurons. *Nature Neuroscience*, 15 (July), 1007–1014

Harris, J. J., & Attwell, D. (2012). The energetics of CNS white matter. *Journal of Neuroscience*, 32(1), 356–371.

Heck, D., & Sultan, F. (2002). Cerebellar structure and function: Making sense of parallel fibers. *Human Movement Science*, 21(3), 411–421.

Hille, B. (1970). Ionic channels in nerve membranes. *Progress in Biophysics & Molecular Biology*, 21, 3–28.

Hille, B. (2001). Evolution and origins. In B. Hille, *Ion channels of excitable membranes* (pp. 693–722). Sunderland, MA: Sinauer.

Hodgkin, A. L., & Huxley, A. F. (1952). A quantitative description of membrane current and its application to conduction and excitation in nerve. *Journal of Physiology*, 117(4), 500–544).

Hsu, A., Tsukamoto, Y., Smith, R. G., & Sterling, P. (1998). Functional architecture of primate cone and rod axons. *Vision Research*, 38, 2539–2549.

Karbowski, J. (2009). Thermodynamic constraints on neural dimensions, firing rates, brain temperature and size. *Journal of Computational Neuroscience*, 27(3), 415–436.

Keynes, R., & Ritchie, J. M. (1965). The movements of labelled ions in mammalian non-myelinated nerve fibres. *Journal of Physiology*, 179(2), 333–367.

Kim, J., Wei, D., & Hoffman, D. A. (2005). Kv4 potassium channel subunits control action potential repolarization and frequency-dependent broadening in rat hippocampal CA1 pyramidal neurones. *Journal of Physiology*, 569, 41–57.

Klein, M., & Kandel, E. R. (1980). Mechanism of calcium current modulation underlying presynaptic facilitation and behavioral sensitization in Aplysia. *Proceedings of the National Academy of Sciences of the USA*, 77(11), 6912–6916).

Kole, M. H. P., Letzkus, J. J., & Stuart, G. J. (2007). Axon initial segment Kv1 channels control axonal action potential waveform and synaptic efficacy. *Neuron*, 55, 633–647.

Korngreen, A., Kaiser, K. M. M., & Zilberter, Y. (2005). Subthreshold inactivation of voltage-gated K^+ channels modulates action potentials in neocortical bitufted interneurones from rats. *Journal of Physiology*, 562 (January), 431–437.

Lass, Y., & Abeles, M. (1975). Transmission of information by the axon: 1. Noise and memory in the myelinated nerve fiber of the frog. *Biological Cybernetics*, 19, 61–67.

Laughlin, S. B. (1989). The reliability of single neurons and circuit design: A case study. In R. Durbin, C. Miall, & G. Mitchison (Eds.), *The computing neuron* (pp. 322–336). New York, NY: Addison-Wesley.

Laughlin, S. B. (2001). Energy as a constraint on the coding and processing of sensory information. *Current Opinion in Neurobiology*, 11 (August), 475–480.

Laughlin, S. B., de Ruyter van Steveninck, R. R., & Anderson, J. C. (1998). The metabolic cost of neural information. *Nature Neuroscience*, 1(1), 36–41.

Llinas, R., Steinberg, I. Z., & Walton, K. (1981). Relationship between presynaptic calcium current and postsynaptic potential in squid giant synapse. *Biophysics Journal*, 33 (March), 323–352.

Llinas, R., Sugimori, M., & Silver, R. B. (1992). Microdomains of high calcium concentration inpresynaptic terminal. *Science*, 256(5057), 677–679.

Llinas, R., Sugimori, M., & Simon, S. M. (1982). Transmission by presynaptic spike-like depolarization in the squid giant synapse. *Proceedings of the National Academy of Sciences of the USA*. 79(7), 2415– 2419.

MacKinnon, R., Cohen, S. L., Kuo, A., Lee, A., & Chait, B. T. (1998). Structural Conservation in Prokaryotic and Eukaryotic Potassium Channels. *Science* 280(5360), 106–109.

Neher, E., & Sakmann, B. (1976). Single-channel currents recorded from membrane of denervated frog muscle fibres. *Nature*, 260,799–802).

Neishabouri, A., & Faisal, A. A. (2014). Axonal noise as a source of synaptic variability. *PloS Computational Biology*, 10(5), e1003615. doi: 10.1371/journal.pcbi.1003615

Niven, J. E., Vahasoyrinki, M., Kauranen, M., Hardie, R. C., Juusola, M., & Weckstrom, M. (2003). The contribution of Shaker K+ channels to the information capacity of Drosophila photoreceptors. *Nature*, 421, 630–634.

Niven, J. E., & Farris, S. (2012). Miniaturization of nervous systems and neurons. *Current Biology*, 22 (May), R323–R329.

Olivares, R., Montiel, J., & Aboitiz, F. 2 (2001). Species differences and similarities in the fine structure of the mammalian corpus callosum.' *Brain Behavior & Evolution*, 57(2), 98–105).

Qian, N., & Sejnowski, T. J. (1989). An electro-diffusion model for computing membrane potentials and ionic concentrations in branching dendrites, spines and axons. *Biological Cybernetics*, 62(1), 1–15.

Ramon y Cajal, S. (1897). Leyes de la morfologia y dinamismo de las celulas nerviosas. *Revista Trimestral Mocrográfica*, 2, 1–12.

Rieke, F., Warland, D., de Ruyter van Steveninck, R. R., & Bialek, W. (1997). *Spikes: Exploring the neural code*. Cambridge, MA: MIT Press.

Rushton, W. A. H. (1951). A theory of the effects of fibre size in medullated nerve. *Journal of Physiology*, 115(1), 101–122).

Sabatini, B. L., & Regehr, W. G. (1997). Control of neurotransmitter release by presynaptic waveform at the granule cell to Purkinje cell synapse. *Journal of Neuroscience*, 17(10), 3425–3435.

Sasaki, T. (2013). The axon as a unique computational unit in neurons. *Neuroscience Research*, 75 (February), 83–88.

Sasaki, T., Matsuki, N., & Ikegaya, Y. (2011). Action-potential modulation during axonal conduction. *Science*, 331(6017), 599–601.

Sengupta, B., Stemmler, M., Laughlin, S. B., & Niven, J. E. (2010). Action potential energy efficiency varies among neuron types in vertebrates and invertebrates. *PLoS Computational Biology*, 6(7). doi: 10.1371/journal.pcbi.1000840

Shadlen, M. N., & Newsome, W. T. (1995). Is there a signal in the noise? *Current Opinion in Neurobiology*, 5(2), 248–50.

Shepherd, G. M., & Harris, K. M. 1 (1998). Three-dimensional structure and composition of CA3->CA1 axons in rat hippocampal slices: Implications for presynaptic connectivity and compartmentalization. *Journal of Neuroscience*, 18, no. 20 (October):, 8300–8310).

Shu, Y., Hasenstaub, A., Duque, A., Yu, Y., & McCormick, D. A. (2006). Modulation of intracortical synaptic potentials by presynaptic somatic membrane potential. *Nature*, 441(7094), 761–765.

Skoyles, J. R. (2014). Skeletal muscle-induced hypoglycemia risk, not life history energy trade-off, links high child brain glucose use to slow body growth. *Proceedings of the National Academy of Sciences of the USA*, 111 (November), E4909

Small, R. K., & Pfenninger, K. H. 1 (1984). Components of the plasma membrane of growing axons. I. Size and distribution of intramembrane particles. *Journal of Cell Biology*, 98(4), 1422–1433).

Softky, W. R., & Koch, C. (1993). The highly irregular firing of cortical cells is inconsistent with temporal integration of random EPSPs. *Journal of Neuroscience*, 13(1), 334–350).

Spencer, A. N., Przysiezniak, J., Acosta-Urquidi, J., & Basarsky, T. A. (1989). Presynaptic spike broadening reduces junctional potential amplitude. *Nature*, 340(6235), 636–638.

Steinmetz, P. N., Manwani, A., Koch, C., London, M., & Segev, I. (2000). Subthreshold voltage noise due to channel fluctuations in active neuronal membranes. *Journal of Computational Neuroscience*, 9(2), 133–148.

Stemmler, M., & Koch, C. (1999). How voltage-dependent conductances can adapt to maximize the information encoded by neuronal firing rate. *Nature Neuroscience*, 2(6), 521–527.

Sugimoto, T., Fukuda, Y., & Wakakuwa, K. (1984). Quantitative analysis of a cross-sectional area of the optic nerve: A comparison between albino and pigmented rats. *Brain Research*, 54(2), 266–274).

Sultan, F. (2000). Exploring a critical parameter of timing in the mouse cerebellar microcircuitry: The parallel fiber diameter. *Neuroscience Letters*, 280(1), 41–44).

Wang, S. S.-H., Shultz, J. R., Burish, M. J., Harrison, K. H., Hof, P. R., Towns, L. C., Wagers, M. W., & Wyatt, K. D. (2008). Functional trade-offs in white matter axonal scaling. *Journal of Neuroscience*, 28(15), 4047–4056).

Waxman, S. G., & Bennett, M. V. L. (1972). Relative conduction velocities of small myelinated and non-myelinated fibres in the central nervous system. *Nature*, 238(85), 217–219).

Weiss, T. F. (1997). *Cellular biophysics* (Vol. 2). Cambridge, MA: MIT Press.

Wheeler, D. B., Randall, A., & Tsien, R. W. (1996). Changes in action potential duration alter reliance of excitatory synaptic transmission on multiple types of Ca2+ channels in rat hippocampus. *Journal of Neuroscience* 16(7), 2226–2237.

White, J. A., Klink, R., Alonso, A., & Kay, A. R. (1998). Noise from voltage-gated ion channels may influence neuronal dynamics in the entorhinal cortex. *Journal of Neurophysiology*, 80, 262–269).

White, J. A., Rubinstein, J. T., & Kay, A. R. (2000). Channel noise in neurons. *Trends in Neuroscience*, 23(3), 131–137.

Williams, R. W., & Chalupa, L. M. 1 (1983). 'An analysis of axon caliber within the optic nerve of the cat: Eevidence of size groupings and regional organization.' *Journal of Neuroscience*, 3, no. (8 August):, 1554–1564).

Wozniak, W., & O'Rahilly, R. (1981). Fine structure and myelination of the developing human vagus nerve. *Acta Anatomica*, 109(3), 218–230.

8

The Central Nervous System of Invertebrates

Volker Hartenstein

8.1 Organizing Principles of Nervous System Architecture

8.1.1 Evolutionary Origin of the Nervous System

All animal taxa except porifera and placozoa have nervous systems consisting of neurons specialized for the repeated conduction of an excited state from receptor sites or other neurons to effectors or other neurons (Bullock & Horridge, 1965). Structural hallmarks of a neuron are the elongated neuronal processes (neurites) projecting from the cell body (soma) and the specialized membrane domains, called synapses, by which the electric activity of one neuron is transmitted to another. Molecularly, neurons are equipped with a multitude of specialized proteins allowing for signal conduction and transmission, including ion pumps and ion channels, transmitter receptors and transporters, and many others.

It was traditionally thought (reviewed in Moroz, 2014) that neurons are monophyletic in origin—that after the "primordial neuron" appeared on the scene, it evolved into all neurons found in present day animal taxa. However, based on current molecular as well as comparative-morphological data, it is equally likely that neurons derived multiple times independently. It is important to note in this context that the molecular building blocks of neurons, such as gated ion channels, neurotransmitters, vesicular transporters and many others, predate the appearance of neurons, and can be found in protists, placozoa, and sponges (Achim & Arendt, 2014; Alié & Manuel, 2010; Burckhardt, 2015; Cai, 2012; Jékely, 2011; Leys & Degnan, 2001; Liebeskind, Hillis, & Zakon, 2011; Renard et al., 2009; Sakarya et al., 2007). Cells with ultrastructural building blocks of neurons, such as specialized motile or sensory cilia or vesicles releasing signaling molecules, are also common in such animals. All these cells had to do in order to evolve into what would be considered a neuron by the above definition would have been to "invent" a neurite, by rearranging certain components of their cytoskeleton. This event could easily have happened multiple times during early animal evolution (Moroz, 2009).

In the simplest case, found in many extant cnidarian and ctenophore taxa, the nervous system constitutes a nerve net (Bullock & Horridge, 1965; Eichinger & Satterlie,

2014; Grimmelikhuijzen & Westfall, 1995; Mackie, 2004; Sakaguchi, Mizusina, & Kobayakawa, 1996; Satterlie & Eichinger, 2014; Watanabe, Fujisawa, Holstein, 2009). Cnidarians are built of two epithelial tissue layers, an outer epidermis (skin) and an inner gastrodermis (gut). There is no separate muscle layer; the epidermal cells contain contractile fibrils responsible for the movement of the body. Sandwiched in between the two tissue layers is a network of neurons, called ganglion cells, that are connected to each other and to the myoepithelial epidermal cells by synapses (see Figure 8.1). Ganglion cells may have two neurites (bipolar neurons), or multiple neurites (multipolar neurons). Synapses at the junction between two ganglion cells typically conduct signals in both directions (non-polarized synapses). However, polarized synapses, which transmit signals in only one direction and which form the prevalent type of synaptic contact in most animals, also exist already in cnidaria (Westfall, Yamataka, & Enos, 1971).

In addition to the ganglion cells, the epidermis and gastrodermis of cnidaria contain sensory neurons that are able to sense various mechanical and chemical stimuli. Sensory neurons are specialized epithelial cells with a modified apical cilium, which serves as the stimulus-receiving apparatus, and a basal neurite that feeds into the nerve net. Long neurites conducting signals away from the soma are referred to as axons; neurites, such as the modified cilium of a sensory neuron, that receive a signal and transmit it to the soma (or, in many cases, directly to the axon) are called dendrites.

Ctenophora (comb jellyfish), also possess a nerve net with sensory neurons and ganglion cells (Bullock & Horridge, 1965; Jager et al., 2011), and were previously united with Cnidaria in the superphylum "Coelenterata." However, several recent studies suggest strongly that Ctenophora split much earlier from the metazoan tree and may have evolved a nervous system independently (Moroz et al., 2014; Moroz, 2015; Ryan et al., 2013).

8.1.2 Central and Peripheral Nervous System

In all bilaterian taxa, the nervous system is formed of two major components, the central nervous system (CNS) and the peripheral nervous system (PNS) (see Figure 8.2). In the nerve net of many cnidarians and ctenophores, regions of higher neuronal density are detected in some body parts that play specialized roles in feeding or locomotion. Examples are the ring ganglion surrounding the mouth opening of polyps and medusae (Garm, Poussart, Parkefelt, Ekström, & Nilsson, 2007; Mackie, 2004), or the ganglia at the base of polyp tentacles, associated with sensory structures called rhopalia (Nakanishi, Hartenstein, & Jacobs, 2009; Skogh et al., 2006). It is likely that from such local densities or ganglia true nerve centers evolved in higher phyla (Jacobs et al., 2007; Koizumi, 2007; Watanabe et al., 2009). Ganglia are formed by two classes of cells, motor neurons and interneurons. Motor neurons are large cells that connect to muscles and glands and thereby directly control body functions. Interneurons form connections within the CNS, integrating the activity of sensory input and motor neuron activity.

The peripheral nervous system (PNS) consists of sensory neurons that are able to sense touch, motion, and sound (mechanoreceptors), position and state of the body (proprioceptors), smell and taste (chemoreceptors), and light (photoreceptors).

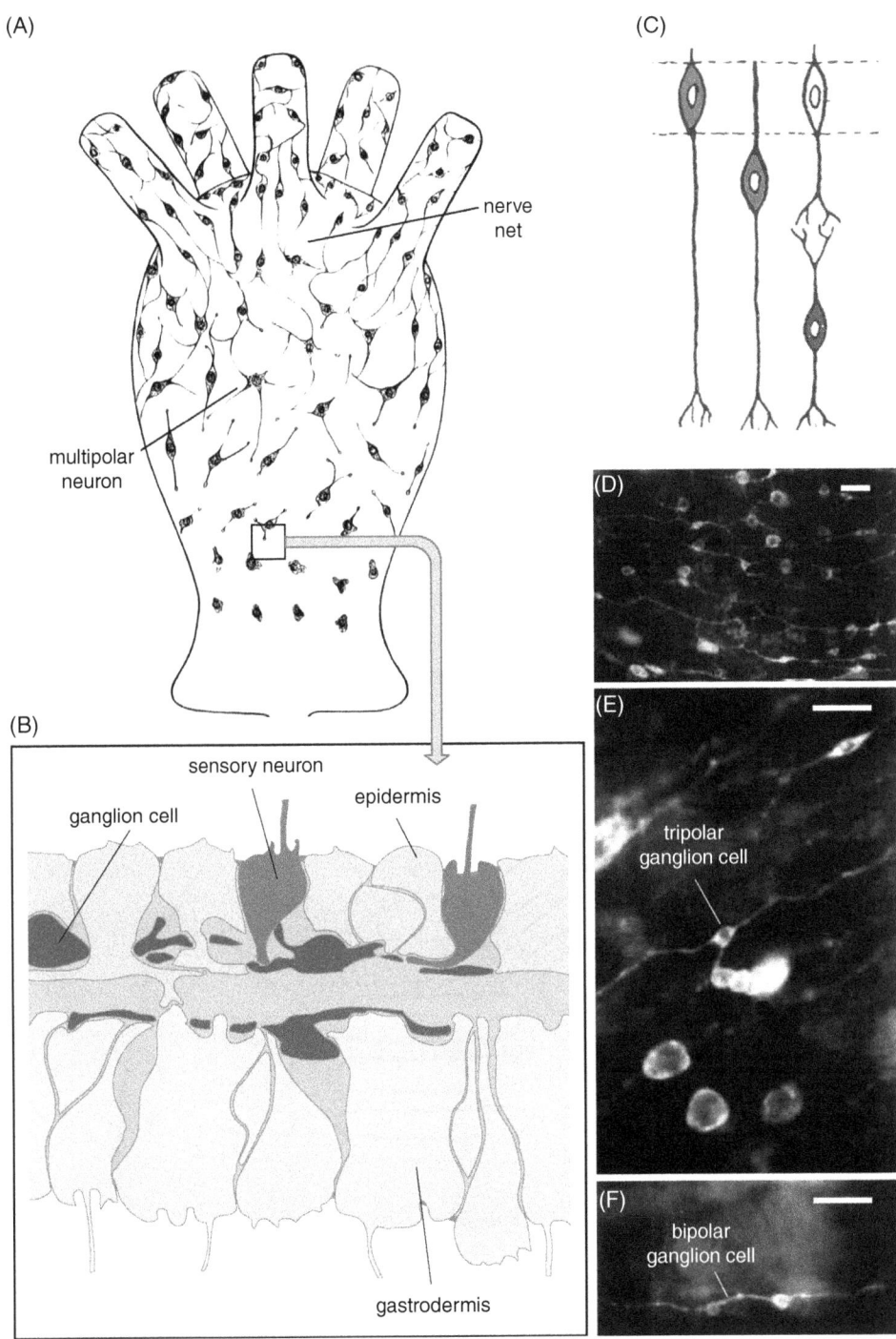

Figure 8.1 Nerve Net in Cnidarians.
(A) Line drawing of polyp, showing multipolar neurons (after). (B) Schematic cross-section of bodywall of cnidarian, showing outer layer (epidermis), inner layer (gastrodermis), sensory neurons, and ganglion cells. (C) Hertwig's hypothesis of evolution of nerve cells. Hanstroem, 1968. Reproduced with permission of Springer. (D–F) Labeling of nerve net of *Hydra viridissima* with antibody CC04. Sakaguchi 1996. Reproduced with permission of John Wiley and Sons. (*See insert for color representation of the figure*).

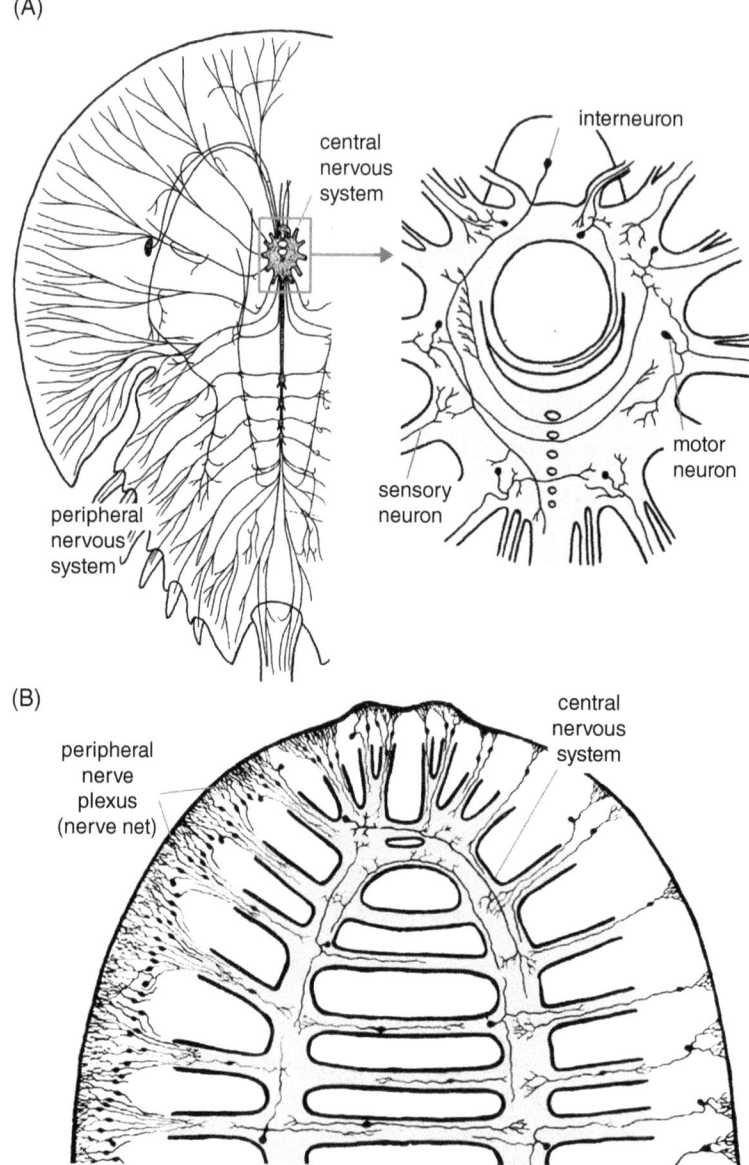

Figure 8.2 Division of Central and Peripheral Nervous System.
(A) Chelicerate *Limulus polyphemus*; (B) Platyhelminth *Bdelloura candida*. Hanstroem 1968. Reproduced with permission of Springer. (*See insert for color representation of the figure*).

Axons of sensory neurons projecting towards the CNS and axons of motor neurons going in the opposite direction form peripheral nerves. Typically, sensory neuronal cell bodies of invertebrates form part of peripherally located sense organs, or sensory ganglia closely attached to sense organs (Bullock & Horridge, 1965). In some instances, cell bodies of sensory neurons are integrated into the central ganglia (for example most sensory neurons in nematodes and plathyhelminths: Ehlers, 1985;

Ward, Thomson, White, & Brenner, 1975; Wright, 1992; Zhu, Li, Nolan, Schad, & Lok, 2011). In these cases, the long dendritic fibers of the sensory neuron project along the peripheral nerves towards the sensory endings (sensory cilia) located in the epidermis or specialized sensory organs.

Part of the peripheral nervous system is associated with (and often developmentally derived from) the gut, providing the intestinal glands and muscles with sensory and motor innervation. This is called the "visceral," "autonomic," or (in most invertebrates) "stomatogastric" nervous system (Barna et al., 2001; Csoknya, Lengvári, Benedeczky, & Hámori, 1992; Hartenstein, 1997; Reuter & Gustafsson, 1985; Selverston, Elson, Rabinovich, Huerta, & Abarbanel, 1998).

8.1.3 Subepithelial, Basiepithelial and Invaginated Nervous Systems

The basic architecture of the central nervous system falls into three main types: subepithelial, basiepithelial, and invaginated. A subepithelial (subepidermal) nervous system consists of a system of ganglia or nerve chords located within the body cavity, that is, separated from the epidermis and muscle layer of the body wall by a basement membrane (see Figure 8.3A). A ganglion has an outer layer of somata (cell bodies) called the rind or cortex, which surrounds a central core of neuropil (axons, dendrites, and synapses). Each neuron sends a single process into the central neuropil. Synapses are formed between these processes; the cell bodies do not generally receive synapses and therefore do not participate in stimulus conduction. The term "nerve chord" describes an elongated array ("bundle") of nerve fibers surrounded by neuronal cell bodies, as opposed to "nerve" (in the peripheral nervous system) or "tract" (in the central nervous system), which simply constitute a bundle of nerve fibers without accompanying cell bodies. The layout of a subepithelial CNS is shown schematically in Figure 8.3A (for the arthropod *Drosophila melanogaster*); it includes the brain, formed by one or several ganglia in the head region arranged around the pharynx, and the ganglia or nerve chords of the trunk. These can be either located all around the dorso-ventral axis (e.g., in platyhelminthes) or concentrated ventrally (e.g., in annelids, arthropods; see §8.2 below). In segmented animals, such as arthropods and annelids, the ganglia are segmentally arranged. Each trunk segment has one corresponding segmental ganglion, and the brain is formed by the fusion of several segmental ganglia.

In a basiepithelial nervous system (see Figure 8.3B), neurons form clusters or layers within the basal epidermis. The defining criterion is the epidermal basement membrane, which in a basiepithelial nervous system includes neuronal cell bodies and their processes. Basiepithelial nervous systems can be distributed fairly diffusely over the body, similar to the nerve net from which they presumably descended (as, for example, in hemichordates), or can form ganglia and chords in the head and trunk of the animal (as in nematodes) (see Figure 8.3B).

The invaginated nervous system, found in vertebrates and, in rudimentary form, in many lower deuterostomes, is thought to be derived from the basiepithelial configuration. Here, the epithelium overlying the neurons separates from the epidermis during early development and forms a continuous neural tube extending along the dorsal surface of the body (see Figure 8.3C). Neurons are organized in layers surrounding the central lumen of the neural tube. The subdivision into cortex and neuropil as described above for a subepidermal nervous system is absent. Instead, neuronal cell bodies and nerve cell processes are intermingled; synapses are formed on processes and cell bodies.

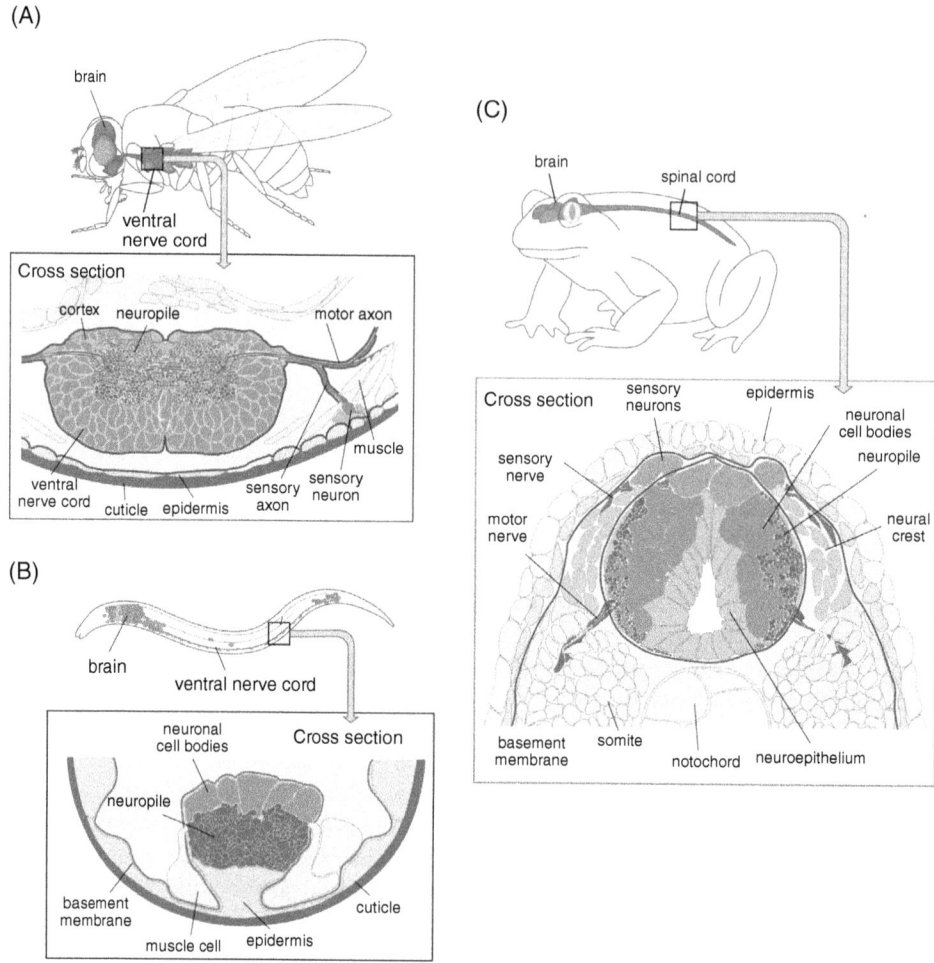

Figure 8.3 Architecture of the Central Nervous System in Bilaterian Animals.
(A) Subepidermal nervous system (arthropod *Drosophila melanogaster*); (B) Basiepithelial nervous system (nematode *Caenorhabditis elegans*); (C) Invaginated nervous system (Chordate *Xenopus laevis*). (*See insert for color representation of the figure*).

It is not clear whether the classification of CNS organization into the types above described has any phylogenetic significance. As outlined in more detail in §8.2, below, most basal protostome and deuterostome phyla possess a basiepithelial nervous system organization (see Figure 8.4). This finding, in conjunction with the assumption that the central nervous system descended from a basiepithelial nerve net, suggests that the basiepithelial organization may be primitive. The term "skin brain" was proposed for the hypothetical nervous system of the bilaterian ancestor (Holland, 2003). However, it is also possible that the basiepithelial and subepithelial types of nervous system organization have no profound phylogenetic implication, but are merely incidental to the overall number/density of neurons, or the embryonic development of the nervous system. In most taxa, the CNS originates from populations of neural progenitors located within the embryonic ectoderm. Frequently, for example, in

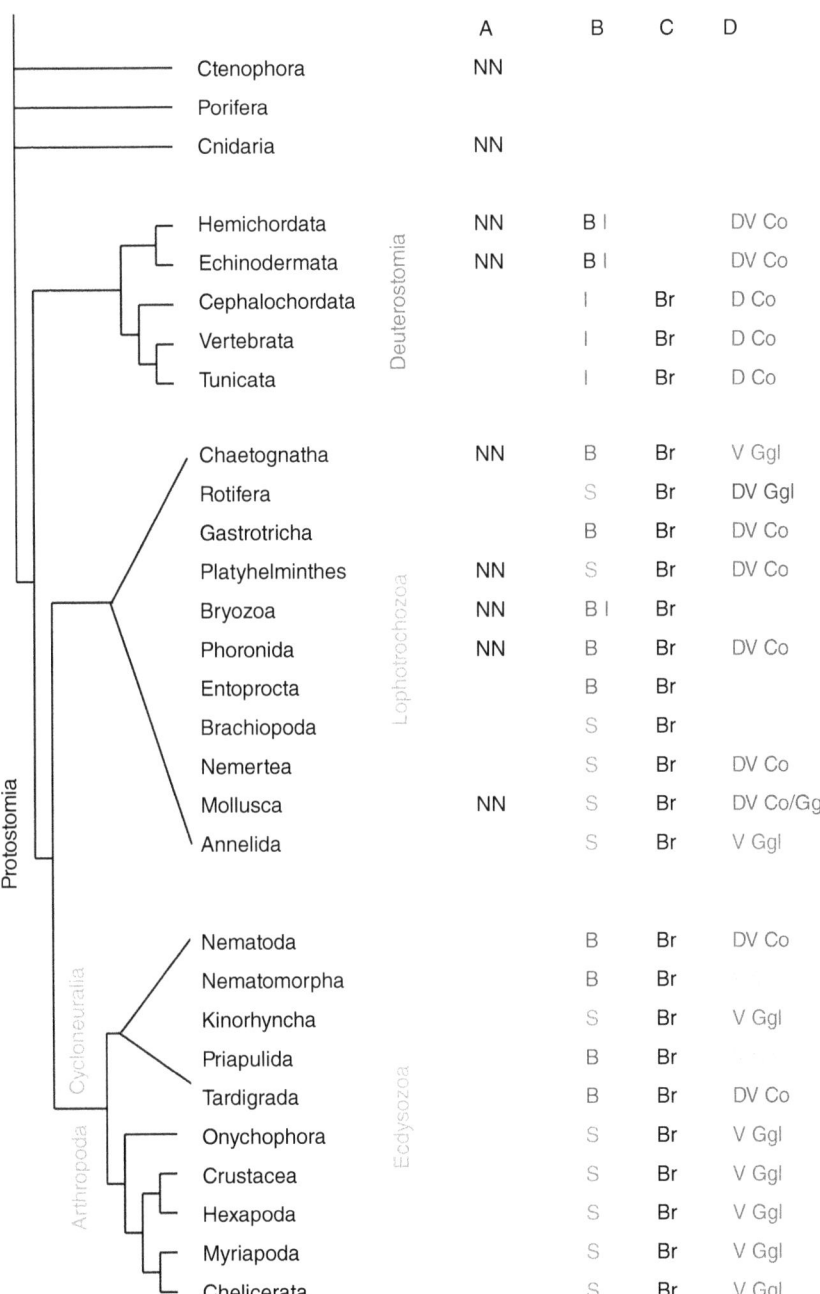

Figure 8.4 Distribution of nervous system characters among major animal phyla. Phylogenetic position of phyla is listed on the left. Column (A) shows presence or absence of peripheral nerve plexus (NN); shown in (B) is position of central nervous system (S subepidermal, B basiepithelial, I invaginated); (C): presence or absence of brain (Br); (D): distribution of nerve cords (Co) or ganglia (Ggl) [D dorsal; V ventral; DV dorsal (or lateral) and ventral].

insects and other arthropod taxa, these progenitors delaminate or invaginate into the body cavity, resulting in a subepithelial neural architecture from early stages onward (Hartenstein & Stollewerk, 2015). In other cases, as described for the annelid *Capitella* sp. (Meyer & Seaver, 2009), neural progenitors first produce a small basiepithelial brain which only secondarily adopts a subepithelial position. It is likely that a similar ontogenetic transition from basiepithelial to subepithelial will become apparent in other animal groups once relevant information on neural development becomes available.

8.1.4 Neuropil Compartmentalization

The multitude of neurites and synapses of the central nervous system form the neuropil. Neural circuits, the elusive objects of study of neurophysiologists and functional neuroanatomists, are embedded in the neuropil. As one of the first steps in approaching neural circuitry, it is important to define within the neuropil discrete compartments formed by neurons which share certain properties. Generally speaking, the interior of a compartment is defined by a high density of terminal neurite branches and their synapses. In many cases, a compartment is subdivided into smaller structurally and functionally distinct modules. On the outside, a compartment is surrounded by a frame of lower synaptic density, caused by the presence of bundles of axons interconnecting different compartments, as well as glial cells (Pereanu, Kumar, Jennett, Reichert, & Hartenstein, 2010; see Figure 8.5A–C).

To introduce the concept of neuropil compartments, a sensory compartment, like the antennal lobe of the insect brain, may serve as an example (see Figure 8.5D). Insects, like most other higher animals, possess special sense organs. The insect antenna serves olfaction, and contains a large number of olfactory sensory neurons whose axons project towards a region of the brain neuropil called the antennal lobe (Stocker, 1994). As a result of the specialized, olfactory input, the neuropil of the antennal lobe acquires properties that make it different from neighboring neuropil domains lacking that input. Olfactory sensory neurons have distinctive, branched axonal endings; they form specialized synapses with the interneurons to which they are connected. Sensory neurons expressing the same olfactory receptor (and thereby sensing the same smell) converge and form a structurally visible unit called "glomerulus" within the antennal lobe (Gao, Yuan, & Chess, 200; Vosshall, Wong, & Axel, 2000: see Figure 8.5D). The interneurons receiving olfactory input project to other domains of the neuropil, which thereby also adopt characteristic structurally and functionally distinct properties. For example, the antennal lobe of the brain of many insects forms a massive projection on to a compartment called the mushroom body, which plays an important role in associative learning (see Figure 8.5D, E).

Compartments like the antennal lobe or mushroom body can be observed in the brains of most animals for which complex sensory organs have been described. An overview of these compartments and the sense organs they are directly or indirectly associated with is provided in §8.2 below. Compartments with a distinctive modular structure and well-defined boundaries are referred to as "structured neuropil" in the literature; they contrast with other "unstructured" regions of the neuropil where compartment boundaries are less evident (see Figure 8.5E). For many invertebrate taxa that have small brains and lack distinctive sensory organs, the entire neuropil has so far remained "unstructured." However, it is likely that, given the use of appropriate markers and applying a high enough resolution to the analysis of individual neurons,

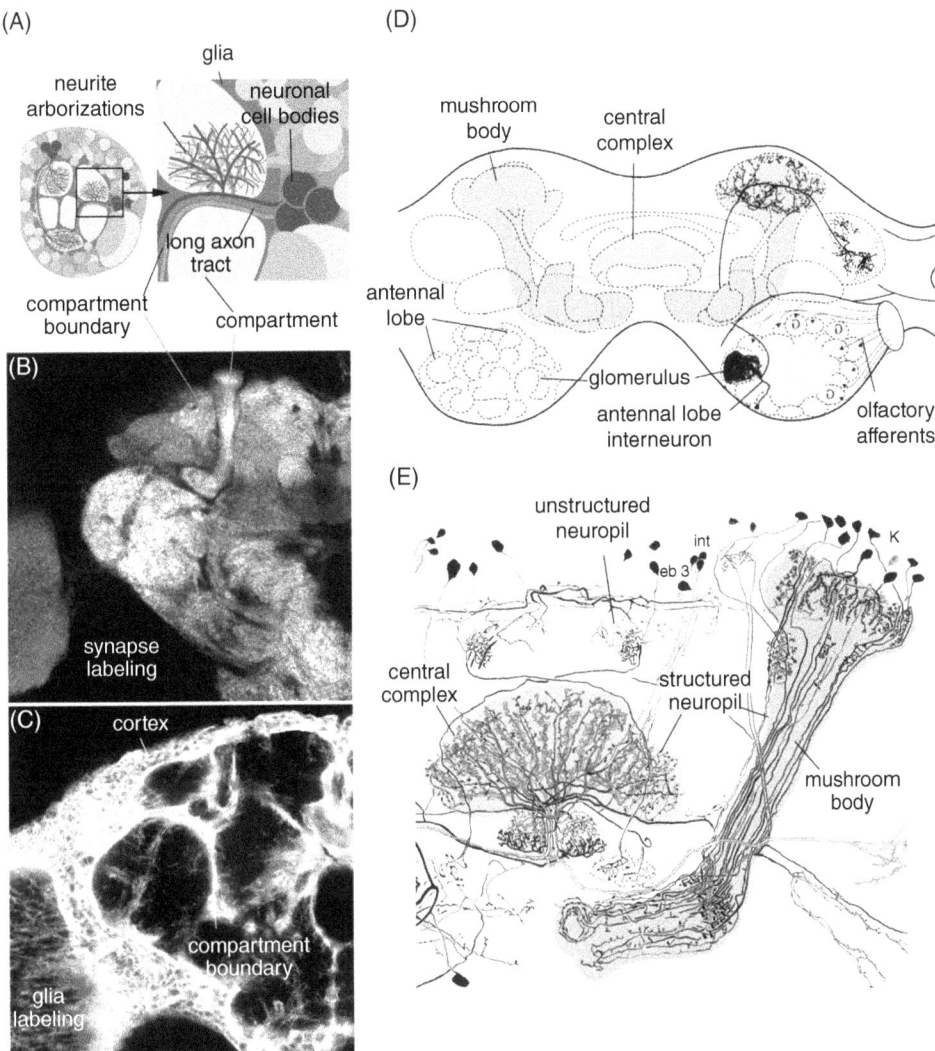

Figure 8.5 Architecture of the Invertebrate Brain Neuropil.
(A) Schematic section of brain, showing relationship between neurons, glia, compartments, and compartment boundaries. (B, C): cross-section of Drosophila brain hemisphere labeled with marker for synapses (B), highlighting compartments, and against glia (C), showing compartment boundaries. (D, E) Line drawings of schematic sections of insect brains, showing examples of Golgi-stained neurons forming structured neuropils (antennal lobe, central complex, mushroom body) and unstructured neuropils. Adapted from Hanstroem 1968. Reproduced with permission of Springer. (*See insert for color representation of the figure*).

compartments will become apparent. For example, the reconstruction of the tiny brain neuropil of the microscopic flatworm *Macrostomum lignano*, containing only a few hundred neurons, has revealed clear compartments that differ in both intrinsic neurite "texture" (branching and orientation of neurites) and connectivity to sensory neurons and nerve cords (see §8.2.2.1 below).

8.2 Invertebrate Nervous Systems: A Brief Comparative Overview

For more than 150 years, neuroanatomists have analyzed the structure of nervous systems of invertebrates using a variety of histochemical techniques to visualize neurons and their fibers. The results of this work has been comprehensively surveyed in two books, the *Vergleichende Anatomie des Nervensystems der wirbellosen Tiere* [Comparative anatomy of the nerve systems of invertebrate animals] (Hanstroem, 1928; reprinted 1968), and Bullock and Horridge's *Structure and function in the nervous system of invertebrates* (1965). Whereas modern techniques have added substantially to our knowledge of the neuroanatomy of many animal species used as neurobiological model systems (cnidarians, arthropods, annelids, molluscs, nematodes, and, to some extent, platyhelminths; for recent review, see Schmidt-Rhaesa, Harzsch, & Purschke, 2016), little has been added to the classical literature dealing with the functional anatomy of the majority of phyla.

As an ordering principle for the following survey of invertebrate neuroanatomy I will use the phylogenetic classification proposed in a number of recent studies (Erwin et al., 2011; see Figure 8.4). Two taxa that branched off the phylogenetic tree at an early stage are placozoa and porifera. Extant members of these groups do not possess a nervous system. All elements of a nervous system are present in two other early branches, Ctenophora and Cnidaria ("diploblastic" animals because of their formation of only two germ layers, ectoderm and endoderm). Here, as described at the beginning of this chapter, sensory neurons and ganglion cells form a peripheral nerve net, with the first appearance of nerve centers associated with the mouth and tentacles.

A significant step towards the potential for higher structural complexity among the multicellular animals took place approximately 700–600 million years ago (MYA) with the appearance of the bilaterian, triploblastic animals. Bilateria are animals with an elongated, bilaterally symmetric body and a third tissue layer, the mesoderm, interposed between the outer ectoderm and inner endoderm. Specialized tissues and organs derived from the mesoderm, including muscle, vascular, excretory, and connective tissue, are thought to have allowed the bilaterian body to move more efficiently, grow to bigger sizes, and occupy new ecological niches. At the same time, neurons became organized into central ganglia and peripheral sensory organs. During the early Cambrian period, 540–515 MYA, bilateria split into different phyla, each with a characteristic body plan. Fossil collections such as the Burgess shale in Canada show that by 515 MYA most currently extant animal phyla had made their appearance ("Cambrian explosion"), implying that the common ancestor of animals as diverse as vertebrates, flies, nematodes and molluscs lived during the early Cambrian period. The early, almost simultaneous appearance of bilaterian taxa is one of the reasons why it is difficult to establish clear phylogenetic relationships between extant phyla. Most of the recently proposed trees assume that bilaterians comprise deuterostomes and protostomes, the latter further subdivided into the "supertaxa" ecdysozoa (molting animals; including arthropods, nematodes and others) and lophotrochozoa (molluscs, annelids, platyhelminths, many others; Aguinaldo et al., 1997; Dunn et al., 2008; Erwin et al., 2011; see Figure 8.4). However, the relationship of phyla within these supertaxa is, in large part, unsettled.

8.2.1 Invertebrate Deuterostomes

The nervous system of invertebrate deuterostomes—echinoderms, hemichordates, cephalochordates, and urochordates—consists of a basiepithelial nerve plexus with local condensations that, in most cases, derive from invaginations of the neuroepithelium (see Figures 8.4 and 8.6). Sensory neurons are distributed over much of the body surface; no large, complex sensory organs, such as image-forming eyes, exist. A neuronal net spreading over the intestine (enteric nervous system) has been described for echinoderms (Garcia-Arraras, Rojas-Soto, Jiménez, Díaz-Miranda, 2001), and is also likely to exist in other invertebrate deuterostomes.

Hemichordates include enteropneusts (acorn worms) and sessile colony-forming pterobranchs (sea angels). Both are characterized by a tripartite body structure: a prosome (proboscis in enteropneusts, tentacles in pterobranchs), mesosome (collar), and metasome (tail) (see Figure 8.6A). Condensations of the basiepithelial nerve layer occur in the collar. In some hemichordates, this region sinks in dorsally (sometimes also ventrally) and forms a tube-like structure (dorsal cord), surrounded by densely packed neurons, similar to the neural tube in vertebrates (Kaul & Stach, 2010; Nomaksteinsky et al., 2009; see Figure 8.6A–C). Distinct neuropil compartments are absent in the hemichordate CNS. However, the dorsal cord of the collar may foreshadow the formation of the vertebrate brain. Thus, in the enteropneust *Saccoglossus*, the anterior-most Hox genes (Hox1, 2), characteristic for hindbrain in vertebrates, are expressed at the boundary between collar and tail (Lowe et al., 2003; Pani et al., 2012). Otx, demarcating forebrain and midbrain in vertebrates, occupies the entire collar region; Engrailed (en), at the vertebrate midbrain-hindbrain boundary, is expressed in the posterior collar.

Like hemichordates, echinoderms include both free-living groups, such as asteroids (star fish) and holothuroids (sea cucumbers), and sessile forms with tentacles (crinoids or sea lillies). Echinoderms are predominantly indirect developers which form from small, bilaterally symmetric larvae. These larvae have a similar three-part body plan to hemichordates, and produce a simple basiepithelial nerve plexus with only a small number of neurons (Burke, 1978; Byrne & Cisternas, 2002; Nakano, Murabe, Amemiya, & Nakajima, 2006). During metamorphosis, the bilateral bodyplan is lost. Larval structures perish, and a five-fold symmetric adult body evolves. Adult echinoderms possess a basiepithelial nerve plexus with condensations around the mouth (circum-oral ring) and along the arms (radial nerve cords) (Cobb & Stubbs, 1981; Mashanov, Zueva, Heinzeller, Aschauer, & Dolmatov, 2010). The cords appear to develop from invaginated epithelial tubes (Märkel & Röser, 1991; Mashanov, Zueva, & Garcia-Arraras, 2007), similar to the collar tube in hemichordates.

Urochordates (tunicates or sea squirts) are predominantly sessile, colonial filter feeders. As larvae they are motile tadpoles that resemble vertebrate larvae in many fundamental aspects. Some tunicates, including the appendicularians (larvaceans), retain a tadpole-like body plan as adults. Tunicate larvae possess gills that function as a straining apparatus for filter feeding, a notochord, and a tripartite neural tube, forming an anterior sensory vesicle, an intermediate visceral ganglion, and a posterior spinal cord (Nicol & Meinertzhagen, 1991; Lacalli, 2001; Sorrentino, Manni, Lane, & Burighel, 2000; see Figure 8.6A, D). The neural tube is formed by an epithelial (ependymal) wall, surrounded by a small number (on the order of 100) of neurons. Given the paucity of neurons, the neuropil forms a loose plexus of nerve bundles surrounding the neuronal somata; specialized neuropil compartments are absent. Neurons associated with the sensory vesicle form the cerebral ganglion. Inserted in

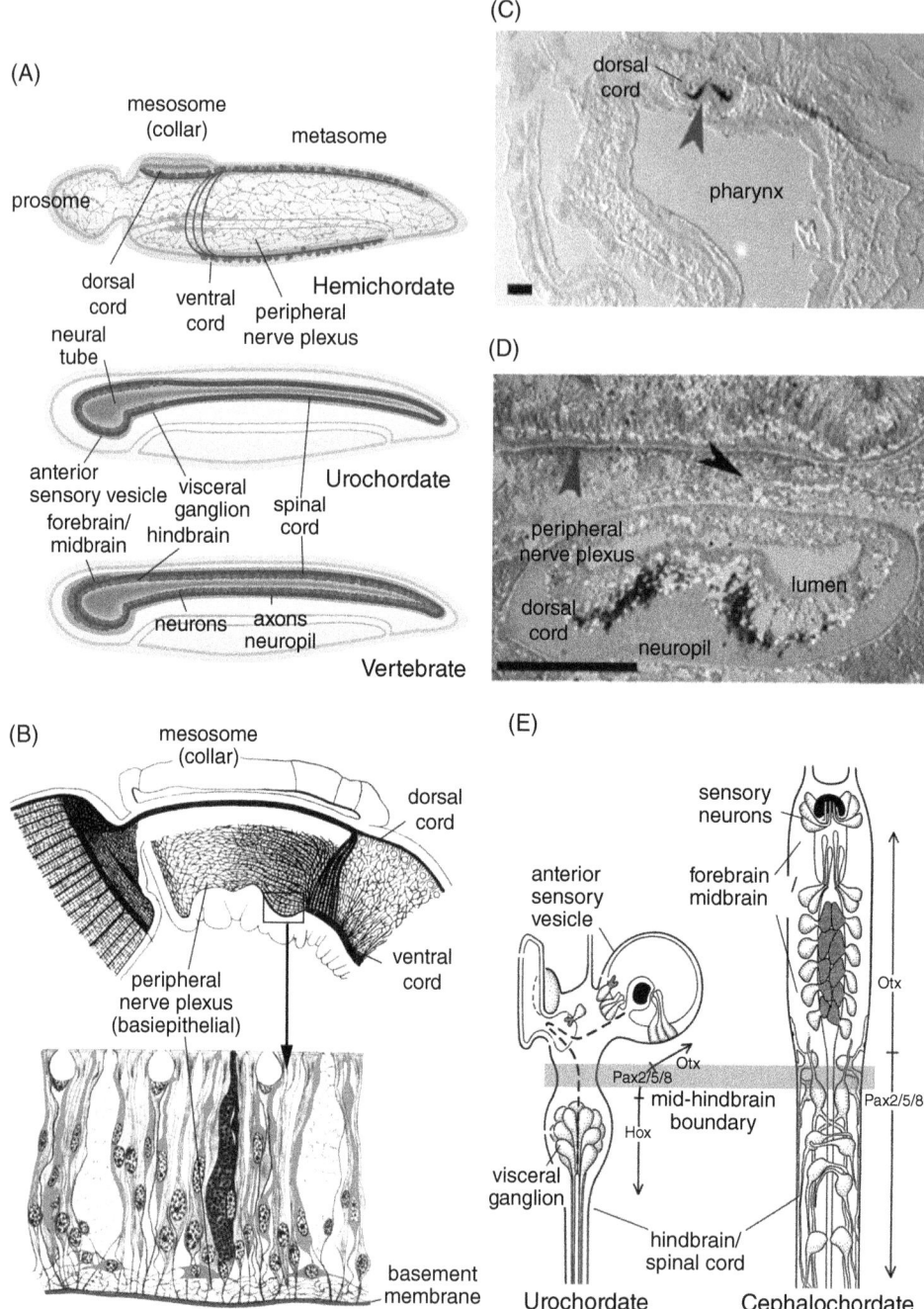

Figure 8.6 The central nervous system of deuterostomes.
(A) Essential characteristics of nervous system architecture in hemichordates, urochordates, and vertebrates, represented as schematic sagittal sections. (B) Line drawing of nervous system of hemichordate *Saccoglossus cambrensis*. Top: nerve net, dorsal cord and ventral cord in collar region. Bottom: cross-section of body wall, showing cytology of basiepithelial nerve plexus. Knight Jones, 1952. Reproduced with permission of the Royal Society. (C, D) Cross-sections of dorsal cord of hemichordate *Ptychodera flava*. Neuronal marker Elav is expressed in ventral

the epithelial wall of the sensory vesicle are groups of photoreceptive and ciliated mechano-receptive sensory neurons, which are considered forerunners of vertebrate eyes and inner ear (Caicci, Burighel, & Manni, 2007; Lacalli, Varona, Arshavsky, Rabinovich, Selverston, 2004; Manni, Caicci, Gasparini, Zaniolo, & Burighel, 2004). Based on such morphological arguments, as well as the expression pattern of patterning genes like the Hox genes (Cañestro, Bassham, & Postlethwait, 2005; Lacalli, 2006; Wada, Saiga, Satoh, & Holland, 1998), the sensory vesicle is homologized with the forebrain/midbrain of vertebrates (see Figure 8.6D). The visceral ganglion, roughly corresponding to the vertebrate hindbrain in regard to gene expression pattern, contains a cluster of neurons with axons that project down the spinal cord—which itself has no neuronal cell bodies attached to it (see Figure 8.6A, D)—and innervate the musculature. The nervous system of cephalochordates (lancelets) is similar in its simplicity and tripartite layout to the tunicate tadpole brain (Lacalli, 1996, 2004, 2006; Lacalli, Holland, & West, 1994; see Figure 8.6A, D).

In urochordates (but not in cephalochordates), the larval neural tube degenerates during metamorphosis and is replaced (through proliferation of progenitors most likely contained within the sensory vesicle) by the cerebral ganglion of the adult sea squirt. Unlike its larval predecessor, the adult ganglion shows attributes of typical non-deuterostome nervous systems, with neuronal somata (measuring in the hundreds) sourrounding a central neuropil (Koyama & Kusunoki, 1993; Manni et al., 1999). No overt similarity to the vertebrate brain is left; however, motor neurons of the cerebral ganglion, projecting fibers towards the muscles of the branchial basket via peripheral nerves, have been likened to hindbrain motor neurons based on gene expression patterns (Dufour et al., 2006).

8.2.2 Lophotrochozoans

Lophotrochozoans combine two major, previously known supra-phyla, the lophophorates and spiralians, as well as several other phyla with unclear phylogenetic position, including chaetognaths, rotifers, and gastrotrichs (see Figure 8.4). Lophophorates are sessile, worm-shaped animals with a conspicuous U-shaped fold, the lophophore, located at their anterior end (see Figure 8.7B). The lophophore bears the multitude of fine tentacles used for filter feeding. Spiralians, including annelids, molluscs, nemertines, and platyhelminths are defined by the spiral mode of cleavage; many species belonging to these phyla also undergo a larval stage known as trochophore (hence the name "lophotrochozoans"). Finally, chaethognaths, rotifers and gastrotrichs are counted as members of the lophotrochozoa based on molecular data.

Figure 8.6 (*Continued*) wall of epithelial dorsal cord (blue; black arrowhead in D). Scale bar: 100 mm. Neuropil of dorsal cord and peripheral nerve plexus is labeled by antibody against Acetylated tubulin (red; red arrowheads). Black arrowhead points at continuous strand of neuropil connecting dorsal cord with peripheral nerve plexus. Nomaksteinsky 2009. Reproduced with permission of Elsevier. (E) Schematic representation of brain of urochordate *Botryllus schlosseri* (left) and cephalochordate *Branchiostoma floridae* (right; both in dorsal view). Expression domains of Otx, Pax2/5/8 and Hox complex demarcatates region considered to be homologous to the vertebrate midbrain-hindbrain boundary (red bar). Lacalli 2001. Reproduced with permission of The Royal Society. (*See insert for color representation of the figure*).

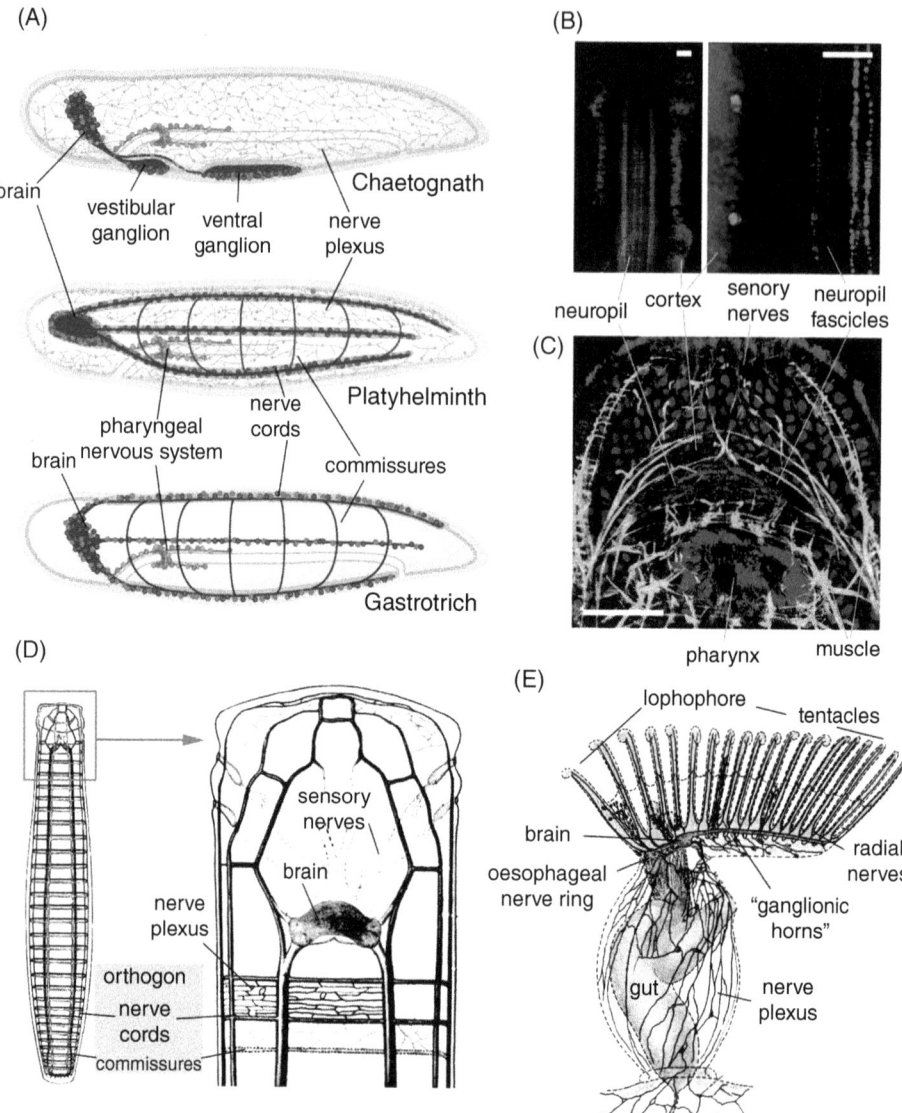

Figure 8.7 The central nervous system of lophotrochozoans.
(A) Essential characteristics of nervous system architecture in chaetognaths, platyhelminths, and gastrotrichs, represented as schematic sagittal sections. (B) Ventral ganglion of chaetognath *Sagitta setosa* (ventral view). Left photograph shows cortex (nuclei of neuronal cell bodies labeled blue) and neuropil (anti-Synapsin, red); right photograph represents higher magnification of neuropil with subset of neurons (anti-RFamide, red) forming distinct longitudinal tracts (scale bars 25 mm). (A, B) Harzsch 2007. Reproduced with permission of BioMed Central Ltd. (C) Brain of platyhelminth *Macrostomum lignano* (dorsal view; scale bar 25 mm). Nerve fibers and cilia of epidermis/pharynx are labeled by anti-Tyrosinated tubulin (red); muscle fibers labeled by phalloidin (green); nuclei of all cells in blue. (D) Line drawing of nervous system of platyhelminth *Bothrioplana semperi* (dorsal view). (E) Line drawing of nervous system of ectoproct (=bryozoan) *Cristatella mucedo* (lateral view). (D, E) Hanstroem, 1968. Reproduced with permission of Springer. (*See insert for color representation of the figure*).

8.2.2.1 Chaetognaths, rotifers, gastrotrichs, flatworms. Chaetognaths (arrow worms) are predatory worm-like animals living among the plankton; rotifers ("wheeled animals") and gastrotrichs ("hairy backs") are microscopic benthic filter feeders. All three groups have a basiepithelial nervous system (see Figure 8.7A). Chaetognaths possess a supraesophageal brain and vestibular ganglion, connected to a large ventral ganglion. These ganglia are considered local condensations of a diffuse basiepithelial nerve net that extends throughout the body (Goto, Katayama-Kumoi, Tohyama, & Yoshida, 1992; Harzsch & Müller, 2007; Rehkämper & Welsch, 1997; Shinn, 1997). Paired eyes and presumed olfactory organs project nerves into the brain neuropil; however, no specialized neuropil compartments have been described.

Platyhelminths (flatworms) represent a large and diverse phylum of animals and include free-living and parasitic groups. All flatworms possess a relatively compact anterior brain organized in a way that resembles a typical invertebrate ganglion (see Figure 8.7A, C; for review, see Bullock & Horridge, 1965; Ehlers, 1985; Hartenstein, 2015; Reuter & Gustavsson, 1985). Nerve cell bodies are arranged in an external layer (cortex), and processes of the predominantly unipolar neurons form a neuropil in the center (see Figure 8.7A, C). Given the relatively small number of neurons (hundreds to a few thousands) and the absence of large sensory organs (see below), the neuropil does not show any overt compartmentalization. However, reproducible patterns of nerve fiber bundles (fascicles) produced by sensory axons entering the brain, or interneuronal connections, can be distinguished (e.g., for *Macrostomum lignano*; Morris, Cardona, De Miguel-Bonet, & Hartenstein, 2007). Non-neuronal cell types, including muscles and glands, penetrate the brain cortex and neuropil.

Issuing posteriorly from the flatworm brain are multiple, paired nerve cords. In most cases, three pairs (dorsal, lateral, and ventral) are distinguished. These longitudinal cords are transversally connected by more or less regularly spaced commissures; longitudinal cords and commissures form the so called "orthogon" (Hanstroem, 1968; see Figure 8.7A, D). Brain and orthogon are subepidermal structures; however, in many instances, a clearly visible basement membrane separating CNS from the overlying body wall is absent. Likewise, glial sheaths forming a cohesive sheath around the brain surface or nerve fiber tracts have not been observed, even though individual cells with glia-like properties may occur (e.g., Biserova, 2008; Koopowitz & Chien, 1974; Sukhdeo & Sukhdeo, 1994).

Aside from the CNS, flatworms possess a peripheral nerve plexus, consisting of nerve cells and sensory cell processes. Mechanosensory and chemosensory neurons are located either in the epidermis or the central nervous system (see Figure 8.7; Ehlers, 1985; Rieger, Tyler, Smith, & Rieger, 1991). A high concentration of these ciliated sensory cells occurs at the anterior tip of the head, where they accompany a diverse array of glandular cells (apical complex). Flatworms have simple eyes and/or statocysts that are located in the brain.

Flatworms have a single gut opening, which was one argument by which previous structure-based cladistic analyses placed them at the very bottom of the bilaterian animals (i.e., closest to cnidarians). A muscular pharynx, sometimes located near the head and sometimes more posteriorly, controls food uptake and egestion and is innervated by neurons forming a dense nerve ring around the pharynx wall (pharyngeal or stomatogastric nervous system).

One group of flatworms, the acoels, lack a gut and pharynx. Their interior is filled with a large syncytium (multinucleated cell), and digestion takes place in the syncytial cytoplasm. Recent molecular phylogenomic studies place the acoels (along with another flatworm taxon, called nemertodermatids) outside the flatworm phylum, and outside the lophotrochozoans altogether; these studies suggest that "acoelomorpha" (acoels and nemertodermatids) represent the sister taxon of all other bilateria (Dunn et al., 2008; Ruiz-Trillo, Riutort, Littlewood, Herniou, & Baguña, 1999). Despite this deep phylogenetic separation, the structure of the central nervous system is similar to that described above for platyhelminths. Notably, and in contrast to some previous claims (e.g., Kotikova & Raikova, 2008; Reuter, Raikova, & Gustafsson, 1998), both types of flatworms have relatively compact anterior brains that consist of a cortex of neural cell bodies and a central neuropil with numerous commissural and longitudinal fiber bundles (Bailly, Reichert, & Hartenstein, 2013; Bery, Cardona, Martinez, & Hartenstein, 2010; Ramachandra, Gates, Ladurner, Jacobs, & Hartenstein, 2002; Semmler, Chiodin, Bailly, Martinez, & Wanninger, 2010).

Gastrotrichs and rotifers contain very small numbers of neurons (100 to a few hundreds). Clusters of neurons in the head form the cerebral ganglion (brain); neurons of the trunk form four paired nerve chords in gastrotrichs, whereas they are clustered in a paired pedal ganglion in rotifers (Clement & Wurdak, 1991; Hochberg, 2009; Hochberg & Litvaitis, 2003; Joffe & Wikgren, 1995; Ruppert, 1991; Teuchert, 1977). Complex sensory organs or neuropil compartments are absent. It is worth noting that the innervation of muscles by motorneurons in these small lophotrochozoans, as well as many other invertebrate taxa (e.g., platyhelminths discussed above) is different from the usual configuration that we know from vertebrates or "higher" invertebrates. There, motor neurons located in the central ganglia send axons towards the muscles, where they form neuro-muscular synapses. Here, in rotifers, gastrotrichs, platyhelminths, nematodes, and other groups, muscle fibers or specialized processes of muscle cells invade the neuropil; neuro-muscular synapses are an integral part of the central nervous system.

8.2.2.2 Lophophorates. Four lophophorate phyla are distinguished: Ectoprocta (Bryozoa: mossy animals); Entoprocta (Kamptozoa: goblet worms); Phoronida (horseshoe worms); Brachiopoda (lamp shells). According to recent molecular analyses (e.g., Dunn et al., 2008; Erwin et al., 2011; Hejnol et al., 2009), these taxa are no longer considered members of a monophyletic group, but have branched off at various positions along the tree (see Figure 8.4). Detailed studies of the adult organization of the nervous system of lophophorates date back to the early 1900s (Hanstroem, 1968); several more recent papers look at the CNS structure of lophophorate larvae (Santagata, 2008, 2011; Wanninger, 2008).

Phoronids have a diffuse nerve net which, at the base of the lophophore, is condensed into a basiepithelial nerve ring. Larvae of phoronids, which share many characteristics with the typical trochophore larvae of other protostomes (annelids, molluscs), have an anterior basiepithelial apical ganglion connected by nerve cords to the ciliated rings and sensory cells of the body (Hay-Schmidt, 1989; Lacalli, 1990; Santagata, 2002; Temereva, 2012; Temereva & Tsitrin, 2014).

The CNS of bryozoa (sessile, colonial filter feeders) consists of a paired ganglion (cerebral ganglion or brain) located close to the mouth; bilateral nerve tracts projecting out of the brain surround the pharynx, thus completing the circum-oral nerve ring typical for most protostomes. Anteriorly, the brain is drawn out into two long processes ("ganglionic horns") that form a U-shaped structure underlying the lophophore (Hanstroem, 1968; Schwaha & Wanninger, 2012; Shunkina, Zaytseva, Starunov, & Ostrovsky, 2014; Weber, Wanninger, & Schwaha, 2014; Figure 8.7E). From these cords emanate many regularly spaced radial nerves that innervate the tentacles of the lophophore. The brain is reported to develop from an invagination of the ectoderm and therefore has an interior lumen, surrounded by inner and outer layers of neuronal cell bodies that enclose a central layer of neuropil (Gruhl & Bartolomaeus, 2008; Weber et al., 2014). Bryozoans possess a peripheral nerve plexus that contacts sensory receptors distributed in the tentacles and all over the body. Complex sensory organs are absent.

The CNS of brachiopods ("lamp shells") and entoprocts ("goblet worms"), containing mostly colonial or solitary, marine filter feeders, consists of subepidermal ganglia. Entoprocts possess only one, paired ganglion located close to the pharynx (Schwaha, Wood, & Wanninger, 2010). In brachiopods, a paired cerebral ganglion and subesophageal ganglion are arranged around the pharynx. Nerves emanating from these ganglia project towards the digestive tract, the tentacles, and stalk (Hanstroem, 1968; Santagata, 2011).

8.2.2.3 Nemertines, annelids, molluscs. Members of these groups of animals can grow to large sizes and inhabit marine, freshwater, and terrestrial environments. They possess complex sensory organs and a central nervous system formed by subepidermal ganglia, containing large numbers of neurons (tens of thousands to millions). Along with the appearance of complex sensory organs and increased neuron number, the brain neuropil exhibits distinct structural/functional compartments.

Nemertines ("ribbon worms") are mostly marine worms. Many are predators; all are characterized by a retractible proboscis involved in prey capture and probing the environment. Nemertines have a paired dorsal (supraesophageal) and ventral (subesophageal ganglion; Beckers, Faller, & Loesel, 2011; Beckers, Loesel, & Bartolomaeus, 2013; Hanstroem, 1968; Turbeville, 1991; Turbeville & Ruppert, 1985; see Figures 8.7, 8.8). These give rise to ventral nerve cords, as well as smaller lateral and dorsal cords. Nerve cords and the commissural tracts interconnecting them, resemble the orthogon defined for flatworms (see above). However, nemertine neural structure is further developed than that of flatworms. Glial layers surround the brain/nerve cords at its outer surface, and separate cortex from neuropil. Aside from ciliary receptors located all over the body, nemertines possess complex, paired sensory organs in their head, including the frontal organ (chemoreception and/or mechanoreception), cerebral organ (probably olfactory in function), and (in some species) eyes. Sensory nerves project into the neuropil of the supraesophageal ganglion. In some species, the dorsolateral part of this ganglion exhibits glomeruli-like neuropil condensations, formed by large masses of densely packed interneurons. We observe here for the first time an example of the structured neuropil compartment, typically seen in conjunction with the appearance of complex sensory organs. Structured neuropil compartments are even more pronounced in the brains of molluscs and annelids.

The diverse phylum of molluscs includes groups, considered phylogenetically basal, which share many characteristics with flatworms, as well as highly derived groups with complex nervous systems. Among the former are the aplacophorans (small, worm-like deep-sea animals) and polyplacophora (chitons). The nervous system of these animals consists of a subepidermal brain ("cerebro-buccal ring") located anterior to the mouth and two paired nerve cords, the lateral pallio-visceral cord, and the ventral pedal cord

190 Volker Hartenstein

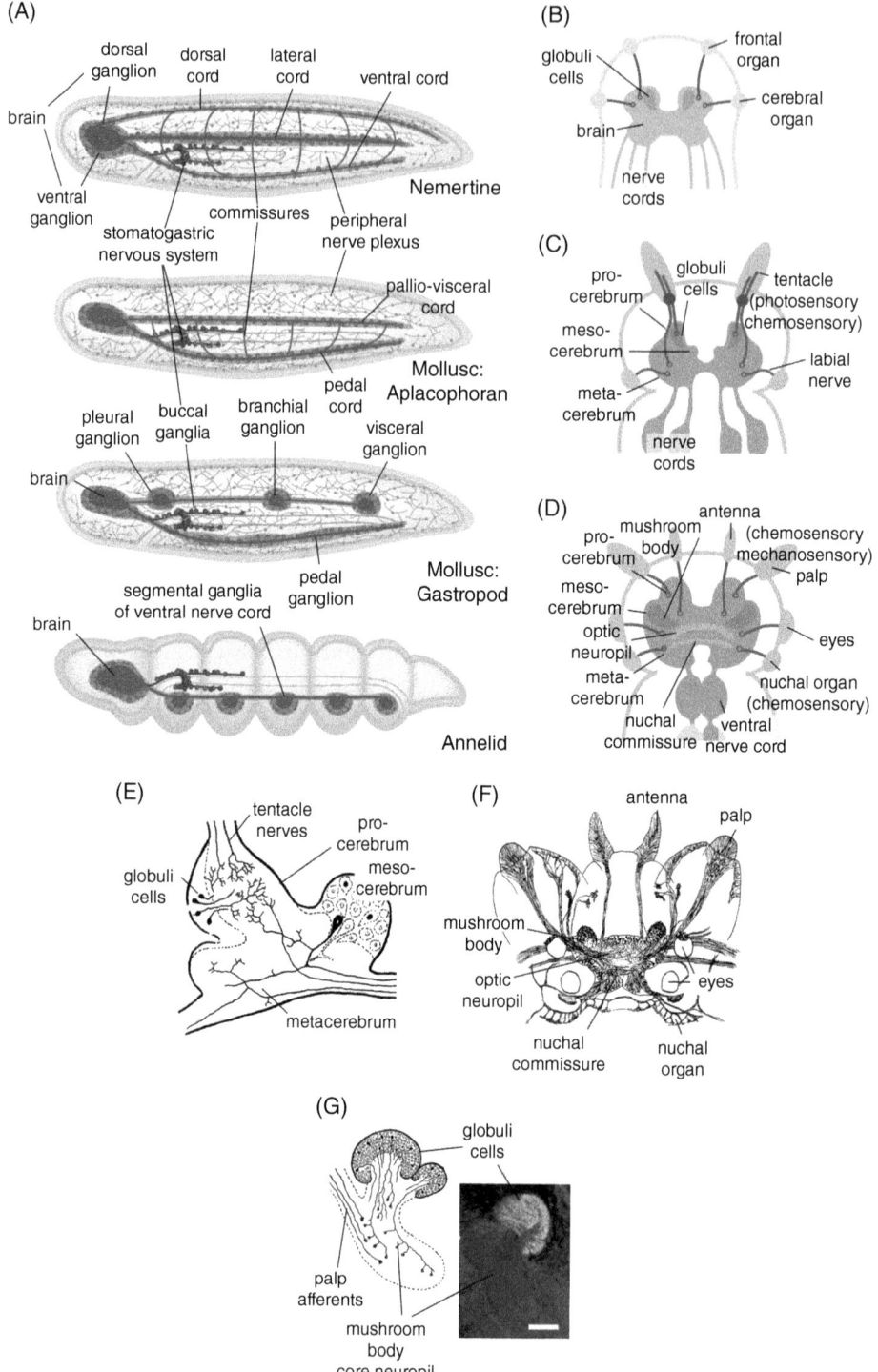

Figure 8.8 (*Continued*)

(Eernisse & Reynolds, 1994; Moroz et al., 1994; Scheltema et al., 1996; Shigeno, Sasaki, & Haszprunar, 2007; see Figure 8.7). Regularly spaced commissures interconnect the cords, bestowing an orthogon-like organization upon the polyplacophoran CNS. Two paired nerves connect the brain to ganglia innervating the foregut (esophageal ganglion) and muscles of the radula (the mollusc-specific toothed tongue used to shred ingested food). Aplacophorans and polyplacophorans have a diffuse peripheral nerve net. Complex sensory organs are absent, and highly structured neuropil compartments have so far not been described.

More highly derived molluscs, including gastropods (snails), bivalves (clams) and cephalopods (squid and octopus), possess subepidermal ganglia and a peripheral diffuse nerve net. The arrangement of ganglia in these animals can be derived from the polyplacophoran condition described above (see Figure 8.7). Gastropods and bivalves have a cerebral ganglion (brain) that gives rise to two pairs of nerve cords. The ventrally located pedal cord innervates the muscular "foot" of gastropods and bivalves (arms in cephalopods); in most species, the cord has condensed into a large ganglion (pedal ganglion) connected to the brain by a fiber bundle, the pedal connective. The lateral, pallio-visceral cord has also condensed into three paired ganglia, the pleural ganglia (innervating the mantle cavity), branchial ganglia (gills) and visceral ganglia (intestinal tract). Nerve tracts connecting the brain to ganglia innervating the pharynx and foregut (buccal ganglia) form the stomatogastric nervous system. In cephalopods, all ganglia are fused into one large "brain" that, in regard to cell number and complexity, surpasses many vertebrate brains. In addition, prominent optic lobes have evolved in conjunction with the large image-forming eyes; intrabrachial ganglia, connected to the pedal ganglion, control movement of the arms.

Gastropods and cephalopods have complex sensory organs; accordingly, the brain neuropil exhibits structured compartments. The gastropod head has paired tentacles, two pairs in most terrestrial gastropods, one pair in many marine opisthobranchs (sea slugs). Tentacles carry eyes, often with lenses and many thousands of photoreceptors. In addition, tentacles act as olfactory organs. Specialized sensory nerves, including the optic nerve and tentacle nerve, carry the axons of these receptors to the cerebral ganglion. The cerebral ganglion of gastropods has three major compartments, called pro-, meso-, and metacerebrum (Chase, 2000; Elekes, 2000; Hanstroem, 1968; see Figure 8.8 schematic). The procerebrum, which receives the optic nerve and tentacle nerve, is formed by large numbers (upwards of 20,000) densely packed neurons ("globuli cells") which form a structured neuropil whose function is the processing and retention of olfactory information. The mesocerebrum, an overtly

Figure 8.8 (*Continued*) The Central Nervous System of Lophotrochozoans. (A) Essential characteristics of nervous system architecture in nemertines, molluscs, and annelids, represented as schematic sagittal sections. (B–D) Schematic representation of head sensory organs and brain of nemertines (B), gastropod molluscs (C), and polychaete annelids (D; all in dorsal view). Sensory nerves in blue, structured neuropil compartments in purple. (E–G) Line drawings of neurons forming structured neuropil compartments. (E) Globuli cells of gastropod *Helix pomata*. (F, G) Sensory nerves, brain and mushroom body of polychaete *Nereis diversicolor*. (F–G) Hanstroem, 1968. Reproduced with permission of Springer. Inset in (G): photograph of mushroom body of *Nereis diversicolor* (scale bar 50 mm). Heuer and Loesel, 2008. Reproduced with permission of Springer. (*See insert for color representation of the figure*).

left-right asymmetric structure, is involved in mating behavior (Chase, 2000). The posteriorly located metacerebrum is connected to the skin of the head, lips, and mouth cavity. It emits the stomatogastric nerves to the buccal ganglia that control feeding behavior.

Annelids (segmented worms) include a wide variety of motile and sessile marine worms (polychaetes, echiurids, sipunculids), as well as leeches and terrestrial earthworms. The central nervous system of annelids consists of subepidermal ganglia (Hanstroem, 1968). The brain (cerebral ganglion, supraesophageal ganglion) is located in the head, anterior to the mouth, a region called prostomium; it receives input from sensory organs of the prostomium. The brain connects to the stomatogastric nervous system and a ventral nerve cord, which consists of a series of segmental ganglia. Segmental ganglia have the tendency to fuse; for example in many species of leeches, the four anterior ganglia have fused into one subesophageal ganglion, the four posterior ganglia into one caudal ganglion. Lateral and dorsal nerve cords, and the peripheral nerve net as seen in molluscs, are absent (see Figures 8.7, 8.8). Segmental ganglia of the ventral nerve cord contain motor neurons and interneurons which control body movement; examples of specific neuronal circuits, which have been elucidated in great detail for leeches, will be discussed in §8.4 of this chapter.

The prostomium of polychaetes possesses several complex sensory organs, including the palps, one or several paired antennae (tentacles), one or more paired eyes, and the nuchal organs. Palps, antennae and nuchal organs bear large arrays of sensory receptors, which are involved in olfaction and mechanoreception (Bullock & Horridge, 1965; Forest & Lindsay, 2008; Golding, 1992; Mill, 1978; see Figure 8.8). The brain, which receives the afferent nerves of these sensory organs exhibits three domains, termed forebrain (procerebrum), midbrain (mesocerebrum), and hindbrain (metacerebrum). The nerve from the palps terminates in the forebrain; eyes and antennae project to the inner part of the midbrain, and the nuchal organ to the hindbrain. The outer part of the midbrain contains large numbers of interneurons which form a structured neuropil, called the mushroom body (Hanstroem, 1968; Heuer & Loesel, 2008; Heuer, Müller, Todt, & Loesel, 2010; see Figure 8.8). It is thought that incoming afferents of all modalities, directly or via interneurons, gain access to the mushroom body. Afferents from the eyes and the nuchal organs also form structured neuropils in the inner midbrain (central optic neuropil; Heuer & Loesel, 2008) and hindbrain (nuchal commissure), respectively. The optic center and nuchal commissure are are "midline neuropils" which cross the midline between left and right brain hemisphere (see Figure 8.8). These midline neuropils, as well as the mushroom body, have been compared to (and may be homologous to) similar structured neuropils observed in arthropods (Heuer et al., 2010; see below). It appears that complex sensory organs and accompanying structured brain neuropils only occur in polychaetes, and have been lost in other annelid taxa, including earthworms or leeches.

8.2.3 Ecdysozoa

The ecdysozoa (animals with a cuticular exoskeleton that is renewed during molts) include two super-phyla, pan-arthropods (likely to be monophyletic) and cycloneuralians (probably para-phyletic; Borner, Rehm, Schill, Ebersberger, & Burmester, 2014; Rota-Stabelli et al., 2010; Telford, Bourlat, Economou, Papillon, & Rota-Stabelli, 2008). Among the pan-arthropods are insects and crustaceans ("tetraconata"), chelicerates (horseshoe crabs, scorpions, spiders, mites), myriapods (centipedes, millipedes), and onychophorans (velvet worms; see Figure 8.4); all of these groups are segmented animals with a subepidermal nervous system forming a complex brain and a ventral chain of segmental ganglia. Tardigrades ("water bears") show many structural similarities to the segmented pan-arthropods, but their inclusion in this clade is unclear. Cycloneuralians are unsegmented, worm-like animals with simple, basiepithelial nervous systems; they include nematodes (round worms) and nematomorphs (hair worms), kinorhynchs, loriciferans, and priapulids.

8.2.3.1 Cycloneuralians and tardigrades. The small, basiepithelial nervous system of cycloneuralians has been studied in great detail for a number of nematodes, most importantly *Ascaris suum* (intestinal roundworm; large parasitic nematode) and *Caenorhabditis elegans* (microscopic nematode living in the soil). Neurons of the nematode CNS form a ring-shaped ganglion that surrounds the pharynx (pharyngeal nerve ring or brain) that emits a ventral, lateral and dorsal nerve cord (Hanstroem, 1968; White, Southgate, Thomson, & Brenner, 1980; Wright, 1991; see Figure 8.9A). The only neuronal cell bodies found in the ventral cord are those of motoneurons innervating the body musculature; posteriorly, neurons controlling mating and defecation form a tail ganglion. As described for platyhelminths and other lower lophotrochozoans, motoneurons have no peripheral axons innervating the muscles; instead muscle cells form processes that travel towards the ventral nerve cord, where they form synapses with motor neurons (see Figure 8.9B).

The pharyngeal nerve ring, or brain, is composed of the anterior ganglion (somata of sensory neurons innervating the simple chemosensory and mechanosensory organs of the head), lateral ganglion, ventral ganglion, and retro-vesicular ganglion. The latter ganglia contain a small number of motoneurons innervating the head and pharynx muscles, and interneurons integrating sensory inputs and synapsing on the motoneurons. The neuropil of the brain and nerve cords is exceedingly small and unstructured, because neurons are low in number (a few hundreds) and are mostly unbranched (White et al., 1980; see Figure 8.9C).

Kinorhynchs are microscopic benthic worms with a basiepithelial nervous system. The brain is formed by clusters of neurons associated with three circular nerve tracts surrounding the pharynx; multiple nerve cords innervate the trunk, but the unpaired ventral cord contains most neurons (Herranz, Pardos, & Boyle, 2013; Kristensen & Higgins, 1991; Neuhaus & Higgins, 2002). Priapulids ("penis worms") are somewhat larger cylindrical worms living in the mud of shallow water; like nematodes and kinorhynchs they possess a basiepithelial nerve ring around the pharynx. A single ventral nerve cord projects into the trunk (Rehkämper, Storch, Alberti, & Welsch, 1989; Rothe & Schmidt-Rhaesa, 2010; Storch, 1991; see Figure 8.9A).

8.2.3.2 Arthropods. Arthropods, representing the largest phylum (by far), are segmented animals with jointed limbs. One distinguishes onychophorans, myriapods, chelicerates, crustaceans, and insects (see Figure 8.4). All of these taxa have subepidermal, ganglionic nervous systems, typically with hundreds of thousands to millions of neurons, and the neuroanatomy of many species has been described in considerable detail (Bullock & Horridge, 1965; Hanstroem, 1968; Strausfeld, 1976). Ganglia form an anterior brain and a ventral nerve cord. A peripheral nerve net is absent. The body of crustaceans and insects, which are more closely related to each other than to

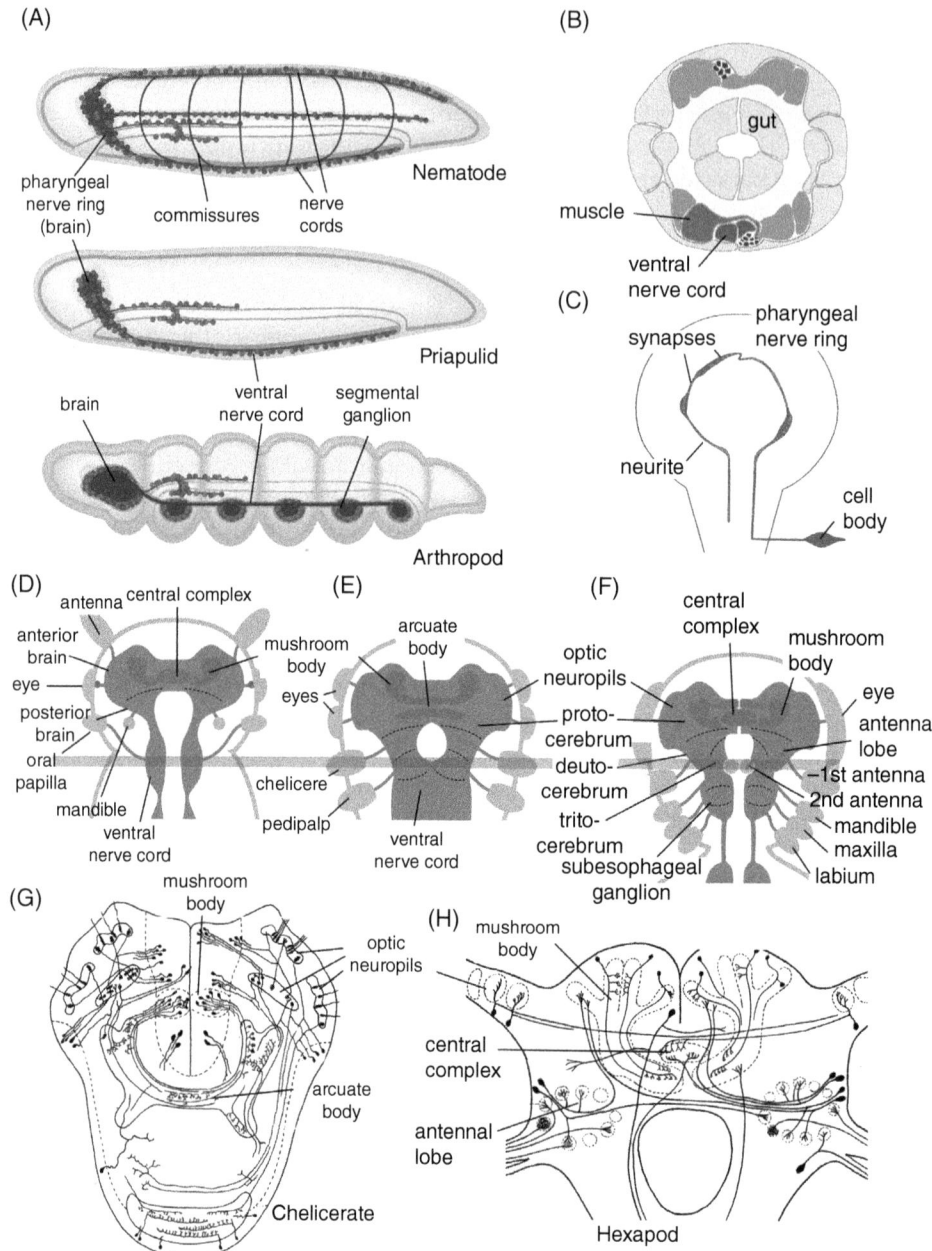

Figure 8.9 The Central Nervous System of Ecdysozoans.
(A) Essential characteristics of nervous system architecture in nematodes, priapulids, and arthropods, represented as schematic sagittal sections. (B) Schematic cross-section of nematode Caenorhabditis elegans, showing basiepithelial dorsal and ventral cord and muscle process forming connection to ventral nerve fiber. (C) Line drawing of representative neuron of nematode C. elegans based on electron microscopic reconstruction. Neuron forms single nerve process with interspersed input and output synapses. (D–F) Schematic representation of head sensory organs and brain of arthropods (D: onychophoran; E: chelicerate; F: hexapod; all in dorsal view). Sensory nerves in blue, structured neuropil compartments in purple.

the remainder of the arthropods, has three domains (tagmata): head, thorax, and abdomen. Each of these domains comprises multiple more or less fused segments. The head is formed by a large anterior, unsegmented component (acron) and five segments flanking the mouth opening; the thorax has eight (most crustaceans) or three (insects) segments; segment number in the abdomen is highly variable. The acron (sometimes considered an additional, "ocular" segment; Schmidt-Ott & Technau, 1992) carries the compound eyes and median ocelli (see Figure 8.9F). The first two segments have chemosensory appendages (two pairs of antennae in crustaceans; one pair of antennae and a gustatory organ in insects); the three segments posterior to the mouth opening (gnathal segments) are equipped with paired appendages (mandible, maxilla, labium) specialized for feeding (see Figure 8.9F). Thoracic segments carry large walking legs in insects and many crustaceans; abdominal segments have no appendages (insects) or small appendages involved in swimming (crustaceans) and other functions.

The brain of insects and crustaceans, according to the segmental organization of the head, is formed by six fused ganglia or "neuromeres" (where separate ganglia are not visible the term "neuromere," which refers to a segmental unit within the nervous system, is preferable). Neuromeres of the acron and two first segments form the supraesophageal ganglion; neuromeres of the three gnathal segments are fused into the subesophageal ganglion (see Figure 8.9F). The anterior neuromere (protocerebrum) consists of large, multi-layered optic lobes which process information from the compound eyes, and a central brain with a number of structured neuropil compartments (Hanstroem, 1968; Krieger et al., 2012; Loesel, Nässel, & Strausfeld, 2002; Strausfeld, 1976; Strausfeld, Sinakevitch, Brown, & Farris, 2009). The most prominent of these are the paired mushroom bodies, crucial for learning and memory (prominent in insects; reduced or absent in crustaceans), and the central complex, which consists of several layers of structured neuropil crossing the midline, and which plays important roles in motor control, spatial orientation/memory, and many other functions (see Figure 8.9F, H). The second neuromere (deutocerebrum) receives olfactory and mechanosensory input from the first antenna and possesses a highly structured neuropil compartment, the antennal lobe. The third neuromere (tritocerebrum) is also predominantly chemosensory, being innervated by the second antenna (crustaceans) and sensory organs located in the mouth cavity and the gut (see Figure 8.9F). The tritocerebrum also connects to the stomatogastric nervous system, formed by a series of peripheral ganglia associated with the intestinal tract.

The structure of the head and brain of Myriapods (centipedes, millipedes) closely resembles that of insects; some neuropil compartments, notably the optic lobe and central complex, are reduced in size (Hanstroem, 1986; Sombke et al., 2012).

Figure 8.9 (*Continued*) Hatched lines indicate boundaries between segmental ganglia. Segmental ganglia of arthropods can be homologized based on anatomical and molecular criteria (for details, see text); green bar registers brains, indicating tritocerebrum (hexapod), pedipalpal ganglion (chelicerate), and ganglion innervating oral papilla (onychophoran) as homologous neuromeres. (G, H) Line drawings of neurons forming structured neuropil compartments. (G) Optic neuropils, mushroom body, and arcuate body of chelicerate. (H) Mushroom body, central complex, and antennal lobe of hexapod *Periplaneta americana*. Hanstroem, 1968. Reproduced with permission of Springer. (*See insert for color representation of the figure*).

The body of chelicerates (horseshoe crabs, spiders, scorpions, mites) has only two tagmata, the cephalo-thorax (fused head and thorax), and abdomen. The cephalo-thorax includes an anterior acron, carrying the eyes, and six segments with appendages involved in chemoreception, feeding, and locomotion. Appendages of the first two segments are the cheliceres ("pincers" for grasping food) and pedipalps (leg-like tactile organs in spiders; claws with pincers in scorpions; see Figure 8.9E); the posterior four segments carry walking legs. Ganglia of the cephalo-thorax form a fused mass, comprising a supraesophageal ganglion and a ventral ganglion. Primitively (e.g., in horseshoe crabs), the supraesophageal ganglion includes only the protocerebrum; ganglia innervating the the cheliceres and pedipalps, as well as the walking legs, form part of the ventral cord (Babu, 1985; Hanstroem, 1968). In spiders and scorpions, the ganglion of the cheliceral segment is incorporated into the supraesophageal ganglion (see Figure 8.9E). Homologizing segments in chelicerates and insects/crustaceans is problematic. Based on neuroanatomical data, the first segment with its cheliceral neuromere corresponds to the insect/crustacean tritocerebrum because its commissural axons cross ventral to the foregut and it is connected to the stomatogastric nervous system (reviewed by Scholtz & Edgecombe, 2006; Weygoldt, 1985). This would mean that a chelicerate counterpart of the first antennal segment/deutocerebrum is missing. However, based on the pattern of Hox gene expression, the cheliceral segment is homologous to the first antennal segment of insects/crustaceans; its neuromere would then correspond to the deuterocerebrum (Damen et al., 1998), and the pedipalpal neuromere would be the tritocerebrum (this interpretation is represented in Figure 8.9E).

In accordance with the presence of large, complex sensory organs, chelicerates also have structured neuropil compartments. They comprise layered optic neuropils, bilateral mushroom bodies, and a central complex (Strausfeld, Weltzien, & Barth, 1993; Strausfeld & Barth, 1993). It is not yet clear in how far, functionally or ontogenetically, these compartments correspond to the structures of the same name in insects. In insects the mushroom bodies receive predominantly input from the antennal lobe and are involved in olfactory learning and memory. The mushroom bodies in spiders are targeted mostly by 2nd order visual interneurons (Strausfeld & Barth, 1993; Strausfeld, Hansen, Li, Gomez, & Ito, 1998; see Figure 8.9E, G). Also, the correspondence between the central complex in insects/crustaceans and spiders (where it is now called arcuate body; Strausfeld, Strausfeld, Loesel, Rowell, & Stowe, 2006) is not clear.

Onychophorans (velvet worms) are terrestrial, segmented, worm-like animals with short limbs on each segment. The head has relatively small cup eyes, a pair of large chemosensory antennae, paired mandibles for mincing food, and oral papillae that produce mucus employed to snare prey. Eyes and antennae, as well as mandibles, are innervated by the supraesophageal ganglion; based on the pattern of innervation, as well as the expression pattern of Hox and other patterning genes, one can subdivide the onychophoran brain into an anterior brain ("protocerebrum") associated with eyes and antennae, and a posterior brain ("deutocerebrum") innervating the mandibles (Mayer et al., 2010; see Figure 8.9D). The oral papillae receive innervation from the first ganglion of the ventral cord. Structured compartments are found in the protocerebrum, and include antennal glomeruli, as well as mushroom bodies and a central complex (Homberg, 2008; Strausfeld et al., 2006).

Tardigrades have a subepidermal CNS that includes a brain (supraesophageal ganglion) and a ventral nerve cord, composed of a pharyngeal (subesophageal) ganglion and four segmental ganglia innervating the segmented body with its stubby appendages (Hanstroem, 1968; Mayer et al., 2013; Persson et al., 2012). Unlike arthropods and onychophorans, tardigrades possess only small numbers (20–30 per ganglion) of neurons in their brain and ventral chord, and complex sensory organs or structured neuropil compartments are absent.

8.3 Morphological Building Blocks of the Invertebrate CNS

8.3.1 Synapses

Synapses are the points of contact between neurons, or between neurons and other target cells, such as muscle, gland, or sensory receptor. Two types of synapses, chemical and electrical, can be distinguished. The ultrastructural criteria of a chemical synapse are vesicles, containing neurotransmitter, and membrane thickenings, corresponding to pre- and postsynaptic protein complexes involved in signal transmission, such as vesicle-docking protein complexes (presynaptic) and receptors/ion channels (postsynaptic). Another presynaptic specialization found in many synapses is the synaptic ribbon, or T-bar (because, in cross section, it adopts the shape of the letter "T"); this organelle is formed by protein complexes involved in the docking of vesicles to the presynaptic membrane (Wichmann & Sigrist, 2010; see Figure 8.10A). Based on their ultrastructural features, various classifications of synapses have been proposed. In his classical electron microscopic investigation, Gray (1959) defined type 1 (asymmetric) and type 2 (symmetric) synapses in the mammalian visual cortex. Type 1 synapses are characterized by a larger diameter of the postsynaptic (compared to presynaptic) membrane thickening. Type 2 synapses have presynaptic and postsynaptic membrane thickenings of equal diameter (symmetric). It was originally proposed that asymmetric synapses are associated with round vesicles and are excitatory, whereas symmetric synapses possessed flattened or irregularly shaped (pleomorphic) synaptic vesicles, and are inhibitory. However it became soon clear that synaptic membrane specializations (e.g., symmetric vs. asymmetric) are unrelated to the type of vesicles present (Ebner & Colonnier, 1975; 1978; see Figure 8.10B), and that functional characteristics of synapses cannot be deduced from *Morphology* ogical observations.

Synapses documented in a wide variety of animal taxa, from "simple" invertebrate groups like platyhelminths to highly derived invertebrates (e.g., insects) show similar charcteristics as vertebrate synapses (Cobb & Pentreath, 1978; Gerschenfeld, 1973). Symmetric synapses appear to prevail (see Figure 8.10C). Synaptic vesicles are round or pleomorphic, clear or electron-dense. In invertebrate species where detailed studies of synapses, neuronal transmitter phenotypes, and synaptic function have been conducted, the majority of synapses have clear vesicles of 30–50nm diameter which contain transmitters such as acetylcholine, GABA, or glutamate. A smaller number of neurons containing peptide transmitters (e.g., insulin-like peptide) form homogenously electron-dense synaptic vesicles of 80–120nm diameter. Synapses with both types of vesicles are frequent (see Figure 8.10D), suggesting that many synapses release more than one transmitter. In addition to the prevailing small/clear and large/dense vesicles, a number of other vesicle types are known from different phyla.

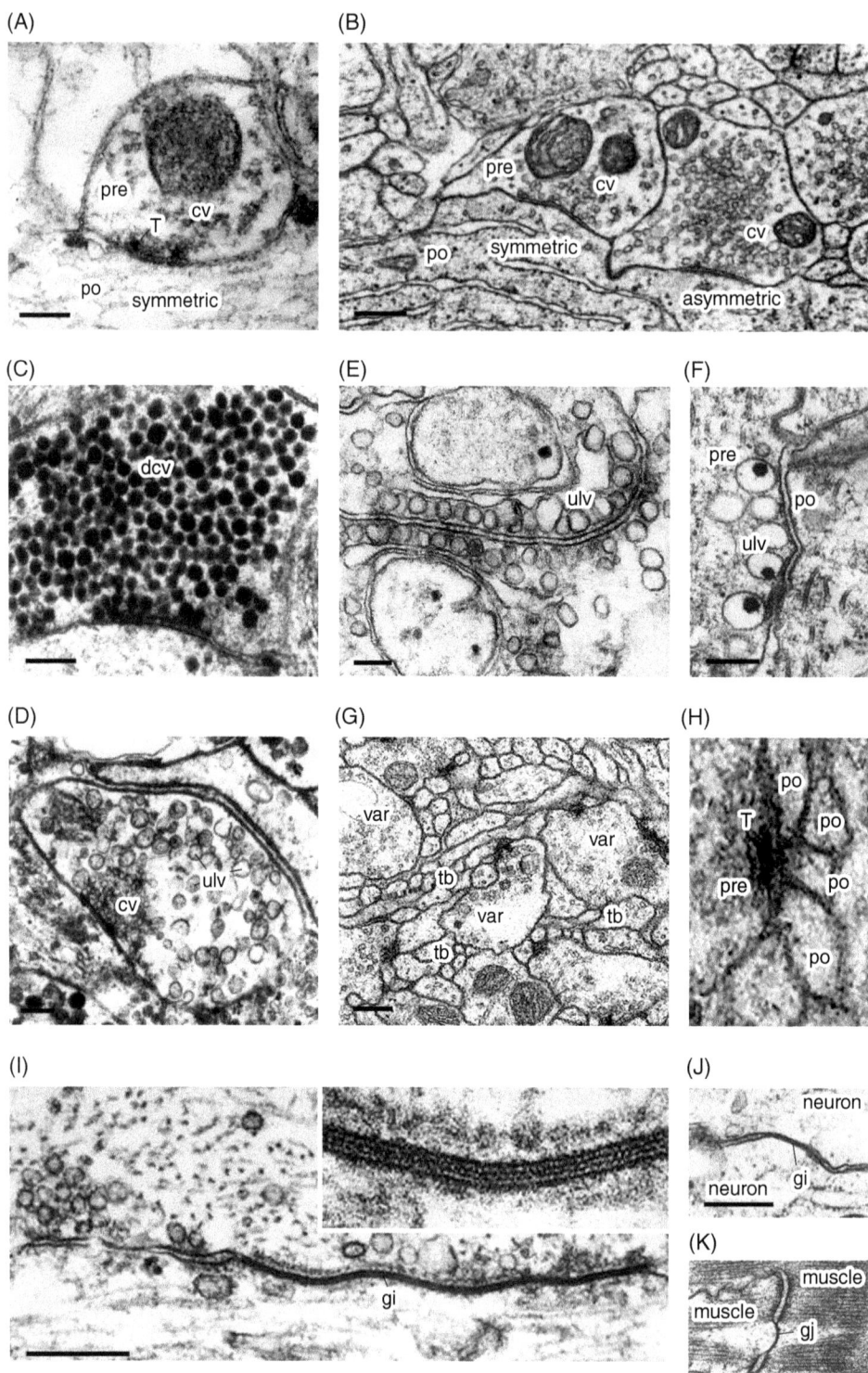

Figure 8.10 (Continued)

An example shown in Figure 8.10D are the "ultra-large" (120–150nm) vesicles, clear or with electron-dense center, shown for a synapse in the *Aplysia* brain (Tremblay, Colonnier, & McLennan, 1979). Synapses with this type of vesicles appears to be the dominant ones found in the two pre-bilaterian taxa with neurons, ctenophores and cnidarians (see Figure 8.10E, F). In these taxa, synapses are typically non polarized possessing vesicles on either side of the synaptic cleft, and thereby transmitting activity in both directions (see Figure 8.10E); however, polarized synapses have been documented (Anderson & Schwab, 1981; Westfall et al., 1971; see Figure 8.10E).

Chemical synapses are often found at the thickened tips of short axon branches, called synaptic boutons. When occurring along the shaft of axons, one speaks of "en passant" synaptic boutons. A synaptic bouton has typically multiple synaptic sites, by which one given presynaptic axon contacts several different postsynaptic dendrites. Individual synaptic sites, defined by a discrete pre/postsynaptic membrane thickening, are relatively uniform in size, measuring 0.2–0.5μm (Cardona et al., 2010; Cobb & Pentreath, 1978; see Figure 8.10G, H). In most cases of invertebrate synapses documented to date, synaptic sites are polyadic, which means that at the same synaptic site, one presynaptic (axonal) element contacts two or more postsynaptic (dendritic) elements (see Figure 8.10G, H). This is different from the typical scenario in vertebrate nervous systems, where synaptic sites connect one axon terminal to one dendrite.

Electrical synapses, or gap junctions, form pores connecting the cytoplasm of neighboring cells, allowing for the passage of ions between these cells. Gap junctions consist of cylindrical arrays of membrane molecules called connexins. Electrical synapses play a widespread role in cell–cell communication during development; in the mature nervous system, they have been documented in many species to occur between some selected neurons (Bennett, 2000, 2006; see Figure 8.10I, J). Electrical synapses between sensory neurons and epithelio-muscle cells form a major component of signal conduction in cnidarians (Mackie, 1970, 2004; Satterlie & Spencer, 1983).

Figure 8.10 (*Continued*) Ultrastructure of Synapses.
(A) Type 2 (asymmetric) synapse from rat cortex (cv clear synaptic vesicles; po postsynaptic element; pre presynaptic element; T T-bar (synaptic ribbon). Gray 1959. Reproduced with permission of John Wiley and Sons. (B) Type 1 (asymmetric) and type 2 (symmetric) synapse from ccerebral cortex of turtle *Pseudemys scripta*. Ebner and Colonnier 1975. Reproduced with permission of John Wiley and Sons. (C, D) Synapses in brain of mollusc *Aplysia californica*: (C) Peptidergic synapse with dense core vesicles (dcv). (D) Mixed synapse with clear vesicles (cv) and ultralarge vesicles (ulv). (C, D) Tremblay 1979. Reproduced with permission of John Wiley and Sons. (E) Non-polarized synapse from pharyngeal nerve net of ctenophore *Beroe ovata*. Hernandez-Nicaise ML 1973. Reproduced with permission of Springer. (F) Polarized synapse from cnidarian *Hydra pseudoligactis*. Westfall 1971. Reproduced with permission of The Rockefeller University Press. (G, H) Polyadic synapses from brain of hexapod Drosophila melanogaster. Note concentration of presynaptic sites on large diameter varicosities (var) of preterminal axonal branches, and abundance of small diameter terminal branches (tb) in (G). (H) shows high magnification of one polyadic synapse. Varicosity with presynaptic site (pre; T denotes T-bar) contacts four postsynaptic (po) terminal branches of dendrites. (I) Gap junction (gj) forming electric synapse in primate neocortex. Bennett 2006. Reproduced with permission of Electroneurobiología. (J, K) Gap junctions between neurons (J) and epithelio-muscular cells (K) of cnidarian *Polyorchis penicillatus*. Bars: 250nm. Satterlie RA 1983. Reproduced with permission of Springer.

8.3.2 Neurons

Neurons come in many different shapes and have many different functions, and it has so far not been possible to relate particular classes of neurons to discrete taxa, or to discern clear evolutionary trends underlying the variety in neuronal architecture encountered in the animal kingdom. The bipolar sensory neurons introduced in §8.1.1—in all likelihood the evolutionarily oldest type of neuron—are modified epithelial cells with a distinct apico-basal polarity. Occuring in all animal taxa, these cells are integrated in epithelia at the body surface or sunken inside the body, and consist of an apical dendrite (a modified cilium, characteristic of typical epithelial cells) and a basal axon that project towards a subepithelial nerve net or the central nervous system. Simple ganglion cells found in the nerve net of cnidarians and the basiepithelial nervous system of many invertebrates are typically unipolar, bipolar or tripolar cells (Sakaguchi et al., 1996), with cell bodies sunken beneath the surface, and fibers forming the basiepithelial plexus (see Figure 8.1).

8.3.2.1 Unipolar vs. multipolar neurons. Ganglion cells of larger, more highly evolved subepidermal nervous systems, such as those encountered in annelids, molluscs, or arthropods, are most often unipolar neurons. Their cell body, located in the cortex at the periphery of the ganglion, sends a single-cell body fiber radially towards the neuropil center, where it branches into primary and higher order neurites (see Figure 8.11; see also examples shown in Figures 8.5, 8.8, 8.9). The unipolar architecture of these neurons is distinctly different from that of vertebrate neurons, which are typically multipolar (see Figure 8.11). The distinction between unipolar and bipolar/multipolar neurons may simply reflect the spatial relationship between neuronal cell body and neuropil. If the cell body is remote from the domain of neurite interactions (as in the case of invertebrate ganglia), it produces a single fiber reaching towards the neuropil, where it branches. The cell body itself typically bears no synapses, and does not participate in signal conduction. On the other hand, cell bodies surrounded by neuropil (as in most compartments of the vertebrate CNS) emit multiple fibers, and also carry synapses themselves. Experimental data suggest that the seemingly fundamental difference between unipolar invertebrate ganglion cells and their multipolar vertebrate counterparts may not have a profound genetic basis. Thus, by experimentally bringing a (normally unipolar) Drosophila motor neuron into a closer proximity to the neuropil, it switched to a multipolar phenotype (Sánchez-Soriano et al., 2005; see Figure 8.11). Furthermore, invertebrate neurons in culture often switch to a multipolar phenotype, even when retaining many other aspects of their normal phenotype (Sánchez-Soriano et al., 2005; see Figure 8.11).

Figure 8.11 Neuronal Architecture: Multipolar vs. Unipolar Neurons.
(A, B) Schematic representation of multipolar motor neuron (vertebrate) and unipolar motor neuron (Drosophila). (C) Unipolar motor neurons (marked by expression of GFP, green) in wild-type Drosophila. Note cell bodies (arrowhead) emitting single-cell body fiber (double arrowhead) towards neuropil (dashed lines) where multiple dendritic branches (curved arrow) are formed by each neuron. Peripheral axon indicated by straight arrow. (D) Expression of activated cdc42 construct variably displaces motor neuronal cell bodies closer towards neuropil. In these cases, multiple dendrites directly branch off the cell body, turning cell into a multipolar neuron. (E, F) Drosophila neurons in culture variably express a bipolar (E) or multipolar (F) phenotype, rather than their normal unipolar phenotype. Scale bars: 10 µm. Soriano 2005. Reproduced with permission of Elsevier. (*See insert for color representation of the figure*).

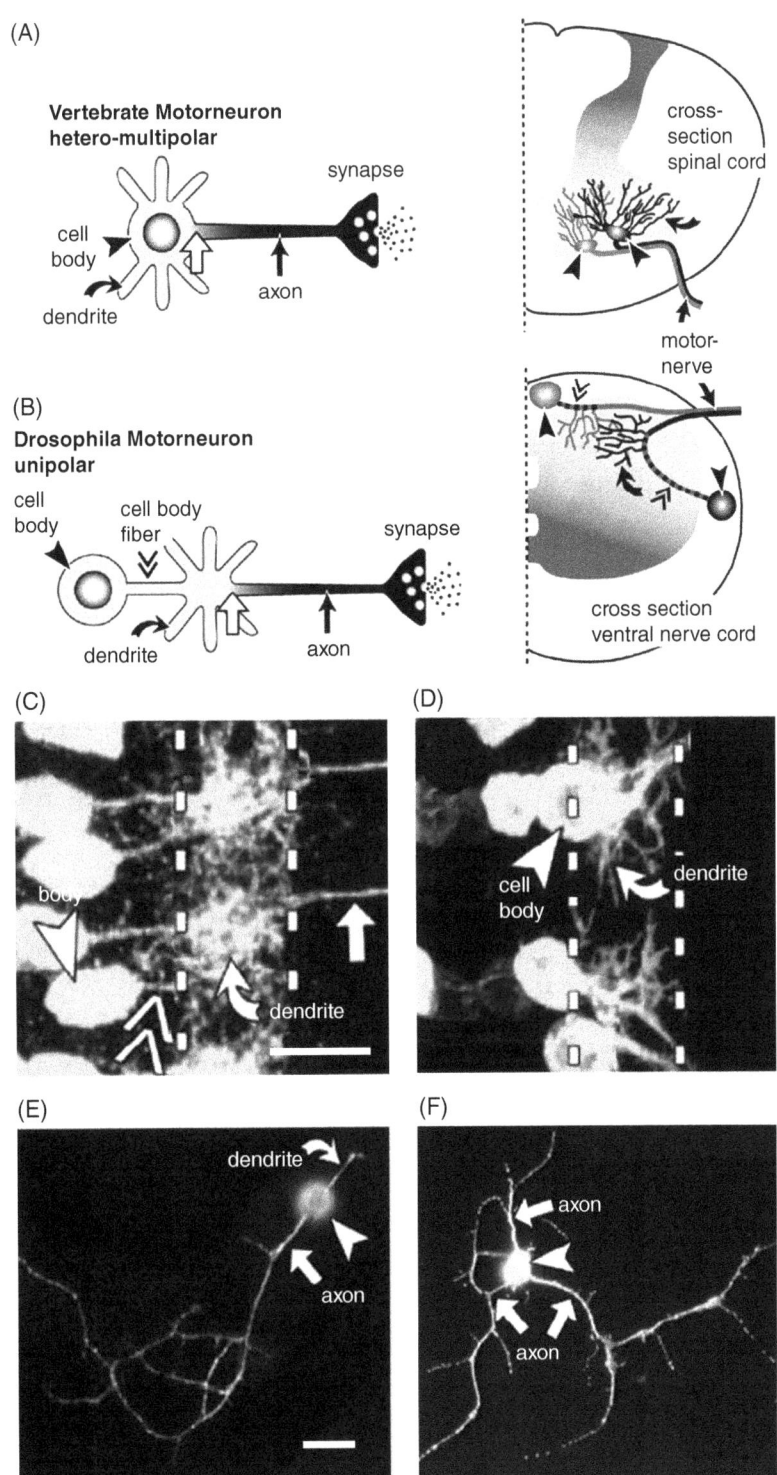

Figure 8.11 (*Continued*)

8.3.2.2 Neurite branching. The size and branching pattern of a neuron is correlated to the overall number of neurons and synaptic contacts made between them. As previously stated, ganglion cells in peripheral nerve nets with a low neuronal density have generally few, unbranched processes. Central neurons of invertebrates possessing a small number of neurons (in the order of 100–1000), such as nematodes, rotifers, or gastrotrichs, typically have one or two unbranched fibers. In the nematode *C. elegans*, where every single neuron and its connectivity has been mapped (a feat achieved for no other animal so far), the large majority of neurons have a single neurite that travels anteriorly or posteriorly in the ring ganglion ("brain") and/or nerve cords (White et al., 1986; see Figure 8.9C). These simple, unbranched neurites provide enough surface area for all synapses formed by the 302 neurons found in the *C. elegans* CNS. Organisms with higher neuron numbers (10,000–1,000,000 in most annelids, molluscs, or arthropods; billions in some vertebrates) require an increase in neuronal surface area, which occurs by branching into secondary and higher order neurites (see examples shown in Figures 8.4, 8.12).

8.3.2.3 Pattern of neurite tree. The geometry of branched central neurons, that is, the pattern in which they distribute neurites within the neuropil, is enormously varied. When referring to this pattern one often speaks of the "neurite tree." Two main types of neurite trees can be distinguished: (1) local neurons, also called amacrine cells ("amacrine" = "no long processes"), which branch more or less evenly within one (small) neuropil domain (see Figure 8.12A, B, C); (2) projection neurons, which interconnect different compartments within the neuropil, and which have two or more neurite trees located in these compartments, and long, unbranched fibers connecting these trees (see Figure 8.12A, D, E). Other examples of projection neurons are the neurons connecting sensory compartments (like the olfactory or visual compartments in arthropods) to higher brain centers (Figures 8.5D, 8.9H), or the motor neurons in most taxa which possess highly branched dendrites in the central nervous system, and send long peripheral axons to muscles or glands.

8.3.2.4 Distribution of synapses. An important—and for the most part unknown—aspect of invertebrate neuron geometry is how input and output synapses are distributed along the neurite tree. In vertebrates, the neurite tree consists of three different domains, the soma, dendrite, and axon (see Figure 8.12A). Dendrites branch off directly from the soma, and, in terms of structure (e.g., cytoskeleton) and function, are similar to the soma (Baas & Yu, 1996; Peters, Palay, & Webster, 1976). Dendrites and soma represent the input domain of the neuron; they carry postsynaptic membrane specializations. By contrast, axons are specialized to conduct axon potentials, and carry presynaptic sites at their branched terminals. As already pointed out, invertebrate neurons differ from this pattern. In cases where combined physiological and anatomical studies have been carried out (and these cases are restricted to the nervous system of insects, crustaceans, molluscs, and annelids), the soma and cell body fiber most often lacks synapses; branches of the neurite tree are either dendritic (i.e., postsynaptic), axonal (i.e., presynaptic), or, in many cases, mixed (both post and presynaptic sites intermingled). It is therefore not easy to use the terms axon or dendrite when referring to neuronal processes in invertebrate neurons. Many local interneurons have neurite trees on which input and output synapses are thoroughly intermingled (see Figure 8.12A, E). On the other hand, many invertebrate projection

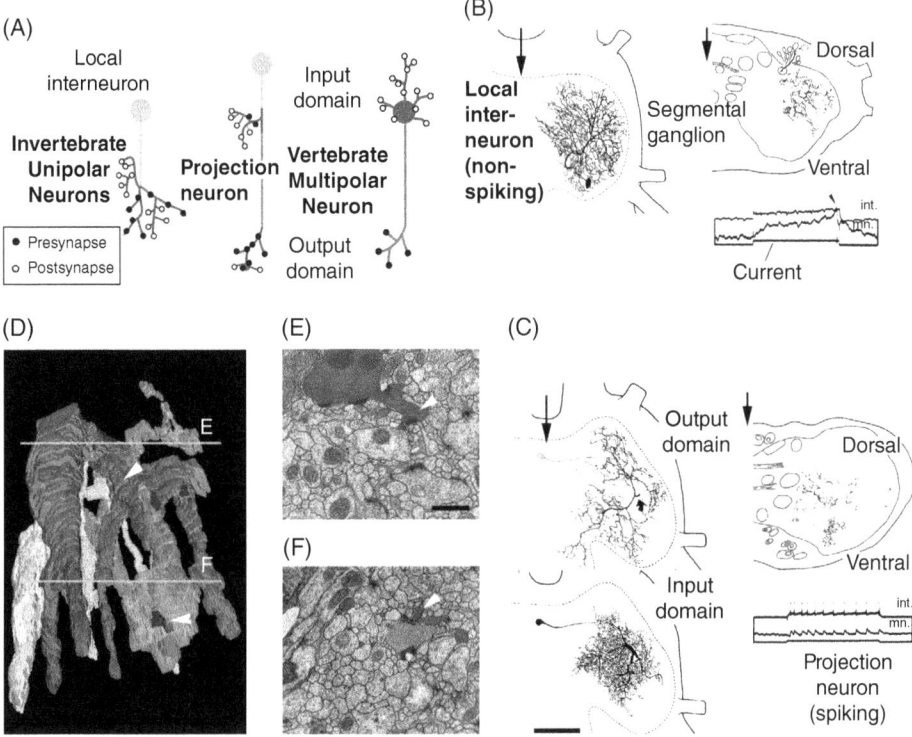

Figure 8.12 Neuronal Architecture: Distribution of Synapses.
(A) Schematic representation of invertebrate local interneuron and projection neuron compared to typical vertebrate neuron. (B) Local non-spiking interneuron in locust segmental ganglion. Left: dorsal view; arrow demarcates midline. Top right: cross-section of hemiganglion, showing distribution of branches of neuron in ventral as well as dorsal domains within neuropil. Bottom right: Physiology of non-spiking neuron. Injection of current (bottom trace) causes depolarization without action potentials in interneuron (int); this in turn leads to slowly increasing depolarization with terminal spike (arrowhead) in postsynaptic motor neuron (mn). Watkins BL 1985. Reproduced with permission of John Wiley and Sons. (C) Spiking interneuron in locust segmental ganglion. Left: dorsal view of ventral (input) domain of neuron (bottom; shaded red) and dorsal (output) domain (top; shaded green); arrow demarcates midline. Top right: cross section of hemiganglion, showing spatial separation of input branches ventrally and output branches dorsally. Bottom right: Physiology of spiking interneuron. Injection of current (bottom trace) causes depolarization and train of action potentials in interneuron (int) and in postsynaptic motor neuron (mn). Siegler MV 1979. Reproduced with permission of John Wiley and Sons. (D) Digitial 3D model of short segments of synaptically connected neurons in Drosophila larval brain rendered in different shades of yellow, blue, and green. Red lines indicates presynaptic sites. Cardona 2010. PLOS Biology. (E, F) cross-section of neuronal fibers shown in (D); level of section indicated by lettered horizontal lines in (D). Note concentraton of presynaptic sites at varicosities (thickenings) of blue fiber (E) and green fiber (F). Varicosity of blue fiber gives off thin branch (white arrowheads); this branch is postsynaptic to green fiber in (F). Scale bars: 200 μm (B, C); 0.5 μm (D–F). (*See insert for color representation of the figure*).

neurons resemble, to some extent, the typical vertebrate neuron. Thus, they often form primarily postsynaptic neurites (dendrites) proximally, close to the soma; a single unbranched axon leads away from the dendrites towards another neuropil compartment, where it ends in multiple terminal branches carrying presynaptic sites. But, even in these cases, output synapses are often found in the predominantly dendritic domain of the neurite tree, and vice versa. This intermingling of input and output has important consequences for the function of neuronal circuits, as will become clearer in §8.4.6 (see below).

8.3.2.5 Spiking vs. non-spiking neurons. The encoding and conduction of signals by a neuron is reflected in several aspects of neuronal geometry. Functionally, invertebrate neurons fall into two classes, spiking neurons and non-spiking neurons (Pearson, 1976; Burrows & Siegler, 1978). Spiking neurons are able to generate an action potential and transmit this potential without decrement along part of their neurite tree, typically along the axon(s) that connect branching sites of the tree. Most projection neurons are spiking; the action potential is typically generated proximally, at the point of origin of the long axon (Burrows & Siegler, 1978; Hoyle & Burrows, 1973), and conducted unidirectionally towards the distal tip of the neurite tree where output synapses are concentrated. Non-spiking neurons do not produce action potentials. They react to stimuli by a graded response that is conducted decrementally along the neurite tree. Many local interneurons are non-spiking; these cells have been estimated to make up 65% of a typical insect ganglion (Watkins, Burrows, & Siegler, 1985; see Figure 8.12, right). Other local interneurons were found to be spiking. One should note that, even though the distinction between local interneurons and projection neurons is generally helpful when describing neuronal connectivity, the use of these terms can be context-dependent. Local interneurons have been studied in great detail in the ventral nerve cord of insects, where they play a crucial role in the formation of central pattern generators and reflex arches (see below). The term "local interneuron" in these cases was chosen because the neurite tree is confined to one segmental ganglion; the term "projection neuron" would be reserved for neurons that interconnect different ganglia. In other words, the neuropil of one ganglion is considered as one single compartment, and a neuron confined to that compartment is deemed "local." In reality, the neuropil of a ganglion can be further subdivided into structural functional subcompartments. For example, the discrete domains within an insect ganglion where different types of sensory afferents (e.g., proprioceptor, touch receptor, chemoreceptor) terminate, or where pools of motor neurons form dendritic branches, represent subcompartments. "Local" neurons that interconnect such subcompartments might be better called "local projection neurons," in particular if input and output synapses are segregated in different parts of their neurite tree. This is the case for the spiking local neurons described for the ventral ganglia of various insects: these neurons possess predominantly dendritic branches, often receiving input from sensory afferents in the ventral neuropil, project a short axon postsynaptic dorsally, and form a second, predominantly axonal tree in the dorsal neuropil that overlaps with dendrites of motor neurons (Watson & Burrows, 1985; see Figure 8.12D, E). By contrast, non-spiking local neurons do not possess a separate dendritic and axonal component, and presumably have mixed input and output synapses distributed over their neurite tree (see Figure 8.12B, C).

8.3.3 Glia

Glia represents the second class of cells of the vertebrate brain, where one distinguishes between two major types, astrocytes and oligodendrocytes. Both are multipolar cells which form sheath-like processes around neuronal cell bodies, neurites and synapses. Astrocytes mainly interact with capillaries, neuronal cell bodies, and synapses, whereas oligodendrocytes form multilayered sheaths (myelin) around axons. In regard to these morphological criteria, glial cells are sparse or non-existent in many "lower" invertebrate phyla, such as coelenterates, flatworms, hemichordates, or tunicates (Hartline, 2011; Radojcic & Pentreath, 1979). Among the invertebrate deuterostomes, only echinoderms were recently reported to possess significant numbers of cells with molecular and (some) structural similarities to vertebrate glia (Mashanov et al., 2010). Thus, many epithelial cells integrated in the nerve cords of echinoderms express markers of glial cells, and show morphological attributes of the radial glial cells of vertebrate embryos. Vertebrate radial glia forms guiding tracks for neuronal precursors and lines channels enclosing early differentiating axon tracts (Campbell & Götz, 2002; Mission, Takahash, & Caviness, 1991; Rakic, 1978); they later differentiate into astrocytes, but also act as neural progenitors and give rise to neural stem cells (Parnavelas & Nadarajah, 2001). It is possible that the radial glia-like cells described for echinoderms—which might also exist in the basiepithelial nerve plexus of other lower deuterostomes, but would be difficult to distinguish morphologically from epidermal/epithelial cells—perform a similar role in the mature nervous system of these animals.

In the protostomes, most of the basal phyla (e.g., lophophorates, most flatworms, most cycloneuralians) lack glia, whereas the highly derived phyla, including nemertines, annelids, molluscs and arthropods, possess a diverse and complex assembly of glial cells. A small number of cells, called cephalic sheath cells and labial sheath cells, has been described in the nematode *C. elegans*; they ensheath the nerve ring and sensory axons entering the nerve ring (Bird, 1971; Chitwood & Chitwood, 1950). For the rest of the nematode nervous system, instead of glial cells, the epidermis and its basement membrane surround neuronal cell bodies and axon tracts.

Glia of the annelid (Baskin, 1971; Kai-Kai & Pentreath, 1981), mollusc (Fernandez, 1966) and arthropod central nervous systems (Edwards & Meinertzhagen, 2010; Freeman, 2015; Hartenstein, 2011) fall into three major classes, defined by location and cell shape (see Figure 8.13). The first class ("surface glia") includes cells that form sheaths around the outer brain surface and peripheral nerves. In insects, where this type of glia has been studied in considerable detail, surface glia are important for establishing a blood–brain barrier, and for the mechanical stabilization of the nervous system. A second type of glia, cortex glia, are confined to the cortex of the ganglia forming the brain and ventral nerve cord. Processes of cortex glia encapsulate neuronal somata. Genetic studies in the fruitfly *Drosophila melanogaster*, demonstrated that, among other functions, cortex glia protect neurons from apoptotic cell death. The third class of glia (neuropil glia) includes different types of cells forming sheaths around the neuropil as a whole, or around smaller assemblies of axons and dendrites. Neuropil glia also contacts synapses, and important functions relating to the re-uptake of neurotransmitter have been attributed to neuropil glia.

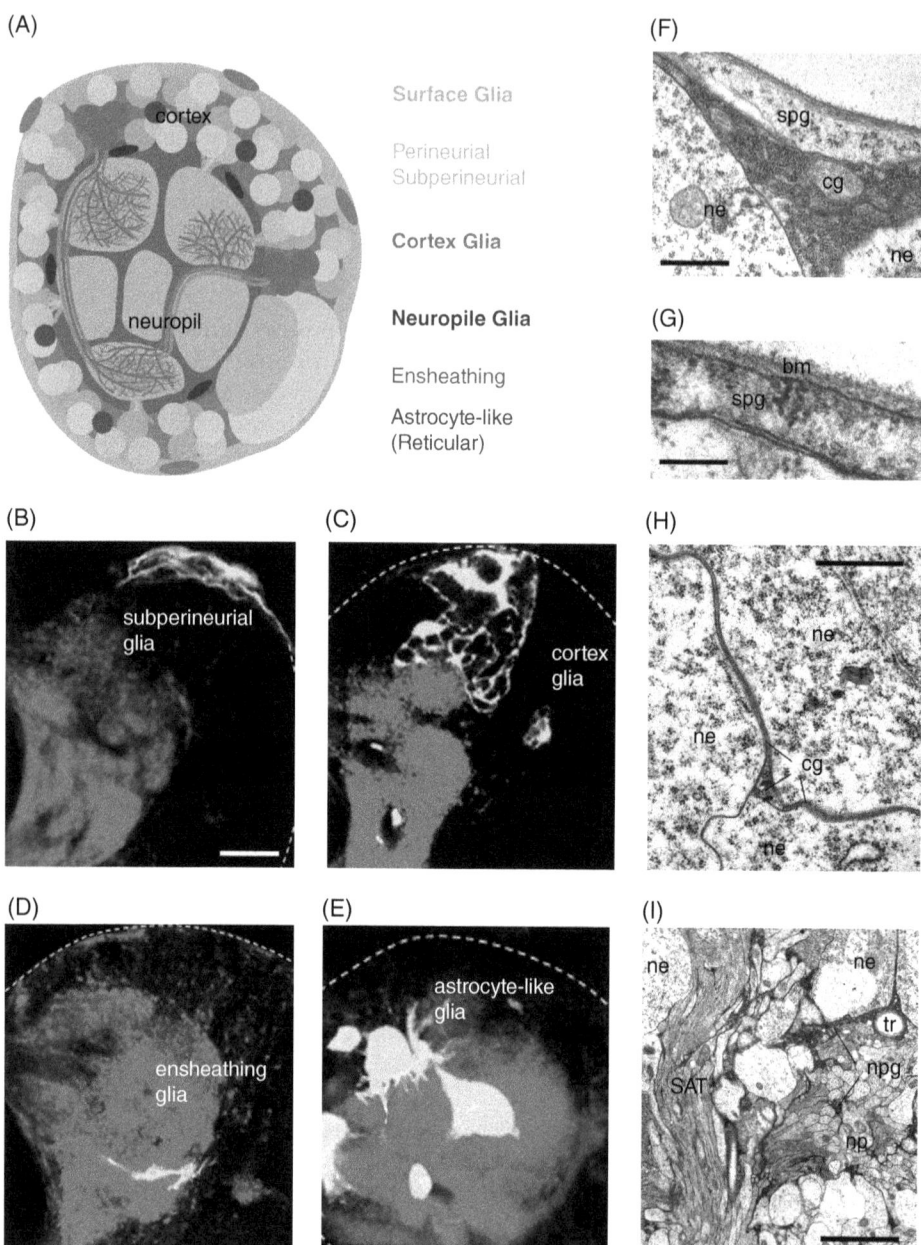

Figure 8.13 Glia in Insect Ganglion.
(A) schematic section of Drosophila brain showing main types of glial cells. (B–E) Z-projections of confocal sections of Drosphila larval brain. Individual glial cells of different types are labeled by expression of a GFP reporter (green); neuropil is labeled by anti-DN-cadherin (red). (F–I) Electron micrographs of parts of cross sections of Drosophila larval brain. (F, G) Brain surface, covered by subperineurial glia (spg) producing basement membrane (bm) and by outer lamella of cortex glia (cg). (H) Cortex, showing thin lamellae of cortex glia (cg) surrounding neuronal cell bodies (ne). (I) Cortex–neuropil boundary, demarcated by neuropil glial sheath (npg). Other abbreviations: np neuropil; SAT secondary axon tract; tr trachea. Scale bars: 40 μm (B–E); 0.5 μm (F); 0.2 μm (G); 1 μm (H); 2 μm (I). (*See insert for color representation of the figure*).

8.4 Neuronal Circuitry and CNS Function: Insights from Invertebrate Nervous Systems

Approximately 50 years ago many neurobiologists interested in analyzing neural circuits turned their attention to invertebrate brains, which are typically smaller and have fewer neurons than vertebrate brains. Most importantly, workers in the field had come to realize that many neurons in various invertebrate systems were unique, that is, could be identified in each specimen, a property that (with very few exceptions) is absent in vertebrate brains. The hope was that, by focusing on carefully chosen parts of invertebrate brains with relatively few, individually recognizable neurons, it might be possible to gain a comprehensive picture of some of the circuits. Significant progress has been made, even though we are far from having a complete physiological/anatomical map of any circuit. Instead, for a good number of behaviors, there exist "partial maps": estimates of how many neurons are involved in a given circuit, and physiological/anatomical characterizations of representative neurons and their synaptic connections forming part of this circuit. The following sections will discuss some well-studied circuits in a number of different invertebrate systems. We will start out by introducing some of the important concepts that structure our understanding of how groups of interconnected neurons control behavior.

8.4.1 Oscillators, Central Pattern Generators, and Command Centers

Many components of invertebrate behavior are stereotyped sequences of movements which are elicited and terminated by defined stimuli. These fixed behaviors are controlled by neuronal circuits called central pattern generators (CPGs). Fixed behaviors can be further broken down into rhythmic contractions of individual muscles or muscle groups, which are controlled by subunits of the CPG called oscillators, or pacemakers (see Figure 8.14). In the simplest case, oscillator and CPG are the same: for example, propulsion of hydromedusae (jellyfish) is effected by rhythmic contractions of all subumbrellar circular muscle fibers, which are controlled by a pacemaker that consists of a small group of ganglion cells with fibers arranged around the margin

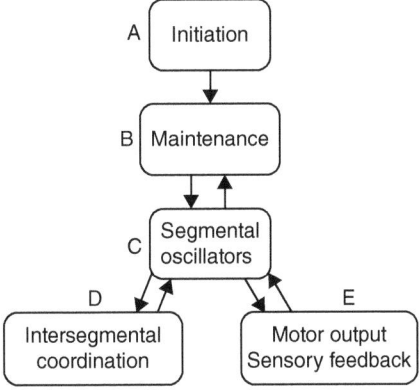

Figure 8.14 Elements of Central Pattern Generator. Mullins 2011. Reproduced with permission of Elsevier.

of the umbrella (see below). In most cases, multiple oscillators, each one responsible for a certain group of muscles, are strung together to form the CPG. Swimming in leech consists of waves of alternating dorsal/ventral contractions of the length musculature of each segment. Each segmental ganglion possesses an oscillator responsible for contracting first the ventral and then the dorsal muscles of the corresponding segment. Oscillators of neighboring ganglia are connected in such a way that they are active in a phase-locked manner, resulting in an output that controls the swimming undulations propelling the animal forward (see below). Here, the CPG consists of the series of segmental oscillators, in addition to the connections responsible for intersegmental coordination. In many insects, more than one oscillator exists per ganglion; for example, the muscles controlling each joint of a leg are driven by a dedicated neuronal oscillator (see below).

The activity of CPGs is affected by input by sensory feedback. Thus, external mechano-receptors (sensitive to touch, movement of air or water) and internal proprio-receptors (sensing stretching of muscles, ligaments, or body wall) feed back onto the CPG and modulate its output. For example, the liquid environment through which a leech moves affects the muscle tone in a dynamic manner; muscle tone is monitored by stretch receptors that influence the intersegmental coordination of the oscillators. In a partially dissected leech preparation that is experimentally "swimming" in air (a low-resistance medium) segmental oscillators are coordinated in such a way that the undulating movements would form a standing wave; in water (high-resistance), contractions of segmental muscles are coordinated to generate a traveling wave (Mullins, Hackett, Buchanan, & Friesen, 2011). In many cases, studied in great detail in insects and crustaceans, the sensory feedback on CPGs is organized in the form of a reflex arch, similar to the canonical spinal reflex in vertebrates: stretch of a given muscle stimulates a proprio-receptor that forms monosynaptic or polysynaptic connections with the motorneuron that contracts that same muscle.

Aside from sensory afferents, CPGs, which are typically organized around the motor neurons directly effecting movement, receive input from higher "command" centers, which are closely associated with sensory neuropils in the brain (see Figure 8.14). Command centers trigger and sustain the activity of CPGs, and modulate CPG output in order to steer movement in an adaptive way towards or away from stimuli.

What types of neurons participate in CPGs and command centers? What types of connections generate their rhythmic impulses? Have nervous systems of different animals solved similar functional problems in similar ways? These and several other questions will be kept in mind in the following when surveying some of the well studied invertebrate circuits.

8.4.2 Swimming Activity in Jellyfish

Swimming in jellyfish is effected by rhythmic contraction of circular epithelio-muscle cells (cnidarians do not possess muscle cells separate from epidermal cells). A central oscillator (pacemaker) controls the swimming rhythm. The location of the pacemaker varies for different cnidarian clades. In Scyphozoa and Cubozoa, pacemaker neurons are associated with specialized sensory complexes, called rhopalia, located at the base of the tentacles (reviewed in Eichinger and Satterlie, 2014). From these rhopalial ganglia, rhythmic electric activity is conducted via a diffuse motor nerve net over the entire umbrella, where it excites, via chemical synapses, the epithelio-muscle cells. The motor nerve net, which appears to use taurin as transmitter, is separate from a second diffuse nerve net, which consists of neurons using the neuropeptide FMRFamide, and which may play a sensory-neuromodulatory role (Eichinger & Satterlie, 2014; see Figure 8.15C).

Organization of the pacemaker system and the conduction of nerve activity is fundamentally different in hydrozoan jellyfish. Here, neurons associated with the control of swimming are concentrated in a nerve ring surrounding the margin of the umbrella, rather than forming a diffuse nerve net covering the entire umbrella. The pacemaker consists of a group of giant bipolar ganglion cells that form part of the inner nerve ring around the margin of the umbrella. Depending on the species, between 2 and 15 of these pacemaker neurons were found (Mackie, 2004; Satterlie, 2002; Satterlie & Spencer, 1983). Axonal diameter of the pacemaker neurons, varying between 4 and 30 µm, far exceeding the size of most axons in cnidaria or higher invertebrate animals. "Gigantism" of axons is a mechanism to increase stimulus conduction velocities, in particular in invertebrate neurons that most often lack the myelin sheath which, in vertebrate neurons, is able to increase conduction velocity. Giant axons are found in neurons that form part of "escape circuits," where certain stimuli evoke a rapid behavioral response that propels the animal away from the stimulus (e.g., the giant fiber system in Drosophila; Koto, Tanouye, Ferrus, Thomas, & Wyman, 1981).

Individual pacemaker neurons in cnidarians produce an endogenous rhythm (Satterlie, 2002). The neuronal resting potential undergoes cyclical changes, and during each phase of depolarization a burst of action potentials is produced. Furthermore, pacemaker neurons are electrically coupled by gap junctions, resulting in a fast spread of electric activity all around the nerve ring. Pacemaker impulses are communicated to the directly adjacent epithelio-muscle cells that border the inner nerve ring via chemical synapses; electric activity then spreads via gap junctions throughout the epithelio-musculature of the umbrella (see Figure 8.15). In addition, radially oriented motorneurons and/or (depending on species) a diffuse nerve net help to spread the electric activity generated by the pacemaker network at the bell margin towards the center of the bell (Mackie, 2004; Mackie & Meech, 2000; Satterlie, 2002; Satterlie, 2008).

The swimming pacemaker of the inner nerve ring receives synaptic input from other, similarly organized neuronal networks located in the outer nerve ring. One of these, the so called "B-system," consists of electrically coupled, spontaneously bursting neurons whose bursting frequency is increased by light-induced input from clusters of photoreceptor cells (ocelli) situated on the tentacles. Through the B-system of neurons, different light levels modulate the bursting rhythm of the pacemaker (Satterlie, 2002; Spencer & Arkett, 1984; see Figure 8.15).

8.4.3 Locomotion in Leeches

Leeches are among the best studied neurophysiological model systems; like other invertebrates lacking limbs (e.g., molluscs, nematodes), they have central nervous systems with small numbers of neurons that are often large and uniquely identifiable. Each ganglion in the leech ventral nerve cord has approximately 400 neurons, and precise maps of these neurons have been generated (Muller, Nicholls, & Stent, 1981; Nicholls & Baylor, 1968; see Figure 8.16A, B). Leeches move by wave-like (undulating) swimming movements, or by crawling (see Figure 8.16C). Both types of

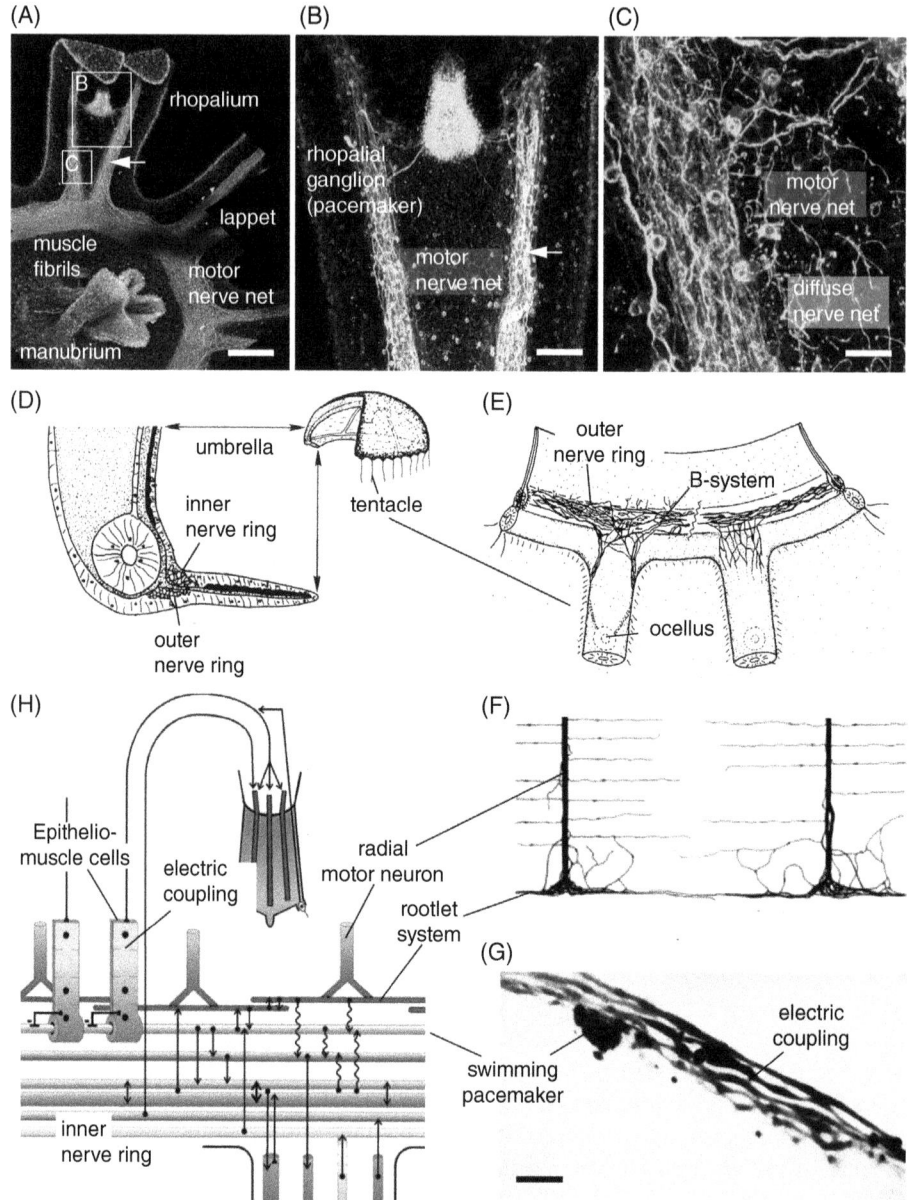

Figure 8.15 Neuronal Circuitry Controlling Swimming in Cnidaria.
(A–C) Neuronal networks in scyphozoan *Aurelia aurita*. Z-projection of horizontal confocal sections of ephyra stage. Myofibrils are labeled by phalloidin in (A) and (B). Neurons forming motor nerve net are labeled by anti-Acetylated tubulin (green) in (A–C); this antibody also binds to cilia at epidermal surface (arrowhead) and manubrium. Note concentration of motor network along radially oriented myofibrils (arrows in A and B), and rhopalia (rhopalial ganglion in B). Neurons forming diffuse nerve net are labeled by anti-FMRFamide (blue) in (C). Nakanishi et al. 2010. (D–F) Line drawings of hydrozoan medusa. (D) Section of umbrella margin, indicating position of inner and outer nerve ring. Satterlie 1983. Reproduced with permission of Springer. (E) detail of umbrella margin, showing outer nerve ring with nerve fibers forming B-system. Spencer AN 1984. Reproduced with permission of J. Exp. Biol.

movement are generated by rhythmic, alternating contraction and relaxation of groups of body muscles. Other movements include food ingestion, as well as reflexive bending in response to touch. The CPGs controlling these behaviors consist of sets of segmentally repeated interneurons and motor neurons, as well as sensory neurons. Motor neurons in each segment comprise 17 pairs of large, unipolar cells (Mullins et al., 2011; Purves & McMahan, 1972; Stuart, 1970; see Figure 8.16D). Dendritic branches extend bilaterally throughout most of each ventral ganglion, and the axon (often in two or more branches) leaves through the segmental nerves to reach the musculature. Sensory input to each ganglion is provided by six large, paired sensory neurons whose somata are located centrally within the ganglion, an exception to the general rule whereby sensory neurons are located peripherally. Sensory dendrites branch widely in partially overlapping fields in the skin (Fett, 1978; Nicholls & Baylor, 1968). Sensory neurons respond to touch (three pairs; T-cells), pressure (two pairs; P-cells) or noxious stimuli (two pairs; N-cells) (see Figure 8.16A). Whereas motor neurons and sensory neurons can be unambiguously identified by retrograde labeling (i.e., injecting a dye into the musculature), interneurons that form part of the CPG are only distinguishable by recording their electric activity, which is temporally linked to the activity of motor neurons and muscles and alters the behavioral rhythm when experimentally stimulated. In this manner, relatively small numbers of interneurons were identified as part of CPGs, for example, 13 interneurons per segment form part of the CPG controlling swimming (Friesen, Poon, & Stent, 1978; Mullins et al., 2011; Weeks, 1982a, 1982b). More recent studies in which neuronal activity was monitored by Ca-sensitive fluorescence (Briggman & Kristan, 2006) demonstrate that the number of neurons involved in swimming and other behaviors is much larger (90 neurons per ganglion in swimming, 188 in crawling; see Figure 8.16E). Interneurons resemble motor neurons in their widespread bilateral branching throughout the ganglion (see Figure 8.16G). Many interneurons project to anteriorly or posteriorly adjacent segments.

The oscillating activity of the CPGs controlling swimming or crawling is generated by recurrent inhibition. Most of the swim interneurons form inhibitory synapses among each other, as do three pairs of motor neurons which are also part of the CPG (Friesen et al., 1978; Mullins et al., 2011; Zheng et al., 2007; see Figure 8.16H). Unlike the swimming pacemaker described for jellyfish, neurons of the leech swimming CPG are not spontaneously active. They react to tonically activating input from higher centers (see below), and, by means of reciprocal inhibition, generate oscillating bursts of action potentials that drive the alternating contractions of dorsal and ventral muscles.

The activity of CPGs is controlled by neurons located in the brain (the supraesophageal and subesophageal ganglion in leech), as well as in the segmental ganglia (see Figure 8.16H, I). One segmentally reiterated neuron, #204, needs to be tonically

---

Figure 8.15 (*Continued*) The Company of Biologists Limited. (F) Radial motor neurons connecting to inner nerve ring. Mackie 2000. Reproduced with permission of J. Exp. Biol. The Company of Biologists Limited. (G) Pacemaker neuron of inner nerve ring of hydrozoan *Aequorea aequorea* injected with fluorescent dye. All pacemaker neurons are labeled because of electric coupling (gap junctions) among these cells. Satterlie RA 1983. Reproduced with permission of Springer. (H) Circuit diagram of neuronal populations forming inner nerve ring of hydrozoa. Arrows indicate synaptic input. Scale bars: 100 µm (A); 50 µm (B, G); 20 µm (C). Mackie 2003. Reproduced with permission of J. Exp. Biol. The Company of Biologists Limited. (*See insert for color representation of the figure*).

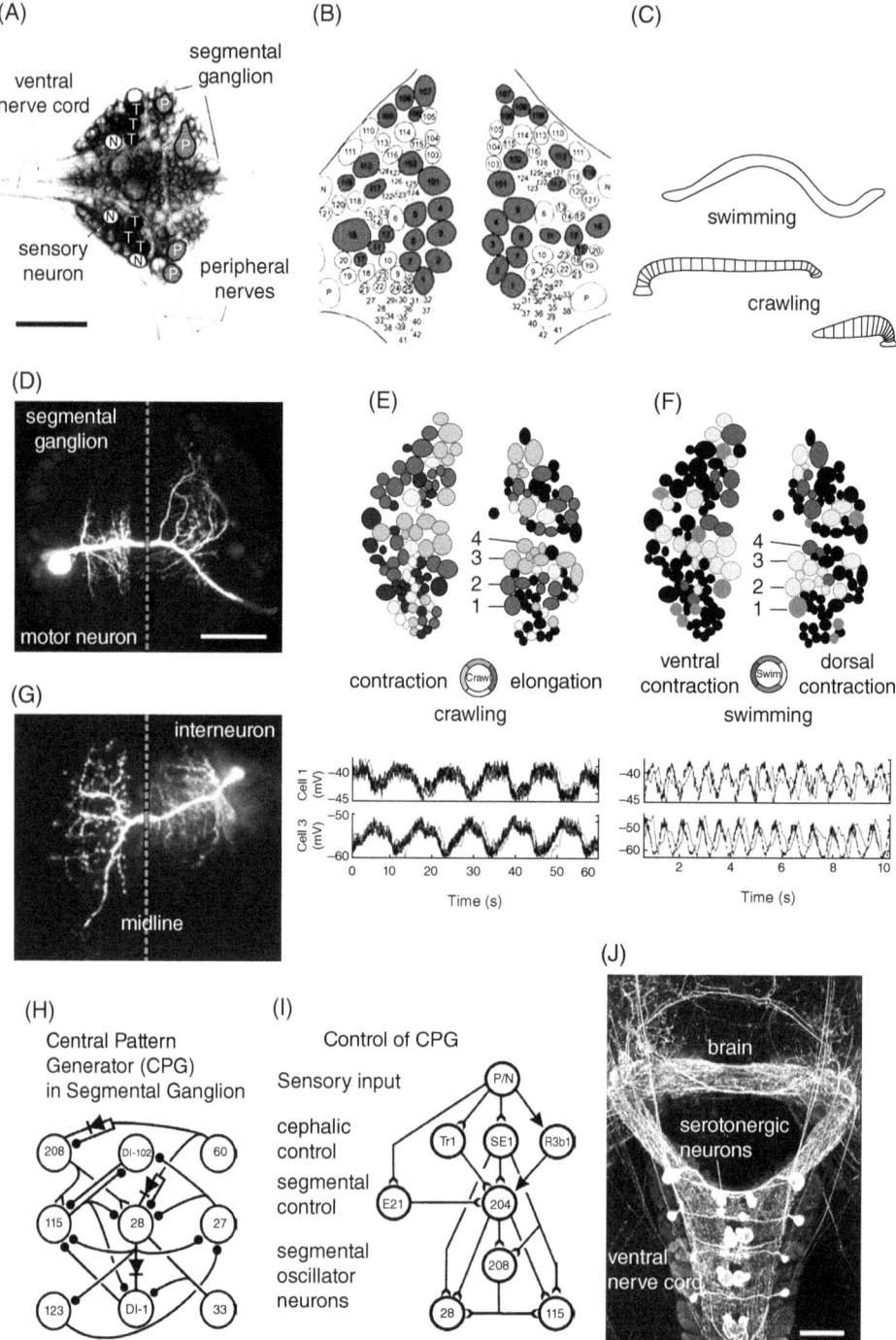

Figure 8.16 Circuitry Controlling Locomotion in Leech.
(A) Microphotograph of segmental ganglion of leech (anterior to the left). Sensory neurons are annotated. Nicholls 1968. Reproduced with permission of J Neurophysiol., The American Physiological Society. (B) Dorsal surface map of leech segmental ganglion (anterior towards the top). (C) Pattern of swimming and crawling in leech. (B, C) Briggman 2006. Reproduced with

active to maintain swimming (Weeks & Kristan, 1978). Similar "activity maintenance" neurons, not yet identified, may exist for the maintenance of crawling. In many ways the activity maintenance cells resemble the reticular formation in the vertebrate brain, which also functions to maintain certain types of behavior (Mullins et al., 2011). Neurons in higher/sensory brain centers are presynaptic to activity maintenance neurons, triggering or inhibiting their bursting activity. Some of these higher neurons, variably called trigger neurons, command neurons, or decision neurons, have been individually identified in leech (Brodfuehrer & Friesen, 1986; Mullins et al., 2011). For example, neuron R3b1 triggers either swimming or crawling, depending on the amount of liquid and the salinity surrounding the animal (Esch, Mesce, & Kristan, 2002). Thus, sensory input reaching the command neurons, as well as the (so far unknown) circuitry among them, selects the activity maintenance system that switches on the appropriate CPG.

Aside from centrally generated, rhythmic nerve cell activity, sensory reflexes play a role in controlling leech locomotion. A well-studied reflex is the bending reflex whereby local stimulation of a P-sensory neuron results in contraction of the muscles close to the stimulus, and relaxation of muscles on the opposite side. P-cells synapse on eight paired and one unpaired interneuron with widespread branching throughout the ganglion, forming contacts with both excitatory and inhibitory motor neurons (Kristan, McGirr, Simpson, 1982; Lockery & Kristan, 1990a, 1990b). Dorsal touch will stimulate excitatory motor neurons for dorsal muscle, as well as inhibitory motor neurons that hyperpolarize (relax) ventral muscle and excitatory motor neurons for ventral muscle. Recent studies showed that aside from this localized "lateral inhibition," P-cell stimulation resulted in an intensity dependent global inhibition of ganglionic interneurons, which adjusted the amplitude (gain) of the response (Baca, Marin-Burgin, Wagenaar, & Kristan, 2008).

It is of great interest to unravel how the neuronal circuits controlling different types of behavior relate to each other. On the one hand, circuits may consist of different, nonoverlapping sets of neurons that are active at different times. In this case we would speak of dedicated circuits. It has become increasingly clear that dedicated circuits are rare; instead, most circuits overlap widely. For example, the above-cited study of Briggman and Kristan (2006) who visualized neuronal activity during swimming and

---

Figure 8.16 (*Continued*) permission of J Neurosci., Society for Neuroscience. (D) Motor neuron labeled by injection of fluorescent dye. Gray-hatched line indicates midline. Mullins 2011. Reproduced with permission of Elsevier. (E, F) Neurons of segmental ganglion active in both crawling (E) and swimming (F). Neuronal activity was monitored in live preparations by Ca-sensor. Neurons active during particular phase of crawling cycle or swimming cycle were color-coded (as indicated in center of panels) and projected on the neuron map (top of panels). Note neurons 1–4 (numbering according to standard map shown in panel B) which are active in both crawling and swimming. Traces at bottom of panels show activity of neurons 1 and 3 during crawling and swimming, respectively. Briggman 2006. Reproduced with permission of J Neurosci., Society for Neuroscience. (G) Segmental interneuron of swimming CPG labeled by injection of fluorescent dye. Hatched grey line indicates midline. Mullins 2011. Reproduced with permission of Elsevier. (H) Central pattern generator in leech segmental ganglion. All neurons shown are interneurons, except DI-1 and DI-102, which are inhibitory motor neurons. (I) Suprasegmental control of CPG. Mullins 2011. Reproduced with permission of Elsevier. (J) Microphotograph of leech brain and anterior segmental ganglia, showing antibody labeled serotonergic neurons. Crisp 2006. Reproduced with permission of J. Exp. Biol. The Company of Biologists Limited. Scale bars: 200 μm (A, D, F); 100 μm (J). (*See insert for color representation of the figure*).

crawling showed that more than 90% of the neurons active during swimming were also active during crawling (see Figure 8.16F). Circuits consisting of neurons that are active in multiple behaviors are characterized as "reorganizing" (or "neuromodulatory multiplexing") circuits. The general mechanism by which the multifunctionality is made possible relies on neuromodulatory input, which changes the phase relationships of CPG neurons or modifies ("gates") the strength of sensory input synapses. For example, during swimming, dorsal and ventral muscles of a given segment always contract in an alternating fashion, whereas they contract simultaneously during the retraction phase of crawling. These different muscular rhythms are reflected in the bursting activity of motor neurons and interneurons. Intracellular recordings have shown directly how the phase relationship between individual motor neurons switches from simultaneous to 180° phase-shifted during the transition from crawling to swimming (Briggman & Kristan, 2008).

Aside from switching the output of multifunctional circuits, neuromodulatory systems of the brain and segmental ganglia affect quantitative aspects of behavior, such the amplitude and speed of swimming movements or the frequency/likelihood with which a given behavior occurs. Neurons producing serotonin, dopamine, and a variety of neuropeptides represent an important component of such neuromodulatory systems. In general, these neurons are large and widely branched and at the same time small in number. The leech serotonergic system includes segmental pairs of giant cells (Retzius cells; cf. Retzius, 1891) and LL cells whose branches pervade the neuropil of the entire brain, subesophageal ganglion and ventral nerve cord (Crisp & Mesce, 2006; see Figure 8.16J). Serotonergic neurons release transmitters both synaptically, like "normal neurons," as well as extrasynaptically (so called volume transmission). Thus, serotonin-containing vesicles are distributed throughout the neuron, including the cell body, and are released upon depolarization of the cell. Serotonin has complex effects on the CPGs and behaviors controlled by them (Brodfuehrer, Debski, O'Gara, & Friesen, 1995). On the one hand, serotonin depolarizes neurons of the swim CPG, which shortens the intervals during which neurons are inhibited, and increases the oscillator frequency. Via feedback from the swim CPG, serotonergic neurons are tonically activated during bouts of swimming. On the other hand, increasing serotonin release in the brain has an inhibitory effect on the swim CPG, reducing the likelihood of swimming, and the length of swim episodes, by decreasing the bursting of the swim activity maintenance system (neuron #204) (Crisp & Mesce, 2006). Serotonin is also the neuromodulator that, during feeding-induced suppression of locomotion, gates the sensory input from the P-cells to its target neurons (Gaudry & Kristan, 2009). The serotonin release during feeding activity acts directly on presynaptic terminals of the P-sensory cells and reduces transmitter release, which silences downstream targets of the P-cells, such as the interneurons evoking the bending reflex discussed previously.

8.4.4 Swimming in Molluscs

Like leeches, molluscs possess central ganglia with small numbers of neurons. Individual neurons are large and uniquely identifiable, which has helped to decipher the circuitry controlling a variety of reflexes and rhythmic behaviors, including swimming (Arshavsky et al., 1998; Newcombe, Sakurai, Lillvis, Gunaratne, & Katz, 2012), feeding (Elliott & Susswein, 2002), and breathing (Haque et al., 2006; Syed & Winslow, 1991). Many marine gastropods swim by undulating movements of their body; others, notably species of the group known as pteropods ("sea butterflies"), have evolved bilateral wing-shaped extensions of their foot which, when moved rhythmically, propel the animal forward. The circuit controlling "wing" movement in several species, among them, *Clione limacina* and *Tritonia diomedea*, has been studied over many years. In these animals, 20–30 motor neurons located in the pedal ganglion (the mollusc counterpart of the annelid ventral nerve cord) form excitatory connections with the muscles responsible for wing movement. Motor neurons include those which elevate the wing, and those which depress it; each cell forms widespread dendritic branches throughout the ipsilateral neuropil of the pedal ganglion (Arshavsky, Beloozerova, Orlovsky, Panchin, & Pavlova, 1985; Arshavsky, Orlovsky, & Panchin, 1985; Getting, 1981, 1983a, 1983b; Getting, Lennard, & Hume, 1980; Satterlie, 1985; see Figure 8.17). Interneurons whose activity correlated with the wing beat

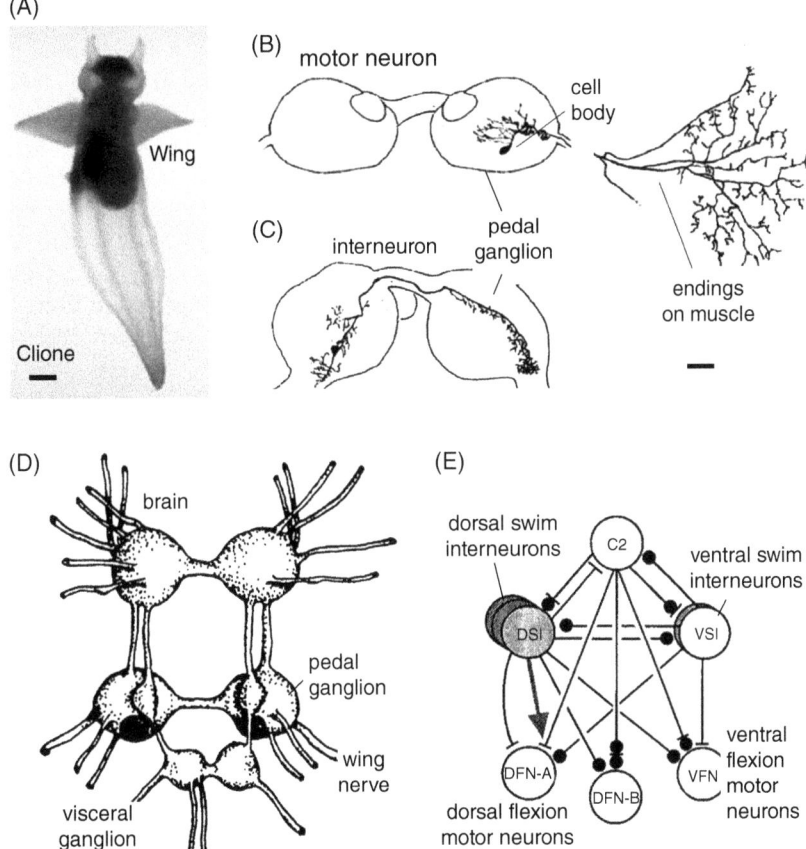

Figure 8.17 Neuronal Circuitry Controlling Swimming in Molluscs.
(A) The nudibranch *Clione limacina*. Levi R 2004. Reproduced with permission of J Neurophysiol., The American Physiological Society. (B) Motor neuron of *Clione* with cell body in pedal ganglion and terminal arborizations in wing muscles. (C) Swim interneuron of *Clione* located in pedal ganglion. (D) Organization of central ganglia of *Clione*. (B–D) Satterlie 1985. Reproduced with permission of J. Exp. Biol. The Company of Biologists Limited. (E) Swim network in *Tritonia diomedea*. Katz PS 1995. Reproduced with permission of J Neurosci. Society for Neuroscience. Scale bars: 5 mm (A); 100 μm (B, C).

were also located in the pedal ganglion, forming dendritic trees on both sides of the ganglion. Interneurons in phase with the upbeat (D-phase interneurons) and with the downbeat (V-phase interneurons) formed two groups of intrinsically bursting cells. On top of this intrinsic activity, the two groups were synaptically interconnected, resulting in a rhythm generator that consisted of two half-centres which exerted strong, short-duration inhibitory and weak long-duration excitatory actions upon each other.

By contrast to the swim CPG in leech in which motor neurons formed an integral part of the rhythm generating circuit, the swim CPG of *Clione* consisted exclusively of interneurons. Thus, following experimental ablation of motor neurons, interneurons of the pedal ganglia continued to produce the swimming rhythm (Arshavsky, Orlovsky, & Panchin, 1985). The same seems to apply for the swim generator in *Tritonia*, where small groups of interneurons located in the brain (cerebral ganglion) generated the oscillation that drives the motor neurons of the pedal ganglion.

As in leech, trigger neurons and modulatory neurons located in the brain (cerebral ganglion) act upstream of the CPG (Satterlie & Norekian, 1995; Panchin, Popova, Deliagina, Orlovsky, & Arshavsky, 1995). Neuromodulation of the swimming circuitry occurs also at two other levels. At the peripheral level, serotonergic neurons directly innervate the musculature and, when stimulated, enhance muscular contractility (Satterlie, 1995). Recent findings demonstrated that neuromodulatory, serotonergic interneurons also form an intrinsic part of the CPG (Katz, 1998). In these neurons, serotonin functions, on the one hand, as a fast-acting transmitter; on the other hand, it tonically increases transmitter release from other interneurons of the CPG.

Molluscs possess an extensive peripheral nerve net which potentially could be involved in the generation and modulation of motor patterns. For example, one might envisage a scenario where "motor neurons" of the central ganglia synapse on neurons of the peripheral net, which in turn activate the musculature. However, at least for the swim circuit of *Clione*, it was shown that the connection between motor neurons and muscles was direct and monosynaptic (Satterlie, 1993).

8.4.5 CPGs of Locomotion in Arthropods

Arthropods move with multi-jointed legs and (in the case of many insects) wings. CPGs controlling walking and flight are considerably more complex than the "simple" oscillators we have encountered so far, because the spatiotemporal precision with which they have to adjust muscle contractions is so great. This is reflected in higher neuron numbers. A typical insect ganglion possesses on the order of 5,000 neurons, compared to the 400 or so neurons contained within a leech or mollusc ganglion. The populations of neurons that are most strongly increased in arthropod ganglia are local premotor interneurons (Burrows, 1996). Motor neurons, on the other hand, are small in number, and can be individually assigned to specific muscle fibers, as in leeches or molluscs discussed above. In view of the high number of interneurons involved, our knowledge of arthropods CPGs and their command systems is rudimentary. Typically, interneurons are identified at the class level, whereby a given class may contain ten or more members (Nagayama & Burrows, 1990; Siegler & Burrows, 1984).

An important consequence of increased neuron numbers in arthropod ganglia is the large neuropil volume. Concomitantly, the neuropil becomes compartmentalized anatomically and functionally (Pflüger, H. J., Bräunig, P., & Hustert, 1988; Strausfeld, 1976; Tyrer & Gregory, 1982). Neurite trees of motor neurons and premotor interneuons occupy the dorsal neuropil. Sensory afferents, dependent on their modality (e.g., stretch, touch, pain, vibration) target the ventral and medial neuropil, and interneurons that mediate reflexes and form part of CPGs obtain a position in between. Groups of interneurons form conspicuous longitudinal and transverse axon bundles (connectives and commissures, respectively) which serve as landmarks to which specific neuropil subcompartments can be related (see Figure 8.18A, B).

Arthropod CPGs have been studied for a number of species and behaviors; well-known examples are walking and flight in locusts. Locusts have two pairs of wings attached to the second (mesothorax) and third (metathorax) thoracic segments. Muscles include (in each of the two segments) five fibers that elevate the wing, and another five that act as wing depressors. In addition to these "power flight muscles," a small muscle acts as "steering muscle." A precise map of motor neurons assigned to these muscles has been assembled (Bentley, 1970; Burrows, 1976; Pflüger, Elson, Binkle, & Schneider, 1986; Tyrer & Altman, 1974). Flight motor neurons have wide, overlapping dendritic trees that occupy a major portion of the dorsal neuropil of one hemi-ganglion (see Figure 8.18A, B). All motor neurons show a spiking, excitatory activity, using glutamate as transmitter at the neuromuscular junction.

Interneurons associated with flight fall into two major groups, local premotor interneurons, which form direct (monosynaptic) connections with motor neurons, and pattern generator interneurons, which are not directly linked to motor neurons and frequently span multiple ganglia (Robertson & Pearson, 1983, 1985). Pattern generator interneurons are functionally defined by their ability to reset the flight rhythm: stimulating a pattern generator interneuron (e.g., neuron #301) during a flight sequence changed the frequency of the wing beat oscillation. Premotor interneurons do not have this ability. It is assumed that only pattern generator interneurons, and not motor neurons or premotor interneurons, form part of the flight CPG. As discussed in previous sections, inhibitory synaptic connections in between CPG interneurons form the basis of their oscillatory output that drives the alternating bursting of premotor neurons/motor neurons ().

The distinction between premotor and pattern generator interneurons can be easily explained when considering how flight in locusts and other insects is controlled. Swift changes in flight direction are caused by shifts of the strength and/or phase of the activation of individual muscles, whereas the frequency of wing beat remains unaltered (Robertson & Pearson, 1983). In other words, two control systems converge on motor neurons: a CPG that provides a stable oscillation determining wing-beat frequency, and a system of afferents from sensory centers that act on specific premotor interneurons and motor neurons, causing short-term changes in motor neuronal output that modify flight direction without affecting wing-beat frequency. This would explain the functional significance of the large number of local premotor interneurons that have been found in ganglia of locusts and other insects.

The analysis of insect ganglia has also helped to direct our attention to a principle that will be of great importance for the understanding of microcircuitry in general: the principle of local neuronal interaction and integration. Three discoveries form the basis of this principle.

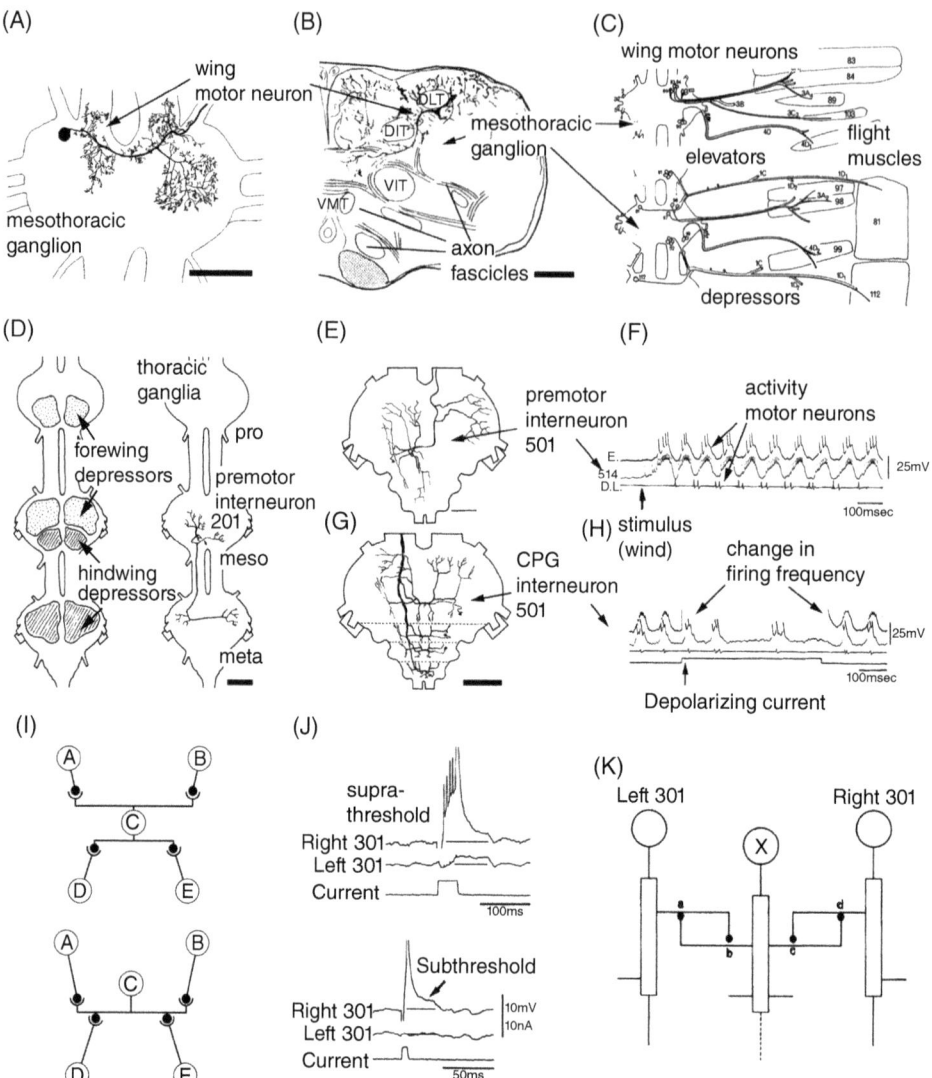

Figure 8.18 Neuronal Circuitry Controlling Flight in Locusts.
(A) Line drawing of mesothoracic motorneuron innervating dorsal longitudinal muscle (indirect depressor of wing; dorsal view). Tyrer 1974. Reproduced with permission of John Wiley and Sons. (B) Cross-section of locust mesothoracic hemiganglion (midline to the left), showing landmark axon fascicles (DIT dorsal intermediate tract; DLT dorsal lateral tract; VIT ventral intermediate tract; VMT ventral medial tract). Branches of motor neuron innervating pleuro-axillary muscle are seen in dorsal neuropil. Pflüger HJ 1986. Reproduced with permission of J. Exp. Biol. The Company of Biologists Limited. (C) Motor neurons innervating flight musculature in locust. Tyrer 1974. Reproduced with permission of John Wiley and Sons. (D) Arborization of premotor interneuron 201(right), whose firing is in phase with motor neurons innervating hindwing depressors, overlaps with neuropil domains innervated by dendrites of these motor neurons (left). (E, F) Premotor interneuron 514, located in metathoracic ganglion (E), is coupled to activity of motor neuron (F). (G, H) Metathoracic CPG interneuron 501 (G) is able to reset firing frequency of motor neurons (D, DL in panel H) when injected with a current. Robertson RM 1983. Reproduced with permission of John Wiley and Sons. (I) Top: Neuron "C" has separate domains for input from "A" and "B," and for output to "D" and "E." Bottom: "C" has mixed input/output domains. (J, K): connectivity of left–right pair of flight interneurons 301. Injection of suprathreshold current in right 301 caused depolarization in itself, as well as contralateral 301 (J, top). Subthreshold current (J, bottom) causes depolarization of itself, but not contralateral 301. (I, J, K) Robertson RM 1988. Reproduced with permission of J Neurosci. Society for Neuroscience. Scale bars: 250 μm (A, D, E, G); 100 μm (B).

1. Much of the conduction of stimuli within the neuropil is mediated by non-spiking local interneurons, which form more than half of the interneuron population in a typical insect ganglion (Siegler & Burrows, 1979);
2. conduction in non-spiking neurons (as well as in large parts of the neurite tree of neurons that do generate action potentials) is attenuated over distance at a rate dependent on neurite diameter: the thicker the neurite, the faster a potential will decrease (Rall, 1981; Siegler, 1984);
3. interneurons have intermingled input and output synapses (Watson & Burrows, 1983).

Taken together, these findings imply a picture of neuronal networks that differs fundamentally from what is usually taken as the basis for modeling circuits. Thus, we are used to the idea that if neurons A and B provide input to C, and C is shown to activate both D and E, then both A and B will affect both D and E. This idea assumes that input and output domains in C are separated, and that its output domain (connecting C to D and E) is equally reached by stimulating A or B (see Figure 8.18I). Instead, assuming that C is a non-spiking interneuron with decremental conduction we often encounter a scenario where C may have input synapses with A in close proximity to output synapses with D, but not E; at the same time, input from B on to C is close to output from C (see Figure 8.18I). Under these conditions, stimulation of A will reach D, but not E, because the postsynaptic potential generated at the synapse A>C will still cause a depolarization at the C>D synapse, resulting in transmitter release and postsynaptic potential in D, but will degrade before it reaches the synapse C>E. In other words: neuron C provides a scaffold of neurites that mediate interactions between neurons A-E, but this scaffold is compartmentalized, such that one compartment provides the link between only a subset of neurons (e.g., A>C>D), but not others.

Actual electrophysiological recordings of interneurons substantiate the significance of local interactions and graded conduction. One of the pattern generator interneurons, a cell called neuron #301, forms indirect disinhibitory interactions with itself, as well as with its contralateral counterpart (Robertson & Reye, 1988). Thus, suprathreshold stimulation of #301 causes a train of action potentials that results in a lasting depolarization of the contralateral #301, and of itself (see Figure 8.18J, top). By contrast, when #301 received a subthreshold current which did not evoke action potentials, the depolarization of the ipsilateral 301 remained in place, but no effect occurred in the contralateral 301 (see Figure 8.18J, bottom). These data support a model where

1. neuron #301 contacts another, inhibitory interneuron ("X" in Figure 8.18K) that provides output synapses onto the ipsilateral #301, as well as its contralateral counterpart (see Figure 8.18K). Stimulating #301 hyperpolarizes (inhibits) X and causes depolarization of both ipsi- and contralateral #301.
2. At subthreshold levels, X conducts activity with decrement; activity reaches the output synapse to ipsilateral #301 and depolarizes this cell, but does not reach the far away synapse between InX and contralateral #301.

8.4.6 Information Processing in the Insect Optic Lobe

Insects and other arthropods have image-forming eyes that are built of many repeated modules, called ommatidia. Each ommatidium contains eight photoreceptor cells (R1–R8; primary sensory cells) which are arranged concentrically, such that photoreceptors R1–R6 (outer photoreceptors) form a circle that surrounds R7 and R8 (inner photoreceptors), located in the center. R1–R6 express a broad spectrum rhodopsin, Rh1, and project their short axons into the distalmost compartment of the brain optic lobe, called lamina (Morante & Desplan, 2004; Wernet & Desplan, 2004; see Figure 8.19A, B). As will be discussed in more detail below, the outer photoreceptors and their target neurons in the lamina form the input channel to a sensory circuit that perceives motion. Photoreceptors R7 and R8 bypass the lamina and project to the next deeper compartment, called medulla. R7 expresses the UV-sensitive rhodopsins Rh3 or Rh4, and R8 expresses either the blue-sensitive Rh5, or the green-sensitive Rh6. Ommatidia with R8 cells that have Rh5 or Rh6 are intermingled; the fly eye, similar to the fovea of the vertebrate eye, forms a mosaic where receptor elements sensitive to different wave lengths are randomly distributed (Wernet & Desplan, 2004; see Figure 8.19C). The inner photoreceptors and their target neurons in the medulla represent the input to a circuit that detects color.

The modular structure of the compound eye is reflected in the neuronal architecture of the lamina and medulla and, to a certain extent, the deeper optic lobe neuropils, called the lobula and lobular plate. Lamina neurons are arranged in stereotypic units, called cartridges, that exactly match the number of ommatidia in the eye. The fly lamina and medulla is one of the few neuropil(s) that at the present moment have been investigated both electron microscopically (Meinertzhagen & O'Neil, 1991; Shaw, 1984; Takemura et al., 2013) and physiologically/genetically (establishing cellular function) to such an extent that we can start to understand neuronal function at the microcircuit level. This will be highlighted in this closing section.

Each lamina cartridge receives input from six photoreceptors, R1–R6, which directly or indirectly contact six different intrinsic lamina interneurons (see Figure 8.19D, E). Five interneurons, L1-L5, are the so called lamina monopolar cells. The single neurite of a monoplar cell penetrates the cartridge vertically and gives off short side branches most of which stay within one cartridge, although some reach the directly adjacent cartridge. The neurite (axon) of the monopolar cells terminate in distinct layers of the next deeper optic neuropil, the medulla (see Figure 8.19E). The sixth lamina interneuron is an amacrine cell (Am) which lacks an axon towards deeper optic layers, but interconnects with its neurite tree multiple neighboring cartridges (see Figure 8.19D). In addition to these six lamina neurons, one cell with dendritic branches in both medulla and lamina (T1) receives R1–R6 input. Finally, at least two medulla interneurons (i.e., neurons with cell bodies and dendritic branches located in the medulla), called C2 and C3, form output synapses in each cartridge (see Figure 8.19D, E).

As a result of the existing synaptic map ("connectome"), as well as the neurotransmitters involved (in so far as known; Hardie, 1989; Kolodziejczyk, Sun, Meinertzhagen, & Nässel, 2008) one can reconstruct the flow of information in the

Figure 8.19 (*Continued*) Reproduced with permission of Springer. (F) Electron micrographs of synapses in Drosophila lamina. Left: synapse of L2 on R6 (feedback) and L4. Right: gap junction between photoreceptors (arrowheads). Bar: 200 nm. Meinertzhagen IA 1991. Reproduced with permission of John Wiley and Sons. (G) Schematic of motion detector formed by pairs of L1 and L2 neurons in neighboring cartridges. Rister J 2007. Reproduced with permission of Elsevier. (H) Schematic of feedback between photoreceptors, amacrine and L2 neuron, mediating gain control circuit. Zheng L 2006. Reproduced with permission of The Rockefeller University Press. (*See insert for color representation of the figure*).

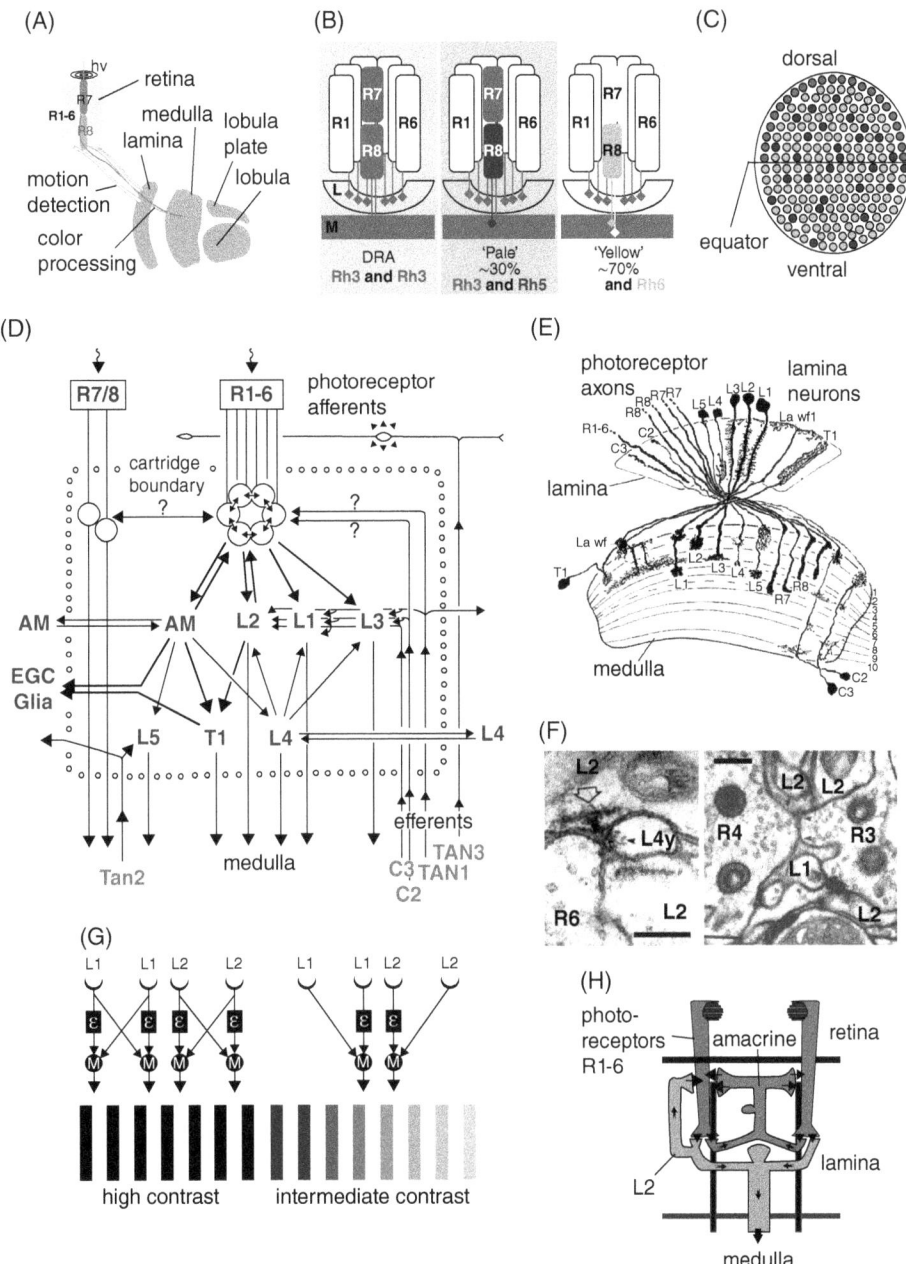

Figure 8.19 Circuitry in the Fly Visual System.
(A) Schematic section of the visual system, showing input from the outer photoreceptors R1–6 (motion and shape detection) and the inner photoreceptors R7/8 (color detection) to the lamina and medulla, respectively. Morante J 2004. Reproduced with permission of Elsevier.
(B) Inner photoreceptors of different ommatidia express different rhodopsins. (C) Mosaic distribution of ommatidia with different rhodopsins. (B, C) Wernet MF 2004. Reproduced with permission of Elsevier. (D) Schematic wiring diagram of dipteran lamina (AM amacrine neuron; C2/3, TAN1-3 medulla neurons projection to lamina; L1-L5 lamina neurons). Shaw 1984. Reproduced with permission of J. Exp. Biol. The Company of Biologists Limited. (E) Line drawings of Golgi-labeled photoreceptors, lamina neurons, and medulla neurons. Fischbach 1989.

lamina and medulla as follows (see Figure 8.19D). R1–R6 upon light-on hyperpolarize L1, L2, L3, and Am. L1–3 forward the signal to the next neuropil, the medulla. Recent evidence suggests that L1/L2 axons from neighboring cartridges converge on certain types of medulla neurons, thereby forming the "arms" of a movement detector network (Rister et al., 2007). L2 and Am also forward the signal to L4 and L5, which in turn join L1–L3 efferents towards the medulla. C2/C3 neurons from the medulla form inhibitory "long" feed back to L1-L3 and photoreceptors. Am and L2 form a "short" feed back onto the photoreceptors (see Figure 8.19D, F); we will discuss below in more detail the functional significance of this peculiar reciprocal connection between sensory input fibers and their target neuron(s), a connection which may actually occur in many sensory neuropils. Finally, a number of cartridge elements, including Am, L4, as well as the epithelial glia form connections to neighboring columns, which may be important to further sharpen the flow of information through a cartridge (lateral inhibition).

A motion detector is a network motif where two or more sequentially activated photoreceptor elements (input channels) converge upon a target element that is tuned (by means of its local interactions) to a specific timing difference between the input channels. The lamina L1/L2 monopolars of neighboring ommatidia (which are sequentially active when confronted with a moving stimulus) represent the input channels of a motion detector whose target element is a medulla neuron. The role of L1 and L2 could be clarified in a set of genetic experiments, which became interpretable in the context of the detailed electron microscopic synaptic map of the lamina sketched out above. It is possible to specifically eliminate L1 or L2, or to separately put back L1 or L2 to fly eyes in which all L neurons are missing (Rister et al., 2007). Furthermore, the ability to detect motion can be monitored by an assay where tethered flies are exposed to moving stripes (dark–light; at different contrasts) which, normally, elicit a defined behavioral output. At a high stimulus contrast, L1 and L2 act redundantly as input channels of the motion detector (see Figure 8.19G, left). At intermediate stimulus contrast (which reflects conditions in the natural environment), L1 and L2 are differentially required for detecting back-to-front motion versus front-to-back motion respectively (see Figure 8.19G, right).

The reciprocal connections between Am/L2 and the photoreceptor input, which form a large fraction of the synapses in the lamina, and most likely in other first order sensory neuropils as well, represent an efficient mechanism of controlling the gain of the R>Am/L2 synapse depending on stimulus intensity (see Figure 8.19H). The model, which was tested using electrophysiological combined with genetic tools (Zheng et al., 2006), predicts that at high light intensity, the hyperpolarization of L2 and Am will be fed back onto the photoreceptor terminals, with the effect that their depolarization decreases, followed by decreased L2/Am hyperpolarization. The result of this microcircuit is a relatively low gain of the R>Am/L2 synapse at high light intensities, and a high gain at low light intensity. Evidence from other systems, in particular the vertebrate retina, suggests similar mechanisms. What is peculiar in the insect lamina, and other insect/invertebrate neuropil compartments, is the fact that reciprocal interactions underlying gain control are spatially compressed to fit into volumes of a few micron cubed; this is made possible because input and output synapses can be placed right next to each other on a neurite branch. Similar circuitry elements in the vertebrate brain requires considerably larger volumes, given that input and output synapses occupy separate compartments of the neuron (i.e., dendrite vs. axon).

References

Achim, K., & Arendt, D. (2014). Structural evolution of cell types by step-wise assembly of cellular modules. *Current Opinion in Genetics and Development*, 27, 102–108.

Aguinaldo, A. M., Turbeville, J. M., Linford, L. S., Rivera, M. C., Garey, J. R., ... Lake, J. A. (1997). Evidence for a clade of nematodes, arthropods and other moulting animals. *Nature*, 29, 489–493.

Alié, A., & Manuel, M. (2010). The backbone of the post-synaptic density originated in a unicellular ancestor of choanoflagellatesw and metazoans. *BMC Evolutionary Biology*, 10, 34. doi: 10.1186/1471-2148-10-34

Anderson P. A. V., & Schwab, W. E. (1981). The organization and structure of nerve and muscle in the jellyfish Cyanea capillata (coelenterata; scyphozoa). *Journal of Morphology*, 170, 383–399.

Arshavsky Y. I., Beloozerova, I. N., Orlovsky, G. N., Panchin, Y. V., & Pavlova, G. A. (1985). Control of locomotion in marine mollusk–Clione limacina. III. On the origin of locomotory rhythm. *Experimental Brain Research*, 58, 273–284.

Arshavsky, Y. I., Orlovsky, G. N., & Panchin, Y. V. (1985). Control of locomotion in marine mollusc--Clione limacina. V. Photoinactivation of efferent neurons. *Experimental Brain Research*, 59, 203–205.

Arshavsky, Y. I., Deliagina, T. G., Orlovsky, G. N., Panchin, Y. V., Popova, L. B., & Sadreyev, R. I. (1998). Analysis of the central pattern generator for swimming in the mollusk Clione. *Annals of the New York Academy of Science*, 860, 51–69.

Baas, P. W., & Yu, W. (1996). A composite model for establishing the microtubule arrays of the neuron. *Molecular Neurobiology*, 12, 145–161.

Babu, K. S. (1985). Patterns of arrangement and connectivity in the central nervous system of arachnids. In F. G. Barth, (Ed.), *Neurobiology of arachnids* (pp. 3–19). Berlin: Springer Verlag.

Baca, S. M., Marin-Burgin, A., Wagenaar, D. A., & Kristan, W. B. Jr. (2008). Widespread inhibition proportional to excitation controls the gain of a leech behavioral circuit. *Neuron*, 57, 276–289.

Bailly, X., Reichert, H., & Hartenstein, V. (2013). The urbilaterian brain revisited: Novel insights into old questions from new flatworm clades. *Development Genes & Evolution*, 223, 149–157.

Barna, J., Csoknya, M., Lázár, Z., Barthó, L., Hámori, J., & Elekes, K. (2001). Distribution and action of some putative neurotransmitters in the stomatogastric nervous system of the earthworm, *Eisenia fetida* (Oligochaeta, Annelida). *Journal of Neurocytology*, 30, 313–325.

Baskin, D. G. (1971). The fine structure of neuroglia in the central nervous system of Nereid polychaetes. *Zeitschrift für Zellforschung und Mikroskopische Anatomie*, 119, 295–308.

Beckers, P., Faller, S., & Loesel, R. (2011). Lophotrochozoan neuroanatomy: An analysis of the brain and nervous system of Lineus viridis(Nemertea) using different staining techniques. *Frontiers in Zoology*, 8, 17. doi: 10,1186/1742-9994-8-17

Beckers, P., Loesel, R., & Bartolomaeus, T. (2013). The nervous systems of basally branching nemertea (Palaeonemertea). *PLoS One* 8, e66137. doi: 10, 1371/journal.pone.0066137

Bennett, M. V. (2000). Seeing is relieving: Electrical synapses between visualized neurons. *Nature Neuroscience*, 3, 7–9.

Bennett, M. V. (2006). Electrical synapses between neurons synchronize gamma oscillations generated during higher level processing in the nervous system. *Electroneurobiología*, 14, 227–250.

Bentley, D. R. (1970). A topological map of the locust flight system motor neurones. *Journal of Insect Physiology*, 16, 905–918.

Bery, A., Cardona, A., Martinez, P., & Hartenstein, V. (2010). Structure of the central nervous system of a juvenile acoel, Symsagittifera roscoffensis. *Development Genes & Evolution*, 220, 61–76.

Bird, A. F. (1971). The structure of nematodes. New York: Academic Press
Biserova, N. M. (2008). Do glial cells exist in the nervous system of parasitic and free-living flatworms? An ultrastructural and immunocytochemical investigation. *Acta Bioligica Hungarica*, 59 Suppl., 209–219.
Borner, J., Rehm, P., Schill, R. O., Ebersberger, I., & Burmester, T. (2014). A transcriptome approach to ecdysozoan phylogeny. *Molecular Phylogenetics & Evolution*, 80, 79–87.
Briggman, K. L., & Kristan, W. B. Jr. (2006). Imaging dedicated and multifunctional neural circuits generating distinct behaviors. *Journal of Neuroscience*, 26, 10925–10933.
Briggman, K. L., & Kristan, W. B. Jr. (2008). Multifunctional pattern-generating circuits. *Annual Review of Neuroscience*, 31, 271–94. doi: 10.1146/annurev.neuro.31.060407.125552.
Brodfuehrer, P. D., & Friesen, W. O. (1986). Initiation of swimming activity by trigger neurons in the leech subesophageal ganglion. I. Output connections of Tr1 and Tr2. *Journal of Comparative Physiology A*, 159, 489–502.
Brodfuehrer, P. D., Debski, E. A., O'Gara, B. A., & Friesen, W. O. (1995). Neuronal control of leech swimming. *Journal of Neurobiology* 27, 403–418.
Bullock, T. H., & Horridge, G. A. (1965). *Structure and function in the nervous systems of invertebrates* (Vol. 1). New York, NY: W. H. Freeman.
Burke, R. D. (1978). The structure of the nervous system of the pluteus larva of Strongylocentrotus purpuratus. *Cell Tissue Research*, 191, 233–247.
Burkhardt, P. (2015). The origin and evolution of synaptic proteins— Choanoflagellates lead the way. *Journal of Experimental Biology*, 218, 506–514.
Burrows, M. (1976). Neural control of flight in the locust. In R. M. Herman, S. Grillner, P. S. G. Stein, & D. G. Stuart (Eds.), *Neural control of locomotion* (pp. 419–438). New York, NY: Plenum Press,.
Burrows, M., & Siegler, M. V. (1978). Graded synaptic transmission between local interneurones and motor neurones in the metathoracic ganglion of the locust. *Journal of Physiology*, 285, 231–255.
Burrows, M. (1996). *The neurobiology of an insect brain*. Oxford: Oxford University Press.
Byrne, M., & Cisternas, P. (2002). Development and distribution of the peptidergic system in larval and adult Patiriella: Comparison of sea star bilateral and radial nervous systems. *Journal of Comparative Neurology*, 451, 101–114.
Cai, X. (2012). Evolutionary genomics reveals the premetazoan origin of opposite gating polarity in animal-type voltage-gated ion channels. *Genomics*, 99, 241–245.
Caicci, F., Burighel, P., & Manni, L. (2007). Hair cells in an ascidian (Tunicata) and their evolution in chordates. *Hearing Research*, 231, 63–72.
Campbell, K., & Götz, M. (2002). Radial glia: Multi-purpose cells for vertebrate brain development. *Trends in Neuroscience*, 25, 235–238.
Cañestro, C., Bassham, S., & Postlethwait, J. (2005). Development of the central nervous system in the larvacean Oikopleura dioica and the evolution of the chordate brain. *Developmental Biology*, 285, 298–315.
Cardona, A., Saalfeld, S., Preibisch, S., Schmid, B., Cheng, A., Pulokas, J., ... Hartenstein, V. (2010). An integrated micro- and macroarchitectural analysis of the Drosophila brain by computer-assisted serial section electron microscopy. *PLoS Biology*, 8(10). doi: 10.1371/journal.pbio.1000502
Cardona, A., Saalfeld, S., Tomancak, P., & Hartenstein, V. (2009). Drosophila brain development: Closing the gap between a macroarchitectural and microarchitectural approach. *Cold Spring Harbor Symposia on Quantitative Biology*, 74, 235–248.
Chase, R. (2000). Structure and function in the cerebral ganglion. *Microscopy Research and Technique*, 49, 511–520.
Chitwood, B.G., & Chitwood, M.B. (1950). *An introduction to nematology*. Baltimore, MD: University Park Press.

Clement, P., & Wurdak, E. (1991). Rotifera. In F. W. Harrison & E. E. Ruppert (Eds.), *Microscopic anatomy of invertebrates* (Vol. 4). *Aschelminthes* (pp. 219–297). New York, NY: Wiley-Liss.

Cobb, J. L., & Pentreath, V. W. (1978). Comparison of the morphology of synapses in invertebrate and vertebrate nervous systems: Analysis of the significance of the anatomical differences and interpretation of the morphological specializations. *Progress in Neurobiology*, 10, 231–252.

Cobb, J. L., & Stubbs, T. R. (1981). The giant neurone system in Ophiuroids. I. The general morphology of the radial nerve cords and circumoral nerve ring. *Cell Tissue Research*, 219, 197–207.

Crisp, K. M., & Mesce, K. A. (2006). Beyond the central pattern generator: amine modulation of decision-making neural pathways descending from the brain of the medicinal leech. *Journal of Experimental Biology*, 209, 1746–1756.

Csoknya, M., Lengvári, I., Benedeczky, I., & Hámori, J. (1992). Immunohistochemical and ultrastructural study of the enteric nervous system of earthworm, Lumbricus terrestris L. *Acta Biologica Hungarica*, 43, 241–251.

Damen, W. G., Hausdorf, M., Seyfarth, E. A., & Tautz, D. (1998). A conserved mode of head segmentation in arthropods revealed by the expression pattern of Hox genes in a spider. *Proceedings of the National Acadamy of Sciences of the USA.*, 95, 10665–10670.

Dufour, H. D., Chettouh, Z., Deyts, C., de Rosa, R., Goridis, C., Joly, J. S., & Brunet, J. F. (2006). Precraniate origin of cranial motoneurons. *Proceedings of the National Academy of Sciences of the USA*, 103, 8727–8732.

Dunn, C. W., Hejnol, A., Matus, D. Q., Pang, K., Browne, W. E., Smith, S. A., ... Giribet, G. (2008). Broad phylogenomic sampling improves resolution of the animal tree of life. *Nature*, 452, 745–479.

Ebner, F. F., & Colonnier, M. (1975). Synaptic patterns in the visual cortex of turtle: An electron microscopic study. *Journal of Comparative Neurology*, 160, 51–79.

Ebner, F. F., & Colonnier, M. (1978). A quantitative study of synaptic patterns in turtle visual cortex. *Journal of Comparative Neurology*, 179, 263–276.

Edwards, T. N., & Meinertzhagen, I. A. (2010). The functional organisation of glia in the adult brain of Drosophila and other insects. *Progress in Neurobiology*, 90(4), 471–497.

Eernisse, D. J., & Reynolds, P. D. (1994). Polyplacophora. In F. W. Harrison & A. Kohn (Eds.), *Microscopic anatomy of invertebrates* (Vol. 5). *Mollusca I* (pp 55–110). New York, NY: Wiley-Liss.

Ehlers, U. (1985). *Das phylogenetische System der Plathelminthes*. Jena: Fischer.

Eichinger, J. M., & Satterlie, R. A. (2014). Organization of the ectodermal nervous structures in medusae: cubomedusae. *Biological Bulletin*, 226, 41–55.

Elekes, K. (2000). Snail nervous system: From classical histology to chemical and molecular neuroanatomy. *Microscopy Research & Technique*, 49, 509–510.

Elliott, C. J., & Susswein, A. J. (2002). Comparative neuroethology of feeding control in molluscs. *Journal of Experimental Biology*, 205, 877–896.

Erwin, D. H., Laflamme, M., Tweedt, S. M., Sperling, E. A., Pisani, D., & Peterson, K. J. (2011). The Cambrian conundrum: Early divergence and later ecological success in the early history of animals. *Science*, 334, 1091–1097.

Esch, T., Mesce, K. A., & Kristan, W. B. (2002). Evidence for sequential decision making in the medicinal leech. *Journal of Neuroscience*, 22, 11045–11054.

Fernandez, J. (1966). Nervous system of the snail Helix aspersa. I. Structure and histochemistry of ganglionic sheath and neuroglia. *Journal of Comparative Neurology*, 127, 157–182.

Fett, M. J. (1978). Quantitative mapping of cutaneous receptive fields in normal and operated leeches, Limnobdella. *Journal of Experimental Biology*, 76, 167–179.

Fischbach, K. F., & Dittrich, A.P.M. (1989). The optic lobe of Drosophila melanogaster. I. A Golgi analysis of wild type structure. *Cell and Tissue Research*, 258, 441-475.

Forest, D. L., & Lindsay, S. M. (2008). Observations of serotonin and FMRFamide-like immunoreactivity in palp sensory structures and the anterior nervous system of spionid polychaetes. *Journal of Morphology*, 269, 544–551.

Freeman, M. R. (2015). *Drosophila* central nervous system glia. *Cold Spring Harbor Perspectives on Biology*. doi: 10.1101/cshperspect.a020552

Friesen, W. O., Poon, M., & Stent, G. S. (1978). Neuronal control of swimming in the medicinal leech. IV. Identification of a network of oscillatory interneurones. *Journal of Experimental Biology*, 75, 25–43.

Gao, Q., Yuan, B., & Chess, A. (2000). Convergent projections of Drosophila olfactory neurons to specific glomeruli in the antennal lobe. *Nature Neuroscience* 3, 780–785.

García-Arrarás, J. E., Rojas-Soto, M., Jiménez, L. B., & Díaz-Miranda, L. (2001). The enteric nervous system of echinoderms: Unexpected complexity revealed by neurochemical analysis. *Journal of Experimental Biology*, 204, 865–873.

Garm, A., Poussart, Y., Parkefelt, L., Ekström, P., & Nilsson, D. E. (2007). The ring nerve of the box jellyfish Tripedalia cystophora. *Cell Tissue Research*, 329, 147–157.

Gaudry, Q., & Kristan, W. B., Jr. (2009). Behavioral choice by presynaptic inhibition of tactile sensory terminals. *Nature Neuroscience*, 12, 1450–1457.

Gerschenfeld, H. M. (1973). Chemical transmission in invertebrate central nervous systems and neuromuscular junctions. *Physiological Reviews*, 53, 1–119.

Getting, P. A. (1981). Mechanisms of pattern generation underlying swimming in Tritonia. I. Neuronal network formed by monosynaptic connections. *Journal of Neurophysiology*, 46, 65–79.

Getting, P. A. (1983a). Mechanisms of pattern generation underlying swimming in Tritonia. II. Network reconstruction. *Journal of Neurophysiology*, 49, 1017–1035.

Getting, P. A. (1983b). Mechanisms of pattern generation underlying swimming in Tritonia. III. Intrinsic and synaptic mechanisms for delayed excitation. *Journal of Neurophysiology*, 49, 1036–1050.

Getting, P. A., Lennard, P. R., & Hume, R. I. (1980). Central pattern generator mediating swimming in Tritonia. I. Identification and synaptic interactions. *Journal of Neurophysiology*, 44, 151–164.

Golding, D. W. 1992. Polychaeta: Nervous system. In F. W. Harrison & S. L. Gardiner (Eds.), *Microscopic anatomy of invertebrates* (Vol. 7, pp. 153–179). New York, NY: Wiley-Liss, Inc.

Goto, T., Katayama-Kumoi, Y., Tohyama, M., & Yoshida, M. (1992). Distribution and development of the serotonin-and RFamide-like immunoreactive neurons in the arrow-worm, Paraspadella gotoi (Chaetognatha). *Cell Tissue Research*, 267, 215–222.

Gray, E. G. (1959). Axosomatic and axodendritic synapses in the cerebral cortex. *Journal of Anatomy*, 93, 420–433.

Grimmelikhuijzen, C. J., & Westfall, J. A. (1995). The nervous systems of cnidarians. *Experientia Supplementum*, 72, 7–24.

Gruhl, A., & Bartolomaeus, T. (2008). Ganglion ultrastructure in phylactolaemate Bryozoa: evidence for a neuroepithelium. *Journal of Morphology*, 269, 594–603.

Hanstroem, B. (1968 [1928]). *Vergleichende Anatomie des Nervensystems der wirbellosen Tiere*. Amsterdam: A Asher.

Haque, Z., Lee, T. K., Inoue, T., Luk, C., Hasan, S. U., Lukowiak, K., & Syed, N. I. (2006). An identified central pattern-generating neuron co-ordinates sensory-motor components of respiratory behavior in Lymnaea. *European Journal of Neuroscience*, 23, 94–104.

Hardie, R. C. (1989). Neurotransmitters in compound eyes. In D. G. Stavenga & R. C. Hardie (Eds.), *Facets of vision* (pp. 235–256). Berlin, Heidelberg: Springer-Verlag.

Hartenstein, V. (1997). Development of the insect stomatogastric nervous system. *Trends in Neuroscience*, 20, 421–427.

Hartenstein, V. (2011). Morphological diversity and development of glia in Drosophila. *Glia*, 59, 1237–1252.

Hartenstein, V., & Stollewerk, A. (2015). The evolution of early neurogenesis. *Developmental Cell*, 32, 390–407.

Hartline, D. K. (2011). The evolutionary origins of glia. *Glia*, 59, 1215–1236.

Harzsch, S., & Müller, C. H. (2007). A new look at the ventral nerve centre of Sagitta: Implications for the phylogenetic position of Chaetognatha (arrow worms) and the evolution of the bilaterian nervous system. *Frontiers in Zoology*, 4, 14. doi: 10.1186/1742-9994-4-14

Hay-Schmidt, A. (1989). The nervous system of the actinotroph larva of Phoronis muelleri (Phoronida). *Zoomorphology*, 108, 333–351.

Hejnol, A., Obst, M., Stamatakis, A., Ott, M., Rouse, G. W., Edgecombe, G. D., ... Dunn, C. W. (2009). Assessing the root of bilaterian animals with scalable phylogenomic methods. *Proceedings of the Royal Society B. Biological Science*, 276, 4261–4270.

Hernandez-Nicaise, M. L. (1973). The nervous system of ctenophores. III. Ultrastructure of synapses. *Journal of Neurocytology*, 2, 249–263.

Herranz, M., Pardos, F., & Boyle, M. J. (2013). Comparative morphology of serotonergic-like immunoreactive elements in the central nervous system of kinorhynchs (Kinorhyncha, Cyclorhagida). *Journal of Morphology*, 274, 258–274.

Heuer, C. M., & Loesel, R. (2008). Immunofluorescence analysis of the internal brain anatomy of Nereis diversicolor (Polychaeta, Annelida). *Cell Tissue Research*, 331, 713–24.

Heuer, C. M., Müller, C. H., Todt, C., & Loesel, R. (2010). Comparative neuroanatomy suggests repeated reduction of neuroarchitectural complexity in Annelida. *Frontiers in Zoology*, 7, 13.

Hochberg, R. (2009). Three-dimensional reconstruction and neural map of the serotonergic brain of Asplanchna brightwellii (Rotifera, Monogononta). *Journal of Morphology*, 270, 430–441.

Hochberg, R., & Litvaitis, M. K. (2003). Ultrastructural and immunocytochemical observations of the nervous system of three macrodasyidan gastrotrichs. *Acta Zoologica*, 84, 171–178.

Holland, N. D. (2003). Early central nervous system evolution: An era of skin brains? *Nature Reviews Neuroscience*, 4, 617–627.

Homberg, U. (2008). Evolution of the central complex in the arthropod brain with respect to the visual system. *Arthropod Structure and Development*, 37, 347–362.

Hoyle, G., & Burrows, M. (1973). Neural mechanisms underlying behavior in the locust Schistocerca gregaria. I. Physiology of identified motorneurons in the metathoracic ganglion. *Journal of Neurobiology*, 4, 3–41.

Jacobs, D. K., Nakanishi, N., Yuan, D., Camara, A., Nichols, S. A., & Hartenstein, V. (2007). Evolution of sensory structures in basal metazoa. *Integrative & Comparative Biology*, 47, 712–723.

Jager, M., Chiori, J. R., Alié, A., Dayraud, C., Quéinnec, E., & Manuel, M. (2011). New insights on ctenophore neural anatomy: Immunofluorescence study in Pleurobrachia pileus (Müller, 1776). *Journal of Experimental Zoology B, Molecular Development & Evolution*, 316, 171–187.

Jékely, G. (2011). Origin and early evolution of neural circuits for the control of ciliary locomotion. *Proceedings of the Royal Society, B. Biological Sciences*, 278, 914–922.

Joffe, B. I., & Wikgren, M. (1995). Immunocytochemical distribution of 5-HT (serotonin) in the nervous system of the gastrotrich Turbanella cornuta. *Acta Zoologica*, 76, 7–9.

Kai-Kai, M. A., & Pentreath, V. W. (1981). The structure, distribution, and quantitative relationships of the glia in the abdominal ganglia of the horse leech, Haemopis sanguisuga. *Journal of Comparative Neurology*, 202, 193–210.

Katz, P. S., & Frost, W. N. (1995). Intrinsic neuromodulation in the Tritonia swim CPG: The serotonergic dorsal swim interneurons act presynaptically to enhance transmitter release from interneuron C2. *Journal of Neuroscience*, 15, 6035–6045.

Katz, P. S. (1998). Neuromodulation intrinsic to the central pattern generator for escape swimming in Tritonia. *Annals of the New York Academy of Science*, 860, 181-188.

Kaul, S., & Stach, T. (2010). Ontogeny of the collar cord: neurulation in the hemichordate Saccoglossus kowalevskii. *Journal of Morphology*, 271, 1240–1259.

Knight-Jones, E. W. (1952). On the nervous system of Saccoglossus cambrensis (Enteropneusta). *Philosophical Transactions of the Royal Society of London B*, 236, 315–354.

Koizumi, O. (2007). Nerve ring of the hypostome in hydra: Is it an origin of the central nervous system of bilaterian animals? *Brain Behavior & Evolution*, 69, 151–159.

Kolodziejczyk, A. M., Sun, X., Meinertzhagen, I. A., & Nässel, D. R. (2008). Glutamate, GABA and acetylcholine signaling components in the lamina of the Drosophila visual system. *PLoS One*, 3, e2110. doi: 10.1371/journal.pone.0002110

Koopowitz, H., & Chien, P. (1974). Ultrastructure of the nerve plexus in flatworms. I. Peripheral organization. *Cell Tissue Research*, 155, 337–351.

Kotikova, E. A., & Raikova, O. L. (2008). Architectonics of the central nervous system of Acoela, Platyhelminthes, and Rotifera. *Journal of Evolutionary Biochemistry and Physiology*, 4(1), 95–108.

Koto, M., Tanouye, M. A., Ferrus, A., Thomas, J. B., & Wyman, R. J. (1981). The morphology of the cervical giant fiber neuron of Drosophila. *Brain Research*, 221, 213–217.

Koyama, H., & Kusunoki, T. (1993). Organization of the cerebral ganglion of the colonial ascidian Polyandrocarpa misakiensis. *Journal of Comparative Neurology*, 338, 549–559.

Krieger, J., Sombke, A., Seefluth, F., Kenning, M., Hansson, B. S., & Harzsch, S. (2012). Comparative brain architecture of the European shore crab *Carcinus maenas* (Brachyura) and the common hermit crab *Pagurus bernhardus* (Anomura) with notes on other marine hermit crabs. *Cell Tissue Research*, 348, 47–69.

Kristan, W. B., Jr, McGirr, S. J., & Simpson, G. V. (1982). Behavioral and mechanosensory neurone responses to skin stimulation in leeches. *Journal of Experimental Biology*, 96, 143–160

Kristensen, R. M., & Higgins, R. P. (1991). Kinorhyncha. In F. W. Harrison & E. E. Ruppert (Eds.), *Microscopic anatomy of invertebrates* (Vol. 4). *Aschelminthes* (pp. 377–404). New York, NY: Wiley-Liss.

Lacalli, T. C. (1990). Structure and organization of the nervous system in the actinotroch larva of Phoronis vancouverensis. *Philosophical Transactions of the Royal Society of London B*, 327, 655–685.

Lacalli, T. C. (1996). Frontal eye circuitry, rostral sensory pathways and brain organization in amphioxus larvae: Evidence from 3D reconstructions. *Philosophical Transactions of the Royal Society of London B*, 351, 243–263.

Lacalli, T. C. (2001). New perspectives on the evolution of protochordate sensory and locomotory systems, and the origin of brains and heads. *Philosophical Transactions of the Royal Society of London B*, 356, 1565–1572.

Lacalli, T. C. (2004). Sensory systems in amphioxus: A window on the ancestral chordate condition. *Brain Behavior & Evolution*, 64, 148–162.

Lacalli, T. C. (2006). Prospective protochordate homologs of vertebrate midbrain and MHB, with some thoughts on MHB origins. *International Journal of Biological Science*, 2, 104–109.

Lacalli, T. C., Holland, N. D., & West, J. E. (1994). Landmarks in the anterior central nervous system of amphioxus larvae. *Philosophical Transactions of the Royal Society of London B*, 344, 165–185.

Levi, R., Varona, P., Arshavsky, Y. I., Rabinovich, M. I., & Selverston, A. I. (2004). Dual sensory-motor function for a molluskan statocyst network. *Journal of Neurophysiology*, 91, 336–345.

Leys, S. P., & Degnan, B. M. (2001). Cytological basis of photoresponsive behavior in a sponge larva. *Biological Bulletin*, 201, 323–338.

Liebeskind, B. J., Hillis, D. M., & Zakon, H. H. (2011). Evolution of sodium channels predates the origin of nervous systems in animals. *Proceedings of the National Academy of Sciences of the USA*, 108, 9154–9159.

Lockery, S. R., & Kristan, W. B., Jr. (1990a). Distributed processing of sensory information in the leech. I. Input-output relations of the local bending reflex. *Journal of Neuroscience*, 10, 1811–1815.

Lockery, S. R., & Kristan, W. B., Jr. (1990b). Distributed processing of sensory information in the leech. II. Identification of interneurons contributing to the local bending reflex. *Journal of Neuroscience*, 10, 1816–1829.

Loesel, R., Nässel, D. R., & Strausfeld, N. J. (2002). Common design in a unique midline neuropil in the brains of arthropods. *Arthropod Structure & Development*, 31, 77–91.

Lowe, C. J., Wu, M., Salic, A., Evans, L., Lander, E., Stange-Thomann, N., ... Kirschner, M. (2003). Anteroposterior patterning in hemichordates and the origins of the chordate nervous system. *Cell*, 113, 853–865.

Mackie, G. O. (1970). Neuroid conduction and the evolution of conducting tissues. *The Quarterly Review of Biology*, 45, 319–332.

Mackie, G. O. (2004). Central neural circuitry in the jellyfish Aglantha: A model "simple nervous system." *Neurosignals*, 13, 5–19.

Mackie, G. O., Marx, R. M., & Meech, R. W. (2003). Central circuitry in the jellyfish Aglantha digitale IV. Pathways coordinating feeding behaviour. *Journal of Experimental Biology*, 206, 2487–2505.

Mackie, G. O., & Meech, R. W. (2000). Central circuitry in the jellyfish Aglantha digitale. III. The rootlet and pacemaker systems. *Journal of Experimental Biology*, 203, 1797–1807.

Manni, L., Caicci, F., Gasparini, F., Zaniolo, G., & Burighel, P. (2004). Hair cells in ascidians and the evolution of lateral line placodes. *Evolution & Development*, 6, 379–381.

Manni, L., Lane, N. J., Sorrentino, M., Zaniolo, G., & Burighel, P. (1999). Mechanism of neurogenesis during the embryonic development of a tunicate. *Journal of Comparative Neurology*, 412, 527–541.

Märkel, K., & Röser, U. (1991). Ultrastructure and organization of the epineural canal and the nerve cord in sea urchins. *Zoomorphology*, 110, 267–279.

Mashanov, V. S., Zueva, O. R., Heinzeller, T., Aschauer, B., & Dolmatov, I. Y. (2007). Developmental origin of the adult nervous system in a holothurian: an attempt to unravel the enigma of neurogenesis in echinoderms. *Evolution & Development*, 9, 244–256.

Mashanov, V. S., Zueva, O. R., Garcia-Arraras, J. E. (2010). Organization of glial cells in the adult sea cucumber central nervous system. *Glia*, 58, 1581–1593.

Mayer, G., Whitington, P. M., Sunnucks, P., & Pflüger, H. J. (2010). A revision of brain composition in Onychophora (velvet worms) suggests that the tritocerebrum evolved in arthropods. *BMC Evolutionary Biology*, 10, 255. doi: 10.1186/1471-2148-10-255

Mayer, G., Kauschke, S., Rüdiger, J., & Stevenson, P. A. (2013). Neural markers reveal a one-segmented head in tardigrades (water bears). *PLoS One*, 8:e59090. doi: 10.1371/journal.pone.0059090

Meinertzhagen, I. A., & O'Neil, S. D. (1991). Synaptic organization of columnar elements in the lamina of the wild type in Drosophila melanogaster. *Journal of Comparative Neurology*, 305, 232–263.

Mill, P. J. (1978). Sense organs and sensory pathways. In P. J. Mill (Ed.), *Physiology of annelids* (pp. 63–113). Academic Press, London.

Meyer, N. P., & Seaver, E. C. (2009). Neurogenesis in an annelid: Characterization of brain neural precursors in the polychaete Capitella sp. I. *Developmental Biology*, 335, 237–252.

Mission, J. P., Takahashi, T., & Caviness, V. S., Jr. (1991). Ontogeny of radial and other astroglial cells in murine cerebral cortex. *Glia*, 4, 138–148.

Morante, J., & Desplan, C. (2004). Building a projection map for photoreceptor neurons in the Drosophila optic lobes. *Seminars in Cell & Developmental Biology*, 15, 137–143.

Moroz, L. L. (2009). On the independent origins of complex brains and neurons. *Brain Behavior & Evolution*, 74, 177–190.

Moroz, L. L. (2014). The genealogy of genealogy of neurons. *Communicative & Integrative Biology*, 7, 6, e993269. doi: 10.4161/19420889.2014.993269

Moroz, L. L. (2015). Convergent evolution of neural systems in ctenophores. *Journal of Experimental Biology*, 218, 598–611.

Moroz, L. L., Kocot, K. M., Citarella, M. R., Dosung, S., Norekian, T. P., Povolotskaya, I. S., ... Kohn, A. B. (2014). The ctenophore genome and the evolutionary origins of neural systems. *Nature*, 510, 109–114.

Moroz, L. L., Nezlin, L., Elofsson, R., Sakhorarov, D. (1994). Serotonin and FMRFamide-immuno-reactive nerve elements in the chiton Lepidopleurus asellus (Mollusca, Polyplacophora). *Cell Tissue Research*, 275, 277–282.

Morris, J., Cardona, A., De Miguel-Bonet, Mdel M., & Hartenstein, V. (2007). Neurobiologyogy of the basal platyhelminth Macrostomum lignano: map and digital 3D model of the juvenile brain neuropile. *Development Genes & Evolution*, 217, 569–584.

Muller, K. J., Nicholls, J. G., & Stent, G. S. (1981). *Neurobiology of the leech*. New York, NY: Cold Spring Harbor.

Mullins, O. J., Hackett, J. T., Buchanan, J. T., & Friesen, W. O. (2011). Neuronal control of swimming behavior: comparison of vertebrate and invertebrate model systems. *Progress in Neurobiology*, 93, 244–269.

Nagayama, T., & Burrows, M. (1990). Input and output connections of an anteromedial group of spiking local interneurons in the metathoracic ganglion of the locust. *Journal of Neuroscience*, 10, 785–794.

Nakanishi, N., Hartenstein, V., & Jacobs, D. K. (2009). Development of the rhopalial nervous system in Aurelia sp.1 (Cnidaria, Scyphozoa). *Development Genes & Evolution*, 219, 301–317.

Nakano, H., Murabe, N., Amemiya, S., & Nakajima, Y. (2006). Nervous system development of the sea cucumber, *Stichopus japonicus*. *Developmental Biology*, 292, 205–212.

Neuhaus, B., & Higgins, R. P. (2002). Ultrastructure, biology, and phylogenetic relationships of kinorhyncha. *Integrative & Comparative Biology*, 42, 619–632.

Newcomb, J. M, Sakurai, A., Lillvis, J. L., Gunaratne, C. A., & Katz, P. S. (2012). Homology and homoplasy of swimming behaviors and neural circuits in the Nudipleura (Mollusca, Gastropoda, Opisthobranchia). *Proceedings of the National Academy of Sciences of the USA*, 109, Suppl. 1, 10669–10676.

Nicholls, J. G., & Baylor, D. A. (1968). Specific modalities and receptive fields of sensory neurons in CNS of the leech. *Journal of Neurophysiology*, 31, 740–756.

Nicol, D., & Meinertzhagen, I. A. (1991). Cell counts and maps in the larval central nervous system of the ascidian Ciona intestinalis (L.). *Journal of Comparative Neurology*, 309, 415–429.

Nomaksteinsky, M., Röttinger, E., Dufour, H. D., Chettouh, Z., Lowe, C. J., Martindale, M. Q., & Brunet, J. F. (2009). Centralization of the deuterostome nervous system predates chordates. *Current Biology*, 19, 1264–1269.

Panchin, Y. V., Popova, L. B., Deliagina, T. G., Orlovsky, G. N., & Arshavsky, Y. I. (1995). Control of locomotion in marine mollusk Clione limacina. VIII. Cerebropedal neurons. *Journal of Neurophysiology*, 73, 1912–1923.

Pani, A. M., Mullarkey, E. E., Aronowicz, J., Assimacopoulos, S., Grove, E. A., & Lowe, C. J. (2012). Ancient deuterostome origins of vertebrate brain signalling centres. *Nature*, 483, 289–94.

Parnavelas, J. G., & Nadarajah, B. (2001). Radial glial cells. are they really glia? *Neuron* 31, 881–884.

Pearson, K. G. (1976). Nerve cells without action potentials. In J. C. Fentress (Ed.), *Simpler networks and behavior* (pp. 99–110). Sunderland, MA.: Sinauer Associates.

Pereanu, W., Kumar, A., Jennett, A., Reichert, H., & Hartenstein, V. (2010). Development-based compartmentalization of the Drosophila central brain. *Journal of Comparative Neurology*, 518, 2996–3023.

Persson, D. K., Halberg, K. A., Jørgensen, A., Møbjerg, N., Kristensen, R. M. (2012). *Journal of Morphology*, 273, 1227–1245.

Peters, A., Palay, S., & Webster, H. E. (1976). *The fine structure of the nervous system.* Philadelphia, PA: W. B. Saunders.

Pflüger, H. J., Bräunig, P., & Hustert, R. (1988). The organization of mechanosensory neuropiles in locust thoracic ganglia. *Philosophical Transactions of the Royal Society of London B*, 321, 1–26.

Pflüger, H. J., Elson, R., Binkle, U., & Schneider, H. (1986). The central nervous organization of the motor neurons to a steering muscle in locusts. *Journal of Experimental Biology*, 120, 403-420.

Purves, D., & McMahan, U. J. (1972). The distribution of synapses on a physiologically identified motor neuron in the central nervous system of the leech. An electron microscope study after the injection of the fluorescent dye procion yellow. *Journal of Cell Biology*, 55, 205–220.

Radojcic, T., & Pentreath, V. W. (1979). Invertebrate glia. *Progress in Neurobiology*, 12, 115–179.

Rakic, P. (1978). Neuronal migration and contact guidance in the primate telencephalon. *Postgraduate Medical Journal*, 54 Suppl. 1, 25–40.

Rall, W. (1981). Functional aspects of neuronal geometry. In A. Roberts & B. M. H. Bush (Eds.), *Neurones without Impulses* (pp. 223–254). Cambridge: Cambridge University Press.

Ramachandra, N. B., Gates, R., Ladurner, P., Jacobs, D., & Hartenstein, V. (2002). Neurogenesis in the primitive bilaterian Neochildia: Normal development and isolation of genes controlling neural fate. *Development Genes & Evolution*, 212, 55–69.

Rehkämper, G., Storch, V., Alberti, G., & Welsch, U. (1989). On the fine structure of the nervous system of Tubiluchus philippinensis (Tubiluchidae, Priapulida). *Acta Zoologica*, 70, 111–120.

Rehkämper, G., & Welsch, U. (1985). On the fine structure of the cerebral ganglion of Sagitta (Chaetognatha). *Zoomorphology*, 105, 83–89.

Renard, E., Vacelet, J., Gazave, E., Lapébie, P., Borchiellini, C., & Ereskovsky, A. V. (2009). Origin of the neuro-sensory system: New and expected insights from sponges. *Integrative Zoology*, 4, 294–308.

Retzius, G. (1891). *Biologische Untersuchungen, Neue Folge II.* Stockholm: Sampson & Wallin.

Reuter, M., & Gustafsson, M. K. (1995). The flatworm nervous system: Pattern and phylogeny. *Experientia Supplementum*, 72, 25–59.

Reuter, M., Raikova, O. I., & Gustafsson, M. K. S. (1998). An endocrine brain? The pattern of FMRF-amide immunoreactivity in Acoela (Platyhelminthes). *Tissue Cell*, 30, 57–63.

Rieger, R. M., Tyler, S., Smith, J. P. S., III, & Rieger, G. E. (1991). Platyhelminthes: Turbellaria. In F. W. Harrison & B. J. Bogitsh (Eds.), *Microscopic anatomy of invertebrates* (Vol.3, pp. 7–140). New York, NY: Wiley-Liss.

Rister, J., Pauls, D., Schnell, B., Ting, C. Y., Lee, C. H., Sinakevitch, I., ... Heisenberg, M. (2007). Dissection of the peripheral motion channel in the visual system of Drosophila melanogaster. *Neuron*, 56, 155–170.

Robertson, R. M., & Pearson, K. G. (1983). Interneurons in the flight system of the locust: Distribution, connections, and resetting properties. *Journal of Comparative Neurology*, 215, 33–50.

Robertson, R. M., & Pearson K. G. (1985). Neural circuits in the flight system of the locust. *Journal of Neurophysiology*, 53, 110–128.

Robertson, R. M., & Reye, D. N. (1988). A local circuit interaction in the flight system of the locust. *Journal of Neuroscience*, 8, 3929–3936.

Rota-Stabelli, O., Kayal, E., Gleeson, D., Daub, J., Boore, J. L., Telford, M. J., ... Lavrov, D. V. (2010). Ecdysozoan mitogenomics: Evidence for a common origin of the legged invertebrates, the Panarthropoda. *Genome Biology & Evolution*, 2, 425–440.

Rothe, B. H., & Schmidt-Rhaesa, A. (2010). Structure of the nervous system in Tubiluchus troglodytes (Priapulida). *Invertebrate Biology*, 129, 39–58

Ruiz-Trillo, I., Riutort, M., Littlewood, D. T., Herniou, E. A., & Baguña, J. (1999). Acoel flatworms: Earliest extant bilaterian Metazoans, not members of Platyhelminthes. *Science*, 283, 1919–1923.

Ruppert, E. E. (1991). Gastrotricha. In F. W. Harrison & E. E. Ruppert (Eds.), *Microscopic anatomy of invertebrates* (Vol. 4). Aschelminthes (pp. 41–109). New York, NY: Wiley-Liss.

Ryan, J. F., Pang, K., Schnitzler, C. E., Nguyen, A. D., Moreland, R. T., Simmons, D. K., ... Baxevanis, A. D. (2013). The genome of the ctenophore Mnemiopsis leidyi and its implications for cell type evolution. *Science*, 342. doi: 10.1146/science.1242592.

Sakaguchi, M., Mizusina, A., & Kobayakawa, Y. (1996). Structure, development, and maintenance of the nerve net of the body column in Hydra. *Journal of Comparative Neurology*, 373, 41–54.

Sakarya, O., Armstrong, K. A., Adamska, M., Adamski, M., Wang, I. F., Tidor, B., ... Kosik, K. S. (2007). A post-synaptic scaffold at the origin of the animal kingdom. *PLoS One* 2, e506. doi: 10.1371/journal.pone.0000506

Sánchez-Soriano, N., Bottenberg, W., Fiala, A., Haessler, U., Kerassoviti, A., Knust, E., ... Prokop, A. (2005). Are dendrites in Drosophila homologous to vertebrate dendrites? *Developmental Biology*, 288, 126–138.

Santagata, S. (2002). Structure and metamorphic remodeling of the larval nervous system and musculature of Phoronis pallida (Phoronida). *Evolution & Development*, 4, 28–42.

Santagata, S. (2008). Evolutionary and structural diversification of the larval nervous system among marine bryozoans. *Biological Bulletin*, 215, 3–23.

Santagata, S. (2011). Evaluating neurophylogenetic patterns in the larval nervous systems of brachiopods and their evolutionary significance to other bilaterian phyla. *Journal of Morphology*, 272, 1153–1169.

Satterlie, R. A., (1985). Swimming in the Pteropod mollusk, Clione limacine. II. Physiology. *Journal of Experimental Biology*, 116, 205–222.

Satterlie, R. A., (1993). Neuromuscular organization in the swimming system of the pteropod mollusc Clione limacina. *Journal of Experimental Biology*, 181, 119–140.

Satterlie, R. A., (1995). Serotonergic modulation of swimming speed in the pteropod mollusc Clione limacina. II. Peripheral modulatory neurons. *Journal of Experimental Biology*, 198, 905–916.

Satterlie, R. A., (2008). Control of swimming in the hydrozoan jellyfish Aequorea victoria: Subumbrellar organization and local inhibition. *Journal of Experimental Biology*, 211, 3467–3477.

Satterlie, R. A., & Eichinger, J. M. (2014). Organization of the ectodermal nervous structures in jellyfish: scyphomedusae. *Biological Bulletin*, 226, 29–40.

Satterlie, R. A., & Norekian, T. P. (1995). Serotonergic modulation of swimming speed in the pteropod mollusc Clione limacina. III. Cerebral neurons. *Journal of Experimental Biology*, 198, 917–930.

Satterlie, R. A., & Spencer, A. N. (1983). Neuronal control of locomotion in Hydrozoan medusae. A comparative study. *Journal of Comparative Physiology*, 150, 195–206.

Scheltema, A. H., Tscherkassky, M., & Kuzirian, A. M. (1994). Aplacophora. In F. W. Harrison & A. J. Kohn (Eds.), *Microscopic anatomy of Invertebrates* (Vol. 5). Mollusca I (pp. 13–54). New York, NY: Wiley-Liss.

Schmidt-Ott, U., & Technau, G. M. (1992). Expression of en and wg in the embryonic head and brain of Drosophila indicates a refolded band of seven segment remnants. *Development*, 116, 111–125.

Schmidt-Rhaesa, A., Harzsch, S., & Purschke, G. (Eds.) (2016). *Structure and evolution of Invertebrate nervous systems*. Oxford: Oxford University Press

Scholtz, G., & Edgecombe, G. D. (2006). The evolution of arthropod heads: Reconciling morphological, developmental and palaeontological evidence. *Development Genes & Evolution*, 216, 395–415.

Schwaha, T., Wood, T. S., & Wanninger, A. (2010). Trapped in freshwater: The internal anatomy of the entoproct Loxosomatoides sirindhornae. *Frontiers in Zoology*, 7, 7. doi: 10.1186/1742-9994-7-7

Schwaha, T., & Wanninger, A. (2012). Myoanatomy and serotonergic nervous system of plumatellid and fredericellid Phylactolaemata (Lophotrochozoa, Ectoprocta). *Journal of Morphology*, 273, 57–67.

Selverston, A., Elson, R., Rabinovich, M., Huerta, R., & Abarbanel, H. (1998). Basic principles for generating motor output in the stomatogastric ganglion. *Annals of the New York Academy of Sciences*, 860, 35–50.

Semmler, H., Chiodin, M., Bailly, X., Martinez, P., & Wanninger, A. (2010). Steps towards a centralized nervous system in basal bilaterians: Insights from neurogenesis of the acoel Symsagittifera roscoffensis. *Development Growth & Differentiation*, 52, 701–713.

Shaw, S. R. (1984). Early visual processing in insects. *Journal of Experimental Biology*, 112, 225–251.

Shigeno, S., Sasaki, T., & Haszprunar, G. (2007). Central nervous system of Chaetoderma japonicum (Caudofoveata, Aplacophora): Implications for diversified ganglionic plans in early molluscan evolution. *Biological Bulletin*, 213, 122–134.

Shinn, G. L. (1997). Chaetognatha. In F. W. Harrison & E. E. Ruppert (Eds.), *Microscopic anatomy of invertebrates, hemichordata, chaetognatha, and the invertebrate chordates* (Vol. 15, pp. 103–220). New York, NY: Wiley-Liss.

Shunkina, K. V., Zaytseva, O. V., Starunov, V. V., & Ostrovsky, A. N. (2014). Sensory elements and innervation in the freshwater bryozoan Cristatella mucedo lophophore. *Doklady Biological Sciences*, 455, 125–128.

Siegler, M. V. (1984). Local interneurons and local interactions in arthropods. *Journal of Experimental Biology*, 112, 253-281.

Siegler, M. V., & Burrows, M. (1979). The morphology of local non-spiking interneurones in the metathoracic ganglion of the locust. *Journal of Comparative Neurology*, 183, 121-47.

Siegler, M. V., & Burrows, M. (1984). The morphology of two groups of spiking local interneurons in the metathoracic ganglion of the locust. *Journal of Comparative Neurology*, 224(4), 463–482.

Skogh, C., Garm, A., Nilsson, D. E., & Ekström, P. (2006). Bilaterally symmetrical rhopalial nervous system of the box jellyfish, *Tripedalia cystophora*. *Journal of Morphology*, 267, 1391–1405.

Sombke, A., Lipke, E., Kenning, M., Müller, C. H., Hansson, B. S., & Harzsch, S. (2012). Comparative analysis of deutocerebral neuropils in Chilopoda (Myriapoda): Implications for the evolution of the arthropod olfactory system and support for the Mandibulata concept. *BMC Neuroscience*, 13, 1–17.

Sorrentino, M., Manni, L., Lane, N. J., & Burighel, P. (2000). Evolution of cerebral vesicles and their sensory organs in an ascidian larva. *Acta Zoologica*, 81, 243–258.

Spencer, A. N., & Arkett, S. A. (1984). Radial symmetry and the organization of central neurones in a Hydrozoan jellyfish. *Journal of Experimental Biology*, 110, 69–90.

Stocker, R. F. (1994). The organization of the chemosensory system in Drosophila melanogaster: A review. *Cell Tissue Research*, 275, 3–26.

Storch, V. (1991). Priapulida. In F. W. Harrison & E. E. Ruppert (Eds.), *Microscopic anatomy of invertebrates* (Vol. 4). *Aschelminthes* (pp. 333–350). New York, NY: Wiley-Liss.

Strausfeld, N. J. (1976). *Atlas of an insect brain*. Berlin: Springer.

Strausfeld, N. J., & Barth, F. G. (1993). Two visual systems in one brain: Neuropils serving the secondary eyes of the spider, Cupiennius salei. *Journal of Comparative Neurology*, 328, 43-62.

Strausfeld, N. J., Hansen, L., Li, Y., Gomez, R. S., & Ito, K. (1998). Evolution, discovery, and interpretations of arthropod mushroom bodies. *Learning and Memory*, 5, 11-37.

Strausfeld, N. J., Sinakevitch, I., Brown, S. M., & Farris, S. M. (2009). Ground plan of the insect mushroom body: Functional and evolutionary implications. *Journal of Comparative Neurology*, 513, 265–291.

Strausfeld, N. J., Strausfeld, C. M., Loesel, R., Rowell, D., & Stowe, S. (2006). Arthropod phylogeny: Onychophoran brain organization suggests an archaic relationship with a chelicerate stem lineage. *Proceedings of the Royal Society B*, 273, 1857–1866.

Strausfeld, N. J., Weltzien, P., & Barth, F. G. (1993). Two visual systems in one brain: Neuropils serving the principal eyes of the spider, Cupiennius salei. *Journal of Comparative Neurology*, 328, 63–75.

Stuart, A. E. (1970). Physiological and morphological properties of motoneurones in the central nervous system of the leech. *Journal of Physiology*, 209, 627–646.

Sukhdeo, S. C., & Sukhdeo, M. V. (1994). Mesenchyme cells in Fasciola hepatica (platyhelminthes): Primitive glia? *Tissue Cell*, 26, 123–131.

Syed, N. I., & Winlow, W. (1991). Respiratory behavior in the pond snail Lymnaea stagnalis. II. A neural network that comprises the central pattern generator (CPG). *Journal of Comparative Physiology*, 169, 557–568.

Takemura, S. Y., Bharioke, A., Lu, Z., Nern, A., Vitaladevuni, S., Rivlin, P. K., … Chklovskii, D. B. (2013). A visual motion detection circuit suggested by Drosophila connectomics. *Nature*, 500, 175–181.

Telford, M. J., Bourlat, S. J., Economou, A., Papillon, D., & Rota-Stabelli, O. (2008). The evolution of the Ecdysozoa. *Philosophical Transactions of the Royal Society of London B. Biological Sciences*. 363, 1529–1537.

Temereva, E. N. (2012). Ventral nerve cord in Phoronopsis harmeri larvae. *Journal of Experimental Zoology B. Molecular & Developmental Evolution*, 318, 26–34.

Temereva, E. N., & Tsitrin, E. B. (2014). Organization and metamorphic remodeling of the nervous system in juveniles of Phoronopsis harmeri (Phoronida): Insights into evolution of the bilaterian nervous system. *Frontiers in Zoology*, 11, 35. doi:10.1186/1742-994-11-35

Teuchert, G. (1977). The ultrastructure of the marine gastrotrich Turbanella cornuta Remane (Macrodasyoidea) and its functional and phylogenetic importance. *Zoomorphologie*, 88, 189–246.

Tremblay, J. P., Colonnier, M., McLennan, H. (1979). An electron microscope study of synaptic contacts in the abdominal ganglion of Aplysia californica. *Journal of Comparative Neurology*, 188, 367–389.

Turbeville, J. M. (1991). Nemertinea. In F. W. Harrison & B. J. Bogitsh (Eds.), *Microscopic anatomy of invertebrates* (Vol. 3). Platyhelminthes and nemertinea (pp. 258–328). New York, NY: Wiley-Liss.

Turbeville J. M., & Ruppert, E. E. (1985). Comparative ultrastructure and the evolution of nemertines. *American Zoology*, 25, 53–71.

Tyrer, N. M., & Altman, J. S. (1974). Motor and sensory flight neurones in a locust demonstrated using cobalt chloride. *Journal of Comparative Neurology*, 157, 117–138.

Tyrer, N. M., & Gregory, G. E. (1982). A guide to the neuroanatomy of locust suboesophageal and thoracic ganglia. *Philosophical Transactions of the Royal Society of London B*, 297, 91–123.

Vosshall, L. B., Wong, A. M., & Axel, R. (2000). An olfactory sensory map in the fly brain. *Cell*, 102, 147–159.

Wada, H., Saiga, H., Satoh, N., & Holland, P. W. (1998). Tripartite organization of the ancestral chordate brain and the antiquity of placodes: Insights from ascidian Pax-2/5/8, Hox and Otx genes. *Development*, 125, 1113–1122.

Wanninger, A. (2008). Comparative lophotrochozoan neurogenesis and larval neuroanatomy: recent advances from previously neglected taxa. *Acta Biologica Hungarica*, 59 Suppl., 127–136.

Ward, S., Thomson, N., White, J. G., & Brenner, S. (1975). Electron microscopical reconstruction of the anterior sensory anatomy of the nematode Caenorhabditis elegans. *Journal of Comparative Neurology*, 160, 313–337.

Watanabe, H., Fujisawa, T., & Holstein, T. W. (2009). Cnidarians and the evolutionary origin of the nervous system. *Development Growth & Differentiation*, 51, 167–183.

Watkins, B. L., Burrows, M., & Siegler, M. V. (1985). The structure of locust nonspiking interneurones in relation to the anatomy of their segmental ganglion. *Journal of Comparative Neurology*, 240, 233–255.

Watson, A. H., & Burrows, M. (1983). The morphology, ultrastructure, and distribution of synapses on an intersegmental interneurone of the locust. *Journal of Comparative Neurology*, 214, 154—169.

Watson, A. H., & Burrows, M. (1985). The distribution of synapses on the two fields of neurites of spiking local interneurones in the locust. *Journal of Comparative Neurology*, 240, 219–232.

Weber, A. V., Wanninger, A., & Schwaha, T. F. (2014). The nervous system of Paludicella articulate—First evidence of a neuroepithelium in a ctenostome ectoproct. *Frontiers in Zoology*, 11, 89. doi: 10.1186/s12983-014-0089-2

Weeks, J. C. (1982a). Segmental specialization of a leech swim-initiating interneuron, cell 205. *Journal of Neuroscience*, 2, 972–985.

Weeks, J. C. (1982b). Synaptic basis of swim-initiation in the leech. II. A pattern generating neuron (cell 208) which mediates motor effects of swim-initiating neurons. *Journal of Comparative Physiology A*, 148, 265–279.

Weeks, J. C., & Kristan, W. B. Jr. (1978). Initiation, maintenance and modulation of swimming in the medicinal leech by the activity of a single neurone. *Journal of Experimental Biology*, 77, 71–88.

Wernet, M. F., & Desplan, C. (2004). Building a retinal mosaic: Cell-fate decision in the fly eye. *Trends in Cell Biology*, 14, 576–584.

Westfall, J. A., Yamataka, S., & Enos, P. D. (1971). Ultrastructural evidence of polarized synapses in the nerve net of Hydra. *Journal of Cell Biology*, 51, 318–323.

Weygoldt, P. (1985). Ontogeny of the arachnid central nervous system. In F. G. Barth (Ed.), *Neurobiology of arachnids* (pp. 20–37). Berlin: Springer.

White, J. G., Southgate, E., Thomson, J. N., & Brenner, S. (1986). The structure of the nervous system of the nematode Caenorhabditis elegans. *Philosophical Transactions of the Royal Society of London B, Biological Sciences*, 314, 1–340.

Wichmann, C., & Sigrist, S. J. (2010). The active zone T-bar–A plasticity module? *Journal of Neurogenetics*, 24, 133–145.

Wright, K. A. (1991). Nematoda. In F. W. Harrison & E. E. Ruppert (Eds.), *Microscopic anatomy of invertebrates* (Vol. 4, pp. 111–195). New York, NY: Wiley.

Wright, K. A. (1992). Peripheral sensilla of some lower invertebrates: The Platyhelminthes and Nematoda. *Microscopy Research Technique*, 22, 285-97.

Zheng, M., Friesen, W. O., & Iwasaki, T. (2007). Systems-level modeling of neuronal circuits for leech swimming. *Journal of Computational Neuroscience*, 22, 21–38.

Zheng, L., de Polavieja, G. G., Wolfram, V., Asyali, M. H., Hardie, R. C., & Juusola, M. (2006). Feedback network controls photoreceptor output at the layer of first visual synapses in Drosophila. *Journal of General Physiology*, 127, 495–510.

Zhu, H., Li, J., Nolan, T. J., Schad, G. A., Lok, J. B. (2011). Sensory neuroanatomy of Parastrongyloides trichosuri, a nematode parasite of mammals: Amphidial neurons of the first-stage larva. *Journal of Comparative Neurology*, 519, 2493–2507.

9

Nervous System Architecture in Vertebrates

Mario F. Wullimann

9.1 Introduction

The vertebrate brain has a distinct conservative bauplan that consists of a limited number of major brain parts and functional systems shared by all extant vertebrate species. Although some elements of this bauplan have been recognized for over a century, there were continued debates about possible additions of anterior brain elements, mostly in the context of a linear "scala naturae" view of brain evolution. Molecular genetic work corroborated many earlier ideas, and brought additional insights into an extended concept of a vertebrate brain bauplan. The current view is that vertebrate brain evolution has occurred by building brains in various lineages based on a fundamental bauplan rather than by sequential addition of new brain parts. During early development, this bauplan is represented by an array of histogenetic units - proliferative zones arranged at the ventricular side of the neuroepithelial wall of the neural tube - from which the adult structures arise by radial or other types of cellular migration and subsequent differentiation. Modern gene expression and functional studies have begun to unravel the complexity of the genetic machinery acting differentially in these histogenetic units. This work has helped to establish firmer conclusions regarding homologies of brain structures arising from particular histogenetic units, while at the same time revealing cases of homoplastic (convergent) evolution of brain characters in vertebrates. Most current debates involve the forebrain, and in particular the telencephalon. In this chapter, after describing the basic bauplan of the vertebrate brain, I will focus on the comparative neural architecture of the forebrain in major vertebrate taxa, in particular of the telencephalon, reviewing functional neuroanatomy, sensory inputs, and premotor/motor outputs, and discussing basal ganglia organization.

9.2 Natural Brain Units: Vesicles/Neuromeres and Longitudinal Columns

What are the morphological and functional units of the brain? In classical embryology the vertebrate brain is said to develop from the anterior portion of the neural tube, which passes from a two-vesicle into a three-vesicle stage (rhombencephalon, mesencephalon, prosencephalon) and then into a five-vesicle stage whereby the prosencephalon or forebrain develops into telencephalon and diencephalon, and the rhombencephalon or hindbrain develops into metencephalon and myelencephalon (see Figure 9.1A and Butler & Hodos, 2005). These five brain parts are also recognized in adult vertebrates (see Figure 9.2). The rediscovery in the late 20th century of the equally old neuromeric paradigm established that the vertebrate rhombencephalon develops from seven to eight transitory neuromeres (called rhombomeres, Rh 1–8; plus possibly three more "cryptorhombomeres"; Marín, Aroca, & Puelles, 2008), and that the prosencephalon develops from at least three neuromeres (prosomeres, P1–P3; Puelles & Rubenstein, 1993, 2003), caudally, plus from a complex most anterior area called the secondary prosencephalon (see Figure 9.1B). Neuromeres are true transverse units of the central nervous system in the sense that they contain all dorsoventral parts of the neural tube at their anteroposterior axial location (see below). More specifically, rhombomeres are morphologically definable entities, with glial intersegmental boundaries and metameric sets of neurons and axonal branching (Holland & Hogan, 1988; Lumsden & Krumlauf, 1996; Wilkinson & Krumlauf, 1990). They furthermore represent compartments with clonal cell restriction (Fraser, Keynes, & Lumsden, 1990; Larsen, Zeltser, & Lumsden, 2001) and differential gene expression (Hunt & Krumlauf, 1992; Oxtoby & Jowett, 1993); neuromeres thus represent a useful topological framework to compare and interpret morphological observations across taxa. However, the vesicle and neuromeric (prosomeric) models are partially in conflict, both in regard to the number of transverse units and to the course of the longitudinal (anteroposterior) axis—and therefore also to the organization of longitudinal columns in the neural tube (Puelles & Rubenstein, 1993).

Three early embryonic vesicles reflect real transverse units. The midbrain–hindbrain boundary (MHB) is uniquely definable in molecular genetic terms as a transverse boundary region and an important signaling center (Brand et al., 1996; Marín & Puelles, 1994; Wurst & Bally-Cuif, 2001). Similarly, the forebrain–midbrain boundary (FMB) has been described as exhibiting local gene expression (Scholpp & Brand, 2003) and acting in clonal cell lineage restriction (Larsen et al., 2001). However, the later division of the rhombencephalon into metencephalon and myelencephalon is mostly epiphenomenonal. In jawed vertebrates, a dorsally located cerebellum and, in both birds and mammals, a ventrally located palliocerebellar relay or pons, characterize the anterior hindbrain (metencephalon); these features distinguish the metencephalic portion of the medulla oblongata from the myelencephalic one. However, because the cerebellum is entirely within rhombomere 1 whereas the pons extends into rhombomere 3/4 (see Figure 9.1; Alonso et al., 2012; Aroca & Puelles, 2005), the metencephalon is not a true transverse brain unit. Similarly, the myelencephalon consists accordingly of unequal numbers of alar rhombomeric units (7) compared to basal plate units (4) and is, thus, also not a real transverse unit. Because a pons is only seen in birds and mammals, the adult rhombencephalon of anamniotes (all craniate taxa except amniotes) is sometimes divided into cerebellum and medulla oblongata (Figure 9.1A). A more severe problem relates to the prosencephalon. In the vesicle model, the telencephalon has been interpreted as lying in front of the diencephalon (Figure 9.1A). However, the existence of two neural tube flexures—one between midbrain and forebrain, the other (which is only pronounced in amniotes) between spinal cord and hindbrain—led to the recognition of the true

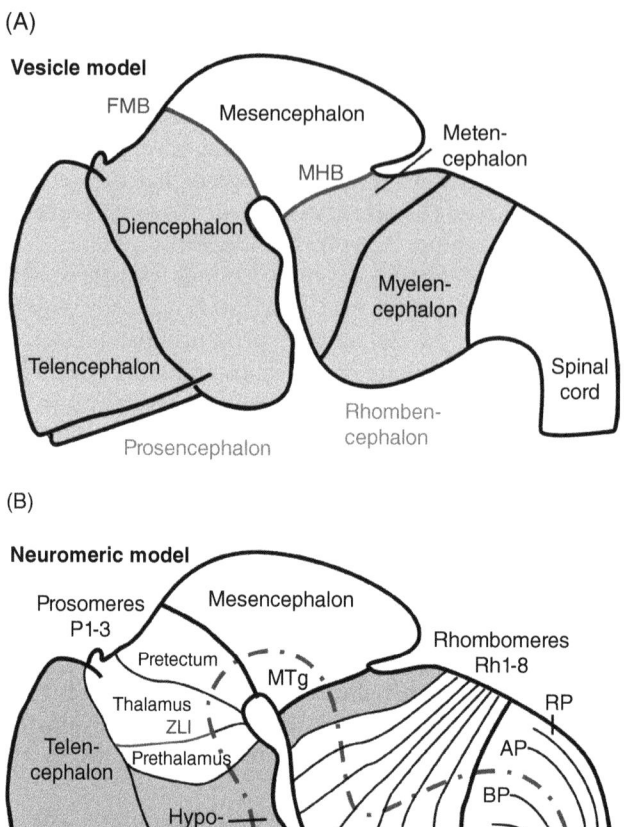

Figure 9.1 Vesicle and Neuromeric Models.
Schematics of embryonic mouse brain at 12.5 days in lateral views. (A) classical vesicle and (B) neuromeric (prosomeric) models. Red interrupted line: anteroposterior axis following hindbrain and forebrain flexures (see text).
Abbreviations: AP alar plate; BP basal plate; FMB forebrain–midbrain boundary; FP floor plate; MHB midbrain–hindbrain boundary; MTg midbrain tegmentum; P1—P6 prosomeres 1—6; Rh1–Rh8 rhombomeres 1—8; RP roof plate; sm somatomotor zone; ss somatosensory zone; vm visceromotor zone; vs viscerosensory zone; ZLI zona limitans intrathalamica. (*See insert for color representation of the figure*).

longitudinal brain axis (Figure 9.1B; red interrupted line). An important consequence of this new axis is that the telencephalon is the dorsal part of the most anterior transverse neural tube unit and that the hypothalamus represents its ventral part, together forming the secondary prosencephalon (Figure 9.1B). The classical diencephalon of the five-vesicle model (Figure 9.1A) is, therefore, not a true transverse unit. Moreover, the remainder of the diencephalon posterior to the hypothalamus forms three more transverse units (prosomeres, P1-3): pretectum (P1), thalamus (P2), and prethalamus (P3, formerly ventral thalamus) (Figure 9.1B) (Puelles & Rubenstein, 1993, 2003). As mentioned above, clonal cell lineage restriction is present at the FMB (Larsen et al. 2001), as well as in the boundary zone of thalamus/prethalamus, i.e., within the zona limitans intrathalamica (ZLI) (Zeltser, Larsen, & Lumsden, 2001), another important transverse signalling region, which separates these two prosomeres as neuromeres (Figure 9.1B). Although clonal cell restriction has been demonstrated more stringently for rhombomeres than prosomeres, approximately 5% of cells also transgress rhombomeric boundaries (Birgbauer & Fraser, 1994). Furthermore, there is selective regulatory gene expression in P1 (*Prox*), P2 (*Gbx2*) and P3 (*Dlx2*) (Larsen et al., 2001). Moreover, early proliferation zones in the zebrafish brain also reveal these three prosomeres, while the secondary prosencephalon follows a more complex proliferation pattern (Wullimann & Mueller, 2004; Wullimann & Puelles, 1999).

What about longitudinal elements of the neural tube? The embryonic longitudinal columns in the spinal cord, that is the floor, basal, alar, and roof plates, represent the prototypical situation for the vertebrate central nervous system. They will develop into dorsal (roof plate) and ventral (floor plate) midline structures, and in particular—going from ventral to dorsal—into functional somatomotor and visceromotor (basal plate), as well as viscerosensory and somatosensory (alar plate) columns of the gray matter receiving sensory and giving rise to motor components of the spinal nerves, respectively (Figure 9.1B). These functional columns seemingly continue into the rhombencephalon and midbrain (Nieuwenhuys, 2011) because up to the midbrain tegmentum there are (basal-plate-derived) somatomotor elements (third nerve oculomotor nucleus) and alar-plate-derived sensory structures (optic tectum/superior colliculus and torus semicircularis/inferior colliculus). Anterior to this level, different paradigms have been used historically to explain the possible continuation of longitudinal zones (Puelles & Rubenstein, 1993; Shimamura, Martinez, Puelles, & Rubenstein, 1997). However, modern gene expression studies demonstrate how the domains of longitudinally expressed genes (such as *sonic hedgehog* and its downstream genes; Ericson et al. 1995; Shimamura, Martinez, Puelles, & Rubenstein, 1997) run in various longitudinal columns by following the neural tube flexures, thus demonstrating directly the new anteroposterior axis mentioned above. In the modern neuromeric model there is now consensus that all four embryonic longitudinal zones and their adult derivatives extend into the forebrain, and that the neural tube has its anterior tip in the area of the optic chiasm. Thus, the neuromeric model of Puelles and Rubenstein (1993, 2003) (Figure 9.1B) integrates approaches to divide the central nervous system into transverse and longitudinal zones and suggests a coherent definition of a vertebrate brain bauplan by proposing a matrix of transverse zones (neuromeres) along the anteroposterior axis which contain a segment of all longitudinal zones (an excellent historical account on these topics is found in Nieuwenhuys, 1998).

The outlined neuromeric model is of great heuristic value for a comparative discussion of the brains of vertebrates, which include jawless (agnathan) lampreys and jawed (gnathostome) vertebrates (see Figure 9.2). The latter include cartilaginous and ray-finned fishes, actinistians (coelacanths), lungfishes, as well as tetrapod amphibians and amniotes (reptiles, birds, mammals). Craniates include additionally the myxinoids (slime eels, hagfishes). An example of a successful application of the neuromeric model is the development and adult location of primary motor nuclei in the vertebrate rhombencephalon which has resulted in improved understanding of developmental intergroup variability in the final adult arrangement of motor nuclei (Gilland & Baker, 2005). The trigeminal motor nucleus is formed and remains located in Rh2/3 in all vertebrates. The gnathostome facial motor nucleus is formed in Rh4 where it remains in adult anurans, while it migrates into Rh5 (and even Rh6) in all other gnathostome groups. This migration is under the control of Rh4 specific expression of the *Hoxb1* gene (Lumsden, 2004). The motor nuclei of the oculomotor nerve (III) in the midbrain tegmentum and of the trochlear nerve (IV) in Rh1 are fixed landmarks in all vertebrate mid- and hindbrains. In contrast, the abducens motor nucleus neurons are located in Rh5/6 in lampreys, birds, and teleosts, but restricted to Rh5 in sharks, frogs, and mammals. Similar analyses exist for primary sensory and other rhombencephalic centers (Aroca & Puelles, 2005; Cambronero & Puelles, 2000; Marín & Puelles, 1995). This and other examples illustrate the power of the neuromeric model for the correct interpretation of both intergroup/interspecific diffences as well as commonalities in neural organization which otherwise would remain in the realm of contentious debate.

9.3 The Ancestral Bauplan of the Adult Craniate Brain

Despite great morphological differences between adult craniate brains (see Figure 9.2), the comparative phyletic method allows to recognize those characters which define its ancestral condition (bauplan or morphotype; Northcutt, 1985; Wicht & Northcutt, 1992). Analysis of adult characteristics, like the developmental ones reviewed above, suggests there was no "terminal addition" of brain regions to the anterior pole of the neuraxis (as envisaged by Ernst Haeckel's recapitulation theory). Rather there was a basic brain bauplan at the outset of craniate evolution which was differentially modified in various taxa (as more generally proposed in pre-Darwinian terms by Karl E. von Baer; see discussion in Butler & Hodos, 2005).

9.3.1 Hindbrain Cranial Nerves

The craniate hindbrain or rhombencephalon is ancestrally associated with the majority of cranial nerves (Figures 9.2 and 9.3 and Table 9.1) and related sense organs (reviews: agnathans: Braun, 1996; teleosts: Wullimann, 1998; cartilaginous fish: Hofmann, 1999). Craniate head development, including that of the cranial nerves and brain, is far more complex than that of the body trunk with respect to interactions of the three

---

Figure 9.2 (*Continued*) Abbreviations: 1 SP first spinal nerve; a anterior cerebellar lobe; A accessory olfactory nerve; al anterior lateral line nerve; BO bulbus olfactorius; c central cerebellar lobe; Ce cerebellum; Di diencephalon; ds dorsal spinal nerve; EG eminentia granularis; H hypothalamus; Ha habenula; Hy hypophysis (pituitary); L lateral pallium (piriform cortex); MO medulla oblongata; MTg midbrain tegmentum; p posterior cerebellar lobe; PC pedunculus cerebri; pl posterior lateral line nerve; Spocc spino-occipital nerve; Tel telencephalon; TeO tectum opticum; vs ventral spinal nerve. 0–XII as in Figure 9.3. (*See insert for color representation of the figure*).

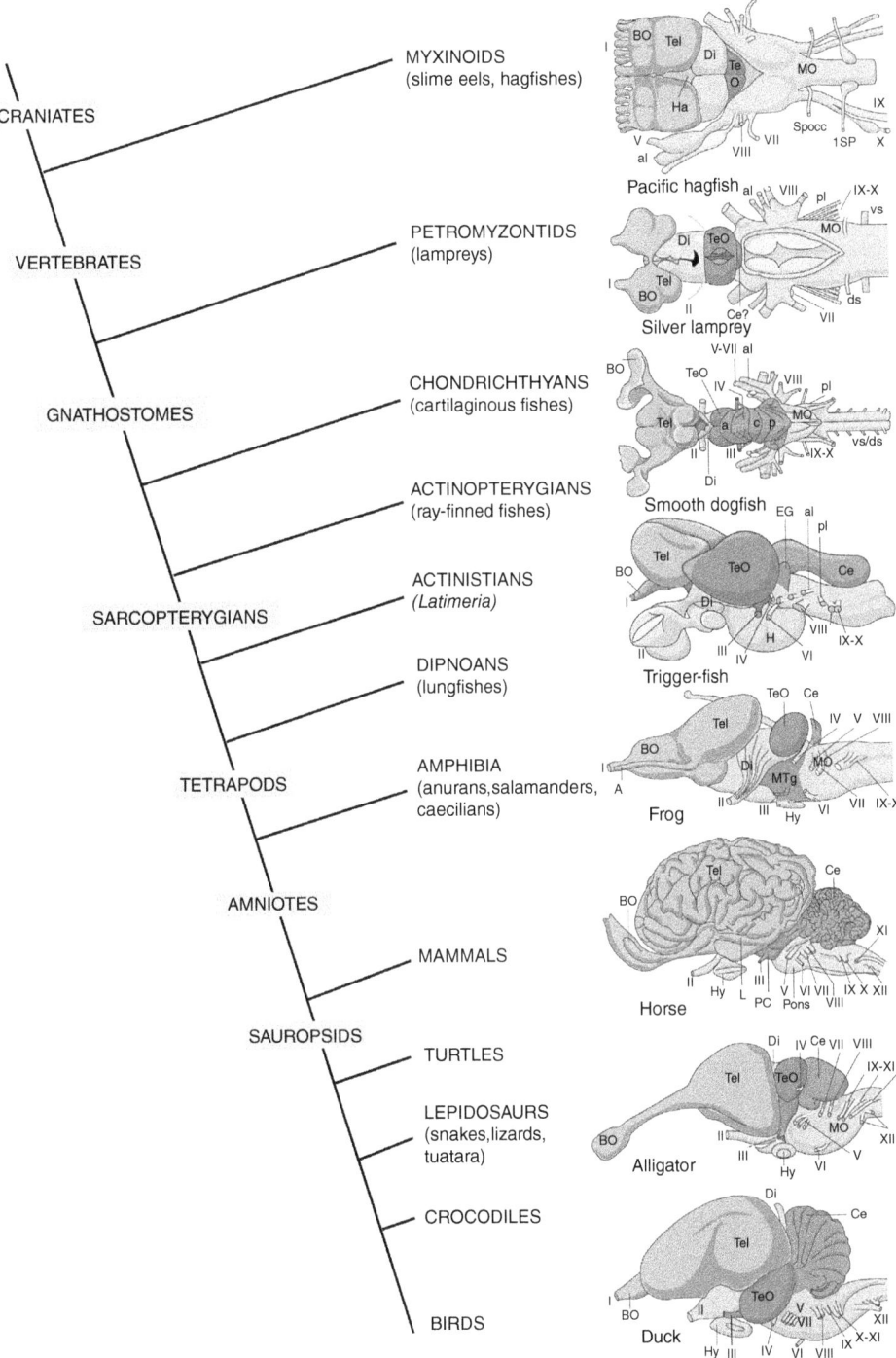

Figure 9.2 Cladogram of Craniate Taxa and Illustrations of Representative Brains.
Note that three terms for nonmonophyletic groups are used in text: agnathans for myxinoids and petromyzontids, reptiles for amniotes except birds/mammals, and anamniotes for all craniates except amniotes.

embryonic germ layers during neurulation ("theory of the new head": Northcutt & Gans, 1983; see also Butler & Hodos, 2005). In particular, a third set of neuroectodermal structures are involved in cranial nerve development in addition to neural tube (somatomotor and preganglionic autonomic nerve components) and neural crest (sensory nerve components, postganglionic autonomic nerve components), both of which are involved in spinal nerve development as well. These third structures, called "placodes," are embryonic epidermal thickenings of neurogenic tissues that give rise to most special sense head organs and—together with the head neural crest—to the cranial sensory ganglia and nerves.

The two most posterior of the 12 classically described human cranial nerves are present in craniates only among tetrapods: they are the hypoglossal nerve (XII, motor innervation of the tongue) and the accessory nerve (XI, motor innervation of two neck muscles). All vertebrates (including lampreys) possess the remaining ten cranial nerves and their primary motor and sensory centers: that is, the olfactory (I), optic (II), oculomotor (III), trochlear (IV), trigeminal (V), abducens (VI), facial (VII), otic (VIII), glossopharyngeal (IX), and vagal (X) nerves (see Table 9.1). The most anterior rhombencephalic cranial nerve (trochlear, IV), together with the abducens (VI) and the oculomotor nerve of the midbrain tegmentum (III), are somatomotor nerves and innervate the six extraocular eye muscles. The branchiomeric cranial nerves innervate one (V/VII/IX) or more (X) visceral arches (mandibular, hyoid, and gill arches or their derivatives; Butler & Hodos, 2005). The trigeminal nerve (V) carries most of the somatosensory information from the head surface (including proprioceptive signals) and oral cavity towards a principal and a spinal trigeminal nucleus. The facial (VII), glossopharyngeal (IX) and vagal nerves (X) all contain a gustatory component which innervates multicellular sensory organs called taste buds. These are located on the tongue in tetrapods and their primary sensory representation is in the medullary solitary tract nucleus. However, taste buds may be located outside of the tongue and oral cavity in fishes (such as goldfish or catfish) where they are always innervated by the facial nerve, whereas the oral cavity is subserved by the glossopharyngeal and vagal nerves. Depending on taste bud density, teleosts have more or less enlarged associated medullary primary sensory facial or vagal centers (FLo/VLo in Figure 9.3A). In gnathostomes, the branchiomeric cranial nerves always have a somatomotor component subserving the jaw musculature (V), the musculature of the hyoid arch (or its derivatives including the facial muscles of mammals, VII), and the gill musculature (or its derivatives, IX/X). Although the situation is basically similar in myxinoids and lampreys, their motor innervation of the trigeminal nerve is not to jaw muscles, but to specialized musculature associated with their convergently evolved feeding apparatus (Kuratani & Ota, 2008); an additional peculiarity unique to myxinoids is that they have (apparently independently evolved) skin taste receptor organs innervated by the trigeminal and spinal nerves (Braun, 1996). Branchiomeric cranial nerves display a viscerosensory and visceromotor (parasympathetic) component related to the innervation of head glands and visceral organs. The otic or vestibulocochlear nerve (VIII) innervates the mechanosensory inner ear (labyrinth) hair cells which subserve the vestibular senses (gravity, torsion); in all gnathostomes it additionally subserves at least a rudimentary sense of hearing (likely without directional hearing, except in amniotes). Details of mammalian cranial nerve components and their innervation sites are summarized in Table 9.1.

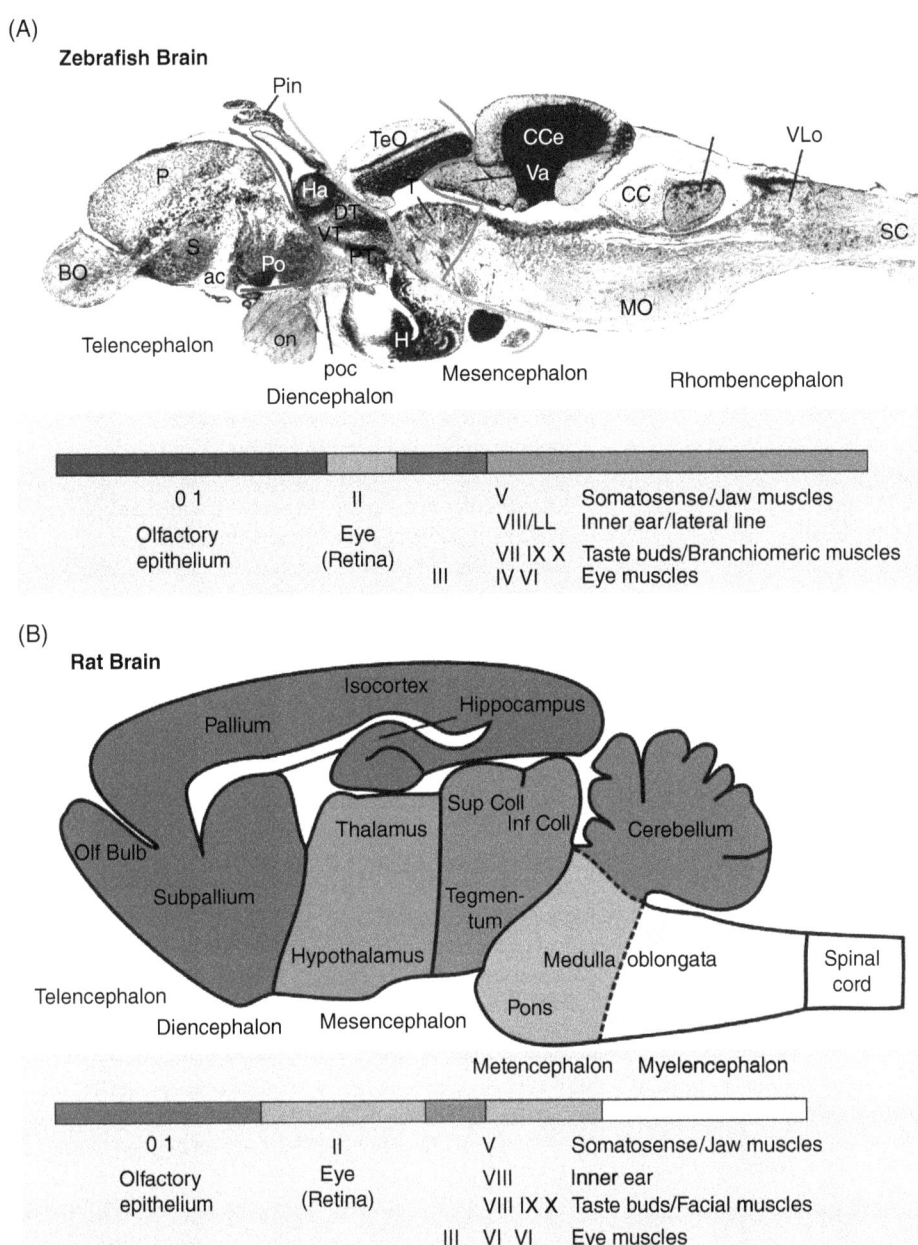

Figure 9.3 Classical Brain Parts and Differences Regarding Presence of Cranial Nerves. Bodian-Nissl stained sagittal adult zebrafish brain section (A) and schema of sagittal adult rat brain section (B) exemplifying classical brain parts and differences regarding presence of cranial nerves between anamniotes (zebrafish) and amniotes (rat).
Abbreviations: 0 nervus terminalis; I nervus opticus; II nervus olfactorius; III nervus oculomotorius; IV nervus trochlearis; V nervus trigeminus; VI nervus abducens; VII nervus facialis; VIII nervus octavus (or vestibulocochlearis); IX nervus glossopharyngeus; X nervus vagus; XI nervus accessorius; XII nervus hypoglossus; ac anterior commissure; BO bulbus olfactorius; CC crista cerebellaris; CCe corpus cerebelli; Corp call corpus callosum; DT dorsal thalamus; FLo facial lobe; Inf Coll inferior colliculus; H hypothalamus; Ha habenula; LL lateral line nerves; on optic nerve; MO medulla oblongata; P pallium; Po preoptic region; Pin pineal; poc postoptic commissure; PT posterior tuberculum; T midbrain tegmentum; S subpallium; SC spinal cord; Sup Coll superior colliculus; TeO tectum opticum; Va valvula cerebelli; VLo vagal lobe; VT ventral thalamus (prethalamus). (*See insert for color representation of the figure*).

Table 9.1 Cranial Nerves in Mammals.

0 N. terminalis	olfactory epithelium VM/SS (?)
I N. olfactorius	olfactory epithelium SS
II N. opticus	retina SS
III N. oculomotorius	ciliary muscle/pupillary sphincter VM
	4 extraocular eye muscles: M obliquus inferior, M rectus superior/inferior/medialis SM
IV N. trochlearis	1 extraocular eye muscle: M obliquus superior SM
V N. trigeminus	skin somatosenses: Face, oral/nasal cavity, tongue, teeth, plus proprioception SS
	jaw muscles and M tensor tympani SM
VI N. abducens	1 extraocular eye muscle: M rectus lateralis SM
VII N. facialis	outer ear/auditory duct/tympanum (SS)
	tongue (taste buds) VS
	sublingual-/submandibular gland VM
	facial musculature, M stapedius SM
VIII N. octavus	inner ear (vestibular senses/hearing) SS
IX N. glossopharyngeus	tongue, pharynx, middle ear SS
	tongue (taste buds) VS
	parotis gland VM
	pharynx SM
X N. vagus	pharynx/larynx, outer ear/auditory duct, tympanum, meninges (dura mater), epiglottis SS
	inner organs, epiglottis (taste buds) VS
	inner organs VM
	pharynx/larynx SM
XI N. accessorius	2 neck muscles: M sternocleidomastoideus, M trapezius SM
XII N. hypoglossus	tongue musculature SM

SS, somatosensory (incl. special senses such as vision, olfaction, hearing, plus proprioception), VS, viscerosensory (incl. gustation); VM, visceromotor (parasympathicus); SM, somatomotor. Note that striated muscles derived from visceral arch musculature (the jaw, hyomandibular and gill arches) are considered somatomotor.

The ancestral vertebrate condition is characterized by some additional cranial nerves. The comparative and embryological study of placodes, their developmental fate, and the adult configuration of cranial nerves and sense organs resulted in a new understanding of the vertebrate head and its evolutionary history, due mostly to the work of Glenn Northcutt and coworkers (review in Northcutt & Bemis, 1993). The so-called lateral line nerves (up to six in gnathostome fishes, Northcutt in Coombs, Görner, & Münz, 1989) innervate multicellular mechanosensory neuromasts (containing hair cells) and electroreceptors on the body surface (Bullock, Bodznick, & Northcutt, 1983). Similar to extraoral taste buds, lateral line organs are always innervated by their cranial nerves and never by spinal nerves. Thus, lateral line nerves are not part of the branchiomeric nerves, as was erroneously assumed in earlier decades.

In fact, lateral line nerves fulfill all criteria for independent cranial nerves, as they have a distinct origin in specific placodes from which their peripheral ganglia and sensory organs develop. Also, distinct central nervous primary sensory nuclei are associated with them (Coombs et al., 1989; Wullimann & Grothe, 2013). Lateral line sensory organs and nerves are present together with an inner ear and octaval nerve in all vertebrates and appear, thus, to be equally old (Northcutt, 1986). Therefore, the inner ear did not evolve by internalization of a part of the superficial lateral line organs, as envisioned previously by the octavolateralis/acousticolateralis hypothesis. Also the concept of a medullary "octavolateral region" with overlapping acoustic and lateral line primary sensory nerve input is likewise clearly flawed. The ancestral vertebrate situation is, apparently, that separate primary sensory nuclei are present for electrosensory, mechanosensory, and auditory hair-cell-related modalities (McCormick, 1992). In sum, the medulla oblongata, together with the spinal cord, links the central nervous system, through the peripheral nervous system (i.e., all sensory and motor nerves described above) to most organ systems of the vertebrate body, enabling it to coordinate functions and enact behavior.

9.3.2 Cerebellum

The first rhombomere of the gnathostome rhombencephalon includes a large, unique dorsal brain part: the cerebellum. It has a three-layered cortex (deep granular, intermediate ganglionic [exhibiting Purkinje cells], and superficial molecular layers) that contains comparable cell types with a similar modular microcircuitry across all jawed vertebrates. The cerebellar afferent mossy fiber inputs (from vestibular nuclei, spinal cord, somatosensory lateral reticular/external cuneate, and accessory optic system nuclei) and efferent connections (to thalamus, nucleus ruber, vestibular nuclei, reticular formation) bear great resemblance among gnathostomes. Additionally, in mammals and birds, a ventrally located pons exists which is a relay center to the cerebellum for input from the mammalian cortex or its homologous area in birds. Furthermore, climbing fiber input from the inferior olive is present in all gnathostome vertebrates. This suggests that a cerebellum is ancestral for gnathostomes with similar functions in motor learning and coordination/execution. However, a palliopontine input and likely also a cerebellothalamopallial loop seems derived for birds and mammals. Interestingly, the cerebellum and its input nuclei share a common ontogenetic origin. The rhombencephalon has a dorsal (rhombic) groove, which is covered by the strongly vascularized neuroepithelial chorioid plexus that generates the cerebrospinal fluid. In gnathostome embryos, the rim of this groove is called the rhombic lip, representing the most dorsal area of the rhombencephalic alar plate. It gives rise, through extensive cellular migration, to all cerebellar projecting systems mentioned, plus the cochlear nuclei, and to the cerebellum itself (review: Wullimann et al., 2011).

9.3.3 Midbrain

The craniate midbrain or mesencephalon displays ancestrally a dorsal (alar plate) region, the tectum mesencephali, which is divided into an optic tectum (visual/multisensory; Figure 9.2) and a torus semicircularis (auditory/lateral line). In mammals, these paired structures are named superior and inferior colliculi (Figure 9.3B), with the former lying anterior to the latter (hence, corpora quadrigemina). However, in the remaining vertebrates, the optic tectum often expands in a dome-like fashion over the torus semicircularis which thus comes to lie posteroventrally to the optic tectum (particularly in birds and teleosts). Although the optic nerve (II) is primarily associated with the diencephalon (see §9.3.4), in most vertebrates its major projection is to the superficial layers of the optic tectum/superior colliculus. The cytoarchitectonic and modular organization of the craniate optic tectum, which is a multilayered cortex, its segregated multimodal input, and the topographical representation of this input (in particular the visual one, plus additional sensory inputs) and output to the reticular formation provide very likely an ancestral neuronal machinery apparently exquisitely designed for integrative orientation tasks, such as object identification and location, and coordinated motor control (details for various taxa: see Nieuwenhuys, ten Donkelaar, & Nicholson, 1998; cartilaginous fishes: Hofmann, 1999; Northcutt, 1978; Smeets, Nieuwenhuys, & Roberts, 1983; ray-finned fishes: Meek & Nieuwenhuys, 1998; Wullimann, 1998). The vertebrate torus semicircularis (including that of lampreys, González, Yáñez, & Anadón, 1999) is recipient to lateral line and/or auditory input via the lateral lemniscus from the respective primary sensory centers in the medulla. Even myxinoids have an ill-defined posterior tectal area receiving bulbar lemniscal input (Wicht & Nieuwenhuys, 1998). The vertebrate ventral (basal plate) midbrain includes the midbrain tegmentum where motor structures are located, such as the oculomotor nerve (III) and its motor nucleus (Figure 9.2). The oculomotor nerve also includes a parasympathetic component (the Edinger–Westphal or accessory oculomotor nucleus, which controls pupillary light reflex). Other important basal midbrain centers in amniotes include the nucleus ruber, periaqueductal gray and dopaminergic ascending systems, that is, the substantia nigra/ventral tegmental area (Puelles, 2007) (see §9.4.2 for situation in anamniotes).

9.3.4 Forebrain

The classic concept of the "diencephalon" or "between-brain" envisaged epithalamus (pineal, habenula), dorsal thalamus (with pretectum), ventral thalamus, and hypothalamus as a dorsoventral series of divisions within the embryonic diencephalic vesicle. In contrast, the neuromeric model (Puelles & Rubenstein, 1993, 2003) established that pretectum, dorsal thalamus (including epithalamus), and ventral thalamus (prethalamus) represent a series of transverse neural tube units arranged from posterior to anterior (prosomeres; see §9.2) with the hypothalamus (basal plate) and telencephalon (alar plate; including preoptic region) lying further rostrally. The posterior tuberculum of fishes develops from the basal plate portions of prosomeres 2 and 3, and the region of the nucleus of the medial longitudinal fascicle from the basal plate of prosomere 1 (Vernier & Wullimann, 2009). Both thalamus/prethalamus and hypothalamus are important integration/control centers present in all craniates. The hypothalamus is the integration center for autonomic-visceral functions (body homeostasis), including the hormonal regulation of bodily functions via the pituitary. The thalamus is in receipt of sensory information from various brain regions, directly via the optic nerve from the retina, as well as through ascending sensory pathways from the midbrain and hindbrain. The functional interrelationship of thalamus and telencephalon will be discussed for each major taxon below. The optic nerve (II) enters the brain between preoptic region and hypothalamus. However, this nerve is in fact a brain tract, since the retina develops from the central rather than the peripheral nervous system. Retinal ganglion cell axons are, nonetheless, traditionally called the optic nerve up to the optic chiasm, and from then to the brain they form the optic tract.

The telencephalon (endbrain) of all craniates has a pallium (all parts of the mammalian cortex, pallial amygdala, and olfactory bulb) and a subpallium (septum and basal ganglia). The craniate amygdala comprises both pallial and subpallial telencephalic regions. The pallium includes multisensory integration centers, particularly pronounced in gnathostomes (details in §9.4 below). The craniate olfactory nerve (I) consists of axons of sensory olfactory epithelial cells (so-called primary sensory cells unlike any other craniate nerve) which project to the olfactory bulb. All other sensory craniate (spinal and cranial) nerves have peripheral ganglia containing pseudounipolar neurons with a dendrite into their sensory organ and a centrally projecting axon. In tetrapods, the olfactory system is divided into a main olfactory epithelium/nerve/bulb and a vomeronasal or accessory epithelium/nerve/bulb; a similar functional subdivision may also exist, at the olfactory receptor cell level within an undivided olfactory epithelium, in ray-finned fishes and, probably, cartilaginous fishes (Eisthen, 2004). A late-discovered cranial nerve, the terminal nerve (0), is also associated with the telencephalon (Kawai, Oka, & Eisthen, 2009; Wirsig-Wiechmann, Wiechmann, & Eisthen, 2002). In contrast to the olfactory nerve, the terminal nerve has ganglion cells (deriving from the cranial neural crest and/or olfactory placode) which have a neurite directed peripherally, into the olfactory epithelium/bulb, and another directed centrally (Schlosser, 2006; von Bartheld, 2004; Whitlock, 2004). These ganglion cells are heterogenous within vertebrate species with respect to location (near olfactory epithelium/bulb/within telencephalon) and to transmitters (gonadotropin-releasing hormone, acetylcholine, glutamate; Edwards, Greig, Sakata, & Elkin, 2007; Kawai et al., 2009; von Bartheld, 2004). The function of the terminal nerve likely involves modulatory/parasympathetic innervation of the olfactory epithelium/bulb (Fujita, Sorensen, Stacey, & Hara, 1991), not pheromone detection (Demski & Northcutt, 1983).

9.3.5 Descending Premotor Systems

Finally, similar descending premotor systems in craniates are in control of the primary motor nuclei of the midbrain, hindbrain, and spinal cord which, altogether, are the sole neurons to directly release motor behaviors. At their simplest, motor behaviors are mediated by central pattern generators: small locally acting neuronal networks in spinal cord or medulla oblongata which organize reflexes (withdraw, cough, sneeze, swallow) or rhythmic movements (walk, chew, breathe; Grillner in Squire et al., 2008). Posture and eye movements involve already larger brainstem–spinal cord networks (e.g., reticulospinal, vestibulospinal, rubrospinal, tectoreticular); these locomotor patterns may be coordinated with one another and adapted to their environmental conditions, even without forebrain control (as evidenced by experiments with decerebrated mammals). However, brainstem premotor systems may be influenced by forebrain centers for more complex adaptive, goal-directed behaviors. For example, hypothalamic centers guide eating, drinking, and aggression-related behaviors (observed even in mammals without cortex; Grillner in Squire et al., 2008). Moreover, in amniotes, telencephalic (pallial/cortical) control over lower level behaviors is possible (from suppression of breathing to skillful learning-dependent movement). Both cerebellum (in all gnathostomes) and basal ganglia (in all vertebrates) play complementary roles in these motor processes. Although long palliospinal and palliopontine pathways originated (independently) exclusively in mammals (pyramidal tract) and birds, all craniates possess various multisynaptic descending control systems. These travel from forebrain, via midbrain and hindbrain, to cranial and spinal nerve motor neurons (those involving the basal ganglia will be detailed in §9.4.1.3). As in amniotes, in cartilaginous (Smeets et al., 1983) and ray finned-fishes (Wullimann, 1998), descending projections to the spinal cord arise from all parts of the reticular formation, the inferior (serotoninergic) raphe, vestibular and sensory trigeminal nuclei, and from the nucleus ruber. An important descending system is present in the nucleus of the medial longitudinal fascicle, an ancestral craniate premotor nucleus located in the basal part of pretectal prosomere 1, which projects to the medulla oblongata and the spinal cord and plays a crucial role in the stereotyped escape reflex (startle response). In mammals, the corresponding interstitial nucleus of the medial longitudinal fascicle plus the interstitial nucleus of Cajal are involved in head and eye movement control (Nieuwenhuys et al., 2008). Also myxinoids and lampreys have descending connections to the spinal cord from all parts of the reticular formation, and from vestibular and sensory trigeminal nuclei (Ronan, 1989). However, both agnathan taxa lack a nucleus ruber, likely because of an absence of pectoral and pelvic fins, the homologs of tetrapod limbs. In all craniates, the optic tectum (superior colliculus) and, in gnathostomes, the cerebellum act on premotor centers, in particular on the reticular formation.

9.3.6 The Agnathan Situation

Most of the above-discussed vertebrate characteristics are also found in myxinoids, and are thus ancestral to all craniates. However, myxinoids lack a cerebellum, and even lampreys lack both *Pax6* expression (required for gnathostome cerebellar development; Murakami, Uchida, Rijli, & Kuratani, 2005; Wullimann et al., 2011) and Purkinje cells (Lannoo & Hawkes, 1997) in the anterior rhombencephalic region (upper rhombic lip) where a small cerebellum was once suspected. However, recent gene expression studies suggest that a lower rhombic lip exists in agnathans (Sugahara et al., 2016). Another difference is that all jawed vertebrates have three semicircular canals in the inner ear labyrinth, whereas myxinoids have one and lampreys two. This may be directly related to the absence of any *Otx* gene expression in the otic cyst of agnathans, an interpretation supported by observations of *Otx1* mutant mice, which lack the horizontal canal (Mazan, Jaillard, Baratte, & Janvier, 2000; Germot et al., 2001). Thus, since both agnathan groups lack many rhombic-lip-derived structures, including a cerebellum, these can be surmised to represent evolutionary novelties of gnathostomes. Moreover, myxinoids—but not lampreys—lack extraocular eye muscles, as well as the associated cranial nerves (III/IV/VI) and their motor nuclei. Finally, myxinoids lack both the terminal nerve and the electrosensory (but not the mechanosensory) component of the lateral line nerves (Braun, 1996; Wicht, 1996; Wicht & Northcutt, 1998). The absence of these neural characters likely represents the ancestral situation of craniates, rather than secondary losses, and highlights the position of myxinoids as the sister group to the vertebrates.

9.4 Comparative Brain Architecture in Craniates

Because the organization of the central nervous system is better known in amniotes than anamniotes, the following discussion will start with a comparison of mammals and birds/reptiles before going into various anamniote taxa.

9.4.1 Comparative Brain Architecture in Amniotes

Based on the shared bauplan of the craniate brain outlined above, which shows essential similarities in primary sensory input (spinal and cranial nerves/primary central nervous projection areas) and in final premotor output to motor nuclei of the brainstem (hindbrain and midbrain), the adult organization of the forebrain shall now be described with emphasis on two focal topics: first, the sensory pathways to telencephalon; second, the motor networks of the basal ganglia.

9.4.1.1 Correlation of brain size and complex cognitive behavior in birds and mammals. Within craniates, birds and mammals have the largest brains relative to body size (discussed in Wullimann & Vernier, 2009b). Moreover, similarly sized avian and mammalian species share equally large brains. Even on superficial inspection, some macroscopical correspondences are apparent (e.g. compare duck and horse brains in Figure 9.2). A notable exception is that the large optic tectum of birds is displaced laterally relative to the mammalian homologue, the superior colliculus. Both mammal and bird brains have a large cerebellum and telencephalon (Northcutt, 2011), the pallium of which overgrows not only the rest of the prosencephalon but also most of the mesencephalon (see Figure 9.2).

In mammals, the telencephalic ventricle separates the isocortex (dorsal pallium) from the basally lying subpallium (which includes the basal ganglia, i.e., striatum and pallidum, plus septum; Figure 9.4A, B). In birds, massive telencephalic brain parts lie subjacent to the dorsal pallium (hyperpallium, formerly called the Wulst) and displace the telencephalic ventricle medially (Figure 9.4D). These neural masses, mesopallium and nidopallium (formerly called the dorsal ventricular ridge or DVR, as are their reptilian homologues), are basally bordered by subpallial structures, the basal ganglia (striatum/pallidum). Up to the mid 20th century, the DVR had been interpreted as an extended basal ganglionic region. This was in line with the then widely accepted view that birds display mostly hard-wired, stereotyped, instinctive behaviors, in contrast to the plastic, learned, and cognitive behaviors of mammals. Because the mammalian basal ganglia were already then known to be involved in the initiation/selection and execution of motor activity (Mink, in Squire et al., 2008), the contemporary historical interpretation of the avian DVR was that birds need larger basal ganglia for the production of complicated, but inflexible, motor acts than do mammals, who instead have a large isocortex dedicated to flexible modification of learned behavior. This view of the DVR turned out to be neurobiologically flawed (Jarvis et al., 2005; Reiner et al., 2004); moreover, stunning reports now demonstrate the learning capabilities and cognitive capacity of birds such as ravens and parrots (Emery & Clayton, 2004). The functional neuroanatomical and developmental findings which parallel this modern view of bird brain and behavior will be discussed below, using recently updated terminology for bird brain regions (Reiner et al., 2004).

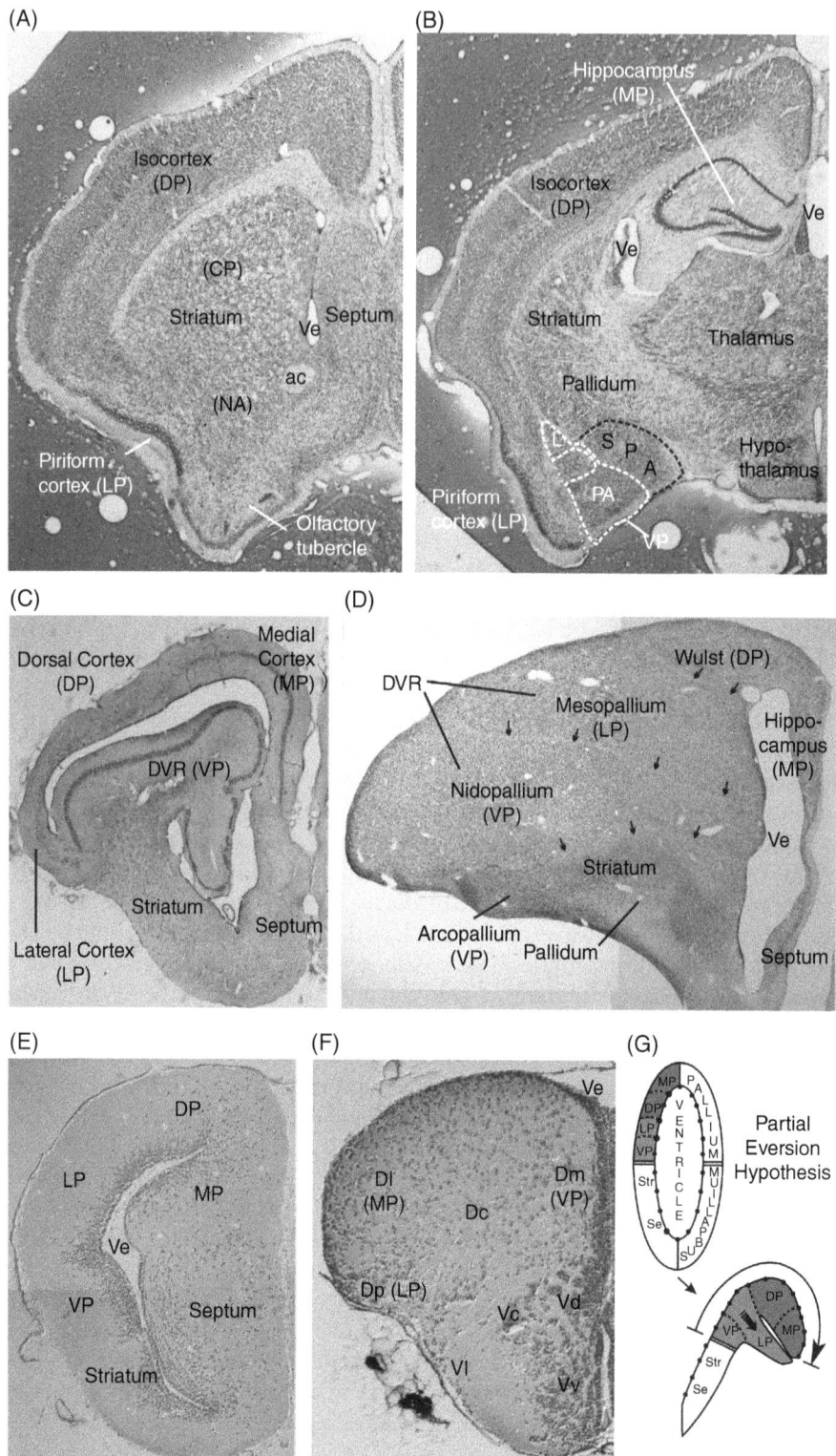

Figure 9.4 (*Continued*)

9.4.1.2 Ascending sensory pathways in birds and mammals. Mammals have two major visual pathways to isocortex originating from retinal ganglion cells (Figure 9.5A). The dominating thalamofugal pathway is associated with detailed perception of visual clues and includes parallel processing streams for color/shape and movement running via the dorsal thalamus (lateral geniculate nucleus) to primary visual cortex (Brodmann area 17). The tectofugal pathway runs via the superior colliculus to the thalamus (pulvinar/lateral posterior nucleus) and from there to secondary visual cortex (areas 18/19). In mammals, this second pathway is associated with unconscious visual processing (blindsight) and is involved in the control of eye movements during object localization (Nieuwenhuys et al., 2008). Birds also have a thalamofugal pathway running to the hyperpallium (Wulst) via the dorsal thalamus (dorsolateral thalamic nucleus), but the dominant avian visual pathway is tectofugal (Engelage & Bischof, 1993; Remy & Güntürkün, 1991; Shimizu & Bowers, 1999), which courses through optic tectum and dorsal thalamus (nucleus rotundus) en route to the entopallium of the DVR (see Figure 9.5B). In lateral-eyed birds, such as pigeons, the tectofugal pathway is the major perceptual processing stream for pattern recognition, innervating separate areas in the nucleus rotundus related to color, size, brightness, and movement perception (Bischof & Watanabe, 1997; Engelage & Bischof, 1993; Wang, Jiang, & Frost, 1993; Watanabe et al., 2008). Furthermore, in zebra finches, entopallial lesions lead to disruption of pattern recognition, while lesions of visual hyperpallium (Wulst) do not (though the latter do result in spatial discrimination deficits) (Watanabe et al., 2011).

The somatosensory systems in birds and mammals also display comparable ascending pathways from primary sensory nuclei of spinal cord/medulla oblongata through thalamus to dorsal pallium (avian hyperpallium (Wulst)/mammalian primary somatosensory cortex) and, in birds, additionally to the DVR (Necker, 1989; Nieuwenhuys et al., 2008; Schneider & Necker, 1989; Wild, 1985, 1987, 1989, 1994; Wild, Arends, & Zeigler, 1984, 1985; Wild & Zeigler, 1996). In the mammalian thalamus, the dorsal column-medial lemniscal system (label 2 in Figure 9.6A) brings fine touch into the lateral and medial ventroposterior nuclei (VPL/VPM) and proprioception into the superior ventroposterior nucleus (VPS). The anterolateral system (label 1 in Figure 9.6A)

Figure 9.4 (*Continued*) Transverse Nissl-Stained Sections through Left Telencephalic Hemispheres of Mouse (*Mus musculus*, A: anterior, B: posterior), Tuatara (*Sphenodon punctatus*, C), Pigeon (*Columba livia*, D), Fire-Bellied Toad (*Bombina orientalis*, E), and Zebrafish (*Danio rerio*, F). Schema shows partial eversion hypothesis (G). Major pallial and subpallial divisions are present in all gnathostome taxa. Arrows in D point to cell-free laminae at boundaries between avian telencephalic divisions. See acknowledgments for origin of sections.

Abbreviations: ac anterior commissure; CP caudate-putamen; Dc central zone of dorsal telencephalic area; Dl lateral zone of dorsal telencephalic area; Dm medial zone of dorsal telencephalic area; Dp posterior zone of dorsal telencephalic area; DP dorsal pallium; DVR dorsal ventricular ridge; L lateral amygdala; LP lateral pallium; MP medial pallium; NA nucleus accumbens; PA pallial amygdala (note fine white dots indicating parts of lateral pallial origin); Se Septum; SPA subpallial amygdala; Str Striatum; Vc central nucleus of ventral telencephalic area; Vd dorsal nucleus of ventral telencephalic area; Ve telencephalic (lateral) ventricle; Vl lateral nucleus of ventral telencephalic area; VP ventral pallium; Vs supracommissural nucleus of ventral telencephalic area; VT ventral thalamus (prethalamus); Vv ventral nucleus of ventral telencephalic area. (*See insert for color representation of the figure*).

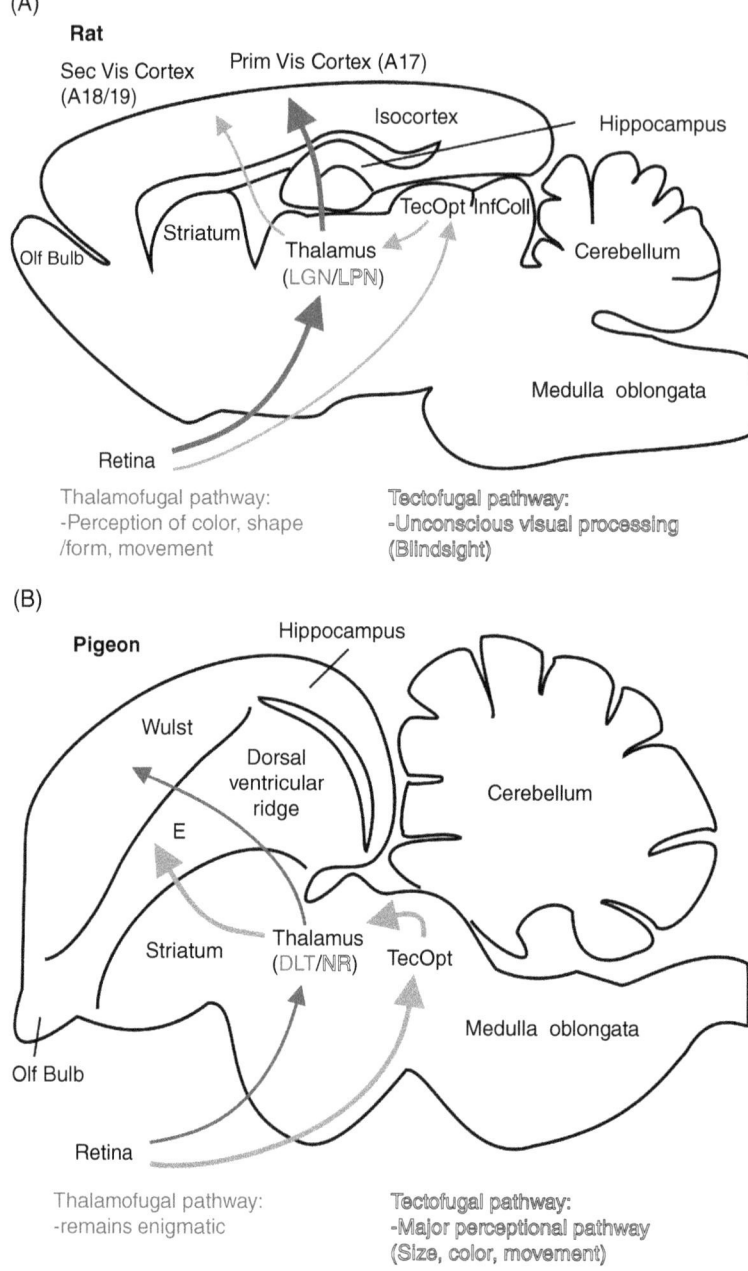

Figure 9.5 Schematics of Sagittal Sections of (A) Rat and (B) Pigeon Brains Showing Ascending Visual Pathways (Rat after Nieuwenhuys et al. 2008, pigeon after Remy & Güntürkün, 1991; Engelage & Bischof, 1993; Shimizu & Bowers, 1999).

Abbreviations: DLT dorsolateral thalamic nuclei/principal optic nuclear complex; E entopallium (old name: ectostriatum); InfColl inferior colliculus; LGN lateral geniculate nucleus; LPN lateral posterior nucleus; NR nucleus rotundus; Olf Bulb olfactory bulb; TecOpt tectum opticum.

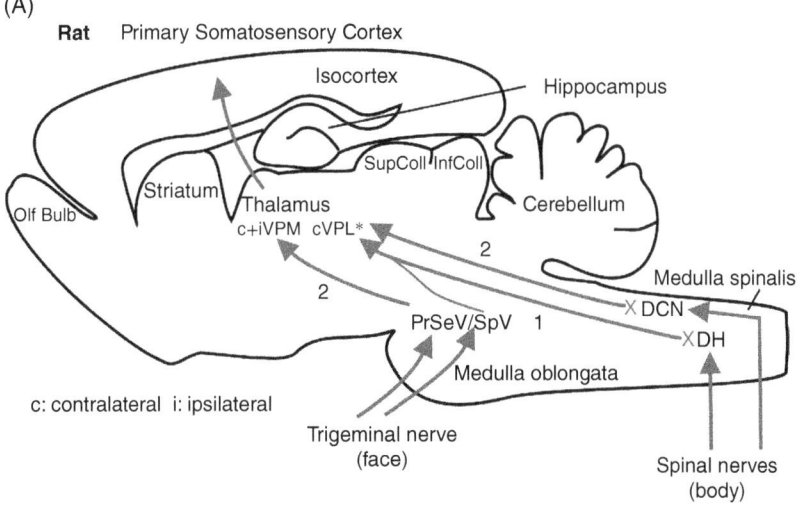

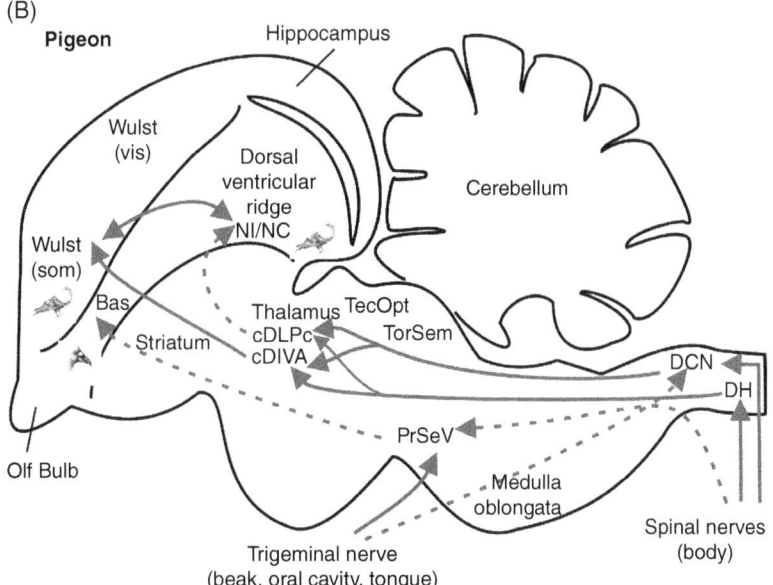

Figure 9.6 Schematics Show Sagittal Sections of (A) Rat and (B) Pigeon Brain with Ascending Somatosensory Pathways (Rat after Nieuwenhuys et al., 2008, pigeon after Necker, 1989; Schneider & Necker, 1989; Wild, 1985, 1987, 1989, 1994; Wild et al., 1984, 1985; Wild & Zeigler, 1996).

* Only nuclei for dorsal column-medial lemniscal system (2; fine touch) shown. See text for nuclei for proprioception (VPS) and for pain/temperature (VMpo, VPI, VPS, MD, and ILN), that is, the anterolateral system (1).

Abbreviations (mammals): DCN dorsal column nuclei (cuneate/gracile); DH dorsal horn of spinal cord; ILN intralaminar nuclei (to cingulum); MD medial dorsal thalamic nucleus (to cingulum); PrSeV principal sensory trigeminal nucleus; SpV descending/spinal sensory trigeminal nucleus; VPL lateral ventroposterior thalamic nucleus (to S1/S2); VPM medial ventroposterior thalamic nucleus (to S1/S2); VPI inferior ventroposterior thalamic nucleus (to S1/S2); VPS superior ventroposterior thalamic nucleus (to S1/S2); VMpo posterior ventromedial thalamic nucleus (to insula); X crossing to contralateral side.

Abbreviations (birds): Bas nucleus basalis rostralis; DCN dorsal column nuclei (cuneate/gracile; plus external cuneate: receives wing and trigeminal information); DLPc dorsolateral posterior thalamic nucleus; caudal part; DIVA nucleus dorsalis intermedius ventralis anterior; NI/NC intermediate/caudal nidopallium; PrSeV: principal sensory trigeminal nucleus (spinal nucleus exists; no ascending connections).

mediates pain and temperature via a spinal relay to posterior ventromedial (VMpo), inferior and superior ventroposterior (VPI/VPS) nuclei, as well as to the medial dorsal thalamic (MD), and intralaminar (ILN) nuclei.

In birds, somatosensory information from the head and beak bypasses the thalamus to arrive, via the principal trigeminal sensory nucleus, directly to a ventral division of the DVR (nucleus basalis; Figure 9.6B), but also ascends, via the thalamic DLPc, to the nidopallium (NI/NC). Thus, while bird somatosensory pathways via DIVA to the hyperpallium (Wulst) appear homologous to mammalian pathways from thalamus to cortex, those via DLPc to NI/NC and more directly to the nucleus basalis do not have a clear mammalian homologue. Also unique to birds is that the body and head have discontinuous pallial representations, with the body represented in the hyperpallium (Wulst) and NI/NC and the head represented in nucleus basalis.

Finally, the avian auditory system has comparable pathways to the thalamus as seen in mammals (Figure 9.7), but distinct primary pallial projections targeting Field L (in the nidopallium) rather than fields in the hyperpallium (Wulst) (summarized in Reiner, Yamamoto, & Karten, 2005). Because of these similarities in the synaptic conformation of ascending sensory pathways to the pallium, two conclusions have been drawn: firstly, that the DVR is not part of the (subpallial) basal ganglia, but rather that the DVR, similar to the hyperpallium (Wulst), is of pallial nature and, secondly, that the DVR, like the hyperpallium, is homologous to corresponding portions of the mammalian primary sensory and additional cortical regions (reviewed in Reiner et al., 2004, 2005). However, only the first conclusion is confirmed in developmental studies, while the second is not (see §9.4.1.4).

9.4.1.3 Basal ganglia in birds and mammals. Comparative work on basal ganglia circuitry and neurochemistry in birds and mammals has greatly contributed to our understanding of amniote telencephalic organization. Isocortical excitatory input reaches two different populations of inhibitory neurons in the mammalian striatum: those associated with the direct and the indirect striatal pathways (Mink in Squire et al., 2008) (Figure 9.8A; note that both descending pallial input to basal ganglia and descending basal ganglia output are omitted from the schematics). The direct pathway starts with GABA/SubstanceP striatal cells and acts to inhibit the reticular substantia nigra and the internal globus pallidus. Since these target structures are also GABAergic, direct-pathway activation results in disinhibition of their target, a glutamatergic dorsal thalamic nucleus, which, in turn, provides excitatory feedback to premotor and motor isocortex. The indirect pathway arises from GABA/enkephalin striatal neurons, and synapses sequentially in the external globus pallidus (GABAergic), subthalamic nucleus (glutamatergic), internal globus pallidus (GABAergic), and thalamus (glutamatergic). Therefore, activation of this indirect pathway has a net inhibitory effect on isocortex. In behaviorally relevant situations, the dopaminergic projections of the substantia nigra pars compacta, in the basal midbrain, release dopamine onto both populations of GABAergic striatal neurons. However, these two neuronal populations carry different dopamine receptors, leading to a net excitatory effect of striatal dopamine on cortex: The D1 receptors on "direct" neurons excite the cells and support the excitatory feedback to cortex; D2 receptors on "indirect" neurons inhibit them, reducing indirect inhibitory feedback to cortex. Release of nigral dopamine, in mammals, thus acts to release planned motor behavior held in the basal ganglia motor loop. The work of Harvey Karten, Anton Reiner, and many colleagues established that a highly comparable neural network is present in birds (Figure 9.8B). The

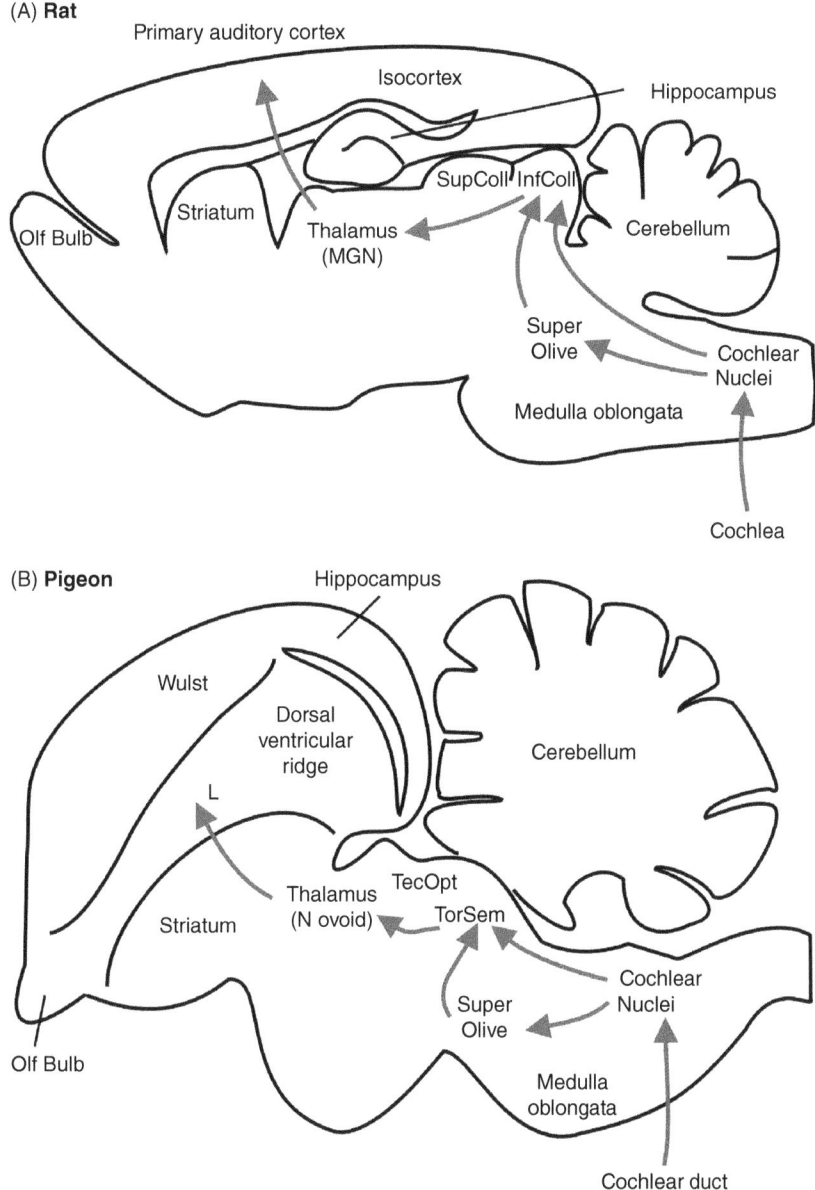

Figure 9.7 Schematics of Sagittal Sections of (A) Rat and (B) Pigeon Brains Showing Ascending Auditory Pathways (Rat after Nieuwenhuys et al., 2008, pigeon after Shimizu & Karten, 1993).
Abbreviations: L Field L, InfColl inferior colliculus, MGN medial geniculate nucleus, N ovoid nucleus ovoidalis, Olf Bulb olfactory bulb, SupColl superior colliculus, SuperOlive superior olivary complex, TecOpt tectum opticum, TorSem torus semicircularis.

functional organization of avian basal ganglia structures has been demonstrated with respect to neuroanatomical location, axonal connections, transmitter characteristics, and motor effects after specific lesioning (Figure 9.8B, see legend for citations). It is

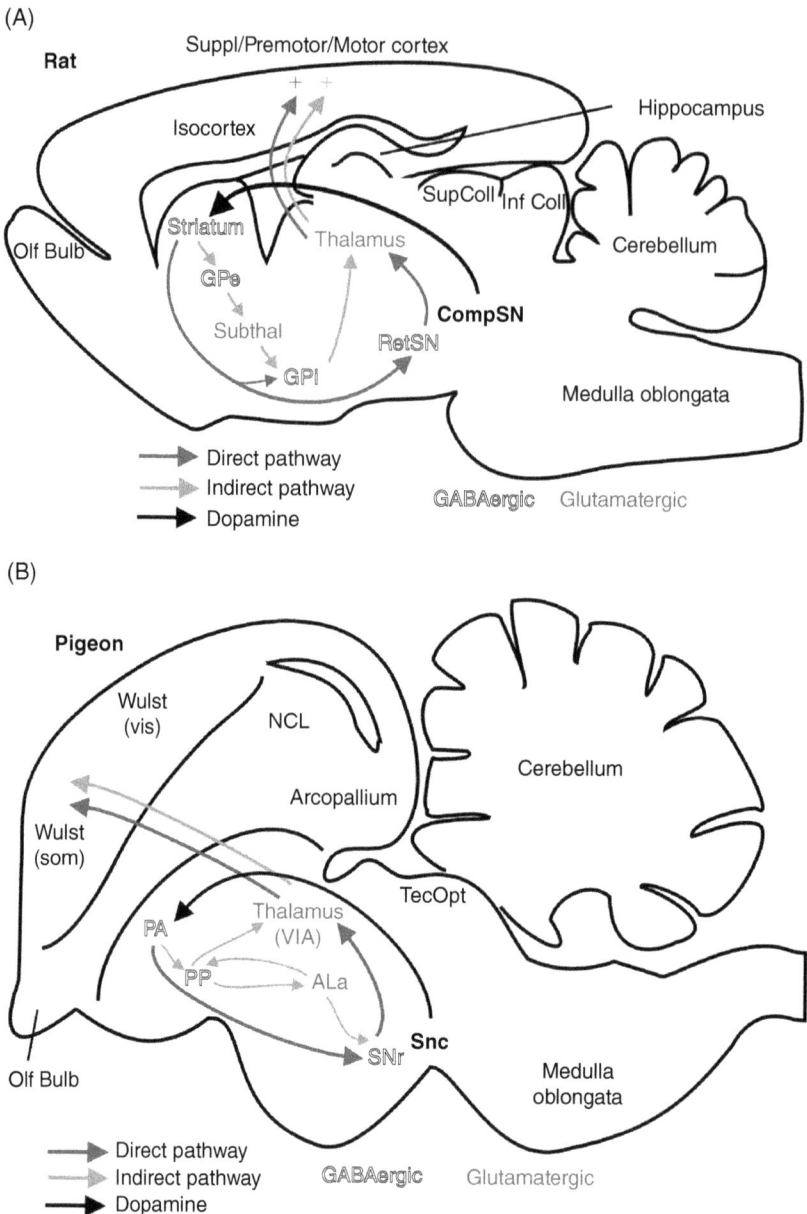

Figure 9.8 Schematics Show Sagittal Sections of (A) Rat and (B) Pigeon Brain with Basal Ganglia Circuitry (Rat after Mink in Squire et al., 2008; pigeon after Jiao et al., 2000; Kröner & Güntürkün, 1999; Reiner, 2002; Veenman, Wild, & Reiner, 1995).
+/- in (A) indicates activation of cortex by both direct and indirect pathways upon dopamine release in striatum. Note that in addition to the dorsal striatopallidal system (somatomotor loop) shown here, ventral striato-pallidal (limbic) loops do exist in all amniotes.
Abbreviations (mammals): CompSN substantia nigra compacta, GPe external globus pallidus, GPi internal globus pallidus, RetSN substantia nigra reticulata Abbreviations (birds): ALa anterior nucleus of ansa lenticularis (=subthalamic nucleus), NCL caudolateral nidopallium, PA paleostriatum augmentatum (=lateral striatum), PP paleostriatum primitivum (=pallidum), SNc substantia nigra compacta, SNr substantia nigra reticulata, VIA: ventrointermediate thalamic area. (*See insert for color representation of the figure*).

now clear that the "paleostriatum," ventral to the DVR in the bird telencephalon, is essentially identical to the mammalian striatum and pallidum. As in mammals, the origin of the activating input to the striatum includes the dorsal pallium (Wulst), which, in turn, receives the glutamatergic feedback from the thalamus. It is therefore not surprising that the long motor pathways descending from the hyperpallium (Wulst) (septomesencephalic tract; Figure 9.9 see legend for citations) bear some resemblance to the mammalian pyramidal tract (e.g. projections to pons and ventral horn of spinal cord). A second motor system descending from avian telencephalon, the occipitomesencephalic tract, carries characteristics of the mammalian amygdalofugal output (see §9.4.1.4), but also contacts medullary premotor centers. This line of work complements the studies on ascending sensory system organization, confirming that the DVR is functionally and neuroanatomically pallial and not part of basal ganglia as once assumed. Further, it supports homology between hyperpallium (Wulst) and isocortex. However, the interpretation that the DVR is homologous to isocortex turned out to be flawed by molecular genetic studies.

9.4.1.4 Comparison of reptiles, birds and mammals: convergence and homology. Based on these comparative studies of neurochemistry and connectivity, a consensus was reached in the late 20th century that the avian DVR and Wulst together are homologous to the mammalian isocortex and that the bird "paleostriatum" is homologous to the mammalian basal ganglia, with the "paleostriatum augmentatum" (PA) homologous to the mammalian striatum proper, and the "paleostriatum primitivum" (PP) homologous to the mammalian pallidum (see Figures 9.4–9); their revised names reflect these homologies. However, mammals and birds are terminal taxa of early-diverging evolutionary amniote lineages (synapsid and diapsid, respectively; Carroll, 1988). Not surprisingly, there are also differences between them, for example in the relative importance of the thalamofugal visual pathway or of the re-entrant pathway of the thalamopallial basal ganglionic loop (Figure 9.8). In mammals, this re-entrant path is the dominant output system of the basal ganglia. In birds, there is a massive descending output of basal ganglia via the substantia nigra pars reticulata and pretectum to the optic tectum and from it to the reticular formation of the medulla oblongata (not shown in Figure 9.8). This descending pathway is equally important as the re-entrant path in the context of avian head, neck and eye movements, while it is more subdued in mammals.

In order to sort out similarities that are based on common descent (homology) versus convergence (homoplasy) in birds and mammals, the situation in reptiles needs to be considered. Extant reptiles show the general organizational plan of the bird brain, in that they have a DVR and, dorsal to it, a relatively small dorsal pallium. Ironically, this structure has long been known as the dorsal cortex, since, unlike in birds, it has three cellular layers. Reptilian medial and lateral pallial divisions are also evident (Figure 9.4C). As far as is known, the ascending sensory pathways follow the outline given above for birds (summarized by ten Donkelaar, 1998b): visual and somatosensory pathways run via the thalamus to the dorsal cortex (thalamofugal), along with a tectofugal visual pathway, and an auditory and somatosensory pathway to the reptilian DVR. Unexpectedly, reptiles (incl. crocodiles) have a third somatosensory pathway which bypasses the thalamus, carrying information from the head to the DVR; this suggests the path is not a specific adaptation to avian beak usage. The general organization of basal ganglia is similar in reptiles and birds, with distinct

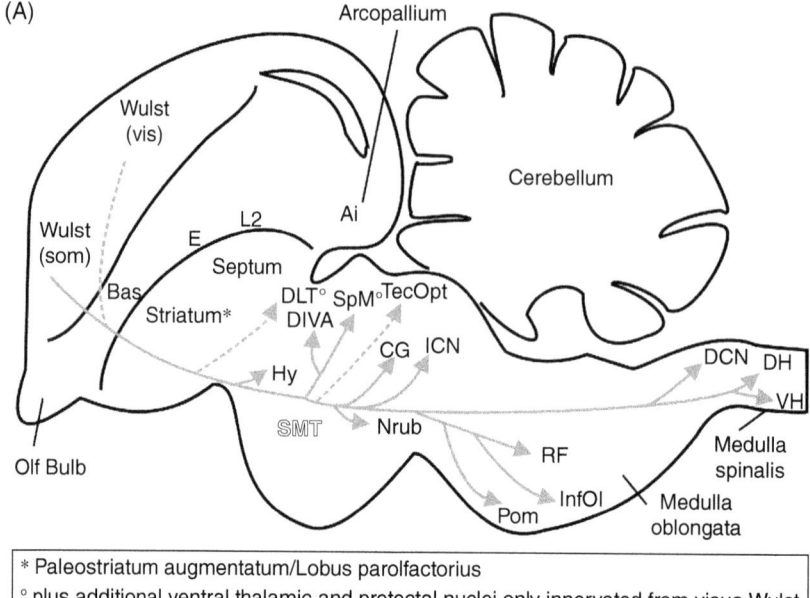

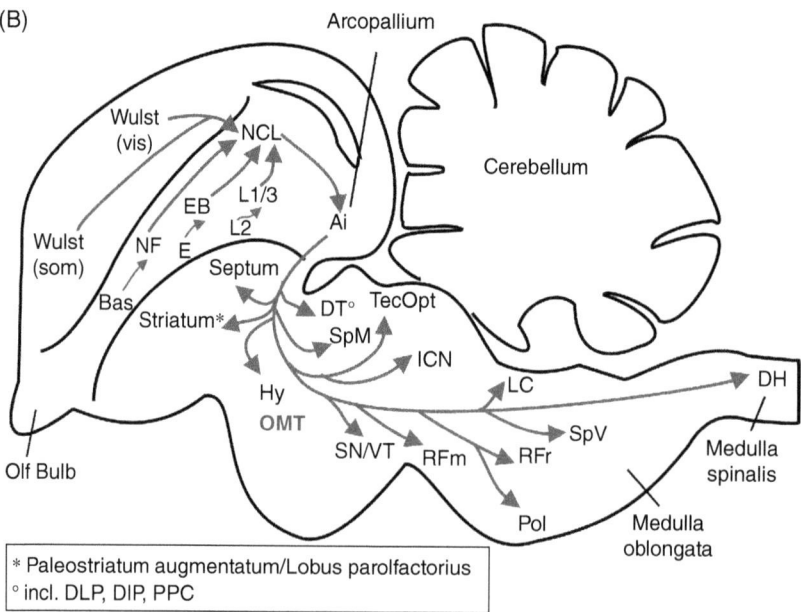

Figure 9.9 Schematics Show Sagittal Sections of Pigeon Brain with Long Descending Telencephalic Pathways.
(A) Septomesencephalic tract including data from pigeon (Karten et al., 1977; Wild, 1992) and zebra/green finch (Wild & Williams, 2000). (B) Occipitomesencephalic tract including data from pigeon (Kröner & Güntürkün, 1999; Miceli, Repérant, Villalobos, & Dionne, 1987; Wild et al., 1985; Zeier & Karten, 1971), zebra finch (Wild & Farabaugh, 1996) and mallard (Dubbeldam, den Boer-Visser, & Bout, 1997).
Abbreviations: Ai intermediate arcopallium; Bas Nucleus basalis rostralis; CG central gray; DCN dorsal column nuclei; DH: dorsal horn; DIP dorsointermediate posterior nucleus; DIVA nucleus dorsalis intermedius ventralis anterior; DLP dorsolateral posterior nucleus; DLT

striatal and pallidal parts containing bird-like distribution of neurotransmitters (striatal GABA/SubstanceP and GABA/Enkephalin, mesencephalic dopamine neuron populations) and axonal connections (Figure 9.10A) (González, Russchen, & Lohman, 1990; Guirado, Dávila, Real, Medina, 1999; Jiao et al., 2000; Reiner, Medina, & Veenman, 1998; Smeets & Medina, 1995). Behavioral experiments in reptiles and birds support similar inhibitory effects of dopamine depletion on motor behavior. Birds and reptiles likewise possess pallial input to striatum both from the DVR (Figure 9.10A) and, as in mammals, from the dorsal pallium.

Important differences exist also between reptilian and avian telencephalic organization. For example, the re-entrant (thalamopallial) pathway to the reptilian dorsal cortex is absent. Thus, the reptilian basal ganglionic output runs via the descending pathway to the optic tectum and medulla oblongata (Figure 9.10A). Also, the long descending connections of the reptilian dorsal cortex are less pronounced and extend less caudally compared to birds (Figure 9.10B) (summarized in ten Donkelaar, 1998b). While similar motor targets are reached in the hypothalamus and thalamus and in the midbrain (torus semicircularis), this is not true for targets in the brainstem or the ventral horn of the spinal cord, which are reached from dorsal pallium via the septomesencephalic tract in birds (and the pyramidal tract of mammals), but not from the reptilian dorsal cortex (Figure 9.10B). Also, the reptilian descending connections from the DVR are more restricted in comparison to those of the bird occipitomesencephalic tract.

A key question is how to interpret the reptilian/avian DVR. Molecular genetic and comparative anatomical developmental studies on telencephalic organization (Bruce & Neary, 1995; Puelles et al., 2000; Striedter, 1997) have shed new light on this issue. The embryonic mammalian pallium forms a dorsal "mantle" on top of deeper-lying subpallium in each telencephalic hemisphere (Figure 9.11) and has four parts: the medial, dorsal, lateral, and the recently distinguished ventral pallium. The mammalian medial pallium develops into the adult hippocampus (Figure 9.4B), necessary for declarative and episodic memory formation (spatial maps, scenes, persons, names etc.). Procedural memory (learned motor programs), by contrast, depends on basal ganglia and cerebellum. The mammalian dorsal pallium develops into isocortex (formerly "neocortex") with its characteristic six layers. It contains the primary sensory (cortical) areas (except those for olfaction) and—depending on the species—extensive association areas with integrative functions, including a prefrontal cortex (Wise, 2008) for complex long-term planning of behavior and, finally, premotor and motor cortices for motor execution. The lateral pallium develops into the olfactory or piriform cortex (Figure 9.4A,B), the main recipient of secondary olfactory input from the olfactory bulb. The lateral pallium also contributes to the pallial amygdala, claustrum, and endopiriform nuclei (Medina et al., 2004). The ventral pallium gives rise to the

---

Figure 9.9 (*Continued*) dor solateral nuclei (=principal optic nuclear complex); DT: dorsal thalamus; E: entopallium (primary visual); EB: entopallial belt; Hy hypothalamus; ICN intercollicular nucleus; LC locus coeruleus; L1;2;3 field L (auditory); NCL caudolateral nidopallium; NF frontal nidopallium; OMT occipitomesencephalic tract; Pol/Pom lateral/medial pontine nucleus; PPC principal precommissural nucleus; RFm mesencephalic reticular formation; RFr rhombencephalic reticular formation (incl. parabrachial nucleus); SMT septomesencephalic tract; SN/VT (dopaminergic) substantia nigra/ventral tegmental area; SpM medial spiriform nucleus; SpV descending/spinal trigeminal nuclei; VH ventral horn. (*See insert for color representation of the figure*).

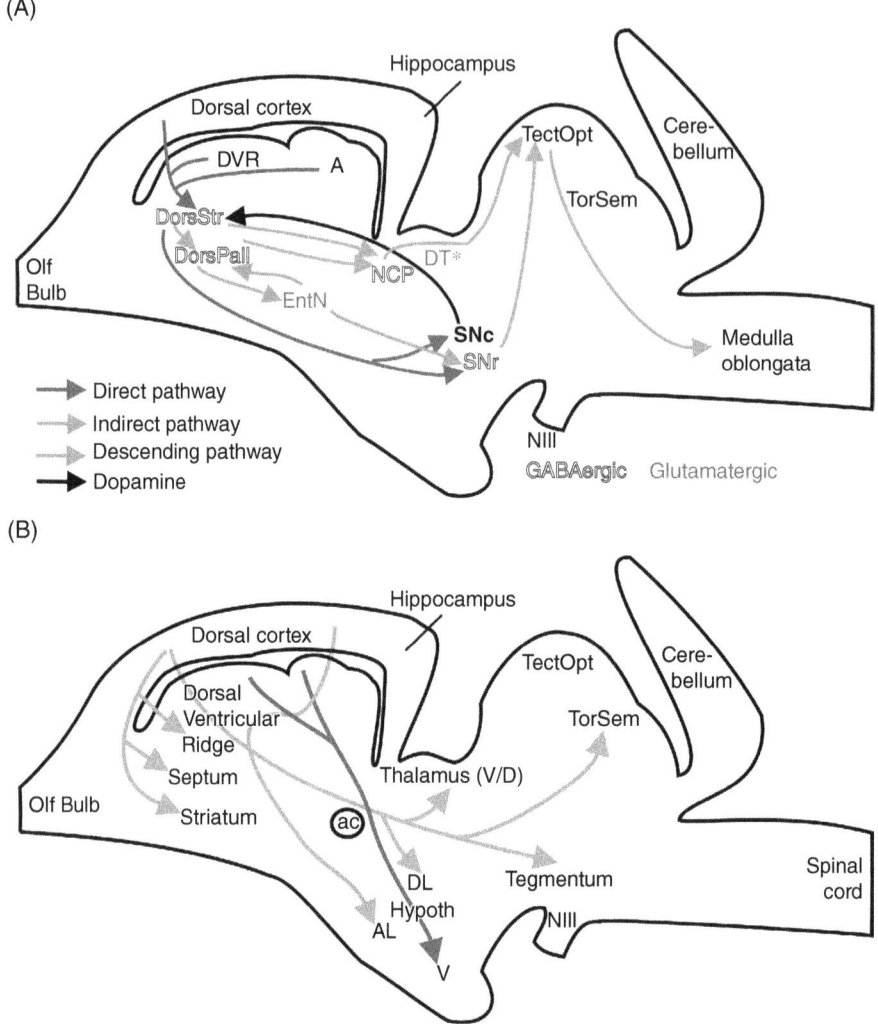

Figure 9.10 Schematics Show Sagittal Sections of Lizard Brain with (A) Basal Ganglia Circuitry (after González et al., 1990; Guirado et al., 1999; Jiao et al., 2000; Reiner et al., 1998; Smeets & Medina, 1995) and (B) Long Descending Telencephalic Pathways (after ten Donkelaar, 1998).
Abbreviations: A arcopallium; DorsPall dorsal pallidum; DorsStr: dorsal striatum DT dorsal thalamus (* not involved in basal ganglia circuitry); DVR dorsal ventricular ridge; EntN (anterior) entopeduncular nucleus; NCP (dorsal) nucleus of posterior commissure; NIII ocolumotor nerve; OlfBulb olfactory bulb; SNc/r: substantia nigra compacta/reticulata; TectOpt tectum opticum; TorSem torus semicircularis. (*See insert for color representation of the figure*).

remaining parts of the pallial amygdala (e.g., the lateral amygdala, see Figure 9.4B), but also of endopiriform region and claustrum. The embryonic mammalian subpallium consists medially of the septum and laterally of the medial and lateral ganglionic eminences (MGE and LGE, respectively, forming the adult pallidum and striatum) (Figure 9.11). The caudal region of both MGE and LGE, the CGE also gives rise to the subpallial amygdala, which lies adjacent to its pallial divisions in adults (Figure 9.4B)

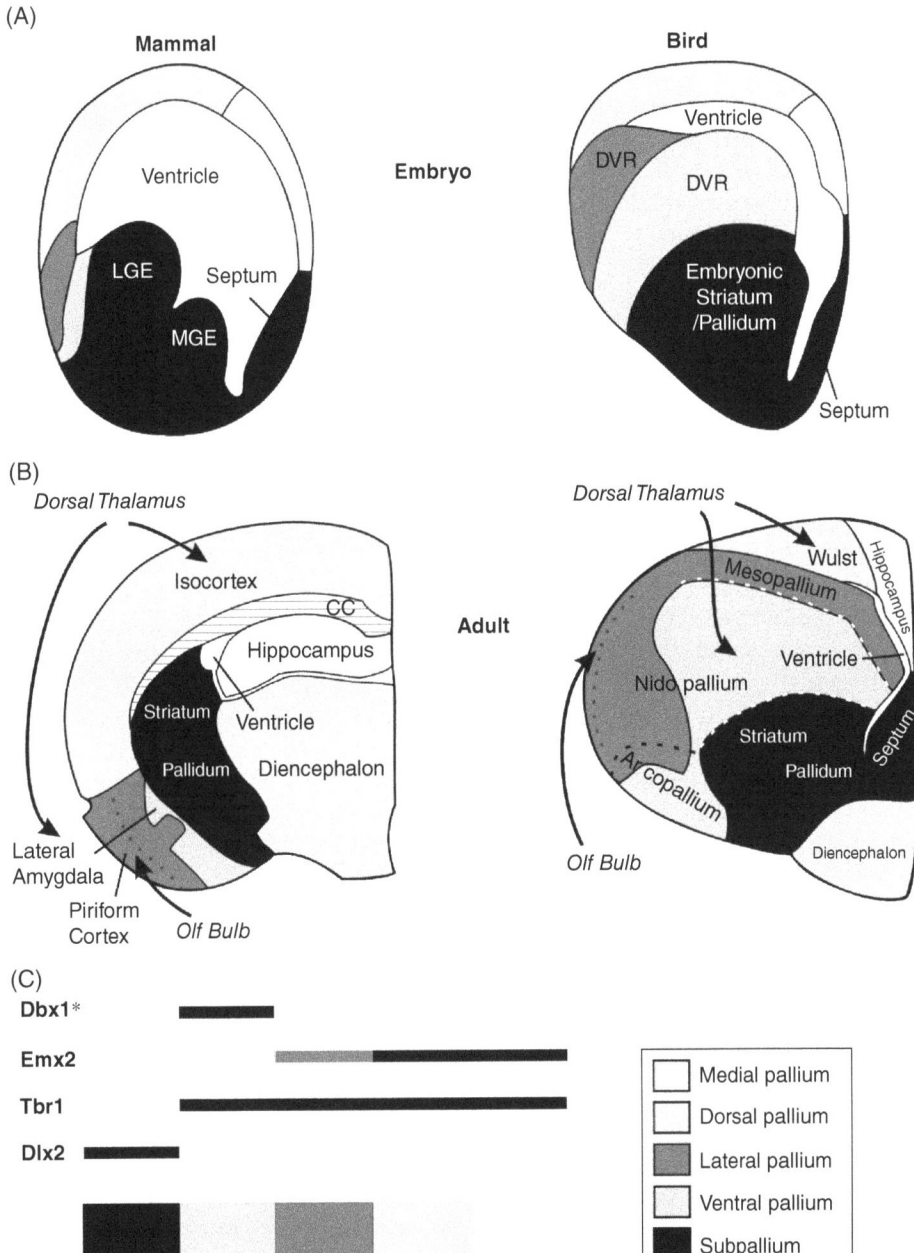

Figure 9.11 Amniote Telencephalic Bauplan and Major Sensory Inputs to Pallium in Mammals and Birds.

Amniote embryonic (A) and adult (B) telencephalic bauplan and major sensory inputs to pallium shown in schematics of the left telencephalic hemisphere in mammals on the left side and birds on the right side. (C) shows some transcription factors critical for telencephalic development and their expression domains in pallial and subpallial regions. * The expression of the *Dbx1* gene has only been described in mammals so far (Medina et al., 2004). Note that there is no direct thalamic input to hippocampus (medial pallium). Adapted from Farries (2001), using data from Martínez-García et al. (2009), Medina et al. (2004) and Puelles et al. (1999, 2000). (*See insert for color representation of the figure*).

(Puelles et al., 2000). Both the pallial and subpallial mammalian amygdala include olfactory and nonolfactory recipient nuclei (for details see Swanson & Petrovich, 1998 and Martínez-García, Novejarque, & Lanuza, 2009). The pallial amygdala, which includes the lateral and basal nuclei, is crucial for fear avoidance (LeDoux, 2000) and has reciprocal connections with isocortex and an amygdalofugal output to the striatum and hypothalamus; it also influences the subpallial central amygdala, the autonomic output structure toward the brainstem.

Gene expression studies in birds and mammals showed the extent of early active transcription factors to be topologically highly selective and consistent between the two taxa. There are regulatory genes with exclusive pallial domains (e.g., *Emx-2* and *Tbr-1*) or subpallial domains (e.g., *Mash1*, *Dlx-2*; see Figure 9.11). This underlies the generation of glutamatergic neurons in the pallium while GABAergic neurons come from the subpallium (see discussion in Osório, Mueller, Rétaux, Vernier, & Wullimann, 2010). Comparative developmental studies show that the DVR originates both from the lateral (mesopallium) and ventral pallium (nidopallium; see Figure 9.4D), but not from dorsal pallium (Martínez-García et al., 2009; Medina et al., 2004; Puelles, Kuwana, Puelles, & Rubenstein 1999; Puelles et al., 2000; Szele et al., 2002; von Frowein, Campbell K, & Götz, 2002). Thus, the ascending thalamic somatosensory, visual, and auditory inputs to the DVR are not homologous to those reaching the primary sensory isocortex of mammals. By contrast, the visual and somatosensory thalamopallial projections to the hyperpallium (Wulst) are homologous to the respective sensory paths and nuclei leading to the mammalian isocortex, because both end in the dorsal pallium and the inputs are present in all amniotes investigated so far. The mammalian lateral amygdala (LA)—like the sensory DVR—develops from the embryonic ventral pallium and in the adult brain receives collateral sensory input from corresponding thalamic nuclei (visual LPN, compare Figure 9.5, and the auditory medial geniculate nucleus, compare Figure 9.7; Doron & LeDoux, 1999) which otherwise mainly project to isocortex. Thus, mammalian lateral amygdala and sensory DVR (nidopallium) may be homologous. Nevertheless, the function of the two structures is apparently very different in birds and mammals: The avian DVR clearly serves for discrimination of fine details (perceptual path), whereas the lateral amygdala has a critical role in fear recognition and expression (LeDoux, 2000). Therefore, the (thalamorecipient) sensory part of the pallial amygdala apparently specialized in different ways in birds and mammals. Moreover, birds have an additional part of the ventral pallium (arcopallium) which, by way of its efferent output to hypothalamus and brainstem, resembles other parts of mammalian pallial (and subpallial) amygdala (see Figures 9A and 11). Thus, it appears that a large part of the ventral (and lateral) pallium in birds and reptiles develops into pallial perception-related structures that may be interpreted as convergent to parts of the primary sensory isocortex in mammals (Martínez-García et al., 2009). Furthermore, the caudolateral nidopallium (NCL; Figure 9.9B) has been said to share functional characteristics of mammalian prefrontal cortex (Güntürkün, 2005). However, its caudal topology and genetics indicate that this is the result of convergence.

This current differentiated picture of amniote telencephalic organization has improved the distinction between convergent and shared amniote characteristics. In particular, developmental biology has corroborated that similarities between DVR and sensory isocortex are due to convergent evolution.[1] Furthermore, a comparison of adult reptiles and birds reveals that the previously mentioned long descending pathways from the dorsal pallium (hyperpallium or Wulst/Isocortex) to brainstem/spinal cord and the re-entrant (thalamopallial) basal ganglia pathway also developed independently in birds and mammals. In contrast, the descending basal ganglia pathways via optic tectum to brainstem and the ascending sensory thalamic inputs to the dorsal pallium are truly ancestral for amniotes and therefore homologous. Unfortunately, a critical question will likely remain unanswered, namely what the situation was in basal synapsid reptiles, which are all extinct. Among them are the ancestors of mammals, and it would be highly exciting to know whether or not they had a DVR and dorsal cortex like extant reptiles and at what point during synapsid evolution an isocortex evolved. Maybe early amniote brains even shared more similarity with today's amphibians than reptiles. In any case, the massive expansion of pallial areas, irrespective of their embryonic ventral/lateral pallial (reptiles/birds) or dorsal pallial (mammals) origin, is a characteristic of extant amniotes. This was interpreted as a consequence of increased demands that the new amniotic lifestyle imposed on the brain. Freed from an amphibian lifestyle through the invention of a yolk-rich egg containing an embryo protected by an amnion that allows for longer embryonic maturation, amniotes developed more complex social behaviors in the context of these more differentiated reproductive and parental care strategies, perhaps including improved strategies for foraging. Such selection pressures might have facilitated an increase in complexity of the amniote forebrain because improved and finer sensory and motor skills may be essential for this new life style (Shimizu, 2001). In any case, no extant amphibian shows a DVR separate from a pallial amygdala and lateral pallium, and the dorsal pallium is relatively small and ill defined: how, then, should we interpret anamniote brain organization?

9.4.2 Comparative Brain Architecture in Anamniotes

As was the case with amniotes above, forebrain organization in anamniotes will be considered on the basis of the previously established bauplan of the craniate brain. Although much less is known about the latter in comparison with amniotes, common patterns have been observed. In particular, the historic "smell brain" theory, which assumed that the telencephalon of fishes is exclusively a secondary olfactory processing structure, turned out to be flawed. In fact, a similar synaptic chain of sensory pathways ascending to the telencephalon is present in all craniates, and a multisensory telencephalon with all major pallial divisions and shared integrative functions now seems to be present in all gnathostomes.

9.4.2.1 Amphibians: Secondarily simplified or ancestral tetrapod condition? Amphibians are the sister group of amniotes (see Figure 9.3) and, from a didactic viewpoint, their telencephalon is ideal for understanding the telencephalon's basic organization. The adult anuran telencephalon consists of two simple evaginated hemispheres, each with a central ventricle (see Figure 9.4E). Histological and neurochemical studies show that the neural walls of the adult amphibian hemispheres contain pallial and subpallial divisions (Marín, Smeets, & González, 1998; Roth & Dicke, 2009) as seen in amniotes, but without the apparent specializations of the adult reptilian/avian or mammalian brains—such as a DVR or a complex isocortex—resembling somewhat the above discussed situation seen in embryonic amniotes (Figure 9.11A). Moreover, the four pallial divisions recognized in amniotes (see §9.4.1), as well as the subpallial septal, pallidal, and striatal divisions, were confirmed in gene expression studies in amphibians (Brox, Puelles, Ferreiro, & Medina, 2003; 2004; Smith-Fernandez, Pieau, Repérant, Boncinelli, & Wassef, 1998). Pallial and subpallial amygdalar divisions were also established, including a subpallial central amygdala (autonomic center) as well as a lateral amygdala (ventral pallium) from which amygdalofugal output projects to the central amygdala, as well as striatum and hypothalamus. However, the amphibian lateral amygdala is multisensory (including olfaction), and does not have additional separate visual, somatosensory, and auditory representations as seen in amniotes (Martínez-García et al., 2009; Moreno & González, 2007).

Regarding sensory pathways, the secondary olfactory tracts reach the frog lateral pallium and, weakly, the dorsal pallium. Visual, auditory, and somatosensory projections reach the medial pallium and, weakly, the dorsal pallium from the anterior dorsal thalamus. Differently from amniotes, these inputs also reach the frog striatum from the posterior dorsal thalamus (summarized in ten Donkelaar, 1998a; Marín et al., 1998; Roth & Dicke, 2009). However, it has been argued that these thalamo-telencephalic projection neurons relay indirect sensory inputs via the ventral thalamus (Roth & Dicke, 2009). In any case, the main recipient of thalamopallial sensory information in amphibians is the medial pallium, the hippocampus homologue, and not the dorsal pallium, which is small, histologically ill-defined, and has overlapping input with both lateral and medial pallia. In addition, while the amphibian dorsal pallium does not have extratelencephalic efferent projections, the medial pallium projects (apart from septum and striatum) to the preoptic region, hypothalamus, thalamus, pretectum, and perhaps optic tectum, though not to the hindbrain and spinal cord (summarized by ten Donkelaar, 1998a). These forebrain characteristics of extant amphibians (lissamphibia) may be ancestral for tetrapods. Alternatively, secondary simplification could be involved, as was shown for the morphology and cell numbers of certain amphibian brain regions (Roth & Dicke, 2009).

Despite the morphological simplicity of adult amphibian brain organization, many details of forebrain organization are surprisingly similar to amniotes, as noted above, including basal ganglia organization (for example separate dorsal and ventral striato-pallidal systems; Marín et al., 1998). Neurochemical and connectional studies have showed, in essence, all major neural elements of the direct and indirect pathways of the somatomotor loop as described above for the reptilian brain, for example, a GABAergic striatum and pallidum (Endepols, Helmbold, & Walkowiak, 2007), a glutamatergic subthalamic nucleus, and a dopaminergic substantia nigra (Marín et al., 1998). The basal ganglia output is, as in reptiles, through the descending pathway via pretectum and optic tectum to medulla oblongata and a re-entrant thalamopallial pathway does not exist. Instead of the strong dorsal pallial input seen in amniotes, the amphibian striatum receives massive sensory-related thalamic input. Despite these differences, the presence of a basal ganglia network in amphibians shows that it is an ancestral characteristic of tetrapods and does not originate in amniotes only.

9.4.2.2 Ray-finned fishes: The parallel universe in vertebrate brain evolution. Ray-finned fishes represent half of all extant vertebrate species with the most derived representatives of their largest taxon, the teleosts, being equally remote from their paleozoic ancestors as today's mammals are from synapsid reptiles. Ray-finned fishes went through subsequent radiations (chondrosteans in the paleozoic, holosteans in the mesozoic) documented today only by few extant species (bichirs, sturgeons, bony gars). The fossil origin of the now-dominant extant teleosts is in the early mesozoic (late Triassic, as it is for mammals, Carroll, 1988), when the now-extinct pholidophorids and leptolepids formed large pelagic swarms. At, and just after, the Cretaceous–Tertiary boundary, a massive teleostean radiation occurred, leading to great species diversity and the invasion of practically every aquatic environment (Wullimann, 1997). For example, the speciose perciform (perch-like fishes) and tetraodontiform (e.g., trigger-fish, Figure 9.3) taxa arose then, at the same geological period as placental mammals. The extent of this radiation is also reflected in teleostean brain diversity (Meek & Nieuwenhuys, 1998; Wullimann, 1998).

Of course teleosts share with other gnathostomes the above-mentioned brain bauplan characteristics. However, some derived developmental changes resulted in massive consequences for adult forebrain organization. Like tetrapods, ray-finned fishes have a telencephalic subpallium ventrally (septum/striatum) and pallium dorsally (Northcutt & Braford, 1980), corroborated by gene expression studies (Figure 9.4F; Mueller, Wullimann, & Guo, 2008; Wulllimann, 2009; Wullimann & Mueller, 2004). However, unlike tetrapods, who have paired evaginated hemispheres with telencephalic divisions that develop around lateral ventricles (compare Figure 9.4E), the actinopterygian telencephalon undergoes eversion, a derived condition in which the medial pallium detaches from the septum and rolls out laterally (Figure 9.4G). A morphocline from basal actinopterygians to teleosts demonstrates the increasing degree of telencephalic eversion, with bichirs, chondrosteans, and gars having the medial pallium positioned increasingly more lateroventrally (see Northcutt 2009; Wullimann & Vernier, 2009a). In teleosts (e.g., zebrafish; Figure 9.4F), telencephalic eversion is most advanced and, consequently, the dorsal surface of the telencephalon is periventricular in nature, covered by a thin neural epithelium which encloses the flattened telencephalic ventricle.

As a result, eversion displaces the medial pallium of bichirs (*Polypterus*) *laterally* to the lateral pallium. Nonetheless, the latter receives the expected dense secondary olfactory input from the olfactory bulb (see Wullimann & Vernier, 2009a); likewise, the everted bichir medial pallium receives its expected sensory (at least visual) input from the posterior tuberculum, a basal diencephalic division (see Figure 9.2A for zebrafish, and below) (Northcutt 2009). This is different from amphibians, where sensory input to the medial pallium arises from the dorsal thalamus (see §9.4.2.1.). Teleosts have a large posterior-tuberculum-derived relay region for ascending sensory systems, called the preglomerular area, which relays various sensory inputs to the pallium. Different teleostean pallial zones (see Figure 9.4 for abbreviations) have variable sensory inputs (Dl: visual, lateral line; Dm: visual, lateral line, auditory, gustatory; see discussion in Northcutt, 2006; Vernier & Wullimann, 2009; Wullimann & Mueller, 2004; Wullimann & Vernier, 2009b). In African mormyrids, segregated bi- and unimodal sensory areas were found in the pallium (mostly Dm; Prechtl et al., 1998).

Lesion experiments involving the lateral pallial zone of goldfish (Dl; Rodríguez, López, Vargas, Broglio, Gómez, & Salas, 2002; Salas, Broglio, & Rodríguez, 2003) lead to deficits in the retention of learned spatial maps resembling the effects of hippocampal dysfunction in mammals, and thus suggest Dl is homologous to the medial pallium of other vertebrates. Complementary lesions of the medial pallial zone (Dm) lead to a specific loss of fear recognition, indicating a homology with the amniote pallial amygdala (ventral pallium). Furthermore, these functional data are in line with detailed connectional studies of these two major pallial zones in goldfish (Northcutt, 2006) and with topological position according to eversion (Figure 9.4F-G). However, the identity of the teleostean lateral (olfactory) pallium remains controversial. Clearly, the densest secondary olfactory input to the teleostean pallium is to the posterior pallial zone (Dp; Figure 9.4F). Assuming that Dp is everted, this would suggest that it is a part of the medial pallium (Nieuwenhuys, 2009) which acquired novel olfactory input during teleost brain evolution. However, the surface of Dp has been interpreted as pial rather than periventricular in nature (Lillesaar, Stigloher, Tannhäuser, Wullimann, & Bally-Cuif, 2009), and to have migrated radially into its peripheral (pial) position, ventrolateral to Dl, from a periventricular proliferation zone between Dm and Dl (Mueller, Dong, Berberoglu, & Guo, 2011). Thus, Dp may *not* be everted and may instead represent the lateral pallium (LP) (Figure 9.4G) (partial eversion hypothesis: Wullimann & Mueller, 2004). The teleostean dorsal pallium has been hypothesized to include portions of Dl and Dm at their merging point, plus the central telencephalic pallial zone (Dc; Wullimann & Mueller, 2004). Alternatively, Dc may represent a exclusively dorsal pallial histogenetic unit with a separate periventricular origin between Dm and Dl (Mueller et al., 2011).

Thus, a plausible comparative neuroanatomical picture of actinopterygian pallial divisions emerges, despite difficulties arising from eversion. Quite possibly, all four teleostean pallial divisions are conserved between fish and tetrapods (Figure 9.4F), and receive sensory information from the diencephalon in addition to secondary olfactory input as discussed above. The major difference from tetrapods is that the teleostean input from the diencephalon arises in the posterior tuberculum (posterior tuberculum/preglomerular complex) rather than the dorsal thalamus, which instead projects mainly subpallially, likely to the striatum (Northcutt, 2006). The preglomerular complex is a peripherally migrated (posterior tubercular) diencephalic structure which contains specific nuclei for ascending sensory modalities (visual, auditory, gustatory, somatosensory, mechanosensory lateral line; Northcutt, 2006) with reciprocal connections to the pallium. The preglomerular complex thus shows many functional correspondences with the sensory dorsal thalamus of tetrapods without being homologous to it, so the extensive sensory input to the pallium in both taxa may have arisen convergently. This may also apply to the suspected dorsal pallium (Dc), as it cannot be discriminated unambiguously in basal actinopterygians (Northcutt, 2009).

The subpallium of actinopterygians includes a septum (ventral nucleus of area ventralis, Vv) and basal ganglia (dorsal nucleus of area ventralis, see; Figure 9.4F; Wullimann & Mueller, 2004; Mueller et al., 2008). The position of the septum—ventral instead of medial to the basal ganglia—is indirectly due to the pallial eversion and the flattening of the lateral ventricles (Figure 9.4F-G). Recent developmental studies in the zebrafish have finally evidenced that striatal and pallidal divisions exist in the early dorsal subpallial area (Mueller et al., 2008). The pallidal part gives rise to GABAergic neurons of the future subpallium as well as of those invading the pallium as seen in tetrapods. In addition, dopaminergic neurons in the posterior tuberculum innervate the teleostean striatum. Since tetrapods also have some dopaminergic neurons which develop in the posterior basal diencephalon, in addition to their majority in the basal midbrain, the former may be considered homologous (Rink & Wullimann, 2001; see Wullimann, 2014 for a current controversy). This strongly indicates that ray-finned fishes, like tetrapods, have a dopaminergic modulation of the telencephalic basal ganglia, but our functional understanding is in its infancy. Beyond dopamine, the distribution of major modulatory transmitter systems has been shown to be highly

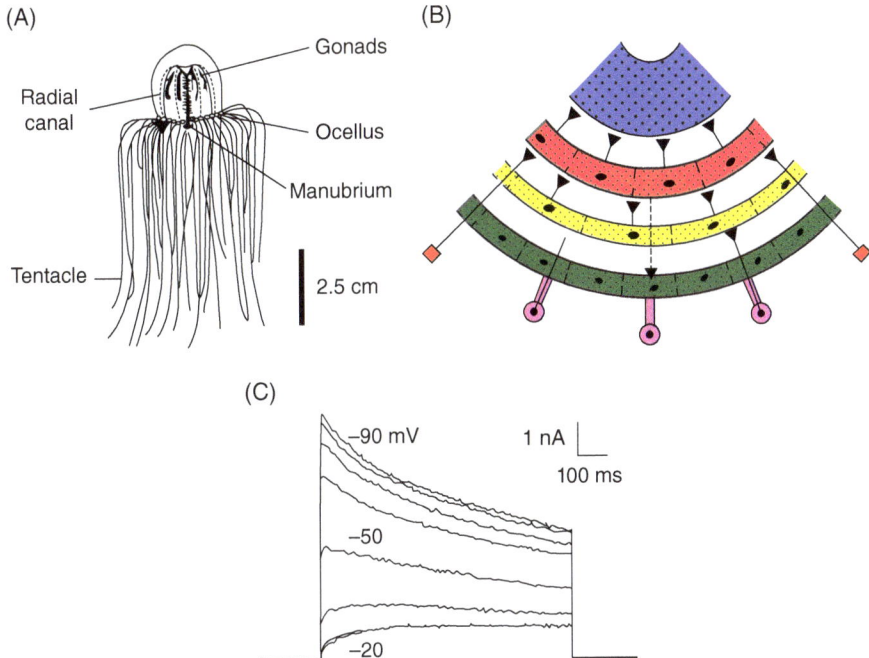

Figure 5.5 Modular Nature of *Polyorchis penicillatus* Nervous System.
(A) Line drawing of a *Polyorchis* specimen at rest. (B) Partial representation of the ring-shaped neuronal networks showing known synaptic contacts. The muscle epithelium (blue) is directly excited by the swim motor neurons (red) in the inner nerve ring. They in turn receive an excitatory synaptic input from neurons of the "B" system (yellow) in the outer nerve ring and from unicellular receptors (orange). Excitatory inputs to the "B" system are shown arising in the ocelli (purple). Also shown is an excitatory pathway from the swim motor neurons to the "O" system (green) in the outer nerve ring; other connections may be present. Excitatory synapses are indicated by filled triangles (▼). All three systems consist of a ring of electrically coupled nerve cells. Adapted from Spencer & Arkett 1984. (C) Inactivating K⁺ currents recorded under voltage clamp in response to test commands to +50 mV. Each test command preceded by a conditioning command lasting 1 s. The superimposed current traces show the effect of the conditioning level (range −90 to −20 mV; 10 mV steps) on the availability of the inactivating current. At −20 mV the inactivating component (IK_{fast}) is absent, leaving IK_{slow}. Adapted from Przysiezniak & Spencer, 1994.

The Wiley Handbook of Evolutionary Neuroscience, First Edition. Edited by Stephen V. Shepherd.
© 2017 John Wiley & Sons, Ltd. Published 2017 by John Wiley & Sons, Ltd.

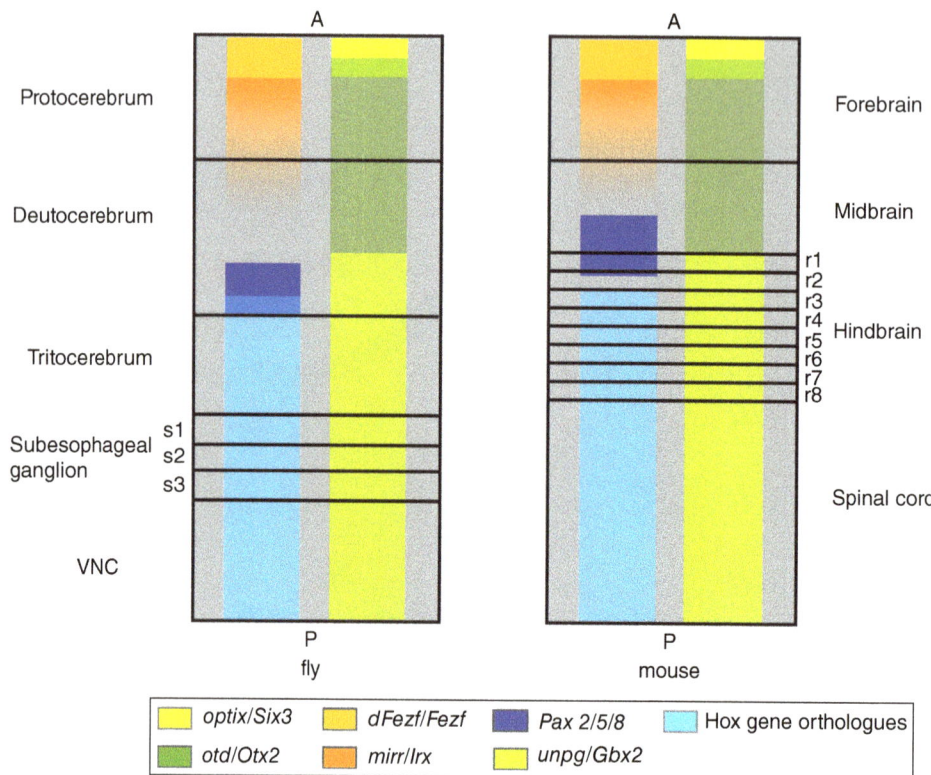

Figure 6.2 Simplified Summary Scheme of the Anteroposterior Order of Conserved Gene Expression in Embryonic CNS Development of Bilaterians.

Dorsoventral patterning is not indicated. Schematic diagram shows the expression of the patterning genes *optix/Six3*, *otd/Otx2*, *dFezf/Fezf*, *mirr/Irx*, *Pax 2/5/8*, *unpg/Gbx2* and Hox gene orthologues in the developing CNS of *Drosophila* and mouse. Expression domains are color-coded. (left) Gene expression in *Drosophila* CNS of embryonic stage 14. Borders of the protocerebral, deutocerebral, tritocerebral, mandibular (s1), maxillary (s2), labial (s3), and ventral nerve cord (VNC) neuromeres are indicated by horizontal lines. (right) Gene expression in mouse CNS of embryonic day 9.5–12.5. Borders of the forebrain, midbrain and the hindbrain and its rhombomeres (r1-r8) are indicated by horizontal lines. In both fly and mouse, an *optix/Six3* expression domain patterns the most anterior CNS region and overlaps with the *otd/Otx2* expression pattern (Steinmetz et al., 2010) which is anterior to the abutting *unpg/Gbx2* expression (Bouillet, Chazaud, Oulad-Abdelghani, Dollé, & Chambon, 1995; Urbach, 2007; Wassarman et al., 1997). In both animals, a *Pax2/5/8-* expression domain is positioned close to the interface between the anterior *otd/Otx2* and the posteriorly abutting *unpg/Gbx2* expression domains (Asano & Gruss, 1992; Hirth et al., 2003; Rowitch & McMahon, 1995). Hox genes orthologues expression follows posteriorly to the *Pax2/5/8* expression domain in both animals (Davenne et al., 1999; Hirth et al., 1998; Lichtneckert & Reichert, 2007). Furthermore, the interface of the relative expression of *dFezf/Fezf* and *mirr/Irx* was reported to be conserved between fly and mouse (Irima et al., 2010; Oliver et al., 1995). Adapted from Lichtneckert and Reichert, 2007.

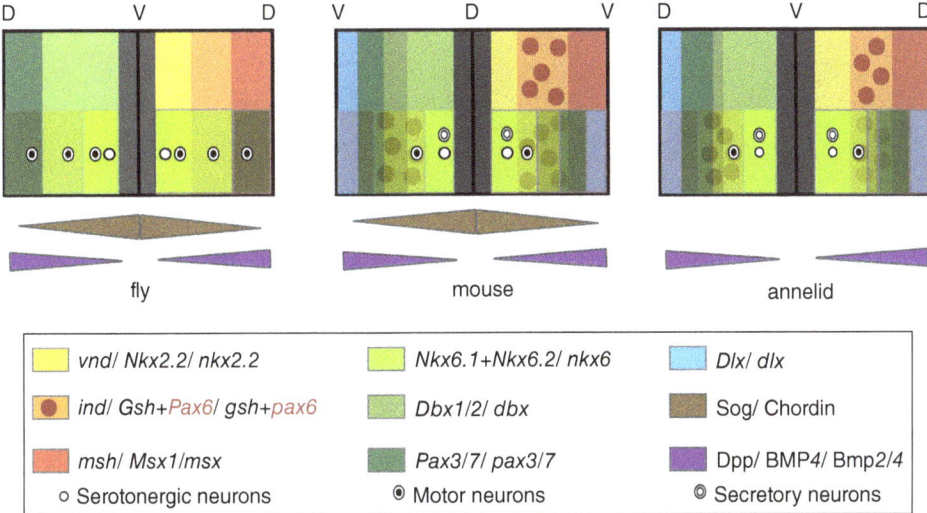

Figure 6.3 Schematic Representation of Examples of Conserved Dorsoventral Genetic Expression Boundaries in a Segmental Part of the Neuroectoderm in Arthropods (Left), Vertebrates (Middle) and Annelids (Right).

The vertebrate neuroectoderm is shown before folding. Anteroposterior patterning is not indicated. The neurogenic region is patterned in a dorsoventral fashion by a set of conserved patterning genes in all three animals, here indicated by color code. Note that the neuroectoderm of each animal is subdivided in two parts at its midline by a black vertical line enabling to show normally overlapping gene expression domains more clearly. At the bottom of the bars the overlap is shown for better comprehension. Within this overlay conserved neuron cell types emerging from this particular region are indicated by different circles (Arendt et al., 2008; Denes et al., 2007; Mizutani & Bier, 2008). The homologous proteins Dpp/ BMP4/ Bmp2/4 (violet) form a dorsoventrally inverted gradient in vertebrates with respect to *Drosophila melanogaster* and *Platynereis dumerilii*. In *Drosophila* and vertebrates, another homologous protein pair, namely Sog/ Chordin (brown) forms an opposing gradient with respect to the Dpp/ BMP4 pattern, where it inhibits Dpp/ BMP4 and therefore enables induction of neurogenesis and with different gradients gives identity to different subdomains of the neuroectoderm (Lichtneckert & Reichert, 2005). The dorsoventral columnar patterning genes are highly conserved between the bilaterian animals (see comparable relative expression domains of *vnd/ Nkx2.2/ nkx2.2* (yellow), *ind/ Gsh/ gsh* (orange), *msh/ Msx1/msx* (red), *Nkx6.1 + Nkx6.2/ nkx6* (light green) in *Drosophila*, mouse and *Platynereis*) (Lichtneckert & Reichert 2007; Seibert, Volland, & Urbach, 2009). In the annelid and the mouse neuroectoderm even more similarities compared to *Drosophila* are apparent, such as the additional *Dbx1/2/ dbx* and *Dlx/ dlx* expression domains, the columnar medial *Pax6* expression (red dots) domain (Mizutani & Bier, 2008), as well as the *Pax3/7* expression which in *Drosophila* is expressed in a strictly segmented fashion (dark green) (Denes et al., 2007).

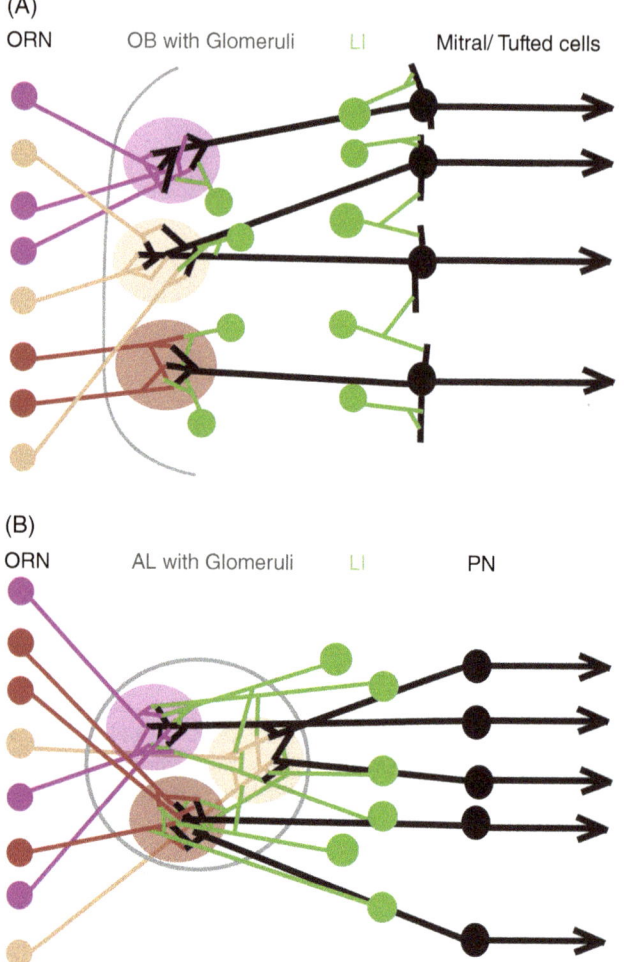

Figure 6.4 General Similarities of Olfactory Circuit Organization in Mammals (A) and Insects (B).
ORN expressing the same olfactory receptor project to the same glomerulus in both animals (expressed olfactory receptor type in neurons is indicated by differently colored neurons). In the glomeruli the ORN connect to the dendrites of the mitral/tufted cells in the mammals (A) or PN in insects (B). In both animals, the sensory information is then transmitted by the mitral/tufted cells or the PN into higher brain centres. Different LI interconnect the information from the various glomeruli and process this olfactory information in fly and mouse. AL, antennal lobe; ORN, olfactory receptor neurons; OB, olfactory bulb; LI, local interneurons; PN, projection neurons. Adapted from Kay & Stopfer, 2006.

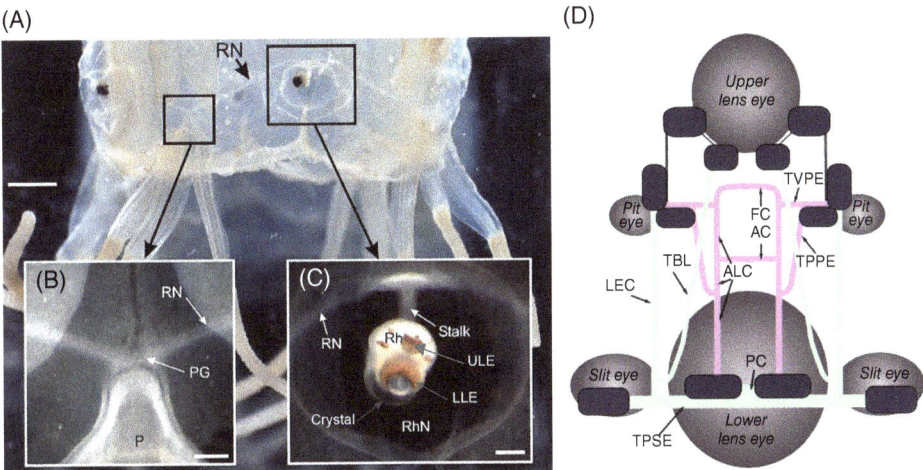

Figure 6.5 Complexity of the CNS of the Cubozoan Jellyfish *Tripedalia cystophora*. (A) The ring nerve RN connects the pedalial ganglion PG (B) with the rhopalia Rh (C) in the central nervous system. (C) the Rh constitute the main sensory structures of cubomedusae. Rh hang within the RhN on a stalk and carry six eyes (only the two lens eyes ULE and LLE are indicated). (A–C) Modified from Garm 2006. Reproduced with permission of Springer Publishing. (D) Schematic overview of commissural connections (light green and violet) between the different neuronal cell groups (dark blue) of the 6 distinct eyes (grey circles) in the rhopalium, indicating the remarkable complexity of this visual and integrating structure. Rh, rhopalium; RhN, rhopalial niche; LLE, large lens eye; ULE, upper lens eye; RN, ring nerve; PG, pedalial ganglion; P, pedalium; AC, anterior commissure; ALC, apical lateral connective; FC, frontal commissure; LEC, lateral exe connective; PC, posterior commissure; TBL, basal lateral tract; TPPE, posterior pit eye tract; TPSE, posterior slit eye tract; TVPE, vertical pit eye tract. Bars indicate 1 mm (A), 100 μm (B, C). Adapted from Parkefelt 2005. Reproduced with permission of John Wiley & Sons, Inc.

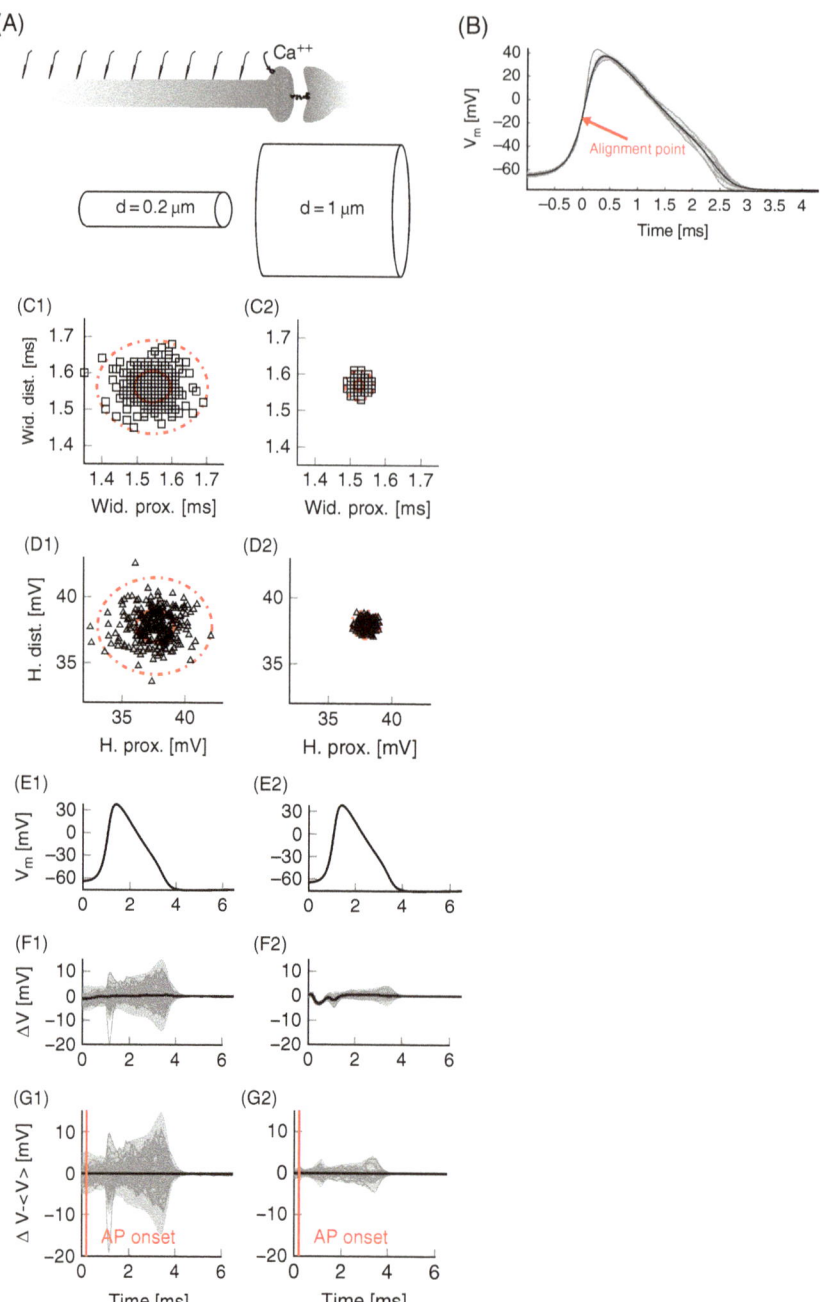

Figure 7.4 Noise Modifies the AP Waveform.
(A) Membrane potential and ionic currents are recorded at regularly placed points along the axon. The subfigures display data for N>200 single APs triggered by identical stimuli and initial conditions in thin squid giant axon model. (B) Due to channel noise, APs triggered in an identical fashion will have different shapes across trials. Here APs in a 0:2 μm diameter axon from 5 trials out of 250 are superimposed. The point at which the membrane potential crossed the half-height line is used to align APs. (C) Distribution of AP width and (D) AP height (red circle, 1SD; dotted circle 3SD). (E) Mean waveform of the AP at the proximal site. (F) Pairwise difference between an AP's shape at the proximal and the distal location. The average difference is plotted in thick black, while the light grey shaded area represents the 3SD range. Grey lines represent sample traces plotted individually. (G) Fluctuations around the mean pairwise difference. The average difference is plotted in thick black (0 by definition), while the light grey shaded area represents the 3SD range. Grey lines represent sample traces plotted individually. Adapted from Neishabouri & Faisal 2014.

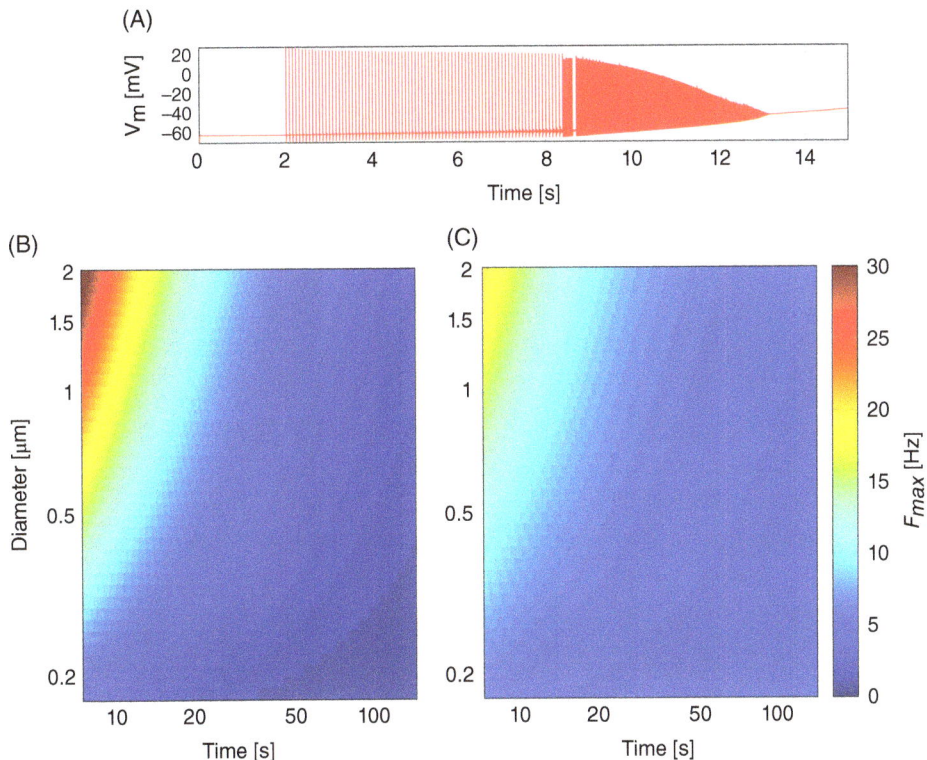

Figure 7.6 Maximum Sustainable Firing Rates.
(A) Firing at too high a firing rate for too long can drive the neuron into a burst which will eventually deplete ionic concentration gradients and lead to cell death. (B) Color map of maximum sustainable firing rates given by fitting a simplified model using parameters from the squid giant axon. Over long time periods, the sustainable firing rate is independent of diameter. (C) For shorter time periods, larger axons can fire at higher rates. The refractory period prevents the axon from firing at more than approx. 40 Hz.

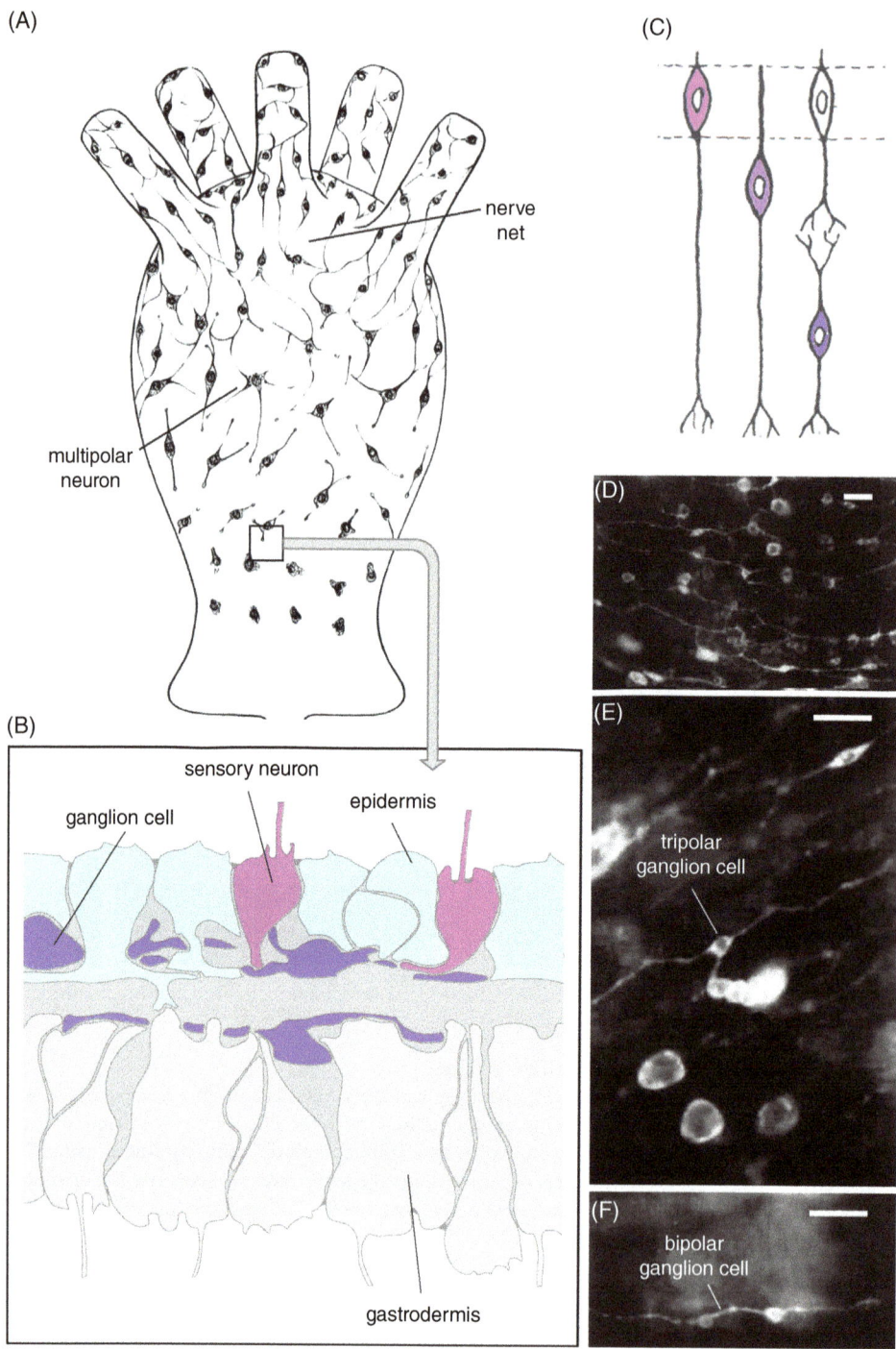

Figure 8.1 Nerve Net in Cnidarians.
(A) Line drawing of polyp, showing multipolar neurons (after). (B) Schematic cross-section of bodywall of cnidarian, showing outer layer (epidermis), inner layer (gastrodermis), sensory neurons, and ganglion cells. (C) Hertwig's hypothesis of evolution of nerve cells. Hanstroem, 1968. Reproduced with permission of Springer. (D–F) Labeling of nerve net of *Hydra viridissima* with antibody CC04. Sakaguchi 1996. Reproduced with permission of John Wiley and Sons.

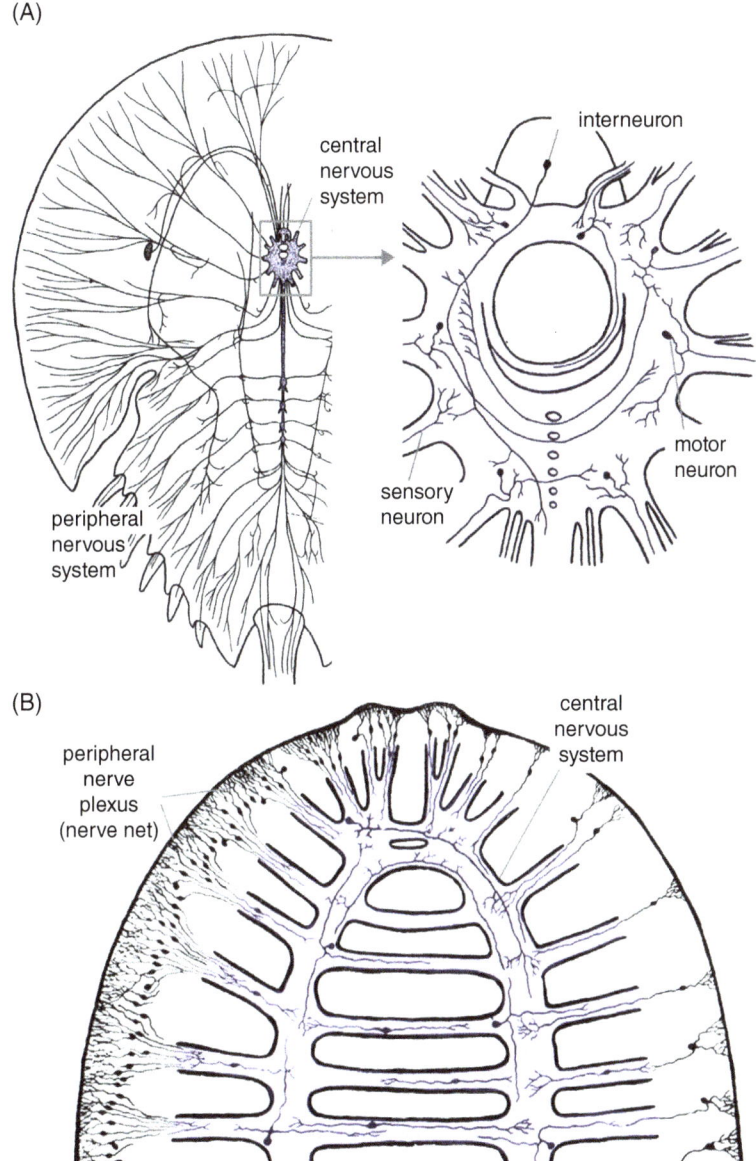

Figure 8.2 Division of Central and Peripheral Nervous System.
(A) Chelicerate *Limulus polyphemus*; (B) Platyhelminth *Bdelloura candida*. Hanstroem 1968. Reproduced with permission of Springer.

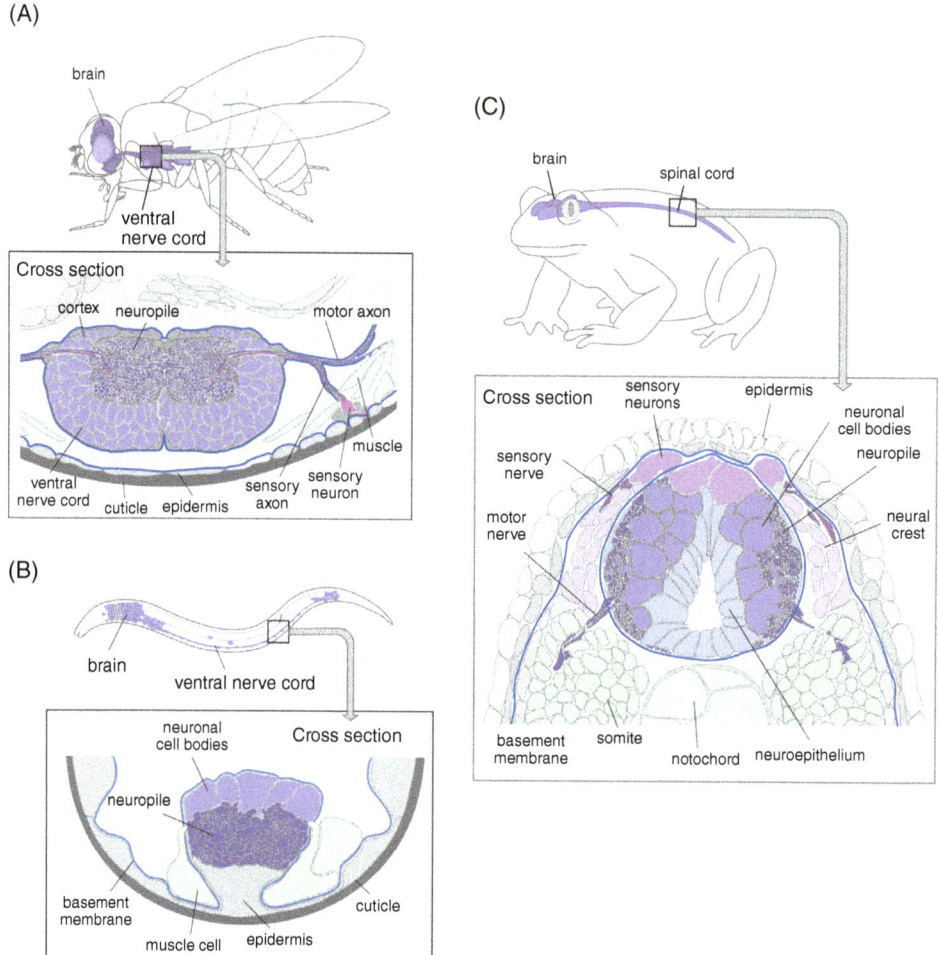

Figure 8.3 Architecture of the Central Nervous System in Bilaterian Animals.
(A) Subepidermal nervous system (arthropod *Drosophila melanogaster*); (B) Basiepithelial nervous system (nematode *Caenorhabditis elegans*); (C) Invaginated nervous system (Chordate *Xenopus laevis*).

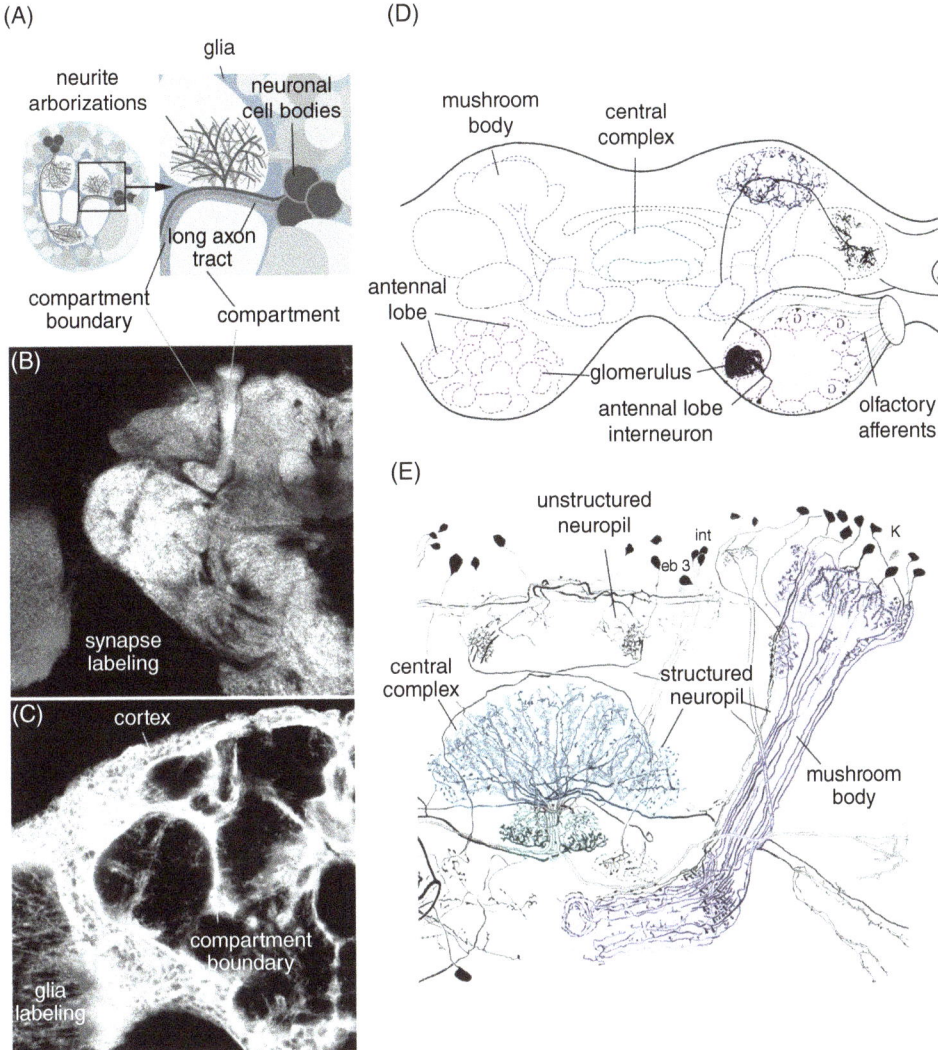

Figure 8.5 Architecture of the Invertebrate Brain Neuropil.
(A) Schematic section of brain, showing relationship between neurons, glia, compartments, and compartment boundaries. (B, C): cross-section of Drosophila brain hemisphere labeled with marker for synapses (B), highlighting compartments, and against glia (C), showing compartment boundaries. (D, E) Line drawings of schematic sections of insect brains, showing examples of Golgi-stained neurons forming structured neuropils (antennal lobe, central complex, mushroom body) and unstructured neuropils. Adapted from Hanstroem 1968. Reproduced with permission of Springer.

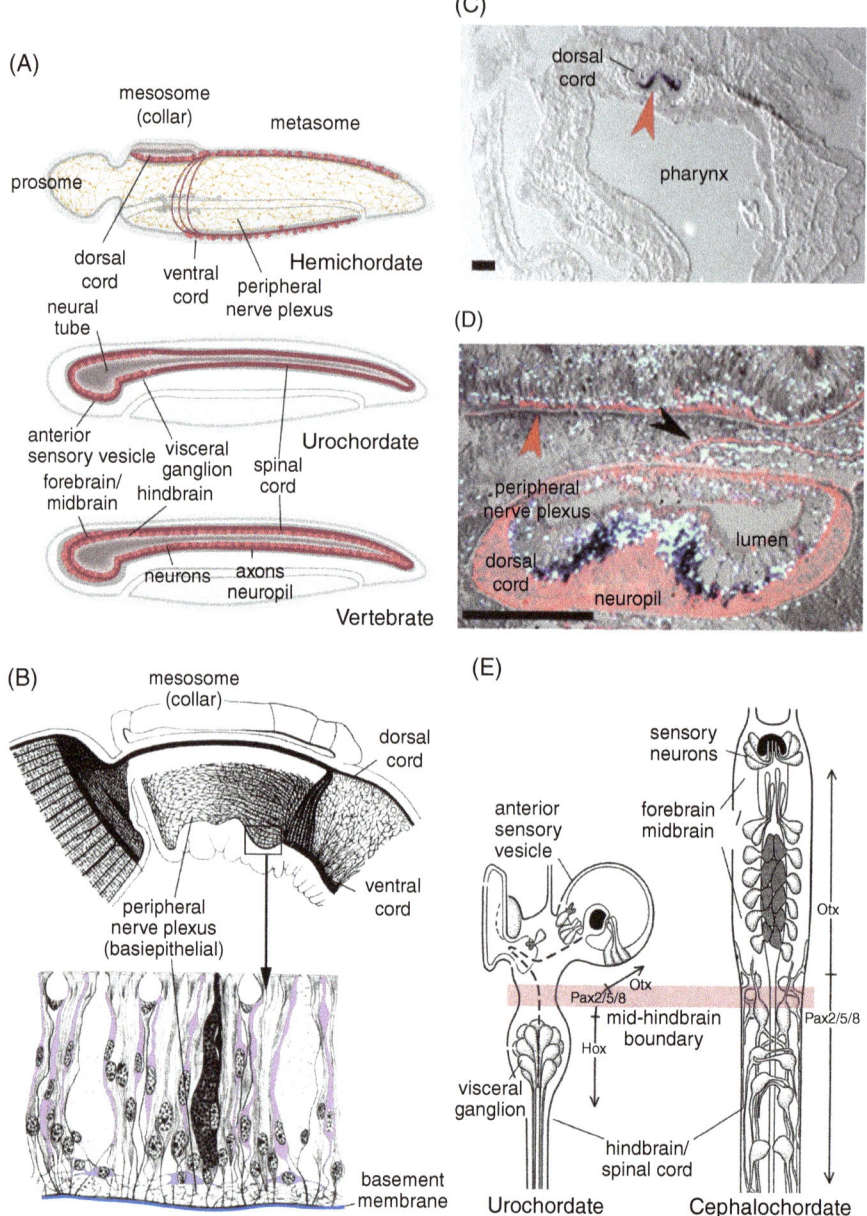

Figure 8.6 The central nervous system of deuterostomes.
(A) Essential characteristics of nervous system architecture in hemichordates, urochordates, and vertebrates, represented as schematic sagittal sections. (B) Line drawing of nervous system of hemichordate *Saccoglossus cambrensis*. Top: nerve net, dorsal cord and ventral cord in collar region. Bottom: cross-section of body wall, showing cytology of basiepithelial nerve plexus. Knight Jones, 1952. Reproduced with permission of the Royal Society. (C, D) Cross-sections of dorsal cord of hemichordate *Ptychodera flava*. Neuronal marker Elav is expressed in ventral wall of epithelial dorsal cord (blue; black arrowhead in D). Scale bar: 100 mm. Neuropil of dorsal cord and peripheral nerve plexus is labeled by antibody against Acetylated tubulin (red; red arrowheads). Black arrowhead points at continuous strand of neuropil connecting dorsal cord with peripheral nerve plexus. Nomaksteinsky 2009. Reproduced with permission of Elsevier. (E) Schematic representation of brain of urochordate *Botryllus schlosseri* (left) and cephalochordate *Branchiostoma floridae* (right; both in dorsal view). Expression domains of Otx, Pax2/5/8 and Hox complex demarcatates region considered to be homologous to the vertebrate midbrain-hindbrain boundary (red bar). Lacalli 2001. Reproduced with permission of The Royal Society.

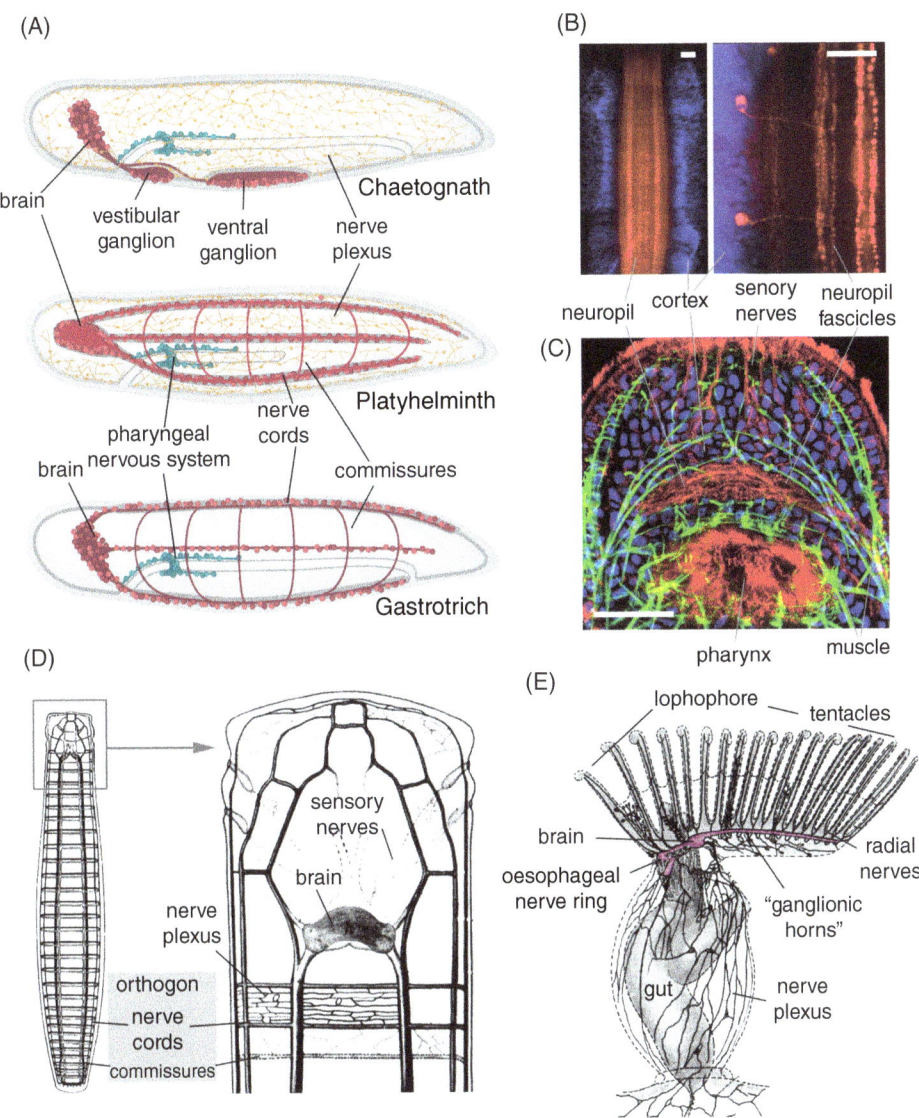

Figure 8.7 The central nervous system of lophotrochozoans.
(A) Essential characteristics of nervous system architecture in chaetognaths, platyhelminths, and gastrotrichs, represented as schematic sagittal sections. (B) Ventral ganglion of chaetognath *Sagitta setosa* (ventral view). Left photograph shows cortex (nuclei of neuronal cell bodies labeled blue) and neuropil (anti-Synapsin, red); right photograph represents higher magnification of neuropil with subset of neurons (anti-RFamide, red) forming distinct longitudinal tracts (scale bars 25 mm). (A, B) Harzsch 2007. Reproduced with permission of BioMed Central Ltd. (C) Brain of platyhelminth *Macrostomum lignano* (dorsal view; scale bar 25 mm). Nerve fibers and cilia of epidermis/pharynx are labeled by anti-Tyrosinated tubulin (red); muscle fibers labeled by phalloidin (green); nuclei of all cells in blue. (D) Line drawing of nervous system of platyhelminth *Bothrioplana semperi* (dorsal view). (E) Line drawing of nervous system of ectoproct (=bryozoan) *Cristatella mucedo* (lateral view). (D, E) Hanstroem, 1968. Reproduced with permission of Springer.

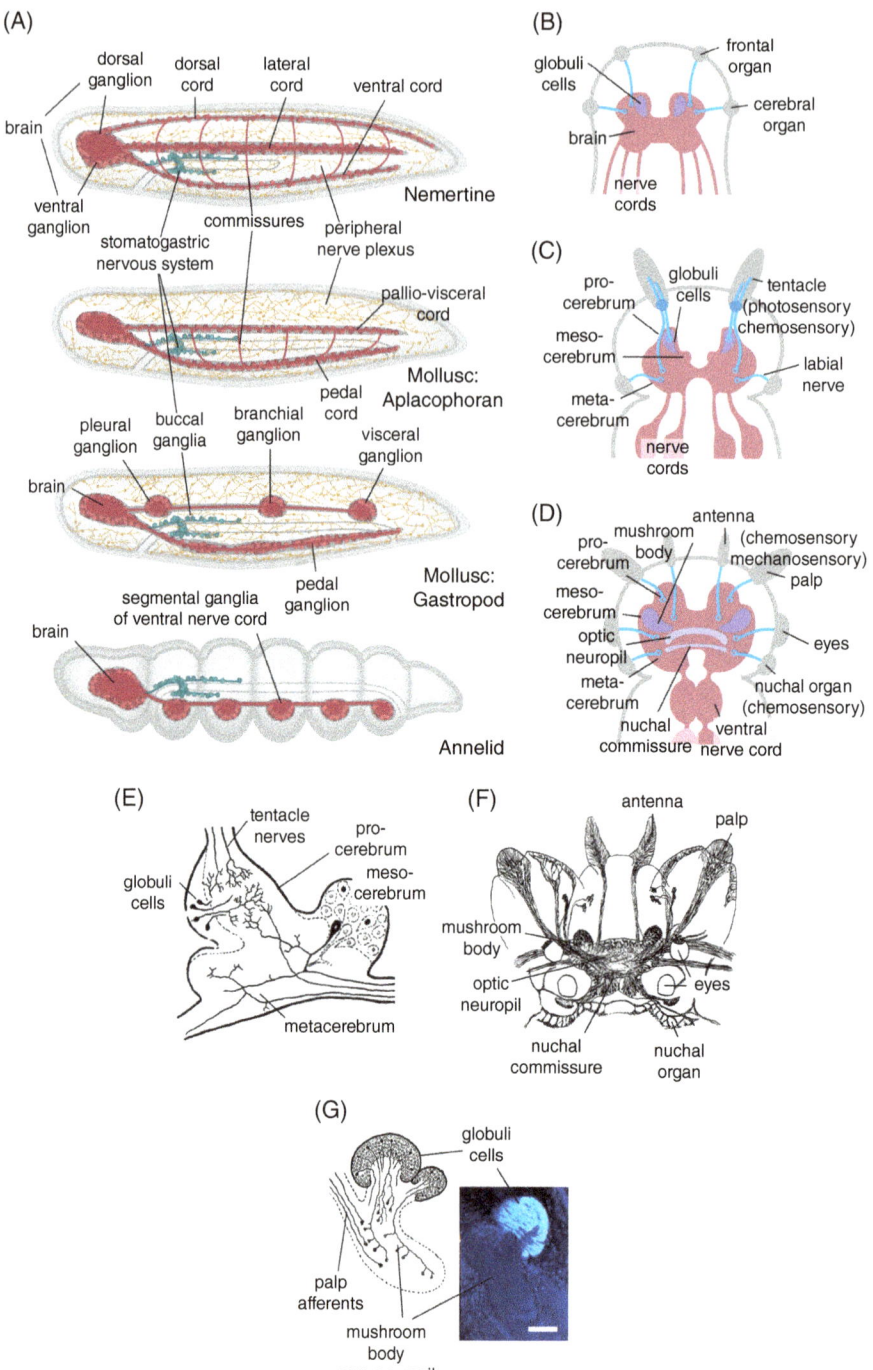

Figure 8.8 The Central Nervous System of Lophotrochozoans.
(A) Essential characteristics of nervous system architecture in nemertines, molluscs, and annelids, represented as schematic sagittal sections. (B–D) Schematic representation of head sensory organs and brain of nemertines (B), gastropod molluscs (C), and polychaete annelids (D; all in dorsal view). Sensory nerves in blue, structured neuropil compartments in purple. (E–G) Line drawings of neurons forming structured neuropil compartments. (E) Globuli cells of gastropod *Helix pomata*. (F, G) Sensory nerves, brain and mushroom body of polychaete *Nereis diversicolor*. (F–G) Hanstroem, 1968. Reproduced with permission of Springer. Inset in (G): photograph of mushroom body of *Nereis diversicolor* (scale bar 50 mm). Heuer and Loesel, 2008. Reproduced with permission of Springer.

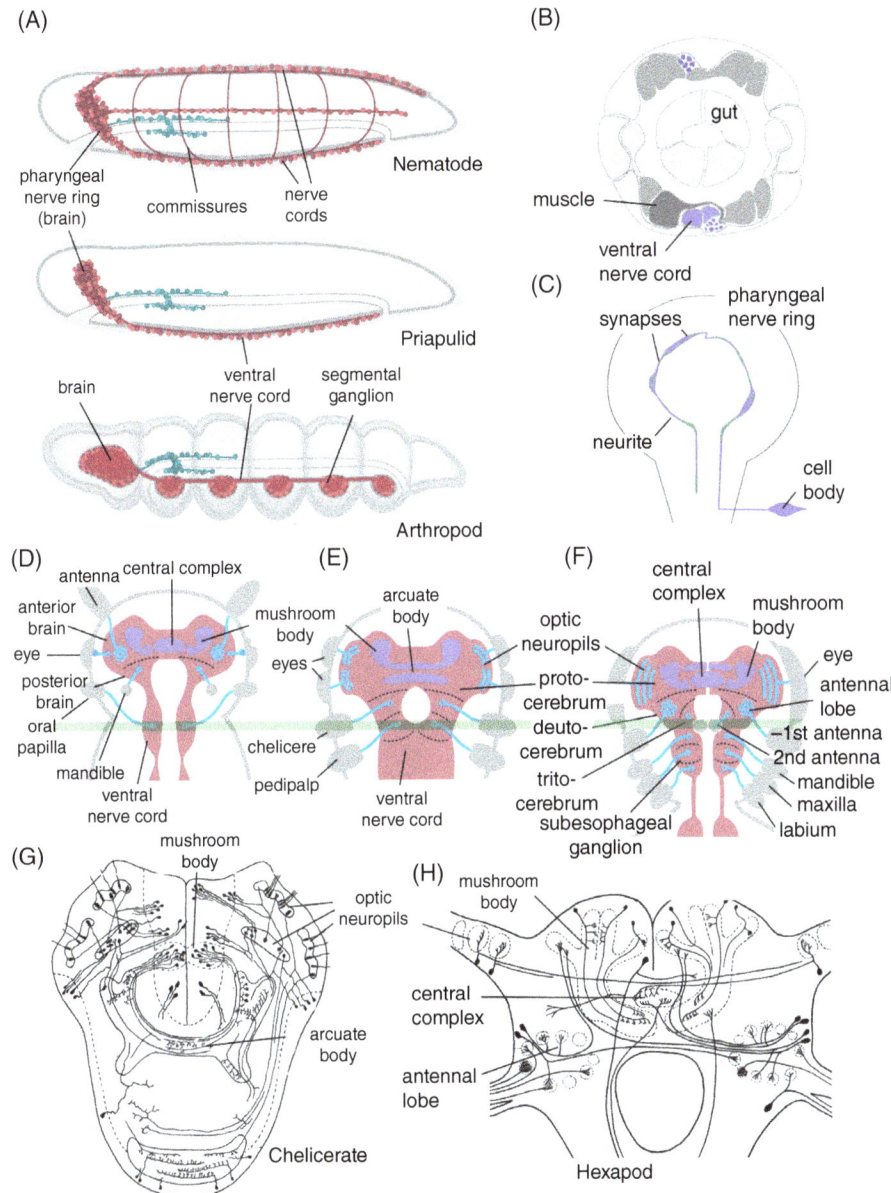

Figure 8.9 The Central Nervous System of Ecdysozoans. (A) Essential characteristics of nervous system architecture in nematodes, priapulids, and arthropods, represented as schematic sagittal sections. (B) Schematic cross-section of nematode Caenorhabditis elegans, showing basiepithelial dorsal and ventral cord and muscle process forming connection to ventral nerve fiber. (C) Line drawing of representative neuron of nematode C. elegans based on electron microscopic reconstruction. Neuron forms single nerve process with interspersed input and output synapses. (D–F) Schematic representation of head sensory organs and brain of arthropods (D: onychophoran; E: chelicerate; F: hexapod; all in dorsal view). Sensory nerves in blue, structured neuropil compartments in purple. Hatched lines indicate boundaries between segmental ganglia. Segmental ganglia of arthropods can be homologized based on anatomical and molecular criteria (for details, see text); green bar registers brains, indicating tritocerebrum (hexapod), pedipalpal ganglion (chelicerate), and ganglion innervating oral papilla (onychophoran) as homologous neuromeres. (G, H) Line drawings of neurons forming structured neuropil compartments. (G) Optic neuropils, mushroom body, and arcuate body of chelicerate. (H) Mushroom body, central complex, and antennal lobe of hexapod *Periplaneta americana*. Hanstroem, 1968. Reproduced with permission of Springer.

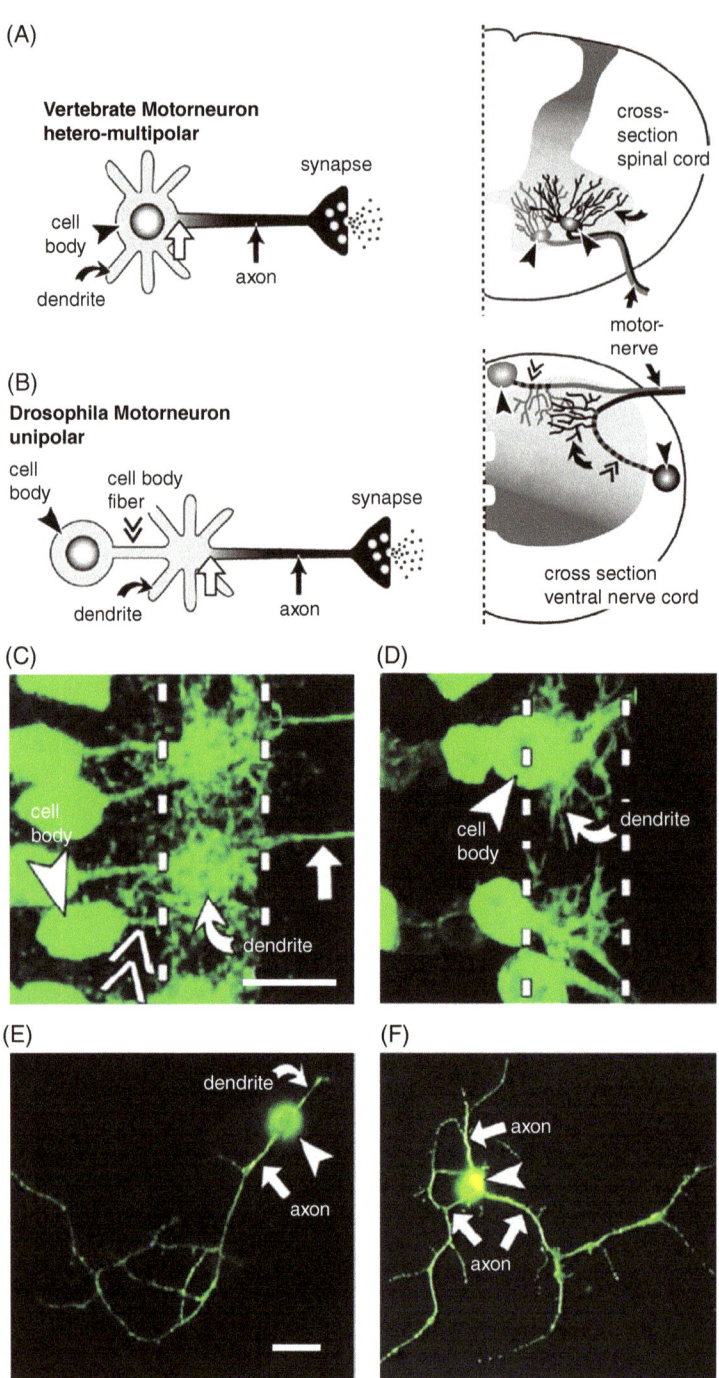

Figure 8.11 Neuronal Architecture: Multipolar vs. Unipolar Neurons.
(A, B) Schematic representation of multipolar motor neuron (vertebrate) and unipolar motor neuron (Drosophila). (C) Unipolar motor neurons (marked by expression of GFP, green) in wild-type Drosophila. Note cell bodies (arrowhead) emitting single-cell body fiber (double arrowhead) towards neuropil (dashed lines) where multiple dendritic branches (curved arrow) are formed by each neuron. Peripheral axon indicated by straight arrow. (D) Expression of activated cdc42 construct variably displaces motor neuronal cell bodies closer towards neuropil. In these cases, multiple dendrites directly branch off the cell body, turning cell into a multipolar neuron. (E, F) Drosophila neurons in culture variably express a bipolar (E) or multipolar (F) phenotype, rather than their normal unipolar phenotype. Scale bars: 10 μm. Soriano 2005. Reproduced with permission of Elsevier.

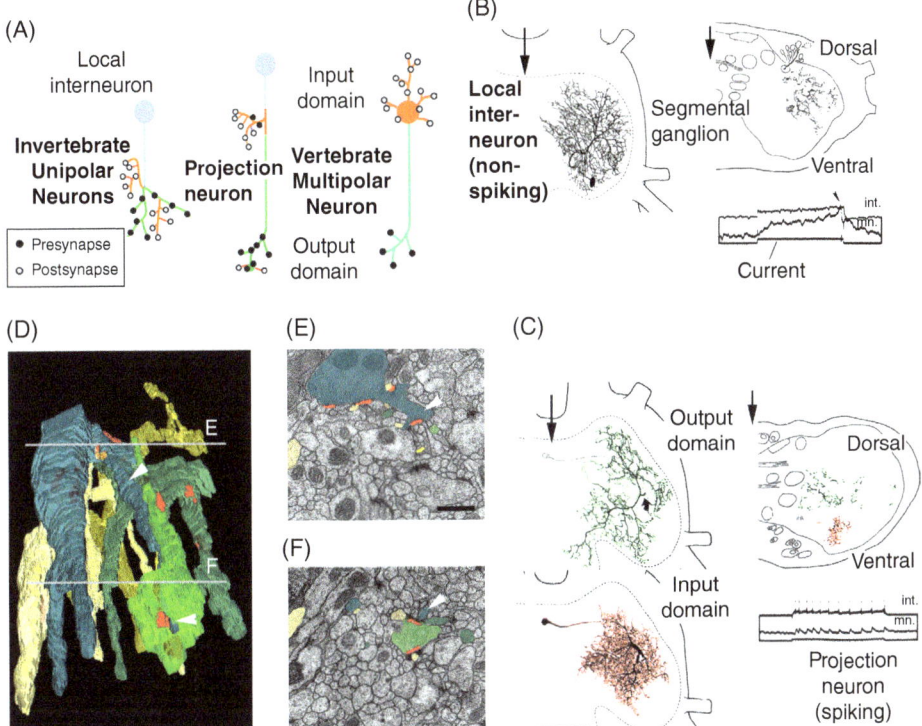

Figure 8.12 Neuronal Architecture: Distribution of Synapses.
(A) Schematic representation of invertebrate local interneuron and projection neuron compared to typical vertebrate neuron. (B) Local non-spiking interneuron in locust segmental ganglion. Left: dorsal view; arrow demarcates midline. Top right: cross-section of hemiganglion, showing distribution of branches of neuron in ventral as well as dorsal domains within neuropil. Bottom right: Physiology of non-spiking neuron. Injection of current (bottom trace) causes depolarization without action potentials in interneuron (int); this in turn leads to slowly increasing depolarization with terminal spike (arrowhead) in postsynaptic motor neuron (mn). Watkins BL 1985. Reproduced with permission of John Wiley and Sons. (C) Spiking interneuron in locust segmental ganglion. Left: dorsal view of ventral (input) domain of neuron (bottom; shaded red) and dorsal (output) domain (top; shaded green); arrow demarcates midline. Top right: cross section of hemiganglion, showing spatial separation of input branches ventrally and output branches dorsally. Bottom right: Physiology of spiking interneuron. Injection of current (bottom trace) causes depolarization and train of action potentials in interneuron (int) and in postsynaptic motor neuron (mn). Siegler MV 1979. Reproduced with permission of John Wiley and Sons. (D) Digitial 3D model of short segments of synaptically connected neurons in Drosophila larval brain rendered in different shades of yellow, blue, and green. Red lines indicates presynaptic sites. Cardona 2010. PLOS Biology. (E, F) cross-section of neuronal fibers shown in (D); level of section indicated by lettered horizontal lines in (D). Note concentraton of presynaptic sites at varicosities (thickenings) of blue fiber (E) and green fiber (F). Varicosity of blue fiber gives off thin branch (white arrowheads); this branch is postsynaptic to green fiber in (F). Scale bars: 200 μm (B, C); 0.5 μm (D–F).

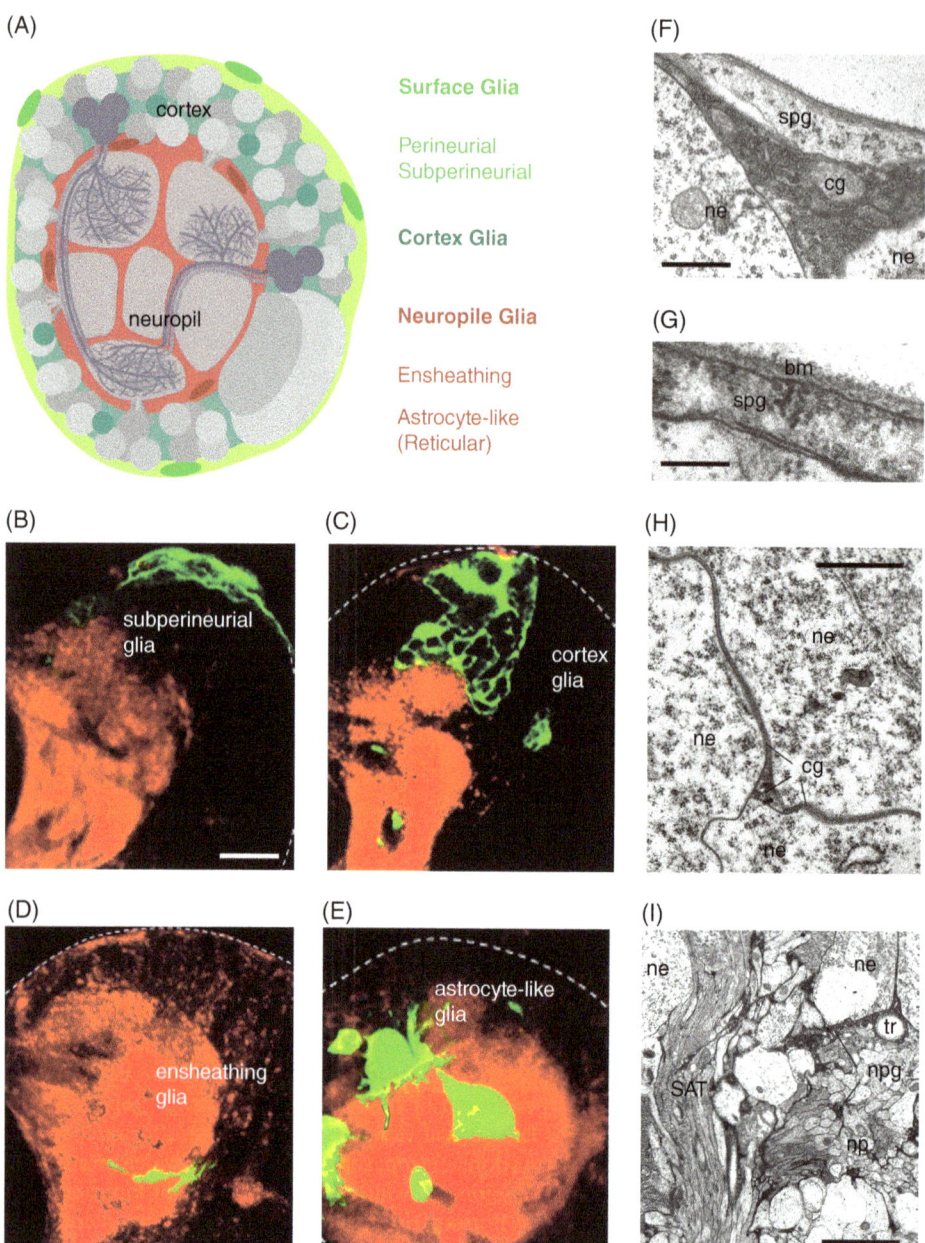

Figure 8.13 Glia in Insect Ganglion.
(A) schematic section of Drosophila brain showing main types of glial cells. (B–E) Z-projections of confocal sections of Drosphila larval brain. Individual glial cells of different types are labeled by expression of a GFP reporter (green); neuropil is labeled by anti-DN-cadherin (red). (F–I) Electron micrographs of parts of cross sections of Drosophila larval brain. (F, G) Brain surface, covered by subperineurial glia (spg) producing basement membrane (bm) and by outer lamella of cortex glia (cg). (H) Cortex, showing thin lamellae of cortex glia (cg) surrounding neuronal cell bodies (ne). (I) Cortex–neuropil boundary, demarcated by neuropil glial sheath (npg). Other abbreviations: np neuropil; SAT secondary axon tract; tr trachea. Scale bars: 40 μm (B–E); 0.5 μm (F); 0.2 μm (G); 1 μm (H); 2 μm (I).

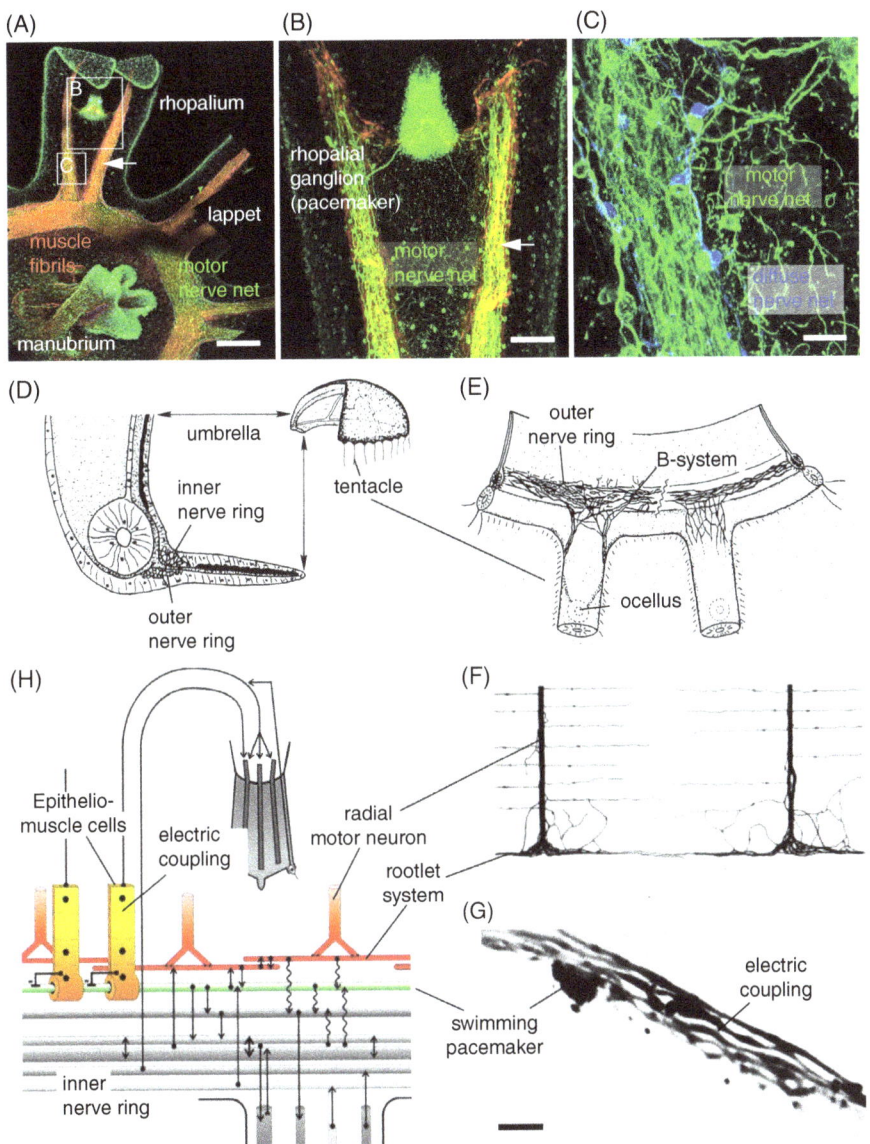

Figure 8.15 Neuronal Circuitry Controlling Swimming in Cnidaria.
(A–C) Neuronal networks in scyphozoan *Aurelia aurita*. Z-projection of horizontal confocal sections of ephyra stage. Myofibrils are labeled by phalloidin in (A) and (B). Neurons forming motor nerve net are labeled by anti-Acetylated tubulin (green) in (A–C); this antibody also binds to cilia at epidermal surface (arrowhead) and manubrium. Note concentration of motor network along radially oriented myofibrils (arrows in A and B), and rhopalia (rhopalial ganglion in B). Neurons forming diffuse nerve net are labeled by anti-FMRFamide (blue) in (C). Nakanishi et al. 2010. (D–F) Line drawings of hydrozoan medusa. (D) Section of umbrella margin, indicating position of inner and outer nerve ring. Satterlie 1983. Reproduced with permission of Springer. (E) detail of umbrella margin, showing outer nerve ring with nerve fibers forming B-system. Spencer AN 1984. Reproduced with permission of J. Exp. Biol. The Company of Biologists Limited. (F) Radial motor neurons connecting to inner nerve ring. Mackie 2000. Reproduced with permission of J. Exp. Biol. The Company of Biologists Limited. (G) Pacemaker neuron of inner nerve ring of hydrozoan *Aequorea aequorea* injected with fluorescent dye. All pacemaker neurons are labeled because of electric coupling (gap junctions) among these cells. Satterlie RA 1983. Reproduced with permission of Springer. (H) Circuit diagram of neuronal populations forming inner nerve ring of hydrozoa. Arrows indicate synaptic input. Scale bars: 100 µm (A); 50 µm (B, G); 20 µm (C). Mackie 2003. Reproduced with permission of J. Exp. Biol. The Company of Biologists Limited.

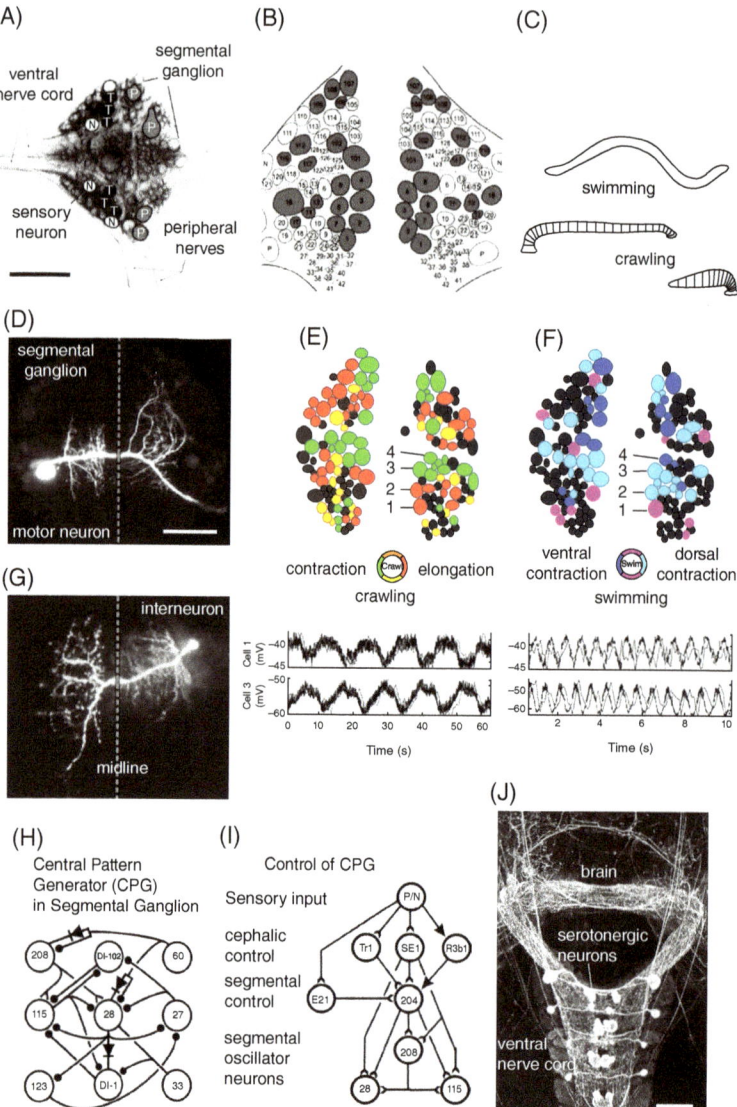

Figure 8.16 Circuitry Controlling Locomotion in Leech.
(A) Microphotograph of segmental ganglion of leech (anterior to the left). Sensory neurons are annotated. Nicholls 1968. Reproduced with permission of J Neurophysiol., The American Physiological Society. (B) Dorsal surface map of leech segmental ganglion (anterior towards the top). (C) Pattern of swimming and crawling in leech. (B, C) Briggman 2006. Reproduced with permission of J Neurosci., Society for Neuroscience. (D) Motor neuron labeled by injection of fluorescent dye. Gray-hatched line indicates midline. Mullins 2011. Reproduced with permission of Elsevier. (E, F) Neurons of segmental ganglion active in both crawling (E) and swimming (F). Neuronal activity was monitored in live preparations by Ca-sensor. Neurons active during particular phase of crawling cycle or swimming cycle were color-coded (as indicated in center of panels) and projected on the neuron map (top of panels). Note neurons 1–4 (numbering according to standard map shown in panel B) which are active in both crawling and swimming. Traces at bottom of panels show activity of neurons 1 and 3 during crawling and swimming, respectively. Briggman 2006. Reproduced with permission of J Neurosci., Society for Neuroscience. (G) Segmental interneuron of swimming CPG labeled by injection of fluorescent dye. Hatched grey line indicates midline. Mullins 2011. Reproduced with permission of Elsevier. (H) Central pattern generator in leech segmental ganglion. All neurons shown are interneurons, except DI-1 and DI-102, which are inhibitory motor neurons. (I) Suprasegmental control of CPG. Mullins 2011. Reproduced with permission of Elsevier. (J) Microphotograph of leech brain and anterior segmental ganglia, showing antibody labeled serotonergic neurons. Crisp 2006. Reproduced with permission of J. Exp. Biol. The Company of Biologists Limited. Scale bars: 200 μm (A, D, F); 100 μm (J).

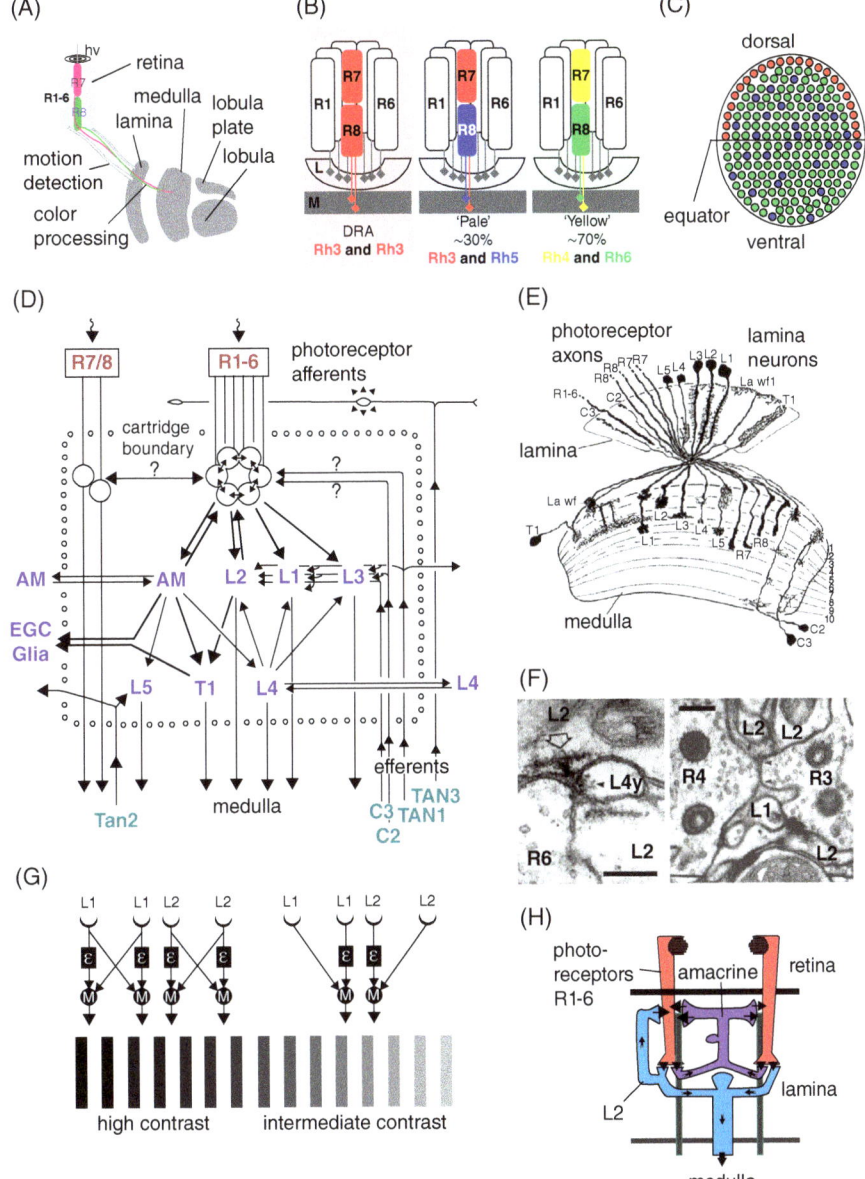

Figure 8.19 Circuitry in the Fly Visual System.
(A) Schematic section of the visual system, showing input from the outer photoreceptors R1–6 (motion and shape detection) and the inner photoreceptors R7/8 (color detection) to the lamina and medulla, respectively. Morante J 2004. Reproduced with permission of Elsevier. (B) Inner photoreceptors of different ommatidia express different rhodopsins. (C) Mosaic distribution of ommatidia with different rhodopsins. (B, C) Wernet MF 2004. Reproduced with permission of Elsevier. (D) Schematic wiring diagram of dipteran lamina (AM amacrine neuron; C2/3, TAN1-3 medulla neurons projection to lamina; L1-L5 lamina neurons). Shaw 1984. Reproduced with permission of J. Exp. Biol. The Company of Biologists Limited. (E) Line drawings of Golgi-labeled photoreceptors, lamina neurons, and medulla neurons. Fischbach 1989. Reproduced with permission of Springer. (F) Electron micrographs of synapses in Drosophila lamina. Left: synapse of L2 on R6 (feedback) and L4. Right: gap junction between photoreceptors (arrowheads). Bar: 200 nm. Meinertzhagen IA 1991. Reproduced with permission of John Wiley and Sons. (G) Schematic of motion detector formed by pairs of L1 and L2 neurons in neighboring cartridges. Rister J 2007. Reproduced with permission of Elsevier. (H) Schematic of feedback between photoreceptors, amacrine and L2 neuron, mediating gain control circuit. Zheng L 2006. Reproduced with permission of The Rockefeller University Press.

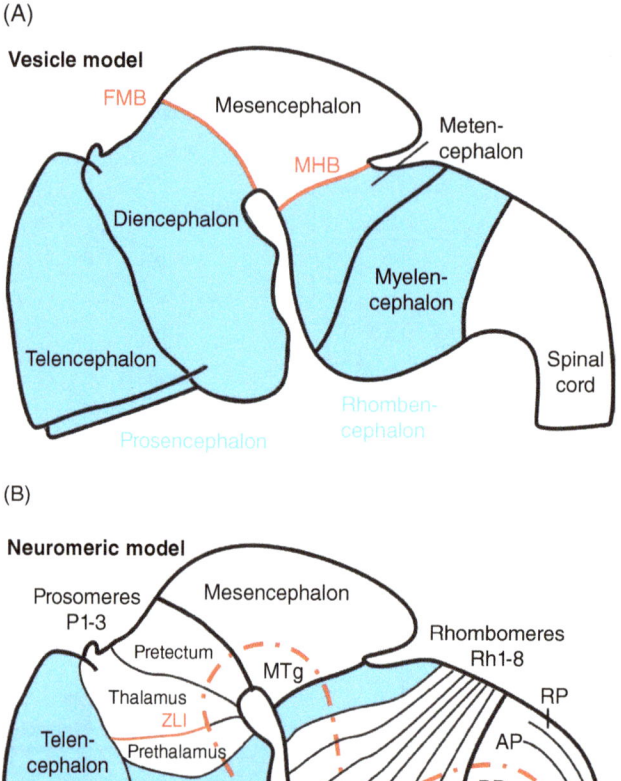

Figure 9.1 Vesicle and Neuromeric Models.
Schematics of embryonic mouse brain at 12.5 days in lateral views. (A) classical vesicle and (B) neuromeric (prosomeric) models. Red interrupted line: anteroposterior axis following hindbrain and forebrain flexures (see text).
Abbrevations: AP alar plate; BP basal plate; FMB forebrain–midbrain boundary; FP floor plate; MHB midbrain–hindbrain boundary; MTg midbrain tegmentum; P1—P6 prosomeres 1—6; Rh1–Rh8 rhombomeres 1—8; RP roof plate; sm somatomotor zone; ss somatosensory zone; vm visceromotor zone; vs viscerosensory zone; ZLI zona limitans intrathalamica.

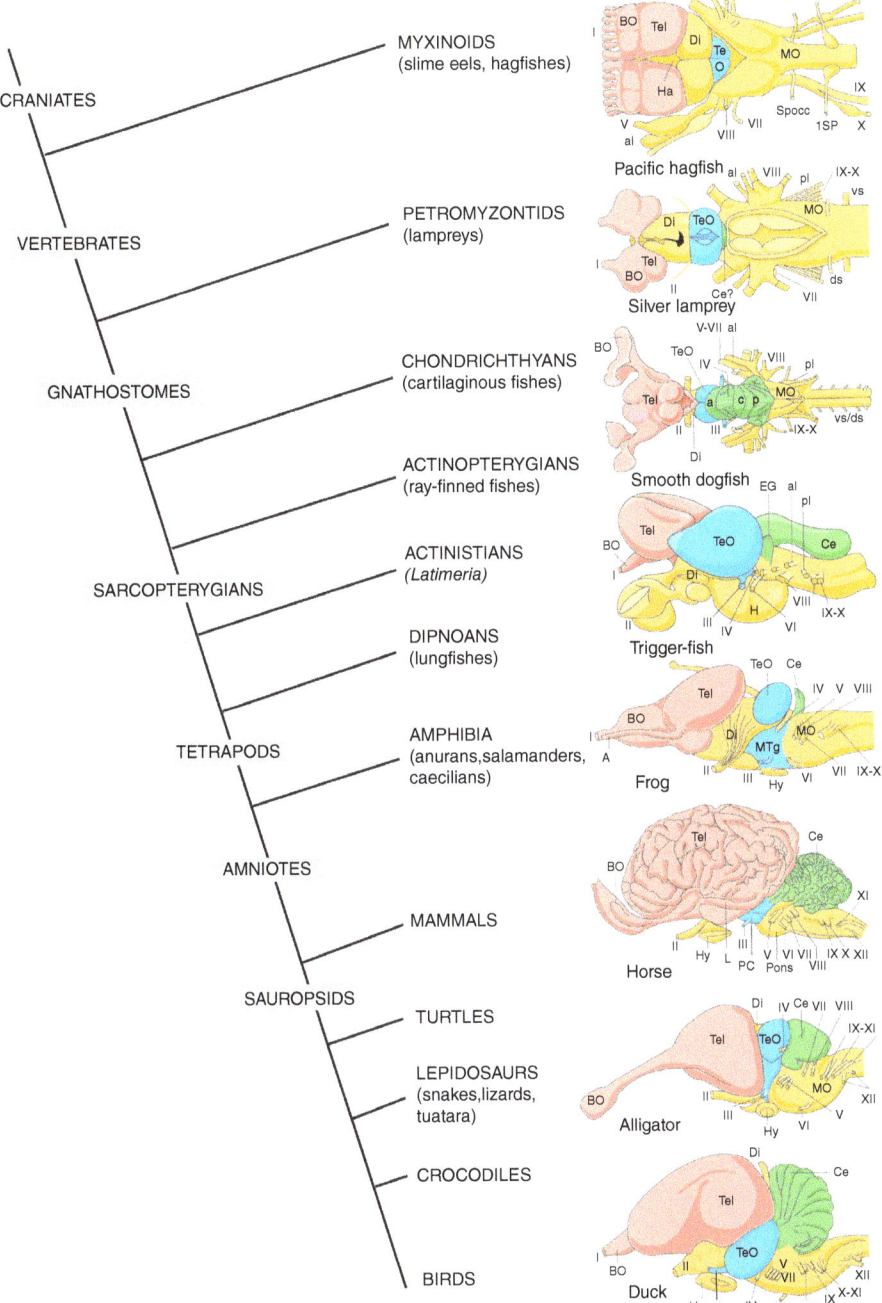

Figure 9.2 Cladogram of Craniate Taxa and Illustrations of Representative Brains.
Note that three terms for nonmonophyletic groups are used in text: agnathans for myxinoids and petromyzontids, reptiles for amniotes except birds/mammals, and anamniotes for all craniates except amniotes.
Abbreviations: 1 SP first spinal nerve; a anterior cerebellar lobe; A accessory olfactory nerve; al anterior lateral line nerve; BO bulbus olfactorius; c central cerebellar lobe; Ce cerebellum; Di diencephalon; ds dorsal spinal nerve; EG eminentia granularis; H hypothalamus; Ha habenula; Hy hypophysis (pituitary); L lateral pallium (piriform cortex); MO medulla oblongata; MTg midbrain tegmentum; p posterior cerebellar lobe; PC pedunculus cerebri; pl posterior lateral line nerve; Spocc spino-occipital nerve; Tel telencephalon; TeO tectum opticum; vs ventral spinal nerve. 0–XII as in Figure 9.3.

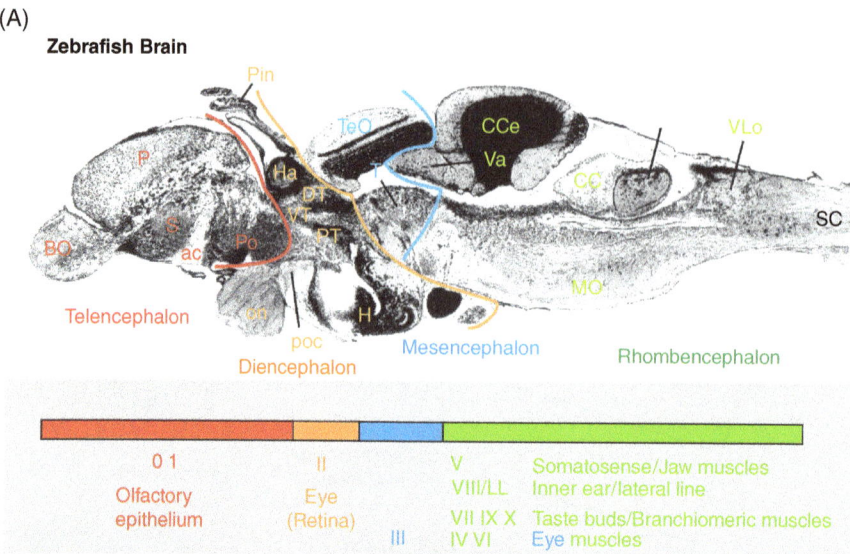

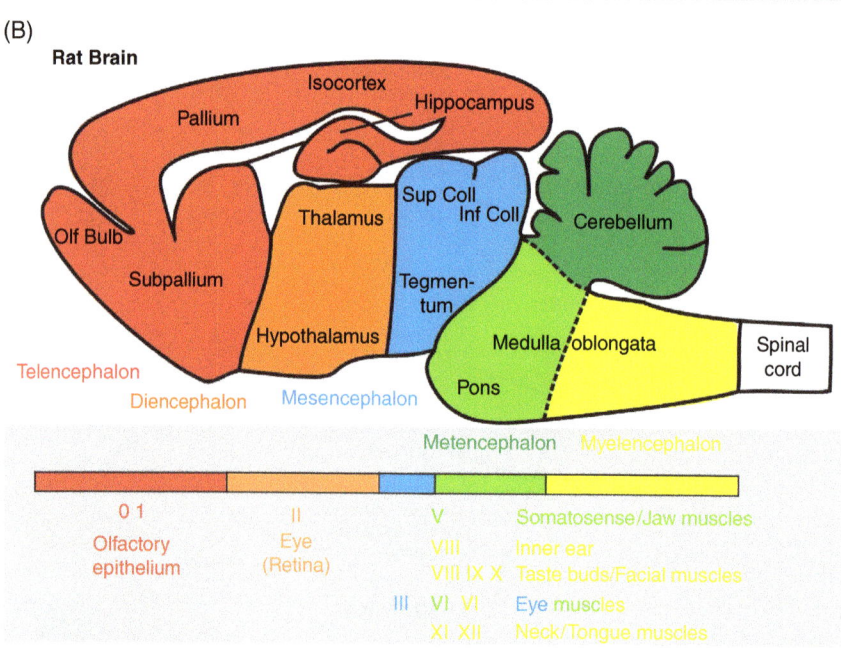

Figure 9.3 Classical Brain Parts and Differences Regarding Presence of Cranial Nerves. Bodian-Nissl stained sagittal adult zebrafish brain section (A) and schema of sagittal adult rat brain section (B) exemplifying classical brain parts and differences regarding presence of cranial nerves between anamniotes (zebrafish) and amniotes (rat).

Abbreviations: 0 nervus terminalis; I nervus opticus; II nervus olfactorius; III nervus oculomotorius; IV nervus trochlearis; V nervus trigeminus; VI nervus abducens; VII nervus facialis; VIII nervus octavus (or vestibulocochlearis); IX nervus glossopharyngeus; X nervus vagus; XI nervus accessorius; XII nervus hypoglossus; ac anterior commissure; BO bulbus olfactorius; CC crista cerebellaris; CCe corpus cerebelli; Corp call corpus callosum; DT dorsal thalamus; FLo facial lobe; Inf Coll inferior colliculus; H hypothalamus; Ha habenula; LL lateral line nerves; on optic nerve; MO medulla oblongata; P pallium; Po preoptic region; Pin pineal; poc postoptic commissure; PT posterior tuberculum; T midbrain tegmentum; S subpallium; SC spinal cord; Sup Coll superior colliculus; TeO tectum opticum; Va valvula cerebelli; VLo vagal lobe; VT ventral thalamus (prethalamus).

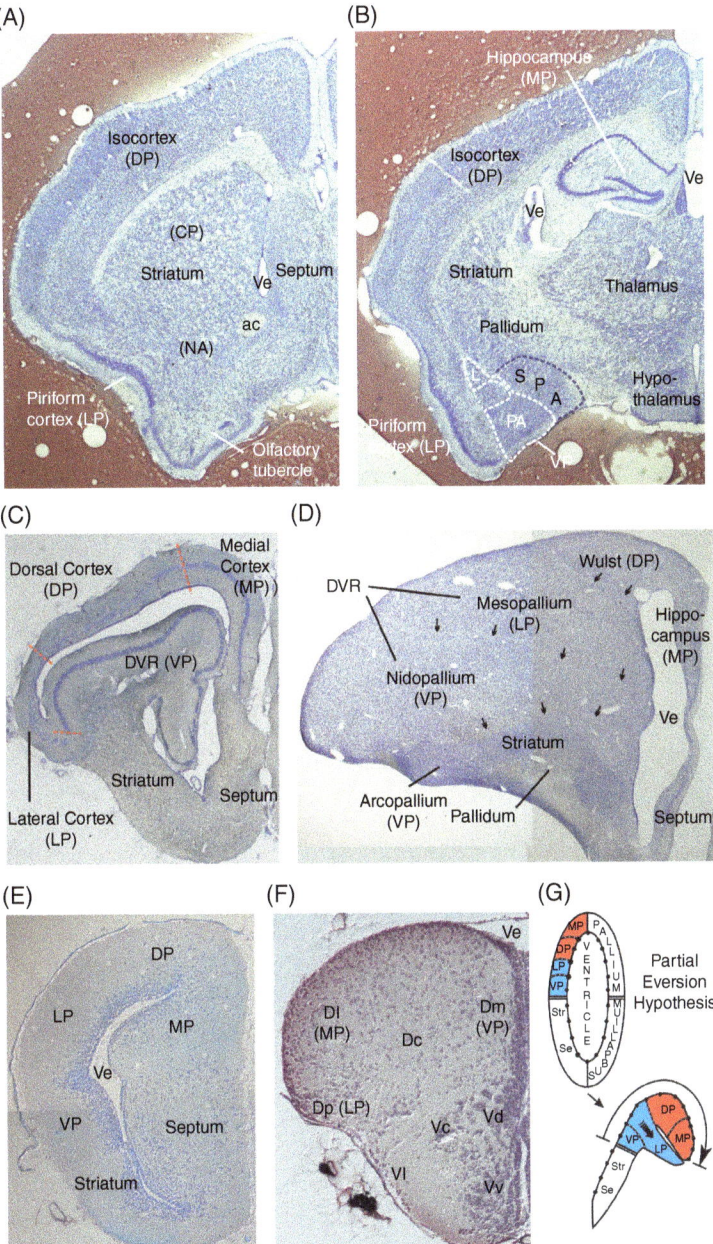

Figure 9.4 Transverse Nissl-Stained Sections through Left Telencephalic Hemispheres of Mouse (*Mus musculus*, A: anterior, B: posterior), Tuatara (*Sphenodon punctatus*, C), Pigeon (*Columba livia*, D), Fire-Bellied Toad (*Bombina orientalis*, E), and Zebrafish (*Danio rerio*, F). Schema shows partial eversion hypothesis (G). Major pallial and subpallial divisions are present in all gnathostome taxa. Arrows in D point to cell-free laminae at boundaries between avian telencephalic divisions. See acknowledgments for origin of sections.

Abbreviations: ac anterior commissure; CP caudate-putamen; Dc central zone of dorsal telencephalic area; Dl lateral zone of dorsal telencephalic area; Dm medial zone of dorsal telencephalic area; Dp posterior zone of dorsal telencephalic area; DP dorsal pallium; DVR dorsal ventricular ridge; L lateral amygdala; LP lateral pallium; MP medial pallium; NA nucleus accumbens; PA pallial amygdala (note fine white dots indicating parts of lateral pallial origin); Se Septum; SPA subpallial amygdala; Str Striatum; Vc central nucleus of ventral telencephalic area; Vd dorsal nucleus of ventral telencephalic area; Ve telencephalic (lateral) ventricle; Vl lateral nucleus of ventral telencephalic area; VP ventral pallium; Vs supracommissural nucleus of ventral telencephalic area; VT ventral thalamus (prethalamus); Vv ventral nucleus of ventral telencephalic area.

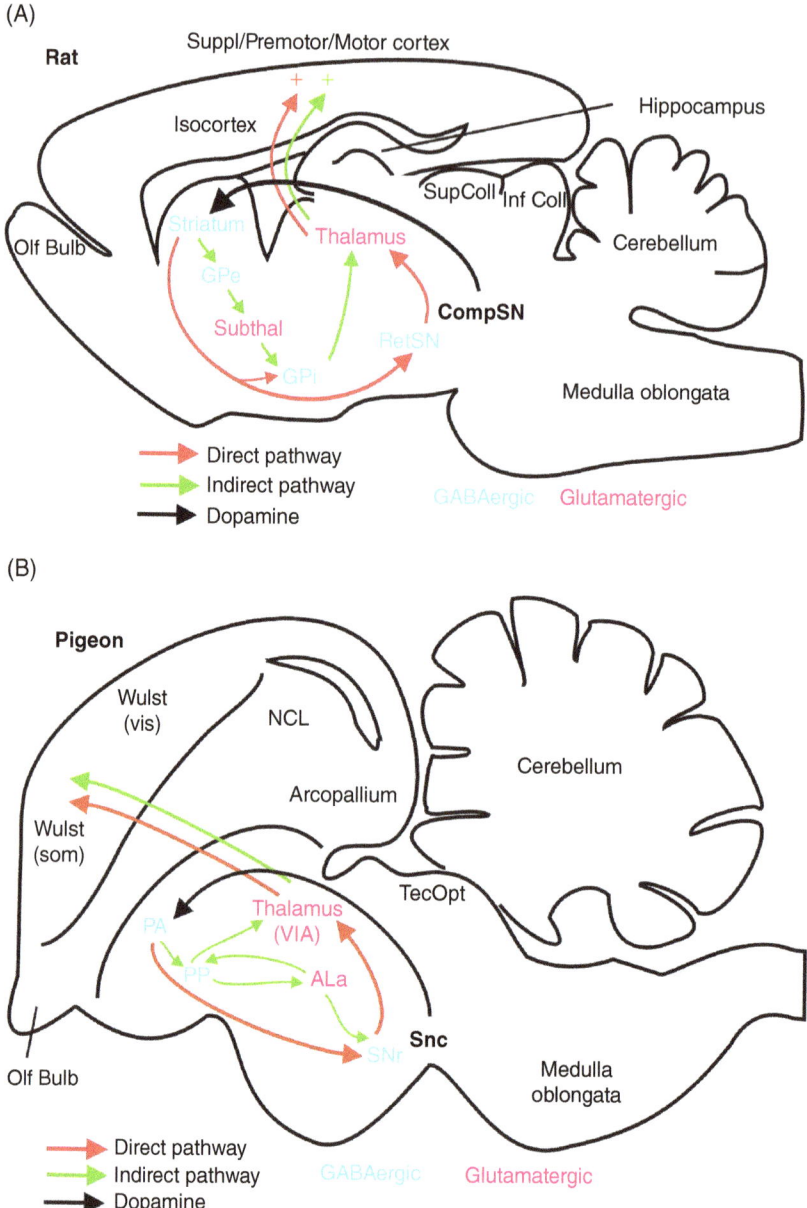

Figure 9.8 Schematics Show Sagittal Sections of (A) Rat and (B) Pigeon Brain with Basal Ganglia Circuitry (Rat after Mink in Squire et al., 2008; pigeon after Jiao et al., 2000; Kröner & Güntürkün, 1999; Reiner, 2002; Veenman, Wild, & Reiner, 1995).
+/− in (A) indicates activation of cortex by both direct and indirect pathways upon dopamine release in striatum. Note that in addition to the dorsal striatopallidal system (somatomotor loop) shown here, ventral striato-pallidal (limbic) loops do exist in all amniotes.
Abbreviations (mammals): CompSN substantia nigra compacta, GPe external globus pallidus, GPi internal globus pallidus, RetSN substantia nigra reticulata Abbreviations (birds): ALa anterior nucleus of ansa lenticularis (=subthalamic nucleus), NCL caudolateral nidopallium, PA paleostriatum augmentatum (=lateral striatum), PP paleostriatum primitivum (=pallidum), SNc substantia nigra compacta, SNr substantia nigra reticulata, VIA: ventrointermediate thalamic area.

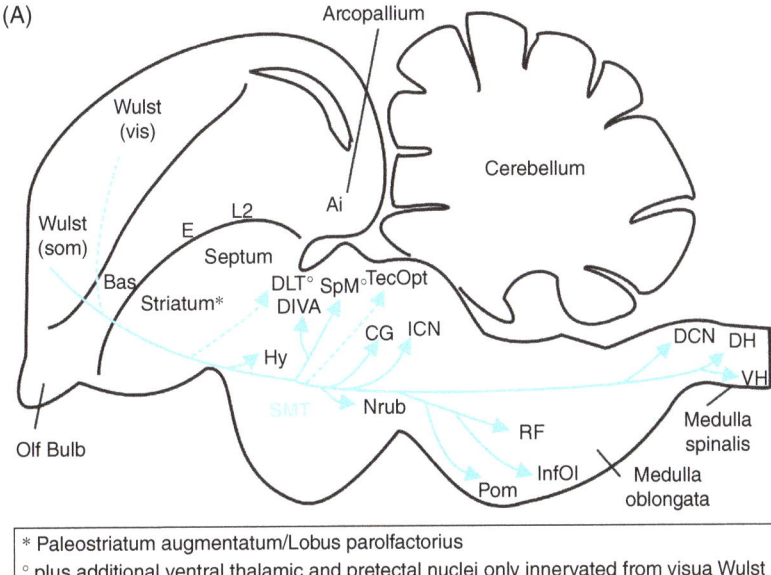

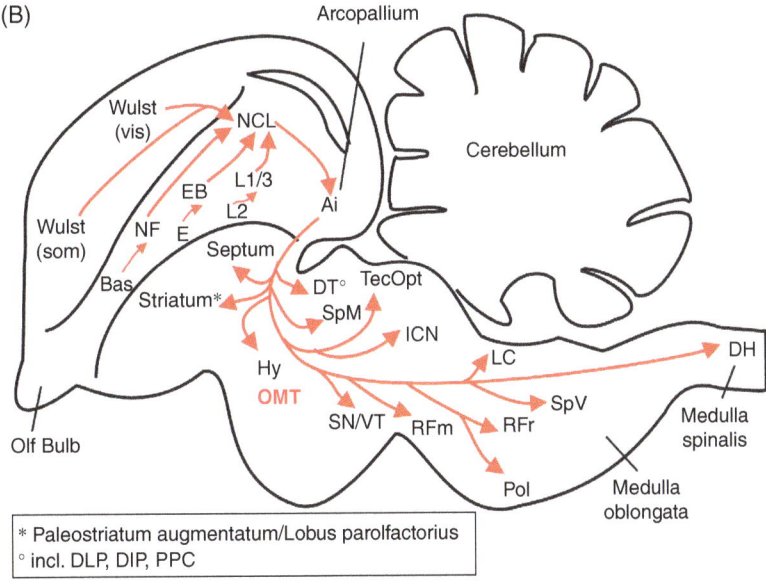

Figure 9.9 Schematics Show Sagittal Sections of Pigeon Brain with Long Descending Telencephalic Pathways.
(A) Septomesencephalic tract including data from pigeon (Karten et al., 1977; Wild, 1992) and zebra/green finch (Wild & Williams, 2000). (B) Occipitomesencephalic tract including data from pigeon (Kröner & Güntürkün, 1999; Miceli, Repérant, Villalobos, & Dionne, 1987; Wild et al., 1985; Zeier & Karten, 1971), zebra finch (Wild & Farabaugh, 1996) and mallard (Dubbeldam, den Boer-Visser, & Bout, 1997).
Abbreviations: Ai intermediate arcopallium; Bas Nucleus basalis rostralis; CG central gray; DCN dorsal column nuclei; DH: dorsal horn; DIP dorsointermediate posterior nucleus; DIVA nucleus dorsalis intermedius ventralis anterior; DLP dorsolateral posterior nucleus; DLT dor solateral nuclei (=principal optic nuclear complex); DT: dorsal thalamus; E: entopallium (primary visual); EB: entopallial belt; Hy hypothalamus; ICN intercollicular nucleus; LC locus coeruleus; L1;2;3 field L (auditory); NCL caudolateral nidopallium; NF frontal nidopallium; OMT occipitomesencephalic tract; Pol/Pom lateral/medial pontine nucleus; PPC principal precommissural nucleus; RFm mesencephalic reticular formation; RFr rhombencephalic reticular formation (incl. parabrachial nucleus); SMT septomesencephalic tract; SN/VT (dopaminergic) substantia nigra/ventral tegmental area; SpM medial spiriform nucleus; SpV descending/spinal trigeminal nuclei; VH ventral horn.

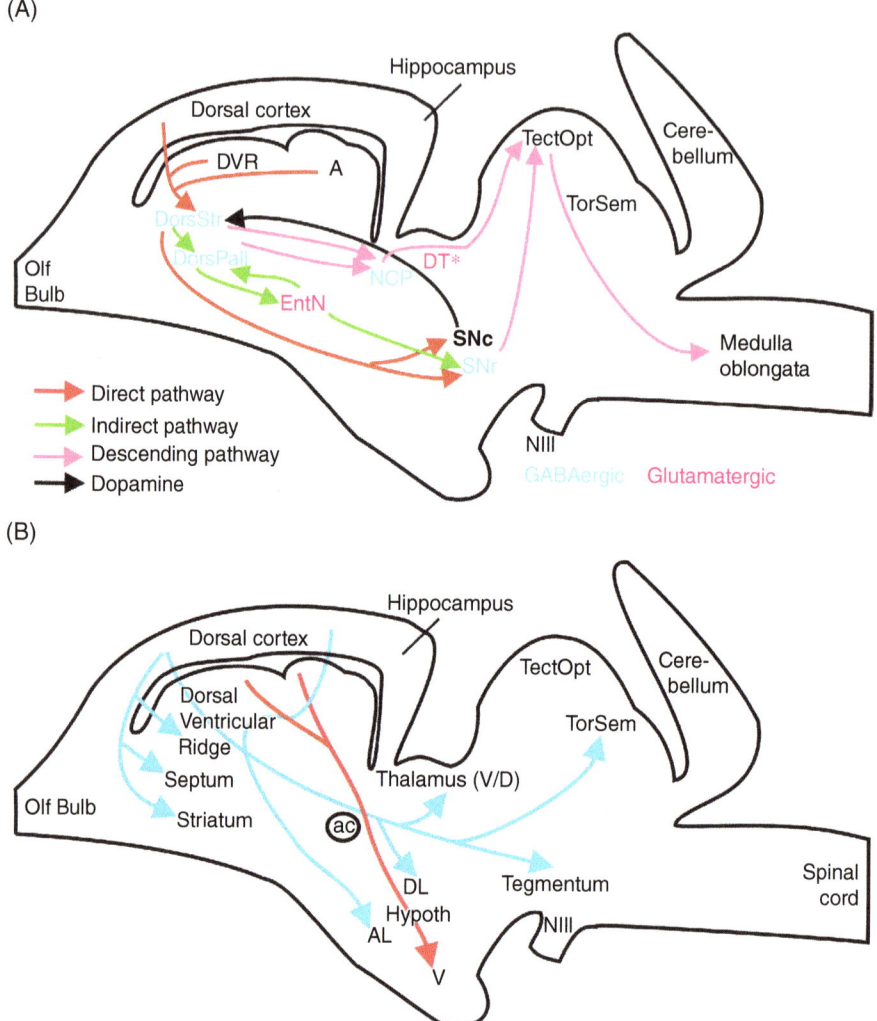

Figure 9.10 Schematics Show Sagittal Sections of Lizard Brain with (A) Basal Ganglia Circuitry (after González et al., 1990; Guirado et al., 1999; Jiao et al., 2000; Reiner et al., 1998; Smeets & Medina, 1995) and (B) Long Descending Telencephalic Pathways (after ten Donkelaar, 1998).

Abbreviations: A arcopallium; DorsPall dorsal pallidum; DorsStr: dorsal striatum DT dorsal thalamus (* not involved in basal ganglia circuitry); DVR dorsal ventricular ridge; EntN (anterior) entopeduncular nucleus; NCP (dorsal) nucleus of posterior commissure; NIII ocolumotor nerve; OlfBulb olfactory bulb; SNc/r: substantia nigra compacta/reticulata; TectOpt tectum opticum; TorSem torus semicircularis.

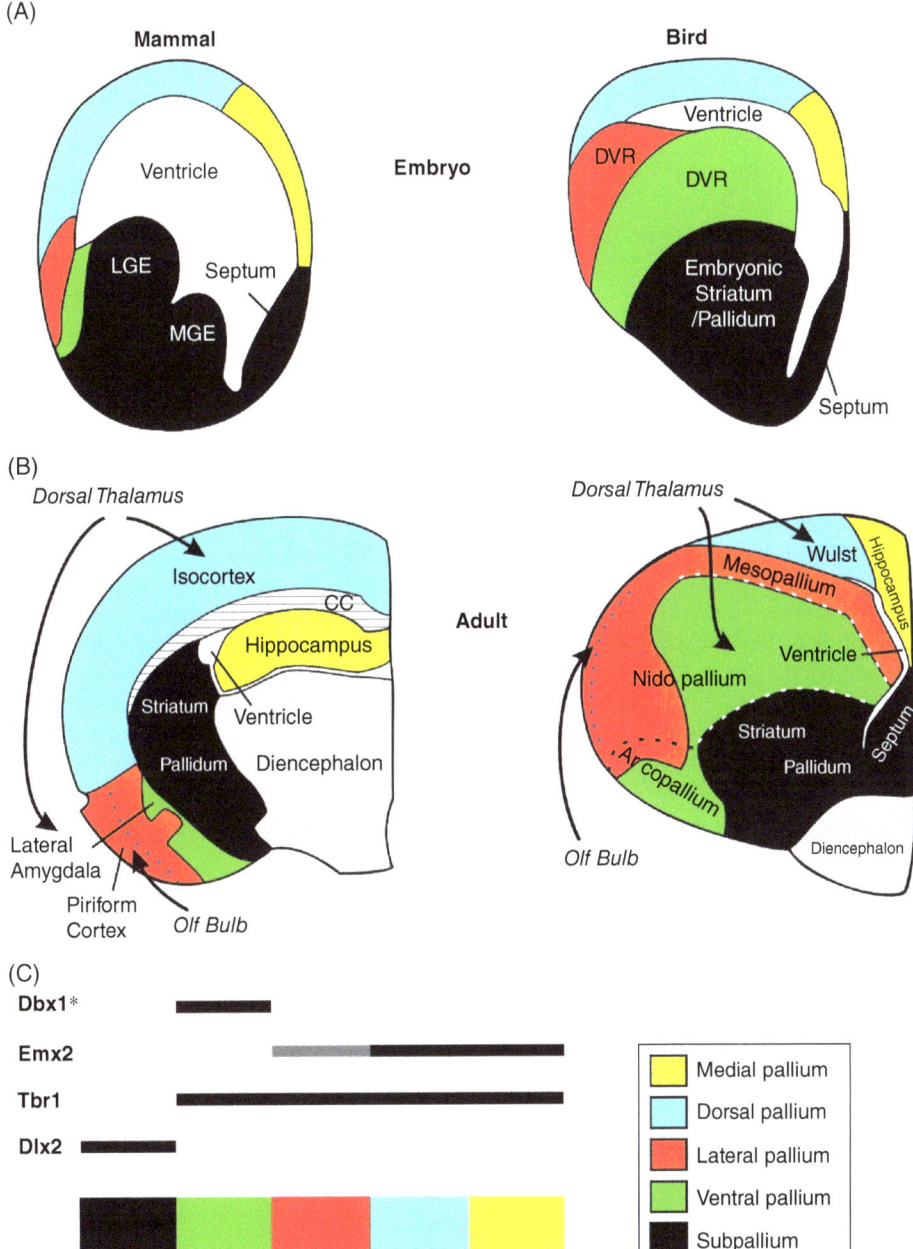

Figure 9.11 Amniote Telencephalic Bauplan and Major Sensory Inputs to Pallium in Mammals and Birds.

Amniote embryonic (A) and adult (B) telencephalic bauplan and major sensory inputs to pallium shown in schematics of the left telencephalic hemisphere in mammals on the left side and birds on the right side. (C) shows some transcription factors critical for telencephalic development and their expression domains in pallial and subpallial regions. * The expression of the *Dbx1* gene has only been described in mammals so far (Medina et al., 2004). Note that there is no direct thalamic input to hippocampus (medial pallium). Adapted from Farries (2001), using data from Martínez-García et al. (2009), Medina et al. (2004) and Puelles et al. (1999, 2000).

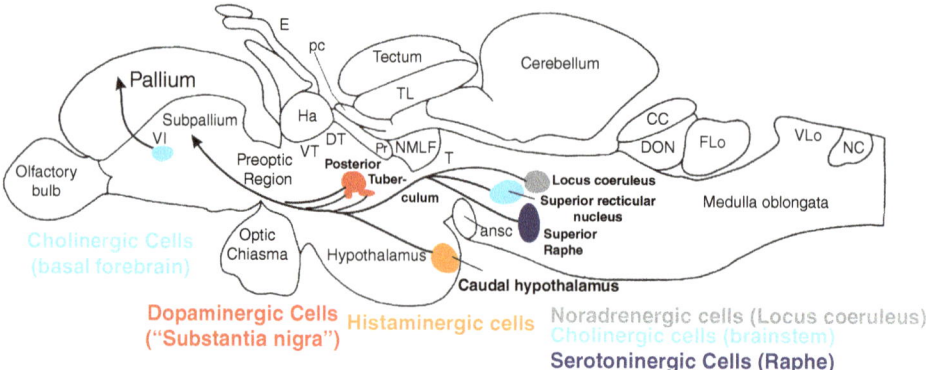

Figure 9.12 Schema Shows Sagittal Section of Adult Zebrafish Brain Depicting Neuromodulatory Systems with Long Ascending Connections to the Telencephalon.

The connections of dopaminergic (Rink & Wullimann, 2001) and serotoninergic (Lillesaar et al., 2009) cells were corroborated by double-label through axonal tracing. The indicated telencephalic connections of the histaminergic hypothalamic (Kaslin & Panula, 2001) and cholinergic brainstem cells (Mueller et al., 2004) are derived from an independent connectional study (Rink & Wullimann, 2004). The innervation of the pallium through Vl was shown for a different teleost species (Murakami, Morita, & Ito, 1983).

Abbreviations: ansc ansulate commissure; CC cerebellar crest; DON descending octaval nucleus; DT dorsal thalamus; E epiphysis; FLo facial lobe; Ha habenula; NMLF nucleus of the medial longitudinal fascicle; pc posterior commissure; Pr pretectum; TL torus longitudinalis; Vl lateral nucleus of ventral telencephalic area; VT ventral thalamus (prethalamus).

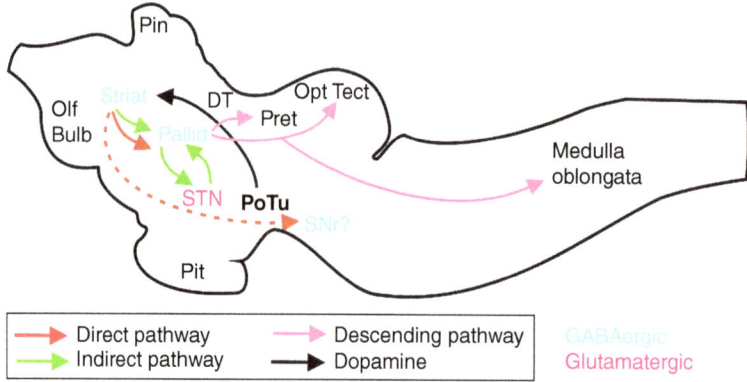

Figure 9.13 Schema Shows Sagittal Section of Lamprey Brain with Basal Ganglia Circuitry. (after Wullimann, 2011, data from Stephenson-Jones et al., 2011.)

Abbreviations: DT dorsal thalamus; OlfBulb olfactory bulb; OptTect optic tectum; Pin pineal. Pit pituitary PoTu posterior tuberculum; Pret pretectum; SNc substantia nigra compacta; SNr substantia nigra reticulata; STN Subthalamic nucleus; Striat striatum; Pallid pallidum.

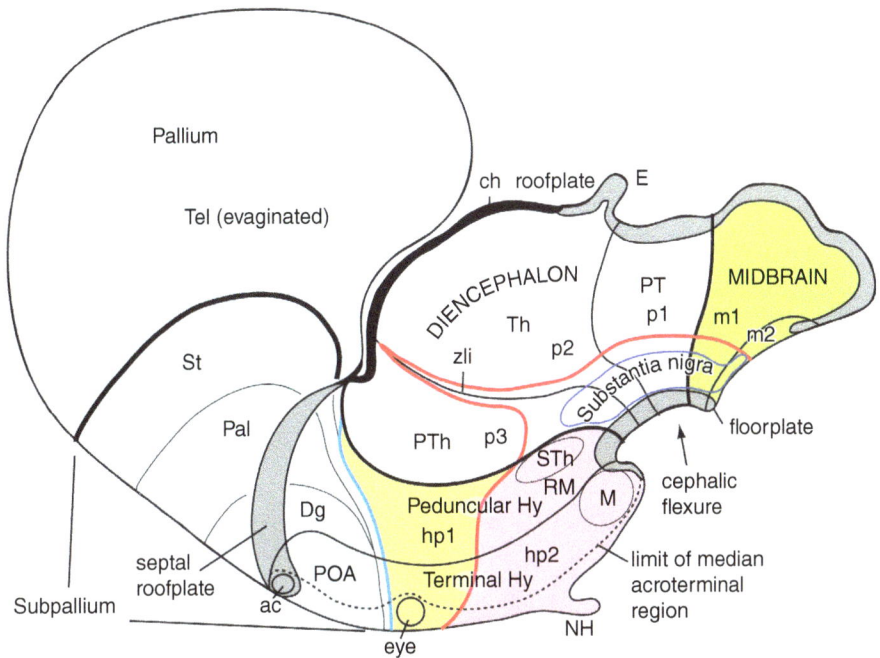

Figure 12.1 Forebrain Prosomeric Model. From Puelles et al., 2012.
The roof and floor plates are shaded gray; note, the choroidal roof (ch) is solid black. The alar–basal boundary plus the *zona limitans intrathalamica* (zli) are represented by the red line; note, this line joins its contralateral counterpart near the eye stalk. There are three proneuromeric regions: rostrally, the secondary prosencephalon comprises the telencephalon (blue) and the eye and hypothalamus (yellow and pink). The largest part of the telencephalon is evaginated, and is thus drawn as seen beyond the midline structures, that is, the septal roof plate and anterior commissure, ac. The telencephalon is delimited from the hypothalamus along a darker blue longitudinal line; the alar–basal boundary separates the hypothalamus into alar (yellow) and basal (pink) parts. The diencephalon appears next in the caudal direction, and is also divided into alar and basal parts (red line, with transverse detour at the zli). Caudally, there is the midbrain (green), also divided into alar and basal parts (red line). Anteroposterior neuromeric subdivisions are separated by transverse black lines that extend from the roof to the floor. The secondary prosencephalon comprises hypothalamic prosomeres, hp2 and hp1, stretched across hypothalamus and telencephalon. Note, the hypothalamus (Hy) is resultantly divided into terminal and peduncular parts; the former terminates in the preoptic area of the telencephalon and the latter expands into the whole evaginated telencephalon. Note, as well, the pallio-subpallial boundary (thick black line) within the hemisphere, and the diverse subpallial subdivisions (St, Pal, Dg, and POA, detailed in Puelles et al., 2013). The acroterminal region is a medial hypothalamic and preoptic locus (part of terminal hypothalamus) where right and left halves of the alar and basal plates are continuous across the midline; the neurohypophysis, NH, is a basal tuberal specialization within this medial domain. The alar domains of the diencephalic prosomeres, labelled p3, p2 and p1, generate respectively the prethalamic (PTh), thalamic (Th) and pretectal (PT) nuclear formations; the epiphysis, E, is a roof plate specialization within p2. Finally, the midbrain subdivides into mesomeres m2 and m1. The m2 segment is bounded posteriorly by the hindbrain isthmus (not shown). Four anatomic landmarks have been added: the mamillary and retromamillary areas (M and RM); the subthalamic nucleus (STh), which originates within RM and migrates dorsalward within the basal plate); and the substantia nigra, which spans across several neuromeres throughout the midbrain and diencephalon.

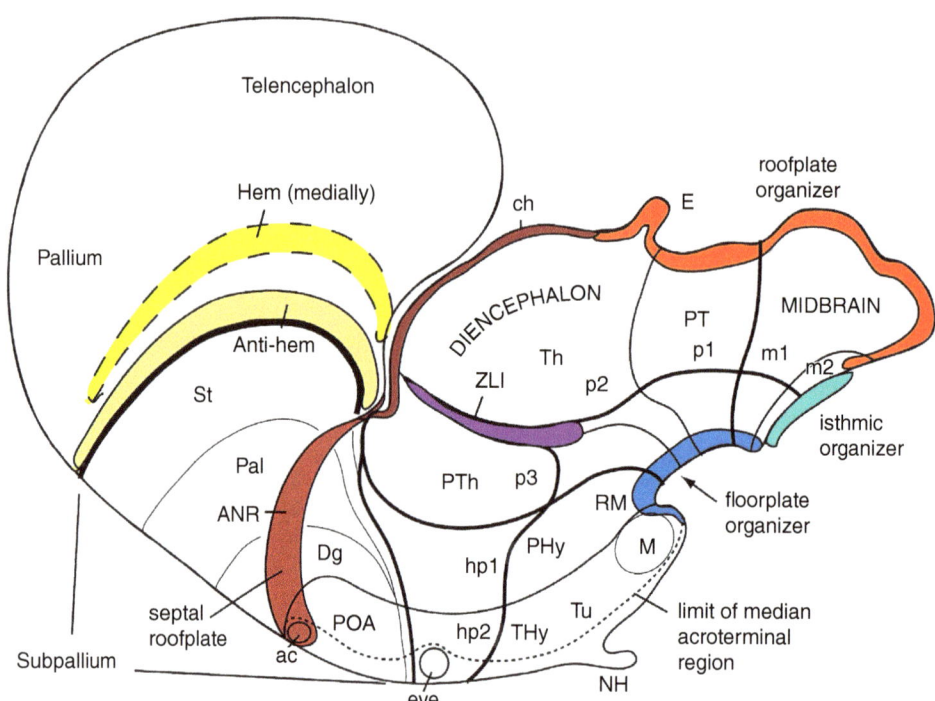

Figure 12.2 Topological Position of Widely Recognized Secondary Organizers, Represented in the Prosomeric Model.

The roof plate organizer in general appears in red, with the darker shade depicting the ANR organizer in the rostral roof plate. The primary floor plate organizer is shown in blue. Note, the alar–basal boundary (compare Figure 12.1) is roughly parallel to both roof and floor organizers (resulting from the equilibrium between dorsoventral antagonistic mechanisms of dorsalization and ventralization across the forebrain wall). The isthmic organizer (cyan), the source of FGF8, is strictly in the rostralmost hindbrain, though its effects also encompass the midbrain. The ZLI (purple) is a transverse ridge between thalamus and prethalamus. The hem (bright yellow) and anti-hem (pale orange) are tertiary organizers which pattern the telencephalic pallium. The anti-hem lies next to the pallio-subpallial boundary (black line), and the hem lies next to the choroidal roof plate tissue on the medial aspect of the hemisphere.

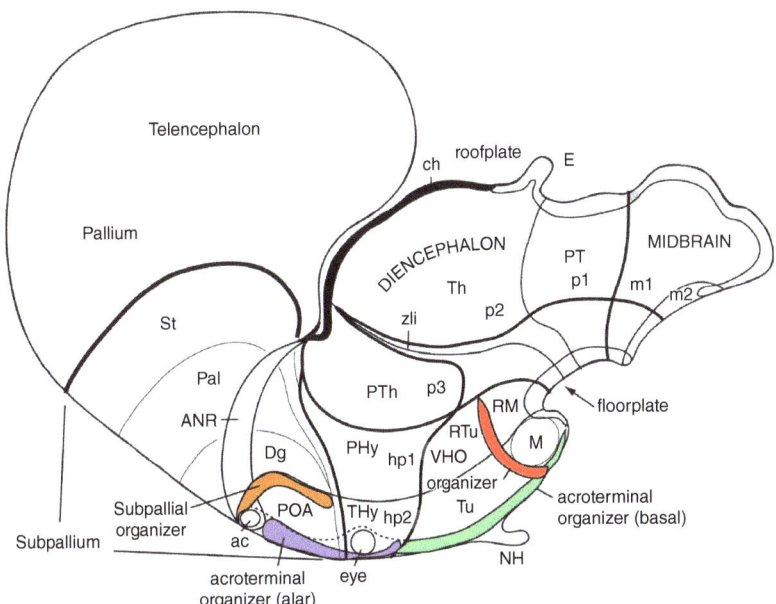

Figure 12.3 Additional Suspected Forebrain Organizers, Placed upon the Prosomeric Model. The SHH-positive subpallial organizer (dark orange shade) lies transversely within the preoptic area next to the hp2/hp1 boundary; note, the boundaries between the diverse subpallial domains are roughly parallel to it. This area is also a source of cells migrating tangentially into the subpallium and pallium (see text). The acroterminal hypothalamic domain is the rostral-most transverse domain in the forebrain and probably has organizer roles, which may be different in its alar and basal moieties (purple and green): Both express *Six3*, and secrete FGFs (see Ferran et al., 2015). The newly postulated ventricular hypothalamic organ (VHO) organizer (red) is a longitudinal domain that separates the main basal territories of the hypothalamus across both hp1 and hp2, namely the M/RM domains from the TU/RTu domains, and probably patterns their different histological fates (see Puelles et al., 2012a).

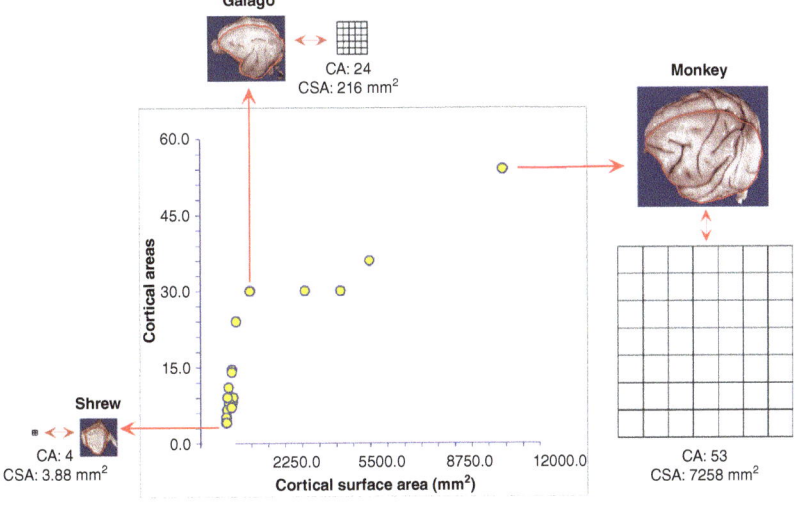

Figure 13.5 The number of cortical areas plotted versus cortex surface area. The approximate total cortical surface area of mapped visual and somatomotor areas (CSA) and number of cortical areas (CA) for the shrew, galago, and macaque are depicted, as an example of small, medium, and large cortex area. Note that the entire cortex of the Galago, comprising 24 areas, could be accommodated within a single cortical area of the rhesus monkey. Reproduced from Finlay et al., 2005.

- Progenitors undergo cell divisions to produce more progenitors or differentiated cortical neurons.

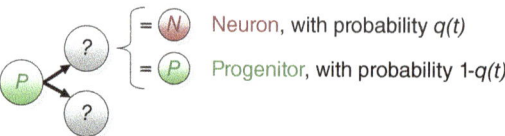

- The relative probability of those outcomes changes during the neurogenetic interval.

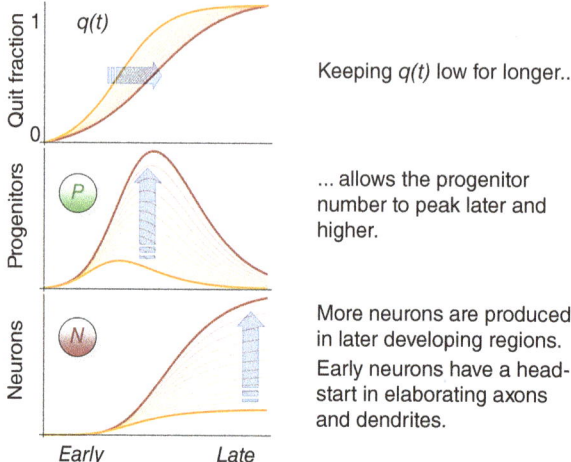

Keeping $q(t)$ low for longer...

... allows the progenitor number to peak later and higher.

More neurons are produced in later developing regions. Early neurons have a head-start in elaborating axons and dendrites.

Figure 13.6 Top: Defining the "quit fraction" as the probability that a daughter cell in the ventricular zone is a differentiated neuron. Bottom: Delaying the rise of the quit fraction has a large effect on the peak size of the precursor pool and the total number of neurons produced.

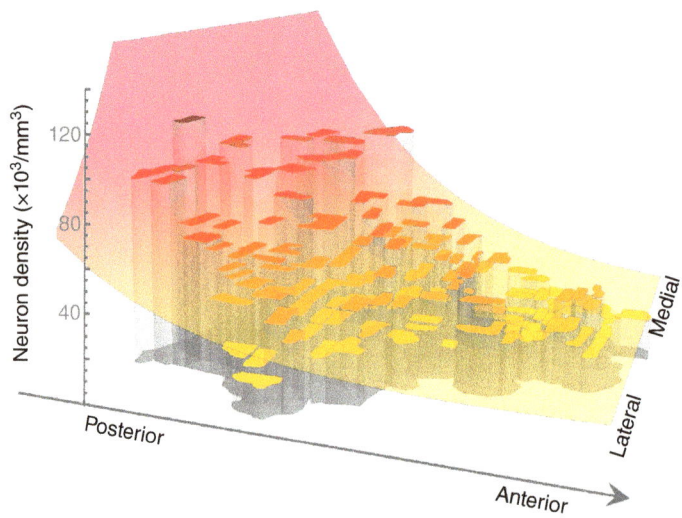

Figure 13.7 The density of neurons was measured in N=141 samples, together comprising the entire (flattened) cortical sheet of a baboon (Collins et al., 2010). Neuron density exhibits marked variation across the cortex of the baboon, the general trend (indicated here by the transparent surface) being to increase along an axis from anterior lateral cortex to posterior medial cortex. Figure redrawn, based on a figure which appeared in Cahalane et al., 2012.

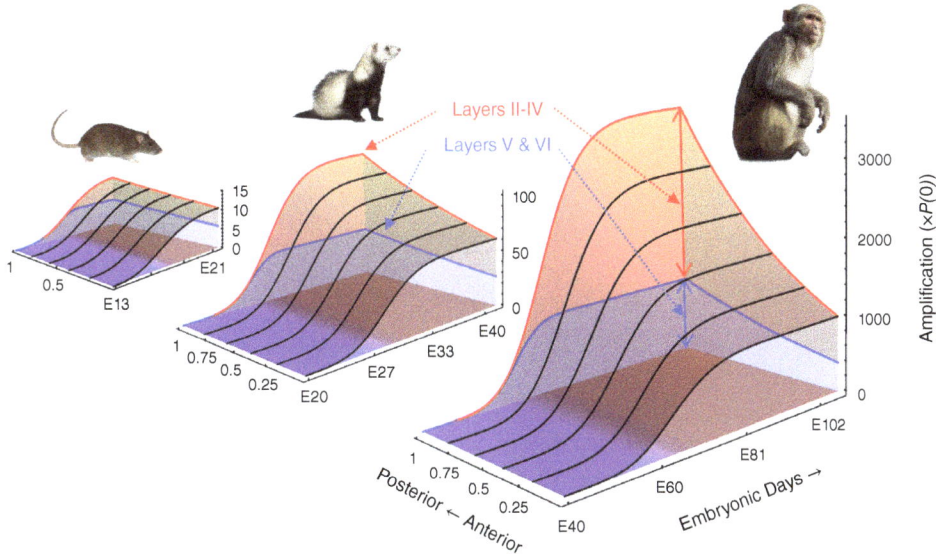

Figure 13.8 Within-cortex gradients in neurogenesis timing lead to pronounced changes in the total number of neurons and in their layer distribution across a large cortex like that of a primate. By contrast, the shorter total duration of neurogenesis and the lesser variation in neurogenesis timing in a small cortex leads to neuronal and layer distributions which are relatively uniform across the cortex. Figure redrawn, based on a figure which appeared in Cahalane et al., 2014.

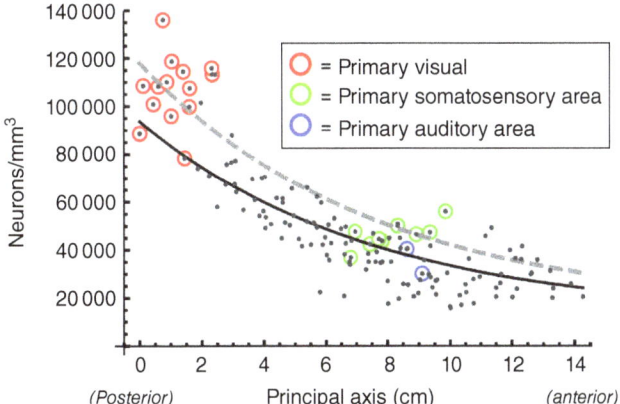

Figure 13.9 Using a two-factor model (location and primary or non-primary area) of neuronal density is better than a location-only model. In the two factor model, primary sensory areas have a neuronal density 26% higher than would a non-primary sensory area at the same location. The origin of the spatial "principal" axis is at the posterior medial pole of the flattened cortex and it extends towards the anterior lateral pole. Figure redrawn, based on an original which appeared in Cahalane et al., 2014.

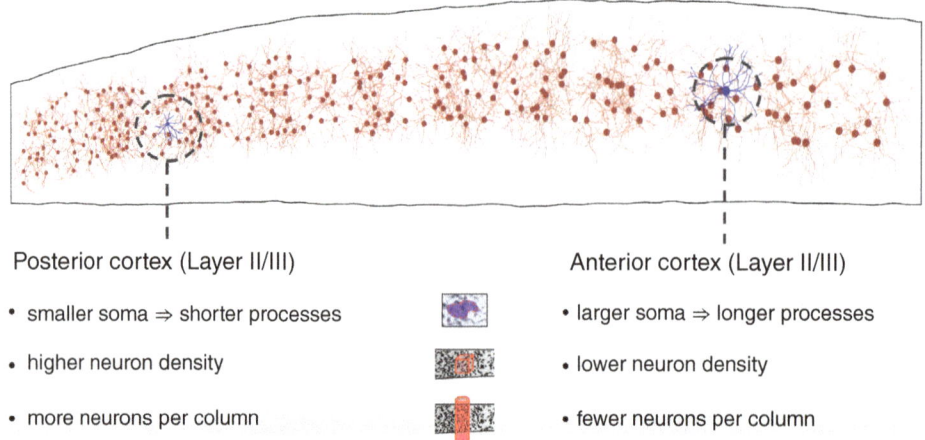

Posterior cortex (Layer II/III)
- smaller soma ⇒ shorter processes
- higher neuron density
- more neurons per column

Anterior cortex (Layer II/III)
- larger soma ⇒ longer processes
- lower neuron density
- fewer neurons per column

Figure 13.10 Schematic summary of the changes in the cortical architecture of layers II and III as implied by increased neuronal density but decreased neuron size along the anterior-to-posterior axis. Figure redrawn, based on an original which appeared in Cahalane et al., 2012.

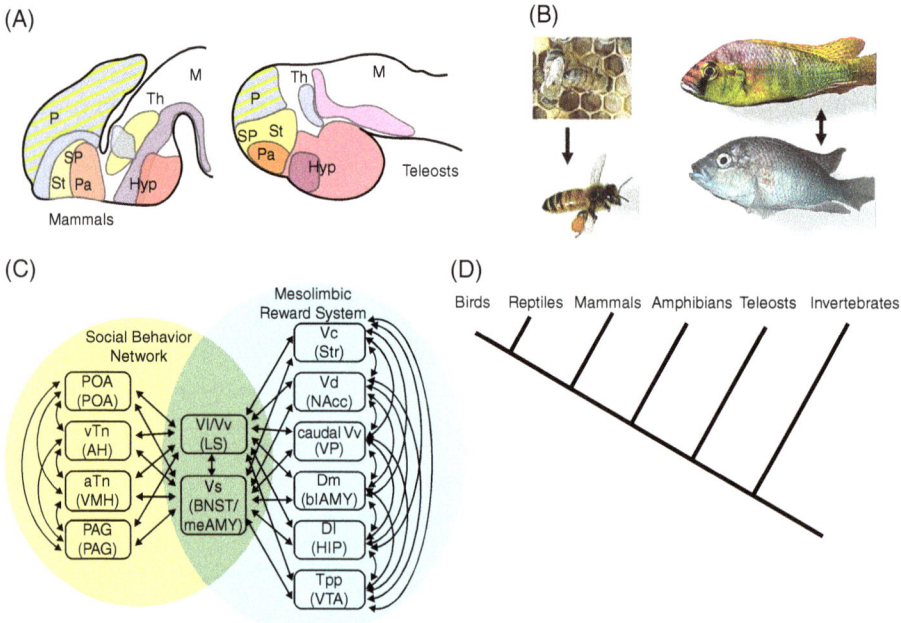

Figure 15.1 Levels and Time Scales of Neural and Behavioral Plasticity.
Understanding how genetic and environmental factors affect neural plasticity requires detailed analyses of development, neural networks, neural connectivity, life history, and evolution. (A) Gene expression patterns of developmental genes (colored regions) are highly conserved and regulate development of homologous brain structures. Adapted from O'Connell 2013. (B) The social decision making network provides a framework for analyzing structurally and functionally homologous brain regions across vertebrates. Adapted from O'Connell & Hofmann 2011b. (C) Life history traits and transitions are highly diverse, but social behavior (e.g., courtship and mating) is present in all kingdoms. Neural activation patterns underlying behavior can be conserved or divergent across species. (D) Phylogenetic relationships between major vertebrate lineages.

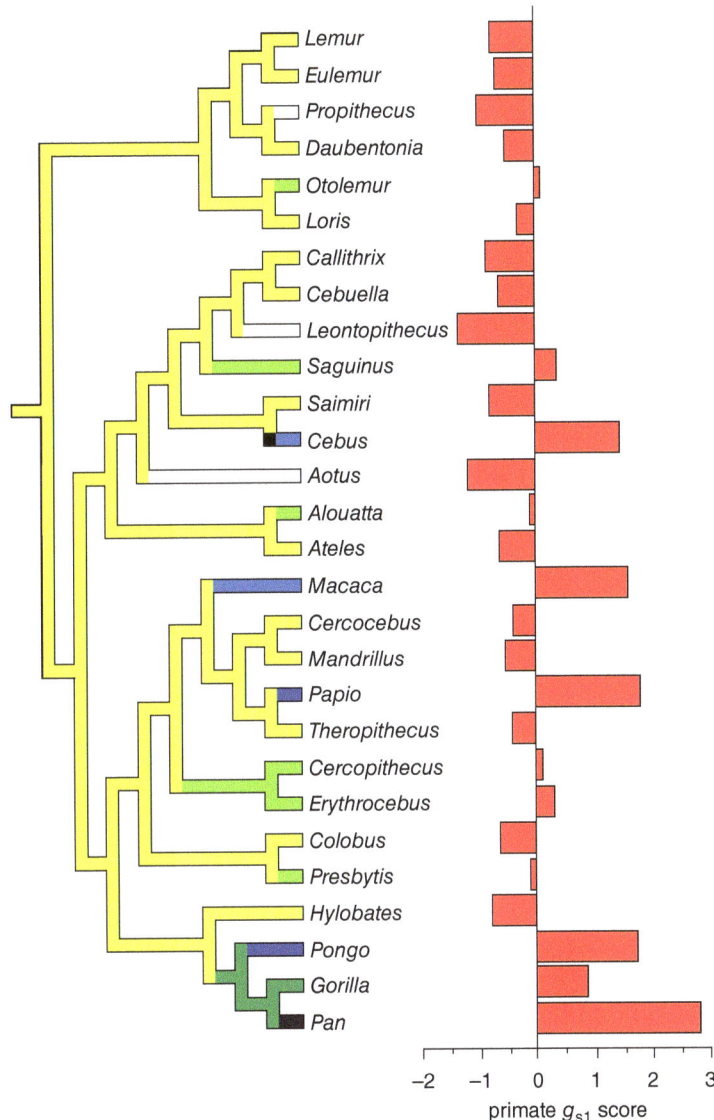

Figure 18.2 Reader et al's (2011) g_s—A Measure of General Intelligence across Primate Species, Mapped on to a Genus-Level Primate Phylogeny. Reader 2011. Reproduced with permission of Kevin Laland.

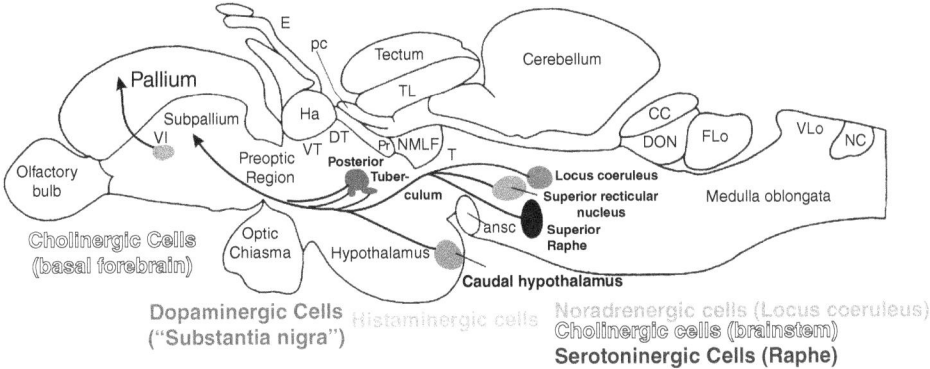

Figure 9.12 Schema Shows Sagittal Section of Adult Zebrafish Brain Depicting Neuromodulatory Systems with Long Ascending Connections to the Telencephalon.
The connections of dopaminergic (Rink & Wullimann, 2001) and serotoninergic (Lillesaar et al., 2009) cells were corroborated by double-label through axonal tracing. The indicated telencephalic connections of the histaminergic hypothalamic (Kaslin & Panula, 2001) and cholinergic brainstem cells (Mueller et al., 2004) are derived from an independent connectional study (Rink & Wullimann, 2004). The innervation of the pallium through Vl was shown for a different teleost species (Murakami, Morita, & Ito, 1983).
Abbreviations: ansc ansulate commissure; CC cerebellar crest; DON descending octaval nucleus; DT dorsal thalamus; E epiphysis; FLo facial lobe; Ha habenula; NMLF nucleus of the medial longitudinal fascicle; pc posterior commissure; Pr pretectum; TL torus longitudinalis; Vl lateral nucleus of ventral telencephalic area; VT ventral thalamus (prethalamus).
(*See insert for color representation of the figure*).

comparable between teleosts and tetrapods (see discussion in Wullimann & Vernier, 2009b), in particular regarding the ascending activating systems towards the telencephalon (see Figure 9.12).

9.4.2.3 Enigmatic cartilaginous fish brains. Cartilaginous fish, being the most ancestral extant gnathostomes, are crucial for understanding vertebrate brain evolution since they form an outgroup to tetrapods and actinopterygians (Figure 9.3). Independent brain enlargement, involving in particular telencephalon and cerebellum, occurs in sharks and rays (Northcutt, 1978), but its neurobiological significance remains enigmatic (discussed in Wullimann & Vernier, 2009b). Cartilaginous fishes have paired evaginated telencephalic hemispheres surrounding lateral ventricles (which clearly represents the ancestral condition, seen also in tetrapods) and a comparable arrangement of medial, dorsal, and lateral pallial divisions dorsal to a subpallium, which displays a medially lying septum and a lateral striatum and area superficialis basalis (Northcutt, 1981). Unfortunately, nothing is known regarding the possible role of the cartilaginous fish medial pallium in spatial orientation, although sensory processing does occur there (see below). In big-brained galeomorph sharks (such as the smooth dogfish, Figure 9.2), a large (presumably dorsal pallial) central nucleus is recognized. Although developmental studies in sharks suggest a similar pattern of subpallial versus pallial divisions as in tetrapods (Carrera, Ferreiro-Galve, Sueiro, Anadón, & Rodríguez-Moldes, 2008), a ventral pallium (and, thus, pallial amygdala) has yet to be identified in any chondrichthyian. In addition to the origin and predominance of GABAergic neurons in the subpallium, little is known regarding the basal ganglia organization in cartilaginous fish. However, unlike actinopterygians, elasmobranchs have large basal midbrain dopaminergic nuclei (substantia nigra pars compacta and ventral tegmental area), possibly convergently with tetrapods (see Wullimann & Vernier, 2009b for details).

More is known on ascending sensory systems and their representation in the prosencephalon (reviewed in Wullimann & Vernier, 2009a, b). Apart from the (largely unknown) gustatory system, most sensory systems in cartilaginous fishes have been demonstrated to reach the dorsal thalamus, that is, vision (retinothalamic and retinotectothalamic pathways), lateral line mechanosensation, somatosensation, and probably audition. In some species, electrosensation and lateral line mechanosensation ascend via the lateral lemniscal system to the lateral posterior tuberculum and dorsal thalamus, respectively, which form relays to the telencephalon (see Wullimann & Vernier, 2009b and Wullimann & Grothe, 2013 for citations).

Regarding the telencephalon, Sven Ebbesson and colleagues (Ebbesson, 1980) discovered that the nurse shark telencephalon (a galeomorph) receives only very restricted secondary olfactory bulb projections to pallial (lateral pallium) and subpallial territories, and that the central pallial nucleus receives substantial contralateral dorsal thalamic input while the lateral part of the dorsal pallium receives ipsilateral dorsal thalamic input of unspecified modality. This input to the central pallial nucleus was later determined by tract tracing and electrophysiology to be multimodal (visual, somatosensory). Furthermore, the medial pallium of the spiny dogfish (a squalomorph) receives multisensory (vision, electrosensory lateral line) dorsal thalamic and posterior tubercular inputs (see Wullimann & Vernier, 2009b; Wullimann & Grothe, 2013, for citations). Although extensive intratelencephalic higher order connections of primary olfactory areas exist in the guitarfish (Hofmann & Northcutt, 2008), the above findings disprove the "smell brain" theory (see §9.4.2) because the ancestral gnathostome condition apparently displays ascending pathways from most (if not all) sensory systems to a multisensory telencephalon. Whether nonolfactory unimodal pallial areas exist in cartilaginous fishes remains an open question. In any case, a comparison of gnathostome anamniotes shows that the medial pallium (hippocampus homologue) is always the main target of ascending sensory information from the thalamus and posterior tuberculum, and that the dorsal pallium becomes dominant in this function only in amniotes.

The data in cartilaginous fishes furthermore suggest that a dual innervation of the dorsal thalamus and posterior tuberculum by at least some ascending sensory systems is the ancestral pattern for gnathostomes. Thus, the fact that some hair cell sensory organs (audition, lateral line mechanosense, maybe vestibular sense) are represented in the dorsal thalamus while others (e.g., electroreception) are relayed in the posterior tubercular region, may be a gnathostome plesiomorphy. Possibly, then, the evolutionary loss of the lateral line system in amniotes may directly explain why the dorsal thalamus becomes the primary diencephalic sensory region.

9.4.2.4 Living agnathans may reveal ancestral forebrain condition. In both agnathan groups (myxinoids and lampreys), subpallial and pallial divisions have been recognized in the telencephalon in classical (see citations in Wullimann & Vernier, 2009b) and in modern developmental studies (reviews: Osório, Mazan, & Rétaux, 2005; Pombal, Àlvarez-Otero, Pérez-Fernández, Solveira, & Mégias, 2011). GABAergic cell production occurs in the subpallium as expected and new gene

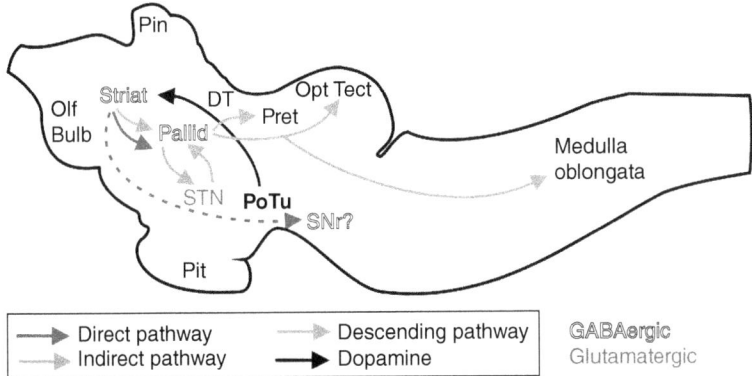

Figure 9.13 Schema Shows Sagittal Section of Lamprey Brain with Basal Ganglia Circuitry. (after Wullimann, 2011, data from Stephenson-Jones et al., 2011.)
Abbreviations: DT dorsal thalamus; OlfBulb olfactory bulb; OptTect optic tectum; Pin pineal. Pit pituitary PoTu posterior tuberculum; Pret pretectum; SNc substantia nigra compacta; SNr substantia nigra reticulata; STN Subthalamic nucleus; Striat striatum; Pallid pallidum. (*See insert for color representation of the figure*).

expression studies indicate the presence of a hitherto undetected pallidum (Sugahara et al., 2016). Morever, a basic basal ganglia circuitry has recently been reported in the lamprey brain with critical features (including an adult pallidum) of the pattern described above for tetrapods (see Figure 9.13; Stephenson-Jones, Samuelsson, Ericsson, Robertson, 7 Grillner, 2011).

The recognition of agnathan pallial subdivisions, as seen in gnathostomes, is controversial. Myxinoids (e.g. Pacific hagfish; Figure 9.2) show a uniform pallium which is organized as a pallial cortex with five distinct neuronal layers (P1-5) throughout (Wicht & Northcutt, 1998). Layers P1 and P5 receive olfactory bulb input, and all layers receive additional dorsal thalamic input of unspecified modality. Reciprocal pallial connections exist with olfactory bulb and dorsal thalamus. Likewise, extensive secondary olfactory projections reach the pallium in petromyzontids (e.g., silver lamprey, Figure 9.2), but these are again complemented by dorsal thalamic input (Polenova & Vesselkin, 1993). Thus, although olfaction is clearly more dominant in the agnathan pallium (and telencephalon) than in the gnathostome pallium, this does not support the "smell brain" theory: Extant agnathans differ greatly from their fossil ancestors in morphology and life habits (Carroll, 1988) and have become very specialized for olfactory orientation, suggesting that olfactory prevalence in the CNS may be a derived trait. Even so, the agnathan telencephalon (including pallium) is multisensory, rather than of exclusive olfactory nature.

9.5 Epilogue

Modern comparative neurobiological work on craniate central nervous system development, functional neuroanatomy/neurochemistry, and behavior has changed fundamentally our understanding of brain evolution. The previously common metaphor and teleological notion of a "scala naturae" leading from fish to man has been replaced by the recognition of a conserved bauplan for brain organization, upon which a large number of adaptive modifications occurred in various lineages—sometimes with puzzling results: the cerebellum appears to have arisen later in evolution than the basal ganglia, and a multisensory, integrative telencephalic pallium likely arose in the ancestral craniate.

Acknowledgments

I thank my colleagues Agustín Gonzáles, Benedikt Grothe, Onur Güntürkün, Fernando Martínez-García, R. Glenn Northcutt, Luis Puelles, Anton Reiner, Georg Striedter, Gerhard Schlosser, Hans ten Donkelaar, and Martin Wild for enlightening discussions. However, the views expressed in this article may not all be shared by these colleagues. Further, all possible errors are of course only attributable to the author. A special thank you goes to Stephen V. Shepherd for excellent editorial work. A series of Nissl-stained forebrain sections used for Figure 9.4 were donated to me by Onur Güntürkün/Ariane Schwarz (pigeon), Alexander Kaiser (mouse), Glenn Northcutt (tuatara), and Gerhard Roth (fire-bellied toad). The author is supported by the Deutsche Forschungsgemeinschaft, Bonn, Germany (SPP Olfaction 1392, Project Wu211/2-1 and 2-2) and the Graduate School for Systemic Neurosciences at the LMU-Munich, Planegg, Germany.

Note

1 Contrary to the view presented in this chapter, there are recent reports that continue to suggest a homology of DVR with mammalian isocortex (Chen, Winkler, Pfenning, & Jarvis, 2013; Jarvis et al., 2012; Karten, 2013). Gene expression studies in the adult (52 genes) and developing bird pallium (16 genes) are interpreted in such a way that layers I to VI of isocortex correspond functionally to various compartments of the bird pallium running from nido- and meso- through hyperpallium. However, only a minority of the investigated genes are developmentally relevant transcription factors reflecting on the developmental origin of a given pallial area. Most other genes may arguably be necessary for functional, and therefore convergent, reasons (for example glutamatergic phenotype maintenance in visual processing or axonal pathfinding). Furthermore, a suspected massive ventrodorsal migration necessary for this explanatory model is still purely hypothetical in these reports. Moreover, transcriptome analyses using 5000 genes (Belgard et al., 2013) are in conflict with these results. However, most recent gene expression studies in mouse and chicken brains (Puelles et al., 2015; 2016) were interpreted as follows: the mammalian lateral pallium/avian mesopallium gives rises to claustrum and insular cortex (and its homologues in birds), and the ventral pallium gives rise to the piriform cortex and most of the pallial amygdala. This would not only make the term lateral pallium obsolete in comparative terms, but also indicate that the avian mesopallium—but not the nidopallium—has a close relationship to isocortex.

References

Alonso, A., Merchán, P., Sandoval, J. E., Sánchez-Arrones, L., Garcia-Cazorla, A., Artuch, R., Puelles, L. (2012). Development of the serotonergic cells in murine raphe nuclei and their relations with rhombomeric domains. *Brain Structure & Function*. doi: 10.1007/s00429-012-0456-8

Aroca, P., & Puelles, L. (2005). Postulated boundaries and differential fate in the developing rostral hindbrain. *Brain Research*, 49, 179–190.

Belgard, T. G., Montiel, J. F., Wang, W. Z., García-Moreno, F., Margulies, E. H., Ponting, C. P., & Molnár, Z. (2013). Adult pallium transcriptomes surprise in not reflecting predicted homologies across diverse chicken and mouse pallial sectors. *Proceedings of the National Academy of Sciences of the USA*, 110, 13150–13155.

Birgbauer, E., & Fraser, S. E. (1994). Violation of cell lineage restriction compartments in the chick hindbrain. *Development*, 120, 1347–1356.

Bischof, H. J., & Watanabe, S. (1997) On the structure and function of the tectofugal visual pathway in laterally eyed birds. *European Journal of Morphology*, 35, 246–254.

Brand, M., Heisenberg, C. P., Jiang, Y. J., Beuchle, D., Lun, K., Furutani-Seiki, M., & Nüsslein-Volhard, C. (1996). Mutations in zebrafish genes affecting the formation of the boundary between midbrain and hindbrain. *Development*, 123, 179–190.

Braun, C. B. (1996). The sensory biology of lampreys and hagfishes: A phylogenetic assessment. *Brain Behavior & Evolution*, 48, 262–276.

Brox, A., Puelles, L., Ferreiro, B., & Medina, L. (2003). Expression of the genes *GAD67* and *Distal-less-4* in the forebrain of *Xenopus laevis* confirms a common pattern in tetrapods. *Journal of Comparative Neurology*, 461, 370–393.

Brox, A., Puelles, L., Ferreiro, B., & Medina, L. (2004). Expression of the genes Emx1, Tbr1, and Eomes (Tbr2) in the telencephalon of Xenopus laevis confirms the existence of a ventral pallial division in all tetrapods. *Journal of Comparative Neurology*, 474, 562–577.

Bruce, L. L., & Neary, T. J. (1995). The limbic system of tetrapods: a comparative analysis of cortical amygdalar populations. *Brain Behavior & Evolution*, 46, 224–234.

Bullock, T. H., Bodznick, D. A., & Northcutt, R. G. (1983). The phylogenetic distribution of electroreception: Evidence for convergent evolution of a primitive vertebrate sensory modality. *Brain Research Reviews*, 6, 25–46.

Butler, A. B., & Hodos, W. (2005). *Comparative vertebrate neuroanatomy*. New York, NY: John Wiley & Sons.

Cambronero, F., & Puelles, L. (2000). Rostrocaudal nuclear relationships in the avian medulla oblongata: a fate map with quail chick chimeras. *Journal of Comparative Neurology*, 427, 522–545.

Carrera, I., Ferreiro-Galve, S., Sueiro, C., Anadón, R., & Rodríguez-Moldes, I. (2008). Tangentially migrating GABAergic cells of subpallial origin invade massively the pallium in developing sharks. *Brain Research Bulletin*, 75, 405–409.

Carroll, R.L. (1988). *Vertebrate paleontology and evolution*. New York, NY: Freeman.

Chen, C.-C., Winkler, C. M., Pfenning, A. R., & Jarvis, E. D. (2013). Molecular profiling of the developing avian telencephalon: Regional timing and brain subdivision continuities. *Journal of Comparative Neurology*, 521, 3666–3701.

Coombs, S., Görner, P., & Münz, P. (1989). *The mechanosensory lateral line*. New York, NY: Springer.

Demski, L., & Northcutt, R. G. (1983). The terminal nerve: A new chemosensory system in vertebrates? *Science*, 220, 435–437.

Doron, N. N., & Ledoux, J. E. (1999). Organization of projections to the lateral amygdala from auditory and visual areas of the thalamus in the rat. *Journal of Comparative Neurology*, 412, 383–409.

Dubbeldam, J. L., den Boer-Visser, A. M., & Bout, R. G. (1997). Organization and efferent connections of the archistriatum of the mallard, *Anas platyrhynchos* L.: An anterograde and retrograde tracing study. *Journal of Comparative Neurology*, 388, 632–657.

Ebbesson, S. O. E. (1980). On the organization of the telencephalon in elasmobranchs. In S. O. E. Ebbesson (Ed.), *Comparative neurology of the telencephalon* (pp. 1–16). New York, NY: Plenum Publishing Corporation.

Edwards, J. G., Greig, A., Sakata, Y., Elkin, D., & Michel, W. C. (2007). Cholinergic innervation of the zebrafish olfactory bulb. *Journal of Comparative Neurology*, 504, 631–645.

Eisthen, H. (2004). The goldfish knows: Olfactory receptor cell morphology predicts receptor gene expression. *Journal of Comparative Neurology*, 477, 341–346.

Emery, N. J., & Clayton, N. S. (2004). The mentalitiy of crows: Convergent evolution of intelligence in corvids and apes. *Science*, 306, 1903–1907.

Endepols, H., Helmbold, F., & Walkowiak, W. (2007). GABAergic projection neurons in the basal ganglia of the green tree frog (*Hyla cinerea*). *Brain Research*, 1138, 76–85.

Engelage, J., & Bischof, H. J. (1993). The organization of the tectofugal pathway in birds: A comparative review. In H. P. Zeigler and H. J. Bischof (Eds.), *Vision, Brain, and Behaviour in Birds* (pp. 137–158). Cambridge, MA: MIT Press.

Ericson, J., Muhr, J., Placzek, M., Lints, T., Jessell, T. M., & Edlund, T. (1995). Sonic hedgehog induces the differentiation of ventral forebrain neurons: A common signal for ventral patterning within neural tube. *Cell*, 81, 747–756.

Farries, M. A. 2001 The oscine song system considered in the context of the avian brain: Lessons learned from comparative neurobiology. *Brain Behavior & Evolution*, 58, 80–100

Fraser, S., Keynes, R., & Lumsden, A. (1990). Segmentation in the chick embryo hindbrain is defined by cell lineage restrictions. *Nature*, 344, 431–435.

Fujita, I., Sorensen, P. W., Stacey, N. E., & Hara, T. J. (1991). The olfactory system, not the terminal nerve, functions as the primary chemosensory pathway mediating responses to sex pheromones in male goldfish. *Brain Behavior & Evolution*, 38, 313–321.

Germot, A., Lecointre, G., Plouhinec, J. L., Le Mentec, C., Girardot, F., & Mazan S. (2001). Structural evolution of *Otx* genes in craniates. *Molecular Biology & Evolution*, 18, 1668–1678.

Gilland, E., & Baker, R. (2005). Evolutionary patterns of cranial nerve efferent nuclei in vertebates. *Brain Behavior & Evolution*, 66, 234–254.

González, A., Russchen, F. T., & Lohman, A. H. M. (1990). Afferent connections of the striatum and the nucleus accumbens in the lizard *Gekko gecko*. *Brain Behavior & Evolution*, 36, 39–58

González, M. J., Yáñez, J., & Anadón, R. (1999). Afferent and efferent connections of the torus semicircularis in the sea lamprey: An experimental study. *Brain Research*, 826, 83–94.

Guirado, S., Dávila, J. C., Real, M. Á., & Medina, L. (1999). Nucleus accumbens in the lizard Psammogrammus algirus: Chemoarchitecture and cortical afferent connection. *Journal of Comparative Neurology*, 405, 15–31.

Güntürkün, O. (2005). The avian "prefrontal cortex" and cognition. *Current Opinion in Neurobiology*, 15, 686–693.

Hofmann, M. H. (1999). Nervous system. In W. C. Hamlett (Ed.), *Sharks, skates, rays: The biology of elasmobranch fishes* (pp. 273–299). Baltimore, MD: Johns Hopkins University Press.

Hofmann, M. H., & Northcutt, R. G. (2008). Organization of major telencephalic pathways in an elasmobranch, the thornback ray *Platyrhinoidis triseriata*. *Brain Behavior & Evolution*, 72, 307–325.

Holland, P. W. H., & Hogan, B. L. M. (1988). Expression of homeobox genes during mouse development: A review. *Genes & Development*, 2, 773–782.

Hunt, P., & Krumlauf, R. (1992). *Hox* codes positional specification in vertebrate embryonic axes. *Annual Review of Cell Biology*, 8, 227–256.

Jarvis, E. D., Güntürkün, O., Bruce, L., Csillag, A., Karten, H., Kuenzel, W., … Butler, A. B. (2005). Avian brains and a new understanding of vertebrate brain evolution. *Nature, Reviews Neuroscience*, 6, 151–159.

Jarvis E. D., Yu, J., Rivas, M. V., Horita, H., Feenders, G., Whitney, O., … Wada, K. (2013). Global view of the functional molecular organization of the avian cerebrum: mirror images and functional columns. *Journal of Comparative Neurology*, 521, 3614–3665.

Jiao, Y., Medina, L., Veenman, L. C., Toledo, C., Puelles, L., & Reiner, A. (2000). Identification of the anterior nucleus of the ansa lenticularis in birds as the homolog of the mammalian subthalamic nucleus. *Journal of Neuroscience*, 20, 6998–7010.

Karten, H. J. (2013). Neocortical evolution: Neuronal circuits arise independently of lamination. *Current Biology*, 23, R12–R15.

Karten, H. J., Hodos, W., Nauta, W. J. H., & Rezvin, A. M. (1977). Neural connections of the "visual Wulst" of the avian telencephalon. Experimental studies in the pigeon (*Columba livia*) and owl (*Speotyto cunicularia*). *Journal of Comparative Neurology*, 150, 253–276

Kaslin, J., & Panula, P. (2001). Comparative anatomy of the histaminergic and other aminergic systems in zebrafish (*Danio rerio*). *Journal of Comparative Neurology*, 440, 342–377.

Kawai, T., & Oka, Y., & Eisthen, H. (2009). The role of the terminal nerve and GnRH in olfactory system neuromodulation. *Zoological Science*, 26, 669–680.

Kröner, S., & Güntürkün, O. (1999). Afferent and efferent connections of the caudolateral neostriatum in the pigeon (*Columba livia*): A retro- and anterograde pathway tracing study. *Journal of Comparative Neurology*, 407, 228–260.

Kuratani, S., & Ota, K. G. (2008). Primitive versus derived traits in the developmental program of the vertebrate head: Views from cyclostome developmental studies. *Journal of Experimental Zoology*, 310B, 294–314.

Lannoo, M. J., & Hawkes, R. (1997). A search for primitive Purkinje cells: Zebrin II expression in sea lampreys (*Petromyzon marinus*). *Neuroscience Letters*, 237, 53–55.

Larsen, C. W., Zeltser, L. M., & Lumsden, A. (2001). Boundary formation compartition in the avian diencephalon. *Journal of Neuroscience*, 21, 4699–4711.

LeDoux, J. (2000). Emotion circuits in the brain. *Annual Review of Neuroscience*, 23, 155–184.

Lillesaar, C., Stigloher, C., Tannhäuser, B., Wullimann, M. F., & Bally-Cuif, L. (2009). Axonal projections of raphe serotonergic neurons in the developing adult zebrafish, *Danio rerio*, visualized by transgenic demonstration of raphe-specific *pet1* expression. *Journal of Comparative Neurology*, 512, 158–182.

Lumsden, A. (2004). Segmentation and compartition in the early avian hindbrain. *Mechanisms of Development*, 121, 1081–1088.

Lumsden, A., & Krumlauf, R. (1996). Patterning the vertebrate neuraxis. *Science*, 274, 1109–1115.

Marín, F., & Puelles, L. (1994). Patterning of the embryonic avian midbrain after experimental inversions: A polarizing activity from the isthmus. *Developmental Biology*, 163, 19–37.

Marín, F., & Puelles, L. (1995). Morphological fate of rhombomeres in quail/chick chimeras: A segmental analysis of hindbrain nuclei. *European Journal of Neuroscience*, 7, 1714–1738.

Marín, F., Aroca, P., & Puelles, L. (2008). *Hox* gene colinear expression in the avian medulla oblongata is correlated with pseudorhombomeric domains. *Developmental Biology*, 323, 230–247.

Marín O, Smeets W. J. A. J., & González, A. 1998 Evolution of the basal ganglia in tetrapods: A new perspective based on recent studies in amphibians. *Trends in Neuroscience*, 23, 487–494.

Martínez-Garcia, F., Novejarque, A., & Lanuza, E. (2009). The evolution of the amygdala in vertebrates, In J. H. Kaas (Ed.), *Evolutionary neuroscience* (pp. 313–392). Amsterdam: Elsevier-Academic Press.

Mazan, S., Jaillard, D., Baratte, B., & Janvier, P. (2000). Otx1 gene-controlled morphogenesis of the horizontal semicircular canal and the origin of the gnathostome characteristics. *Evolution & Development*, 2, 186–193.

McCormick, C. A. (1992). Evolution of central auditory pathways in anamniotes. In D. B. Webster, R. R. Fay, & A. N. Popper (Eds.), *The evolutionary biology of hearing* (pp. 323–350). New York, NY: Springer.

Medina, L., Legaz, I., González, G., De Castro, F., Rubenstein, J. L. R., & Puelles, L. (2004). Expression of *Dbx1*, *Neurogenin2*, *Semaphorin 5A*, *Cadherin 8*, *Emx1* distinguish ventral lateral pallial histogenetic divisions in the developing mouse claustroamygaloid complex. *Journal of Comparative Neurology*, 474, 504–523.

Meek, J., & Nieuwenhuys, R. (1998). Holosteans and teleosts. In R. Nieuwenhuys, H. J. ten Donkelaar, & C. Nicholson (Eds.), *The central nervous system of vertebrates* (pp. 759-937). Berlin: Springer.

Miceli, D., Repérant, J., Villalobos, J., & Dionne, L. (1987). Extratelencephalic projections of the avian visual Wulst. A quantitative autoradiographic study in the pigeon *Columba livia*. *Journal für Hirnforschung*, 28, 45–57.

Moreno, N., & González, A. (2007). Evolution of the amygdaloid complex in vertebrates, with special reference to the anamnio-amniotic transition. *Journal of Anatomy*, 211, 151–163.

Mueller, T., Wullimann, M. F., & Guo, S. (2008). Early teleostean basal ganglia development visualized by zebrafish *Dlx2a*, *Lhx6*, *Lhx7*, *Tbr2 (eomesa)*, and *GAD67* gene expression. *Journal of Comparative Neurology*, 507, 1245–1257.

Mueller, T., Dong, Z., Berberoglu, M. A., & Guo, S. (2011). The dorsal pallium in zebrafish, Danio rerio (Cyprinidae, Teleostei). *Brain Research*, 1381, 95–105.

Murakami, T., Morita, Y., & Ito, H. (1983). Extrinsic and intrinsic fiber connections of the telencephalon in a teleost, *Sebastiscus marmoratus*. *Journal of Comparative Neurology*, 216, 115–131.

Murakami, Y., Uchida, K., Rijli, F. M., & Kuratani, S. (2005). Evolution of the brain developmental plan: Insights from agnathans. *Developmental Biology*, 280, 249–259.

Necker, R. (1989). Cells of origin of spinothalamic, spinotectal, spinoreticular, and spinocerebellar pathways in the pigeon by the retrograde transport of horseradish peroxidase. *Journal für Hirnforschung*, 30, 33–43.

Nieuwenhuys, R. (1998). Morphogenesis and general structure. In R. Nieuwenhuys, H. J. ten Donkelaar, & C. Nicholson (Eds.), *The central nervous system of vertebrates* (pp. 159–226). New York, NY: Springer.

Nieuwenhuys, R. (2009). The forebrain of actinopterygians revisited. *Brain Behavior & Evolution*, 73, 229–252.

Nieuwenhuys, R. (2011). The structural, functional, and molecular organization of the brainstem. *Frontiers in Neuroanatomy*, 5, 33. doi: 10.3389/fnana.2011.00033

Nieuwenhuys, R., ten Donkelaar, H. J., & Nicholson, C. (Eds.) (1998). *The central nervous system of vertebrates*. New York, NY: Springer.

Nieuwenhuys, R., Voogd, J., & van Huijzen, C. (2008). *The human central nervous system* (4th ed.). New York, NY: Springer.

Northcutt, R. G. (1978). Brain organization in the cartilaginous fishes. In E. S. Hodgson & R. F. Mathewson (Eds.), *Sensory biology of sharks, skates, rays*, (pp. 117–193). Arlington, VA: Office of Naval Research, Department of the Navy.

Northcutt, R. G. (1981). Evolution of the telencephalon in nonmammals. *Annual Review of Neuroscience*, 4, 301–350.

Northcutt, R. G. (1985). The brain and sense organs of the earliest vertebrates: Reconstruction of a morphotype. In R. E. Foreman, A. Gorbman, J. M. Dodd, & R. Olsson (Eds.), *Evolutionary biology of primitive fishes* (pp. 81–112). Plenum Press, New York.

Northcutt, R. G. (1986). Evolution of the octavolateralis system: Evaluation and heuristic value of phylogenetic hypotheses. In R. W. Ruben, T. R. van der Water, & E. W. Rubel (Eds.), *The biology of change in otolaryngology* (pp. 3–14). Amsterdam: Elsevier.

Northcutt, R. G. (2006). Connections of the lateral and medial divisions of the goldfish telencephalic pallium. *Journal of Comparative Neurology*, 494, 903–943

Northcutt, R. G. (2009). Phylogeny of nucleusl medianus of the posterior tuberculum in ray-finned fishes. *Integrative Zoology*, 4, 134–151.

Northcutt, R. G. (2011). Evolving large and complex brains. *Science*, 332, 926–927.

Northcutt, R. G., & Bemis, W. W. (1993). Cranial nerves of the coelacanth *Latimeria chalumnae* (Osteichthyes: Sarcopterygii: Actinistia) and comparison with other craniata. *Brain Behavior & Evolution*, 42(Suppl. 1), 1–76.

Northcutt, R. G., & Braford, M. R., Jr. (1980). New observations on the organization and evolution of the telencephalon of actinopterygian fishes. In S. O. E. Ebbesson (Ed.), *Comparative neurology of the telencephalon* (pp. 41–98). New York, NY: Plenum Press.

Northcutt, R. G., & Gans, C. (1983). The genesis of neural crest epidermal placodes: A reinterpretation of vertebrate head origins. *Quarterly Review of Biology*, 58, 1–28.

Osório, J., Mazan, S., & Rétaux, S. (2005). Organisation of the lamprey (Lampetra fluviatilis) ambryonic brain from LIM-homeodomain, Pax and hedgehog genes. *Developmental Biology*, 288, 100–112.

Osório, J., Mueller, T., Rétaux, S., Vernier, P., & Wullimann, M. F. (2010). Phylotypic expression of the bHLH genes Neurogenin2, NeuroD, and Mash1 in the mouse embryonic fore-brain. *Journal of Comparative Neurology*, 518, 851–871.

Oxtoby, E., & Jowett, T. (1993). Cloning of the zebrafish krox-20 gene (krx-20) its expression during hindbrain development. *Nucleic Acids Research*, 21, 1087–1095.

Polenova, O. A., & Vesselkin, N. P. (1993). Olfactory and nonolfactory projections in the river lamprey (*Lampetra fluviatilis*) telencephalon. *Journal für Hirnforschung*, 34, 261–279.

Pombal, M. A., Àlvarez-Otero, R., Pérez-Fernández, J., Solveira, C., & Mégias, M. (2011). Development and organization of the lamprey telencephalon with special reference to the GABAergic system. *Frontiers in Neuroanatomy*, 5, 20. doi: 10.3389/fnana.2011.00020

Prechtl, J. C., von der Emde, G., Wolfart, J., Karamürsel, S., Akoev, G. N., Andrianov, Y. N., & Bullock, T. H. (1998). Sensory processing in the pallium of a mormyrid fish. *Journal of Neuroscience*, 18, 7381–7393.

Puelles, E. (2007). Genetic control of basal midbrain development. *Journal of Neuroscientific Research*, 85, 3530–3534.

Puelles, L., Ayad, A., Sandoval, J. E., Alonso, A., Medina, L., & Ferran, J. L. (2016) Selective expression of the orphan nuclear receptor Nr4a2 identifies the claustrum homolog in the avian mesopallium: impact on sauropsidian/mammalian pallium comparisons. *Journal of Comparative Neurolology*, 524, 665–703.

Puelles, L., Kuwana, E., Puelles, E., & Rubenstein, J. L. R. (1999). Comparison of the mammalian and avian telencephalon from the perspective of gene expression data. *European Journal of Morphology*, 37, 139–150.

Puelles, L., Kuwana, E., Puelles, E., Bulfone, A., Shimamura, K., Keleher, J., ... Rubenstein, J. L. R. (2000). Pallial subpallial derivatives in the embryonic chick and mouse telencephalon, traced by the expression of the genes Dlx-2, Emx-1, Nkx-2.1, Pax-6, Tbr-1. *Journal of Comparative Neurology*, 424, 409–438.

Puelles, L., Medina, L., Borello, U., Legaz, I., Teissier, A., Pierani, A., & Rubenstein, J. L. (2015) Radial derivatives of the mouse ventral pallium traced with Dbx1-LacZ reporters. *Journal of Chemical Neuroanatomy*. doi:10.1016/j.jchemneu.2015.10.011

Puelles, L., & Rubenstein, J. L. R. (1993). Expression patterns of homeobox and other putative regulatory genes in the embyonic mouse forebrain suggests a neuromeric organization. *Trends in Neurosciences*, 16, 472–479.

Puelles, L., & Rubenstein, J. L. R. (2003). Forebrain gene expression domains the evolving prosomeric model. *Trends in Neurosciences*, 26, 469–476.

Reiner, A. (2002). Functional circuitry of the avian basal ganglia: implications for basal ganglia organization in stem amniotes. *Brain Research Bulletin*, 57, 513–528

Reiner, A., Medina, L., & Veenman, C. L. (1998). Structural and functional evolution of the basal ganglia in vertebrates. *Brain Research Reviews*, 28, 235–285.

Reiner, A., Perkel, D. J., Bruce, L. L., Butler, A. B., Csillag, A., Kuenzel, W., ... Avian Brain Nomenclature Consortium (2004). Revised nomenclature for avian telencephalon and some related brainstem nuclei. *Journal of Comparative Neurology*, 473, 377–414.

Reiner, A., Yamamoto, K., & Karten, H J. (2005). Organization and evolution of the avian forebrain. *The Anatomical Record*, 287A, 1080–1102.

Rink, E., & Wullimann, M. F. (2001). The teleostean (zebrafish) dopaminergic system ascending to the subpallium (striatum) is located in the basal diencephalon (posterior tuberculum). *Brain Research*, 889, 316–330.

Rink, E., & Wullimann, M. F. (2004). Connections of the ventral telencephalon (subpallium) in the zebrafish (Danio rerio). *Brain Research*, 1011, 206– 220.

Remy, M., & Güntürkün, O. (1991). Retinal afferents to the tectum opticum and nucleus opticus principalis thalami in the pigeon. *Journal of Comparative Neurology*, 305, 57–70

Rodríguez, F., López, J. C., Vargas, J. P., Broglio, C., Gómez, Y., & Salas, C. (2002). Spatial memory and hippocampal pallium through vertebrate evolution: Insights from reptiles and teleost fish. *Brain Research Bulletin*, 57, 499–503.

Ronan, M. (1989). Origins of the descending spinal projections in petromyzontid and myxinoid agnathans. *Journal of Comparative Neurology*, 281, 54–68.

Roth, G., & Dicke, U. (2009). Evolution of the amphibian nervous system. In J. H. Kaas (Ed.), *Evolutionary neuroscience* (pp. 169–232). Amsterdam: Elsevier-Academic Press.

Salas, C., Broglio, C., & Rodriguez, F. (2003). Evolution of forebrain spatial cognition in vertebrates: Conservation across diversity. *Brain Behavior & Evolution*, 62, 72–82

Schlosser, G. (2006). Induction and specification of cranial placodes. *Developmental Biology*, 294, 303–351.

Schneider, A., & Necker, R. (1989). Spinothalamic projections in the pigeon. *Brain Research*, 484, 139–149.

Scholpp, S., & Brand, M. (2003). Integrity of the midbrain region is required to maintain the diencephalic-mesencephalic boundary in zebrafish no isthmus/*pax2.1* mutants. *Developmental Dynamics*, 228, 313–322.

Shimamura, K., Martinez, S., Puelles, L., & Rubenstein, J. L. R. (1997). Patterns of gene expression subdivide the embyonic forebrain into transverse longitudinal zones. *Developmental Neuroscience*, 19, 88–96.

Shimizu, T. (2001). Evolution of the forebrain in tetrapods. In G. Roth & M. F. Wullimann (Eds.), *Brain evolution cognition* (pp. 135–184). Berlin/New York: Spektrum Akademischer Verlag/Wiley.

Shimizu, T., & Bowers, A. N. (1999). Visual circuits of the avian telencephalon: Evolutionary implications. *Behavioural Brain Research*, 89, 183–191.

Shimizu, T., & Karten, H. J. (1993). The avian visual system and the evolution of the neocortex. In H. P. Zeigler & H. J. Bischof (Eds.), *Vision, brain, and behavior in birds* (pp 103–114). Cambridge, MA: MIT Press.

Smeets, W. J. A. J., & Medina, L. (1995). The efferent connections of the nucleus accumbens in the lizard *Gekko gecko*. *Anatomy & Embryology*, 191, 73–81.

Smeets, W. J. A. J., Nieuwenhuys, R., & Roberts, B. L. (1983). *The central nervous system of cartilaginous fishes*. New York, NY: Springer.

Smith-Fernandez, A., Pieau, C., Repérant, J., Boncinelli, E., & Wassef, M. (1998). Expression of the *Emx-1* and *Dlx-1* homeobox genes define three molecularly distinct domains in the telencephalon of the mouse, chick, turtle and frog embryos. *Development*, 125, 2099–2111.

Squire, L. R., Bloom, F. E., Spitzer, N. C., du Lac, S., Ghosh, A., & Berg, D. (2008). *Fundamental neuroscience* (3rd ed.). New York, NY: Academic Press.

Stephenson-Jones, M., Samuelsson, E., Ericsson, J., Robertson, B., & Grillner S (2011). Evolutionary conservation of the basal ganglia as a common vertebrate mechanism for action selection. *Current Biology*, 21, 1081–1091.

Striedter, G. F. (1997). The telencephalon of tetrapods in evolution. *Brain Behavior & Evolution*, 49, 179–213.

Sugahara, F., Pascual-Anaya, J., Oisi, Y., Kuraku, S., Aota, S-i., Adachi, N., ... Kuratani, S. (2016) Evidence from cyclostomes for complex regionalization of the ancestral vertebrate brain. *Nature*, 531(7392), 97–100. doi: 10.1038/nature16518

Swanson, L. W., & Petrovich, G. D. (1998). What is the amygdala? *Trends in Neuroscience*, 21, 323–331.

Szele, F. G., Chin, H. K., Rowlson, M. A., & Cepko, C. L. (2002). Sox-9 and cDachshund-2 expression in the developing chick telencephalon. *Mechanisms of Development*, 12, 179–182.

ten Donkelaar, H. (1998a). Anurans. In R. Nieuwenhuys, H. ten Donkelaar, & C. Nicholson (Eds.), *The central nervous system of vertebrates* (pp. 1151–1314). Berlin: Springer Verlag.

ten Donkelaar, H. (1998b). Reptiles. In R. Nieuwenhuys, H. ten Donkelaar, & C. Nicholson (Eds.), *The central nervous system of vertebrates* (pp. 1315–1524). Berlin: Springer Verlag.

Veenman, C. L., Wild, J. M., & Reiner, A. (1995). Organization of the avian "corticostriatal" projection system: A retrograde and anterograde pathway tracing study in pigeons. *Journal of Comparative Neurology*, 354, 87–126.

Vernier, P., & Wullimann, M. F. (2009). The posterior tuberculum. In M. D. Binder, N. Hirokawa, & U. Windhorst (Eds.), *Encyclopedia of neuroscience* (Vol 2, pp. 1404–1413). New York, NY: Springer.

von Bartheld, C. (2004). The terminal nerve and its relation with extrabulbar "olfactory" projections: Lessons from lampreys and lungfishes. *Microscopy Research & Technique*, 65, 13–24.

von Frowein, J., Campbell, K., & Götz, M. (2002). Expression of *Ngn*1, *Ngn*2, *Cash*1, *Gsh*2 and *Sfrp*1 in the developing chick telencephalon. *Mechanisms of Development*, 110, 249–252.

Wang, Y. C., Jiang, S., & Frost, B. J. (1993). Visual processing in pigeon nucleus rotundus: Luminance, color, motion, and looming subdivisions. *Visual Neuroscience*, 10, 21–30.

Watanabe, S., Mayer, U., & Bischof H.-J. (2008). Pattern discrimination is affected by entopallial but not by hippocampal lesions in zebra finches. *Behavioural Brain Research*, 190, 201–205

Watanabe, S., Mayer, U., & Bischof H.-J. (2011). Visual Wulst analyses "where" and entopallium analyses "what" in the zebra finch visual system. *Behavioural Brain Research*, 222, 51–56.

Whitlock, K. E. (2004). Development of the nervus terminalis: Origin and migration. *Microscopy Research & Technique*, 65, 2–12

Wicht, H. (1996). The brains of lampreys and hagfishes: Characteristics, characters, comparisons. *Brain Behavior & Evolution*, 48, 248–261.

Wicht, H., & Nieuwenhuys, R. (1998). Hagfishes (Mixinoidea). In R. Nieuwenhuys, H. J. ten Donkelaar, & C. Nicholson (Eds.), *The central nervous system of vertebrates* (pp. 497–549). New York, NY: Springer.

Wicht, H., & Northcutt, R.G. (1992). The forebrain of the pacific hagfish: A cladistic reconstruction of the ancestral craniate forebrain. *Brain Behavior & Evolution*, 40, 25–64.

Wicht, H., & Northcutt, R. G. (1998). Telencephalic connections in the Pacific hagfish (*Eptatretus stouti*), with special reference to the thalamopallial system. *Journal of Comparative Neurology*, 395, 245–260.

Wild, J. M. (1985). The avian somatosensory system. I. Primary spinal afferent input to the spinal cord and brainstem in the pigeon (*Columba livia*). *Journal of Comparative Neurology*, 240, 377–395.

Wild, J. M. (1987). The avian somatosensory system: Connections of regions of body representation in the forebrain of the pigeon. *Brain Research*, 412, 205–223.

Wild, J. M. (1989). Avian somatosensory system: II. Ascending projections of the dorsal column and external cuneate nuclei in pigeon. *Journal of Comparative Neurology*, 287, 1–18.

Wild, J. M. (1992). Direct and indirect "cortico"-rubral and rubro-cerebellar cortical projections in the pigeon. *Journal of Comparative Neurology*, 326, 623–636.

Wild, J. M. (1994). Visual and somatosensory inputs to the avian song system via nucleus uvaeformis (Uva) and a comparison with the projections of a similar thalamic nucleus in a nonsongbird, *Columba livia*. *Journal of Comparative Neurology*, 349, 512–535.

Wild, J. M., Arends, J. J. A., & Zeigler, H. P. (1984). A trigeminal sensorimoor circuit for pecking, grasping and feeding in the pigeon (*Columba livia*). *Brain Research*, 300, 146–151.

Wild, J. M., Arends, J. J. A., & Zeigler, H. P. (1985). Telencephalic connections of the trigeminal system in the pigeon (*Columba livia*): A trigeminal sensorimotor circuit. *Journal of Comparative Neurology*, 234, 441–464.

Wild, J. M., & Farabaugh, S. M. (1996). Organization of afferent and efferent projections of the nucleus basalis prosencephali in a passerine *Taeniopygia guttata*. *Journal of Comparative Neurology*, 365, 306–328.

Wild, J. M., & Williams, M. N. (2000). Rostral Wulst in passerine birds. I. Origin, course, and terminations of an avian pyramidal tract. *Journal of Comparative Neurology*, 416, 429–450

Wild, J. M., & Zeigler, H. P. (1996). Central projections and somatotopic organisation of trigeminal primary afferents in pigeon (*Columba livia*). *Journal of Comparative Neurology*, 368, 136–152.

Wilkinson, D. G., & Krumlauf, R. (1990). Molecular approaches to the segmentation of the hindbrain. *Trends in Neuroscience*, 13, 335–339.

Wirsig-Wiechmann, C. R., Wiechmann, A. F., & Eisthen, H. L. (2002). What defines the nervus terminalis? Neurochemical, developmental, and anatomical criteria. *Progress in Brain Research*, 141, 45–58

Wise, S. P. (2008). Forward frontal fields: Phylogeny and fundamental function. *Trends in Neuroscience* 12, 599–608.

Wullimann, M. F. (1997). Major patterns of visual brain organization in teleosts and their relation to prehistoric events and the paleontological record. *Paleobiology*, 23, 101–114

Wullimann, M. F. (1998). The central nervous system. In D. H. Evans (Ed.), *Physiology, of fishes* (pp. 245–282). Boca Raton: CRC Press.

Wullimann, M. F. (2009). Secondary neurogenesis and telencephalic organization in zebrafish and mice. *Integrative Zoology*, 4, 123–133.

Wullimann, M. F. (2011). Basal ganglia: Insights into origins from lamprey brains. *Current Biology*, 21, R497–500.

Wullimann, M. F. (2014). Ancestry of basal ganglia circuits: New evidence in teleosts. *Journal of Comparative Neurology*, 522, 2013–2018.

Wullimann, M. F., & Grothe, B. (2013). The central nervous organization of the lateral line system. In S. Coombs, H. Bleckmann, A. N. Popper, & R. R. Fay (Eds.), *The lateral line* (Springer handbook of auditory research 48, pp. 195–251). New York, NY: Springer.

Wullimann, M. F., & Mueller, T. (2004). Teleostean mammalian forebrains contrasted: Evidence from genes to behavior. *Journal of Comparative Neurology*, 475, 143–162.

Wullimann, M. F. Mueller, T., Distel, M., Babaryka, A., Grothe, B., & Köster, R. F. (2011). The long adventurous journey of rhombic lip cells in jawed vertebrates: A comparative developmental analysis. *Frontiers in Neuroanatomy*, 5, 27. doi: 10.3389/fnana.2011.00027

Wullimann, M. F., & Puelles, L. (1999). Postembryonic neural proliferation in the zebrafish forebrain and its relationship to prosomeric domains. *Anatomical Embryology*, 199, 329–348.

Wullimann, M. F., & Vernier, P. (2009a). Evolution of the telencephalon in anamniotes. In M. D. Binder, N. Hirokawa, U. Windhorst (Eds.), *Encyclopedia of neuroscience* (Vol 2, pp. 1424–1431). New York, NY: Springer.

Wullimann, M.F., & Vernier, P. (2009b). Evolution of the nervous system in fishes. In J. H. Kaas (Ed.), *Evolutionary neuroscience* (pp. 147–168). Amsterdam: Elsevier-Academic Press.

Wurst, W., & Bally-Cuif, L. (2001). Neural plate patterning: Upstream downstream of the isthmic organizer. *Nature Reviews Neuroscience*, 2, 99–108.

Zeier, H., & Karten, H. J. (1971). The archistriatum of the pigeon: organization of afferent and efferent connections. *Brain Research*, 31, 313–326.

Zeltser, L. M., Larsen, C. W., & Lumsden, A. (2001). A new developmental compartment in the forebrain regulated by Lunatic fringe. *Nature Neuroscience*, 4, 683–684.

10

Neurotransmission—Evolving Systems

Michel Anctil

10.1 Introduction

Connecting with others is a deeply entrenched characteristic of our social self and we routinely see this played out in the animal world. In the living world at large we witness the trappings of connectivity in the form of signals carrying meanings that are not always clear to us. At the cellular level we have become accustomed to linking the emission of a chemical signal with a cascade of reactions. We regard this as the *modus operandi* of signaling systems, the linchpin of cellular communication.

Trams (1981) theorized that signaling systems evolved in five steps of increasing complexity: (1) basic "chemoception" first present in unicellular organisms; (2) open-loop communication, in which a molecule is widely broadcast and becomes recognized at a distance by an individual, as in chemotaxis; (3) closed-loop communication, in which two-way chemical information transfer occurs between cells of a multicellular organism; (4) communications networks in which signaling is diversified through multiple molecular channels allowing more complex excitation–response processes; and (5) symbolic logic exchanges, in which hierarchically layered neural networks mediate aggression, courtship, and other complex behaviors. In this scheme, the emergence of neurons and neurotransmission fits in the third step.

The activity of a cell's signaling systems can be controlled by a number of external factors. In neurotransmission, the transfer of a message from a neuron to another neuron, or from a neuron to an effector cell, is a controlling factor that can activate or modulate the ongoing activity of a signaling pathway in the post-junctional cell. Constructing a network of such cells is considered the most efficient way of managing the interaction of a multicellular organism with its environment. While nonneuronal cells are able to communicate with each other and to transmit information, neurons do these things better owing to two key features. First, they make specialized contacts (synapses) with each other or with other cells that ensure speedy transmission. Second, convergence or divergence of synapses to and from neurons leads to integration of multiple signals and improves the sophistication of final response outputs.

Early research on the nervous system focused, not surprisingly, on the mammalian brain or on vertebrate models. Although Ramon y Cajal and other pioneers identified contacts between neurons in histological preparations, experimentalists were hard at work to give substance to the concept of chemical neurotransmission. In spite of accumulating evidence that neurotransmitters are released by nerves and bind to receptors to effect responses in postsynaptic cells, some researchers posed stiff resistance to the concept and proposed instead that neurotransmission occurs by current flowing from one neuron to another. With the advent of electron microscopy and more sophisticated electrophysiological techniques, the existence of both chemical and electrical synapses was established and accepted. The roles of exocytosis in chemical synapses and of gap junctions in electrical synapses became basic to our understanding of neurotransmission at the cellular level.

Just as neuroscientists became comfortable with the notion of chemical synapses functioning with a limited number of small neurotransmitters binding each to a class of specific postsynaptic receptors, the discipline of comparative neurobiology emerged and forced a revision of the paradigm. Similarly, the notion of neurotransmitters and hormones operating at two separate, mutually exclusive levels was belied by the discoveries of comparative endocrinology. That neurons can synthesize and release hormones, thereby bridging the nervous and endocrine systems, was discovered thanks to pioneering investigations by Ernst and Berta Scharrer on fish and insects in the 1920s and 1930s (Sawin, 2003). These studies led to the identification of similar neurosecretory cells in the mammalian hypothalamo-hypophyseal complex and to experimental evidence of neurohormones accumulating in nerve endings and released into the circulatory system (Bargmann & Scharrer, 1951).

From the discovery that many neural secretions are peptides to the realization that these same peptides directly influence neurons (the concept of neuropeptides) there passed only a short time, spanning the late 1960s and early 1970s (de Wied, 1984; Strand, 2007). The techniques developed to identify the first peptides acting as neurohormones (for example, high-performance liquid chromatography coupled with radioimmunoassay) were also applied to identify peptides in the invertebrate and vertebrate CNS (Scharrer, 1990). As the realization sank in that peptide families could be found in a variety of neurons throughout the nervous system, evidence emerged that these peptides were released via similar mechanisms to classical neurotransmitters. On the postsynaptic side, it became increasingly clear that the diversity of neuropeptide receptors surpassed that of the neuropeptides themselves. However, the functional role of neuropeptides in neurotransmission has proved difficult to pinpoint. While a modulatory role as cotransmitter was demonstrated for neuropeptides in several studies, their involvement as bona fide frontline neurotransmitters cannot be dispelled, as we shall see below.

The concept of co-transmission originated from observations of co-localization of ATP with norepinephrine in sympathetic neurons (Burnstock, 1976). Soon after, numerous studies documented the co-release of transmitters, usually a neuropeptide with a classical neurotransmitter or another peptide, and much of what we know of the functional implications of co-transmission came from studies on invertebrate experimental models (Kupfermann, 1991). It is now conceivable that the majority of synapses in any nervous system are multitransmitters and it has become apparent that synapses that involve two classical neurotransmitters but no neuropeptide are much less frequent in both vertebrates and invertebrates (Miller, 2009; Trudeau, 2004).

Comparative neurobiology provided relevant animal models amongst invertebrates and lower vertebrates (providing large tractable neurons) that helped illuminate fundamental features of neurotransmission at the cellular level. But comparative studies are also critical to efforts aimed at tracing the origin and evolution of neurotransmission and, more specifically, transmitters. The bulk of data accumulated so far on transmitter systems in a variety of phyla was obtained using the tools of biochemical analysis and immunohistochemistry combined with electrophysiological monitoring techniques that proved successful in assessing the range of transmitters and receptors available to support the behavioral repertoire proper to each animal grade. With the advent of molecular tools and the recent opportunities for genomic analysis, a new era of evolutionary neuroscience is unleashed, opening up the field for more targeted gene expression studies and phylogenetic analyses of neurotransmitter/receptor systems.

These developments led to the views on the evolution of neurotransmitters that serve as the basis for the following narrative. Figure 10.1 summarizes some hypothetical phylogenetic steps that led to electrical and chemical neurotransmission.

10.2 Unicellulars and Neurotransmitters: The Concept of Biomediators

Our view of neurotransmitters as unique to neurons was challenged at about the same time as our view of hormones as unique to endocrine tissues. In this case, the leadership did not come from the comparative neurobiology and endocrinology community, but from researchers in the medical field. It started with reports of the detection of insulin-related peptides and conventional neurotransmitters in fungi and the ciliated protozoan *Tetrahymena* in the early 1980s (Le Roith, Shiloach, Roth, & Lesniak, 1980; Roth et al., 1985).

Although these findings threatened to alter the consensual view of what constitutes a neurotransmitter, evidence of autocrine or paracrine signaling pathways for the putative transmitters was often absent or sketchy. Sometimes the case for the putative transmitter was merely its biochemical detection or its effect on some activity by the unicellular. It was seldom clear whether the substance was released by the cell as a step in a signaling pathway or as an inactive by-product of some separate activity. Also, the putative transmitters must be released outside the cell for signaling to occur. Does that involve release mechanisms proper to metazoan synapses, such as exocytosis? Exocytosis of dense-cored vesicles was visualized in detail in *Paramecium*, in which vesicle docking and membrane fusion appear to involve proteins similar to those associated with synapses of multicellular animals (Plattner & Kissmehl, 2003). However, key proteins such as SNARE and V_0 components have yet to be detected, so the jury is still out as to how strongly the release mechanisms in paramecia resemble the presynaptic machinery of metazoans.

More recent evidence supports the view that substances acting as fast neurotransmitters in many metazoan synapses are likely candidates for regulatory function in protozoan development, ciliary locomotion, and intercellular communication. This applies particularly to acetylcholine (ACh) and the amino acids glutamate and GABA. Acetylcholine and its biosynthetizing enzyme, choline acetyltransferase (ChAT), were detected in the ciliate *Paramecium* (Horiushi et al., 2003; Kawashima et al., 2007). Expression levels of ChAT were high in mature individuals but absent in immature individuals; blocking acetylcholinesterase activity led to a significant reduction of cell pairing during mating (Delmonte Corrado et al., 2001; Trielli, Politi, Falugi, & Delmonte Corrado, 1997). Thus, the case is strong for a role of ACh in *Paramecium* conjugation.

Although glutamate is not known to accumulate in ciliates, highly specific L-glutamate binding sites were reported on the cilia membranes of *Paramecium* (Preston & Usherwood, 1988a), and L-glutamate was found to induce a transient membrane hyperpolarization and to act as a chemoattractant for swimming paramecia (Preston & Usherwood, 1988b). The reason for their attractiveness to ciliates is that release of glutamate by bacteria serves as a signal that food is at hand (Van Houten, 1978). In addition, the action of glutamate is mediated by a fast increase in intracellular cAMP (Yang, Braun, Plattner, Purvee, & Van Houten, 1997), suggesting that the hallmarks of glutamatergic signaling are present and functioning in paramecia. More recently, a candidate gene for a NMDA-like receptor was identified when its down-regulation by RNA interference caused the loss of the chemoresponse of paramecia to L-glutamate (Valentine, Yano, & Van Houten, 2008). Furthermore, searching the genome survey sequence (GSS) and EST transcripts of another ciliate, *Tetrahymena*, for receptor proteins yielded a putative gene sequence matching NMDA receptors (Orias, 2002).

In contrast to glutamate, GABA was reported to be present in paramecia and released in the external medium by KCl-induced depolarization (Delmonte Corrado et al., 2002; Ramoino et al., 2003). Pharmacological experiments indicated that GABA modulates swimming in paramecia through two signaling pathways. One pathway appears to promote forward swimming through GABA-A receptors and Ca^{2+} entry (Bucci, Ramoino, Diaspro, & Usai, 2005) whereas the other pathway promotes ciliary reversal and backward swimming through GABA-B receptors, G protein activation and inhibition of Ca^{2+} entry (Ramoino et al., 2003). Although no attempt was made to back these studies with evidence of GABA binding sites on the ciliate's membranes, the evidence for the presence of conventional GABAergic pathways in ciliates is strong and supported by reading matchups to GABA receptors in a GSS of the ciliate *Tetrahymena* (Orias, 2002).

With these and other reports of the presence of ACh and biogenic amines in microorganisms and plants (see Freestone & Lyte, 2008; Murch, 2006 for reviews), it becomes more and more tempting to conclude that classical neurotransmitters should be considered as multifunctional substances universally involved in physiological processes in all living organisms (see also Chapter 4). Roshchina (2010) proposed that, in view of this ubiquity, calling these substances "biomediators" instead of neurotransmitters or neurohormones is more accurately descriptive of their status in biological signaling. She advances the hypothesis that such biomediators were first involved as participants in chemotaxis through their secretion in the environment as attractants or repellents to other cells searching for food or mating partner. This would have served as the basis for cell-to-cell communication in a fundamental ecological community or ecosystem. Because chemotaxis involves a ligand-receptor relationship, this type of basic cellular signaling would have more likely carried over without much ado in multicellular organisms, by simple inheritance from unicellulars.

10.3 Sponges: The Trappings of Neurotransmission without the Neurons

Although there is disagreement over the phylogenetic position of sponges (Edgecombe et al., 2011), phylogenomic studies (Philippe et al., 2009) and the recent sequencing of the genome of a sponge from the Great Barrier Reef (Srivastava et al., 2010) support the view that sponges are a sister group to the Eumetazoa (Cnidaria, Placozoa, and Bilateria). Therefore, sponges stand at a pivotal position in the evolution of cell-to-cell signaling in multicellulars, and especially concerning neurotransmission. However, sponges lack neurons and, as a result, share with unicellulars the ambiguity of determining how "neuroid," to use Parker's (1919) expression, their modes of cell-to-cell communication are. Recent discussions have revolved around the hypothesis that sponges, being able to detect external stimuli and produce coordinated effector responses, possess sensory cells and semi-specialized epithelial cells (myocytes) capable of transmitting chemical signals (Nickel, 2010; Renard et al., 2009). A similar scenario holds for the small, obscure phylum Placozoa (see Figure 10.1), but, as little is known about transmitters in the group except for evidence of aminergic-like elements in the genome (Srivastava et al., 2008), the focus of this section will remain on sponges.

For a start, do sponges, as multicellular assemblies, possess the building blocks of neurotransmission? The answer appears to be a qualified yes. The sequenced genome of the demosponge *Amphimedon queenslandica* reveals a large number of genes involved in the assembly and functioning of both pre- and postsynaptic elements in

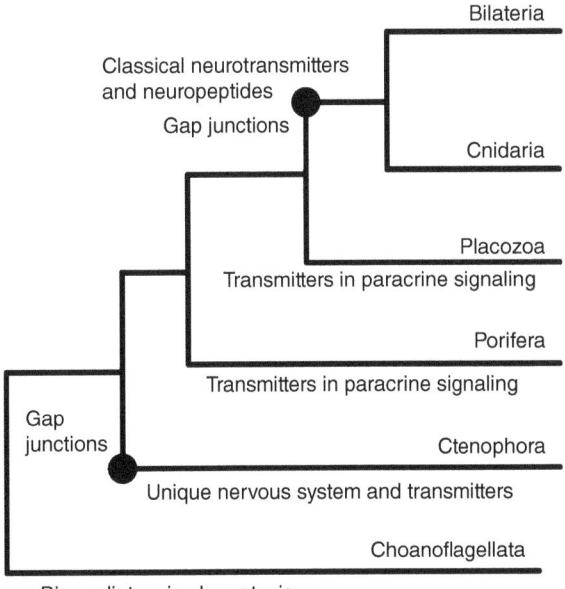

Figure 10.1 Proposed evolutionary steps for the emergence of transmitter systems.
The phylogenetic relationships are based on the tree proposed by Moroz et al. (2014). The two nodes represent independently evolved traits in Ctenophore and the Cnidaria/Bilateria lineages (homoplasy).

eumetazoans (Srivastava et al., 2010). These include scaffolding molecules that manage the deployment of synaptic vesicles, calcium channels, and signaling machinery on the presynaptic side, and proteins (ion channels and receptors) on the postsynaptic side. In addition, *A. queenslandica* possesses genes that encode proteins involved in the regulation of synaptic vesicle exocytosis by calcium influx (synaptophysin, synaptotagmin, SV2). Calcium influx appears to be implicated in sponges because it was demonstrated that contractile responses to transmitters cannot be propagated (presumably from cell to cell) in calcium-free media (Elliott & Leys, 2010). However, the lack of a homolog of RIMS, a scaffold for the facilitation of synaptic vesicle fusion in exocytosis, casts doubt on the potential ability of sponges to effect synapse-like transmission. As for the other sequences, a cautionary note is in order as to the reliability of the relatively superficial annotations associated with the disclosure of genome sequences. It has been the experience of this author that closer, manual inspections of the sequences sometimes reveal erroneous annotations in the genome, as hits on a protein family do not necessarily convert into a bona fide match (Anctil, 2009). Given this caveat, some of the claims for the presence of synapse genes in this sponge may turn out to be invalid after a more thorough analysis. This can apply to other gene families discussed further below.

If expression of synaptic proteins remains to be demonstrated in sponges, are classical transmitters such as those found in unicellulars present in sponges and processed through some synaptic-like machinery? Early histochemical investigations reported the presence of acetylcholinesterases and biogenic amines in sponges (Lentz, 1966), but the poor reliability of the techniques used and the lack of supporting evidence for the involvement of these substances in signal processing have cast doubt on the claims (Mackie, 1990). However, recent studies have provided convincing physiological evidence for a role of the amino acid transmitters glutamate and GABA in coordinating behavior in sponges. Two laboratories independently reported propagated contractions in sponges induced by glutamate (Elliott & Leys, 2010; Ellwanger & Nickel, 2007) and their inhibition by GABA (Elliott & Leys, 2010), in keeping with the traditional opposite actions of these transmitters in higher animals. In addition, Elliott and Leys detected by HPLC relatively large amounts of glutamate in sponge tissues, although no glutamate or GABA released in the medium could be detected after KCl-induced depolarization of those tissues. A third transmitter, nitric oxide (NO), was found to induce contractions and modulate the amplitude and frequency of endogenous rhythmic contraction waves (Ellwanger & Nickel, 2006). Nitric oxide synthase activity was visualized in cells of the osculum and in pinacocytes, considered to be the substrates for coordinated responses in the demosponge *Ephydatia muelleri* (Elliott & Leys, 2010). A common target in NO signaling pathways of eumetazoans is cyclic guanosine monophosphate (cGMP), and NO was shown to increase cGMP production in the osculum and pinacoderm of *E. muelleri* (Elliott & Leys, 2010).

These findings are supported by (1) the cloning in the sponge *Geodia cydonium* of a metabotropic glutamate receptor with apparent mixed selectivity for glutamate and GABA (Perovic, Krasko, Prokic, Müller, & Müller, 1999) and (2) the presence of sequences matching metabotropic glutamate receptors and the PDZ domain of a glutamate receptor interacting protein in the *A. queenslandica* genome (Srivastava et al., 2010). In contrast, no GABA receptor sequence was identified, which begs the possibility of a nonspecific action by GABA on sponge contractility, a suspicion strengthened by the conflicting results of Ellwanger and Nickel (2007) and Elliott and Leys (2010) with GABA. On the other hand, the presence of a NOS sequence in the genome is consonant with the nitrergic signaling pathway uncovered by Elliott and Leys (2010). The genome also revealed sequences apparently related to dopamine and serotonin receptors. While it is premature to label these sequences as such, in view of the difficulty of finding close orthologues of specific biogenic amine receptor subfamilies even in cnidarians (Bouchard, Ribeiro, Dubé, & Anctil, 2003; Bouchard, Ribeiro, Dubé, Demers, & Anctil, 2004), a report of serotonin immunoreactivity in cells of a sponge larva and settled juvenile (Weyrer, Rutzler, & Rieger, 1999) encourages further inquiry.

In addition to all these candidate transmitters, many sequences related to neuropeptide production and secretion were identified in the genome of *A. queenslandica* (Srivastava et al., 2010). These include peptidases that cleave precursor chains to produce active peptides (proprotein convertases, an aminopeptidase, a carboxypeptidase, and cathepsins), a gene for post-translational C-terminal amidation (PAM) and another for N-terminal pyrolation, and finally two genes encoding proteins involved in the calcium-dependent secretion of neuropeptides. Despite this complex machinery, no neuropeptide precursor homologue was detected in the sponge genome. As Baggerman, Liu, Wets, and Schoofs (2005) have noted, it is notoriously difficult to hunt for neuropeptide precursor proteins in genomes when there is no biochemically isolated peptide or cloned sequences at hand, as is the case with sponges. Even when some are available, as in the sea anemone *Nematostella vectensis*, locating the homologues in the genome is a painstaking process (Anctil, 2009). However, the presence in the sponge genome of G-protein-coupled receptors (GPCR) belonging to the LGR/hormone, opsin/prostanoid and gonadotropin-releasing hormone/neurotensin/somatostatin receptor families should serve to encourage efforts to uncover neuropeptide-like transmitters in the sponge genome. A recent report of sponge larval settlement triggered by the cnidarian neuropeptide GLWamide (Whalan, Webster, & Negri, 2012) is a step in the right direction.

From the evidence reviewed above, it is clear that sponges use chemical substances known to act as neurotransmitters in neurons of eumetazoans in order to coordinate their behavior, and that they do so through paracrine channels of transmission in the absence of neurons (Elliott & Leys, 2010; Nickel, 2010). Their "management style" of signal transmission appears to be a mere upgrade from that of unicellulars like paramecia. However, the sponge's tissue integration, as opposed to a loose community of unicellulars, ensures that cells are more proximal to each other and that relatively tight coordination of activities by chemical transmission can be achieved through a chain of epithelial cells. Leys and Riesgo (2012) have recently argued that true functional epithelia appeared first in sponges. This is where the innovation in transmission stands—not in the chemical identity of the transmitters, which is carried over from the unicellulars and which may include acetylcholine, glutamate, and GABA.

10.4 Cnidarians: Neurotransmission Enters the Stage

Recent analyses of basal metazoan relationships based on molecular phylogeny have been mired in controversy. One such analysis claiming to use more appropriate methods and larger sampling of genes proposes that Ctenophora and Cnidaria, which share the possession of nervous systems and muscle cells, are sister groups in a clade (the revived Coelenterata) which branched early from the bilaterian line (Philippe et al., 2009). On that basis nervous systems would have originated in a common ancestor of these groups. In addition, it was previously assumed that extant ctenophores are more derived forms than cnidarians (Bridge, Cunningham, Schierwater, DeSalle, & Buss, 1992).

However, the recent sequencing of the genome of the ctenophore *Pleurobrachia bachei* has allowed new insights into the set of neural genes of this phylum that challenge any view of cnidarians and ctenophores sharing close affiliations (Moroz et al., 2014). Quite to the contrary, this new analysis leads to the surprising conclusion that ctenophores constitute "the most basal animal lineage" and that sponges are "the sister taxon to remaining metazoans" (Moroz et al., 2014, p. 109). Ctenophores lack many of the neural genes associated with cnidarians and bilateral animals, suggesting that neurons arose independently in ctenophores and in the cnidarian/bilaterian lineage. What is particularly relevant to this chapter is the lack in ctenophores of nonpeptidergic neurotransmitter systems with the possible exception of glutamate, thus reinforcing the notion that these are cnidarian/bilaterian innovations. Because the implication of the study is that cnidarians constitute the sister clade of bilateral animals and that the nervous system of ctenophores was an evolutionary dead end, in the following account cnidarians will serve as representatives of the basal nervous system shared with the lineage of the bilaterians. All extant cnidarian species examined so far possess neurons organized in assemblies ranging from diffuse, two-dimensional nerve nets to condensed, ganglion-like "nervous centers" (Mackie, 2004; Satterlie, 2011). For more on how they got there—presumably from a basal ancestor with a level of tissue organization between sponge and cnidarian grades—refer to Chapters 5 and 6 of this volume. This initial emergence of neurons in Metazoa set the stage for a remodeling of cell-to-cell signal transmission through the coemergence of specialized contact zones, known as synapses.

Electrical synapses—gap junctions—make their first appearance in Cnidaria and Ctenophora, where they link epithelial cells or neurons to propagate behavioral control signals (Mackie, 2004; Mackie & Passano, 1968; Spencer, 1981; Spencer & Satterlie, 1980). While this mode of signal transmission is important as an adjunct to chemical transmission, it has only been morphologically and functionally demonstrated in one cnidarian class, Hydrozoa (but see Germain & Anctil, 1996; Mire, Nasse, & Venable-Thibodeaux, 2000).

Contrary to sponges, where traces of synaptic machinery must be sniffed out by genomic analysis, in cnidarians the synapses observed by electron microscopy possess most of the familiar features seen in invertebrate and vertebrate synapses (see Westfall, 1970, 1996 for review). Due to their epithelial position, sensory cells tend to make unidirectional synaptic contacts with neurons of the nerve nets (Holtmann & Thurm, 2001), and cnidarian neuromuscular synapses are likewise unidirectional (Spencer, 1982). But transmission is nonpolarized in the sheet-like nerve nets, with many bidirectional, symmetrical synapses found between neurons. Based on reconstruction of serial sections from the jellyfish *Cyanea capillata*, these early versions of synapses are characterized by a single layer of a few large synaptic vesicles abutting the terminal membrane, an arrangement mirrored on the opposite side of the terminal (Anderson & Grünert, 1988). Electrophysiological analysis of these jellyfish synapses revealed that postsynaptic potentials are indeed generated alternately in the opposite neuron, so that each synaptic partner can play a presynaptic role at one moment and a postsynaptic one just a few milliseconds later (Anderson, 1985). These synapses display unusually high depolarization thresholds for transmission to avoid continuous "babbling" between the two partners and loss of meaningful signal transmission.

Conventional electrophysiological approaches, such as those used by Anderson (1985), allow chemical transmission only to be indirectly monitored from the postsynaptic side. To *directly* record presynaptic events such as exocytosis in real time, amperometry with carbon-fiber microelectrodes was used to record currents generated by electro-oxidizable monoamines released from isolated neurons of the sea pansy *Renilla koellikeri* (Gillis & Anctil, 2001). Depolarization-evoked transmitter release displayed the fast kinetics and calcium sensitivity typical of exocytotic events in neurons of higher invertebrates (Bruns & Jahn, 1995; Chen & Ewing, 1995; Chen, Gutman, Zerby, & Ewing, 1996). Many larger exocytotic events appeared as multiples of a subunit spike, and therefore suggested a quantal mode of secretion as classically defined for mammalian neuromuscular synapses by Boyd and Martin (1956), but the skewed distribution of spikes in favour of the lowest (subunit) amplitude contradicted the classical model (Gillis & Anctil, 2001).

With the hindsight of more recent data from mammalian cell models, it is likely that the fewer, larger spikes (≥ 5 pA) recorded in sea pansy neurons represent the complete emptying of a vesicle (all-or-none exocytosis) whereas the higher frequency spikes of small amplitude (<5 pA) reflect the partial discharge of transmitter through the fusion pore (kiss-and-run exocytosis). These mechanisms have been observed in chromaffin cells (Henkel, Kang, & Kornhuber, 2001) which, like the sea pansy, contain catecholamines stored in large dense-cored vesicles; they are also seen in rat dopaminergic neurons (Staal, Mosharov, & Sulzer, 2004). Even the strong bursts of fusion pore flickers seen in chromaffin cells, an indicator of secretion "hot spots" (Henkel et al., 2001), are present in the majority of the recorded monoaminergic neurons of the sea pansy (Gillis & Anctil, 2001). Staal et al. (2004) proposed that kiss-and-run exocytosis increases the longevity of a synaptic vesicle, a distinct advantage in synapses where synaptic vesicles are low in number such as in cnidarian synapses. It is clear from these observations that sophisticated delivery of transmitter at the synapse was an early step of neuronal evolution that underwent little, if any, change up to mammals.

With synapses fully operational in Cnidaria, we may ask to what extent substances reach their full potential as neurotransmitters. The question is legitimate, insofar as declaring a candidate to be a neurotransmitter requires that the substance meets stringent criteria: synthesis in presynaptic neuron, presence in presynaptic terminal vesicles, release upon stimulation (depolarization), binding to specific postsynaptic receptors and production of a fast, reversible response in the target cell. While an impressive array of transmitters satisfy some of the criteria in Cnidaria, as documented in the comprehensive review by Kass-Simon and Pierobon (2007), very few have been shown to meet all of them, likely due in part to difficulties in finding tractable experimental models at the cellular level. Genomic approaches, such as the survey of sequences matching those of transmitter systems in the genome of the starlet sea anemone *N. vectensis* (Anctil, 2009), may help to focus efforts on cloning and expressing candidate genes, thereby establishing neurotransmitters on more solid ground, but this will require substantial investments from many researchers.

Let us first examine which of the candidate transmitters have the best case for a neurotransmitter role in Cnidaria. Three candidate classes include peptides, amines, and amino acids. The RFamide-related peptides (RFaPs), a major neuropeptide family, are strong contenders. Not only are they present exclusively in neurons of all the species examined so far from all cnidarian classes, but neurons containing them form nerve nets and appear to be more numerous than any other class of chemically identified neurons. The precursors of these peptides contain multiple copies of the peptide (up to 36 copies in the sea pansy), thus leading to high concentrations of the bioactive peptide in the terminals. One of them, Antho-RFamide, was localized by immuno-electron microscopy in synaptic vesicles of hydra and sea anemones (Koizumi, Wilson, Grimmelikhuijzen, & Westfall, 1989; Westfall & Grimmelikhuijzen, 1993), but synaptic release has not been investigated. In addition, it was demonstrated that *Hydra* RFamides were ligands for a *Hydra* peptide-gated ion channel (Golubovic et al., 2007), thus qualifying them as fast neurotransmitters, a highly unusual role for neuropeptides which, as a rule, bind to metabotropic receptors, usually rhodopsin-like G-protein-coupled receptors (GPCRs). The only other report of a neuropeptide-gated ion channel relates to a snail (Cottrell, 1997; Cottrell, Jeziorski, & Green, 2001) and also involves a RFaP, thus suggesting that these peculiar ionotropic receptors are evolutionarily ancient. There is physiological evidence for a role of RFaPs as neuromuscular transmitters in cnidarians (Anctil & Grimmelikhuijzen, 1989; McFarlane, Graf, & Grimmelikhuijzen, 1987), but these actions appear to be mediated by as-yet unidentified metabotropic receptors.

Biogenic amines are the next best contenders. There are numerous reports of the presence and actions of biogenic amines in Cnidaria (see Kass-Simon & Pierobon, 2007 for review), but none provide multiple converging lines of evidence except in one anthozoan, the sea pansy *R. koellikeri*. The enzyme machinery for the synthesis of dopamine and norepinephrine appears to be present and expressed in neurons (Anctil, Hurtubise, & Gillis, 2002; Pani & Anctil, 1994) and even in nerve terminals of a sea anemone (Westfall, Elliott, MohanKumar, & Carlin, 2000), although genomic evidence from the starlet sea anemone suggests that the enzymatic pathways differ from those of higher invertebrates and vertebrates (Anctil, 2009). Serotonin and norepinephrine immunoreactivities were detected in sea pansy neurons (Umbriaco, Anctil, Descarries, 1990; Pani, Anctil, & Umbriaco, 1995), but only in a sea anemone has serotonin been localized in synaptic vesicles (Westfall et al., 2000). Norepinephrine was released from sea pansy tissues by KCl-induced depolarization in a calcium-dependent manner (Pani et al., 1995) and neuronal catecholamine release was confirmed at the cellular level by Gillis and Anctil (2000) as mentioned earlier. Adrenergic and serotonergic receptors were characterized by analysis of membrane binding sites in the sea pansy (Awad & Anctil, 1993; Hajj-Ali & Anctil, 1997) in relation to their roles in bioluminescence control and modulation of peristaltic contractions in this species (Anctil, 1989; Anctil, Boulay, & LaRivière, 1982). Two receptors for biogenic amines were also cloned from the sea pansy (Bouchard et al., 2003, 2004), but none correspond to those characterized by membrane binding and in fact their ligand is unknown at this point. Numerous orthologues of biogenic amine receptors, transporters for biogenic amine re-uptake and aminergic vesicular transporters were also found in the genome of the sea anemone *N. vectensis* (Anctil, 2009).

The third contenders are the amino acids taurine/β-alanine. In a search for the fast transmitter involved in the bidirectional synapses of the previously-mentioned jellyfish nerve net (Anderson, 1985), taurine immunoreactivity was localized in motor nerve net neurons (Carlberg, Alfredsson, Nielsen, & Anderson, 1995). Anderson and Trapido-Rosenthal (2009) have since reported that both taurine and β-alanine induce depolarizations and conductance changes in nerve-net neurons that are consistent with the excitatory postsynaptic potentials recorded during synaptic activity. They also found that both amino acids were released by depolarizing nerve-net neurons, but so were other amino acids, and the calcium sensitivity of release could not be demonstrated (Anderson & Trapido-Rosenthal, 2009). In addition, it is unknown if these amino acids are stored in synaptic vesicles, and nothing is known about their receptors. While this is the best existing model to identify a fast transmitter at an identified synapse, there is still a lot of work ahead to confirm these amino acids as cnidarian neurotransmitters.

Other amino acids were enlisted as candidate transmitters in Cnidaria. Ionotropic receptors for glutamate, GABA and glycine were characterized by pharmacological analysis of binding sites from membranes of *Hydra vulgaris* (Concas et al., 1998; Grosvenor, Bellis, Kass-Simon, & Rhoads, 1992; Pierobon et al., 1995, 2001, 2004). Receptors for all three putative transmitters are represented in transcripts of the genome of *N. vectensis*, including metabotropic receptors (Anctil, 2009). Vesicular transporters for glutamate and for GABA were also spotted in the genome, suggesting that these amino acids have the potential to be stored in synaptic vesicles. There are a number of reports pointing to a role for these transmitters in the control of nematocyst discharge (Kass-Simon & Scappaticci, 2004), pacemaker activity in motor systems (Kass-Simon, Pannaccione, & Pierobon, 2003; Ruggieri, Pierobon, & Kass-Simon, 2004), and the feeding response (Pierobon et al., 1995, 2001, 2004).

As in sponges, there is no convincing evidence for a transmitter role of ACh in cnidarians, but the detection of several transcripts in the genome of the starlet sea anemone for choline acetyltransferase (ChAT), acetylcholinesterase (AChase) and nicotinic receptors (Anctil, 2009) may revive the search for cholinergic transmission. Another potential transmitter, adenosine triphosphate (ATP), was shown to activate muscles (Hoyle, Knight, & Burnstock, 1989) and to modulate sensory transmission (Watson, Venable, Hudson, & Repass, 1999), but ACh has no such bolstering support as a cnidarian neurotransmitter. However, the presence of two orthologues in the genome of *N. vectensis* with strong homology with vertebrate P2X purinergic receptors (Anctil, 2009), suggests that ATP may play a role as a signaling molecule, though its action as a neurotransmitter remains to be examined. The last of the small transmitters for which evidence is available is nitric oxide (NO). Nitrergic neurons were visualized by NADPH-diaphorase histochemistry in various cnidarian species and NO was reported to modulate feeding in hydra, swimming in the jellyfish *Aglantha digitale* and peristaltic contractions in *R. koellikeri* (Anctil, Poulain, & Pelletier, 2005; Colasanti, Venturini, Merante, Musci, & Lauro, 1997; Cristino, Guglielmotti, Cotugno, Musio, Santillo, 2008; Moroz, Meech, Sweedler, & Mackie 2004). An orthologue each of nitric oxide synthase (NOS) and NOS binding protein was found in the genome of the starlet sea anemone (Anctil, 2009), which puts the role of NO as a neurotransmitter on even firmer ground.

While there is persuasive evidence that a few small transmitters are active at synapses in cnidarians, there is even better evidence that neuropeptides are distinctly more widespread in cnidarian neurons (Grimmelikhuijzen, Williamson, & Hansen, 2004). While there is evidence that small transmitters are also present in non-neuronal cells

(Anctil & Carette, 1994; Anctil & Ngo Minh, 1997; Umbriaco et al., 1990), all peptide transmitters examined so far are exclusively localized in neurons. The RFaP family mentioned earlier is more widespread in cnidarian neurons than any other peptide family, to the point where FMRFamide immunoreactivity was used to visualize the general organization of cnidarian nervous systems (Anderson, Moosler, & Grimmelikhuijzen, 1992; Grimmelikhuijzen, 1985; Grimmelikhuijzen & Graff, 1986; Marlow, Srivastava, Matus, Rokhsar, & Martindale, 2009; Pernet, Anctil, & Grimmelikhuijzen, 2004). In fact, RFaPs constitute one of the largest and most diversified peptide families in the animal world and are considered very ancient peptides that may have given rise to other amidated peptide families (Vandingenen et al., 2004). The ubiquity of RFaPs in what is considered one of the first evolved nervous systems certainly supports this proposal.

That the phylum Cnidaria was a laboratory for the early evolution of neuropeptides is also supported by the presence in cnidarian neurons of several peptide families which, unlike RFaPs, are not found in the rest of the animal kingdom and probably represent evolutionary dead ends in the sense that they vanished in the bilaterian lineage. These are the KAamide, RIamide, RNamide, RPamide, RWamide and LWamide families (Carstensen, Rinehart, McFarlane, & Grimmelikhuijzen, 1992; Carstensen et al., 1993; Graff & Grimmelikhuijzen, 1988a, 1988b; Grimmelikhuijzen et al., 1990; Leitz, Morand, & Mann, 1994; Nothacker, Rinehart, & Grimmelikhuijzen, 1991; Nothacker, Rinehart, McFarlane, & Grimmelikhuijzen 1991). Antho-RWamide immunoreactivity was found in synaptic vesicles of neuromuscular synapses (Westfall, Sayyar, Elliott, & Grimmelikhuijzen, 1995) and peptides from all but one of these families are known to act on muscles (Carstensen et al., 1992; McFarlane, Anderson, Grimmelikhuijzen, 1991; McFarlane, Reinscheid, Grimmelikhuijzen, 1992). Only the LWamides stand out, as regulators of metamorphosis of planula larvae into mature adults (Leitz, 1998). Precursors for all these peptide families (except the KAamides) were found in the genome of *N. vectensis*, in addition to numerous unclassifiable neuropeptide receptors on which these uniquely cnidarian peptides may bind (Anctil, 2009).

The genome of the starlet sea anemone provides also tantalizing glimpses of other neuropeptide families that are commonly found in bilaterian animals. These include tachykinin-related, galanin-related, gonadotropin-releasing hormone (GnRH)-related, vasopressin/oxytocin-related and melanocortin-related peptides (Anctil, 2009). However, there is only limited, if any, immunohistochemical or physiological evidence for their role as neurotransmitters in examined cnidarian species. The best contenders are the GnRH family, two members of which were partially isolated and purified in the sea pansy (Anctil, 2000), mirroring the two GnRH octapeptides predicted from a precursor transcript in the genome, and the vasopressin/oxytocin-related family, two members of which were identified in *Hydra* (Morishita, 2003) and three have been predicted from the sea anemone genome. Immunoreactivity for both peptide families was found in neurons, and roles in modulation of muscle activity (GnRH: Anctil, 2000) and in neuronal differentiation (vasopressin/oxytocin: Takahashi et al., 2000) were reported. Their receptors have yet to be characterized, but analysis of transcripts of GnRH-like receptors in the genome of *N. vectensis* suggests that these receptors are derived from ancestral forms that may have given rise to specific GnRH and vasopressin/oxytocin receptors (Anctil, 2009).

10.5 Neurotransmission Comes of Age

The survey of cnidarian neurotransmission gives the impression that the synaptic infrastructure and the neurotransmitter systems were fully deployed before bilaterian animals evolved and, therefore, that little evolving was left to take place during animal evolution in the last 500 million years. In the following account we ask how accurate that impression is.

The internal structure and basic functioning of the synapse, with its full array of organelles, does not deviate from its cnidarian framework in the many bilaterians that have been examined, but the complexity of postsynaptic arrangements increases sharply in most synapses of higher invertebrates where the release site juts against multiple postsynaptic elements to form dyad, triad, and even tetrad modules (Meinertzhagen, 2010). This evolutionary trend, which allows for divergent transmission, did not continue in vertebrates except in retinal circuits where such modules are integral to visual processing. All the other vertebrate synapses have presynaptic release sites abutting a single postsynaptic element such as a dendrite. Other features of synaptic organization where bilaterian evolution has made inroads, in comparison to cnidarians, are increases in the number and packing density of synapses and the specialized means by which synapses participate in neural networks that inform behavior (Meinerthagen, 2010), as discussed in the previous chapters.

If synaptic morphology and physiology are largely conserved, have transmitter systems diversified during bilaterian evolution? There is no straightforward answer to this question. Tables 10.1 and 10.2 provide a tally of gene orthologues for classical neurotransmitters and neuropeptides based on genome database mining. In addition, Figures 10.2 and 10.3 summarize major phylogenetic shifts that may account for the data computed in Tables 10.1 and 10.2, respectively.

Before addressing what insights we may gain by comparing these numbers amongst the representatives of the different phyla, it is advisable to take heed of possible pitfalls in doing this. Whenever no genes can be found coding for a transmitter-related protein, it does not follow necessarily that the transmitter element is absent: the sequence may just have changed considerably or there may have been misleading genome annotations. Also, associating a receptor sequence with a specific ligand on the basis of signature residues and motifs is a perilous and, at best, tentative exercise. There are examples of predicted assignations of receptor identity that turned out, when functionally expressed, to bind to a different category of ligands (Staubli et al., 2002) or to have no known ligand and instead display constitutive activity (Bouchard et al., 2003). This is an especially critical issue with neuropeptide receptors, too few of which have been formally identified by functional expression. For this reason, no assignation of receptor category was attempted in Table 10.2.

Table 10.1 shows the number of transcripts matching receptors for different categories of classical neurotransmitters and the key enzyme for the production of nitric oxide (NOS). The most striking trends in the table are the late emergence of metabotropic acetylcholine receptors and the pattern of gene losses suggested by the absence of purinoreceptors, histamine receptors, and nitric oxide synthase in the ecdysozoan superclade (Arthropoda and Nematoda), and the absence of metabotropic GABA receptors in the echinoderm *Strongylocentrotus purpuratus*. In contrast, gene amplification is reflected by the unusually large number of transcripts for ionotropic acetylcholine receptors in the nematode *Caenorhabditis elegans*, and for GABA ionotropic receptors and serotonin

Table 10.1 Number of Transcripts Matching Different Neurotransmitter Receptors and Nitric Oxide Synthase in the Genome of Species Representing Various Phyletic Levels in Metazoa.

	Ionotropic AChR	Metabotropic AChR	Ionotropic GluR	Metabotropic GluR	GABA A/B receptors	Glycine receptors	P2X purino-receptors	DA/5-HT receptors	+Histamine receptors	Nitric oxide synthase
Sea anemone	12	0	11	8	11/4	1	2	13*	10	1
Sea slug	16	0	15	3	5/1	2	2	3/6	2	2
Nematode	56	3	10	3	4/2	3	0	3/5	0	0
Fruit fly	12	1	11	2	4/3	4	0	5/4	0	1
Sea urchin	12	4	?	2	1/0	2	1	7/4	3	5
Human	17	5	16	8	19/4	6	10	6/20	4	3

* All these receptors had mixed matches for both dopamine and serotonin receptors.
Data culled from the following sources: Anctil (2009), sea anemone; Moroz et al. (2006), sea slug, nematode, fluit fly and human; Burke et al. (2006), sea urchin.

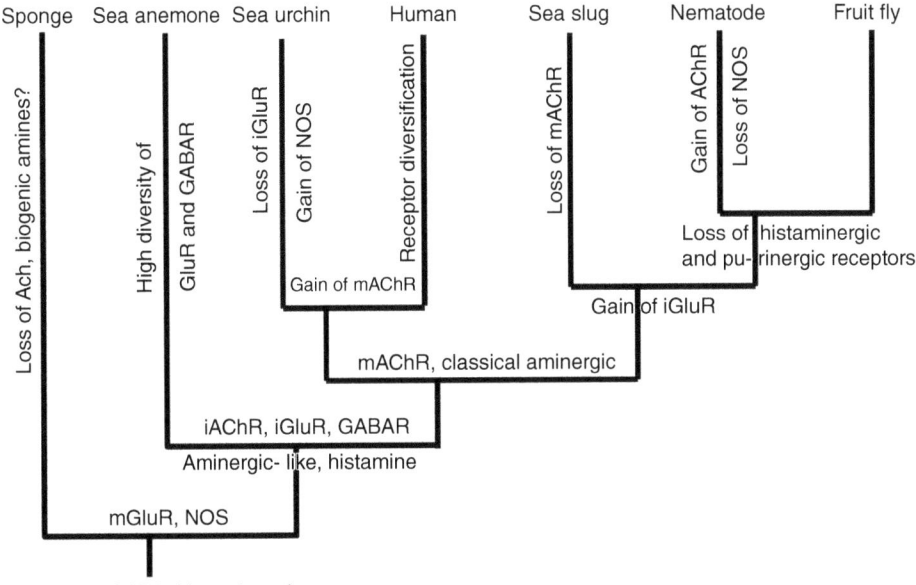

Figure 10.2 Guideposts for phylogenetic trends of classical neurotransmitter systems based on information in Table 10.1.
This and the tree in Fig. 10.3 reflect current views on the phylogenetic relationships of major metazoan phyla. Ach, acetylcholine; AChR, cholinergic receptor; GABA, gamma-aminobutyric acid; GABAR, GABAergic receptor; Glu, glutamate; GluR, glutamate receptor; iAChR ionotropic cholinergic receptor; iGluR, ionotropic glutamate receptor; mAChR, muscarinic cholinergic receptor; mGluR, metabotropic glutamate receptor; NOS, nitric oxide synthase.

receptors in the human genome. Modest diversification is apparent for the glycine, dopamine, and serotonin receptors; otherwise the suite of transmitter receptors in place in cnidarians was largely conserved in bilaterians.

Neuropeptides and neurohormones predicted from precursor genes are computed in Table 10.2. Many peptide families are lacking for several taxa, but the absence of data here may reflect just how notoriously difficult it is to sniff out the signature peptide sequences embedded in precursor genes. In spite of this hurdle, a few insights can be gained from the available information. For a start, each species of the represented phyla are endowed with similar numbers of shared neuropeptide families, even though the families represented may differ between phyla. The only exception is the sea urchin where 4 families are listed, and this is probably due to inadequate attempts to screen the genome for gene precursors. The RFamide family of peptides, which is the most important in the sea anemone both in the number of neurons and the variety of functions it regulates, have diversified even more in the bilaterian phyla to reach a climax in *C. elegans*, where 75 transcripts were tallied. Again, sea urchin RFamides remain to be discovered, as RFamide-related receptors were spotted in the genome (Burke et al., 2006). Insulin-related peptides and the somatostatin/allatostatin superfamily show a similar pattern of diversification, again with evidence of considerable gene expansion in the nematode and the mollusc *Lottia gigantea*. The data for the GnRH and vasopressin/oxytocin families confirm that, contrary to prevailing views of

Table 10.2 Number of Transcripts Matching Different Neuropeptide and Neurohormone Families in the Genome of Species Representing Various Phyletic Levels in Metazoa.

	RFamide and NPY family	RIamide family	RPamide family	RWamide family	GLWamide family	CCK/ gastrin family	Somatostatin/ allatostatin	SALMFamide family	PDF-related family
Sea anemone	4	1	4	2	5	x	x	x	x
Annelid	6	x	x	1	x	x	2	x	1
Mollusk	11	3	x	x	x	x	26	x	1
Nematode	75	2	5	x	3	2	8	12	x
Insect	11	x	x	2	x	?	10	3	1
Sea urchin	x	x	x	x	x	x	x	7	x
Human	13	x	x	x	x	6	6	x	x

	Tachykinin family	Galanin family	GnRH/ vasopressin family	Melanocortin family	Insulin- related	Allatotropin family	Glycoprotein family	Phylum specific[b]	Neuropeptide GPCRs
Sea anemone	2	2	5	3	2	x	x[a]	1	89
Annelid	1	x	5	x	6	1	6	8	?
Mollusk	4	x	2	x	4	1	2	40	~37
Nematode	x	x	x	x	40	8	1	80	~50
Insect	6	x	1	x	7	1	2	31	49
Sea urchin	x	x	2	x	2	x	5	?	37
Human	5	2	6	11	3	x	6	83	88

x, not found
?, unable to confirm
[a] seven receptor transcripts found, but no candidate ligand
[b] peptide family unique to the species representing a phylum or superclade (Lophotrochozoa or Ecdysozoa)
Anctil (2009), sea anemone; Veenstra (2011), annelid; Veenstra (2010), mollusc; Nathoo, Moeller, Westlund, and Hart (2001) and Li and Kim (2008), nematode; Hewes and Taghert (2001), insect; Burke et al. (2006), sea urchin; Burbach (2010) and Vassilatis et al. (2003), human.

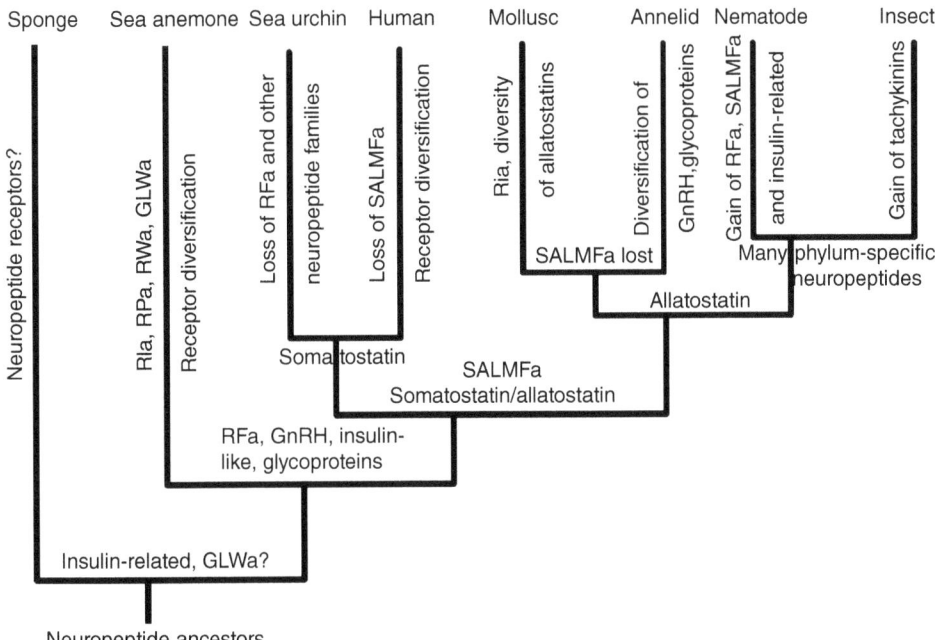

Figure 10.3 Guideposts for Phylogenetic Trends of Some Neuropeptide Families Based on Information in Table 10.2.
GLWa, GLWamides; GnRH, gonadotropin-releasing hormone/vasopressin/oxytocin peptide superfamily; RFa, RFamides; RIa, RIamides; RPa, RPamides; RWa, RWamides; SALMFa, SALMFamides.

the past, these peptide families are not unique to the chordates and show orthologues in every genome examined except that of *C. elegans*, thus providing yet another example of gene loss in this group.

There are a few peptide families represented in the sea anemone that appear to be absent in most bilaterians, including vertebrates (Table 10.2). These are relatively obscure families such as the suite of non-RFaP R(X)amides and the GLWamides and their functional significance in the few bilaterians that possess them is unknown. The galanin and melanocortin families are represented only in the sea anemone and human genomes in Table 10.2. However, the predicted galanin-like peptides of the sea anemone are much shorter than the human counterparts and the melanocortin precursor for the α-MSH-like peptides of the sea anemone is not related to the POMC precursor of the vertebrates (Anctil, 2009), thus suggesting that the two sets of peptides arose independently in cnidarians and vertebrates. Table 10.2 also shows that numerous neuropeptides in the mollusc (*L. gigantea*), ecdysozoan and human genomes belong to families that are unique to these clades and represent evolutionary dead-ends.

The tally of neuropeptide GPCRs shown for each genome in Table 10.2 can also be instructive. For most phyla the numbers range approximately between 40 and 50 such receptors, but in the sea anemone and human genomes the numbers are much higher. The count in the sea anemone is suspiciously high if we take into account the ratio of predicted peptide to receptor ratio (1:3 versus 1.6:1 in human), especially as several neuropeptides belong to the same family and may bind to one and the same receptor.

Many of the sea anemone neuropeptide GPCRs (39) could not be assigned to any of the ligand families (Anctil, 2009), and in fact showed more homology with each other than with specific ligand families. The same can be said for many of the sea anemone biogenic amine GPCRs, especially the surprisingly large number of melatonin-like receptors. Such receptors were described as "surreal-GPCRs" in a genomic analysis of sea urchin GPCRs (Raible et al., 2006).

Rhodopsin-based GPCRs are highly versatile membrane proteins that can adapt to light signals and chemoreception as well as to cell-to-cell signaling, including neurotransmission. This versatility creates ambiguity, in that some of the chemicals used as cues for chemoreception can also be used in neurotransmission, as observed in unicellulars earlier in this chapter. Raible et al. (2006) argued that the unusually large repertoire of chemosensory GPCRs in the sea urchin reflects retention of the ancestral molecular machinery for chemoreception such as seen in unicellulars, followed by multiple and independent gene expansions. We can argue that similar events occurred in cnidarians, which share with echinoderms radial symmetry, relatively decentralized nervous systems and a heavy reliance on chemoreception for detection of chemical cues from preys or predators. It implies that fewer GPCRs are directly involved in neurotransmission in such phyla than the numbers would lead us to deduce. The challenge ahead will be to functionally express as many of these GPCRs as possible to tease out the receptors relevant to neurotransmission from those that are hypothesized to represent chemosensory GPCRs.

Another issue with evolutionary implications is the dilemma of co-evolution of ligands and receptors. It takes two to tango, so how the two partners, as they undergo changes in the course of evolutionary timelines, coordinate themselves in order to keep the signaling going? Bridgham, Carroll, and Thornton (2006, p. 97) put it this way: "If the hormone [or neurotransmitter] is not yet present, how can [natural] selection drive the receptor's affinity for it? Conversely, without the receptor, what selection pressure could guide the evolution of the ligand?"

To answer the first question, these authors traced the ancestral corticoid receptor from which the vertebrate mineralocorticoid (MR) and the glucocorticoid (GR) receptors descended by gene duplication. The hormone aldosterone selectively activates MR but not GR, which has cortisol for ligand. And yet, aldosterone also activates the ancestral corticoid receptor even though aldosterone is not available in the ancestor (lamprey or hagfish). They found that only two amino acid substitutions introduced in the ligand-binding domain of the ancestral receptor were necessary to confer a GR-like phenotype (and the attendant loss of aldosterone sensitivity). Thus a new functional pairing of ligand and receptor can occur by recruitment of an old receptor.

To answer the second question, Thornton, Need, and Crews (2003) and Keay, Bridgham, and Thornton (2006) used the example of the classical estrogen receptors, that is, those that once activated by estrogen translocate to the nucleus where they act as DNA-binding transcription factors. Estrogen receptors are absent in cnidarians (Reitzel & Tarrant, 2009), nematodes, and arthropods, and when they are present in invertebrates, such as in molluscs, they are not activated by estrogens but are instead constitutively active (Thornton, Need, & Crews, 2003; Keay et al., 2006). And yet, estrogens are present in all these phyla and are known to exert specific actions in cnidarians and molluscs (D'Aniello et al., 1996; Pernet & Anctil, 2002; Tarrant, Atkinson, & Atkinson, 2004). Bridgham et al. (2006) proposed that the reproductive role of estrogens in invertebrates is mediated by ancient signaling pathways other than transcriptional factors, in order to keep estrogen production active in the phyletic lines until the ancestral classical estrogen receptor emerged later in evolution. Thus the ancestral steroid receptor's pairing with estrogen may have evolved by exploitation of an even more ancient ligand, estrogen. One can logically predict that similar evolutionary scenarios were involved in the co-evolution of neurotransmitters and their receptors, but confirmatory data have yet to emerge.

10.6 The Role of Glia in Neurotransmission

From structural (ensheathment) and metabolic support, glial cells have graduated to even more pivotal roles in the functioning of the nervous system of both vertebrates and invertebrates (Murai & van Meyel, 2007). One of these roles is to support and modulate synaptic activity, to the point of bidirectional signaling between neurons and glial cells (Coles & Abbott, 1996). Using the highly tractable squid giant axon model, Evans, Reale, Merzon, and Villegas (1992) found that activation of the squid axon releases glutamate which binds to mGluRs on the Schwann glial cells, leading to ACh release from the Schwann cell. The released ACh feeds back via nicotinic autoreceptors on the Schwann cell, setting in motion cascading events leading to the glial cell's hyperpolarization. In addition, these events are modulated by octopamine and neuropeptides (Evans, Reale, & Villegas, 1986). This level of complexity in neuron–glia signaling is repeated in other invertebrate and vertebrates models (Domingues, Taylor, & Fern, 2010; Fields, 2010; Murai & van Meyel, 2007). In addition to the role of glia in affecting synaptic activity through the release of neuromodulators, another important way that glia assist neurotransmission is through neurotransmitter uptake, processing and recycling (Bringmann et al., 2009).

Based on observations of *C. elegans*, Heiman and Shaham (2007) suggested that the ancestral glial cells interacted with neurons in far more limited ways than in the complex nervous systems of higher animals. However, where glial cells come short, other cell types take over roles carried out by glia in higher animals. Hartline (2011) explored how far back in phylogeny one must travel to encounter the earliest examples of glial cells, defined as "non-neuronal cells closely associated with neurons and not present outside of the nervous system." He concluded that glia were an innovation of basal bilaterians, as acoelomorphs appear to possess poorly developed glia or protoglia. These protoglia may have derived from basiectodermal or epithelial support cells, phagocytes or developmental guidepost cells. In phyla with nervous systems basal to bilaterians such as cnidarians, glia are absent and other cell types must substitute if necessary. It has long been considered that epithelial (basiectodermal) nerve nets do not need glia (Bullock & Horridge, 1965), and axon ensheathment is unneeded in synaptic nerve nets with short neurites. However, there are exceptional cases in cnidarians where nerve nets extend away from epithelia and cross wide, gelatinous spaces sandwiched between the epithelia, called the mesoglea. In some anthozoans the mesogleal nerve net is closely associated with a superimposed net of large stellate amoebocytes that partially enwrap neurons (Buisson, 1970; Buisson & Franc, 1969; Satterlie, Anderson, & Case, 1980). In the sea pansy these amoebocytes take up extracellular catecholamines (Anctil, Germain, &

LaRivière, 1984), and they are norepinephrine- (Pani et al., 1995) and dopamine β-hydroxylase-immunoreactive (Anctil et al., 2002), as are nearby neurons. Anthozoan amoebocytes are regarded as multifunctional cells (Chapman 1966; Fautin & Mariscal 1991), exhibiting phagocytic and other activities usually associated with the hemocytes of higher invertebrates. This can be construed as an early example of cells exhibiting glial-like functions, and clearly one of these functions is neurotransmitter uptake and processing.

10.7 Conclusion

Molecules that act as neurotransmitters/neuromodulators/neurohormones in animals with nervous systems can be traced in aneural multicellular organisms such as sponges and plants as well as in unicellular organisms such as bacteria and protozoans. Among these molecules the amino acids glutamate and GABA, the biogenic amines, and the cholinergic system are supported by sound evidence. The evidence for neuropeptide families is less compelling except for insulin-related peptides. They evolved as "biomediators" involved in autoregulatory functions and in cell-to-cell communication during the pre-nervous period. This culminated in the relatively complex signaling seen in protozoan chemotaxis and sponge behavioral coordination.

With the emergence of nervous systems in prebilaterian animals, the synapse developed as a more efficient method of cell-to-cell communication with the potential to subtend the coordination of more sophisticated sensorimotor activities and complex behavior than in sponges. While the exocytotic machinery responsible for transmitter release on the presynaptic side already existed in unicellular organisms, the postsynaptic elements across the narrow junctional cleft and the molecular cross-talk between the two sides of the synapse had to be introduced in the common ancestor of cnidarians and ctenophores.

In contrast to aneural organisms, in which the biomediator role of classical small molecules such as acetylcholine, glutamic acid, GABA and biogenic amines is better established, the emergence of nervous systems is accompanied by the rise and diversification of neuropeptides as dominant transmitters at the cnidarian synapse, and as major players in the modulation of central nervous system activities in bilaterian animals. The genomic records suggest a bumpy history of diversification and gene loss in the course of the evolution of neurotransmitter systems.

Major challenges lay ahead before we come to a better understanding of what drove the evolution of neurotransmission beyond the generalities above. We have not yet addressed the evolutionary steps leading to the extant complex systems except for the issue of ligand-receptor pairing, a "chicken-or-egg"-like causality dilemma. While the sequencing of genomes from animals representing key branches in the phylogenetic tree has helped us get a better picture of what is available as neurotransmitter systems, we have far to go before a clear picture of what systems are functionally implemented emerges, especially in basal metazoans. This is a particularly critical issue in the case of receptors where sequence matches and the presence of signature motifs are not reliable harbingers of correct ligand identification. It will take ambitious projects of cloning and expressing candidate genes to get beyond that hurdle and to fashion as accurate a story of evolving neurotransmitter systems as possible.

References

Anctil, M. (1989). Modulation of a rhythmic activity by serotonin via cyclic AMP in the coelenterate *Renilla köllikeri*. *Journal of Comparative Physiology*, B159, 491–500.

Anctil, M. (2000). Evidence for gonadotropin-releasing hormone-like peptides in a cnidarian nervous system. *General & Comparative Endocrinology*, 119, 317–328.

Anctil, M. (2009). Chemical transmission in the sea anemone *Nematostella vectensis*: A genomic perspective. *Comparative Biochemistry & Physiology, Part D, Genomics & Proteomics*, 4, 268–289.

Anctil, M., Boulay, D., & LaRivière, L. (1982). Monoaminergic mechanisms associated with control of luminescence and contractile activities in the coelenterate, *Renilla köllikeri*. *Journal of Experimental Zoology*, 223, 11–24.

Anctil, M., & Carette, J.-P., (1994). Glutamate immunoreactivity in non-neuronal cells of the sea anemone *Metridium senile*. *The Biological Bulletin*, 187, 48–54.

Anctil, M., Germain, G., & LaRivière, L., (1984). Catecholamines in the coelenterate *Renilla köllikeri*. Uptake and radioautographic localization. *Cell Tissue Research*, 238, 69–80.

Anctil, M., & Grimmelikhuijzen, C. J. P., (1989). Excitatory action of the native neuropeptide Antho-RFamide on muscles in the pennatulid *Renilla köllikeri*. *General Pharmacology*, 20, 381–384.

Anctil, M., Hurtubise, P., & Gillis, M. A., 2002. Tyrosine hydroxylase and dopamine-β-hydroxylase immunoreactivities in the cnidarian *Renilla koellikeri*. *Cell Tissue Research*, 310, 109–117.

Anctil, M., & Ngo Minh, C., (1997). Neuronal and non-neuronal taurine-like immunoreactivity in the sea pansy, *Renilla koellikeri* (Cnidaria, Anthozoa). *Cell Tissue Research*, 288, 127–134.

Anctil, M., Poulain, I., & Pelletier, C., (2005). Nitric oxide modulates peristaltic muscle activity associated with fluid circulation in the sea pansy *Renilla koellikeri*. *Journal of Experimental Biology*, 208, 2005–2017.

Anderson, P. A. V. (1985). Physiology of a bidirectional, excitatory, chemical synapse. *Journal of Neurophysiology*, 53, 821–835.

Anderson, P. A. V., & Grünert, U., 1988. Three-dimensional structure of bidirectional, excitatory chemical synapses in the jellyfish *Cyanea capillata*. *Synapse*, 2, 606–613.

Anderson, P. A. V., Moosler, A., & Grimmelikhuijzen, C. J. P. (1992). The presence and distribution of Antho-RFamide-like material in scyphomedusae. *Cell Tissue Research*, 267, 67–74.

Anderson, P. A. V., & Trapido-Rosenthal, H. G. (2009). Physiological and chemical analysis of neurotransmitter candidates at a fast excitatory synapse in the jellyfish *Cyanea capillata* (Cnidaria, Scyphozoa). *Invertebrate Neuroscience*, 9, 167–173.

Awad, E. W., & Anctil, M., (1993). Identification of ß-like adrenoceptors associated with bioluminescence in the sea pansy *Renilla koellikeri* (Cnidaria, Anthozoa). *Journal of Experimental Biology*, 177, 181–200.

Baggerman, G., Liu, F., Wets, G., & Schoofs, L. (2005). Bioinformatic analysis of peptide precursor proteins. *Annals of the New York Academy of Sciences*, 1040. 59–65.

Bargmann, W., & Scharrer, E. (1951). The site of origin of the hormones of the posterior pituitary. *American Scientist*, 39, 255–259.

Bouchard, C., Ribeiro, P., Dubé, F., & Anctil, M. (2003). A new G protein-coupled receptor from a primitive metazoan shows homology with vertebrate aminergic receptors and displays constitutive activity in mammalian cells. *Journal of Neurochemistry*, 86, 1149–1161.

Bouchard, C., Ribeiro, P., Dubé, F., Demers, C., & Anctil, M., (2004). Identification of a novel aminergic-like G protein-coupled receptor in the cnidarian, *Renilla koellikeri*. *Gene*, 341, 67–75.

Boyd, I. A., & Martin, A. R. (1956). The end-plate potential in mammalian muscle. *Journal of Physiology (London)*, 132, 74–91.

Bridge, D., Cunningham, C. W., Schierwater, B., DeSalle, R., & Buss, L. W. (1992). Class-level relationships in the phylum Cnidaria: Evidence from mitochondrial genome structure. *Proceedings of the National Academy of Sciences of the USA*, 89, 8750–8753.

Bridgham, J. T., Carroll, S. M., & Thornton, J. W. (2006). Evolution of hormone-receptor complexity by molecular exploitation. *Science*, 312, 97–101.

Bringmann, A., Pannicke, T., Biedermann, B., Francke, M., Iandiev, I., Grosche, J., ... Reichenbach, A. (2009). Role of retinal glial cells in neurotransmitter uptake and metabolism. *Neurochemistry International*, 54, 143–160.

Bruns, D., & Jahn, R. (1995). Real-time measurement of transmitter release from single synaptic vesicles. *Nature*, 377, 62–65.

Bucci, G., Ramoino, P., Diaspro, M., & Usai, C. (2005). A role for $GABA_A$ receptors in the modulation of *Paramecium* swimming behavior. *Neuroscience Letters*, 386, 179–183.

Buisson, B. (1970). Les supports morphologiques de l'intégration dans la colonie de *Veretillum cynomorium* Pall. (Cnidaria, Pennatularia). *Zoomorphology*, 68, 1–36.

Buisson, B., & Franc, S. (1969). Structure et ultrastructure des cellules mésenchymateuses et nerveuses intramésogléennes de *Veretillum cynomorium* Pall. (Cnidaire Pennatulidae). *Vie Milieu*, 20, 279–292.

Bullock, T. H., & Horridge, G. A. (1965). *Structure and function in the nervous systems of invertebrates* (Vol. 1). San Francisco and London: Freeman & Co.

Burbach, J. P. H. (2010). Neuropeptides from concept to online database www.neuropeptides.nl. *European Journal of Pharmacology*, 626, 27–48.

Burke, R. D., Angerer, L. M., Elphick, M. R., Humphrey, G. W., Yaguchi, S., Liang, S., ... Thorndike, M. C. (2006). A genomic view of the sea urchin nervous system. *Developmental Biology*, 300, 434–460.

Burnstock, G. (1976). Do some nerve cells release more than one transmitter? *Neuroscience*, 1, 239–248.

Carlberg, M., Alfredsson, K., Nielsen, S.-O., & Anderson, P. A. V. (1995). Taurine-like immunoreactivity in the motor nerve net of the jellyfish *Cyanea capillata*. *The Biological Bulletin*, 188, 78–82.

Carstensen, K., Rinehart, K. L., McFarlane, I. D., & Grimmelikhuijzen, C. J. P. (1992). Isolation of Leu-Pro-Pro-Gly-Pro-Leu-Pro-Arg-Pro-NH_2 (Antho-RPamide), an N-terminally protected, biologically active neuropeptide from sea anemones. *Peptides*, 13, 851–857.

Carstensen, K., McFarlane, I. D., Rinehart, K. L., Hudman, D., Sun, F., & Grimmelikhuijzen, C. J. P. (1993). Isolation of < Glu-Asn-Phe-His-Leu-Arg-Pro-NH_2 (Antho-RPamide II), a novel, biologically active neuropeptide from sea anemones. *Peptides*, 14, 131–135.

Chapman, G. (1966). The structure and function of the mesoglea. In W. J. Rees (Ed.), *The cnidaria and their evolution* (pp. 147–168). New York, NY: Academic Press.

Chen, G., & Ewing, A. G. (1995). Multiple classes of catecholamine vesicles observed during exocytosis from the *Planorbis* cell body. *Brain Research*, 701, 167–174.

Chen, G., Gutman, D. A., Zerby, S. E., & Ewing, A. G. (1996). Electrochemical monitoring of bursting exocytotic events from the giant dopamine neuron of *Planorbis corneus*. *Brain Research*, 733, 119–124.

Colasanti, M., Venturini, G., Merante, A, Musci, G., & Lauro, G. M. (1997). Nitric oxide involvement in *Hydra vulgaris* very primitive olfactory-like system. *Journal of Neuroscience*, 17, 493–499.

Coles, J. A., & Abbott, N. J. (1996). Signaling from neurones to glial cells in invertebrates. *Trends in Neuroscience*, 19, 358–362.

Concas, A., Pierobon, P., Mostallino, M. C., Porcu, P., Marino, G., Minei, R., & Biggio, G. (1998). Modulation of γ-aminobutyric acid (GABA) receptors and the feeding response by neurosteroids in *Hydra vulgaris*. *Neuroscience*, 85, 979–988.

Cottrell, G. A. (1997). The first peptide-gated ion channel. *Journal of Experimental Biology*, 200, 2377–2386.

Cottrell, G. A., Jeziorski, M., & Green, K. (2001). Location of a ligand recognition site of FMRFamide-gated Na channels. *FEBS Letters*, 489, 71–74.

Cristino, L., Guglielmotti, V., Cotugno, A., Musio, C., & Santillo, S., (2008). Nitric oxide signaling pathways at neural level in invertebrates: Functional implications in cnidarians. *Brain Research*, 1199, 148–158.

D'Aniello, A., Di Cosmo, A., Di Cristo, C., Assisi, L., Botte, V., & Di Fiore, M. M. (1996). Occurrence of sex steroid hormones and their binding proteins in *Octopus vulgaris*. *Biochemical & Biophysical Research Communications*, 227, 782–788.

Delmonte Corrado, M. U., Ognibene, M., Trielli, C., Politi, H., Passalacqua, M., & Falugi, C. (2002). Detection of molecules related to the GABAergic system in a single-cell eukaryote, *Paramecium primaurelia*. *Neuroscience Letters*, 329, 65–68.

Delmonte Corrado, M. U., Politi, H., Ognibene, M., Angelini, C., Trielli, C., Ballarini, P., & Falugi, C. (2001). Synthesis of the signal molecule acetylcholine during the developmental cycle of *Paramecium primaurelia* (Protista, Ciliophora) and its possible role in conjugation. *Journal of Experimental Biology*, 204, 1901–1907.

De Wied, D. (1984). The neuropeptide concept. *Maturitas*, 6, 217–223.

Domingues, A. M. de J., Taylor, M., & Fern, R. (2010). Glia as transmitter sources and sensors in health and disease. *Neurochemistry International*, 57, 359–366.

Edgecombe, G. D., Giribet, G., Dunn, C. W., Hejnol, A., Kristensen, R. M., Neves, R. C., ... Sorensen, M.V. (2011). Higher-level metazoan relationships: Recent progress and remaining questions. *Organisms Diversity & Evolution*, 11, 151–172.

Elliott, G. R. D., & Leys, S.P. (2010). Evidence of glutamate, GABA and NO in coordinating behavior in the sponge, *Ephydatia muelleri* (Demospongiae, Spongillidae). *Journal of Experimental Biology*, 213, 2310–2321.

Ellwanger, K., & Nickel, M. (2006). Neuroactive substances specifically modulate rhythmic body contractions in the nerveless metazoon *Tethya wilhelma* (Demospongiae, Porifera). *Frontiers in Zoology*, 3, 7.

Ellwanger, K., & Nickel, M. (2007). GABA and glutamate specifically induce contractions in the sponge *Tethya wilhelma*. *Journal of Comparative Physiology*, A193, 1–11.

Evans, P. D., Reale, V., Merzon, R. M., & Villegas, J. (1992). N-methyl-D-aspartate (NMDA) and non-NMDA (metabotropic) type glutamate receptors modulate the membrane potential of the Schwann cell of the squid giant nerve fibre. *Journal of Experimental Biology*, 173, 229–249.

Evans, P. D., Reale, V., & Villegas, J. (1986). Peptidergic modulation of the membrane potential of the Schwann cell of the squid giant nerve fibre. *Journal of Physiology (London)*, 379, 61–82.

Fautin, D. G., & Mariscal, R. N. (1991). Cnidaria: Anthozoa. In F. W. Harrison & J. A. Westfall (Eds.) *Microscopic anatomy of invertebrates, Vol. 2: Placozoa, porifera, cnidaria, and ctenophora* (pp. 267–358). New York, NY: Wiley-Liss.

Fields, R. D. (2010). Release of neurotransmitters from glia. *Neuron Glia Biology*, 6, 137–139.

Freestone, P. P., & Lyte, M. (2008). Microbial endocrinology: Experimental design issues in the study of interkingdom signaling in infectious disease. *Advances in Applied Microbiology*, 64, 75–105.

Germain, G., & Anctil, M. (1996). Evidence for intercellular coupling and connexion-like protein in the luminescent endoderm of *Renilla koellikeri* (Cnidaria, Anthozoa). *The Biological Bulletin*, 191, 353–366.

Gillis, M.-A., & Anctil, M. (2001). Monoamine release by neurons of a primitive nervous system: an amperometric study. *Journal of Neurochemistry*, 76, 1774–1784.

Golubovic, A., Kuhn, A., Williamson, M., Kalbacher, H., Holstein, T. W., Grimmelikhuijzen, C. J. P., & Gründer, S. (2007). A peptide-gated ion channel from the freshwater polyp *Hydra*. *Journal of Biological Chemistry*, 282, 35098–35103.

Graff, D., & Grimmelikhuijzen, C. J. P. (1988a). Isolation of <Glu-Ser-Leu-Arg-Trp-NH_2, a novel neuropeptide from sea anemones. *Brain Research*, 442, 354–358.

Graff, D., & Grimmelikhuijzen, C. J. P. (1988b). Isolation of <Glu-Gly-Leu-Arg-Trp-NH$_2$ (Antho-Rwamide II), a novel neuropeptide from sea anemones. *FEBS Letters*, 239, 137–140.

Grimmelikhuijzen, C. J. P. (1985). Antisera to the sequence Arg-Phe-amide visualize neuronal centralization in hydroid polyps. *Cell Tissue Research*, 241, 171–182.

Grimmelikhuijzen, C. J. P., & Graff, D. (1986). Organization of the nervous systems of physonectid siphonophores. *Cell Tissue Research*, 246, 463–479.

Grimmelikhuijzen, C. J. P., Rinehart, K. L., Jacob, E., Graff, D., Reinscheid, R. K., Nothacker, H.-P., & Staley, A. L. (1990). Isolation of L-3-phenyllactyl-Leu-Arg-Asn-NH$_2$ (Antho-RNamide), a sea anemone neuropeptide containing an unusual amino-terminal blocking group. *Proceedings of the National Academy of Sciences of the USA*, 87, 5410–5414.

Grimmelikhuijzen, C. J. P., Williamson, M., & Hansen, G. N. (2004). Neuropeptides in cnidarians. In I. Fairweather (Ed.), *Cell signaling in prokaryotes and lower metazoans* (pp. 115–139). Dordrecht, Netherlands: Kluwer Academic Publishing.

Grosvenor, W., Bellis, S. L., Kass-Simon, G., & Rhoads, D.E. (1992). Chemoreception in hydra: Specific binding of glutathione to a membrane fraction. *Biochimica et Biophysica Acta*, 1117, 120–125.

Hajj-Ali, I., & Anctil, M., 1997. Characterization of a serotonin receptor in the cnidarian *Renilla koellikeri*: A radiobinding analysis. *Neurochemistry International*, 31, 83–93.

Hartline, D. K. (2011). The evolutionary origins of glia. *Glia*, 59, 1215–1236.

Heiman, M. G., & Shaham, S. (2007). Ancestral roles of glia suggested by the nervous system of *Caenorhabditis elegans*. *Neuron Glia Biology*, 3, 55–61.

Henkel, A. W., Kang, G., & Kornhuber, J. (2001). A common molecular machinery for exocytosis and the kiss-and-run mechanism in chromaffin cells is controlled by phosphorylation. *Journal of Cell Science*, 114, 4613–4620.

Hewes, R. S., & Taghert, P. H. (2001). Neuropeptides and neuropeptide receptors in the *Drosophila melanogaster* genome. *Genome Research*, 11, 1126–1142.

Holtmann, M., & Thurm, U. (2001). Mono- and oligo-vesicular synapses and their connectivity in a cnidarian sensory epithelium (*Coryne tubulosa*). *Journal of Comparative Neurology*, 432, 537–549.

Horiushi, Y., Kimura, R., Kato, N., Fujii, T., Seki, M., Endo, T., ... & Kawashima, K. (2003). Evolutional study on acetylcholine expression. *Life Science*, 72, 1745–1756.

Hoyle, C. H. V., Knight, G. E., & Burnstock, G. (1989). Actions of adenylyl compounds in the pedal disc of the cnidarian *Actinia equina*. *Comparative Biochemistry & Physiology, Part C. Comparative Pharmacology*, 94C, 111–114.

Kass-Simon, G., Pannaccione, A., & Pierobon, P. (2003). GABA and glutamate receptors are involved in modulating pacemaker activity in hydra. *Comparative Biochemistry Physiology*, A136, 329–342.

Kass-Simon, G., & Pierobon, P. (2007). Cnidarian chemical neurotransmission, an updated overview. *Comparative Biochemistry Physiology*, A146, 9–25.

Kass-Simon, G., & Scappaticci, A. A., (2004). Glutamatergic and GABAergic control in the tentacle effector systems of *Hydra vulgaris*. *Hydrobiologia*, 530/531, 67-71.

Kawashima, K., Misawa, H., Moriwaki, Y., Fujii, Y. X., Fujii, T., Horiushi, Y., ... Kamekura, M. (2007). Ubiquitous expression of acetylcholine and its biological functions in life forms without nervous systems. *Life Science*, 80, 2206–2209.

Keay, J., Bridgham, J. T., & Thornton, J. W. (2006). The *Octopus vulgaris* estrogen receptor is a constitutive transcriptional activator: Evolutionary and functional implications. *Endocrinology*, 147, 3861–3869.

Koizumi, O., Wilson, J. D., Grimmelikhuijzen, C. J. P., & Westfall, J. A. (1989). Ultrastructural localization of RFamide-like peptides in neuronal dense-cored vesicles in the peduncle of *Hydra*. *Journal of Experimental Zoology*, 249, 17–22.

Kupfermann, I., (1991). Functional studies of cotransmission. *Physiological Reviews*, 71, 683-732.

Leitz, T., (1998). Metamorphosin A and related compounds – a novel family of neuropeptides with morphogenic activity. *Annals of the New York Academy of Sciences*, 839, 105–110.

Leitz, T., Morand, K., & Mann, M., (1994). Metamorphosin A: A novel peptide controlling development of the lower metazoan *Hydractinia echinata* (Coelenterata, Hydrozoa). *Developmental Biology*, 163, 440–446.

Lentz, T. L. (1966). *The cell biology of hydra*. Amsterdam: North-Holland Publishing Co.

Le Roith, D., Shiloach, J., Roth, J., & Lesniak, M. A. (1980). Evolutionary origins of vertebrate hormones: Substances similar to mammalian insulins are native to unicellular eukaryotes. *Proceedings of the National Academy of Sciences of the USA*, 77, 6184–6188.

Leys, S. P., & Riesgo, A. (2012). Epithelia, an evolutionary novelty of metazoans. *Journal of Experimental Zoology (Molecular & Developmental Evolution)*, 318B, 438–447.

Li, C., & Kim, K. (2008). Neuropeptides. In E. M. Jorgensen & J.M. Kaplan (Eds.), *Wormbook* (pp. 1–36). The *C. elegans* Research Community, Wormbook, doi/10.1895/wormbook.1.142.1. Retrieved from http://www.wormbook.org

Mackie, G. O. (1990). The elementary nervous system revisited. *American Zoologist*, 30, 907–920.

Mackie, G. O. (2004). Epithelial conduction: Recent findings, old questions, and where do we go from here? *Hydrobiologia*, 530/531, 73–80.

Mackie, G. O., & Passano, L. M. (1968). Epithelial conduction in hydromedusae. *Journal of General Physiology*, 52, 600–621.

Marlow, H. Q., Srivastava, M., Matus, D. Q., Rokhsar, D., & Martindale, M. Q. (2009). Anatomy and development of the nervous system of *Nematostella vectensis*, an anthozoan cnidarian. *Developmental Neurobiology*, 69, 235–254.

McFarlane, I. D., Graf, D., & Grimmelikhuijzen, C. J. P. (1987). Excitatory actions of Antho-RFamide, an anthozoan neuropeptide, on muscles and conducting systems of the sea anemone *Calliactis parasitica*. *Journal of Experimental Biology*, 133, 157–168.

McFarlane, I. D., Anderson, P. A. V., & Grimmelikhuijzen, C. J. P. (1991). Effects of three anthozoan neuropeptides, Antho-RWamide I, Antho-RWamide II, and Antho-RFamide, on slow muscles from sea anemones. *Journal of Experimental Biology*, 156, 419–431.

McFarlane, I. D., Reinscheid, R. K., & Grimmelikhuijzen, C. J. P. (1992). Opposite actions of the anthozoan neuropeptide Antho-RNamide on antagonistic muscle groups in sea anemones. *Journal of Experimental Biology*, 164, 295–299.

Meinertzhagen, I. A. (2010). The organization of invertebrate brains: cells, synapses and circuits. *Acta Zoologica (Stockholm)*, 91, 64–71.

Miller, M. W. (2009). Colocalization and cotransmission of classical neurotransmitters: An invertebrate perspective. In R. Gutierrez (Ed.), *Co-existence and co-release of classical neurotransmitters* (pp. 243–261). New York, NY: Springer Science + Business Media.

Mire, P., Nasse, J., & Venable-Thibodeaux, S. (2000). Gap junctional communication in the vibration-sensitive response of sea anemones. *Hearing Research*, 144, 109–123.

Morishita, F., Nitagai, Y., Furukawa, Y., Matsushima, O., Takahashi, T., Hatta, M., … Koizumi, O. (2003). Identification of a vasopressin-like immunoreactive substance in hydra. *Peptides*, 24, 17–26.

Moroz, L. L., Meech, R. W., Sweedler, J. V., & Mackie, G. O., (2004). Nitric oxide regulates swimming in the jellyfish *Aglantha digitale*. *Journal of Comparative Neurology*, 471, 26–36.

Moroz, L. L., Edwards, J. R., Puttanveettil, S. V., Kohn, A. B., Ha, T., Heyland, A., … Kandel, E. R. (2006). Neuronal transcriptome of *Aplysia*: Neuronal compartments and circuitry. *Cell*, 127, 1453–1467.

Moroz, L. L., Kocot, K. M., Citarella, M. R., Dosung, S., Norekian, T. P., Povolotskaya, I. S., … Bruders, R., (2014). The ctenophore genome and the evolutionary origins of nervous systems. *Nature*, 510, 109–115.

Murai, K. K., & van Meyel, D. J. (2007). Neuron-glial communication at synapses: Insights from vertebrates and invertebrates. *Neuroscientist*, 13, 657–666.

Murch, S. J. (2006). Neurotransmitters, neuroregulators and neurotoxins in plants. In F. Baluska, S. Mancuso, & D. Volkman (Eds.), *Communication in plants—Neuronal aspects of plant life* (pp. 137–151). Berlin: Springer-Verlag.

Nathoo, A. N., Moeller, R. A., Westlund, B. A., & Hart, A. C. (2001). Identification of neuropeptide-like protein gene families in *Caenorhabditis elegans* and other species. *Proceedings of the National Academy of Sciences of the USA*, 98, 1400–1405.

Nickel, M. (2010). Evolutionary emergence of synaptic nervous systems: What can we learn from the non-synaptic, nerveless Porifera? *Invertebrate Biology*, 129, 1–16.

Nothacker, H.-P., Rinehart, K. L., & Grimmelikhuijzen, C. J. P. (1991). Isolation of L-3-phenyllactyl-Phe-Lys-Ala-NH$_2$ (Antho-KAamide), a novel neuropeptide from sea anemones. *Biochemical & Biophysical Research Communications*, 179, 1205–1211.

Nothacker, H.-P., Rinehart, K. L., McFarlane, I. D., & Grimmelikhuijzen, C. J. P., (1991). Isolation of two novel neuropeptides from sea anemones: The unusual, biologically active L-3-phenyllactyl-Tyr-Arg-Ile-NH$_2$ and its des-phenyllactyl fragment Tyr-Arg-Ile-NH$_2$. *Peptides*, 12, 1165–1173.

Orias, E. (2002). Sequencing the *Tetrahymena thermophila* genome. A White Paper submitted to the National Human Genome Research Institute. Retrieved from https://www.genome.gov/pages/research/sequencing/seqproposals/tetrahymena_genome.pdf

Pani, A. K., & Anctil, M. (1994). Evidence for biosynthesis and catabolism of monoamines in the sea pansy *Renilla koellikeri* (Cnidaria). *Neurochemistry International*, 25, 465–474.

Pani, A. K., Anctil, M., & Umbriaco, D., 1995. Neuronal localization and evoked release of norepinephrine in the cnidarian *Renilla koellikeri*. *Journal of Experimental Zoology*, 272, 1–12.

Parker, G. H. (1919). *The elementary nervous system*. Philadelphia, PA: Lippincott.

Pernet, V., & Anctil, M., (2002). Annual variations and sex-related differences of estradiol-17b levels in the anthozoan *Renilla koellikeri*. *General and Comparative Endocrinology*, 129, 63–68.

Pernet, V., Anctil, M., & Grimmelikhuijzen, C. J. P. (2004). Antho-RFamide-containing neurons in the primitive nervous system of the anthozoan *Renilla koellikeri*. *Journal of Comparative Neurology*, 472, 208–220.

Pernet, V., Galino, V., Galino, G., & Anctil, M. (2002). Variations of lipid and fatty acid contents during the reproductive cycle of the anthozoan *Renilla koellikeri*. *Journal of Comparative Physiology*, B172, 455–465.

Perovic, S., Krasko, A., Prokic, I., Müller, I., & Müller, W. E. G. (1999). Origin of neuronal-like receptors in Metazoa: Cloning of a metabotropic glutamate/GABA-like receptor from the marine sponge *Geodia cydonium*. *Cell Tissue Research*, 296, 395–404.

Philippe, H., Derelles, R., Lopez, P., Pick, K., Borchiellini, C., Bouty-Esnault, N., ... Manual, M. (2009). Phylogenomics revives traditional views on deep animal relationships. *Current Biology*, 19, 706–712.

Pierobon, P., Concas, A., Santoro, G., Marino, G., Minei, R., Pannaccione, A., ... Biggio, G., (1995). Biochemical and functional identification of GABA receptors in *Hydra vulgaris*. *Life Science*, 56, 1485–1497.

Pierobon, P., Minei, R., Porcu, P., Sogliano, C., Tino, A., Marino, G., ... Concas, A., (2001). Putative glycine receptors in *Hydra*: A biochemical and behavioral study. *European Journal of Neuroscience*, 14, 1659–1666.

Pierobon, P., Sogliano, C., Minei, R., Tino, A., Porcu, P., Marino, G., ... Concas, A., (2004). Putative NMDA receptors in *Hydra*: A biochemical and functional study. *European Journal of Neuroscience*, 20, 2598–2604.

Plattner, H., & Kissmehl, R. (2003). Dense-core secretory vesicle docking and exocytotic membrane fusion in *Paramecium* cells. *Biochimica et Biophysica Acta*, 1641, 183–193.

Preston, R. R., & Usherwood, P. N. R. (1988a). Characterization of a specific L-[^{3}H]glutamic acid binding site on cilia isolated from *Paramecium tetraurelia*. *Journal of Comparative Physiology*, B158, 345–351.

Preston, R. R., & Usherwood, P. N. R. (1988b). L-Glutamate-induced membrane hyperpolarization and behavioral responses in *Paramecium tetraurelia*. *Journal of Comparative Physiology*, A164, 75–82.

Raible, F., Tessmar-Raible, K., Arboleda, E., Kaller, T., Bork, P., Arendt, D., & Arnone, M. I. (2006). Opsins and clusters of sensory G-protein-coupled receptors in the sea urchin genome. *Developmental Biology*, 300, 461–475.

Ramoino, P., Fronte, P., Beltrame, F., Diaspro, A., Fato, M., Raiteri, L., … Usai, C. (2003). Swimming behavior regulation by $GABA_B$ receptors in *Paramecium*. *Experimental Cell Research*, 291, 398–405.

Reitzel, A.M., & Tarrant, A.M., (2009). Nuclear receptor complement of the cnidarian *Nematostella vectensis*: phylogenetic relationships and developmental expression patterns. *BMC Evolutionary Biology*, 9, 230.

Renard, E., Vacelet, J., Gazave, E., Lapébie, P., Borchiellini, C., & Ereskovsky, A. V. (2009). Origin of the neuro-sensory system: New and expected insights from sponges. *Integrative Zoology*, 4, 294–308.

Roshchina, V. V. (2010). Evolutionary considerations of neurotransmitters in microbial, plant, and animal cells. In M. Lyte & P.P.E. Freestone (Eds.), *Microbial endocrinology* (pp. 17–52). New York, NY: Springer Science+Business Media.

Roth, J., Le Roith, D., Collier, E. S., Weaver, N. R., Watkinson, A., Cleland, C. F., & Glick, S. M. (1985). Evolutionary origins of neuropeptides, hormones, and receptors: Possible applications to immunology. *Journal of Immunology*, 135, S816–S819.

Ruggieri, R. D., Pierobon, P., & Kass-Simon, G., (2004). Pacemaker activity in hydra is modulated by glycine receptor ligands. *Comparative Biochemistry Physiology*, A 138, 193–202.

Satterlie, R. A. (2011). Do jellyfish have central nervous systems? *Journal of Experimental Biology*, 214, 1215–1223.

Satterlie, R. A., Anderson, P. A. V., & Case, J. F. (1980). Colonial coordination in anthozoans: Pennatulacea. *Marine Behaviour & Physiology*, 7, 25–46.

Sawin, C. T. (2003). Berta and Ernst Scharrer and the concept of neurosecretion. *The Endocrinologist*, 13, 73–76.

Scharrer, B. (1990). The neuropeptide saga. *American Zoologist*, 30, 887–895.

Spencer, A. N. (1981). The parameters and properties of a group of electrically coupled neurones in the central nervous system of a hydrozoan jellyfish. *Journal of Experimental Biology*, 93, 33–50.

Spencer, A. N. (1982). The physiology of a coelenterate neuromuscular synapse. *Journal of Comparative Physiology*, A148, 353–363.

Spencer, A. N., & Satterlie, R. A. (1980). Electrical and dye coupling in an identified group of neurons in a coelenterate. *Journal of Neurobiology*, 11, 13–19.

Srivastava, M., Begovic, E., Chapman, J., Putnam, N. H., Hulsten, U., Kawashima, T., … Rokhsar, D. S. (2008). The Trichoplax genome and the nature of placozoans. *Nature*, 454, 955–960.

Srivastava, M., Simakov, O., Chapman, J., Fahey, B., Gauthier, M. E. A., Mitros, T., … Jackson, D. J. (2010). The *Amphimedon queenslandica* genome and the evolution of animal complexity. *Nature*, 466, 720–727.

Staal, R. G. W., Mosharov, E. V., & Sulzer, D. (2004). Dopamine neurons release transmitter via a flickering fusion pore. *Nature Neuroscience*, 7, 341–346.

Staubli, F., Jorgensen, J. T. D., Cazzamali, G., Williamson, M., Lenz, C., Sondergaard, L., … Grimmelikhuijzen, C. J. P. (2002). Molecular identification of the adipokinetic hormone receptors. *Proceedings of the National Academy of Sciences of the USA*, 99, 3446–3451.

Strand, F. L. (2007). Neuropeptides. In D. Sibley, I. Hanin, M. Kuhar & P. Skolnik (Eds.), *Handbook of contemporary neuropharmacology* (pp. 669–704). New York, NY: John Wiley & Sons.

Takahashi, T., Koizumi, O., Ariura, Y., Romanovitch, A., Bosch, T.C.G., Kobayashi, Y., … Fujisawa, T. (2000). A novel peptide, Hym-355, positively regulates neuron differentiation in *Hydra*. *Development*, 127, 997–1005.

Tarrant, A. M., Atkinson, M. J., & Atkinson, S. (2004). Effects of steroidal estrogens on coral growth and reproduction. *Marine Ecology Progress Series*, 269, 121–129.

Thornton, J. W., Need, E., & Crews, D. (2003). Resurrecting the ancestral steroid receptor: Ancient origin of estrogen signaling. *Science*, 301, 1714–1717.

Trams, E. G. (1981). On the evolution of neurochemical transmission. *Differentiation*, 19, 125–133.

Trielli, F., Politi, H., Falugi, C., & Delmonte Corrado, M. U. (1997). Presence of molecules related to the cholinergic system in *Paramecium primaurelia* (Protista, Ciliophora) and possible role in mating pair formation: An experimental study. *Journal of Experimental Zoology*, 279, 633–638.

Trudeau, L.-E. (2004). Glutamate co-transmission as an emerging concept in monoamine neuron function. *Journal of Psychiatry Neuroscience*, 29, 296–310.

Umbriaco, D., Anctil, M., & Descarries, L., 1990. Serotonin-immunoreactive neurons in the cnidarian *Renilla koellikeri*. *Journal of Comparative Neurology*, 291, 167–178.

Valentine, M., Yano, J., & Van Houten, J. L. (2008). Chemosensory transduction in paramecium. *Japanese Journal of Protozoology*, 41, 1–7.

Vandingenen, A., Mertens, I., Meeusen, T., Cerstiaens, A., Jonckheere, H., Luyten, W., & Schoofs, L. (2004). FMRFamide-related peptide G-protein-coupled receptors throughout the animal kingdom. *Recent Research Developments in Biophysics & Biochemistry*, 4, 117–143.

Van Houten, J. L. (1978). Two mechanisms of chemotaxis in *Paramecium*. *Journal of Comparative Physiology*, A127, 167–174.

Vassilatis, D. K., Hohmann, J. G., Zeng, H., Li, F., Ranchalis, J. E., Mortrud, M. T., ... Gaitanaris, M. J. (2003). The G protein-coupled receptor repertoires of human and mouse. *Proceedings of the National Academy of Sciences of the USA*, 100, 4603–4609.

Veenstra, J. A. (2010). Neurohormones and neuropeptides encoded by the genome of *Lottia gigantea*, with reference to other mollusks and insects. *General & Comparative Endocrinology*, 167, 86–103.

Veenstra, J. A. (2011). Neuropeptide evolution: Neurohormones and neuropeptides predicted from the genomes of *Capitella teleta* and *Helobdella robusta*. *General & Comparative Endocrinology*, 171, 160–175.

Watson, G. M., Venable, S., Hudson, R. R., & Repass, J. J. (1999). ATP enhances repair of hair bundle in sea anemones. *Hearing Research*, 136, 1–12.

Westfall, J. A. (1970). Ultrastructure of synapses in a primitive coelenterate. *Journal of Ultrastructure Research*, 32, 237–246.

Westfall, J. A. (1996). Ultrastructure of synapses in the first-evolved nervous systems. *Journal of Neurocytology*, 25, 735–746.

Westfall, J. A., Elliott, S. R., MohanKumar, P. S., & Carlin, R. W. (2000). Immunocytochemical evidence for biogenic amines and immunogold labelling of serotonergic synapses in tentacles of *Aiptasia pallida* (Cnidaria, Anthozoa). *Invertebrate Biology*, 119, 370–378.

Westfall, J. A., & Grimmelikhuijzen, C. J. P. (1993). Antho-RFamide immunoreactivity in neuronal synaptic and nonsynaptic vesicles of sea anemones. *The Biological Bulletin*, 185, 109–114.

Westfall, J. A., Sayyar, K. L., Elliott, C. F., & Grimmelikhuijzen, C. J. P., 1995. Ultrastructural localization of Antho-RWamides I and II at neuromuscular synapses in the gastrodermis and oral sphincter muscle of the sea anemone *Calliactis parasitica*. *The Biological Bulletin*, 189, 280–287.

Weyrer, S., Rutzler, K., & Rieger, R. (1999). Serotonin in Porifera? Evidence from developing *Tedania ignis*, the Caribbean fire sponge (Demospongiae). *Memoirs of the Queensland Museum*, 44, 659–665.

Whalan, S., Webster, N. S., & Negri, A. P. (2012). Crustose coralline algae and a cnidarian neuropeptide trigger larval settlement in two coral reef sponges. *PLoS ONE*, 7, e30386. doi 10.1371/journal.pone.0030386

Yang, W. Q., Braun, C., Plattner, H., Purvee, J., & Van Houten, J. L. (1997). Cyclic nucleotides in glutamate chemosensory signal transduction of *Paramecium*. *Journal of Cell Science*, 110, 2567–2572.

11
Neural Development in Invertebrates
Roger P. Croll

11.1 Overview of Invertebrate Development

Invertebrates comprise the vast majority of animal species on Earth and exhibit a diverse range of body forms which arise through different developmental programs. Furthermore, many invertebrates possess complex life cycles with larvae often differing greatly from adults in morphology, habitat, and behavior. In fact, some invertebrates have multiple sequential body forms which bear little resemblance to one another and, at first glance, may be mistaken for entirely different species. As if all this were not enough to thoroughly confuse any reader seeking a simple description of neural development in invertebrates, some of these animals can develop into identical adult forms by alternative methods (e.g., through sexual reproduction or through asexual budding).

Experience with our own species, common pets, and perhaps a vertebrate-oriented research career might imbue many neuroscientists with a view of early animal life as a period of development sheltered by gestation and parental care from which a smaller individual, which is more-or-less similar to the adult, may emerge to complete its final growth in preparation for later reproductive activities. Such a perspective is misleading for understanding the development of invertebrates. While a few invertebrates may undergo transitional periods of development sequestered from many of the demands of daily life (e.g., during the pupal stages of insects), most invertebrates are free-living throughout almost the entirety of ontogeny. These animals must grow, sometimes by enormous proportions, and undergo dramatic changes in body shape while their nervous systems continue to respond appropriately to stimuli and produce the motor programs that underlie feeding, locomotion, and escape. Furthermore, often radical transitions in body form must be accomplished quickly (e.g., within the course of a tidal cycle; Hadfield, 2000).

To understand the development of invertebrates and their nervous systems one must also jettison the idea that each developmental stage is simply a rudimentary and imperfect intermediate form in a continuous linear path toward the construction of a mature, reproductive adult. Instead, it must be remembered that each stage of the free-living larva is exposed to selective pressures which shape its form and function. Thus larval stages can and do evolve very differently from the adult stages. In extreme cases, a completely new larval form may be inserted into a developmental plan, permitting growth and dispersal of the species without competition for resources with the reproductive adult. The literature thus refers to primary and secondary larvae, contrasting larvae representative of ancestral forms with those that evolved subsequently (Raff & Byrne, 2006). Significant debates have involved not only questions of phylogeny, but also whether comparisons are most appropriate between larval or adult stages.

A final cornerstone of this chapter is a recommendation as to the level of analysis and perspective that a reader should bring to invertebrate neural development. As mentioned in other chapters, an important driving force in such studies is the comparison of homologous gene networks. However, knowledge of genes themselves provides little insight into how the nervous system controls the physiology and behavior of a developing organism. Furthermore, in many invertebrates, our current knowledge consists only of the existence or nonexistence of certain well-conserved genes—and perhaps their spatiotemporal expression patterns—while their contextual function remains largely unexplored. The focus of this chapter will therefore be dedicated to descriptions of the cellular origins, anatomy, transmitter contents, and, where known, the actual functions of the different components of the developing nervous systems.

Even with the exclusion of molecular biology, however, full coverage of invertebrate neural development in a single chapter is utterly impossible. The reader is therefore also referred to other sources for background literature. As a first step for readers with only a limited background outside the mammals or even the vertebrates, Pechenik (2005) offers a thoroughly readable (in fact, enjoyable) introduction to invertebrate biology, while a more detailed overview is offered by Brusca and Brusca (2003). For a glimpse at the intricacy, diversity, and beauty of the different larval forms, the reader is referred to Young, Sewell, & Rice (2002). For more focused overviews of invertebrate development, larval biology, and evolution of larval forms, the reader should consult Gilbert and Raunio (1997), Hall and Wake (1999), and Nielsen (2001).

A recurring theme of this chapter will be the (often uncertain) evolutionary relationships between different animal groups. The following sections will be organized into a scheme which groups the simplest animals, the diploblasts, and divides the more complex bilaterian invertebrates among three clades: Lophotrochzoa, Ecdysozoa, and Deuterostomia.

11.2 Basal Diplobalastic Metazoa with Nervous Systems

Together with their roughly radial symmetry, the defining characteristic of the diplobastic phyla, Cnidaria and Ctenophora, is their possession of only two germ layers, an outer ectoderm and inner endoderm. The layers are then generally separated by a loose extracellular matrix forming the mesoglea.

11.2.1 Cnidaria (Including Jellyfish, Corals, Anemones)

The Cnidaria are illustrative of many of the complexities encountered in comparative studies of neural development across the animal kingdom. As a sister clade to the Bilateria, the Cnidaria have long been the subject of close scrutiny with regards to the structure and function of the adult nervous system. However, while these animals are relatively simple in terms of overall structure, their life cycles are complex and large differences exist across the phylum. In the general developmental plan of Cnidaria, planula larvae metamorphose into the sessile polyp stage which then divides asexually to produce either additional polyps or numerous swimming medusa stage individuals (see Figure 11.1A). The medusae provide the next generation of planulae through sexual reproduction. There are, however, large variations in this plan. In the Sciphoza (true jellyfish) and Cubozoa (box jellyfish), the medusa stage dominates, while in the Hydrozoa (hydroids), the polyp stage tends to be the most noticeable. In the Anthozoa (anemones and corals) the medusa stage is lost entirely. In addition, the early development of Cnidaria is variable, with cleavage ranging from radial to irregular, unequal, or superficial, eventually giving rise to a blastula which, in turn, undergoes mixed patterns of ingression, invagination, delamination, and epiboly during gastrulation (Martin, 1997). Such variations make lineage studies difficult to generalize. Given such large differences in development it is not surprising that findings from different species seem contradictory, although broad generalities are emerging.

Neurogenesis commences late in, or shortly after, gastrulation and produces two broad categories of nerve cells. Sensory cells appear to transdifferentiate from epitheliomuscle cells and remain in situ to span the epithelium from the mesogleal surface

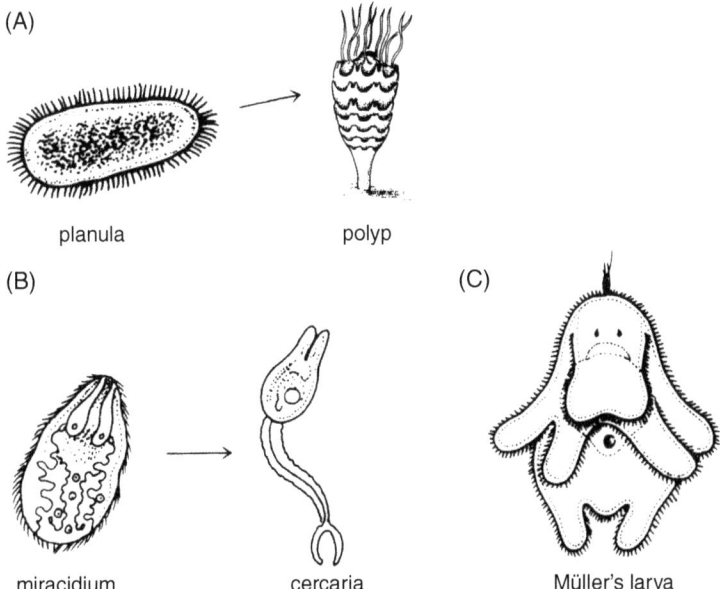

Figure 11.1 Larval Forms of Cnidaria and Flatworms.
(A) The planula larva is typical of the cnidaria, some of which (i.e., the Scyphozoa) also transit through a scyphisoma polyp stage before assuming an adult medusa form. (B) The miracidium and cercaria stages are found in different hosts of parasitic trematode flatworms. (C) The Müller's larva is one of the forms found in polyclad Turbellaria flatworms. Modified from Pechenik 2005. Reproduced with permission of McGraw-Hill Companies, Inc.

to the free surface of the planula, where they bear ciliate or microvillus endings (Martin, 1988; Piraino et al., 2011). Some sensory cells also appear dark and some light under electron microscopy (EM), and a portion contain detectable levels of RFamide neuropeptides (Martin, 1988). Ganglion cells, on the other hand, differentiate from a type of pluripotent stem cell known as the interstitial cell (i-cell), although this progenitor cell type may be missing in some cnidarians (Marlow, Srivastava, Matus, Rokhsar, & Martindale, 2009). Extensive migration may occur before the ganglion cells differentiate to form both the somata and the processes located at the base of the epithelium and constituting an extensive network adjacent to the mesoglea. The ganglion cells thus appear to function as interneurons with connections to sensory cells and other ganglion cells and with possible release of transmitters (e.g., catecholamines, serotonin, taurine, and GABA in addition to RFamide and GLWamide peptides) into the mesoglea. Some processes of the ganglion cells have also been reported to project to the outer surface of the planula, suggesting a possible sensory function for some of these cells as well (Martin, 1988, 1992, 1997).

Although the nervous system of the planula is often described as a diffuse network, an anterior/posterior polarity is commonly reported. For example, Martin (1992) reported concentrations of RFamide containing cells at the anterior (aboral) end, a finding that has been confirmed in other species (Marlow et al., 2009; Piraino et al., 2011). Tightly packed, elongated sensory cells appear to form a discrete apical organ, which may be replete with an apical tuft of cilia in some species (Marlow et al., 2009; Yuan, Nakanishi, Jacobs, & Hartenstein, 2008). Piraino et al. (2011) suggest that polarization may be most extreme in species with planulae that crawl in wormlike fashion rather than swim like most species.

This concentration of peptides at the anterior end of the swimming larvae disappears following settlement and metamorphosis, which involves a fundamental change from swimming to sessile forms with the attachment of the previous anterior end of the larva to the substrate. The previously posterior end of the animal then forms tentacles around the mouth/anus, which in the polyp, as in the planula, is the sole opening into and out of the gastric cavity. The polyp has a nervous system reminiscent of the planula with a diffuse nerve net made of sensory and ganglion cells, although neuronal elements are concentrated into a distinct ring around the mouth (oral ring) and in the tentacles. While neurons can be observed in regions of the endoderm in late stage planulae, they become much more numerous in the polyp.

11.2.2 Ctenophores (Comb Jellies)

The ctenophores have long been recognized as basal metazoans, with body plans and molecular characteristics that provide insight into the first nervous system—nonetheless, in comparison to the cnidaria, they remain poorly studied. It has been estimated that the roughly 200 known species of ctenophores represent only about 25% of the total number in nature, with the majority of undescribed species residing deep in the oceans (Martindale & Henry, 1997) and unsampled by traditional collection through trawling (Pechenik, 2005). Even studying the basic biology of species that dwell near shore is hampered by difficulties in rearing and maintaining specimens in aquaria and by the delicacy of their tissues, which often disintegrate when placed in common fixatives (personal observations). Hence, little is known about the adult nervous system (Jager et al., 2011) and still less about development.

Early cleavage in ctenophores follows neither the spiral nor radial patterns commonly observed in other animals and provides one of the most striking examples of determinate mosaic development, with experimental separation of individual blastomeres at the two cell stage producing two nearly perfect half animals. The eight cell stage consists of four E and four M blastomeres with the first micromeres produced from these cells (the e1 and m1 micromeres) being the source of the nerve net, which begins to form shortly after gastrulation by epiboly (Martindale & Henry, 1997).

The typical larva of the ctenophores, called a "cydippid," has eight comb rows and two tentacles. Larvae possess a nerve net and apical organ similar to the adult, but details are lacking. Cydippid larvae closely resemble the adult form in species such as *Pleurobrachia*, and one could argue that there is no true metamorphosis (Martindale & Henry, 1997). In other species, however, tentacles are lost, lobes grown, or the bodies changed from globular to wormlike. With the paucity of information currently available regarding neural development in this important phylum, the field is clearly wide open for future study.

11.3 Lophotrochozoa

The Lophotrochozoa form the largest clade of protostome animals in terms of numbers of phyla. The name signifies an amalgamation of animal groups characterized by either lophophorate or trochophore larval types (see below); modern molecular evidence also includes Platyzoa. While evidence indicates lophotrochozoans are monophyletic (Philippe, Lartillot, & Brinkmann, 2005), they can differ even with regard to such basic characteristics as early cleavage patterns (i.e., spiral and nonspiral), and relationships between subgroups are matters of constant discussion (Giribet, 2008; Philippe et al., 2005).

11.3.1 Platyzoa

The Platyzoa include the platyhelminthes, gastrotrichs, rotifers and assorted minor phyla, but only the platyhelminths are discussed below. While inroads are being made into understanding the adult nervous systems of other phyla (Hochberg, 2007; Hochberg & Lilley, 2010), little is known about their neural development.

11.3.1.1 Platyhelminthes (flatworms). Because of their simple body plan, together with other special features such as the lack of a coelum, the platyhelminthes had long been considered as a basal sister group to the rest of the Bilateria. However, molecular studies and re-evaluations of early cleavage patterns now favor a position for this group within the Lophotrochozoa (Ellis & Fausto-Sterling, 1997; Giribet, 2008; Rawlinson, 2010). Hypotheses regarding the form of the ancestral flatworm larvae are complicated by phylogenetic uncertainty (Egger et al., 2009; Rawlinson, 2010). Additionally, life histories of the different groups of flatworm are extremely complicated, with three of the four classes showing dramatic adaptations to parasitic lifestyles. For example, many trematodes (flukes) and cestodes (tapeworms) possess sequential larval forms corresponding to different animal hosts (Figure 11.1B). Even within the turbellarians, which have free-living species and appear to best represent the ancestral forms, development can be highly variable; some species hatch as planktonic larvae (for example, as eight-lobed Müller's larvae (Figure 11.1C), four-lobed Götte's larvae, or larvae with other numbers of lobes), others retain features of larval forms embryonically, but hatch as juvenile benthic worms, while yet others manifest few if any larval characteristics as embryos before hatching as juveniles.

A final complication hindering comparisons of ontogeny of different flatworms lies in the special characteristics of the eggs and corresponding cleavage patterns (Ellis & Fausto-Sterling, 1997). One group of flatworms (archoophoran) produces oocytes which show early spiralian cleavage (Boyer, Henry, & Martindale, 1998; Younossi-Hartenstein, Jones, & Hartenstein, 2001) but are surrounded by numerous yolk-rich nurse cells. The dividing embryonic blastomeres eventually form a consolidated mass of cells which moves toward the surface of the embryo. At this point, cells in deep layers become postmitotic neural and muscle precursors and begin to differentiate. Cells at the surface form an epithelium which then differentiates into the epidermis. Thus, in the neoophorans, the formation of separate epithelial germ layers through gastrulation, which characterize the ontogeny of most other animals groups (including other flatworms), is absent. Despite these differences between archoophoran and neoophoran flatworms, Younossi-Hartenstein et al. (2001) point out that neurogenesis in both of these groups shares the feature that progenitors of central neurons are not derived from a pre-existing neurectoderm (see below), but instead are internalized before commitment to a neural fate. Both developmental programs produce similar embryonic nervous systems.

The central nervous system of the embryo and larva consists of an anterior brain that can include between 30 and 50 neurons, depending on the species and presumably related to lifestyle demands upon the young organism (Younossi-Hartenstein & Hartenstein, 2000). Eyes are often situated directly on the brain. Axons from neurons in the brain project posteriorly in 2–3 pairs of connectives. Contralaterally projecting axons form the regularly spaced commissures, which produces the central nervous system characteristic of flatworms. Some fibers within the connectives of the central nervous system contain serotonin or FMRFamide immunoreactivity in a free-living turbellarian larva, but other fibers (labeled with antibodies against acetylated tubulin) apparently contain neither of these transmitters. Indeed, several histological and immunocytochemical studies have revealed extensive larval nervous systems in parasitic flatworms by targeting neurotransmitters such as acetylcholine, monoamines, nitric oxide and a variety of peptides (Gustafsson & Terenina, 2003; Kemmerling, Cabrera, Campos, Inestrosa, & Galanti, 2006).

In addition to the anterior brain and posterior connectives which characterize all flatworm larvae and juveniles, the free-living forms also possess a peripheral nervous system. Lacalli (1982, 1983) used electron microscopy to first describe the detailed ultrastructure of a Müller's larva and identify the regularly spaced sensory neurons with axons closely associated with the various ciliary bands which generate locomotor drive and feeding currents for the larvae. Lacalli emphasized the independent natures of the intraepithelial peripheral nervous system and the subepithelial central nervous system, a finding that has largely been confirmed by anatomical studies on other flatworm larvae (Rawlinson, 2010). Lacalli's suggestion that the peripheral sensory neurons differentiate *in situ* from the surrounding epithelium has also been confirmed (Younossi-Hartenstein et al., 2000; Younossi-Hartenstein & Hartenstein, 2000), as has the suggestion that the apical organ in free-living larvae appears to be composed of central ciliated cells surrounded by flask-shaped glandular cells, and is thus nonneural, although it may still mediate a sensory function through innervation from the brain (Rawlinson, 2010; Younossi-Hartenstein, Ehlers, & Hartenstein, 2000; Younossi-Hartenstein & Hartenstein, 2000).

Metamorphosis in flatworms is gradual with the resorption of lobes and the apparent loss of the intraepithelial sensory neurons and apical organ (Rawlinson, 2010). The number of posterior connectives also generally increases to yield the nervous system typical of the adult flatworm (see Chapter 8).

11.3.2 Lophophorata and Endoprocta

Problematic phylogenetic relationships of many invertebrate groups impede our understanding of the evolution of the nervous system throughout the animal kingdom. The lophophorates have been particularly difficult. These animals have been grouped together based primarily on morphological features of the adults and larvae. However, this group exhibits a bewildering array of embryological and morphological characteristics that have defied even broad categorization into protostomes or deuterostomes. For example, zygote cleavage has long been described as basically radial, and their regulated development is typical of deuterostomes; likewise, Nielsen (1987) has argued that the position and beat direction of cilia generating feeding currents are characteristic of deuterostomes. In sharp contrast, the fact that the mouth of phoronids originates from the blastopore—the defining characteristic of protostomes—argues for inclusion in the latter category. Molecular evidence now places the Lophophorata not only within the protostomes, but close to other Lophotrochozoa (Helmkampf, Bruchhaus, & Hausdorf, 2008), although these findings leave unresolved the mismatch between genetic and morphological/developmental characteristics.

11.3.2.1 Phoronids (horseshoe worms). Cleavage of the phoronids is biradial and by the 16- or 32-cell stage a small blastocoel has already formed. Gastulation occurs by invagination of the vegetal pole thus resulting in the formation of the germ layers which give rise to the actinotroch larva (Figure 11.2A) typical of the phylum (Zimmer, 1997). Many of the basic features of the larval nervous system were described using light microscopy from the end of 19th century through the middle of the 20th century (reviewed by (Hay-Schmidt, 1989). Details of the ultrastructure of the larval nervous system were provided at the EM level by Hay-Schmidt, (1989) and Lacalli (1990) and subsequent immunocytochemical studies provided visualization of discrete populations of neurons containing catecholamines, serotonin or FMRFamide-related peptides (Hay-Schmidt, 1990a, 1990b, 1990c; Santagata & Zimmer, 2002). Those descriptions of the larval nervous system emphasized its intraepithelial nature, thus suggesting that early neurogenesis involves only the differentiation of neurons from surrounding epithelial cells without the need for internalization of neural progenitor cells, as often seen later in other animals.

Some differences occur between species in details of the larval nervous system (Hay-Schmidt, 1989; Lacalli, 1990; Santagata & Zimmer, 2000), but, generally, cell bodies of the nervous system reside within or surrounding the apical ganglion. Up to four types of ganglion cells, including putative sensory cells and neurons, can be identified based on ultrastructural criteria. Serotonergic cells appear early in development, forming a U- or V-shaped configuration around the central core of neuropile.

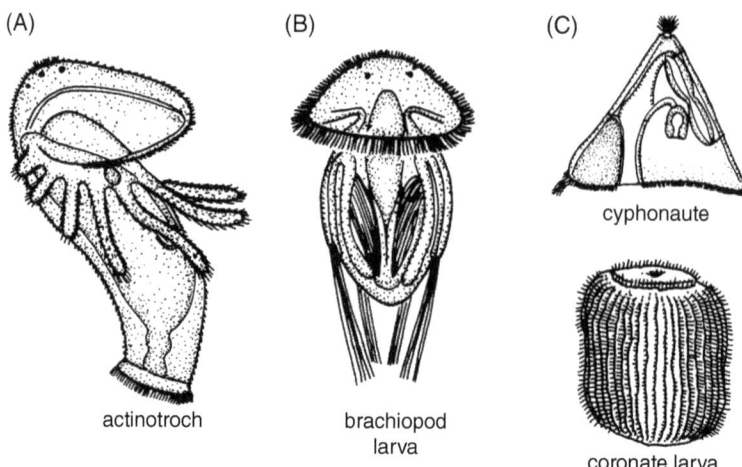

Figure 11.2 Larval Forms of Lophophores.
(A) Actinotroch larvae are typical of the phonorids. (B) Brachiopod also possess a unique larval form. (C) Both cyphonaute and coronate larvae are found in the Bryozoa. Modified from Pechenik 2005. Reproduced with permission of McGraw-Hill Companies, Inc.

Later-appearing cells can occupy more central positions. Several catecholamine-containing cells also appear early and encircle the neurons containing serotonin. Fibers from all these cells contribute axons to the main nerves, which emanate from the apical organ to innervate the rest of the surrounding epistome and from there the rest of the larva and especially the tentacles with their ciliary bands. FMRFamide-related peptides are also abundant in the nerves, although somata are only sparsely labeled immunocytochemically. Rows of specialized epithelial cells that were innervated and presumed to be sensory were also found along the tentacles (Hay-Schmidt, 1989). As the larva reaches competence for metamorphosis, a secondary neuropile forms under the hood organ which also contains putative sensory cells.

Metamorphosis is striking and profound in phoronids, whereby the entire larva turns inside out, apparently through muscular contractions orchestrated by the nervous system (Santagata, 2002). The cells of the apical ganglion undergo cell death, muscles disappear and the gut and tentacles are remodeled in ways that vary between species to produce the adult form.

11.3.2.2 Brachiopods (lamp shells). The brachiopods are generally thought to be closely related to the phoronids and accordingly their early development is very similar, although since the blastopore closes completely during development, its relationship to the mouth is unclear. The articulate brachiopods produce simple lobed larvae, which do not feed and exist for only a few hours before settlement, yet nonetheless may possess complex musculature and reflexes. These larvae have eye spots and respond specifically to environmental cues for settlement (Zimmer, 1997). As the larvae become competent to metamorphose, they develop prominent flask-shaped and presumably sensory serotinergic cells in the apical organ; these disappear during metamorphosis (Altenburger & Wanninger, 2010). The inarticulate brachiopods, on the other hand, have larvae (Figure 11.2B) which can swim and feed for several weeks before settlement using cilia located along well-developed tentacles. The larval nervous systems of *Lingula* and *Glottidia* have been described by Hay-Schmidt, (1992) using EM and immunocytochemistry, and consist of a "ventral" division comprising both the apical organ located within a medial tentacle (Luter, 1996) and the innervation to the lophophore and a "dorsal" division which comprises both the ventral ganglion and the innervation of the body musculature associated with the shells unique to this taxon. The former division compares to the larval nervous system seen in phoronids, whereas the latter is suggested to have evolved only within the brachiopods (Hay-Schmidt, 1992).

11.3.2.3 Bryozoa (ectoprocta, or moss animals). The bryozoans are a strange group of animals which, nonetheless, share enough morphological, developmental, and molecular similarities to other Lophophorata to classify them as distant relatives to phoronids and brachiopods (Hausdorf, Helmkampf, Nesnidal, & Bruchhaus, 2010).

Bryozoans have biradial cleavage, and by around the 64-cells stage they gastrulate through the internalization of the vegetal tier of cells. The tier of eight cells at the animal pole eventually produces the neurons of the apical disc. Some bryozoans, however, exhibit a very unusual form of reproduction (polyembryony), which is poorly studied in its early stages but which involves a single zygote producing several (hundreds of) larvae. More usually, byozoans produce either nearly spherical coronate larvae, which are brooded and hence are short-lived and nonfeeding, or triangular, shelled cyphonaute larvae (Figure 11.2C) which can swim and feed in the water column for several weeks before settlement (Zimmer, 1997).

The larval nervous system has now been studied in several species, and demonstrates numerous similarities over a wide variety of body forms (Santagata, 2008). One consistent component is the numerous neurons in the neural plate in the center of the apical disc and several more cells in surrounding regions. Unlike many other invertebrate larvae, there is no neuropil underlying the apical cells (but see Wanninger, Koop, & Degnan, 2005); instead, the axons from these cells project to a nerve nodule underlying the pyriform organ near the extreme anterior end of the larva. Nerves projecting from the nexus also connect to other regions of the body and can thus provide innervation to the ocelli, balancer cells, and various other putative sensory organs. Linking form to function, Santagata (2008) noted that innervation of regions involved in feeding were more developed in planktotrophic larvae while the fields of sensory cells in the apical organ were more developed in lecitrophic (yolk-feeding) larvae. Pires and Woollacott (1997) report the coronate larvae of *Bugala* are positively phototactic when they are first released but then become negatively phototactic as they prepare to settle: Dopamine, which is detectable in the larvae, prolongs the period of positive phototaxis; serotonin, present in cells in the apical organ and along the edge of the ciliated corona (Gruhl, 2009), causes a rapid shift to negative phototaxis.

The processes mediating metamorphosis vary greatly, as might be expected by the range of larval and adult forms in this phylum, but can involve eversion, inversion, or degeneration of major body regions, including the larval nervous system. While dramatic, these processes must be viewed in a broader context of the rest of the life cycle. The larva metamorphoses into an ancestrula, which founds (by budding) a colony of hundreds or thousands of individuals. The ways that identical adult nervous systems can be derived from either sexual or asexual reproduction are unknown, but are encountered in colonial animals (e.g., corals and colonial ascidians) from other phyla.

One additional and remarkable feature of the bryozoan life history warrants mention. The entire body, except for the thin outer wall of individual bryozoans (or "cystid"), periodically degenerates and is expelled as a "brown body." A new inner body (the polypide), including the tentacles, gut, musculature, and nervous system, is subsequently regenerated. This process, called polypide replacement, occurs either in response to unfavorable environmental conditions or due to accumulating metabolic wastes in animals that lack nephridia. As mentioned by Zimmer (1997), the ability of bryozoans to regenerate all body tissues from just a few cell types challenges long-held concepts of triploblastic germ layer specification through gastrulation.

11.3.2.4 Entoprocta (Kamptozoa). Like the lophophorates, entoprocts are characterized by a ring of ciliated tentacles which surround the mouth and are used for both feeding and respiration. Entoprocts are distinct in that their anus lies within this ring, and that the direction of water currents generated by ciliary beating is opposite to that in lophophorates. Molecular evidence has been limited, since these animals are not often represented in large-scale phylogenetic analyses (but see Hausdorf et al., 2010). Hence, these animals are sometimes hypothesized to have closest affinities with bryozoans or molluscs, with much evidence based on morphology and development.

Distinctly unlike the lophophorates, the cleavage of entoprocts is spiral and determinate. Moreover, the swimming larvae of some entoprocts have a trochophore-like form. Hay-Schmidt, (2000) provided an initial description of the larval nervous system of an entoproct, but more complete descriptions were later provided by Wanninger, Fuchs, & Haszprunar (2007), who examined developing serotonin-like immunoreactivity and Haszprunar and Wanninger (2008), who employed EM. Both studies reported larval nervous systems with similarities to basal molluscs. Specifically, the creeping-type larva of *Loxosomella* was reported to have a complex apical organ containing about eight central and an equal number of peripheral serotonergic neurons. Two ventral and two lateral nerve cords projecting posteriorly from the cerebral ganglion were also reported, as seen in some molluscs and polychaetes. Another similarity was a prototrochal nerve ring, although it was pointed out that the arrangement of ciliary bands on molluscs and entoprocts did not support direct homologies. These types of similarities were sufficient for both studies to suggest strong affinities between the molluscs and the entoprocts.

Many entoprocts, like the bryozoans, form colonies through asexual budding, and Wanninger, Fuchs, Bright, & Funch (2006) reported that neurons expressed serotonin and FMRFamide like immunoreactivity very early during bud formation. The fact that no connections were detected between the nervous systems of the adult and the bud suggests the new nervous system differentiates independently, but further details are lacking.

11.3.3 Trochozoa

This taxa's characteristic trochophore larva (see Figure 11.3) possesses several ciliary bands and an apical organ associated with its own ciliary tuft. Occasional eye spots and tentacles can be found in the episphere, which is demarked caudally by the prototroch, a ciliary band which generates locomotion. The caudal end of the larva (the hyposphere) contains another ciliary band, the metatroch, a ventral mouth between the prototroch and the metatroch, and a caudal-most anus, often associated with a third

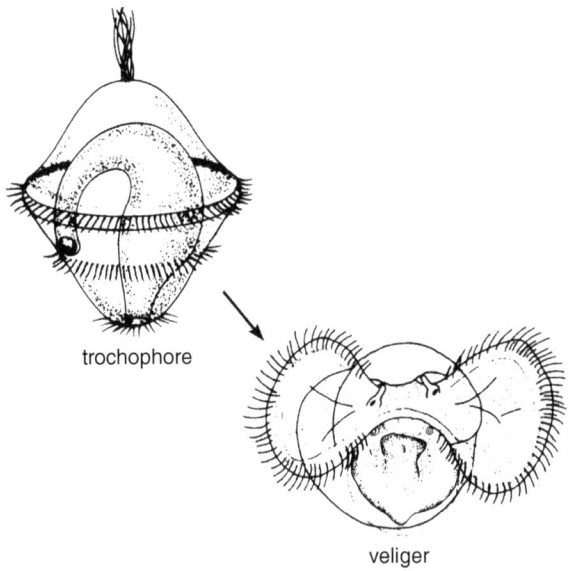

Figure 11.3 The Trochophore Larva Characteristic of the Trochozoa.
Some molluscs, such as the polyplacophora (chitons) and Scaphopoda (tusk shells) metamorphose from the trochophore into a juvenile stage while the gastropods and bivalves generally have an intermediary veliger stage. Polychaete annelids also have an elongated late larval stage referred to as a metatrochophore. Modified from Pechenik 2005. Reproduced with permission of McGraw-Hill Companies, Inc.

ciliary band, the telotroch. Some trochozoans (e.g., the polychaete annelids) also possess an additional longitudinal band of cilia, the neurotroch, along what will become the ventral surface. Many trochozoans metamorphose directly from the trochophore into a juvenile form, while others display intermediate forms, such as the veligers (Figure 11.3) of gastropods and bivalves or the elongated and already-segmented metatrochophore of the polychaetes. The similarity of larval structures and development across this group, and particularly between the annelids and molluscs, have generated interest in possible common functions of organs (e.g. Voronezhskaya & Khabarova, 2003; Nielsen, 2005) and developmental mechanisms (Voronezhskaya & Ivashkin, 2010).

11.3.3.1 Annelids (segmented worms). Neural development has been studied extensively in two groups of annelids, the leeches and the polychaetes, which differ profoundly from each other in several aspects of ontogeny.

The leeches are a highly derived group of annelids which develop within their egg capsules to hatch as miniature, sexually immature versions of the adults. They have thus lost the larval trochophore stage. The large and identifiable cells of the adult CNS, however, have been widely studied and offer well-defined endpoints for developmental investigations. Thus leeches illustrate the development of single cells from their first specification as neurons to the point that they attain their adult phenotypes with stereotyped positions, sizes, transmitter and receptor complements, and axonal morphologies. Furthermore, the development of behavior can be studied in terms of the assemblages of mediating circuitry. These advantages served as incentives propelling leeches into being among the earliest modern subjects of neural development (Stent, Kristan, Torrence, French, & Weisblat, 1992).

As in other annelids, cleavage in leeches is highly stereotyped (Stent et al., 1992). After three cleavages the zygote has produced four macromeres (A,' B,' C,' D') and the first quartet of micromeres (a,' b', c,' d'). The anterior supraesophageal ganglion is derived from this and other early quartets of micromeres. Subsequent divisions of the D' macromere produce bilateral pairs of large stem cells which are referred to as M, N, O, P, (or two roughly equivalent O/P) and Q teloblasts. Asymmetric divisions of these teloblasts produce longitudinal bandlets of n, o, p, and q ectodermal blast cells which constitute the germinal plate and eventually form the ventral nerve cord. Underlying mesodermal m blast cells primarily become muscles. Intracellular dye injections have demonstrated that all four ectodermal teloblasts, and even the mesodermal teloblast, contribute to the segmental ganglia of the ventral nerve cord. The majority of neurons in the segmental ganglia derive from the n bandlet, while the o, p, and q bandlets produced greater proportions of peripheral neurons. The relationship between developmental ancestry and different neuronal attributes is complicated, however, and Stent et al. (1992) concluded that, despite substantial migration of neuroblasts and immature neurons, ancestry correlated best with final positions of neurons in the nervous system and not with other neuronal attributes, which appear to be determined by cell-to-cell interactions.

While the development of the adult annelid nervous system (e.g., anterior ganglia and ventral nerve cord) has been best studied in the leeches, recent research adds a comparative perspective by also describing neural development in the polychaetes. Some information is complementary. For example, Denes et al. (2007) described cellular events and gene expression patterns during the formation of the ventral nerve cord from a neurectoderm in *Playnereis* and described many similarities with development of CNS in vertebrates. Meyer and Seavers (2009) provided further details of neurogenesis by focusing on cellular events during neural differentiation and internalization of neuronal precursors that form the anterior brain (cerebral ganglia) of another polychaete, *Capitella*.

However, in contrast to the leeches, the polychaetes have a distinct trochophore stage and a larval nervous system which permits a free-living mode of existence at this early period of development. Lacalli (1984) made pioneering EM studies and described the larval nervous system as appearing much earlier than the first rudiments of the adult brain and nerve cord and as being composed of two parts. The pretrochal component comprises the apical organ and a few other neurons in the episphere and innervates the prototroch, the anteriormost of the locomotor ciliary bands. The apical organ itself was described as comprising about 16 neuronal and nonneuronal cells, most of which possessed ciliary surface specializations. The second component comprises neurons innervating the mouth regions and metatroch. Lacalli (1984) emphasized that this larval nervous system was largely independent of the later-developing adult nervous system, although he also noted that the cerebral commissure of the adult nervous system became intimately associated with the apical organ and that the apical organ may play a role in organizing the development of the commissural fibers.

More recent immunocytochemical studies have revealed additional cells, added details to the earlier descriptions, and showed that many of the larval neurons exhibit serotonin or FMRFamide-like immunoreactivity. These studies have permitted a fuller view of the developmental sequence in the assemblage of the larval nervous system, but have also revealed needs for additional studies of underlying developmental processes. For instance Voronezhskaya, Tsitrin, & Nezlin (2003) and McDougall, Chen, Shimeld, & Ferrier (2006) demonstrated that the very first neurons to develop in the larvae of different species of polychaetes were not located in the apical region, as would have been consistent with earlier work on other trochophore larvae, but at the posterior end of the larva. The functions of these serotonin-containing posterior cells, and numerous other peripheral neurons, are generally unknown and warrant further investigation. In addition, Hay-Schmidt, (1995) viewed later neural development as a transformation of the larval nervous system into the adult nervous system, whereas Voronezhskaya et al. (2003) argued that certain larval neurons formed a framework or scaffolding of pioneering fibers upon which the adult nervous system was built, before disappearing with the rest of the larval nervous system at metamorphosis. Another important contribution of later immunocytochemical studies is that they have examined a diversity of species and demonstrate major differences in details of early neuronal development in the polychaetes, supporting the notion that the polychaetes are themselves diverse and perhaps paraphyletic (Brinkmann & Wanninger, 2009).

Although details differ, studies on larval polychaete nervous systems have consistently revealed a surprisingly complex and extensive nervous system in what was once viewed as a simple larval form. These studies have also provided information on neural circuitry underlying larval behaviors such as phototaxis (Jékely et al., 2008) and vertical migrations in the water column (Conzelmann et al., 2011) and might therefore additionally suggest potential functions of the larval nervous systems in other phyla of Trochozoa.

Near the time of metamorphosis, the trochophore or elongated metatrochophore larvae of polychaetes contains a brain (the cerebral or supraesophageal ganglion, which, as in the leech, derives from early micromeres, primarily the first quartet) and often 1–3 ganglia of nerve cord (derived primarily from later divisions of the D blastomere). After metamorphosis, the body elongates by sequential additions of posterior segments during postlarval life. The timing of neurogenesis in the nerve cord is thus very different between the leech and polychaetes. Specifically, the germinal plate is laid down by continuous and fairly rapid divisions of the blast cells derived from the D blastomere in the embryonic leech. In polychaetes, however, the segmental ganglia are laid down one by one over a more extended time, and are derived from teloblasts which remain active in a proliferative zone, pygidium, near the anus.

11.3.3.2 Echiura (spoon worms). Echiura constitute a small taxon of only about 150 species which burrow into the mud in shallow marine habitats. Their common name derives from the flattened proboscis with which they feed. The echiurans have generally either been classified as a separate phylum closely related to the annelids or as a subtaxon within the annelids. Recent molecular evidence favors the latter relationship (Struck et al., 2007).

Cleavage is spiral in echiurans, and an annelid cross (rotated 45° from the typical cross of molluscan and sipunculan development) forms around the apical rosette by the 48-cell stage. Gastrulation occurs either by proliferation of animal micromeres and epiboly or by invagination of the vegetal plate. As with annelids, the trochophore larvae of echiurans possess bands of cilia arranged into a prototroch and telotroch, and also neurotroch and metatroch in some species (Pilger, 1997).

Different species of echiurans possess early larval nervous systems associated with the pharynx and the various ciliary bands (Hessling, 2002; Hessling & Westheide, 2002). No evidence for an apical organ was reported, and the first neurons appeared in circumesophageal ganglia. The focus of these studies was the detailed description of ventral nerve cord development, since the lack of demonstrated ventral nerve cord segmentation had been a major argument against including echiruans among annelids. Hessling (2002) and Hessling and Westheide (2002) exploited immunocytochemistry and confocal microscopy to show ventral nerve cord segmentation in these animals during larval stages with serial reiteration of serotonergic and FMRFamidergic elements and tubulin-rich axonal tracks. Furthermore, as in annelids, the nerve cord is initially a bilaterally paired structure which develops in an anterior-to-posterior sequence. These features are lost or obscured in the adult (Pilger, 1997; Hessling, 2002).

A final noteworthy feature of echiuan development involves the profound sexual dimorphism that occurs in some species (Pilger, 1997): The adult males in these animals live inside the reproductive tracts of the females, upon which they depend for nutrition and protection. Accordingly, post-metamorphic development of females is associated with an enlarged proboscis and trunk and corresponding changes in the nervous system. In contrast, post-metamorphic development of males appears to involve only elongation of the trunk, without corresponding changes to the nervous system (Hessling & Westheide, 2002).

11.3.3.3 Sipuncula (peanut worms). Sipunculans are wormlike animals that have variously been aligned with either annelids or molluscs. Cleavage is spiral, and when the embryos reach the 48-cell stage, their apical plate forms a molluscan cross pattern, even though recent molecular evidence favors a closer relationship with annelids (Wanninger, Koop, Bromham, Noonan, & Degnan, 2005). The fates of the early blast cells appear consistent with those of other spirilians (Pilger, 1997). Gastrulation can occur through the processes of epiboly, invagination or both. Although some direct developing species lack any larval form, most sipunculans develop through either a single trochophore phase or a trochophore followed by an prolonged pelagophera stage (Rice, 1975).

Wanninger, Koop, Bromham, et al. (2005) originally reported that one sipunculan, *Phascolion*, possessed no serotonergic neurons in the larval nervous system and no obvious signs of segmentation in either the early nervous system or musculature. That species, however, has an unusually reduced larval stage; examining *Phascolosoma*, Kristof, Wollesen, & Wanninger (2008) found an extensive larval nervous system innervating the prototroch, apical organ, and mouth region, in addition to the anlage of the adult brain. In contrast to the earlier study, they also reported that the early larval nervous system exhibited a metameric organization with repeated FMRFamide and serotonin immunoreactivity in both neuronal cell bodies and commissures, thus resembling neural development in annelids; in late-stage larvae, many neurons disappear or migrate, and these signs of segmentation are lost. During metamorphosis, which can last several weeks, the metatrochal cilia are lost. Lacking propulsive force for locomotion, the larva sinks to the bottom and assumes the benthic life of the juvenile and adult. The basic features of the peptidergic component of the nerve cord survive metamorphosis and serve as a foundation for subsequent postlarval development (Wanninger, Koop, Bomham et al., 2005). Notably, even though the early larval nervous system of certain sipunculans exhibit metameric organization, the adult peanut worm shows no sign of segmentation.

11.3.3.4 Molluscs. Molluscs are the second largest phylum in the animal kingdom in terms of number of species, and their possession of a classical trochophore larva (see Figure 11.3) places them firmly within the Lophotrochozoa *sensu stricto*. Their spiral cleavage pattern of development has been studied for over a century (Conklin, 1897) and, like other spiralians, they derive their anterior nervous system (larval apical and adult cerebral ganglia) from the first quartets of micromeres, and the more posterior portions of the nervous system develop with at least some contributions from the 2D blastomere. Recently, however, Hejnol, Martindale, and Henry (2007) have added important details and corrections to the earlier fate maps, showing which specific cells of the apical organ arise from individual micromeres and demonstrating that the second quartet of micromeres also contributes components to the posterior portions of the adult nervous system.

An extensive literature describes the larval nervous system across a wide variety of molluscs. The cells of the apical organ are among the first neural elements to be detected during development, appearing shortly after gastrulation. Detailed morphology of various cells of the apical organ were first described using EM (Bonar, 1978; Chia & Koss, 1984; Marois & Carew, 1997a; Page & Parries, 2000; Page, 2002) and more recently using immunocytochemistry. The number and arrangement of vase-shaped, presumably sensory, apical cells are similar across the molluscs. For example, three serotonergic cells often first appear as the larvae enter the trochophore stage (Croll, Jackson, & Voronezhskaya, 1997; Kempf, Page, & Pires, 1997; Marois & Carew, 1997a, 1997b; Page & Parries, 2000; Voronezhskaya, Nezlin, Odintsova, Plummer, & Croll, 2008) while many gastropods, bivalves, scaphopods and polyplacophorans have additional vase-shaped cells which contain catecholamines, neuropeptides and nitric oxide synthase (Croll, 2006; Croll & Voronezhskaya, 1996; Dickinson & Croll, 2003; Hens, Fowler, & Leise, 2006; Kempf, Chun, & Hadfield, 1992; Voronezhskaya & Elekes, 2003; Voronezhskaya, Hiripi, Elekes, & Croll, 1999). The apical organs of molluscs also commonly contain cells that have been suggested to be interneurons (Croll, 2006; Friedrich, Wanninger, Bruckner, & Haszprunar, 2002; Kempf et al., 1997; Voronezhskaya, Tyurin, & Nezlin, 2002).

Neurites projecting from the apical cells end in underlying neuropil, or project to peripheral regions of the veliger. Serotonergic varicosities lie in close proximity to ciliated cells of the velum (Croll, 2006; Croll et al., 1997; Dickinson & Croll, 2003; Marois & Carew, 1997c; Voronezhskaya, Nezlin, Odintsova, & Plummer, 2008) and Braubach, Dickinson, Evans, and Croll (2006) showed that serotonergic input enhances ciliary beating on the velum, thus increasing both locomotion and feeding of gastropod larvae. The apical organ has also been suggested to be involved in the transduction of an environmental signal which initiates metamorphosis (Couper & Leise, 1996; Froggett & Leise, 1999; Hadfield, Meleshkevitch, & Boudko, 2000; Hens et al., 2006; Leise & Hadfield, 2000), although other evidence argues for alternative or additional roles for this structure (Kuang, Doran, Wilson, Goss, & Goldberg, 2002; Voronezhskaya, Khabarova, & Nezlin, 2004; Wanninger & Haszprunar, 2003).

Posterior neurons immunoreactive for FMRFamide can also be observed in the early trochophore larvae and have axons that project anteriorly to the apical organ

(Croll & Voronezhskaya, 1996; Dickinson, Nason, & Croll, 1999; Dickinson et al., 2000; Dickinson & Croll, 2003). Although such cells appear to be a general feature of gastropod larvae (Cummins, Tollenaere, Degnan, & Croll, 2011), their presence in other molluscs is less consistent (Friedrich et al., 2002; Voronezhskaya et al., 2002; Wanninger & Haszprunar, 2003), and lateral, pretrochal cells might play analogous roles in bivalve larvae (Voronezhskaya et al., 2008). One such role could be to innervate larval muscles (Dyachuk & Odintsova, 2009; Evans, Dickinson, & Croll, 2009; Page, 1997; Wanninger et al., 1999). Indeed, Braubach et al. (2006) reported that both applications of FMRFamide increased the frequency of velar contractions. Another potential role for the early peptidergic cells involves the pioneering of pathways to scaffold adult nervous system development (Croll & Voronezhskaya, 1996; Voronezhskaya & Ivashkin, 2010).

A final category of neurons known to develop early in molluscan ontogeny are the catecholaminergic cells that appear around the mouth of both gastropod and bivalve larvae (Croll, Boudko, & Hadfield, 2001; Croll et al., 1997; Dickinson & Croll, 2003; Dickinson, Croll, & Voronezhskaya, 2000; Voronezhskaya et al., 1999), perhaps suggesting control of different aspects of feeding. Many catecholamine-containing cells and axons have also been found associated with ciliary bands along the velum and may have an inhibitory effect on locomotion and feeding (Beiras & Widdows, 1995; Braubach et al., 2006).

During the mid-larval stage, many populations of neurons continue to grow in number and complexity, but this period is also marked by the first appearance of structures that persist into adulthood (e.g., eyes, tentacles, foot, and various retractor muscles) and of the ganglia that form the adult central nervous system. Mid-larval life thus marks a time when the early larval and the developing adult nervous systems must operate in concert to produce the coordinated final output upon which the free-living larvae depend for survival. Several previous investigations have focused on the development of the ganglia to understand the origins of identifiable neurons in the adult. Such studies used EM and radioactive birth-dating techniques (Jacob, 1984; Kandel, Kriegstein, & Schacher, 1981; Schacher, Kandel, & Woolley, 1979) to confirm reports that ganglia derive from specific proliferative zones in the body wall (Raven, 1966). After the initial ganglionic anlagen separate from the proliferative zones, they continue to generate postmitotic neurons that migrate inward to the developing ganglia (McAllister, Scheller, Kandel, & Axel, 1983).

During metamorphosis, in preparation for the adult's benthic lifestyle, the bands of cilia used for larval locomotion and feeding and the nerve cells that innervate them are lost. The posterior peptidergic neurons and the apical organ disappear during or soon after metamorphosis in numerous species (Dickinson & Croll, 2003; Lin & Leise, 1996; Marois & Carew, 1997b), as do the posterior, peptidergic cells (Croll & Voronezhskaya, 1996; Dickinson & Croll, 2003), apparently via programmed cell death (Gifondorwa & Leise, 2006; Marois & Carew, 1997b). However, metamorphosis also involves the gain of structures and neurons, such as the buccal muscles and ganglia used by adult gastropods for feeding. Finally, in addition to dramatic losses and gains of entire organs and their innervation, metamorphosis also involves numerous changes in the form and functions of organs, such as the foot, which initially serves merely as an anchor for the operculum and the insertion point for large muscles. By late larval stages, it has gained a sensory role, with swimming larvae "sampling" the substrate with their foot in preparation for settlement. With the completion of metamorphosis, the foot becomes the primary organ for locomotion in gastropods. All these changes must be accompanied by changes in the nervous system.

Alert readers will note that, while the discussions above cover a wide variety of molluscs, the cephalopods have been omitted. Comparisons are difficult. The large yolk distorts early developmental processes seen in other molluscs, permitting cephalopods to develop directly inside the egg capsule, and resulting in a juvenile (paralarval) stage hatchling which lacks characteristically molluscan larval features. Furthermore, the large yolk also impedes detailed histology of earlier stages. Nonetheless, significant progress is now being made: Shigeno, Tsuchiya, and Segawa (2001) reviewed the scattered early literature on cephalopod neural development and provided new information on the squid, *Sepioteuthis*. Baratte and Bonnaud (2009) subsequently demonstrated that peripheral sensory cells which appear to contain catecholamines, and which may therefore be homologous to similar cells in other molluscs, are detectable in early stages of brain development. Other recent studies have examined the development of both peptidergic (FMRFamide) and serotonergic cells in cephalopods (Aroua, Andouche, Martin, Baratte, & Bonnaud, 2011; Wollesen, Cummins, Degnan, & Wanninger, 2010; Wollesen, Degnan, & Wallinger 2010).

A final aspect of neural development in molluscs involves their tremendous postlarval growth. For example, large sea slugs can show a dramatic 10^7-fold increase in volume between metamorphosis and adulthood (Croll, 2009). Some bivalves, like the giant clam, attain gigantic proportions through postlarval growth; the giant squid's postlarval growth is proportionately even greater. Obviously, the nervous system must change to accommodate this growth (cf. Chapter 7). Gastropods appear to have their full, or nearly full, complement of large identifiable neuronal somata early in their juvenile phase, but these cells grow disproportionately, becoming increasingly differentiated from neighboring somata (Croll & Chiasson, 1989); these giant cells also exhibit increasing polyploidy (Chase & Tolloczko, 1987). The large size of these cells presumably supports the increased metabolic demands of expanded axons innervating expanding target fields (Gillette, 1991). In addition, the nervous systems of gastropods also possess clusters of smaller cells which grow more modestly in diameter during development, but which also increase in numbers (Croll & Chiasson, 1989). The degree to which changes in cell size and cell number contribute to the growth of the nervous system of molluscs that do not possess giant identifiable cells is unknown. Finally, the postlarval nervous system must not only increase in size to accommodate growing target tissues, but as the mollusc reaches sexual maturity, new organs appear and new behaviors must be generated, often with the addition of new neurons (Croll & Chiasson, 1989; McAllister et al., 1983).

11.4 Ecdysozoa

The Ecdysozoa comprise only eight phyla but contain far more species than the Lophotrochozoa and Deuterostomes combined (Telford, Bourlat, Economou, Papillon, & Rota-Stabelli, 2008). They share the major characteristic of molting their cuticle during development, but also share several other features such as a general lack of locomotor cilia or primary larval stages. Despite the large number of species, all Ecdysozoa possess only one of two basic adult body plans. They either have a worm-like body with an anterior introvert or proboscis and terminal mouth (the Nematoda,

Nematomopha, Priapulida, Kinohyncha, Loricifera) or segmented body and appendages (the Arthropoda, Tarigrada, Onychophora). Recent evidence suggests that the Ecdysozoa form a monophyletic group (Dunn et al., 2008; Telford et al., 2008). This review will focus on the best known examples of each body type in terms of nematodes or arthropods. As pointed out by Telford et al. (2008), the two characteristic body types of Ecdysozoa happen to be represented by two of the best studied organisms in the entire animal kingdom: the nematode *Caenorhabditis elegans* and the fruit fly *Drosophila melanogaster*.

11.4.1 Nematodes (Round Worms)

The position of the nematodes in phylogeny has long been problematic. Although both molecular and morphological evidence now places them within the Ecdysozoa (Dunn et al., 2008; Telford et al., 2008), their exact relationship within this group remains unresolved (Dorris, De Ley, & Blaxter, 1999). The parasitic nature of many nematodes, amply demonstrated in text books with graphic depictions of heart, hook, ring, and guinea worm infestations, has both shaped the evolution of the phylum in general and driven much early study. For example, a long-standing "model nematode" was an ascarid intestinal parasite, which contributed much of what we know regarding early cleavage in round worms (Schierenberg, 1997). This perspective is often lost in the present day, with the domination of nematode research by work on *C. elegans*, which exists in nature as a free-living worm in the soil.

Nematodes possess a unique, uneven cleavage pattern. The first cleavage produces a larger AB blast cell which is the ancestor of the hypodermis and most neurons of the adult. The smaller cell, P_1, undergoes subsequent unequal cleavages with each producing larger cells (EMS, C, D) which establish other somatic lines and smaller cells (P2, P3, and P4) which eventually produce the germ line. This uneven cleavage is accompanied by chromatin diminution by which large portions of the DNA are degraded in somatic cell lines, with only germ lines possessing the full genetic complement of the fertilized egg (Schierenberg, 1997).

Gastrulation begins at the 24–28 cell stage as the daughters of the E founder cell (itself a daughter of the EMS cell) migrate toward the center of the embryo to produce the gut. This migration is followed shortly thereafter by internalization of the anterior MS cell, which produces the pharynx and a portion of the body musculature, and the more posterior C and D cells, which produce the rest of the body muscles (Schierenberg, 1997). Finally, differentiating neuroblasts from the remaining ectoderm on the surface move inward to give rise to the neurons of the ventral nerve cord and concentrations of additional neurons at the anterior and posterior ends of the cord. (The MS lineage also gives rise to a minor, more anterior set of neurons around the pharynx.) Proliferation slows in the second half of embryogenesis as the embryo elongates and cells begin to differentiate (Chalfie, 1984).

Embryogenesis is rapid in *C. elegans*: The late embryo is capable of coordinated movement and, by the time of hatching, the main body plan of the adult is already established with the majority of neurons residing in the anterior brain and a secondary population forming a string along the ventral midline, contributing to the ventral nerve cord. Subsequent postembryonic development involves gradual changes in body size and cell numbers, without drastic metamorphosis. The postembryonic development that does occur is triggered when the hatchling begins to feed and involves the resumption of divisions by several blast cells which were set aside at the end of embryogenesis. During the first of four larval stages, new classes of motor neurons are generated and the synaptic connections of other motor neuron classes are reorganized (Chalfie, 1984); subsequent larval stages (demarcated by molting) involve few additional changes. By the adult stage, hermaphrodites have a total of 959 somatic nuclei, of which 302 are neuronal, while adult males have a total of 1031 somatic nuclei, of which 381 are neuronal (White, Southgate, Thomson, & Brenner, 1986). The lineages of each of these cells are invariant and offer a fruitful model for the study of widespread developmental mechanisms; for example, its simplicity and genetic manipulability permitted powerful insights into cell-to-cell interactions mediating determinate lineages (Horvitz, Sternberg, Greenwald, Fixsen, Ellis, 1983). Another striking feature of *C. elegans* development is that a large number of cells appear to be generated only to die shortly thereafter. In fact, of the 671 nuclei generated in the embryo, 113 undergo programmed death during development (Horvitz, 2003; Horvitz et al., 1983); study of these apoptotic pathways helped elucidate their role in development across Metazoa. Other studies on *C. elegans* revealed mechanisms of cell migration and axonal pathfinding which proved equally generalizable (Wadsworth & Hedgecock, 1992; 1996). In fact, it is hard to overstate the importance of *C. elegans* to our understanding both of basic developmental mechanisms and of how evolution of those mechanisms has shaped the nervous system across phylogenies—all this *despite* the fact that the exact placement of the nematode nervous system in phylogeny remains unclear.

11.4.2 Arthropods

The arthropods are the best studied invertebrate phylum with regard to the cellular and molecular processes and mechanisms underlying their neural development. In fact, many of the key concepts of neural development in general, and indeed, of development as a whole, have come from studies of the arthropods. For example, cell-to-cell interactions, such as those first examined between differentiating photoreceptor cells of the compound eye (Zipursky, 1989) and the roles of pioneer fibers of the peripheral and central nervous system, have proven applicable across a wide spectrum of animals (Raper & Mason, 2010). Genes involved in specifying neural fate or body position along anterior/posterior and dorsal/ventral axes are amongst the many discoveries first made through studies of insect nervous system development. Comparisons of the expression patterns and actions of these genes serve as a basis for many of the inferences regarding the nervous system evolution described throughout the rest of this volume.

Because excellent reviews of insect neural development have appeared over the years (Harris & Hartenstein, 2008; Hartenstein, Spindler, Pereanu, & Fung, 2008), only an overview will be provided here. Briefly, following cellularization of the initially syncytial blastoderm, a primordium, referred to as the germ band, is formed through proliferation and aggregation of blastomeres. Gastrulation then involves the invagination and formation of a mesodermal tube at the ventral midline of the germinal band, leaving the ectoderm remaining on the surface of the now metamerically-organized embryo (Schwalm, 1997). Adoption of a neural fate for the developing ventral nerve cord occurs within the ventral neurogenic region of the ectoderm. Through a process of *Notch/Delta* lateral inhibition between neighbors, single cells within clusters become neural progenitor cells or "neuroblasts" which then delaminate into the embryo in distinct waves. The neuroblasts number about 30 per hemisegment with 3-6 neuroblasts in each of seven anteroposterior rows (Stollewerk & Chipman, 2006), a pattern observed in all insects examined so far. Subsequent asymmetric divisions of the neuroblasts produce a number of ganglion mother cells, which then divide only once more, giving rise to the neurons and glial cells of the ventral nerve cord. Because the mitotic spindle of the neuroblast is oriented perpendicular to the embryonic surface, each neuroblast produces a stack of ganglion mother cells and eventually neurons; the oldest cells lie deepest within the developing ganglion and furthest from the neuroblast, which remains close to the surface of the neurectoderm. Axons from neurons of a single lineage typically fasciculate to produce tracts which eventually form the connective, commissures and peripheral nerves of the nervous system.

Neurogenesis in the anteriormost (procephalic) regions of the insect nervous system differs in certain details from the process observed in the ventral nerve cord. First, the neurons develop from lateral rather than ventral ectodermal regions. Second, the arrangement of the approximately 100 neuroblasts which eventually generate each side of the supraesophageal brain are less stereotypically arranged than in each hemisegment of the developing nerve cord. Also, the cell divisions before and after neuroblast differentiation are less strictly oriented so that some neuroblasts remain within the neurectoderm, with subsequent delamination only of the ganglion mother cells (Technau & Urbach, 2004; Urbach, Schnabel, Technau, 2003).

Development of peripheral sensory cells in the insect embryo involves similar processes to those for central neurons, with clusters of cells in the ectoderm first expressing proneural genes and eventual selection of single cells in the cluster as a sensory organ progenitor (SOP) cells through *Notch/Delta* feedback inhibition. In the case of mechanosensory sensillia, the SOP divides twice more to produce a single neuron and an inner sheath cell from one lineage and two supportive cuticular cells from the other (Ghysen, Damblychaudiere, Jan, & Jan, 1993; Jan & Jan, 1994).

By the time of hatching, the organism contains both the foundations of what will become the central and peripheral nervous systems of the adult and also all the neural circuitry needed for larvae to behave appropriately for their own survival. Neuroblasts giving rise to the central nervous system have arrested mitosis at this stage, but they begin dividing again later in larval development and during the pupal stage. These later developing central neurons are largely destined to become interneurons, while the larval motor neurons redefine their peripheral projections, withdrawing from larval muscles and sprouting connections to newly developing adult muscles. The majority of larval sensory neurons die at metamorphosis, but new arrays of sensory cells then develop from the imaginal discs which give rise to the adult structures after metamorphosis. For many insects, metamorphosis involves the generation of thousands of new sensory cells, including the chemosensory cells of the antennae and photoreceptors of the eye. By the end of metamorphosis, most neurogenesis ceases, though some continues in structures such as the mushroom bodies (Tissot & Stocker, 2000).

In addition to providing a thorough overview of neural development within class Insecta, the arthropods have provided insight into the variability between classes within a single phylum. For example, many details of neuroblast differentiation differ across Arthropod taxa (Stollewerk & Simpson, 2005). In crustaceans (shrimp, crabs, lobsters, barnacles) the neuroblasts do not delaminate from the neurectoderm but remain at the embryonic surface while generating ganglion mother cells. In chelicerates (spiders, mites) and myriapods (millipedes, centipedes) small groups of contiguous cells rather than single neuroblasts invaginate from the surface and provide the origins for the neurons which eventually populate the ganglia of the ventral nerve cord (Stollewerk & Chipman, 2006; Stollewerk & Simpson, 2005). However, despite these differences in neural progenitor cells, they share striking similarities in number (roughly 30 cells or groups) and position (generally arranged as seven rows with 3–6 elements per row) in each hemisegment. The various arthropods also differ with regard to differentiation of neuroblasts within hemisegments and in anteroposterior progression (Stollewerk & Simpson, 2005).

Given the wealth of information regarding neural development in arthropods, major problems arise when relating this literature to that of early development of other animal groups. First, the evolutionary relationships within the arthropods themselves have proved difficult to unravel (Stollewerk & Simpson, 2005; Truman & Riddiford, 1999). Furthermore, the most striking developmental features of insects like *Drosophila* are its larval stage and metamorphosis into an adult, yet these are derived and not ancestral traits. Instead, the earliest true insects, represented by the extant bristletails and silverfish, appear to have been ametabolous (without metamorphosis). Truman and Riddiford (1999) have hypothesized that the larval stage of holometabolous (metamorphic) insects can be best understood as a stage intercalated between the hatching (prenymph) and the first instar (nymph) stage of heterometabolous insects that show a more gradual change in body form during development. Comparisons between adult and larval forms become even more complicated when

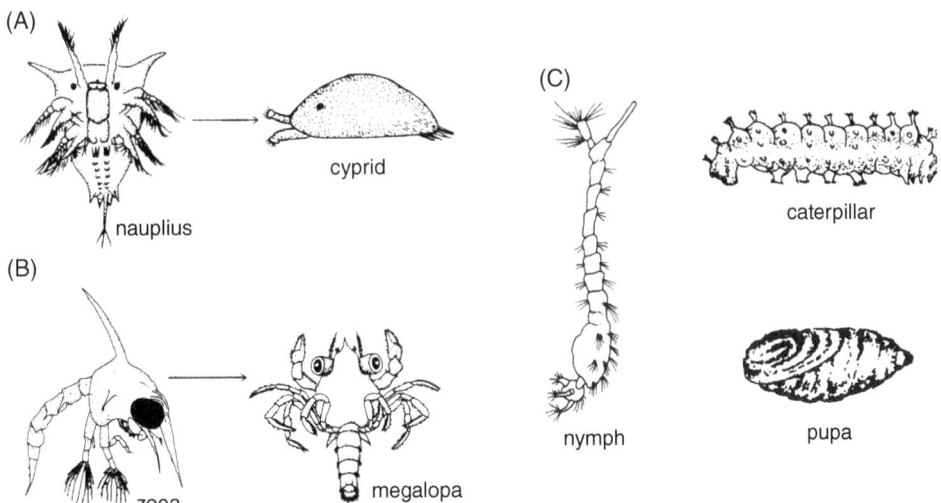

Figure 11.4 Arthropod Larvae and Other Developmental Stages.
(A) The nauplius and cyprid larvae are characteristic of the crustacea. (B) Zoea and megalopa larvae are also found in certain crustacea. (C) The nymph stage is a characteristic insect development, although many insects also possess a secondary larva stage, in the form of a caterpillar, grub, or maggot which transits through a pupal stage before becoming an adult. Modified from Pechenik 2005. Reproduced with permission of McGraw-Hill Companies, Inc.

considering insects and other groups such as the crustaceans, with which they appear to be closely related. Nauplius larva (Figure 11.4A) are characteristic of crustacean development and have been argued to better represent the ancestral condition of those groups (Dahms, 2000) than does the adult stage of many of its members. For instance, the nervous systems of naupilus larva (Semmler, Wanninger, Hoeg, & Scholtz, 2008) and the postlarval cypris stage of the barnacle (Harrison & Sandeman, 1999) are more complex and extensive than the adult stage, which appears to degenerate as adults adopt a sedentary existence. Other groups of crustacea employ alternative intermediate larval stages (zoea and megalopa, see Figure 11.4B) and have also been studied extensively (Beltz, Helluy, Ruchhoeft, & Gammill, 1992). As discussed in §11.7, these comparative studies have fuelled discussion of ancestral modes of neural development in arthropods, particular regarding anterior regions (Harzsch, 2004; Semmler et al., 2008; Vilpoux, Sandeman, & Harzsch, 2006).

11.4.3 Onychophora (Velvet Worms)

Another approach to understanding the phylogenetic origins of the arthropod nervous system has been to examine the developing nervous systems of other panarthropoda, in particular Onychophora. Recent analyses of molecular, morphological and fossil evidence all favor close relationships between the Arthropoda, Terigrada and Onychophora (Dunn et al., 2008; Regier et al., 2010; Rota-Stabelli et al., 2010).

The velvet worms are terrestrial animals that inhabit primarily tropical and subtropical regions of the southern hemisphere. Their nervous system is similar to that of the arthropods in that it consists of a concentration of anterior neurons and neuropil and a pair of posterior ventral cords with transverse commissures. There is debate, however, whether the nervous system of the velvet worms is truly segmentally organized, especially in the posterior regions, or whether seemingly repeated congregations of somata and axons are simply a reflection of the innervation of regularly spaced lobopod feet (Strausfeld, Strausfeld, Stowe, Rowell, & Loesel, 2006; Whitington & Mayer, 2011). An unsegmented nervous system could either arise through regressive evolution from a segmented panarthropod common ancestor, or could indicate that the common ancestor more closely resembled unsegmented wormlike organisms, as in other phyla.

Mechanisms and cellular details of Onychorphoran neurogenesis are still incompletely understood, although they have been studied for well over a century (Anderson, 1973). More recently Eriksson and Stollewerk (2010) reported several differences in neural development between Onychorphora and the arthropods. Perhaps the most notable difference involves the fact that there is neither morphological nor molecular evidence in velvet worms for the distinct patterning of neural precursor cells, as present in arthropod development of distinct brain compartments. Conversely, however, Whitington and Mayer (2011) concluded that neurogenesis in these animals was similar to that observed in insects in one crucial feature, which may therefore be an ancestral pananthropoda trait: Specifically, the central nervous systems in both groups arise from progenitor cells which are internalized by ingression of single cells within the neurectoderm. This developmental trait contrasts with the continued external positions of neural progenitor cells and only subsequent inward migration of clusters of postmitotic neurons seen in myriapods and chelicertates.

11.5 Deuterostomia

The deuterostomes comprise the final clade of Metazoa. They were classically defined as the group of animals in which the blastopore became the anus rather than the mouth of the developing gut, thus distinguishing them from the protostomes. Other classical characteristics of deuterostomes have included radially symmetric cleavage and the ciliary beating patterns on the larvae. The deuterostomes consist of only four extant phyla: Echinodermata, Hemichordata, Chordata and Xenoturbellida, with only the first three groups reviewed here.

11.5.1 Echinoderms

Echinoderms have long played important roles in experimental embryology dating from early studies providing initial and powerful demonstrations of regulated development (Driesch, 1892, as cited by Wray, 1997)). These animals also provide an important evolutionary perspective, as a major non-chordate phylum of deuterostomes. Another advantage of the echinoderms is the wide range of developmental variations both in early cleavage patterns (Wray, 1997) and also larval forms (Raff & Byrne, 2006), thus providing a focus for studies on evolution of developmental programs. Finally, these animals exhibit a fascinating metamorphosis whereby adult tissues form as imaginal rudiments in the late larva. When metamorphosis does occur, much of the larval tissue is simply discarded as the bilaterally symmetric larvae is transformed into the pentamerically radial adult (Wray, 1997).

Cleavage in most echinoderms, including the sea stars (Asterozoa) and sea cucumbers (Holothuroidea), is radial and equal, but in urchins (Echinoidea), the vegetal cells divide unevenly at the fourth cleavage to yield distinct tiers of mesomeres at the animal pole, micromeres at the vegetal pole, and macromeres positioned equatorially. These blast cells are not to be confused or homologized with the micromeres and macromeres in, for example, molluscs: in these echinoderms, the mesomeres are the origins of the ectoderm from which the larval nervous system is derived (Wray, 1997).

Gastrulation occurs through a thickening of the vegetal pole into a plate which then invaginates, accompanied by involution of surrounding cells to form the primitive archenteron. In many species, the gut differentiates into its component regions shortly after gastrulation thereby allowing the ensuing larvae to begin feeding. However, many species have larvae that never feed or that start feeding significantly later during their development (Wray, 1997).

Echinoderm larvae also vary greatly over a range, from the doliolaria larvae of the sea lilies (Crinioidea), to the pluteus larvae of urchins and brittle stars (Ophiuroidea), to the auricularia larvae of sea cucumbers, to the initial bipinnaria and subsequent brachiolaria larvae of sea stars (Figure 11.5A-D). Each larval stage is substantially different from both the forms representing other classes and from the adult forms into which they metamorphose.

The existence and function of a echinoderm larval nervous system was inferred by Strathmann (1975), who studied swimming and feeding behaviors which depend upon coordinated activity of cilia in various bands around the larva and of muscular contractions around the mouth. Building on earlier descriptions (Ryberg, 1977), Burke and colleagues (Burke, 1983a, 1983b; Burke, Brand, & Bisgrove, 1986; Bisgrove & Burke, 1987) laid the foundations for our understanding of the

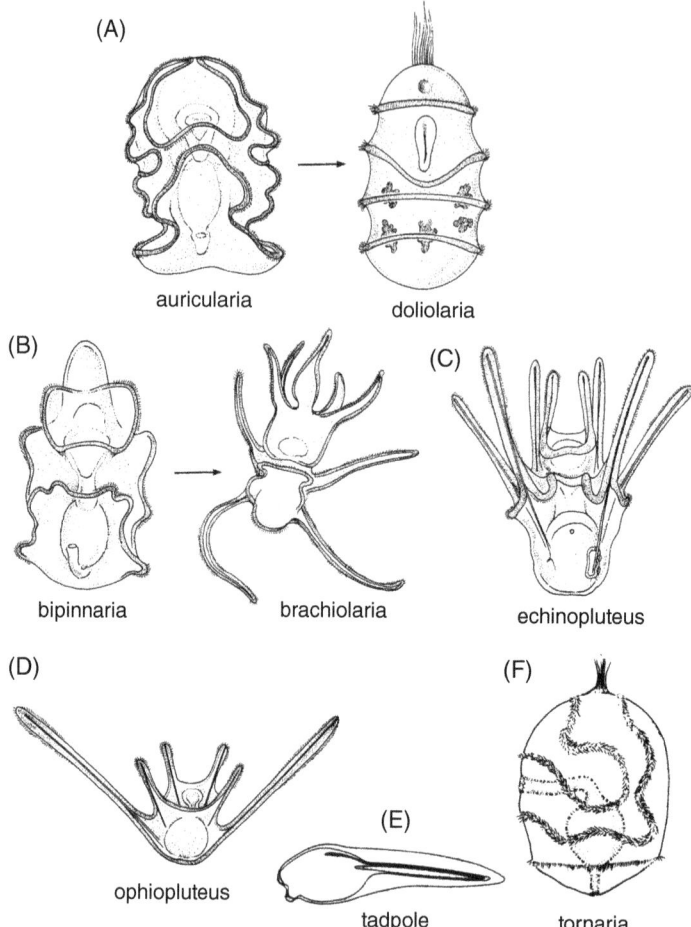

Figure 11.5 Deuterostome Larvae.
(A) Holothurian echinoderms (sea cucumbers) transit through auricularia and doliolaria larval forms. (B) Asteroidea (sea stars) transit through bipinnaria and brachiolaria larval forms. (C) Echiopluteus larvae are typical of echinoid echinoderms (sea urchins). (D) Ophiopluteus are typical of ophiuroid echinoderms (brittle stars). (E) Tadpole larvae are characteristic of tunicates (F) Tornaria larvae are characterisitic of the hemichordates. Modified from Pechenik 2005. Reproduced with permission of McGraw-Hill Companies, Inc.

larval nervous systems of echinoderms through comprehensive descriptions of the different forms of larvae. These studies generally employed histochemical stains for different transmitter types coupled with EM. Chee and Byrne (1999) recently added finer detail to these descriptions through the use of confocal microscopy. More recently an antibody for synaptotagmin has provided a more general stain for the larval neurons (Burke et al., 2006), thus completing the picture sketched by more selective stains.

The first neurons appear at the animal pole at or before the completion of gastrulation. Serotonin-containing cells are generally first detected at this time, scattered near the anterior end of the larva. In some larvae these cells appear to migrate extensively as they become increasingly associated with the complicated band(s) of ciliated cells which surround the oral field and rim the edges of the arms and folds of the larvae. While serotonin-containing cells and axons can be observed along the lengths of the ciliary bands, concentrations of cells often form ganglia in lateral and anterior regions of the larvae, and another ganglion also forms in the lower lip. Catecholaminergic cells are similarly located along ciliary bands and lower lip, whereas GABA-like immunoreactivity tends to be mostly concentrated in the lip and contributes innervation to muscles surrounding the esophagus. SALMFamide peptides are also associated with the various bands of ciliary (Byrne, Cisternas, & Koop, 2001).

However, just as larval forms vary greatly between classes of echinoderms, so too can the larval nervous systems. For example, the degree of consolidation of anterior cells into an apical ganglion with a distinct neuropil varies between different echinoderms. Also, since the ciliary bands are used not only for swimming but also for feeding, the extent of their innervation, along with that of the mouth, correlates with whether the larvae of a particular species are lecithotrophic or planktotrophic. By contrast, structures involved with control of settlement and metamorphosis—processes common to all such larvae—are more conserved. Thus, for instance, innervation of the attachment complex is comparable in feeding and non-feeding species of sea stars (Byrne et al., 2001). Pharmacological experiments confirm neuronal functions inferred by innervation patterns: depletion of serotonin rendered larvae incapable of swimming (Yaguchi & Katow, 2003); conversely, applications of transmitters have been shown to directly affect ciliary beating (Wada, Mogami, & Baba, 1997).

By mid to late larval stages, the imaginal rudiment of the adult body begins to grow, generally on the left side (Wray, 1997). The larva continues to swim and feed as the rudiment develops. Overt metamorphosis can occur quickly, whereby the pentamerically organized adult rudiment everts from the bilaterally organized larvae, the tissues of which then degenerate. The adult nervous system appears at the time that the larval tissues are degenerating and thus there appears to be little overlap between the two (Chia & Burke, 1978). Some elements, such as the serotonergic cells which are prevalent during larval stages, are lost in the adult (Chee & Byrne, 1999; Cisternas & Byrne, 2003). However, some neural elements, such as certain peptidergic cells surrounding mouth, appear to persist through metamorphosis thus suggesting that they may contribute to the oral nervous system of the adult.

11.5.2 Hemichordata (Acorn Worms)

The hemichordates share characteristics with both the echinoderms and the chordates. The dipleurula larval form resembles the auricularia larvae of echinoderms, and as in those animals, the nervous system develops into a diffuse, intraepithelial network in the adults (Lowe et al., 2003; Miyamoto, Nakajima, Wada, & Saito, 2010). In other regards the hemichordates more closely resemble the chordates, with which they share sets of pharyngeal gill slits and a bilateral body plan throughout life. Recent molecular evidence favors grouping the hemichordates and echinoderms as sister groups to one another and more distantly related to the chordates (Gerhart, Lowe, & Kirschner, 2001).

Like other deuterostomes, the hemichordates exhibit radial cleavage and the future body plan is established early. The first cleavage establishes the plane of bilateral symmetry whereas the second cleavage establishes the dorsal and ventral halves of the body. The first quartet of cells at the animal pole is destined to become the anterior ectoderm including progenitors for the anterior nervous system. While progeny of the posterior quartet invaginate during gastrulation, some remain external and form the posterior ectoderm, thus contributing to that part of the nervous system (Gerhart et al., 2001).

Development through subsequent larval forms can vary widely between different hemichordates. Some start feeding shortly after hatching and may then live in the water column for months before settlement and metamorphosis. Other species possess a lecithotrophic larval stage, while still others, such as the well studied enteropneust, *Saccoglossus*, are direct developers. As expected, morphological features are elaborated or lost and the timing of development differs according to which life style is adopted by what are referred to as the tornaria larvae of this phylum (Nielsen & Hay-Schmidt, 2007). As a further complication, postembryonic stages are often named (Müller, Heider, Metschnikoff, Krohn, and Agassiz) as the larva progresses toward metamorphic competence (Nielsen & Hay-Schmidt, 2007).

Some species hatch as roughly spheroidal larvae with incomplete ciliary bands and before the mouth opens. By this point, nerve cells marked with antibodies to synaptotagmin and/or serotonin are already present in association with ciliary bands in the anterior end just outside the apical region. Over the next several days, the larva elongates, adds more ciliated cells to various bands, and starts feeding. As the larva adds a posterior ring of cilia, this band takes over the major function of locomotion leaving more anterior bands to specialize in feeding functions. Development is also marked by increasing numbers of nerve cells associated with the various bands, the mouth region, and a distinct apical organ (Miyamoto et al., 2010; Nielsen & Hay-Schmidt, 2007). In addition to serotonin, catecholamines and peptides are also present in the larval nervous system. Nezlin and Yushin (2004) pointed out that the concentration of cells within the apical organ and the formation of a well-developed, underlying neuropile, along with details of the locations of serotonergic and FMRFamidergic cells, were different from what one normally encounters in echinoderm larvae and suggested that such differences may indicate that these phyla are not as closely related as often assumed.

About halfway through larval development, additional neurons begin to appear on the dorsal midline and thus mark the first appearance of cells that will form the dorsal nerve cord of the adult. Additional cells appear through later stages of larval development and indicate the co-existence of both the larval and adult nervous systems in these pre-metamorphic stages. As metamorphosis begins, the various ciliary bands of the larva begin to disappear, as do the nerve cells associated with them (Miyamoto et al., 2010; Nielsen & Hay-Schmidt, 2007). The apical organ also disappears during metamorphosis. Metamorphosis is a time when apoptotic activity can be observed in numerous cells of the transforming larva (Miyamoto et al., 2010), and yet, during this time new neurons are also added, particularly to the epithelium in the proboscis and collar, thus forming the nervous system observed in the adult stage (Miyamoto et al., 2010).

11.6 Invertebrate Chordates

11.6.1 Cepahalochrodates (Lancelets or Amphioxus)

There are 25–30 species of cephalochordates which burrow into the loose substrate of shallow waters in temperate and tropical marine environments worldwide (Whittaker, 1997). Evidence now favors a closer phylogenetic relationship between tunicates and vertebrates with the cephalochordates basal to the two, but Koop and Holland (2008) argue that the cephalochordates probably better represent ancestral vertebrates than do the ascidians, reinforcing a view that has existed for many years (Whittaker, 1997). Like the tunicates, cephalochordates have a simpler and smaller genome than the vertebrates since neither group has undergone the two rounds of whole-genome duplication which characterize vertebrates. However, cephalochordates, like vertebrates, have indeterminant cleavage, and both contrast to the tunicates, which have greatly simplified body plans with dramatic reductions of total cell numbers.

Gastrulation in the cephalochordates occurs by invagination of the flattened vegetal pole to form a cap-shaped embryo with a broad archenteron (Whittaker, 1997). Further invagination of the dorsal lip of the blastopore internalizes cells that will form the notochord. The late gastrula is somewhat elongated with a neural plate forming on the flattened dorsal surface. The embryo further elongates and hatches as a ciliated neurula.

Neurulation begins with the rounding up and enclosure of the neural plate in a manner similar to that which occurs in vertebrates (Koop & Holland, 2008; Whittaker, 1997). As the neural plate begins to fold, the ectoderm along its lateral edges detaches and migrates medially to reform a continuous dorsal surface over the embryo. During the time of neural tube development, the notochord is forming between the dorsal neural tube and the roof of the archenteron. One large difference between neurulation in cephalochordates and vertebrates, however, is that the tissue along the lateral edges of the neural plate forms only ectoderm and no other cell types. Thus cephalochordates lack a neural crest (and also cranial placodes), which characterizes neural development in vertebrates. Another difference in development involves the underlying notochord which extends all the way to the anterior end of these animals, thus giving them the characteristic of being pointed at both ends, for which they are sometimes named (amphioxus).

Differentiation of neurons begins at the early neurula stage, and by the time that the somites develop, clusters of neurons (as indicated by expression of *Hu/elav*) can be observed near the underlying intersomite boundaries at the anterior end of the animal. Large numbers of putative sensory cells also appear to differentiate in the epithelial ectoderm by the early neurula stage (Satoh, Wang, Zhang, & Satoh, 2001).

The first modern study of the neuroanatomy of cephalochordate larvae was conducted by Bone (1959) but 3D reconstructions based on serial EM sections (Lacalli, 2004; Wicht & Lacalli 2005) provide much of the basis for our current knowledge of the organization of the larval nervous system. The anterior end consists of the slightly enlarged cerebral vesicle; the anterior end of the cerebral vesicle itself consists of the frontal eye, which contains photoreceptors and associated interneurons, and a putative balance organ (Lacalli, 1996). The posterior end of the cerebral vesicle consists of the lamellar bodies, which also appear to contain photoreceptors and have been argued to be a homologue of the vertebrate pineal gland. The region ventral to the lamellar bodies contains numerous neurons associated with infundibular secretory cells. Other tightly packed neurons that are ventral to the central canal possess caudally-directed axons. From this region the nerve cord makes a gradual transition to that typical of more posterior regions responsible for motor control of different swimming behaviors. The anteriormost set of motor neurons and associated interneurons comprise a region named the Primary Motor Centre (PMC) (Lacalli, 2003; Lacalli & Kelly, 2003). There, three sets of Large Paired Neurons (LPNs) receive extensive sensory input from a variety of sources including, the nearby interneurons, and make connections onto motor neurons which in turn project to the ventral and dorsal compartments. These compartments provide connections to muscles controlling fast and slow components of swimming, respectively. Cell types found in the PMC also appear elsewhere in the nerve cord, but detailed descriptions exist only for those in the anteriormost somites.

In addition to the central nervous system, Lacalli, Gilmour and Kelly (1999) also examined the peripheral nervous system, focusing on oral innervation of the larvae. They suggested that numerous Type II secondary sensory receptors (axonless epithelial cells resembling vertebrate taste receptors or hair cells) on oral spines might mediate a coughing response used to eject debris from the oral region. These sensory cells synapse upon the axons of primary sensory neurons with both central and peripheral cells bodies. An alternative approach to examining the peripheral nervous system has employed vital staining of sensory cells with the lipophilic dye, DiI. Holland and Yu (2002) used this approach to describe the early appearance of Type I sensory receptors, which have epithelial somata, ciliated dendrites, and axons which then project into the nerve cord.

The nerve cord increases many fold in diameter during the larval period, and comes to contain both a much greater number and diversity of axons (Wicht & Lacalli, 2005). In accordance with this finding, Holland and Yu (2002) described a continuous increase in the numbers of peripheral sensory cells, with concentrations at the rostral and caudal ends of the larva. At the time of metamorphosis, additional cell types appear including new populations of Type II cells, as well as sensory cells in the ventral pits.

The most dramatic feature of metamorphosis in some cephalochordates involves a shift from a highly asymmetric larva to a roughly symmetric juvenile. The first signs of bilateral asymmetry are seen as left/right shifts in the alignments of intersomite boundaries starting at the early larval stage (Whittaker, 1997). By the time that metamorphosis begins, the larvae of *Branchiostoma* are highly asymmetric with a single row of eight gill slits on the right and the mouth on the left. During metamorphosis a second row of gill slits is generated and the mouth migrates to a medial and more caudal position. Paris et al. (2008) have demonstrated the involvement of thyroid-hormone-like receptors in metamorphosis and their presence in the endostyle, strengthening earlier suggestions that the cephalochordate endostyle is homologous to the vertebrate thyroid.

11.6.2 Tunicates (Ascidians or Sea Squirts)

Although adult ascidians are soft-bodied filter feeders with little resemblance to any vertebrate, the short-lived tadpole larva possesses not only a notochord but a nervous system which shares many features with vertebrates (Meinertzhagen, Lemaire, & Okamura, 2004).

Two species of solitary ascidians, *Ciona intestinalis* and *Halocynthia rortzi*, have been investigated in detail. They possess about 2600 to 3000 cells, respectively, in the entire tadpole larva; of that total, the CNS contains less than 400, of which only about 100 are neurons (Imai & Meinertzhagen, 2007a). It has not yet been determined if all cells are specified as distinct individuals, although variation in total numbers appears to be small. Small cell numbers appear to be common in other urochordates as well, with larvaceans like *Oikopleura* having fewer neurons than the ascidians, but with salps having more (Bone, Pulsford, & Amoroso, 1985; Lacalli & Holland, 1998).

Because of the largely determinate development of ascidians, they have long been a favorite subject of lineage studies (Conklin 1905, as cited by Jeffery & Swalla, 1997). More modern studies (Cole & Meinertzhagen, 2004) revealed that all neurons are apparently derived through only 10–14 cell divisions from the first cleavage, with the posterior and anterior regions of the CNS originating from different sets of blastomeres.

The CNS of the ascidian tadpole larva is organized into: (1) an anterior sensory vesicle, (2) an intermediate visceral ganglion, and (3) a posterior dorsal nerve cord. The sensory vesicle contains the majority of cells in the CNS, of which many are the sensory cells and interneurons associated with a single anterior-and-left otolith and a single posterior-and-right ocellus. The visceral ganglion houses about 45 motor neurons, categorized into at least four subtypes which are responsible for generating locomotor behaviors comprising tail flicks and phototactic swimming. The caudal cord is made up of mostly ependymal cells lining the hollow central canal (Meinertzhagen et al., 2004; Imai & Meinertzhagen, 2007a). A slender neck region containing only six cells has also been identified between the sensory vesicle and visceral ganglion. In addition, ascidians possess a peripheral nervous system consisting of sensory cells from the apical papilla, apical and rostral trunk epidermal neurons, and dorsal and ventral caudal epidermal neurons (Imai & Meinertzhagen, 2007b). Like the cephalochordates, but unlike the vertebrates, the ascidians possess peripheral neural somata around the body. The various neurons of ascidians contain (and presumably use) a number of transmitters including dopamine, GABA, GnRH, acetylcholine, serotonin, and possibly several other substances (Tsutsui, Yamamoto, Ito, & Oka, 1998; Meinertzhagen et al., 2004).

Several homologies with vertebrate brain structures have been suggested in tunicates, as in the cephalochordates. Original suggestions were often based purely on tissue organization and development, but many cases have been strengthened by more recent findings of gene expression patterns. For example, ascidian ocellus and vertebrate eye have long been suggested to be homologs with shared photoreceptive function. In fact, the development of the central nervous system as a whole shares striking similarities with its development in vertebrates (Meinertzhagen et al., 2004).

While the development of the larval nervous system has been the subject of much study, the metamorphic development of the adult nervous system remains poorly understood. Much of the larval nervous system, however, appears to degenerate through programmed cell death (Jeffery, 2002). Another interesting aspect of adult nervous system development involves the use of alternative pathways in colonial ascidians. The founding individual zooid of a colony generally develops through the pathway for neural development described above for other tunicates. This zooid, however, forms a colony through asexual budding. The nervous systems of adult zooids appear to be similar whether derived from sexual or asexual reproduction (Tiozzo, Murray, Degnan, De Tomaso, & Croll, 2009). Presumably many of same genetic and cellular processes are involved in these so-called embryogenic and blastogenic pathways of neural development, but their regulation must occur by different mechanisms.

11.7 Conclusions

As illustrated repeatedly in this chapter, major advances have been made in understanding the development of nervous systems across a wide variety of invertebrates. Most of these advances have come only recently, first with the introduction of electron microscopy and then through application of immunocytochemistry and confocal microscopy. In recent decades, we have discovered surprisingly complex nervous systems in the early life stages of invertebrates. We have also begun to build sufficient knowledge to make meaningfully comparisons of developing nervous systems across phyla, but these views and comparisons remain limited.

One technical limitation to these types of studies involves the fact that, while immunocytochemistry and confocal microscopy provide relatively quick and easy means for visualizing nervous systems, the vast majority of work to date has exploited a very small number of markers for neurons: Only a fraction of each nervous system has been viewed. For example, markers against specific neurotransmitters have been favored in these studies because they reveal discrete subsets of neurons which can then be described in detail. Moreover, transmitter type can be incorporated into the character sets used to homologize nervous systems, or even individual neurons, across animals. In addition, determination of the transmitter contents of cells with known innervation immediately suggests the function of these cells; hypotheses can then be tested pharmacologically for changes in behavior/physiology. The vast majority of comparative studies of larval nervous systems have employed antibodies against only a very few neurotransmitters (i.e., serotonin and FMRFamide-like peptides). Many more transmitters are likely to be used by larval nervous systems and many more cells are likely to be revealed through use of antibodies against those transmitters. Currently, we simply do we do not know what we cannot see. It has proven difficult to raise antibodies against classical, small-molecule transmitters. In studies on vertebrates, the transmitter contents of neurons are often inferred by targeting proteins involved in synthesis/packaging of transmitters (e.g., choline acetyltransferase or vesicular glutamate transporter). Few of the presently available antibodies against these targets, however, have wide cross-species reactivity, especially in distant clades. In fact, this same limitation also applies to potential pan-neuronal markers, such as synaptic proteins, which might label all or most neurons regardless of transmitter phenotype. Hence, antibodies against different tubulins are often used to provide overall views of developing nervous systems, although these antibodies generally only stain axons and not cell bodies, and neural elements are often obscured by cilia. More complete views will be produced once a wider palette of markers is developed.

Another major obstacle to the construction of a generalized, comprehensive framework for understanding the evolution of neural development comes from the sheer number and diversity of invertebrates in the world. In this regard, we are faced more than ever with the long standing challenge of deciding whether it is better to focus research efforts on a small number of model species or to dedicate more resources to the examination of a wider range of organisms. Certainly work on *C. elegans* and *Drosophila* has led to important advances in our understanding of developmental neurobiology, and these advances only came about because of the early availability of whole genome sequences and a large background literature for these animals. But it must also be recognized that laboratory animals are often chosen precisely because they do *not* represent ancestral groups. Instead, they are chosen because they are easy to raise in the lab and have short generation times. They are not, therefore, representative of the large segment of the animal kingdom that has complex life cycles and often spends long periods of time as swimming and feeding larvae. Fortunately, recent work on a wider range of organisms promises better representation of the diversity in the animal kingdom while still permitting focus on shared animals and techniques across labs. For example, several research groups are now working on *Mnemiopsis* and *Pleurobrachia* in ctenophores, *Nematostella*, *Hydra*, and *Clathia* in cnidaria, and *Dugesia* and *Macrostomum* in flatworms, and are beginning to provide a wider understanding of developmental mechanisms employed by more basal taxa in the Metazoa. Although leeches dominated early work on neural development in annelids, recent work on polychaetes such as *Platynereis*, *Capitella*, and *Pomatoceros* supplement this research with examples of neural development through primary larval forms. *Aplysia*, *Ilyanassa*, *Crepidula* provide robust and readily available larvae for understanding early neural development in gastropods, but cephalopods like *Sepia* and bivalves like *Mytilus* are now beginning to round out our views of early nervous systems in molluscs. The diversity of echinoderms defies characterization based on studies of a single species, although *Strongylocentrotus* is the best studied developmental model in this taxon. Finally, *Saccoglossus*, *Brachiostoma*, and *Ciona* have provided focused insights into the development of invertebrate chordates. Of course, this list provides only a partial enumeration of some of the more important organisms which are now contributing to our view of evolving nervous systems. It does, however, offer a healthy alternative to what, at first, was an overreliance on comparisons between a fly, a worm, and a mouse to understand the evolution of developmental mechanisms in the animal kingdom.

With the wider use of different animals, comes also a need for better resolution of the evolutionary relationships between the different invertebrate groups. Great strides have been made with molecular phylogenies, confirming many long-suspected relationships while challenging others. One limitation to these types of studies has often been the lack of key basal groups, which can be rare or technically difficult to assay. However, even well-studied and available groups such as insects and nematodes have proven to be surprisingly resistant to placement into a larger phylogeny. Here, also, more study is needed (and perhaps new approaches).

It must be remembered, however, that, while molecular work has recently provided substantial increases in our understanding of phylogeny, the bedrock for comprehension of evolutionary relationships comes from morphological comparisons between animals at different stages of development. Indeed the very names that demarcate the animal kingdom into protostomes, deuterostomes, lophotrochozoa, etc., derive from anatomical observations of different animal groups during ontogeny; much of that work involved comparisons of the larval stages which have been the focus of this chapter. But such comparisons raise difficult issues with regard to the evolution of the nervous system. Specifically, it is now evident that both the larval and postlarval stages of a wide range of animals possess nervous systems which can be distinct from one another. They develop at different times, originate from different cell lineages, and many, if not all, components of the early larval nervous system can disappear during metamorphosis. Furthermore, as larval stages are lost or gained through evolution, it has not always been clear which components of the nervous system, or even which stages of development, are being represented in comparisons between animals.

Several examples offered in this chapter suggest a possible clarification of this issue. For example, an apical organ and surrounding neurons are common components of the earliest larval nervous system of Lophotrochozoa, and are derived from early micromeres. Although the cerebral ganglion develops later, it is similarly derived from such micromeres. In contrast, the more posterior regions of what will become the adult nervous system develop only later in larval life and are derived at least in part from later divisions of the macromeres. Available evidence suggests that cellular and genetic details of neurogenesis also differ between anterior and posterior nervous systems in Lophotrochozoa (Denes et al., 2007; Meyer & Seaver, 2009). Even in direct-developing leeches, which have lost most other features of ancestral annelid larvae stages, the anterior (cerebral ganglion) and posterior (ventral nerve cord) components of the nervous system continue to form via different developmental programs and apparently reflect vestiges of the ancestral larval and postlarval nervous systems, respectively (Stent et al., 1992). When expanding comparisons to the Ecdysozoa, insects have also lost the primary larval stage, and syncytial development obscures the cellular origins of different components of the nervous system. And yet, anterior ganglia are still derived from a different developmental program than the posterior nerve cord (Hartenstein et al., 2008; Technau & Urbach, 2004). Furthermore, the development of the ventral nerve cord in insects manifests striking similarities to that of the ventral nerve cord in annelids (Denes et al., 2007). Thus, the anterior and posterior nervous systems of insects may reflect vestiges of the early larval and later postlarval nervous systems, respectively, of the common ancestor to all extant protostomes, even though most other traces of those early stages have been lost or obscured in evolution in these animals. It is also tempting to speculate that, given the similarities between insect, annelid, and vertebrate neural development, the last common ancestor to both the protostomes and deuterostomes also possessed larval and postlarval stages and that vestiges of the nervous systems of each of these stages are evident in the development across an even wider spectrum within the Animal Kingdom. For example, although the rooting of the Bilateria within the phylogenetic tree is still not clear (Peterson & Eernisse, 2001; Philippe et al., 2011; Sempere, Martinez, Cole, Baguna, & Peterson, 2007), comparisons of neural development across the animal kingdom suggest that the ancestral bilaterian first elongated its postlarval bauplan and generated the nervous system for the extended posterior portion of the body by processes still evident in the development of the nervous systems of present day animals from all major clades of the higher Metazoa (Arendt, Denes, Jékely, & Tessmar-Raible, 2008).

Acknowledgments

This review was largely written during a sabbatical leave at the Observatoire Océanologique de Villefranche-sur-Mer in 2011–2012. Thanks are extended to the people there for their hospitality, support, and friendship. Funding was provided by a grant from the Natural Sciences and Engineering Research Council of Canada.

References

Altenburger, A., & Wanninger, A. (2010). Neuromuscular development in *Novocrania anomala*: Evidence for the presence of serotonin and a spiralian-like apical organ in lecithotrophic brachiopod larvae. *Evolution & Development*, 12, 16–24.

Anderson, D. T. (1973). *Embryology and Phylogeny in Annelids and Arthropods.* Oxford: Pergamon Press.

Arendt, D., Denes, A. S., Jékely, G., Tessmar-Raible K. (2008). The evolution of nervous system centralization. *Philosophical Transactions of the Royal Society B,* 363, 1523–1528.

Aroua, S., Andouche, A., Martin, M, Baratte, S., & Bonnaud L. (2011). FaRP cell distribution in the developing CNS suggests the involvement of FaRPs in all parts of the chromatophore control pathway in *Sepia officinalis* (Cephalopoda). *Zoology (Jena),* 114, 113–122.

Baratte, S., & Bonnaud, L. (2009). Evidence of early nervous differentiation and early catecholaminergic sensory system during *Sepia officinalis* embryogenesis. *Journal of Comparative Neurology,* 517, 539–549.

Beiras, R, & Widdows, J. (1995). Effects of the neurotransmitters dopamine, serotonin and norepinephrine on the ciliary activity of mussel (*Mytilis edulis*) larvae. *Marine Biology,* 122, 597–603.

Beltz, B. S., Helluy, S. M., Ruchhoeft, M. L., & Gammill, L. S. (1992). Aspects of the embryology and neural development of the American lobster. *Journal of Experimental Zoology,* 261, 288–297.

Bisgrove, B. W., & Burke, R. D. (1987). Development of the nervous system of the pluteus larva of *Strongylocentrotus droebachiensis. Cell Tissue Research,* 248, 335–343.

Bonar, D. B. (1978). Ultrastructure of a cephalic sensory organ in the larvae of the gastropod *Phestilla sibogae* (Aeolidacea, Nudibranchia). *Tissue Cell,* 10, 153–165.

Bone, Q. (1959). The central nervous system in larval acraniates. *Quarterly Journal of Microscopic Science,* 100, 509–527.

Bone, Q., Pulsford, A. L., & Amoroso, E. C. (1985). The placenta of the salp (Tunicata, Thaliacea). *Placenta* 6, 53–64.

Boyer, B. C., Henry, J. J., & Martindale, M. Q. (1998). The cell lineage of a polyclad turbellarian embryo reveals close similarity to coelomate spiralians. *Developmental Biology,* 204, 111–123.

Braubach, O. R., Dickinson, A. J., Evans, C. C., & Croll, R. P. (2006). Neural control of the velum in larvae of the gastropod, *Ilyanassa obsoleta. Journal of Experimental Biology,* 209, 4676–4689.

Brinkmann, N., & Wanninger, A. (2009). Neurogenesis suggests independent evolution of opercula in serpulid polychaetes. *BMC Evolutionary Biology,* 9, 270.

Brusca, R., & Brusca, G. (2003). *Invertebrates.* Sunderland, MA: Sinauer Associates, Inc.

Burke, R. D. (1983a). Development of the larval nervous system of the sand dollar, *Dendraster excentricus. Cell Tissue Research,* 229, 145–154.

Burke, R. D. (1983b). The structure of the larval nervous system of *Pisater ochraceus* (Echinodemata: Asteroidea). *Journal of Morphology,* 178, 23–35.

Burke, R. D., Brand, D. G., & Bisgrove, B. W. (1986). Structure of the nervous system of Auricularia larva of *Parasticopus californicus. Biological Bulletin,* 170; 450–460.

Burke, R. D., Osborne, L., Wang, D., Murabe, N., Yaguchi, S., & Nakajima, Y. (2006). Neuron-specific expression of a synaptotagmin gene in the sea urchin *Strongylocentrotus purpuratus. Journal of Comparative Neurology,* 496, 244–251.

Byrne, M., Cisternas, P., & Koop, D. (2001). Evolution of larval form in the sea star genus *Patiriella*: Conservation and change in the larval nervous system. *Development, Growth & Differ* 43, 459–468.

Chalfie, M. (1984). Neuronal development in *Caenorhabditis elegans. Trends in Neuroscience,* 7, 197–202.

Chase, R, & Tolloczko, B. (1987). Evidence for differential DNA endoreplication during the development of a molluscan brain. *Journal of Neurobiology,* 18, 395–406.

Chee, F., & Byrne, M. (1999). Development of the larval serotonergic nervous system in the sea star *Patiriella regularis* as revealed by confocal imaging. *Biological Bulletin,* 197, 123–131.

Chia, F., & Koss, R. (1984). Fine structure of the cephalic sensory organ in the larva of the nudibranch *Rostanga pulchra* (Mollusca, Opisthobranchia, Nudibranchia). *Zoomorphology*, 104, 131–139.

Chia, F. S., & Burke, R. D. (1978). Echinoderm metamorphosis: fate of larval structures. In F. S. Chia & M. E. Rice (Eds.), *Settlement and metamorphosis of marine invertebrate larvae* (pp 219–234). Amsterdam: North-Holland, Elsevier.

Cisternas, P. A., & Byrne, M. (2003). Peptidergic and serotonergic immunoreactivity in the metamorphosing ophiopluteus of Ophiactis resiliens (Echinodermata, Ophiuroidea). *Invertebrate Biology*, 122, 177–185.

Cole, A. G., & Meinertzhagen, I. A. (2004). The central nervous system of the ascidian larva: Mitotic history of cells forming the neural tube in late embryonic *Ciona intestinalis*. *Developmental Biology*, 271, 239–262.

Conklin, E. G. (1897). The embryology of *Crepidula*. *Journal of Morphology*, 13, 1–230.

Conzelmann, M., Offenburger, S. L., Asadulina, A., Keller, T., Munch, T. A., & Jékely, G. (2011). Neuropeptides regulate swimming depth of Platynereis larvae. *Proceedings of the National Academy of Sciences of the USA*, 108, E1174–1183.

Couper, J. M., & Leise, E. M. (1996). Serotonin injections induce metamorphosis in larvae of the gastropod mollusc *Ilyanassa obsoleta*. *Biological Bulletin*, 191, 178–186.

Croll, R. P. (2006). Development of embryonic and larval cells containing serotonin, catecholamines, and FMRFamide-related peptides in the gastropod mollusc *Phestilla sibogae*. *Biological Bulletin*, 211, 232–247.

Croll, R. P. (2009). Developing nervous systems in molluscs: navigating the twists and turns of a complex life cycle. *Brain Behavior & Evolution*, 74, 164–176.

Croll, R. P., Boudko, D. Y., & Hadfield, M. G. (2001). Histochemical survey of transmitters in the central ganglia of the gastropod mollusc, *Phestilla sibogae*. *Cell Tissue Research*, 305, 417–432.

Croll, R. P., & Chiasson B. J. (1989). Post-embryonic development of serotonin-like immunoreactivity in the central nervous system of the snail, *Lymnaea stagnalis*. *Journal of Comparative Neurology*, 280, 122–142.

Croll, R. P., Jackson D. L., & Voronezhskaya, E. E. (1997). Catecholamine-containing cells in larval and postlarval bivalve molluscs. *Biological Bulletin*, 193, 116–124.

Croll, R. P., Voronezhskaya, E. E. (1996). Early elements in gastropod neurogenesis. *Developmental Biology*, 173, 344–347.

Cummins, S. F., Tollenaere, A., Degnan, B. M., & Croll, R. P. (2011). Molecular analysis of two FMRFamide-encoding transcripts expressed during the development of the tropical abalone *Haliotis asinina*. *Journal of Comparative Neurology*, 519, 2043–2059.

Dahms, H. U. (2000). Phylogenetic implications of the Crustacean nauplius. *Hydrobiologia* 417, 91–99.

Denes, A. S., Jékely, G., Steinmetz, P. R., Raible, F., Snyman, H., Prud'homme, B., … Arendt, D. (2007). Molecular architecture of annelid nerve cord supports common origin of nervous system centralization in bilateria. *Cell*, 129, 277–288.

Dickinson, A. J., Croll, R. P., & Voronezhskaya, E. E. (2000). Development of embryonic cells containing serotonin, catecholamines, and FMRFamide-related peptides in *Aplysia californica*. *Biological Bulletin*, 199, 305–315.

Dickinson, A. J., & Croll, R. P. (2003). Development of the larval nervous system of the gastropod *Ilyanassa obsoleta*. *Journal of Comparative Neurology*, 466, 197–218.

Dickinson, A. J., Nason, J., & Croll, R. P. (1999). Histochemical localization of FMRFamide, serotonin and catecholamine in embryonic *Crepidula fornicata* (Prosobranchia: Gastropoda). *Zoomorphology*, 119, 49–62.

Dorris, M., De Ley, P., & Blaxter, M. L. (1999). Molecular analysis of nematode diversity and the evolution of parasitism. *Parasitology Today*, 15, 188–193.

Dunn, C. W., Hejnol, A., Matus, D. Q., Pang, K., Browne, W. E., Smith, S. A., ... Giribet G. (2008). Broad phylogenomic sampling improves resolution of the animal tree of life. *Nature*, 452, 745–749.

Dyachuk, V., & Odintsova, N. (2009). Development of the larval muscle system in the mussel *Mytilus trossulus* (Mollusca, Bivalvia). *Development, Growth & Differentiation*, 51, 69–79.

Egger, B., Steinke, D., Tarui, H., De Mulder, K., Arendt, D., Borgonie, G., ... Ladurner P. (2009). To be or not to be a flatworm: The acoel controversy. *PLoS One* 4:e5502. doi: 10.1371/journal.pone.0005502

Ellis, C. H. J, & Fausto-Sterlingm A. (1997). *Platyhelminths, the flatworms*. Sunderland, MA: Sinauer Associates, Inc.

Eriksson, B. J., & Stollewerk, A. (2010). The morphological and molecular processes of onychophoran brain development show unique features that are neither comparable to insects nor to chelicerates. *Arthropod Structure & Development*, 39, 478–490.

Evans, C. C., Dickinson, A. J. G., & Croll, R. P. (2009). Major muscle systems in the larval caenogastropod, *Ilyanassa obsoleta*, display different patterns of development. *Journal of Morphology*, 270, 1219–1231.

Friedrich, S., Wanninger, A., Bruckner, M., & Haszprunar, G. (2002). Neurogenesis in the mossy chiton, *Mopalia muscosa* (Gould) (Polyplacophora): Evidence against molluscan metamerism. *Journal of Morphology*, 253, 109–117.

Froggett, S. J., & Leise, E. M. (1999). Metamorphosis in the marine snail *Ilyanassa obsoleta*, yes or NO? *Biological Bulletin*, 196, 57–62.

Gerhart, J., Lowe, C., & Kirschner, M. (2001). *Hemichordates: Development*. New York, New York: John Wiley & Sons, Ltd.

Ghysen, A., Damblychaudiere, C., Jan, L. Y., & Jan, Y. N. (1993). Cell interactions and gene interactions in peripheral neurogenesis. *Genes & Development*, 7, 723–733.

Gifondorwa, D. J., & Leise, E. M. (2006). Programmed cell death in the apical ganglion during larval metamorphosis of the marine mollusc *Ilyanassa obsoleta*. *Biological Bulletin*, 210, 109–120.

Gilbert, S. F., & Raunio, A. M., (Eds.). (1997). *Embryology: Constructing the organism*. Sunderland, MA: Sinauer Associates, Inc.

Gillette, R. (1991). On the significance of neuronal giantism in gastropods. *Biological Bulletin*, 180, 234–240.

Giribet, G. (2008). Assembling the lophotrochozoan (=spiralian) tree of life. *Philosophical Transactions of the Royal Society B*, 363, 1513–1522.

Gruhl, A. (2009). Serotonergic and FMRFamidergic nervous systems in gymnolaemate bryozoan larvae. *Zoomorphology* 128, 135–156.

Gustafsson, M. K. S., & Terenina, N. B. (2003). Nitric oxide and its target cells in cercaria of *Diplostomum chromatophorum*: a histochemical and immunocytochemical study. *Parasitology Research*, 89, 199–206.

Hadfield, M. G. (2000). Why and how marine-invertebrate larvae metamorphose so fast. *Semin Cell Developmental Biology*, 11, 437–443.

Hadfield, M. G., Meleshkevitch, E. A., & Boudko, D. Y. (2000). The apical sensory organ of a gastropod veliger is a receptor for settlement cues. *Biological Bulletin*, 198, 67–76.

Hall B. K., & Wake, M. H., (Eds.). (1999). *The origin and evolution of larval forms*. San Diego, London, Boston: Academic Press.

Harris, W. A., & Hartenstein, V. (2008). Cellular determination. In L. Squire, D. Berg, & F. E. Bloom (Eds.), *Fundamental neuroscience* (3rd ed., pp. 321–349). Burlington, MA: Academic Press.

Harrison, P. J. H., & Sandeman, D. C. (1999). Morphology of the nervous system of the barnacle cypris larva (*Balanus amphitrite* Darwin) revealed by light and electron microscopy. *Biological Bulletin*, 197, 144–158.

Hartenstein, V., Spindler, S., Pereanu, W., & Fung, S. (2008). The development of the Drosophila larval brain. *Advances in Experimental Medicine & Biology*, 628, 1–31.

Harzsch, S. (2004). Phylogenetic comparison of serotonin-immunoreactive neurons in representatives of the chilopoda diplopoda, and chelicerata: Implications for arthropod relationships. *Journal of Morphology*, 259, 198–213.

Haszprunar, G., & Wanninger, A. (2008). On the fine structure of the creeping larva of *Loxosomella murmanica*: additional evidence for a clade of Kamptozoa (Entoprocta) and Mollusca. *Acta Zoologica (Stockholm)*, 89, 137–148.

Hausdorf, B., Helmkampf, M., Nesnidal, M. P., & Bruchhaus, I. (2010). Phylogenetic relationships within the lophophorate lineages (Ectoprocta, Brachiopoda and Phoronida). *Molecular Phylogenetics & Evolution*, 55, 1121–1127.

Hay-Schmidt, A. (1989). The nervous system of the actinotroch larva of *Phoronis muelleri* (Phoronida). *Zoomorphology*, 108, 333–351.

Hay-Schmidt, A. (1990a). Catecholamine-containing, serotonin-like, and FMRFamide-like immunoreactive neurons and processes in the nervous system of the early actinotroch larva of *Phoronis vancouverensis* (Phoronida)—Distribution and development. *Canadian Journal of Zoology*, 68, 1525–1536.

Hay-Schmidt, A. (1990b). Distribution of catecholamine-containing, serotonin-like and neuropeptide FMRFamide-like immunoreactive neurons and processes in the nervous system of the actinotroch larva of *Phoronis muelleri* (Phoronida). *Cell Tissue Research*, 259, 105–118.

Hay-Schmidt, A. (1990c). Distribution of catecholamine-containing, serotonin-like and neuropeptide FMRFamide-like immunoreactive neurons and processes in the nervous system of the early actinotroch larva of *Phoronis vancouverensis* (Phoronida): distribution and development. *Canadian Journal of Zoology*, 68, 1525–1536.

Hay-Schmidt, A. (1992). Ultrastructure and immunocytochemistry of the nervous system of the larvae of *Lingula anatina* and *Glottidia Sp* (Brachiopoda). *Zoomorphology*, 112, 189–205.

Hay-Schmidt, A. (1995). The larval nervous system of *Polygordius lacteus* Schiender, 1868 (Polygordiidae, Polychaeta): Immunocytochemical data. *Acta Zoologica (Stockholm)*, 76, 121–140.

Hay-Schmidt, A. (2000). The evolution of the serotonergic nervous system. *P Roy Soc B-Biol Sci* 267, 1071–1079.

Hejnol, A., Martindale, M. Q., & Henry, J. Q. (2007). High-resolution fate map of the snail *Crepidula fornicata*: The origins of ciliary bands, nervous system, and muscular elements. *Developmental Biology*, 305, 63–76.

Helmkampf, M., Bruchhaus, I., & Hausdorf, B. (2008). Phylogenomic analyses of lophophorates (brachiopods, phoronids and bryozoans) confirm the Lophotrochozoa concept. *Proceedings of the Royal Society B, Biological Sciences*, 275, 1927–1933.

Hens, M. D., Fowler, K. A., & Leise, E. M. (2006). Induction of metamorphosis decreases nitric oxide synthase gene expression in larvae of the marine mollusc *Ilyanassa obsoleta* (Say). *Biological Bulletin*, 211, 208–211.

Hessling, R. (2002). Metameric organisation of the nervous system in developmental stages of Urechis caupo (Echiura) and its phylogenetic implications. *Zoomorphology*, 121, 221–234.

Hessling, R., & Westheide, W. (2002). Are Echiura derived from a segmented ancestor? Immunohistochemical analysis of the nervous system in developmental stages of *Bonellia viridis*. *Journal of Morphology*, 252, 100–113.

Hochberg, R. (2007). Comparative immunohistochemistry of the cerebral ganglion in Gastrotricha: an analysis of FMRFamide-like immunoreactivity in *Neodasys cirritus* (Chaetonotida), *Xenodasys riedli* and *Turbanella cf. hyalina* (Macrodasyida). *Zoomorphology*, 126, 245–264.

Hochberg, R., & Lilley, G. (2010). Neuromuscular organization of the freshwater colonial rotifer, Sinantherina socialis, and its implications for understanding the evolution of coloniality in Rotifera. *Zoomorphology*, 129, 153–162.

Holland, N. D., & Yu, J. K. (2002). Epidermal receptor development and sensory pathways in vitally stained amphioxus (*Branchiostoma floridae*). *Acta Zoologica (Stockholm)*, 83, 309–319.
Horvitz, H. R. (2003). Worms, life, and death (Nobel lecture). *Chembiochem*, 4, 697–711.
Horvitz, H. R., Sternberg, P. W., Greenwald, I. S., Fixsen, W., & Ellis, H. M. (1983). Mutations that affect neural cell lineages and cell fates during the development of the nematode *Caenorhabditis elegans*. *Cold Spring Harbor Symposia on Quantitative Biology*, 48 Pt. 2, 453–463.
Imai, J. H., & Meinertzhagen, I. A. (2007a). Neurons of the ascidian larval nervous system in *Ciona intestinalis*: I. Central nervous system. *Journal of Comparative Neurology*, 501, 316–334.
Imai, J. H., & Meinertzhagen, I. A. (2007b). Neurons of the ascidian larval nervous system in *Ciona intestinalis*: II. Peripheral nervous system. *Journal of Comparative Neurology*, 501, 335–352.
Jacob, M. H. (1984). Neurogenesis in *Aplysia californica* resembles nervous system formation in vertebrates. *Journal of Neuroscience*, 4, 1225–1239.
Jager, M., Chiori, R., Alié, A., Dayraud, C., Quéinnec, E., & Manuel, M. (2011). New insights on ctenophore neural anatomy: Immunofluorescence study in *Pleurobrachia pileus* (Müller, 1776). *Journal of Experimental Zoology, B Molecular Development & Evolution*, 316B, 171–187.
Jan, Y. N., & Jan, L. Y. (1994). Genetic control of cell fate specification in *Drosophila* peripheral nervous system. *Annual Review of Genetics*, 28, 373–393.
Jeffery, W. R. (2002). Programmed cell death in the ascidian embryo: modulation by FoxA5 and Manx and roles in the evolution of larval development. *Mechanisms of Development*, 118, 111–124.
Jeffery, W. R., & Swalla, B. J. (1997). Tunicates. In S. F. Gilbert & A. M. Raunio (Eds.), *Embryology: Constructing the organism*. Sunderland, MA: Sinauer Associates, Inc.
Jékely, G., Colombelli, J., Hausen, H., Guy, K., Stelzer, E., Nedelec, F., & Arendt, D. (2008). Mechanism of phototaxis in marine zooplankton. *Nature*, 456, 395–399.
Kandel, E. R., Kriegstein, A., & Schacher, S. (1981). Development of the central nervous system of *Aplysia* in terms of the differentiation of its specific identifiable cells. *Neuroscience*, 5, 2033–2063.
Kemmerling, U., Cabrera, G., Campos, E. O., Inestrosa, N. C., & Galanti, N. (2006). Localization, specific activity, and molecular forms of acetylcholinesterase in developmental stages of the cestode *Mesocestoides corti*. *Journal of Cell Physiology*, 206, 503–509.
Kempf, S. C., Chun, G. V., & Hadfield, M. G. (1992). An immunocytochemical search for potential neurotransmitters in larvae of *Phestilla sibogae* (Gastropoda, Opisthobranchia). *Comparative Biochemistry & Physiology*, 101C, 299–305.
Kempf, S. C., Page, L. R., & Pires, A. (1997). Development of serotonin-like immunoreactivity in the embryos and larvae of nudibranch mollusks with emphasis on the structure and possible function of the apical sensory organ. *Journal of Comparative Neurology*, 386, 507–528.
Koop, D., & Holland, L. Z. (2008). The basal chordate amphioxus as a simple model for elucidating developmental mechanisms in vertebrates. *Birth Defects Research C, Embryo Today* 84, 175–187.
Kristof, A., Wollesen, T., & Wanninger, A. (2008). Segmental mode of neural patterning in sipuncula. *Current Biology*, 18, 1129–1132.
Kuang, S., Doran, S. A., Wilson, R. J., Goss, G. G., & Goldberg, J. I. (2002). Serotonergic sensory-motor neurons mediate a behavioral response to hypoxia in pond snail embryos. *Journal of Neurobiology*, 52, 73–83.
Lacalli, T. C. (1982). The nervous system and ciliary band of Muller's larva. *Proc Roy Soc Lond B* 217, 37–58.

Lacalli, T. C. (1983). The brain and central nervous system of Muller larva. *Can Journal of Zoology*, 61, 39–51.

Lacalli, T. C. (1984). Structure and organization of the nervous system in the trochophore larva of *Spirobranchus*. *Philosophical Transactions of the Royal Society B*, 306, 79–&.

Lacalli, T. C. (1990). Structure and organization of the nervous system in the actinotroch larva of *Phoronis vancouverensis*. *Philosophical Transactions of the Royal Society B*, 327, 655–685.

Lacalli, T. C. (1996). Frontal eye circuitry, rostral sensory pathways and brain organization in amphioxus larvae: Evidence from 3D reconstructions. *Philosophical Transactions of the Royal Society B*, 351, 243–263.

Lacalli, T. C. (2003). Ventral neurons in the anterior nerve cord of amphioxus larvae. II. Further data on the pacemaker circuit. *Journal of Morphology*, 257, 212–218.

Lacalli, T. C. (2004). Sensory systems in amphioxus: a window on the ancestral chordate condition. *Brain Behavior & Evolution*, 64, 148–162.

Lacalli, T. C., Gilmour, T. H. J., & Kelly, S. J. (1999). The oral nerve plexus in amphioxus larvae: function, cell types and phylogenetic significance. *Proceedings of the Royal Society of London B, Biological Sciences*, 266, 1461–1470.

Lacalli, T. C., & Holland, L. Z. (1998). The developing dorsal ganglion of the salp *Thalia democratica*, and the nature of the ancestral chordate brain. *Philosophical Transactions of the Royal Society B*, 353, 1943–1967.

Lacalli, T. C., & Kelly, S. J. (2003). Ventral neurons in the anterior nerve cord of amphioxus larvae. I. An inventory of cell types and synaptic patterns. *Journal of Morphology*, 257, 190–211.

Leise, E. M., & Hadfield, M. G. (2000). An inducer of molluscan metamorphosis transforms activity patterns in a larval nervous system. *Biological Bulletin*, 199, 241–250.

Lin, M. F., & Leise, E. M. (1996). Gangliogenesis in the prosobranch gastropod *Ilyanassa obsoleta*. *Journal of Comparative Neurology*, 374, 180–193.

Lowe, C. J., Wu, M., Salic, A., Evans, L., Lander, E., Stange-Thomann, N., ... Kirschner M. (2003). Anteroposterior patterning in hemichordates and the origins of the chordate nervous system. *Cell*, 113, 853–865.

Luter, C. (1996). The median tentacle of the larva of *Lingula anatina* (Brachiopoda) from Queensland, Australia. *Australian Journal of Zoology*, 44, 355–366.

Marlow, H. Q., Srivastava, M., Matus, D. Q., Rokhsar, D., & Martindale, M. Q. (2009). Anatomy and development of the nervous system of *Nematostella vectensis*, an anthozoan cnidarian. *Developmental Neurobiology*, 69, 235–254.

Marois, R., & Carew, T. J. (1997a). Fine structure of the apical ganglion and its serotonergic cells in the larva of *Aplysia californica*. *Biological Bulletin*, 192, 388–398.

Marois, R., & Carew, T. J. (1997b). Ontogeny of serotonergic neurons in *Aplysia californica*. *Journal of Comparative Neurology*, 386, 477–490.

Marois, R., & Carew, T. J. (1997c). Projection patterns and target tissues of serotonergic cells in larval *Aplysia californica*. *Journal of Comparative Neurology*, 386, 491–506.

Martin, V. J. (1988). Development of nerve cells in hydrozoan planulae .1. Differentiation of ganglionic Cells. *Biological Bulletin*, 174, 319–329.

Martin, V. J. (1992). Characterization of a RFamide-positive subset of ganglionic cells in the hydrozoan planular nerve net. *Cell Tissue Research*, 269, 431–438.

Martin, V. J. (1997). Cnidarians, the jellyfish and hydras. In S. F. Gilbert & A. M. Raunio (Eds.), *Embryology: Constructing the Organism*. Sunderland, MA, USA: Sinauer Associates, Inc. p 537.

Martindale, M. Q., & Henry, J. (1997). Ctenophorans, the comb jellies. In S. F. Gilbert & A. M. Raunio (Eds.), *Embryology: Constructing the organism*. Sunderland, MA: Sinauer Associates, Inc.

McAllister, L. B., Scheller, R. H., Kandel, E. R., & Axel, R. (1983). *In situ* hybridization to study the origin and fate of identified neurons. *Science*, 222, 800–808.

McDougall, C., Chen, W. C., Shimeld, S. M., & Ferrier, D. E. (2006). The development of the larval nervous system, musculature and ciliary bands of *Pomatoceros lamarckii* (Annelida): Heterochrony in polychaetes. *Frontiers in Zoology*, 3, 16. doi: 10.1186/1742-9994-3-16

Meinertzhagen, I. A., Lemaire, P., & Okamura, Y. (2004). The neurobiology of the ascidian tadpole larva: Recent developments in an ancient chordate. *Annual Review of Neuroscience*, 27, 453–485.

Meyer, N. P., & Seaver, E. C. (2009). Neurogenesis in an annelid: characterization of brain neural precursors in the polychaete *Capitella* sp. I. *Developmental Biology*, 335, 237–252.

Miyamoto, N., Nakajima, Y., Wada, H., & Saito, Y. (2010). Development of the nervous system in the acorn worm *Balanoglossus simodensis*: Insights into nervous system evolution. *Evolution & Development*, 12, 416–424.

Nezlin, L. P., & Yushin, V. V. (2004). Structure of the nervous system in the tornaria larva of *Balanoglossus proterogonius* (Hemichordata : Enteropneusta) and its phylogenetic implications. *Zoomorphology*, 123, 1–13.

Nielsen, C., (1987). Structure and function of metazoan ciliary bands and their phylogenetic significance. *Acta Zoologica*, 68, 205–262.

Nielsen, C., (2001). *Animal evolution: Interrelationships of the living phyla* (2nd ed.). New York, NY: Oxford University Press.

Nielsen, C., (2005). Larval and adult brains. *Evolution & Development*, 7, 483–489.

Nielsen, C., & Hay-Schmidt, A. (2007). Development of the enteropneust Ptychodera flava: Ciliary bands and nervous system. *Journal of Morphology*, 268, 551–570.

Page, L. R. (1997). Larval shell muscles in the abalone *Haliotis kamtschatkana*. *Biological Bulletin*, 193, 30–46.

Page, L. R. (2002). Apical sensory organ in larvae of the patellogastropod *Tectura scutum*. *Biological Bulletin*, 202, 6–22.

Page, L. R., & Parries, S. C. (2000). Comparative study of the apical ganglion in planktotrophic caenogastropod larvae: ultrastructure and immunoreactivity to serotonin. *Journal of Comparative Neurology*, 418, 383–401.

Paris, M., Escriva, H., Schubert, M., Brunet, F., Brtko, J., Ciesielski, F., ... Laudet V. (2008). Amphioxus postembryonic development reveals the homology of chordate metamorphosis. *Current Biology*, 18, 825–830.

Pechenik, J. A. (2005). *Biology of invertebrates* (5th ed.). Boston, Montreal: McGraw Hill.

Peterson, K. J., & Eernisse, D. J. (2001). Animal phylogeny and the ancestry of bilaterians: inferences from morphology and 18S rDNA gene sequences. *Evolution & Development*, 3, 170–205.

Philippe, H., Brinkmann, H., Copley, R. R., Moroz, L. L., Nakano, H., Poustka, A. J., ... Telford, M. J. (2011). Acoelomorph flatworms are deuterostomes related to Xenoturbella. *Nature*, 470, 255–258.

Philippe, H., Lartillot, N., & Brinkmann, H. (2005). Multigene analyses of bilaterian animals corroborate the monophyly of ecdysozoa, lophotrochozoa, and protostomia. *Molecular Biology & Evolution*, 22, 1246–1253.

Pilger, J. F. (1997). Spinculans and echinurans. In S. F. Gilbert & A. M. Raunio (Eds.), *Embryology: Constructing the organism*. Sunderland, MA: Sinauer Associates, Inc. p 167–188.

Piraino, S., Zega, G., Di Benedetto, C., Leone, A., Dell'Anna, A., Pennati, R., ... Reichert H. (2011). Complex neural architecture in the diploblastic larva of *Clava multicornis* (Hydrozoa, Cnidaria). *Journal of Comparative Neurology*, 519, 1931–1951.

Pires, A., & Woollacott, R. M. (1997). Serotonin and dopamine have opposite effects on phootaxis in larvae of the bryozoan *Bugula neritina*. *Biological Bulletin*, 192, 399–409.

Raff, R. A., & Byrne, M. (2006). The active evolutionary lives of echinoderm larvae. *Heredity* 97, 244–252.

Raper, J., & Mason, C. (2010). Cellular strategies of axonal pathfinding. *Cold Spring Harbor Perspectives in Biology*, 2. doi: 10.1101/chsperspect.a001933

Raven, C. P. (1966). *Morphogenesis: The Analysis of Molluscan Development*. Oxford: Pergamon Press.

Rawlinson, K. A. (2010). Embryonic and post-embryonic development of the polyclad flatworm *Maritigrella crozieri*; implications for the evolution of spiralian life history traits. *Frontiers in Zoology*, 7, 12.

Regier, J. C., Shultz, J. W., Zwick, A., Hussey, A., Ball, B., Wetzer, R., ... Cunningham, C. W. (2010). Arthropod relationships revealed by phylogenomic analysis of nuclear protein-coding sequences. *Nature*, 463, 1079–1083.

Rice, M. E. (1975). Observations on the development of six species of Caribbean Sipuncula with a review of development in the phylum. In M. E. Rice, & M. Todorovic (Eds.), *Proceedings of the international symposium on Sippuncula and Echiura* (pp. 141–160). Belgrade: Naucno Delo Press.

Rota-Stabelli, O., Kayal, E., Gleeson, D., Daub, J., Boore, J. L., Telford, M. J., ... Lavrov, D. V. (2010). Ecdysozoan mitogenomics: Evidence for a common origin of the legged invertebrates, the Panarthropoda. *Genome Biology & Evolution*, 2, 425–440.

Ryberg, E. (1977). Nervous system of early echinopluteus. *Cell Tissue Research*, 179, 157–167.

Santagata, S. (2002). Structure and metamorphic remodeling of the larval nervous system and musculature of *Phoronis pallida* (Phoronida). *Evolution & Development*, 4, 28–42.

Santagata, S. (2008). Evolutionary and structural diversification of the larval nervous system among marine bryozoans. *Biological Bulletin*, 215, 3–23.

Santagata, S, & Zimmer, R. L. (2000). Muscle, neural, and epithelial tissue variation between larval and presumptive juvenile structures in different species of actinotroch larvae (Phoronida). *American Zoologist*, 40, 1197–1197.

Santagata, S., & Zimmer, R. L. (2002). Comparison of the neuromuscular systems among actinotroch larvae: systematic and evolutionary implications. *Evolution & Development*, 4, 43–54.

Satoh, G., Wang, Y., Zhang, P., & Satoh, N. (2001). Early development of amphioxus nervous system with special reference to segmental cell organization and putative sensory cell precursors: A study based on the expression of pan-neuronal marker gene Hu/elav. *Journal of Experimental Zoology*, 291, 354–364.

Schacher, S., Kandel, E. R., & Woolley, R. (1979). Development of neurons in the abdominal ganglion of *Aplysia calfornica* I. Axosomatic synaptic contacts. *Developmental Biology*, 71, 176–190.

Schierenberg, E. (1997). Nematodes, the roundworms. In S. F. Gilbert & A. M. Raunio (Eds.), *Embryology: Constructing the organism*. Sunderland, MA: Sinauer Associates, Inc.

Schwalm, F. E. (1997). Arthropods: The insects. In S. F. Gilbert & A. M. Raunio (Eds.), *Embryology: Constructing the organism*. Sunderland, MA: Sinauer Associates, Inc.

Semmler, H., Wanninger, A., Hoeg, J. T., & Scholtz, G. (2008). Immunocytochemical studies on the naupliar nervous system of *Balanus improvisus* (Crustacea, Cirripedia, Thecostraca). *Arthropod Structure & Development*, 37, 383–395.

Sempere, L. F., Martinez, P., Cole, C., Baguna, J., & Peterson, K. J. (2007). Phylogenetic distribution of microRNAs supports the basal position of acoel flatworms and the polyphyly of Platyhelminthes. *Evolution & Development*, 9, 409–415.

Shigeno, S., Tsuchiya, K., & Segawa, S. (2001). Embryonic and paralarval development of the central nervous system of the loliginid squid *Sepioteuthis lessoniana*. *Journal of Comparative Neurology*, 437, 449–475.

Stent, G. S., Kristan, W. B., Torrence, S. A., French, K. A., & Weisblat, D. A. (1992). Development of the leech nervous system. *International Review of Neurobiology*, 33, 109–193.

Stollewerk, A., & Chipman, A. D. (2006). Neurogenesis in myriapods and chelicerates and its importance for understanding arthropod relationships. *Integrative & Comparative Biology*, 46, 195–206.

Stollewerk, A., & Simpson, P. (2005). Evolution of early development of the nervous system: a comparison between arthropods. *Bioessays*, 27, 874–883.

Strathmann, R. R. (1975). Larval feeding in echinoderms. *American Zoologist*, 15, 717–730.

Strausfeld, N. J., Strausfeld, C. M., Stowe, S., Rowell, D., & Loesel, R. (2006). The organization and evolutionary implications of neuropils and their neurons in the brain of the onychophoran *Euperipatoides rowelli*. *Arthropod Structure & Development*, 35, 169–196.

Struck, T. H., Schult, N., Kusen, T., Hickman, E., Bleidorn, C., McHugh, D., & Halanych, K. M. (2007). Annelid phylogeny and the status of Sipuncula and Echiura. *BMC Evolutionary Biology*, 7, 57.

Technau, G. M., & Urbach, R. (2004). Neuroblast formation and patterning during early brain development in *Drosophila*. *Bioessays*, 26, 739–751.

Telford, M. J., Bourlat, S. J., Economou, A., Papillon, D., & Rota-Stabelli, O. (2008). The evolution of the Ecdysozoa. *Philosophical Transactions of the Royal Society B*, 363, 1529–1537.

Tiozzo, S., Murray, M., Degnan, B. M., De Tomaso, A. W., & Croll, R. P. (2009). Development of the neuromuscular system during asexual propagation in an invertebrate chordate. *Developmental Dynamics*, 238, 2081–2094.

Tissot, M., & Stocker, R. F. (2000). Metamorphosis in *Drosophila* and other insects: the fate of neurons throughout the stages. *Progress in Neurobiology*, 62, 89–111.

Truman, J. W., & Riddiford, L. M. (1999). The origins of insect metamorphosis. *Nature*, 401, 447–452.

Tsutsui H, Yamamoto N, Ito H, & Oka Y. (1998). GnRH-immunoreactive neuronal system in the presumptive ancestral chordate, *Ciona intestinalis* (Ascidian). *General & Comparative Endocrinology*, 112, 426–432.

Urbach, R., Schnabel, R., & Technau, G. M. (2003). The pattern of neuroblast formation, mitotic domains and proneural gene expression during early brain development in *Drosophila*. *Development*, 130, 3589–3606.

Vilpoux, K., Sandeman, R., & Harzsch, S. (2006). Early embryonic development of the central nervous system in the Australian crayfish and the Marbled crayfish (*Marmorkrebs*). *Development Genes & Evolution*, 216, 209–223.

Voronezhskaya, E. E., & Elekes, K. (2003). Expression of FMRFamide gene encoded peptides by identified neurons in embryos and juveniles of the pulmonate snail *Lymnaea stagnalis*. *Cell Tissue Research*, 314, 297–313.

Voronezhskaya, E. E., Hiripi, L., Elekes, K., & Croll, R. P. (1999). Development of catecholaminergic neurons in the pond snail, *Lymnaea stagnalis*: I. Embryonic development of dopamine-containing neurons and dopamine-dependent behaviors. *Journal of Comparative Neurology*, 404, 285–296.

Voronezhskaya, E. E., & Ivashkin, E. G. (2010). Pioneer neurons: A basis or limiting factor of lophotrochozoa nervous system diversity? *Russian Journal of Developmental Biology*, 41, 337–346.

Voronezhskaya, E. E., & Khabarova, M. Y. (2003). Function of the apical sensory organ in the development of invertebrates. *Doklady Biological Sciences*, 390, 231–234.

Voronezhskaya, E. E., Khabarova, M. Y., & Nezlin, L. P. (2004). Apical sensory neurones mediate developmental retardation induced by conspecific environmental stimuli in freshwater pulmonate snails. *Development*, 131, 3671–3680.

Voronezhskaya, E. E., Nezlin, L. P., Odintsova, N. A., Plummer, J. T., & Croll, R. P. (2008). Neuronal development in larval mussel *Mytilus trossulus* (Mollusca: Bivalvia). *Zoomorphology*, 127, 97–110.

Voronezhskaya, E. E., Tsitrin, E. B., & Nezlin, L. P. (2003). Neuronal development in larval polychaete *Phyllodoce maculata* (Phyllodocidae). *Journal of Comparative Neurology*, 455, 299–309.

Voronezhskaya, E. E., Tyurin, S. A., & Nezlin, L. P. (2002). Neuronal development in larval chiton *Ischnochiton hakodadensis* (Mollusca: Polyplacophora). *Journal of Comparative Neurology*, 444, 25–38.

Wada, Y., Mogami, Y., & Baba, S. A. (1997). Modification of ciliary beating inn sea urchin larvae induced by neurotransmitters: Beat-plane rotatioin and control of frequency fluctuation. *Journal of Experimental Biology*, 200, 9–18.

Wadsworth, W. G., & Hedgecock, E. M. (1992). Guidance of neuroblast migrations and axonal projections in *Caenorhabditis elegans*. *Current Opinion in Neurobiology*, 2, 36–41.

Wadsworth, W. G., & Hedgecock, E. M. (1996). Hierarchical guidance cues in the developing nervous system of *C. elegans*. *Bioessays*, 18, 355–362.

Wanninger, A., Fuchs, J., Bright, M., & Funch, P. (2006). Immunocytochemistry of the neuromuscular systems of *Loxosomella vivipara* and *Loxosomella parguerensis* (Entoprocta: Loxosomatidae). *Journal of Morphology*, 267, 866–883.

Wanninger, A., Fuchs, J., & Haszprunar, G. (2007). Anatomy of the serotonergic nervous system of an entoproct creeping-type larva and its phylogenetic implications. *Invertebrate Biology*, 126, 268–278.

Wanninger, A., & Haszprunar, G. (2003). The development of the serotonergic and FMRFaminergic nervous system in *Antalis entalis* (Mollusca, Scaphopoda). *Zoomorphology*, 122, 77–85.

Wanninger, A., Koop, D., Bromham, L., Noonan, E., & Degnan, B. M. (2005). Nervous and muscle system development in *Phascolion strombus* (Sipuncula). *Development Genes & Evolution*, 215, 509–518.

Wanninger, A., Koop, D., Degnan, B. M. (2005). Immunocytochemistry and metamorphic fate of the larval nervous system of *Triphyllozoon mucronatum* (Ectoprocta: Gymnolaemata: Cheilostomata). *Zoomorphology*, 124, 161–170.

Wanninger, A., Ruthensteiner, B., Lobenwein, S., Salvenmoser, W., Dictus, W. J., & Haszprunar, G. (1999). Development of the musculature in the limpet Patella (Mollusca, Patellogastropoda). *Development Genes & Evolution*, 209, 226–238.

White, J. G., Southgate, E., Thomson, J. N., & Brenner, S. (1986). The structure of the nervous system of the nematode *Caenorhabditis elegans*. *Philosophical Transactions of the Royal Society B*, 314, 1–340.

Whitington, P. M., & Mayer, G. (2011). The origins of the arthropod nervous system: Insights from the Onychophora. *Arthropod Structure & Development*, 40, 193–209.

Whittaker, J. R. (1997). Cephalochordates, the lancelets. In S.F. Gilbert & A.M. Raunio (Eds.), *Embryology: Constructing the organism*. Sunderland, MA: Sinauer Associates, Inc. p 365–381.

Wicht, H., & Lacalli, T. C. (2005). The nervous system of amphioxus: Structure, development, and evolutionary significance. *Canadian Journal of Zoology*, 83, 122–150.

Wollesen, T., Cummins, S. F., Degnan, B. M., & Wanninger, A. (2010). FMRFamide gene and peptide expression during central nervous system development of the cephalopod mollusk, *Idiosepius notoides*. *Evolution & Development*, 12, 113–130.

Wollesen, T., Degnan, B. M., & Wanninger, A. (2010). Expression of serotonin (5–HT) during CNS development of the cephalopod mollusk, *Idiosepius notoides*. *Cell Tissue Research*, 342, 161–178.

Wray, G. A. (1997). Echinoderms. In S. F. Scott & A.M. Raunio (Eds.), *Embryology: Constructing the organism* (pp. 309–364). Sutherland, MA: Sinauer Associates.

Yaguchi, S., & Katow, H. (2003). Expression of tryptophan 5-hydroxylase gene during sea urchin neurogenesis and role of serotonergic nervous system in larval behavior. *Journal of Comparative Neurology*, 466, 219–229.

Young, C. M., Sewell, M. A., Rice, M. E., (Eds.). (2002). *Atlas of marine invertebrate larvae*. New York: Academic Press.

Younossi-Hartenstein, A., Ehlers, U., & Hartenstein, V. (2000). Embryonic development of the nervous system of the rhabdocoel flatworm *Mesostoma lingua* (Abildgaard, 1789). *Journal of Comparative Neurology*, 416, 461–474.

Younossi-Hartenstein, A., & Hartenstein, V. (2000). The embryonic development of the polyclad flatworm *Imogine mcgrathi*. *Development Genes & Evolution*, 210, 383–398.

Younossi-Hartenstein, A., Jones, M., & Hartenstein, V. (2001). Embryonic development of the nervous system of the temnocephalid flatworm *Craspedella pedum*. *Journal of Comparative Neurology*, 434, 56–68.

Yuan, D., Nakanishi, N., Jacobs, D. K., & Hartenstein, V. (2008). Embryonic development and metamorphosis of the scyphozoan *Aurelia*. *Development Genes & Evolution*, 218, 525–539.

Zimmer, R. L. (1997). The lophophorate phyla: Phoronida, brachiopoda, and bryozoa. In S. F. Gilbert & A. M. Raunio (Eds.), *Embryology: Constructing the organism*. Sunderland, MA: Sinauer Associates, Inc.

Zipursky S. L. (1989). Molecular and genetic analysis of *Drosophila* eye development: Sevenless, bride of sevenless and rough. *Trends in Neuroscience*, 12, 183–189.

12

Forebrain Development in Vertebrates

The Evolutionary Role of Secondary Organizers

Luis Puelles

12.1 Introduction

The present chapter explores the advantages of studying the evolution of vertebrate forebrain structures from the perspective provided by secondary organizers. The nature of these transient developmental mechanisms will be covered in some detail. I emphasize from the beginning that the rationale for this approach depends on specific assumptions about the forebrain's fundamental morphologic organization, or bauplan, since this provides the background for detailed analysis of developmental patterns and their consequences. The bauplan of vertebrate brains is captured schematically by morphological models, which must be tested for validity across phylogeny. I shall comment in detail about the alternative brain models that are presently available, pointing out the improvements offered by the recently updated prosomeric model (Puelles, 2013; Puelles, L., Martínez-de-la-Torre, Bardet, & Rubenstein, 2012). One aim of this essay is to explain conceptual difficulties that arise from the use of obsolete models.

In this framework, those aspects of morphology and structure which are not constrained by the bauplan (e.g., growth of a smaller or larger cerebral cortex) represent what changes in forebrain evolution. Instead, the conserved bauplan specifies a set of invariant topological and morphological features which seem to be heavily buttressed against genetic change (Puelles & Ferran, 2012; Puelles & Medina, 2002). To properly establish this concept, I describe our best current forebrain model, contrast it with competing models, and detail its application to forebrain development and evolution in early vertebrates.

12.2 The Prosomeric Model of Brain Regionalization

12.2.1 The Prosomeric Model and Secondary Organizers

The prosomeric model (see Figure 12.1) defines the longitudinal and transverse dimensions of the developing neural plate, closed neural tube, and eventual adult brain, and traces topologically the positions and fates of relevant anteroposterior (AP)

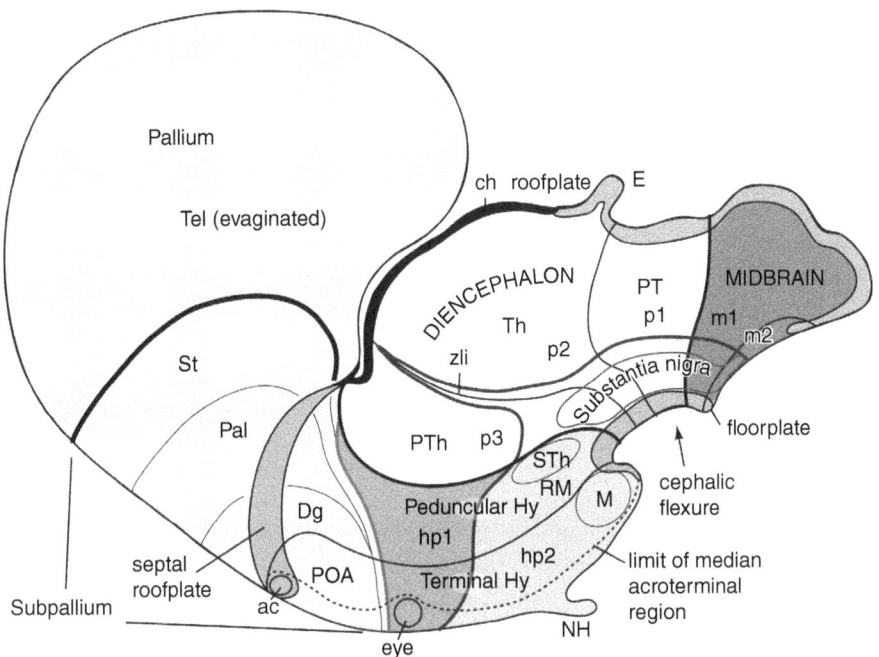

Figure 12.1 Forebrain Prosomeric Model. From Puelles et al., 2012.
The roof and floor plates are shaded gray; note, the choroidal roof (ch) is solid black. The alar-basal boundary plus the *zona limitans intrathalamica* (zli) are represented by the red line; note, this line joins its contralateral counterpart near the eye stalk. There are three proneuromeric regions: rostrally, the secondary prosencephalon comprises the telencephalon (blue) and the eye and hypothalamus (yellow and pink). The largest part of the telencephalon is evaginated, and is thus drawn as seen beyond the midline structures, that is, the septal roof plate and anterior commissure, ac. The telencephalon is delimited from the hypothalamus along a darker blue longitudinal line; the alar–basal boundary separates the hypothalamus into alar (yellow) and basal (pink) parts. The diencephalon appears next in the caudal direction, and is also divided into alar and basal parts (red line, with transverse detour at the zli). Caudally, there is the midbrain (green), also divided into alar and basal parts (red line). Anteroposterior neuromeric subdivisions are separated by transverse black lines that extend from the roof to the floor. The secondary prosencephalon comprises hypothalamic prosomeres, hp2 and hp1, stretched across hypothalamus and telencephalon. Note, the hypothalamus (Hy) is resultantly divided into terminal and peduncular parts; the former terminates in the preoptic area of the telencephalon and the latter expands into the whole evaginated telencephalon. Note, as well, the pallio-subpallial boundary (thick black line) within the hemisphere, and the diverse subpallial subdivisions (St, Pal, Dg, and POA, detailed in Puelles et al., 2013). The acroterminal region is a medial hypothalamic and preoptic locus (part of terminal hypothalamus) where right and left halves of the alar and basal plates are continuous across the midline; the neurohypophysis, NH, is a basal tuberal specialization within this medial domain. The alar domains of the diencephalic prosomeres, labelled p3, p2 and p1, generate respectively the prethalamic (PTh), thalamic (Th) and pretectal (PT) nuclear formations; the epiphysis, E, is a roof plate specialization within p2. Finally, the midbrain subdivides into mesomeres m2 and m1. The m2 segment is bounded posteriorly by the hindbrain isthmus (not shown). Four anatomic landmarks have been added: the mamillary and retromamillary areas (M and RM); the subthalamic nucleus (STh), which originates within RM and migrates dorsalward within the basal plate); and the substantia nigra, which spans across several neuromeres throughout the midbrain and diencephalon. (*See insert for color representation of the figure*).

and dorsoventral (DV) territories, respectively called "neuromeres" and "longitudinal zones," and their respective boundaries. Importantly, the model recognizes a *bent* longitudinal brain axis (due to differential growth), and also attends to the elastic framework of radial glial processes in the neural wall. The boundaries that divide the latter into longitudinal and transverse domains can be identified roughly using morphological methods, but are revealed more precisely by molecular mapping ones (genoarchitecture). There are four primary longitudinal zones (floor, basal, alar, and roof plates; each of them comprises distinct microzones) and 19 segmental, transversally delimited brain neuromeric units (5 in the forebrain, 2 in the midbrain, and 12 in the hindbrain), positioned serially along the AP dimension of the supra-spinal neural tube (Orr, 1887; Puelles, 2013). Whereas the number of these units is conserved among vertebrate brains, the number of comparable spinal neuromeres adapts to the number of vertebrae, and is therefore variable. Neuromeres emerge as molecular/histogenetic subdivisions of earlier protovesicles known as "proneuromeres". The classically-defined forebrain, located rostral to the midbrain, is composed of two such proneuromeres, "secondary prosencephalon" and "diencephalon" (the sum of which gives the "primary prosencephalon"); these further subdivide into prosencephalic neuromeres called "prosomeres". According to the recently updated prosomeric model (Puelles, L., Martínez-de-la-Torre et al., 2012; Puelles, 2013), the secondary prosencephalon subdivides further into two hypothalamo-telencephalic prosomeres, hp1 and hp2, while the diencephalon subdivides into three diencephalic prosomeres, p1-p3 (note, prosomeres are numbered from caudal to rostral). Modern molecular developmental studies suggest that the classical midbrain, comprising two mesencephalic neuromeres or "mesomeres," m1 and m2 (reviewed in Puelles, E., Martínez-de-la-Torre, Watson, & Puelles, L., 2012; Puelles, L., Harrison, Paxinos, & Watson, 2013; note, mesomeres are numbered rostrocaudally), should be considered as a third forebrain proneuromere rather than as a part of the brainstem. As a result, the current molecularly defined forebrain comprises three proneuromeres divided into seven neuromeres and directly abuts the hindbrain, with its 12 rhombomeres, named r0 (isthmus) to r11 (Figure 12.1).

The orthogonal longitudinal and transverse boundaries define a checkerboard framework of distinguishable areas: a primary set of positionally and molecularly diverse domains. These have been variously characterized molecularly (at least in part), and their specific anatomical fates have been experimentally determined via "fate mapping" at neural plate and neural tube stages. These primary domains each generate, over time, a three-dimensional neural complex in the brain wall (including derivatives destined to migrate tangentially) via neurogenesis, gliogenesis and subsequent differentiation and stratification. They, accordingly, have been referred to as *migration areas* (Bergquist & Källén, 1954) or, possibly better, as *histogenetic areas* (Puelles, Amat, & Martínez de la Torre, 1987) or *fundamental morphologic units* (Nieuwenhuys and Puelles, 2016).

These primary histogenetic areas undergo secondary AP and DV regionalization that subdivide them into definitive progenitor domains under the planar influence of *secondary organizers*. These are specific neuroepithelial loci that transiently release diffusible morphogens, whose role is to epistatically modify surrounding neural fates within a given spatial range (in contrast, *primary organizers* are early extraneural sources of signals that induce undifferentiated ectodermal tissue to acquire a neural character, though other effects are possible, as well). The signaling of the secondary organizers precedes a definitive partition of the early histogenetic areas into distinct progenitor subdomains called *histogenetic microzones*. Each acquires a characteristic molecular identity: a combinatorial profile of active and inactive genes which dictate subsequent local production of unique neuronal (and glial) derivatives, as well as the local proliferative pattern and the chemoarchitectonic environment, which, in turn, influence cell migration, axonal navigation, and synaptogenesis. Differential differentiation of glial cells within each domain has been scarcely studied in terms of chemical epitopes, but likely parallels neuronal regionalization as regards some molecular phenotypes (Redies, Treubert-Zimmermann, & Luo, 2003). This is especially true for radial glia, which generally provide a physical, molecularly decorated referential framework for neuronal positioning and axonal navigation (representing so-called "natural coordinates"; Nieuwenhuys, 1998a, 2009a). These radial landmarks are used both by migrating neurons to find their final positions in the mantle zone, and by axons to navigate towards their synaptic targets and sources of neurotrophic support. As a result of this developmental regionalization, characteristic types of neurons are produced from the molecularly-specified mother cells occupying each histogenetic microzone. A progenitor's molecular profile is a combination of several hundred active and inactive developmental genes. This allows neighboring progenitor domains to share many genetic determinants, leading to similar but not identical histogenetic features. The overall continuity of areal and zonal boundaries envisioned in the prosomeric model help explain how shared DV patterning mechanisms can provide to the longitudinal zones their columnar properties, which share similarities across neuromeres at various rostrocaudal positions. Phenomena occurring at distant parts of the brain primordium can thus be explained as particular cases of a more general developmental phenomenon.

An important feature of the revised prosomeric model (see Figure 12.1; Puelles 2013; Puelles, L., Martínez-de-la-Torre et al., 2012; Puelles & Rubenstein, 2015) is that the whole forebrain is now thought to be epichordal. Previous versions of the model held the rostral hypothalamo-telencephalic portion—the secondary prosencephalon—to be prechordal (e.g., Puelles, 1995, 2001a; Puelles et al., 1987; Puelles & Rubenstein, 1993, 2003).

Assuming a general validity for such a model in craniates (see Chapter 9), the forebrain in agnathans (myxinoids and lampreys) and gnathostomes can be compared with respect to an invariant set of orthogonally-demarcated progenitor domains. It seems that most evolutionary divergences in forebrain structure occur in anatomic neighborhoods whose development is influenced by secondary organizers. This suggests a causally-relevant parallel evolution of genomic and developmental control mechanisms. This viewpoint has not been sufficiently explored previously (but see, e.g., Robertshaw & Kiecker, 2012), and is the take-home message of the present essay.

The differences in forebrain structure observed across evolutionary radiations may be interpreted as revealing variant manifestation of the vertebrate neural Bauplan. This can be explained either by basal insufficiency of given organizer signals (i.e., non-production of the signals, or lack of appropriate receptor functions) with ulterior evolutionary implementation as functionality emerges, or by variant dimensioning of patterning effects occurring at specific areas due to changes in other variables, such as surface growth. This approach has benefitted in recent times from the sort of analysis allowed by comparative genoarchitectonics (Puelles & Ferran, 2012), since the presence and extent of both secondary organizers and progenitor domains (with derivatives) can be assessed using selected genetic markers extracted from a known molecular profile.

Comparison of the forebrain of agnathans with that of anamniote and amniote gnathostomes reveals that only the gnathostomes show the full mammalian bauplan (see Chapter 9, this volume).

12.2.2 Alternatives to the Prosomeric Model

Unfortunately, the present-day literature has not settled on which morphological model best represents the neural bauplan, particularly as regards the forebrain. Two main discrepant viewpoints are represented—referred to as the columnar and neuromeric paradigms—which differ in their conception of how the brain axis courses through the forebrain, leading to 90-degree discrepancies in their description of forebrain anatomical relationships. The columnar paradigm has been prevalent during the last 100 years. However, descriptive and experimental molecular data accrued during the last 30 years have triggered a fundamental paradigm shift in favor of the neuromeric paradigm. The function-oriented *columnar* forebrain model of Herrick (1910, 1933, 1948), as well as the morphologically-oriented *columnar* model of Kuhlenbeck (1927, 1973), both mistook sulcal accidents present in the ventricular surface as fundamental boundaries. These models have been superseded by the ontogenetic *segmental* or *neuromeric* paradigm, inspired by the embryologic work of Orr (1887) and a succession of others (Bergquist & Källén, 1954; Coggeshall, 1964; Gribnau & Geijberts, 1985; Keyser, 1972; Puelles et al., 1987; Rendahl, 1924; Tello, 1934; Vaage, 1969; von Kupffer, 1906; Ziehen, 1906).

The prosomeric model (Martínez, Puelles, E., Puelles, L., & Echevarria, 2012; Puelles, E., Martínez-de-la-Torre et al. 2012; Puelles, L., 1995, 2001a, 2009, 2013; Puelles, L., & Rubenstein, 1993, 2003, 2015; Puelles, L., Martínez, Martínez-de-la-Torre, & Rubenstein, 2004; Puelles, L., Martínez-de-la-Torre, Paxinos, Watson, & Martínez, 2007; Puelles, L., Martínez-de-la-Torre et al., 2012; Puelles, L., Watson, Martínez-de-la-Torre M, & Ferran, 2012; Rubenstein, Martínez, Shimamura, & Puelles, 1994; Rubenstein, Shimamura, Martínez, & Puelles, 1998; Shimamura, Hartigan, Martínez, Puelles, & Rubenstein, 1995; Shimamura & Rubenstein, 1997) is a widely tested and continuously revised update of the neuromeric paradigm. This model integrates advances in morphological description, genoarchitectonics, fate mapping and causal mechanisms of development such as patterning, neurogenesis, and histogenesis to underpin the axial and other constitutive elements of the model (longitudinal and transverse partitions). The prosomeric model emphasizes progressive regionalization of the brain wall via intrinsic position-dependent genetic fate specification, with consequent region-specific proliferative, neurogenetic, histogenetic, and connective mechanisms that transform the regionalized brain primordium into the adult form. In all vertebrates, characteristic sets of differentiated AP and DV regions emerge within a topologically invariant morphological framework, which is itself based on evolutionarily conserved (constrained) patterns of causal molecular cell–cell interaction. Note that functional considerations so common in the columnar conceptions are purposefully left aside: function is regarded as a neural epiphenomenon that emerges late, largely in the postnatal period, involving adaptations of the interactive potential of available structures to modes of survival that are efficient in the context of the slowly-changing ecosystem. This does not mean I do not support the usual simple heuristic interpreting the ventral brain as "motor," and the dorsal brain as "sensory," where this correlation applies (mainly spinal cord, hindbrain, and midbrain). My point is that morphological analysis essentially deals, for instance, with explaining how motoneurons and some associated interneurons happen to be produced and established permanently where they do, that is, ventrally (with causal mechanisms preceding their ulterior synaptic connection with muscles, and therefore taking effect before the motor function possibly appears). At early stages of brain development there is no function to speak of, only prospective functions. The mentioned heuristic works best in the postnatal functioning organism, as a simplified rule of thumb indicating which functions have essential hubs at specific positions. Referral to dorsal "sensory" functions more widely implies "analyzer" and "world modeling" functions if we include forebrain levels. However, there emerge many other functions which are neither "sensory" nor "motor" (e.g., is the function of the hypothalamic histaminergic neurons, present in the "motor" hypothalamic basal plate, truly motor?). Such qualifications also apply partly even at brainstem and spinal cord levels.

Let me argue this point more fully, for clarity. Fundamental morphological structure and multiple neuronal types are established developmentally via intrinsic genetic mechanisms, irrespective of prospective neural functions, essentially drawing on information contained in the genome (though the generic contribution of some epigenetic effects should be acknowledged). In a further step, neuronal synaptic connectivity is initially implemented under intrinsic developmental control via complex axonal navigational, synaptogenetic and trophic mechanisms. These fundamental relays include bidirectional functional connection of sensory and motor brain neurons with diverse body parts and (in a sense) with the peripheral world via predetermined motor and sensory nerves and receptor/effector specializations. Biological functions eventually emerge (partly prenatally) via energy-spending dynamic interactions of the diverse structural components among themselves and with the external world, and adaptive fitness (survival) results under normal conditions. Of course, subsequent activity refines structural aspects of brain connectivity, particularly the number, precision and stability of correlative synaptic data mappings, via neural plasticity-mediated learning and adaptation. Most neuronal circuits in charge of biological functions are not limited spatially to the derivatives of a single developmental unit of the neural tube; the normal case instead involves complex relays across many brain regions. It may be deduced, accordingly, that single developmental and morphological units do not have a specific function *per se*. This establishes in developmental theory the primacy of form-analytic and causal morphological analyses over behavioral-functional considerations. The latter's relevance is strictly to fully established *neural systems* at a relatively late period of maturation. Functional capabilities emerge from a complex of preformed structural relationships, which then can be refined and optimized through functional interaction with the surrounding world.

A topological approach, implicit in the application of any morphological model to real brains, is specially needed in comparative neuroanatomy because the brain primordium irregularly thickens, bulges, results compressed, and bends upon itself in flexures and fissures as it develops into adult form. Such morphogenetic deformations are a result of differentially regulated growth phenomena occurring at the multiple histogenetic areas placed side-by-side along the neural tube wall (if growth was homogeneous the overall shape would not change). Moreover, such areal and regional deformations vary quantitatively and/or qualitatively when compared among different vertebrate lineages (for example, hypothalamic lobulations and pallial eversion are characteristic of teleost fishes, not being present in other forms). Differential growth causes much of the apparent variation in brain shape across taxa, without significant changes in the fundamental areal and zonal organization of the neural primordium (Puelles & Medina, 2002).

Abstraction of these differences within the topology embedded in the prosomeric model thus emphasizes the shared fundamental structural pattern, which is comparatively consistent in terms of conserved cell populations, nerves, central pathways and functions, irrespective of species-specific shape variations (see Chapter 9, this volume). The basic areal framework of gene expression patterns observed during forebrain development is largely shared among practically all vertebrates studied so far. This implies potent attractor and canalization (genomic and morphogenetic) mechanisms operating during early development, and underlines the constrained causal conditions through which topologically-invariant brain structure is diversely elaborated in different taxa.

The *columnar model* (Herrick, 1910) originated in a scenario in which the increasing power of microscopes, the availability of better staining methods, and emerging experimental neuroanatomy techniques (e.g., retrograde and anterograde neuronal degeneration) demanded an improved conceptual framework for subdividing brain anatomy. There was scarcely any histogenetic brain embryology at that time, as most work in this field was concerned with the shapes of neural primordia, rather than their histogenesis in terms of cells. Accordingly, there was practically no causal understanding of how brain structure is built. This explains why essential embryologic criteria, though recognized from a theoretical standpoint, were not used in practice in sketching the basic structure of brains, while attention was instead focused on apparent adult brain function. At the end of the 19th century, a number of authors investigating the brainstem and cranial nerves noted a characteristic correlation of sensory and motor neuronal populations (and correlative cranial nerve roots) with longitudinal cell columns, separated more or less distinctly by ventricular sulci (Gaskell, 1889; Herrick, 1910; His, 1904; Johnston, 1902; review in Kuhlenbeck, 1973). The functional insight thus obtained had a large impact. Herrick (1910) exploited the conjecture that subdivisions analogous to these brainstem columns might also extend into the forebrain, where some ventricular sulci were also observable, aiming to illuminate by this extrapolation the obscure functional roles of the diencephalon, hypothalamus and telencephalon. This led to Herrick's *columnar forebrain model*, which places emphasis in columnar regions of the adult forebrain separated by supposedly longitudinal sulci. These columns were assumed to be one-to-one comparable to known brainstem columns, thus extending rostralward the corresponding functional implications. This model largely eschewed differential areal histogenetic properties, transverse neuromeric patterns (already known at the time; reviewed by von Kupffer, 1906; Ziehen, 1906) and any developmental causal considerations. Unfortunately, such functional illumination of the forebrain was obtained at the price of accepting that observable forebrain ventricular sulci were in fact *longitudinal* limiting landmarks, analogous to brainstem sulci. Kuhlenbeck (1927) worked in parallel to Herrick in attaching longitudinal morphological significance to the forebrain sulci, though there were important differences. Kuhlenbeck examined embryonic material and purposefully abstained from functionalistic assumptions; thus, Kuhlenbeck's approach was form-analytic rather than functional.

Amusingly, though his ideas were seen as underpinning Herrick's concept, Herrick did not approve of Kuhlenbeck's form-analytic "support" (Herrick, 1933, 1948).

The procedure used to define forebrain sulci as being longitudinal was tendentious, and led Herrick (1933, 1948) and Kuhlenbeck (review in 1973) to implicitly redefine the longitudinal axis of the brain as it was known at the time. Whereas a forebrain axis curved at the cephalic flexure and ending in the hypothalamus had been previously recognized and widely accepted in the anatomic nomenclature (His 1892, 1893b, 1895, 1904; even in Herrick, 1910; in modern times, cf. Puelles, 2013; Puelles, L. et al., 1987; Puelles, Martínez-de-la-Torre et al., 2012; Puelles & Rubenstein, 2015; see particularly Hauptmann, Söll, & Gerster, 2002), the columnar model abandoned the straightforward morphologic implications of the cephalic flexure (Figure 12.1), and assumed instead a straight (or progressively straightened) forebrain length axis that ended in the telencephalon. Curiously, no explicit statement or discussion about this altered conceptualization of the forebrain axis appeared in the writings of these authors—in his recapitulative book, Herrick (1948, p. 21) said the altered neuraxis was "controversial ... but convenient ," showing he was aware the change lacked supporting evidence.

Interestingly, this *ad hoc* axial change ran against the prior conclusions of several neuroembryologists (Bergquist, 1932; His, 1892, 1893a, 1893b, 1895, 1904; Orr, 1887; Rendahl, 1924; von Kupffer, 1906; Ziehen, 1906) and some comparative neuroanatomists (Johnston, 1902, 1909; Kappers, 1929, 1947). Nevertheless, the apparent functional insights offered by the columnar model generated nearly universal adherence throughout the 20th century, extending even into present times. As a consequence, most neuroanatomical interpretations of the vertebrate forebrain in scientific articles, treatises and textbooks are couched in columnar terms, explicitly or implicitly (e.g., Swanson, 2012). In the long run, however, the "functional light" supposedly provided by the columnar sulcal approach has become dimmer as anatomical, developmental, and physiological data on the forebrain accrued. Nobody thinks nowadays that brainstem function directly explains forebrain function. Forebrain ventricular sulci have been shown to be unreliable for purposes of histogenetic or genoarchitectonic subdivision across species (Nieuwenhuys & Bodenheimer, 1966; Puelles & Rubenstein, 1993). It was realized that, in general, brain ventricular sulci are mechanical deformations, distantly epiphenomenal to the primary developmental molecular mechanisms that actually define longitudinal and transversal patterns in the brain wall. For example: notochord-related ventralization induces via SHH signaling the molecularly distinct basal plate zone, which produces specific cell types such as motor neurons; their similar accumulation along a series of neuromeres eventually causes a longitudinal ventricular bulge, which generates in its turn the appearance of a longitudinal sulcus, the precise course of which depends on multiple variables across a series of distinct developmental units, and, more importantly, is not insured against evolutionary variation by any known genomic mechanism— indeed, how should genes specify a sulcus?

Meanwhile, we have learned much about how brains develop, including the conserved cellular mechanisms that give rise to longitudinal and transversal fields within the rostral neural tube, analogously to what happens more caudally in the hindbrain (even hindbrain columns are now understood as a late consequence of early neuromeric patterning phenomena; review in Puelles, 2013; see Tomás-Roca et al., 2016). It has become increasingly clear that the columnar model is not capable of accounting for the multitude of contemporaneous gene expression data and position-related causal mechanisms, whose morphologic interpretation seems meaningless within the columnar paradigm (reviewed in Puelles, L., Martínez-de-la-Torre et al., 2012; Puelles & Rubenstein, 2015). The forebrain longitudinal dimension is widely acknowledged to be underpinned by a *dorsoventral* patterning mechanism, and the resulting longitudinal zonal pattern of the neural tube literally bends at the cephalic flexure as was originally described by His (1892, 1893a, 1893b, 1895, 1904) and as was actually observed with gene markers by Hauptmann et al. (2002); the true forebrain axis is, therefore, *orthogonal* to the arbitrary columnar model's axis, drawn straight across the diencephalon and hypothalamus into the telencephalon. The developmentally defined forebrain axis clearly does not end in the telencephalon, but in the hypothalamic acroterminal domain (see Hauptmann et al., 2002; Puelles, L., Martínez-de-la-Torre et al., 2012; Puelles & Rubenstein, 2015).

Accordingly, most of Herrick's "longitudinal" diencephalic and telencephalic "columns" (and the separating sulci) turn out to be *transversal*. This conclusion wreaks havoc on neuroanatomical terminology. For instance, the columnar "dorsal thalamus" and "ventral thalamus" are not truly dorsal and ventral with respect to the neuraxis, but *caudal* and *rostral*. The alternative noncolumnar terms "thalamus" and "prethalamus" (introduced by Puelles & Rubenstein, 2003), are therefore preferred (Figure 12.1). Most neuroanatomists are not aware that the cerebral peduncle courses *dorsoventrally* through the hypothalamus, and then bends 90 degrees around the subthalamic nucleus to enter its ulterior *longitudinal* course through diencephalon, midbrain and brainstem (Puelles, L., Martínez-de-la-Torre et al., 2012; Puelles & Rubenstein, 2015). The "hypothalamus" is actually a "hypotelencephalon," since it lies wholly *rostral* to thalamus and prethalamus, rather than "under" them (Figure 12.1), and we now know, from fate mapping of the neural plate, that the telencephalic hemispheres are in fact evaginated derivatives of the dorsal hypothalamus (Cobos, Shimamura, Rubenstein, Martínez, & Puelles, 2001; Fernández-Garre, Rodríguez-Gallardo, Gallego-Díaz, Alvarez, & Puelles, 2002, Inoue, Nakamura S, & Osumi, 2000; Puelles, L., Martínez-de-la-Torre et al., 2012; Puelles & Rubenstein, 2015; Rubenstein et al., 1998). This had been accurately recognized by His a century ago (1893b, 1895, 1904), but his pioneering observations gained little traction among his contemporaries and were swept aside by the columnar model followers. We probably will need several generations to resolve all the terminological problems raised by this 100-years-long conceptual blunder.

Contrariwise, by focusing on development rather than function, neuromeric models allow us to explore the brains of all vertebrates molecularly, structurally, and causally, and thus serve as a solid morphologic basis upon which sophisticated postnatal functional analysis is possible.

12.2.3   Detailed Morphology of the Prosomeric Forebrain

The current version of the prosomeric model (Figure 12.1; based on Puelles, 2013; Puelles, L., Martínez-de-la-Torre et al., 2012; Puelles & Rubenstein, 2015) expands upon the simpler version described in Chapter 9 of this volume. Topologically, longitudinal and transversal aspects of brain structure are given equal weight, with due consideration of both the axial bending of the developing neural tube at the cephalic flexure, and its rostral ending at the terminal wall, within the hypothalamic region. The secondary prosencephalon is now conceived as being primarily epichordal, rather than prechordal, due to the recent genoarchitectonic discovery of a small hypothalamic floor plate, related at early neural tube stages to the rostral tip of the floor-inducing notochord (Puelles, L., Martínez-de-la-Torre et al., 2012; Puelles & Rubenstein, 2015; see also the *Allen Developing Mouse Brain Atlas*, delineated by the present author, at www.developingmouse.brain-map.org).

The hypothalamic *floor plate* appears subdivided into retromamillary (RM) and mamillary (M) portions, lying at the midline of the corresponding basal regions (peduncular and terminal basal hypothalamus, respectively, within hp1 and hp2). The left and right *basal plates* directly meet at the rostral midline, forming the acroterminal basal plate (Figure 12.1): this includes the median retrochiasmatic and tuberal basal regions, the median eminence, the neurohypophysis, and the tuberomamillary area (which seem to lie "in front" of, but should be conceived more correctly as topologically *dorsal* to, the mamillary area (Puelles, L., Martínez-de-la-Torre et al., 2012)).

As regards the rostral end of the *roof plate*, it was shown experimentally to correlate with the median bed of the anterior commissure, ascribed to the septopreoptic region (Figure 12.1; Cobos et al., 2001; see also Flames et al., 2007; Puelles, 2013; Puelles, L., Martínez-de-la-Torre et al., 2012 and Puelles & Rubenstein, 2015 for molecular background and discussion). This means that the terminal lamina, and the preoptic and optic chiasm regions lying "under" the anterior commissure, are ascribed in the model to the acroterminal median alar plate lying topologically ventral to the roof plate (the bilateral *alar plates* meet here, like the basal plates underneath; Puelles, L., Martínez-de-la-Torre et al., 2012; Puelles, 2013).

The primary prosencephalon (which includes the prospective diencephalon) divides into the secondary prosencephalon (hypothalamus plus telencephalon), rostrally, and the diencephalon proper, caudally (Figure 12.1). Early versions of the prosomeric model postulated a tentative subdivision of the secondary prosencephalon into three prosomeres (Puelles & Rubenstein, 1993). However, doubts arose later (Puelles and Rubenstein, 2003) about how the transverse parts of the hypothalamus continued dorsally into the telencephalon (Puelles & Rubenstein, 2003; Wullimann & Puelles, 1999). In Chapter 9, Wullimann stays at this cautious position, which leaves the secondary prosencephalon unsegmented. However, I think that a satisfactory solution has recently become available to the problem of the hypothalamo-telencephalic transition, as a result of understanding better the roof and floor zones as well as the courses of tracts at the front of the brain (Bardet, 2007; Ferran, Puelles, & Rubenstein, 2015; Flames et al., 2007; Pombal, Megias, Bardet, & Puelles, 2009; Puelles, 2013; Puelles et al., 2007; Puelles, L., Martínez-de-la-Torre et al., 2012 Puelles et al., 2013; Puelles & Rubenstein, 2015). We accordingly postulate that the secondary prosencephalon divides into two prosomeres, identified in caudorostral order as hypothalamic prosomeres hp1 and hp2 (Figure 12.1; Nieuwenhuys, 2009b; Pombal et al., 2009; Puelles, 2013; Puelles, L., Martínez-de-la-Torre et al., 2012a; Puelles et al., 2013; Puelles & Rubenstein, 2015). This new terminology aims to evade confusion with the earlier, differently defined rostral prosomeric units (p4–p6) of Puelles and Rubenstein (1993); we now regard that subdivision as obsolete. The newly postulated hypothalamic prosomeres both have hypothalamic and telencephalic portions (Figure 12.1).

The hypothalamic part of hp1 is named the *peduncular hypothalamus* (PHy), thus departing from columnar-flavored terms for the sake of clarity; the new term emphasizes the typical character of this territory as the bed of the peduncular medial and lateral forebrain bundles. PHy lies caudally and borders diencephalic prosomere p3, the prethalamus. The neuromere hp1 continues dorsally into the entire evaginated *telencephalon*, encompassing both its pallial and subpallial components (note this leaves out the preoptic region). The hp1 roof plate contains the median septum (septal roof plate), is crossed by the large telencephalic commissures (hippocampal and callosal), and extends into the caudally attached telencephalic portion of the choroidal tela (which continues into the diencephalon). As already mentioned, the hp1 floor plate lies at the retromamillary area of PHy, rostral to the diencephalic floor and caudal to the mamillary floor.

The hypothalamic part of hp2 is named the *terminal hypothalamus* (THy), to emphasize its position at the axial end of the forebrain, departing from columnar tradition. It lies in front of PHy and rostrally has a distinct median *acroterminal domain*, where the bilateral alar and basal plates are continuous from side to side (Ferran et al., 2015; Puelles, L., Martínez-de-la-Torre et al., 2012; Puelles & Rubenstein, 2015). As observed above, this is topologically the *rostralmost* alar and basal neighborhood (in contrast to the misleading columnar convention that it represents the hypothalamic/diencephalic *floor*). The acroterminal domain displays uniquely specialized median structures: the terminal lamina, optic chiasm, suprachiasmatic nucleus, median anterobasal area, and neurohypophysis, with the surrounding median eminence and arcuate tuberal area. In addition to THy, hp2 also includes dorsally the telencephalic (subpallial) preoptic area—this is the classic nonevaginated, or *unpaired telencephalon*. Dorsally, the hp2 roof plate lies at the septo-preoptic rostralmost part of the roof plate, and contains the bed of the anterior commissure. The latter is well known to be embedded within the adult median preoptic nucleus, which also extends ventrally along the acroterminal lamina terminalis. The hp2 floor plate is mamillary (i.e., is represented by the midline septum separating the pair of basal mamillary bodies). Note as well that the eye vesicles develop out of the THy alar plate, possibly within its alar acroterminal domain.

On the other hand, the diencephalon proper, which represents the caudal part of the classic forebrain, appears structured anteroposteriorly into the three conventional diencephalic prosomeres (termed p1-p3 in caudorostral order; Puelles, L. et al., 2014a,b; Puelles, L., 2013; Puelles, L., Martínez-de-la-Torre et al., 2012; Puelles, Watson et al. 2012; Puelles & Rubenstein, 2003). These have not changed in the updated model (Figure 12.1), and in essence can be traced back via Puelles et al. (1987) to Rendahl's original conception (Rendahl, 1924). The major respective derivatives are in the alar plate, represented by the pretectum (alar p1), the thalamus plus the habenular area (alar p2) and the prethalamus plus the prethalamic eminence (alar p3). The diencephalic roof plate shows rostrally choroidal structure (p3 and rostral p2), and displays caudally the pineal gland (caudal p2) and the posterior commissure (p1). The diencephalic basal plate forms the so-called pre-rubral tegmentum, which shares with the midbrain counterpart a number of characteristics, including the development of meso-diencephalic dopaminergic cell populations (substantia nigra and ventral tegmental area; Puelles, L., Watson et al., 2012).

Finally, the revised prosomeric model incorporates the midbrain into the forebrain, due to its comparable genoarchitectonic patterns. This region is composed of two neuromeres, the m1 and m2 mesomeres (Figure 12.1); note these are numbered rostrocaudally (Palmgren, 1921). The m1 is large and contains the major mesencephalic structural derivatives (oculomotor nucleus and red nucleus in the basal plate, and tectal gray, superior colliculus and inferior colliculus in the alar plate; Puelles, E., et al. 2012; Puelles, L., 2013; Puelles, 2016). The m2 is a narrow caudal midbrain neuromere found caudal to both the inferior colliculus and the oculomotor and red nuclei; it is molecularly distinct and has its own set of derivatives (but no motor neurons); its more extensive alar part is known as the *preisthmus*, since it lies just in front of the hindbrain isthmic nuclei. The preisthmic (caudal) m2 region develops the cuneiform complex of nuclei, previously wrongly thought to lie ventral to the inferior colliculus. A distinct m2 derivative in the basal plate is the conventional A8 group of dopaminergic neurons; there is also a median population of dorsal raphe serotonergic neurons within basal m2 (Alonso et al., 2012; Puelles, E., et al. 2012; Puelles, L. 2013).

12.3 Secondary Organizers and Forebrain Topology

Use of the prosomeric model to localize corresponding forebrain features and developmental processes across vertebrate taxa allows appropriate field homologies to be defined within a shared bauplan. This approach also causes apparent differences to become salient as quantitative or qualitative variations on the bauplan, thus aiding evolutionary interpretation (Puelles & Medina, 2002). Understanding such similarities and differences beyond the rather crude analysis level of the classic (but obsolete) "five vesicle brain" requires consideration of patterning entities known as *secondary organizers*, which are responsible for detailed brain regionalization. In the present context, we only need minimal attention to the extrinsic *primary organizers*, which are responsible for neural induction (e.g., the node, the rostral visceral endoderm, the notochord, and the prechordal plate). The primary organizer signals collectively induce nervous tissue to emerge within non-neural ectoderm, and may also initiate differentiation of the early neural plate along the AP and DV dimensions. The role of secondary organizers is then to modify, subdivide, and specialize the crudely delimited neural regions obtained via primary induction. This is the process known as *brain regionalization*, which begins at neural plate stages, but essentially progresses to culmination once the neural tube has closed.

So far, five loci are widely accepted as being secondary organizers in the rostral neural tube (Figure 12.2): the floor plate (ventralizing organizer), roof plate (dorsalizing organizer), anterior neural ridge (ANR, a dorsalizing organizer restricted to the rostral forebrain), zona limitans intrathalamica (ZLI, a mid-diencephalic AP organizer), and isthmic boundary (a meso-rhombencephalic AP organizer); each is described in more detail below. The *hem* and *anti-hem* pallial regions represent important *tertiary organizers* appearing within the telencephalic hemisphere (see Figure 12.2). They are causally relevant particularly for the pallium; potential effects on the subpallium—particularly of the anti-hem—have not yet been explored (Assimacopoulos, Grove, & Ragsdale, 2003; Puelles, 2011; Shimogori, Banuchi, Ng, Strauss, & Grove 2004).

There may exist additional, unexplored secondary organizers, for instance between diencephalon and midbrain, or between hypothalamus and diencephalon, unless these boundaries merely represent transverse limits between the ranges of action of the already-known secondary organizers. The acroterminal hypothalamic domain, for example, credibly represents a rostral AP secondary organizer, possibly subdivided

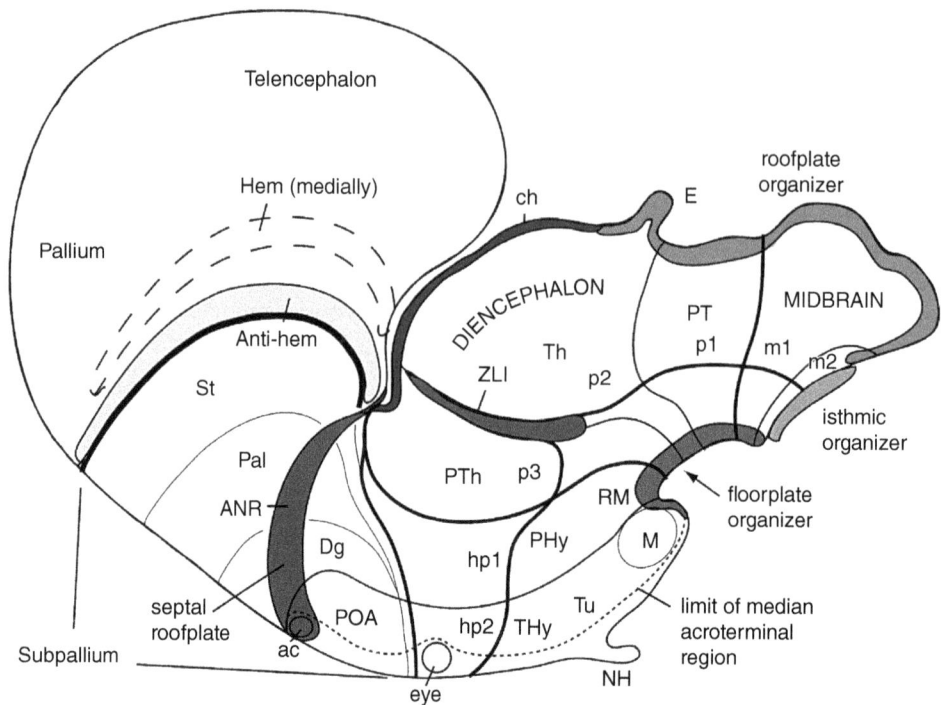

Figure 12.2 Topological Position of Widely Recognized Secondary Organizers, Represented in the Prosomeric Model.

The roof plate organizer in general appears in red, with the darker shade depicting the ANR organizer in the rostral roof plate. The primary floor plate organizer is shown in blue. Note, the alar–basal boundary (compare Figure 12.1) is roughly parallel to both roof and floor organizers (resulting from the equilibrium between dorsoventral antagonic mechanisms of dorsalization and ventralization across the forebrain wall). The isthmic organizer (cyan), the source of FGF8, is strictly in the rostralmost hindbrain, though its effects also encompass the midbrain. The ZLI (purple) is a transverse ridge between thalamus and prethalamus. The hem (bright yellow) and anti-hem (pale orange) are tertiary organizers which pattern the telencephalic pallium. The anti-hem lies next to the pallio-subpallial boundary (black line), and the hem lies next to the choroidal roof plate tissue on the medial aspect of the hemisphere. (*See insert for color representation of the figure*).

dorsoventrally into distinct alar and basal parts (Figure 12.3; Ferran et al., 2015; Puelles & Rubenstein, 2015). This possibility is also explored in more detail below.

Basically, when we refer to secondary neural organizers we are dealing with circumscribed neuroepithelial subregions, usually linear in shape and either longitudinal or transverse in position, which early on acquire a peculiar gene expression profile associated with production and secretion of diffusible morphogenetic signals. The latter, usually, are members of a handful of morphogen protein families (e.g., Sonic Hedgehog, SHH; Bone Morphogenetic Proteins, BMPs; Wingless-Int proteins, WNTs; Fibroblast Growth Factors, FGFs) or their corresponding functional antagonists (e.g., the WNT-antagonizing Secreted Frizzled-Related Peptides, SFRPs; and other anti-WNT, anti-SHH, or anti-BMP products). There is the added complexity that several morphogens may be released at the same site, or at parallel, closely-adjacent sources, and these molecules potentially have diverse diffusion ranges, so that the

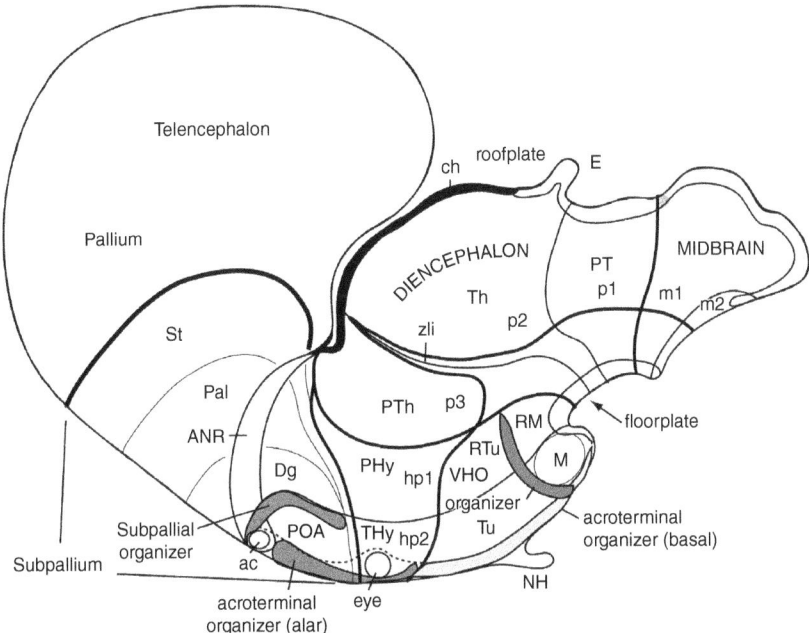

Figure 12.3 Additional Suspected Forebrain Organizers, Placed upon the Prosomeric Model. The SHH-positive subpallial organizer (dark orange shade) lies transversely within the preoptic area next to the hp2/hp1 boundary; note, the boundaries between the diverse subpallial domains are roughly parallel to it. This area is also a source of cells migrating tangentially into the subpallium and pallium (see text). The acroterminal hypothalamic domain is the rostral-most transverse domain in the forebrain and probably has organizer roles, which may be different in its alar and basal moieties (purple and green): Both express *Six3*, and secrete FGFs (see Ferran et al., 2015). The newly postulated ventricular hypothalamic organ (VHO) organizer (red) is a longitudinal domain that separates the main basal territories of the hypothalamus across both hp1 and hp2, namely the M/RM domains from the TU/RTu domains, and probably patterns their different histological fates (see Puelles et al., 2012a). (*See insert for color representation of the figure*).

overall effects are compounded and difficult to localize unambiguously across the time axis (see Verbeni et al., 2013). Interestingly, the 5–6 neuroepithelial secondary organizers cited above tend to be sites where neurogenesis is retarded, reduced, or absent, and these loci may show either low or high proliferative activity.

The graded distribution of signals secreted from secondary organizer sites epigenetically affects the intrinsic (genetic) differentiation patterns taking place in the surrounding immature neural wall under the control of the local complement of co-expressed transcription factors. The latter collectively act upon enhancer and repressor DNA regulatory sites to progressively change the local gene expression pattern (Davidson, 2006; Puelles & Ferran, 2012). Genomic fate specification effects intrinsically progress over developmental time, but are modified, in a position-dependent way, by effect of the superimposed gradients of organizer morphogens in the extracellular matrix. The morphogens act through their respective receptors in neuroepithelial cells (and perhaps also in postmitotic neurons), triggering various cytoplasmic signaling cascades leading to genetic regulation. Similar effects are obtained across the field of cells lying within a similar range of morphogen concentration or distance from the organizer (see Gurdon et al., 1999), but steplike differential genomic sensitivity to distinct levels of the morphogen concentration gradient often results in a small series of discrete cellular fields, progressively more distant from the secondary organizer, which each react differently to the signaling (up-regulating or down-regulating specific transcription factors). This causes new regionalization patterns that subdivide the primary unitary fields. There is, likewise, a temporal range for organizer effects, and the latter frequently are opposed, restricted, or refined by antagonistic signals, or conditioned by intervening growth dynamics, before the results are fixed as a new pattern of molecular boundaries with corresponding fate specifications (Gurdon et al., 1999; Meinhardt, 2008).

The diffusing morphogens normally do not provide specific developmental instructions, but instead act as *positional selector mechanisms*. Recipient cells at different distances from the organizer "choose," in accordance with the local morphogen signal strength, among a variety of developmental options they had *ab initio* due to their intrinsically-active genetic program (Saka & Smith, 2007; see Verbeni et al., 2013 for a novel mechanistic perspective on this process). For instance, the whole hindbrain alar plate down to the medullo-spinal boundary is able to build a cerebellum (i.e., there is an overall cerebellogenic potential in the hindbrain [Martínez, Marín, Nieto, & Puelles, 1995]; note the whole alar midbrain, as well as the alar diencephalon caudal to the zona limitans also have cerebellogenic potential), but normally only a rostral hindbrain part that receives a particular amount of FGF8 signal from the isthmic organizer enters this histogenetic pathway. Thus, the cerebellum normally arises at the level of the caudal isthmus and rhombomere 1. Rostral isthmic territory, lying nearer to the isthmic signal source, receives a higher amount of signal and produces characteristic isthmic nuclear structures quite different from those of the cerebellum. Alar plate tissue within the more caudal rhombomere 2 receives very little or no isthmic organizer signal and produces cochleo-vestibular nuclei within its alar plate, rather than a cerebellum. Further back, in the medulla, other specialized fates appear caudal to the cochlear complex, though the tissue remains capable of building a cerebellum. A morphogenetic gradient signal originating at the midbrain–hindbrain boundary is thus read out positionally as differentially active sets of genes, designating alternative developmental fates within a territory that was initially molecularly homogeneous and equipotential. This is the basis for *regionalization*, a fundamental developmental mechanism that is potentially subject to evolutionary variation and operates under the control of a complex genetic network (Davidson, 2006).

Due to the linear shape of the organizer sites (Figure 12.2), organizer signals diffuse through the immature brain wall in two main directions, orthogonal to the organizer length axis. On either side of the organizer, each recipient field must interpret the local morphogen gradient with respect to its unique sensitivity to the diffused signal, as defined by the respective previous differential genetic background (Gurdon et al., 1999; Meinhardt, 2008; Saka & Smith, 2007; Verbeni et al., 2013). If the signal source lies along the brain midline (floor or roof plates), effects are symmetric and pattern the DV dimension of the neural wall. When the organizer is transversally positioned, effects occur along the AP dimension and are usually different rostral than caudal to the organizer; this is due to the previous AP patterning of molecular states rostral and caudal to the organizer. For example, comparable levels of isthmic signal lead to caudal midbrain fates within the midbrain, and rostral hindbrain fates within the hindbrain, and comparable zona limitans signals extending rostralward or caudalward variously produce prethalamic versus thalamic fates in the diencephalon.

The linear shape of the organizer sites may result from direct induction by a linear signal source; for example, the neural floor plate is induced vertically by the underlying notochord, a midline mesodermal rod. More frequently, though, linear neural organizers emerge at molecular boundaries between two pre-existent areas with different and mutually-opposed (antagonistic) molecular profiles (i.e., showing cross-repressive mutual interaction between differentially-expressed genes). For instance, the roof plate organizer forms bilaterally at the neural/extraneural boundary of the neural plate, prior to fusion of both roof halves at the midline during neurulation. Similarly, the ZLI or mid-diencephalic organizer arises at the transverse boundary between neuromeres p2 and p3, and the isthmic organizer appears at the midbrain/hindbrain molecular boundary even before the latter is visible as a constriction (Figure 12.2). Apparently, the cells exposed to balanced contradictory signals at certain interfaces develop genomic regulation cascades that transform them into signaling organizer entities. Since all primary neuroepithelial domains are areal, their mutual boundaries are linear.

Obviously, organizer-creating phenomena are subject to evolution. It is not yet clear whether the cited antagonistic boundary mechanism is a special or a general consequence of molecular boundary formation. There exist many molecular boundaries at field interfaces which apparently do not become sources of diffusing signals. The known secondary organizer sites may be special cases in which the neighboring fields generate particular sorts of mutual interaction causing organizer function to emerge. Alternatively, it might be speculated that all molecular boundaries necessarily develop some organizer properties, but we have recognized such effects only at places where they are particularly consequential. This issue will become clearer in the future. In any case, linear molecular boundaries emerging in the neuroepithelium seem to have the evolutionary potency to become secondary organizers, and are therefore capable of introducing novelty in terms of new regional subdivisions in patterned neural morphogenesis.

12.3.1 Floor Plate

The earliest secondary organizers are the floor plate and the roof plate. Both are longitudinal and extend throughout the length of the neural tube; some data suggest they may become themselves regionalized into molecularly distinct sectors at later stages. Both domains are apparent at neural plate stages.

The prospective floor plate is the median neural strip lying exactly over the notochord. The latter is a cord-like condensation of axial mesoderm laid down in rostro-caudal order by the nodal gastrulation mechanism as it dwindles caudalwards from its earliest locus. Prospective notochordal cells seem to be nonmotile (Saucedo & Schoenwolf, 1994) and initially form a primordium, the undifferentiated cephalic process, which differentiates and matures in situ to form the notochord proper (this used to be wrongly conceived as a result of rostralward migration of notochordal cells, under the misconception that the nodal structure was fixed in position). As the node moves caudally, new notochordal cells are added. The caudally elongating notochord directly underlies the emergent neural plate midline (prospective floor), derived from a primarily rostral neuroepithelial cell clone that elongates caudally along the neural midline. Secreted notochordal SHH signals induce the floor plate to differentiate as a molecularly distinct median longitudinal zone that likewise expresses the *Shh* gene

(Ruiz i Altaba, 1998). After a short period of close apposition between notochord and neural floor plate, both structures separate as a consequence of neural tube morphogenesis (e.g., axial bending) and mesenchymal differentiation around the notochord, leading to development of the basis of the cranium. Presomitic and somitic para-axial mesodermal cells condense around the notochord to build the basal chondrocranium and the bodies of vertebrae.

The notochord is shorter anteroposteriorly that the overlying neural plate. This occurs because the planar neural induction effects, initially spreading all around the tip of the node, project circumferentially within a radius of some 200-300 micrometers beyond the nodal semicircular outline (Fernández-Garre et al., 2002), whereas the notochord only forms directly underneath the node. This explains why the neural floor plate ends rostrally over the notochordal tip, which lies under the prospective mamillary pouch of the hypothalamus (Bartezko & Jacob, 2002; His, 1892; Jurand, 1974; Puelles, L., Martínez-de-la-Torre et al., 2012; Puelles & Rubenstein, 2015). A number of references in the literature have misidentified this point, generally placing it more caudally (as far back as the isthmus). This is due to lack of proper analysis of the early stages in which axial mesoderm and neural floor are in close contact.

The product of the *Shh* gene is the important secreted morphogen, sonic hedgehog (SHH), one of whose effects on the overlying neural floor cells is to homeotically activate its own gene, so that the floor plate itself becomes a source of diffusible SHH and thus the floor plate organizer (Figure 12.2). This is partly responsible, in its turn, for the specification of the longitudinal zone adjacent to the floor plate as the basal plate (both the floor and basal plates are absent in *Shh* loss-of-function mutants). In the extended forebrain (i.e., including the midbrain), homeotic induction of *Shh* then expands from the floor plate into the basal plate neuroepithelium of the closing neural tube—this is in contrast to the hindbrain and spinal cord, where *Shh* expression remains restricted to the floor plate. This regional peculiarity of the forebrain basal plate seems due to an early prechordal effect (see García-Calero, Fernández-Garre, Martínez, & Puelles, 2008), and correlates with the expression domain of the gene *Otx2*, which equilibrates antagonistically with hindbrain *Gbx2* expression at the nascent isthmic organizer. In the midbrain/forebrain territory the floor plate organizer is thus soon transformed into a more extensive and complex *floor-basal organizer* (compare basal plate in Figure 12.1; note, the basal plate domain is most extensive dorsoventrally within the hypothalamus, another possible prechordal effect). From these median and paramedian ventral sites, the secreted SHH protein diffuses bilaterally dorsalward, generating what is known as a *ventralizing* organizing effect on the patterning of the symmetric lateral walls of the neural tube. Activation of the gene *Gli3* at long range (dorsally) seems to repress, and thus limit, the dorsalward homeotic expansion of the basal *Shh*-expressing domain (Puelles & Martínez, 2013; Martinez-Ferre & Martinez, 2012; Martinez-Ferre et al., 2013). Selective expression of transcription factor genes such as *Nkx2.9*, *Nkx2.2*, and *Ptc*, among others, occurs in a thin liminal band just outside, or partly overlapping, the *Shh*-positive ventral basal domain, directly downstream of high-level SHH signaling. At forebrain and midbrain levels, this effect roughly positions the prospective longitudinal alar-basal limit (red line in Figure 12.1); note, the liminal band later also surrounds what seems a dorsalward transversal extension of the Shh basal expression domain, forming gradually a vertical spike which is the emergent ZLI, the mid-diencephalic organizer (compare Figure 12.2). The issue whether the conceptually linear alar–basal boundary falls above or below the *Nkx2.2*-positive band remains controversial and perhaps is of merely academic value (Ferran, Sánchez-Arrones, Sandoval, & Puelles, 2007; Ferran et al., 2009; Hauptmann et al., 2002; Martínez et al., 2012; Puelles, E., Rubenstein, & Puelles L., 2001; Puelles, E., et al., 2012; Puelles, L., 2013; Puelles, L., Martínez-de-la-Torre et al., 2012). Since the whole neural tube is epichordal, according to the updated map of the rostral tip of the floor plate (and notochord) introduced in the revised prosomeric model (Puelles & Rubenstein, 2015), the floor plate ventralization mechanism that differentiates the basal from the alar plate is common to the entire brain, irrespective of neuromeric topography.

12.3.2 Roof Plate

The roof plate likewise emerges as a secondary organizer before a neural tube roof exists as such. Its early primordium is formed bilaterally at the interface between the neurally-induced (neural) ectoderm and the surrounding nonneural ectoderm. It is accordingly represented at the peripheral border of the neural plate as a band that is concentrically internal relative to the prospective neural crest. It is unclear whether the neural crest anlage itself participates in the antagonistic mechanism forming the *roof plate organizer* (Sanchez-Arrones, Ferrán, Rodríguez-Gallardo, & Puelles, 2009; Sánchez-Arrones, Stern, Bovolenta, & Puelles, 2012). In this regard, it should be kept in mind that the neural crest only extends rostrally so far as the prospective epiphysis (see Figure 12.1). Interestingly, only the rostral roof plate extending from septum to epiphysis later expresses *Fgf8*, building the anterior neural ridge organizer (ANR, Figure 12.2), which represents a specialized part of the roof plate organizer (Echevarría, Vieira, Gimeno, & Martínez, 2003; Puelles, 2011; Sanchez-Arrones et al., 2009, 2012; Vieira et al., 2010). As would be expected, due to this antagonistic neural/non-neural causal background, the roof plate reaches its rostral end where the right and left plate borders meet at the anterior neural ridge. As neurulation proceeds, and the neural tube closes (excluding the crest material), the right and left roof primordia first meet somewhere over the neural canal (midbrain?) and progressively fuse together rostralwards and caudalwards, forming the roof plate proper all the way to the front end, which fate maps to the septopreoptic bed of the prospective anterior commissure (Cobos et al., 2001; Eagleson & Harris, 1990; Inoue et al., 2000; Puelles et al., 1987). The neural roof plate and the overlying equally fused median nonneural ectoderm both are general sources of signaling proteins of the BMP (bone morphogenetic protein) and WNT (wingless-int) families. These diffuse ventralwards into the lateral wall of the neural tube, and jointly exert on it a *dorsalizing* influence (but note that some members of these morphogen families are independently produced at ventral sources, where they may have alternative functional roles).

As a result of the functional antagonism of the floor and roof organizers, the intervening lateral wall of the neural tube is patterned into dorsalized roof and alar plates and ventralized basal and floor plates. This process extends caudally into the spinal cord. Note the specific forebrain distribution of these primary patterning features contradicts the columnar model, since the resulting alar-basal boundary (marked by the *Nkx2.2*-positive liminal band) does not enter the telencephalon (which is wholly dorsalized, irrespective of its own subdivision into pallium and subpallium) and divides neatly the hypothalamus, including its acroterminal median domain, into alar and basal moieties. The latter singular rostral territory behaves, in terms of DV patterning, as if it were a part of the lateral neural wall, producing a unique instance of radial symmetry at the tip of the node and of the derived notochord; this locus is further affected singularly by locally superimposed vertical patterning from the prechordal plate, and possibly transforms into a secondary organizer itself, releasing FGFs (see below). These special properties of the rostralmost forebrain and the corresponding molecular background are not accounted for in the columnar model, which regards the hypothalamic and preoptic median areas as a prechordal extension of the floor plate (e.g., Swanson, 2012).

12.3.3 The Anterior Neural Ridge

As mentioned above, the rostralmost part of the forebrain roof plate acquires early on a distinct molecular profile, and therefore can be characterized as a separate secondary organizer, the anterior neural ridge (ANR; Figure 12.2). This distinction apparently arises due to locally restricted vertical signaling coming from the anterior visceral endoderm during gastrulation via released FGF8 among other possible morphogens (Sanchez-Arrones et al., 2009, 2012; Thomas & Beddington, 1996; see also data on the "rostral first row" cells of Wilson and Houart, 2004 in zebrafish). The ANR eventually produces several members of the Fibroblast Growth Factor protein family, notably FGF8, FGF15 and FGF17. I hold that the signals from this source are topologically comparable to dorsalizing signals elsewhere, since they spread ventralwards from a rostral sector of the roof plate (Figure 12.2); however, in the literature the ANR is most often conceived as a rostralizing signal source. This is likely due to conceptual confusion caused by the columnar model, which wrongly places the ANR at the rostral end of the neural tube (rather than of the roof plate). A modified dorsalizing role accords with the observed joint release, at this site, of some BMPs and WNTs, which are generally roof plate signals, apart from the FGFs resulting from homeotic induction of *Fgf* genes elicited by the anterior visceral endoderm. The ANR clearly participates in the closure of the rostral neuropore, certifying this as a straightforward roof domain. Neuropore closure causes the bilateral halves of the early rostromedian ANR to fuse together at the roof of the closed rostral forebrain. This readily observable topography underlines the longitudinal and roof-like topology of the ANR in the closed neural tube. The fusion of the two earlier halves of the ANR may indeed enhance its efficiency as an organizer at closed tube stages, since its signaling strength is effectively doubled once both ANR halves find the whole forebrain within their signal diffusion range. In the closed tube, median expression of *Fgf8* and other members of the FGF family at the rostral roof plate marks the longitudinal extent of the definitive ANR, which reaches uninterruptedly from the level of the anterior commissure bed in the rostralmost septo-preoptic roof area back to a caudal end in the roof of diencephalic prosomere 2, close to the epiphysis (Crossley, Martínez, Ohkubo, & Rubenstein, 2001; Garcia-Lopez, Vieira, Echevarria, & Martínez, 2004; Garcia-Lopez, Pombero, & Martínez, 2009; Martinez-Ferre & Martínez, 2009, 2012; Puelles, 2011).

Early FGF8 signaling from the ANR at open neural plate stages probably is important for dorsalizing the prospective telencephalon and eye fields during early regionalization of the forebrain alar plate (recall that the telencephalon lies topologically dorsal—i.e., peripheral—to the eye field in the plate). There is little analysis yet of the role of ANR Fgf8 signals in the differential regionalization of the alar part of the hypothalamus (yellow in Figure 12.1), but, interestingly, some mouse Fgf8 hypomorphs show a reduction of vasopressin and oxytocin neurons of the paraventricular area developing in this territory (Brooks, Chung, & Tsai, 2010; McCabe et al., 2011; see alar hypothalamic subdivisions in Puelles, 2013; Puelles, L., Martínez-de-la-Torre et al., 2012; Puelles & Rubenstein, 2015). At later stages, in the closed neural tube, the ANR is also important for details of telencephalic and diencephalic patterning (Martinez-Ferre & Martínez, 2012; Shimamura & Rubenstein, 1997; Vieira et al., 2010). The topologic position of the evaginated telencephalic vesicle in the closed neural tube remains, counterintuitively, *ventral* to the ANR, since the vesicle develops out of the rostral forebrain alar plate and the ANR lies at the roof plate. This topology is disguised in tetrapods by the massive evagination of the telencephalic vesicle, which falsely makes the hemispheres seem to be dorsal to the roof plate. Thus, ANR signals acting upon the telencephalon need to be interpreted topologically as dorsalizing signals. Remarkably, the maximally dorsalized part of the telencephalic field builds up the septum (a region usually misunderstood within columnar tradition as being ventral, due to the overall forebrain topographic deformation created by the cephalic flexure): the septum notably contains in its median roof-derived part the telencephalic roof plate and the corresponding commissures. The least dorsalized part of the telencephalon would be the amygdala and its transition into the preoptic and hypothalamus areas (extended amygdala). In contrast, pallium and subpallium occupy similar intermediate levels of dorsalization and become differentially patterned due to separate AP effects (subpallium lying topologically *rostrally* within the telencephalic field relative to the pallium; Cobos et al., 2001). Similar dorsalizing effects are exerted by the caudal/diencephalic part of the ANR upon the prethalamus (p3; for example, formation of the prethalamic eminence, a dorsal alar prethalamic subregion) and thalamus (p2; for example, formation of the habenular or epithalamic subregion, dorsal to the thalamic mass; Martinez-Ferre & Martínez, 2009; Puelles & Martínez, 2013).

12.3.4 The Prechordal Plate

The prechordal plate organizer is a transient embryonic formation that technically represents a *primary* rather than a secondary organizer, since its signals originate from the migrating mesendodermal cells of the prechordal plate; it is included here for its relevance to forebrain evolution. Note that the prechordal plate of vertebrates, rather than a fixed cell population analogous to the notochord (as sometimes implied), is a dynamic stream of mesodermal or mesendodermal cells that actively migrate out of the nodal tip in a median dorsally-directed fan-shaped array at the beginning of gastrulation. Traditionally, the prechordal plate has been held to be involved in neural induction at rostral forebrain levels, though any such early function remains difficult to dissociate from primary nodal effects and subsequent anterior visceral endodermal effects. The prechordal mesendodermal cells actively advance along the median rostral terminal wall of the neural plate, that is, the prospective acroterminal region. It is usually assumed that they spread "rostral" to the tip of the notochord. However, as I observed above, there is no neural plate lying topologically "rostral" to the notochordal tip: the forebrain neuroepithelium that lies beyond the tip of the notochord represents the prospective *terminal wall* of the forebrain, which stretches *ventrodorsally* between the rostral end of the floor plate and the rostralmost roof plate (i.e., the acroterminal domain, Figure 12.1; Puelles, L., Martínez-de-la-Torre et al., 2012;

Puelles & Rubenstein, 2015). Such topologic considerations force a reinterpretation of the relationship of the migrating rostromedian prechordal tissue to the forebrain vesicle, highlighting that the direction of this migration is *dorsalward*, in terms of closed neural tube topology. The prechordal cells first migrate out from the node when it is placed directly over the rostral tip of the prospective notochord (start of gastrulation), and thereafter they sequentially move dorsally in front first of the median hypothalamic basal plate (infundibular area) and later of the alar parts of the terminal wall (optic chiasm), up into the neighborhood of the terminal lamina, also an alar telencephalic subregion (i.e., the prospective median preoptic area). Eventually, they reach the overlying rostral end of the roof plate (the prospective bed of the anterior commissure).

The continuous signaling of the prechordal cells throughout this migration acts sequentially upon the neighboring neuroepithelium. Experimental data suggest that the small distances existing at the earliest stages of the migration causes that the earliest prechordal plate signals at basal plate level spread all the way from prospective mamillary to prospective isthmic border levels. The prechordal plate cells thus first contribute to differential molecular specification of the midbrain, diencephalic and hypothalamic *basal plate* domains, ending with the specification of the "rostralmost basal" mamillary bodies, which do not form in the absence of prechordal tissue (Garcia-Calero et al., 2008). As the prechordal cells then proceed dorsally along the acroterminal domain, their signals first enlarge the basal hypothalamus (tuberal region, with infundibulum and neurohypophysis as median specializations), then segregate bilaterally the initially cyclopean alar median eye field into separate eye primordia, and finally control the separation of the telencephalic evagination as two hemispheres (note, the latter are *alar plate* domains; additional extraneural effects include induction of the median adenohypophysial placode and the bilateral separation of the olfactory placodes). Partial or total failure of these sequential effects leads to the various holoprosencephalic/ciclopic syndromes. Additional dynamic patterning effects are probably exerted by the prechordal cells during their migration, though these have hardly been analyzed so far. For instance, prechordal signals may contribute directly or indirectly (via secondary acroterminal organizers, Figure 12.3) to the AP patterning of the basal and alar hypothalamus, namely to its subdivision into prosomeric domains hp1 and hp2 (Figure 12.1; Puelles, L., Martínez-de-la-Torre et al., 2012). Some prechordal cells reaching the mesenchymal neighborhood next to the prospective preoptic area move backward bilaterally along the hp1/hp2 boundary, where they apparently induce a preoptic transverse linear domain that strongly expresses *Shh* (see above; Bardet, Ferran, Sanchez-Arrones, & Puelles, 2010; Flames et al., 2007). This *Shh*-positive preoptic domain can be identified as a topologically transverse secondary *subpallial organizer* (Figure 12.3), described below.

12.3.5 The Subpallial Organizer

An independent transverse alar source of SHH signals is established after neural tube closure at the caudal boundary of the preoptic area, where it borders the evaginated telencephalic hemisphere (telencephalic hp2/hp1, Figure 12.3; Bardet et al., 2010; Flames et al., 2007). Vertical prechordal plate signals are needed for the appearance of these bilateral preoptic patches of *Shh* expression (Garcia-Calero et al., 2008). Mappings of prechordal cell markers (e.g., the gene *goosecoid*) show that the median stream of prechordal cells eventually bifurcates into bilateral streams at the end of their median dorsalward migration, and that these bilateral streams roughly correspond in position to the emerging preoptic patches of induced *Shh* expression (Bardet et al., 2010; Izpisua-Belmonte, De Robertis, Storey, & Stern, 1993). Note that here, *Shh* (widely understood as a floor or basal plate marker gene) is expressed selectively within the telencephalic alar plate. Since the preoptic area represents a rostromedian nonevaginated part of the telencephalic subpallium, the resulting bilateral linear site of SHH production contacts the evaginated part of the subpallium across the hp1/hp2 interneuromeric boundary. A number of preoptic cells born in this domain (some also expressing *Shh*) migrate tangentially across this border, incorporating into both subpallial and pallial parts of the hemispheres, where they always appear in the mantle zone, in contrast with the preoptic area, where *Shh* expression occurs in the ventricular zone (Bardet et al., 2010; Gelman et al., 2011). I hold that the preoptic borderline expression of *Shh* represents a novel secondary organizer. This preoptic secondary organizer is strategically placed to exert an AP patterning influence primarily on the subpallium, and can accordingly be defined as the *subpallial organizer*. It is topologically transverse within the prosomeric model, since it parallels the boundary between hp2 and hp1. Interestingly, it is restricted to the telencephalic sector of the local hp2 alar plate, extending dorsalward all the way to the bed of the anterior commissure in the roof plate; in contrast, it does not extend into the underlying hypothalamic alar plate (Flames et al., 2007). The patterning effects of SHH diffusing out of the subpallial organizer would be expected to be different at the preoptic subpallial area, rostral to the postulated organizer, versus the subpallial region of the evaginated hemisphere, lying caudally (it is now a fact established by neural plate fate-mapping in several species that the preoptic region is the rostralmost part of the subpallium; Cobos et al., 2001; Eagleson & Harris, 1999; Inoue et al., 2000; Puelles et al., 2013; Sanchez-Arrones et al., 2009). The medial ganglionic eminence (MGE), which is subdivided into parallel preoptic, diagonal, and pallidal domains (Figure 12.1), is the immediate hemispheric neighbor of the subpallial organizer, whereas the striatum, formed within the lateral ganglionic eminence (LGE), lies beyond the MGE, next to the pallium, farther from the subpallial organizer and close to the pallial anti-hem (Figures 12.1–3). At long range, and in early stages of hemispheric development, the signals of the subpallial organizer may reach even farther caudally into the prospective pallium.

It is widely ignored in the literature that intrinsic patterning of the telencephalon into subpallial and pallial parts has been demonstrated by fate mapping at neural plate stages to involve essentially AP patterning within a dorsal part of the rostral alar plate (see, e.g., Cobos et al., 2001; Sánchez-Arrones et al., 2009, 2012). The mistaken notion often found in the literature that the pallium lies dorsal to the subpallium (though topologically it lies *caudal* to it) is yet another misleading result of the columnar model, due to its arbitrary and unrealistic extension of the basal plate into the telencephalon (e.g., Swanson, 2012; discussion in Puelles & Rubenstein, 2015). The fact that *Shh*-positive cells produced in the subpallial organizer migrate tangentially into subpallial and pallial domains of the telencephalon (Bardet et al., 2010; Gelman et al., 2011) raises the possibility that early direct patterning via diffused SHH at neural plate stages is secondarily prolonged via continued SHH secretion by the tangentially migrating preoptic neurons. According to the present topologic definition, the subpallial organizer is not expected to act significantly upon the ventrally-placed alar hypothalamus (including the eye fields). The restriction of its range of influence to preoptic and strio-pallidal subpallial domains represents part of the modern rationale for considering the preoptic area a part of the telencephalic subpallium, against its obsolete columnar interpretation as "rostral" hypothalamus (Medina & Abellán, 2012; Puelles, L., Martínez-de-la-Torre et al., 2012; Puelles & Rubenstein, 2003).

12.3.6 Potential Organizers in the Hypothalamus

The whole alar and basal acroterminal domain (probably under early influence of the rostral visceral endoderm and prechordal plate) is characterized by precocious and persistent *Six3* expression from the early neural plate stages onward; this transcription factor reportedly has indirect antagonistic effects upon caudalizing WNT signaling (Lagutin et al., 2003) and is one of the genes implicated in holoprosencephalic syndromes.

In the hypothalamus there are also sources of FGFs, which have not yet been investigated as potential secondary organizers (see Ferran et al., 2015). A bilateral medial *Fgf8*-positive spot appears likewise at the acroterminal origin of the optic stalks (not shown); this locus may be suspected of acting upon the eye field derivatives and other nearby specialized subregions of the acroterminal domain (e.g., the vascular organ of the lamina terminalis, or the suprachiasmatic nuclei).

In addition, the acroterminal midline along the basal hypothalamus (median tuberomamillary and tuberal areas) expresses *Fgf8*, *Fgf10* and *Fgf18* (e.g., Ferran et al., 2015; Parkinson, Collins, Dufresne, & Ryan, 2010). Note that here, in the absence of floor tissue, these FGFs overlap basal plate expression of *Shh*, and seem required for the indirect patterning of the adenohypophysis. This molecularly-distinct acroterminal basal domain is perhaps a result of prechordal mesendodermal induction, though Rathke's pouch (the adenohypophysial primordium), which secondarily contacts it, and is influenced itself by this area, may be also relevant for its distinct histogenetic pattern. Interestingly, the surrounding tuberal/tuberomamillary basal neighborhood selectively down-regulates *Shh* expression secondarily (Manning et al., 2006). This topologically transverse median FGF signal source may be conceived as a potential *basal acroterminal organizer* (Figure 12.3; see also Puelles & Rubenstein, 2015). Its FGF signals spread caudally bilaterally along the AP dimension of the forebrain basal plate, starting at its rostralmost acroterminal end (compare Figure 12.1). FGFs diffusing from this source probably participate in large-scale rostralizing anti-WNT effects, which usually are reported in the literature as originated from the ANR (the best-known source of forebrain FGF8; see above). These AP signals also may be involved in intrahypothalamic anteroposterior patterning within the basal plate (hp1 and hp2 neuromeres, or differentiation of the acroterminal median eminence and neurohypophysis; see Haddad-Tivolli et al., 2015; Figure 12.1). Early prechordal plate induction is known to be necessary for the differentiation of the mamillary anlage within hp2 (García-Calero et al., 2008).

There is also a longitudinal tuberal neuroepithelial locus which may represent a basal hypothalamic secondary organizer (Figure 12.3; see also Puelles, L., Martínez-de-la-Torre et al., 2012). This ventral tuberal band is represented at mature stages by the so-called *ventricular hypothalamic organ* (VHO), a sulcal ependymal specialization that develops longitudinally along the boundary between the molecularly distinct ventral tuberal/retrotuberal areas and the underlying peri-mamillary/peri-retromamillary regions. Its plausibility as an organizer is suggested by the observation that this locus selectively expresses *Wnt8* at early embryonic stages, which codes for a secreted morphogen of the Wnt family that is also involved in ZLI *signaling* (Garda, Puelles, Rubenstein, & Medina, 2002; Puelles & Martínez, 2013). Though the VHO is usually described only in non-mammalian vertebrates, a mammalian homologue was recently identified by means of corresponding molecular markers (Puelles, L., Martínez-de-la-Torre et al., 2012). Planar WNT8 signals diffusing from the VHO organizer would be able to influence the DV pattern of the whole basal hypothalamus (the perimamillary and mamillary domains ventrally, as well as the tuberal and retrotuberal domains dorsally).

12.3.7 The *Zona Limitans Intrathalamica*

A well-recognized transverse secondary organizer is found in the diencephalic alar plate at a site known as the *zona limitans intrathalamica* (ZLI; Figures 12.1, 12.2). This is a transverse neuroepithelial ridge expressing *Shh*, formed exclusively at alar plate levels, precisely at the inter-neuromeric boundary between the prethalamus (p3) and the thalamus (p2). The ZLI name was coined by Rendahl (1924), a pioneer of neuromeric forebrain models, who first identified this transverse ventricular ridge in avian, reptilian and mammalian embryos. He characterized it histologically as a cell-poor palisade separating two diencephalic neuromeric fields (p3/p2, which he called the rostral and caudal parencephalon, respectively; his name for p1, still used occasionally, was synencephalon). The ZLI is alternatively known as the mid-diencephalic organizer (MDO).

Although the tranversal ZLI ventricular ridge was long disregarded in mainstream columnar neuroanatomy (or wrongly interpreted as a longitudinal entity), various contemporaneous genoarchitectonic and experimental studies have corroborated both its clearcut transversal orientation (Figure 12.2) and its importance for understanding diencephalic patterning (Echevarría et al., 2003; Kiecker & Lumsden, 2004; Martínez et al., 2012; Martinez-Ferre & Martínez, 2009, 2012; Martinez-Ferre, Navarro-Garberi, Bueno, & Martínez, 2013; Puelles et al., 1987; Puelles & Martínez, 2013; Puelles & Rubenstein, 1993; Rubenstein et al., 1994; Scholpp & Lumsden, 2010; Scholpp, Wolf, Brand, & Lumsden, 2006; Vieira et al., 2010). The ZLI apparently emerges developmentally as a result of early antagonistic rostrocaudal molecular phenomena (antagonism of rostral *Fezf2*-expressing areas versus caudal *Irx3*-expressing territories, among other causes). These antecedents establish the subsequent transversal site in the diencephalic alar plate where the ZLI will later emerge. Next, there occurs homeotic expansion of roof plate *Wnt8* expression ventrally along the transverse prospective-ZLI site. This, apparently, creates conditions permissive for local homeotic upwards expansion of floor/basal markers (primarily *Shh*, but with other ventral markers following suit; see Martínez et al., 2012; Puelles et al., 2004), apparently due to local repression by WNT8 of *Gli3*, the main antagonist of *Shh* (Martinez-Ferre & Martínez, 2012; Martinez-Ferre et al., 2013; Puelles & Martínez, 2013). Simultaneously, this resultant *Wnt8/Shh*-expressing territory is compressed anteroposteriorly by the repressing effects of other rostral and caudal determinants (e.g., *Lfng, Lrrn1*; García-Calero, Garda, & Puelles, 2006; Kiecker & Lumsden, 2004). Expression of *Shh* at the center of the *Wnt8*-positive neuroepithelial ridge is evoked gradually from ventral to dorsal, jointly with downregulation of local mitogenic activity, creating the definitive ZLI transverse ridge. This process does not involve any cell movements; it is strictly a molecular repatterning of the local alar plate, irrespective of the fact the ZLI itself shows clonal restriction properties. At the end of this process, the ZLI appears as a long *Shh*-positive spike, crossing ventrodorsally most of the alar plate, orthogonally to the longitudinal alar-basal boundary (Figures 12.1, 12.2). Thereafter, its signals (including at least SHH and some WNTs, such as WNT8 and WNT3a) diffuse rostrally and caudally along the AP dimension of the alar diencephalon. The ZLI has been shown experimentally to generate differential patterning effects via SHH in p3 (prethalamus) and p2 (thalamus). It is as yet unresolved whether the pretectum (p1), which lies caudal to the thalamus, falls within the range of ZLI patterning, though it shows a tripartite molecular and histogenetic subdivision aligned with the diencephalic AP axis (Ferran et al., 2007, 2008, 2009). Some patterning phenomena, such as the partial downregulation of the initially widespread alar expression of *Pax6* and *Pax7*, which starts at the ZLI and progresses caudalwards, suggest a range of signals that include the pretectum, with the possible exception of the dorsocaudal pretectum and the habenular region, where these markers are not downregulated. The analogous *Pax6*-downregulating effect occurring within alar p3 stops short of the prethalamic boundary with the hypothalamus, leaving a *Pax6*-positive rostral field where the reticular nucleus arises. The hypothalamus itself, placed far rostrally to the ZLI, in front of p3, seems wholly unaffected by ZLI signaling.

12.3.8 The Isthmic Organizer

As mentioned above, it is appropriate to consider the midbrain lying caudal to the diencephalon as the caudalmost part of the forebrain, rather than as a rostral part of the brainstem, as has been done traditionally (probably due to the midbrain origin of the oculomotor nerve). Part of the evidence in favor of its ascription to the forebrain is represented by various gene patterns resulting from DV patterning, which are clearly shared by the midbrain and the more rostral parts of the forebrain, but not by the hindbrain (this includes the secondary homeotic induction of *Shh* expression in the basal plate, and resultant ventralizing cascade inducing the *Nkx2.2*-expressing band along the alar-basal boundary); both patterns stop at the isthmus. In addition, some early neural genes (e.g., *Six3*) which are activated initially in the rostral neural plate, have been shown by fate-mapping to become soon expressed as far caudally as the prospective isthmic boundary, irrespective that secondary down-regulation later restricts their territories to more rostral domains (see Sánchez-Arrones et al., 2009). I also mentioned above that the prospective midbrain is included within the very early range of influence of the prechordal plate, in contrast to the hindbrain (Garcia-Calero et al., 2008). Finally, expression of the important early rostral neural gene *Otx2* stabilizes at the isthmo-mesencephalic boundary, thus placing the whole extended forebrain under its control.

This interpretation places the caudalmost forebrain region under the caudalizing influence of the *isthmic organizer*, the first-identified one among known secondary organizers (see reviews in Echevarría et al., 2003; Martínez, 2001; Vieira et al., 2010). It emerges via antagonistic effects at the transverse boundary between rostral *Otx2* expression and caudal *Gbx2* expression. At neural plate stages, this antagonism first occurs caudally within the prospective spinal cord (i.e., *Otx2* expression initially extends down to prospective spinal levels in the neural plate, though the caudal tip of the spinal cord expresses *Gbx2*). However, the *Otx2/Gbx2* interface sweeps rostrally, due to *Gbx2* dominance in their mutual antagonism, ultimately coming to a definitive equilibrium at the isthmus. This molecular interface soon develops apposed transverse neuroepithelial rings (these do not reach the floor plate) expressing either *Wnt1* (rostrally, in caudalmost midbrain) or *Fgf8* (caudally, in the isthmus, the rostralmost hindbrain). Note some authors, confusingly, do not recognize the isthmus as a separate hindbrain neuromere (in the sense of His [1893b, 1904] and the prosomeric model; see Puelles [2013]), and enclose it instead within a plurineuromeric "r1" entity, causing some confusion in the reported analysis of this organizer. Apparently the mesencephalic WNT1 signal has a very limited diffusion range, since this protein becomes fixed to the intercellular matrix; it has mainly mitogenic and antineurogenic effects, causing observable protracted proliferation and growth at the caudal midbrain in most vertebrates, notably teleost fishes (where growth continues in the adult). The main long-range signal spreads from the isthmic side of the organizer and seems to be FGF8, complemented by other FGFs (see references above). The organizer's area of influence extends caudally to the r1/r2 border (roughly including isthmic, prepontine and cerebellar formations within its range) and rostrally to the diencephalo-mesencephalic border (i.e., including the whole midbrain, that is, mesomeres m1 and m2; Puelles, 2013). However, as mentioned above, the whole diencephalon caudal to the ZLI (p2 and p1) and the whole hindbrain down to the medullo-spinal border (r0-r11) can respond to appropriate levels of FGF8 signaling with the production of an ectopic cerebellum (Martínez et al., 1995).

12.4 The Early Evolution of the Chordate Forebrain

I next examine how these organizers apply to early vertebrate forebrain evolution. For detailed phylogeny and comparative anatomy, see Chapter 9 in this volume.

12.4.1 Early Chordates

A forebrain vesicle, the so-called "sensory" vesicle observed at the rostral end of the neural tube, can be distinguished in ascidian larvae (urochordates or "sea squirts"; Nakamura et al., 2012). Ascidians are presently regarded, on grounds of genomic similarity, to represent the closest ancestors of vertebrates, irrespective whether they retain or not the most basal morphology (Pani et al., 2012). What we know about the genoarchitectonic profile of the ascidian "sensory vesicle" suggests it largely has a regionally diversified hypothalamic character (Moret et al., 2005). There is no recognizable telencephalon, though, but there is a single eye primordium—a pigmented patch—in the median rostral wall of the sensory vesicle (Meinertzhagen & Okamura, 2001). Once thought primordial to a pineal eye, the conclusive hypothalamic molecular character of this vesicle instead suggests an optic latent homology for the pigmented patch (Meinertzhagen & Okamura, 2001). The forebrain of larval amphioxus (cephalochordates or "lancelets"), which belongs to a separate pre-vertebrate lineage, is very much similar (Lacalli, 2008; Nieuwenhuys 1998b, 2002).

This rather simple composition of the rostral forebrain in acraniates reveals differential patterning effects, which possibly arise from isolated primary neural induction implemented solely by the node and the notochord, in the apparent absence of telencephalogenic and oculogenic organizers equivalent to vertebrate anterior visceral endoderm or AVE (primary) and ANR (secondary, see §12.3.3). Interestingly, ascidians have rostral endoderm that lies underneath the front of the neural tube, well beyond the notochordal tip (Nakamura, Terai, Okubo, Hotta, & Oka, 2012), but this does not seem to operate as a primary organizer, since no telencephalic primordium is distinguishable (see §12.3.4). AVE signaling is known to be required early on, in vertebrates, for extra-nodal neural induction of the telencephalic field at the rostral rim of the rostral neural plate, where neural-inducing nodal and notochordal signals are apparently too weak (Sanchez-Arrones et al., 2012; Thomas & Beddington, 1999; Wilson & Houart, 2004). This AVE-induced neural region becomes the primary median telencephalic field, which further develops at its dorsal border (roof plate zone) the secondary ANR organizer (§12.3.3 and Figure 12.2; Fernández-Garre et al., 2002; Sanchez-Arrones et al., 2009).

Moreover, it seems that, in ascidians, there is no distinct rostromedially migrating endomesodermal tissue equivalent to the prechordal plate (Nakamura et al., 2012). It may be speculated, though, that the so-called "notochordal prolongation" found "rostral" to the *Amphioxus* forebrain represents a variant precursor of the prechordal plate (§12.3.4).

Recent data from our laboratory in larval *Amphioxus*, in collaboration with the group of J. García-Fernández in Barcelona, suggest that the rostral part of the forebrain vesicle shows AP and DV molecular patterns suggestive of a general hypothalamic nature (with a median undeveloped eye patch and no telencephalic derivatives), but it is not possible to distinguish molecularly between potential mesencephalic and diencephalic territories in the caudal part of the vesicle. There is, accordingly, also no indication of a mid-diencephalic organizer (ZLI, §12.3.7) in *Amphioxus*. Nevertheless, the entire forebrain vesicle is neatly molecularly divided dorsoventrally into distinct floor, basal and alar longitudinal zones (probably a signaling roof plate organizer is present, since otherwise forebrain alar dorsalization would not be possible). A distinct, apparently latent, isthmic boundary (§12.3.8) appears caudally to the small, and ill-defined diencephalo-mesencephalic primordium. An isthmic organizer proper may be nevertheless absent in the neural tube of acraniates, or at least in cephalochordates (none of the characteristic markers is present), though presence and partial efficacy of a patterning stage antecedent to its formation is indicated by the fact that the relatively unpatterned caudal forebrain is nonetheless well-delimited from the relatively unpatterned hindbrain in both urochordates and cephalochordates, as is also the case (in molecular terms) in various invertebrates. The isthmic boundary itself would thus have arisen earlier in evolution than its role as an organizer mediating the regionalization of adjoining midbrain and hindbrain territories.

12.4.2 Early Craniates and Vertebrates

This analysis suggests that the separate primary induction mechanisms of the anterior visceral endoderm and the prechordal plate first evolved in agnathan vertebrates. Indeed, myxinoids (hagfish) and lampreys, the most primitive extant craniates, clearly possess paired telencephalic vesicles and paired eyes distinct from the hypothalamus (the main schematic features of the secondary prosencephalon). These agnathan craniates also clearly show standard gross parts of the forebrain caudal to hypothalamus, which we identify as straightforward diencephalic and midbrain proneuromere homologs when compared to correlative elements in gnathostomes (Figure 12.1; Nieuwenhuys, 2009b; Pombal & Puelles, 1999; Pombal et al., 2009; Puelles, L., Martínez-de-la-Torre et al., 2012; Puelles, L., Watson et al., 2012). Moreover, the full set of prosomeres and mesomeres distinguished in the prosomeric model is present in agnathans, implying that acroterminal, mid-diencephalic and isthmic secondary organizers must be efficiently active (Pombal et al., 2009). Evaginated eye development occurs also in agnathans essentially like in gnathostomes.

Classical morphologists nevertheless noted the small size of the telencephalon of myxinoids and lampreys, in which the olfactory bulb represents about one half of the whole hemispheric mass (Nieuwenhuys et al., 1998b, 2002; Pombal et al., 2009; Puelles, 2001b), though there is already a clear division between pallial and subpallial telencephalic regions (Martínez-de-la-Torre et al., 2011; Pombal et al., 2009), bespeaking of the existence of an incipient AP patterning of the telencephalic field (subpallial organizer?). In any case, the bilateral hemispheres and optic vesicles of agnathans imply an effective prechordal repression of these tissular fates at the rostral alar midline (thus preventing holoprosencephaly). Although there exists a clearcut rostral subpallium distinct from the caudal pallium, various data suggest that the subpallial organizer (§12.3.6) is not fully functional, at least in lampreys (Murakami et al., 2001, 2002; Myogin et al., 2001). Indeed, no *Shh* expression is observed at the preoptic area in early lamprey larvae, and there is no resultant induction of *Nkx2.1* in the pallidal, diagonal and preoptic areas (Murakami et al., 2001, 2002; Sugahara et al., 2011; but see Sugahara, Murakami, Adachi, & Kuratani, 2013). The MGE is thus apparently absent in lampreys, possibly due to the lack of a complete subpallial organizer (Sugahara et al., 2011). However, recent data suggest that a standard subpallium is indeed defined molecularly in myxine larvae and may be merely heterochronically retarded in its developmental regionalization in lampreys (Sugahara, F. et al., 2016). On the whole, these results suggest that the differential patterning of the subpallium versus pallium within the agnathan telencephalic field implies some novel relevant signals, perhaps due to the AVE, or the prechordal plate (which would act more strongly over the subpallium than over the pallium).

What, then, is the status of the lamprey subpallium? Present-day accounts indicate agnathan subpallium comprises a *Dlx*-positive telencephalic field which is clearly separated from the hypothalamic and prethalamic *Dlx* expression domains (Martínez-de-la-Torre et al., 2011; Murakami et al., 2001, 2002; Myogin et al., 2001; Sugahara et al., 2011). There is already evidence that *Dlx*-positive (GABAergic) subpallial neurons migrate tangentially into the lamprey pallium, a shared phenomenon well studied in gnathostomes. This pattern resembles the general case for vertebrate subpallium, but in larval lampreys this field does not clearly subdivide further into molecularly distinct preoptic, diagonal, pallidal and striatal subdomains (but see reference above to recent contrasting data of Figure 12.1; Sugahara et al., 2016; Puelles et al., 2013). Note that these subdivisions are arranged along the topologic AP dimension of the forebrain (if we imagine them as flattened into neural plate relationships), and may require the SHH-mediated patterning exerted by the subpallial organizer (§12.3.5 and Figure 12.3). The subpallium found in lampreys therefore can be understood as a field-homolog, or as an "unfinished" or latent version of the vertebrate subpallium (i.e., a variant from the more normal myxinoid case). Its neuronal populations possibly manifest a general striatal character, according to their molecular signature (*Dlx+*, *Nkx2.1-*. *Shh-*, etc.; see the comparable case of the *Nkx2.1* loss-of-function mouse phenotype in Sussel, Marin, Kimura, & Rubenstein, 1999). However, the complete set of striatal neuronal types observed in vertebrates includes tangentially migrated elements that originate from the preoptic, diagonal and pallidal subpallial domains. In their absence, the agnathan subpallium cannot contain a true functional analog of the vertebrate striatal intrinsic circuitry, and its participation in forebrain "basal ganglia-like" functions is accordingly unclear, despite some suggested generalizations (Pombal et al., 1997). Further research should elucidate whether this patterning difference in the lamprey subpallium is permanent or transient, due to insufficiency of prechordal plate SHH signaling at the subpallial organizer locus, or to absence of appropriate signal receptors or signal cascades. There is the new possibility to study whether the induction of the subpallial organizer is merely heterochronically retarded in these animals (Sugahara et al., 2016), which may or may not have an effect on the differentiated neuronal phenotypes.

The apparent functional incompleteness or retarded action of the subpallial organizer may relate, as well, to the relatively small size of the telencephalic vesicle in these taxa, since SHH is known to have a widespread mitogenic effect over the whole telencephalon (Dahmane et al., 2001). On the other hand, the extremely reduced size of the lamprey pallium, leaving aside the relatively massive olfactory bulb, also suggests absence or insufficiency of the hem and anti-hem tertiary organizers (§12.3.6).

The hem is formed in mammals at the border between the choroidal roof plate and the adjacent alar pallium, which is fated to become hippocampal cortex. It is thus essentially longitudinal and dorsal in topological position, but is one step removed from the roof plate proper, being an alar subregion. It may be described as a longitudinal alar organizer. It is known to influence the whole pallium, though most importantly the prospective hippocampus or medial pallium, which lies next to it and controls expression of Emx genes, also in lampreys (see Murakami et al., 2001, 2002). The hem is also an important source of Cajal-Retzius neurons that migrate tangentially into layer I of the general cortex (review in Puelles, 2011). The anti-hem represents the ventral pallium sector (Puelles et al., 2000) that borders the striatum across the pallio-subpallial boundary (some authors confusingly misidentify the anti-hem *itself* as the "pallio-subpallial boundary", but the boundary is lineal, whereas the ventral pallium distinctly represents a thin pallial subregion that releases anti-WNTs and other morphogens). There are antagonistic gene effects operating across this boundary that establish, early on, the anti-hem site. Curiously, the anti-hem is wider caudally (encompassing a sizeable part of the amygdala) than rostrally (Bielle et al., 2005). The mutually antagonistic effects exerted by the hem and anti-hem organizers, plus the earlier effects due to the ANR and the roof plate, serve to regionalize the pallium. Puelles et al. (2000) compared a set of pallial gene markers between chicken and mouse embryos, and postulated a pallial model with four parts (ventral, lateral, dorsal and medial pallium sectors), which was initially thought to be valid at least for tetrapods. This pattern in general has later been widely corroborated in gnathostome vertebrates, which have essentially comparable pallial regions. For example, all gnathostomes show distinct olfactory and hippocampal cortical sectors, derived respectively from the ventral and medial pallium; these domains form a peripheral continuous ring around a central island composed of dorsal and lateral pallium sectors; Puelles, 2014 subsequently updated this model (Puelles, 2001b, 2014; Puelles & Rubenstein, 2003). Such conserved fundamental molecular and histogenetic pallial partitioning, however, does not exclude the existence of large histogenetic differences. It is well known that only mammals develop isocortex, a stereotyped six-layered derivative of the dorsal pallium; in contrast, sauropsids (birds and reptiles) develop a small and primitive sort of dorsal cortex (the so-called Wulst), jointly with a massive *dorsal ventricular ridge*, derived from ventropallial nuclear masses accumulating deep to the olfactory cortex (Puelles et al., 2007). The molecularly distinct lateral pallium sector, that always lies intercalated between the dorsal (isocortical) and ventral pallium (olfactory) sectors, was recently revealed to contain a transitional claustro-insular complex, again composed of a deep nucleus (the claustrum) and a superficial layered cortex, though the latter is less complex than the isocortex; apparent homolog lateropallial structures were recently identified in the avian brain (Puelles, 2014; Puelles et al., 2016).

This cardinal partitioning of pallium by the above-mentioned organizers has not yet been examined in any agnathans. Doubts have been expressed that their pallial regionalization is equivalent to that of gnathostomes (Martínez-de-la-Torre et al., 2011; Pombal et al., 2009; Puelles, 2001b). Most of the caudal hemispheric mass of the lamprey receives input from the olfactory bulb, suggesting that it may largely represent the peripheral ring of ventral pallium plus medial pallium (both of which receive olfactory input in many gnathostomes). It is presently very unclear (and controversial) whether a central island containing lateral and dorsal pallial sectors comparable to those distinguished in tetrapods is present in agnathans. If the absence of these domains is corroborated, it would imply that the hem and anti-hem organizers are somehow non-functional or hypofunctional in the agnathan forms. Obviously, the easy recognition of a pallial olfactory bulb in agnathans implies a separate patterning mechanism, possibly dependent on primary inducing signals from the olfactory placodes (possibly retinoic acid), or from the olfactory nerve afferents themselves.

These schematic comparative considerations illustrate the convenience to apply the logic of secondary organizers and their differential patterning effects to agnathan larvae and other problematic instances in the field of comparative neuroanatomy, in an effort to understand the early steps of telencephalic evolution, as well as the evolutionary divergence of the forebrain between ancestral chordates and present-day vertebrates. For instance, the new set of potential hypothalamic organizers might be used to try to understand the homology relationships of the hypothalamic lobules found in teleost fishes with hypothalamic subdivisions observed in tetrapods. In the present chapter, I have aimed to depict a preliminary general framework for this line of work, while pointing out the difficulties created by the traditional nomenclature based on unsupported aged assumptions, and the consequent unproductive ways of thinking that result from obsolete models of forebrain development (e.g., the columnar model). These aspects handicap modern causal morphologic thinking, resisting the strong impulse represented otherwise by comparative genomic analysis and brain genoarchitectonic studies. The prosomeric model tries to avoid these problems by faithfully adapting its tenets to any novel emergent details of vertebrate forebrain development (see Puelles & Rubenstein, 2015).

Acknowledgments

This work was funded by the Spanish Ministry of Science and Innovation grants BFU2008-04156 and BFU2014-57516-P, and SENECA Foundation contract 19904/GERM/15 to L.P.

References

Alonso, A., Merchán, P., Sandoval, J. E., Sánchez-Arrones, L., Garcia-Cazorla, A., Artuch, R., ... Puelles, L. (2012). Development of the serotonergic cells in murine raphe nuclei and their relations with rhombomeric domains. *Brain Structure & Function*, 218, 1229–1277.

Assimacopoulos, S., Grove, E. A., & Ragsdale, C. W. (2003). Identification of a Pax6-dependent epidermal growth factor family signaling source at the lateral edge of the embryonic cerebral cortex. *Journal of Neuroscience*, 23, 6399–6403.

Bardet, S. M. (2007). *Organización morfológica y citogenética del hipotálamo del pollo sobre base de mapas moleculares* (Unpublished doctoral thesis). Department of Neuroscience, University of Murcia, Spain.

Bardet S. M., Ferran J. L., & Sanchez- Arrones, L., & Puelles, L. (2010). Development of Shh expression at the chicken telencephalic stalk. *Frontiers in Neuroanatomy*, 4, 1–16.

Barteczko, K., & Jacob, M. (2002). The morphology of the rostral notochord in embryos of *Ichthyophis kohtaoensis* (*Amphibia, Gymnophiona*) is comparable to that of higher vertebrates. *Anatomical Embryology (Berlin)*, 205, 99–112.

Bergquist, H. (1932). Zur Morphologie des Zwischenhirns bei niederen Vertebraten. *Acta Zoologica (Stockholm)*, 13, 57–303.

Bergquist, H., & Källén, B. (1954). Notes on the early histogenesis and morphogenesis of the central nervous system in vertebrates. *Journal of Comparative Neurology*, 100, 627–659.

Bielle, F., Griveau, A., Narboux-Nême, N., Vigneau, S., Sigrist, M., Arber, S., ... Pierani, A. (2005). Multiple origins of Cajal-Retzius cells at the borders of the developing pallium. *Nature Neuroscience*, 8, 1002–1012.

Brooks, L. R., Chung, W. C., & Tsai, P. S. (2010). Abnormal hypothalamic oxytocin system in fibroblast growth factor 8-deficient mice. *Endocrine*, 38(2), 174–180. doi: 10.1007/s12020-010-9366-9.

Cobos, I., Shimamura, K., Rubenstein, J. L. R., Martínez, S., & Puelles, L. (2001). Fate map of the avian anterior forebrain at the four-somite stage, based on the analysis of quail-chick chimeras. *Developmental Biology*, 239, 46–67.

Coggeshall, R. E. (1964). Study of diencephalic development in the albino rat. *Journal of Comparative Neurology*, 37, 241–270.

Crossley, P. H., Martínez, S., Ohkubo, Y., Rubenstein, J. L. (2001). Coordinate expression of *Fgf8, Otx2, Bmp4*, and *Shh* in the rostral prosencephalon during development of the telencephalic and optic vesicles. *Neuroscience*, 108, 183–206.

Dahmane, N., Sánchez, P., Gitton, Y., Palma, V., Sun, T., Beyna, M., ... Ruiz i Altaba, A. (2001). The Sonic Hedgehog-Gli pathway regulates dorsal brain growth and tumorigenesis. *Development*, 128, 5201–5212.

Davidson, E. H. (2006). *The regulatory genome. Gene regulatory networks in development and evolution*. Burlington, MA: Academic Press.

Eagleson, G. W., & Harris, W. A. (1990). Mapping of the presumptive brain regions in the neural plate of *Xenopus laevis*. *Journal of Neurobiology*, 21, 427–440.

Echevarria, D., Vieira, C., Gimeno, L., & Martínez, S. (2003). Neuroepithelial secondary organizers and cell fate specification in the developing brain. *Brain Research Review*, 43, 179–191.

Fernández-Garre, P., Rodríguez-Gallardo, L., Gallego-Díaz, V., Alvarez, I. S., & Puelles, L. (2002). Fate map of the chicken neural plate at stage 4. *Development*, 129, 2807–2822.

Ferran, J. L., Dutra de Oliveira, E., Sánchez-Arrones, L., Sandoval, J. E., Martínez-de-la-Torre, M., & Puelles, L. (2009). Geno-architectonic analysis of regional histogenesis in the chicken pretectum. *Journal of Comparative Neurology*, 517, 405–451.

Ferran, J. L., Puelles, L., & Rubenstein J. L. R. (2015). Molecular codes defining rostrocaudal domains in the embryonic mouse hypothalamus. *Frontiers in Neuroanatomy*. doi: 10.3389/fnana.2015.00046.

Ferran, J. L., Sánchez-Arrones, L., Sandoval, J. E., & Puelles, L. (2007). A model of early molecular regionalization in the chicken embryonic pretectum. *Journal of Comparative Neurology*, 505, 379–403.

Ferran, J. L., Sánchez-Arrones, L., Bardet, S. M., Sandoval, J., Martínez-de-la-Torre, M., & Puelles, L. 2008. Early pretectal gene expression pattern shows a conserved anteroposterior tripartition in mouse and chicken. *Brain Research Bulletin*, 75, 295–298.

Flames, N., Pla, R., Gelman, D. M., Rubenstein, J. L. R., Puelles, L., & Marín, O. (2007). Delineation of multiple subpallial progenitor domains by the combinatorial expression of transcriptional codes. *Journal of Neuroscience*, 27, 9682–9695.

García-Calero, E., Garda, A.-L., & Puelles, L. (2006). The gene *Lrrn1* marks the prospective site of the zona limitans thalami in the early embryonic chicken diencephalon. *Gene Expression Patterns*, 6, 879–885.

García-Calero, E., Fernández-Garre, P., Martínez, S., & Puelles, L. (2008). Early mamillary pouch specification in the course of prechordal ventralization of the forebrain tegmentum. *Developmental Biology*, 320, 366–377.

Garcia-Lopez, R., Vieira, C., Echevarria, D., & Martínez, S. (2004). Fate map of the diencephalon and the zona limitans at the 10-somites stage in chick embryos. *Developmental Biology*, 268, 514–530.

Garcia-Lopez, R., Pombero, A., Martínez, S. (2009). Fate map of the chick embryo neural tube. *Development, Growth & Differentiation*, 51, 145–165.

Garda, A. L., Puelles, L., Rubenstein, J. L. R., & Medina, L. (2002). Expression patterns of *Wnt8b* and *Wnt7b* in the chicken embryonic brain suggest a correlation with forebrain patterning centers and morphogenesis. *Neuroscience*, 113, 689–698.

Gaskell, W. H. (1889). On the relations between the structure, function and origin of the cranial nerves; Together with a theory on the origin of the nervous system of vertebrates. *Journal of Physiology*, 10, 153–211.

Gelman, D., Griveau, A., Dehorter, N., Teissier, A., Varela, C., Pla, R., ... Marín O. (2011). A wide diversity of cortical GABAergic interneurons derives from the embryonic preoptic area. *Journal of Neuroscience*. 31, 16570–16580.

Gribnau, A. A. M., & Geijsberts, L. G. M. (1985). Morphogenesis of the brain in staged *Rhesus* monkey embryos. *Advances in Anatomy, Embryology & Cell Biology*, 91, 1–69.

Gurdon, J. B., Standley, H., Dyson, S., Butler, K., Langon, T., Ryan, K., ... Zorn A. (1999). Single cells can sense their position in a morphogen gradient. *Development*, 126, 5309–5317.

Haddad-Tovolli, R. P. F., Zhang, Y., Zhou, X., Theil, T., Puelles, L., Blaess. S., & Alvarez-Bolado, G. (2015). Differential requirements for Gli2 and Gli3 in the regional specification of the mouse hypothalamus. *Frontiers in Neuroanatomy*. doi: 10.3389/fnana.2015.00034

Hauptmann, G., Söll, I., & Gerster, T. (2002). The early embryonic zebrafish forebrain is subdivided into molecularly distinct transverse and longitudinal domains. *Brain Research Bulletin*, 57, 371–375.

Herrick, C. J. (1910). The morphology of the forebrain in amphibia and reptilia. *Journal of Comparative Neurology*, 20, 413–547.

Herrick, C. J. (1933). Morphogenesis of the brain. *Journal of Morphology*, 54, 233–258.

Herrick, C. J. (1948). *The brain of the tiger salamander, Amblystoma tigrinum*. Chicago, IL: The University of Chicago Press.

His, W. (1892). Zur allgemeinen Morphologie des Gehirns. *Archiv für Anatomie und Entwickelungsgeschichte (Anatomische Abteilung des Archivs für Anatomie u. Physiologie)*, 2, 346–383.

His, W. (1893a). Über das frontale Ende des Gehirnrohrs. *Archiv für Anatomie und Entwickelungsgeschichte (Anatomische Abteilung des Archivs für Anatomie u. Physiologie)*, 3, 157–171.

His, W. (1893b). Vorschläge zur Eintheilung des Gehirns. *Archiv für Anatomie und Entwickelungsgeschichte (Anatomische Abteilung des Archivs für Anatomie u. Physiologie)*, 3, 172–179.

His, W. (1895). Die anatomische Nomenclatur. Nomina Anatomica. *Archiv für Anatomie und Entwickelungsgeschichte (Anatomische Abteilung des Archivs für Anatomie u. Physiologie)* Suppl.Vol. 1895, 1–180 [+Taf.I, II].

His, W. (1904). *Die Entwicklung des menschlichen Gehirns während der ersten Monate*. Leipzig: Hirzel.

Inoue, T., Nakamura, S., & Osumi, N. (2000). Fate mapping of the mouse prosencephalic neural plate. *Developmental Biology*, 219, 373–383.

Izpisúa-Belmonte JC, De Robertis EM, Storey KG, Stern CD. (1993). The homeobox gene *goosecoid* and the origin of organizer cells in the early chick blastoderm. *Cell*, 74, 645–659.

Johnston, J. B. (1902). An attempt to define the primitive functional divisions of the central nervous system. *Journal of Comparative Neurology*, 12, 87–106.

Johnston, J. B. (1909). The morphology of the forebrain vesicle in vertebrates. *Journal of Comparative Neurology*, 19, 457–539.

Jurand, A. (1974). Some aspects of the development of the notochord in mouse embryos. *Development*, 32, 1–33.

Kappers, C. U. A. (1929). *The evolution of the nervous system in invertebrates, vertebrates and man*. Haarlem: Bohn.

Kappers, C. U. A. (1947). *Anatomie comparée du systeme nerveux, particulièrement de celui des mammifères et de l'homme*. Haarlem: Bohn; Paris: Masson.

Keyser, A. (1972). The development of the diencephalon of the Chinese hamster; An investigation of the validity of the criteria of subdivision of the brain. *Acta Anatomica (Basel)*, Suppl.59-1 ad. 83, 1–178.

Kiecker, C., & Lumsden, A. (2004). Hedgehog signaling from the ZLI regulates diencephalic regional identity. *Nature Neuroscience*, 7, 1242–1249.

Kuhlenbeck, H. (1927). *Vorlesungen über das Zentralnervensystem der Wirbeltiere*. Jena, Gustav Fischer.

Kuhlenbeck, H. (1973). *The central nervous system of vertebrates*. Vol. 3, part II: Overall morphological pattern. Basel: Karger.

Lacalli, T. C. (2008). Basic features of the ancestral chordate brain: a protochordate perspective. *Brain Research Bulletin*, 75,319–323.

Lagutin, O. V., Zhu, C. C., Kobayashi, D., Topczewski, J., Shimamura, K., Puelles, L., … & Oliver, G. (2003). *Six3* repression of Wnt signaling in the anterior neuroectoderm is essential for vertebrate forebrain development. *Genes & Development*, 17, 368–379.

Manning, L., Ohyama, K., Saeger, B., Hatano, O., Wilson, S. A., Logan, M., & Placzek, M. (2006). Regional morphogenesis in the hypothalamus: A BMP-Tbx2 pathway coordinates fate and proliferation through Shh downregulation. *Developmental Cell*, 11, 873–885.

Martínez, S. (2001). The isthmic organizer and brain regionalization. *International Journal of Developmental Biology*, 45, 367–3771.

Martínez, S., Marín, F., Nieto, M. A., & Puelles, L. (1995). Induction of ectopic *engrailed* expression and fate change in avian rhombomeres: Intersegmental boundaries as barriers. *Mechanisms of Development*, 51, 289–303.

Martínez, S., Puelles, E., Puelles, L., & Echevarria, D. (2012). Molecular regionalization of developing neural tube. In C. Watson, G. Paxinos, & L. Puelles (Eds.), *The mouse nervous system* (pp. 2–18). Amsterdam: Academic Press/Elsevier.

Martínez-de-la-Torre, M., Pombal, M. A., & Puelles, L. (2011). Distal-less-like protein distribution in the larval lamprey forebrain. *Neuroscience*, 178, 270–284.

Martinez-Ferre, A., & Martínez, S. (2009). The development of the thalamic motor learning area is regulated by Fgf8 expression. *Journal of Neuroscience*, 29, 13389–13400.

Martinez-Ferre, A., & Martínez, S. (2012). Molecular regionalization of the diencephalon. *Frontiers in Neuroscience*. May 25. doi: 10.3389/fnins.2012.00073.

Martinez-Ferre, A., Navarro-Garberi, M., Bueno, C., & Martínez, S. (2013). Wnt signal specifies the intrathalamic limit and its organizer properties by regulating Shh induction in the alar plate. *Journal of Neuroscience*, 33, 3967–3980.

McCabe, M. J., Gaston-Massuet, C., Tziaferi, V., Gregory, L. C., Alatzoglou, K. S., Signore, M., ... & Dattani, M. T. (2011). Novel FGF8 mutations associated with recessive holoprosencephaly, craniofacial defects, and hypothalamo-pituitary dysfunction. *Journal of Clinical Endocrinology & Metabolism*, 96, 1709–1718. doi: 10.1210/jc.2011-0454

Medina, L., & Abellán, A. (2012). Subpallial structures. In C. Watson, G. Paxinos, & L. Puelles (Eds.), *The mouse nervous system* (pp. 173–220). Amsterdam: Academic Press/Elsevier.

Meinertzhagen, I. A., & Okamura, Y. (2001). The larval ascidian nervous system: The chordate brain from its small beginnings. *Trends in Neuroscience*, 24, 401–410.

Meinhardt, H. (2008). Models of biological pattern formation: From elementary steps to the organization of embryonic axes. *Current Topics in Developmental Biology*, 81, 1–63.

Moret, F., Christiaen, L., Deyts, C., Blin, M., Vernier, P., & Joly, J.-S. (2005). Regulatory gene expressions in the ascidian ventral sensory vesicle: evolutionary relationships with the vertebrate hypothalamus. *Developmental Biology*, 277, 567–579.

Murakami, Y., Ogasawara, M., Satoh, N., Sugahara, F., Myojin, M., Hirano, S., & Kuratani, S. (2002). Compartments in the lamprey embryonic brain as revealed by regulatory gene expression and the distribution of reticulospinal neurons. *Brain Research Bulletin*, 57, 271–275.

Murakami, Y., Ogasawara, M., Sugahara, F., Hirano, S., Satoh, N., & Kuratani, S. (2001). Identification and expression of the lamprey Pax6 gene: Evolutionary origin of the segmented brain of vertebrates. *Development*, 128, 3521–3531.

Myojin, M., Ueki, T., Sugahara, F., Murakami, Y., Shigetani, Y., Aizawa, S., ... & Kuratani, S. (2001). Isolation of Dlx and Emx gene cognates in an agnathan species, *Lampetra japonica*, and their expression patterns during embryonic and larval development: conserved and diversified regulatory patterns of homeobox genes in vertebrate head evolution. *Journal of Experimental Zoology*, 291, 68–84.

Nakamura, M. J., Terai, J., Okubo, R., Hotta, K., & Oka, K. (2012). Three-dimensional anatomy of the Ciona intestinalis tailbud embryo at single-cell resolution. *Developmental Biology*, 372, 274–284.

Nieuwenhuys, R. (1998a). Morphogenesis and general structure. In R. Nieuwenhuys, H. J. ten Donkelaar, & C. Nicholson (Eds.), *The central nervous system of vertebrates* (Vol. 1, pp. 159–228). Berlin: Springer Verlag.

Nieuwenhuys, R. (1998b). Amphioxus. In R. Nieuwenhuys, H. J. ten Donkelaar, & C. Nicholson (Eds.), *The central nervous system of vertebrates* (Vol. 1, pp. 365–396). Berlin: Springer Verlag.

Nieuwenhuys, R. (2002). Deuterostome brains: Synopsis and commentary. *Brain Research Bulletin*, 57, 257–270.

Nieuwenhuys, R. (2009a). Analysis of the structure of the brain stem of mammals by means of a modified D'Arcy Thompson procedure. *Brain Structure & Function*, 214, 79–85.

Nieuwenhuys, R. (2009b). The structural organization of the forebrain: A commentary on the papers presented at the 20th Annual Karger Workshop "Forebrain Evolution in Fishes". *Brain Behavior & Evolution*, 74, 77–85.

Nieuwenhuys, R., & Bodenheimer, T. S. (1966). The diencephalon of the primitive bony fish *Polypterus* in the light of the problem of homology. *Journal of Morphology*, 118, 415–449.

Orr, H. J. (1887). Contribution to the embryology of the lizard. *Journal of Morphology*, 1, 311–372.

Palmgren, A. (1921). Embryological and morphological studies on the midbrain and cerebellum of vertebrates. *Acta Zoologica*, 2, 1–94.

Pani, A. M., Mullarkey, E. E., Aronowicz, J., Assimacopoulos, S., Grove, E. A., & Lowe, C. J. (2012). Ancient deuterostome origins of vertebrate brain signalling centres. *Nature*, 483, 289–294.

Parkinson, N., Collins, M. M., Dufresne, L., & Ryan, A. K. (2010). Expression patterns of hormones, signaling molecules, and transcription factors during adenohypophysis development in the chick embryo. *Developmental Dynamics*, 239, 1197–1210.

Pombal, M. A., El Manira, A., & Grillner, S. (1997). Organization of the lamprey striatum—Transmitters and projections. *Brain Research*, 766, 249–254.

Pombal, M. A., Megias, M., Bardet, S. M., & Puelles, L. (2009). New and old thoughts on the segmental organization of the forebrain in lampreys. *Brain Behavior & Evolution*, 74, 7–19.

Pombal, M. A., & Puelles, L. (1999). A prosomeric map of the lamprey forebrain based on calretinin immunocytochemistry, Nissl stain and ancillary markers. *Journal of Comparative Neurology*, 414, 391–422.

Puelles, E., Martínez-de-la-Torre, M., Watson, C., & Puelles, L. (2012). Midbrain. In C. Watson, G. Paxinos, & L. Puelles (Eds.), *The mouse nervous system* (pp. 337–359). Amsterdam: Academic Press/Elsevier.

Puelles, E., Rubenstein, J. L. R., & Puelles, L. (2001). Chicken Nkx6.1 expression at advanced stages of development identifies distinct brain nuclei derived from the basal plate. *Mechanisms of Development*, 102, 279–282.

Puelles, L. (1995). A segmental morphological paradigm for understanding vertebrate forebrains. *Brain Behavior & Evolution*, 46, 319–337.

Puelles, L. (2001a). Brain segmentation and forebrain development in amniotes. *Brain Research Bulletin*, 55, 695–710.

Puelles, L. (2001b). Thoughts on the development, structure and evolution of the mammalian and avian telencephalic pallium. *Philosophical Transactions of the Royal Society B, Biological Sciences*, 356, 1583–1598.

Puelles, L. (2009). Forebrain development: Prosomere model. In G. Lemle (Ed.), *Developmental neurobiology*, (pp. 95–99), London; Elsevier/Academic Press.

Puelles, L. (2011). Pallio-pallial tangential migrations and growth signalling: New scenario for cortical evolution? *Brain Behavior & Evolution*, 78, 108–127.

Puelles, L. (2013). Plan of the developing vertebrate nervous system relating embryology to the adult nervous system (prosomere model, overview of brain organization). In J. L. R. Rubenstein & P. Rakic (Eds.), *Comprehensive developmental neuroscience: Patterning and cell type specification in the developing CNS and PNS* (pp. 187–209). Amsterdam: Academic Press.

Puelles, L. (2014). Development and evolution of the claustrum. In J. Smythies, V. S. Ramachandran, & L. Edelstein (Eds.), *Functional neuroanatomy of the claustrum* (pp. 119–176). New York, NY: Academic Press.

Puelles, L. (2016). Comments on the limits and structure of the mammalian midbrain. *Anatomy*. doi:10.2399/ana.15.045.

Puelles, L., Amat J. A., & Martínez de la Torre, M. (1987). Segment-related, mosaic neurogenetic pattern in the forebrain and mesencephalon of early chick embryos. I. Topography of AChE-positive neuroblasts up to stage HH18. *Journal of Comparative Neurology*, 266, 147–268.

Puelles, L., Ayad, A., Sandoval, J. E., Alonso, A., Medina, L., & Ferran J. L. (2016). Selective early expression of the orphan nuclear receptor Nr4a2 identifies the claustrum homolog in the avian mesopallium; Impact on sauropsidian/mammalian pallium comparisons. *Journal of Comparative Neurology*, 524, 665–703.

Puelles, L., Fernández, B., & Martinez-de-la-Torre, M. (2014). Neuromeric landmarks in the rat midbrain, diencephalon and hypothalamus, compared with acetylcholinesterase histochemistry. In G. Paxinos (Ed.), *The Rat Nervous System*, (4th ed., pp. 25–43), New York, NY: Academic Press/Elsevier.

Puelles, L., & Ferran, J. L. (2012). Concept of genoarchitecture and its genomic fundament. *Frontiers in Neuroanatomy*. doi:10.3389/fnana.2012.00047.

Puelles, L., Harrison, M., Paxinos, G., & Watson, C. (2013). A developmental ontology for the mammalian brain based on the prosomeric model. *Trends in Neuroscience*, 36, 570–578.

Puelles, L., Kuwana, E., Puelles, E., Bulfone, A., Shimamura, K., Keleher, J., … Rubenstein, J. L. (2000). Pallial and subpallial derivatives in the embryonic chick and mouse telencephalon, traced by the expression of the genes *Dlx-2*, *Emx-1*, *Nkx-2.1*, *Pax-6*, and *Tbr-1*. *Journal of Comparative Neurology*, 424, 409–438.

Puelles, L., & Martínez, S. (2013). Patterning of the diencephalon. In J. L. R. Rubenstein & P. Rakic (Eds.), *Comprehensive developmental neuroscience: Patterning and cell type specification in the developing CNS and PNS* (pp. 151–172). Amsterdam: Academic Press.

Puelles, L., Martínez, S., Martínez-de-la-Torre, M., & Rubenstein, J. L. R. (2004). Gene maps and related histogenetic domains in the forebrain and midbrain. In G. Paxonis (Ed.), *The rat nervous system* (3rd ed., pp. 3–24). San Diego, CA: Academic Press.

Puelles, L., Martínez, S., Martínez-de-la-Torre, M., & Rubenstein, J. L. R. (2014). Gene maps and related histogenetic domains in the forebrain and midbrain. In G. Paxinos (Ed.), *The Rat Nervous System*, (4th ed., pp. 3–24), New York, NY: Academic Press/Elsevier.

Puelles, L., Martínez-de-la-Torre, M., Bardet, S., & Rubenstein, J. L. R. (2012). The hypothalamus. In C. Watson, G. Paxinos, & L. Puelles (Eds.), *The mouse nervous system* (pp. 221–312). London: Academic Press/Elsevier.

Puelles, L., Martínez-de-la-Torre, M., Paxinos, G., Watson, C., & Martínez S. (2007). *The chick brain in stereotaxic coordinates: An atlas featuring neuromeric subdivisions and mammalian homologies*. San Diego: Academic Press/Elsevier.

Puelles, L., & Medina, L. (2002). Field homology as a way to reconcile genetic and developmental variability with adult homology. *Brain Research Bulletin*, 57, 243–255.

Puelles, L., & Rubenstein, J. L. R. (1993). Expression patterns of homeobox and other putative regulatory genes in the embryonic mouse forebrain suggest a neuromeric organization. *Trends in Neuroscience*, 16, 472–479.

Puelles, L., & Rubenstein, J. L. R. (2003). Forebrain gene expression domains and the evolving prosomeric model. *Trends in Neuroscience*, 26, 469–476.

Puelles, L., & Rubenstein, J. L. R. (2015). A new scenario of hypothalamic organization: Rationale of new hypotheses introduced in the updated prosomeric model. In G. Alvarez-Bolado, V. Grinevich, & L. Puelles (Eds.), *Development of the hypothalamus* (pp. 10–32). Lausanne: Frontiers Media.

Puelles, L., Watson, C., Martínez-de-la-Torre, M., & Ferran, J. L. (2012). Diencephalon. In C. Watson, G. Paxinos, & L. Puelles (Eds.), *The mouse nervous system* (pp. 313–336). London: Academic Press/Elsevier.

Redies, C., Treubert-Zimmermann, U., & Luo, J. (2003). Cadherins as regulators for the emergence of neural nets from embryonic divisions. *Journal of Physiology, Paris*, 97, 5–15.

Rendahl, H. (1924). Embryologische und morphologische Studien über das Zwischenhirn beim Huhn. *Acta Zoologica (Stockholm)*, 5, 241–344.

Robertshaw, E., & Kiecker, C. (2012). Phylogenetic origins of brain organisers. *Scientifica (Cairo)*, 2012, 1–14. doi: 10.6064/2012/475017

Rubenstein, J. L. R., Martínez, S., Shimamura, K., & Puelles, L. (1994) The embryonic vertebrate forebrain: The prosomeric model. *Science*, 266, 578–580.

Rubenstein, J. L. R., Shimamura, K., Martínez, S., & Puelles, L. (1998). Regionalization of the prosencephalic neural plate. *Annual Review of Neuroscience*, 21, 445–478.

Ruiz i Altaba, A. (1998). Combinatorial *Gli* gene function in floor plate and neuronal inductions by Sonic hedgehog. *Development*, 125, 2203–2212.

Saka, Y., & Smith, J. C. (2007). A mechanism for the sharp transition of morphogen gradient interpretation in Xenopus. *BMC Developmental Biology*, 7, 47. doi: 10.1186/1471-213X-7-47

Sánchez-Arrones, L., Ferrán, J. L., Rodríguez-Gallardo, L., & Puelles, L. (2009). Incipient forebrain boundaries traced by differential gene expression and fate mapping in the chick neural plate. *Developmental Biology*, 335, 43–65.

Sánchez-Arrones, L., Stern, C. D., Bovolenta, P., & Puelles, L. (2012). Sharpening of the anterior neural border in the chick by rostral endoderm signalling. *Development*, 139, 1034–1044.

Saucedo, R. A., & Schoenwolf, G. C. (1994). Quantitative analyses of cell behaviors underlying notochord formation and extension in mouse embryos. *The Anatomical Record*, 239, 103–112.

Scholpp, S., & Lumsden, A. (2010). Building a bridal chamber: Development of the thalamus. *Trends in Neuroscience*, 33, 373–380.

Scholpp, S., Wolf, O., Brand, M., & Lumsden, A. (2006). Hedgehog signaling from the zona limitans intrathalamica orchestrates patterning of the zebrafish diencephalon. *Development*, 133, 855–864.

Shimamura, K., Hartigan, D. J., Martínez, S., Puelles, L., & Rubenstein, J. L. R. (1995). Longitudinal organization of the anterior neural plate and neural tube. *Development*, 121, 3923–3933.

Shimamura, K., & Rubenstein, J. L. (1997). Inductive interactions direct early regionalization of the mouse forebrain. *Development*, 124, 2709–2718.

Shimogori, T., Banuchi, V., Ng, H. Y., Strauss, J. B., & Grove, E. A. (2004). Embryonic signaling centers expressing BMP, WNT and FGF proteins interact to pattern the cerebral cortex. *Development*, 131, 5639–5647.

Sugahara, F., Aota, S., Kuraku, S., Murakami, Y., Takio-Ogawa, Y., Hirano, S., & Kuratani, S. (2011). Involvement of Hedgehog and FGF signalling in the lamprey telencephalon: Evolution of regionalization and dorsoventral patterning of the vertebrate forebrain. *Development*, 138, 1217–1226.

Sugahara, F., Murakami, Y., Adachi, N., & Kuratani, S. (2013). Evolution of the regionalization and patterning of the vertebrate telencephalon: What can we learn from cyclostomes? *Current Opinion in Genetics & Development*, 23, 475–483. doi: 10.1016/j.gde.2013.02.008.

Sugahara, F., Pascual-Anaya, J., Oisi, Y., Kuraku, S., Aota, S., Adachi, N., Takagi, W., Hirai, T., Sato, N., Murakami, Y., Kuratani, S. (2016). Evidence from cyclostomes for complex regionalization of the ancestral vertebrate brain. *Nature*, 531, 97–100. doi: 10.1038/nature16518. Epub 2016 Feb 15.

Sussel, L., Marin, O., Kimura, S., & Rubenstein, J. L. R. (1999). Loss of *Nkx2.1* homeobox gene function results in a ventral to dorsal molecular respecification within the basal telencephalon: Evidence for a transformation of the pallidum into the striatum. *Development*, 126, 3359–3370.

Swanson, L. W. (2012). *Brain architecture* (2nd ed.). Oxford: Oxford University Press.

Tello, J. F. (1934). Les différentiations neurofibrillaires dans le prosencephale de la souris de 4 a 15 millimétres. *Travaux du Laboratoire des Recherches Biologiques de l'Université de Madrid*, 29, 339–396.

Thomas, P., & Beddington, R. (1996). Anterior primitive endoderm may be responsible for patterning the anterior neural plate in the mouse embryo. *Current Biology*, 6, 1487–1496.

Tomás-Roca, L., Corral-San-Miguel, R., Aroca, P., Puelles, L., & Marín, F. (2016). Cryptorhombomeres of the mouse medulla oblongata, defined by molecular and morphological features. *Brain Structure and Function*, 221, 815–838

Vaage, S. (1969). The segmentation of the primitive neural tube in chick embryos (*Gallus domesticus*). A morphological, histochemical and autoradiographical investigation. *Ergebnisse der Anatomie und Entwicklungsgeschichte*, 41, 3–87.

Verbeni, M., Sánchez, O., Mollica, E., Siegl-Cachedenier, I., Carleton, A., Guerrero, I., ... Soler, J. (2013). Morphogenetic action through flux-limited spreading. *Physics of Life Reviews*, 10, 457–475.

Vieira, C., Pombero, A., Garcia-Lopez, R., Gimeno, L., Echevarría, D., & Martínez, S. (2010). Molecular mechanisms controlling brain development: An overview of neuroepithelial secondary organizers. *International Journal of Developmental Biology*, 54, 7–20.

Von Kupffer, C. (1906). Die Morphogenie des Zentralnervensystems. In O. Hertwig (Ed.), *Handbuch der vergleichenden und experimentellen Entwicklungslehre der Wirbeltieren* (Vol. 2, Part 3, pp. 1–272). Jena: Fischer.

Wilson, S. W., & Houart, C. (2004). Early steps in the development of the forebrain. *Developmental Cell*, 6, 167–181.

Wullimann, M. F., & Puelles, L. (1999). Postembryonic neural proliferation in the zebrafish forebrain and its relationship to prosomeric domains. *Anatomical Embryology*, 329, 329–348.

Ziehen, T. (1906). Die Morphogenie des Zentralnervensystems der Säugetiere. In O. Hertwig (Ed.), *Handbuch der vergleichenden und speziellen Entwicklungslehre der Wirbeltiere* (Vol. 2, 273–368). Jena: Fischer.

13

Brain Evolution and Development
Allometry of the Brain and Arealization of the Cortex
Diarmuid J. Cahalane and Barbara L. Finlay

13.1 Introduction

The first images and descriptions of brains have an unusual leverage on how we pose further questions about the brain. The principal images seen in introductions to neuroscience are sagittal views of the human brain, outlining the large divisions from forebrain to hindbrain derived from the "primitive swellings" of the neural tube, and the lateral view showing some version of Brodmann's original subdivisions of the cortex with its numbered areas and lobes (Figure 13.1, bottom). Such first impressions underlie the concepts of brain structure and function that we form as individual researchers and the conversations we have as a research community.

The first studies of brain evolution enumerated these divisions and described their allometry. Studies of brain function attempt to map adaptive behavior, the mechanisms of perception and action, onto the same regions and areas delimited in those early images. However, the existence of an easily accessible set of morphological subdivisions of the brain is no guarantee at all that those subdivisions have identically corresponding developmental mechanisms on which evolution may act independently (cf. Chapter 12, this volume). How different might our current understanding of the brain be if our first view of it had been its wiring diagram (Figure 13.1, right), or the gradients of gene expression in the developing brain and the initial segmentation those gradients imply (Figure 13.1, bottom)?

The subdivisions drawn up by the first descriptive neuroanatomists were exalted by physical anthropologists studying brain evolution. The anthropologists applied the *prima facie* reasonable idea of "proper mass" (Jerison, 1973; Stephan, Baron, & Frahm, 1988) which says that the brain of an individual species should devote more of its volume to those sensory and motor capacities most important to it. Insofar as brain segments embody these particular capacities, therefore, to select an animal on the basis of some particular capacity is to select on the volume of its associated brain segment—this is the basis of the "mosaic" view of brain evolution (Barton & Harvey, 2000; Striedter, 2005). This discussion is not to suggest that the divisions and strategies employed by the first brain researchers were poorly chosen or necessarily

---

The Wiley Handbook of Evolutionary Neuroscience, First Edition. Edited by Stephen V. Shepherd.
© 2017 John Wiley & Sons, Ltd. Published 2017 by John Wiley & Sons, Ltd.

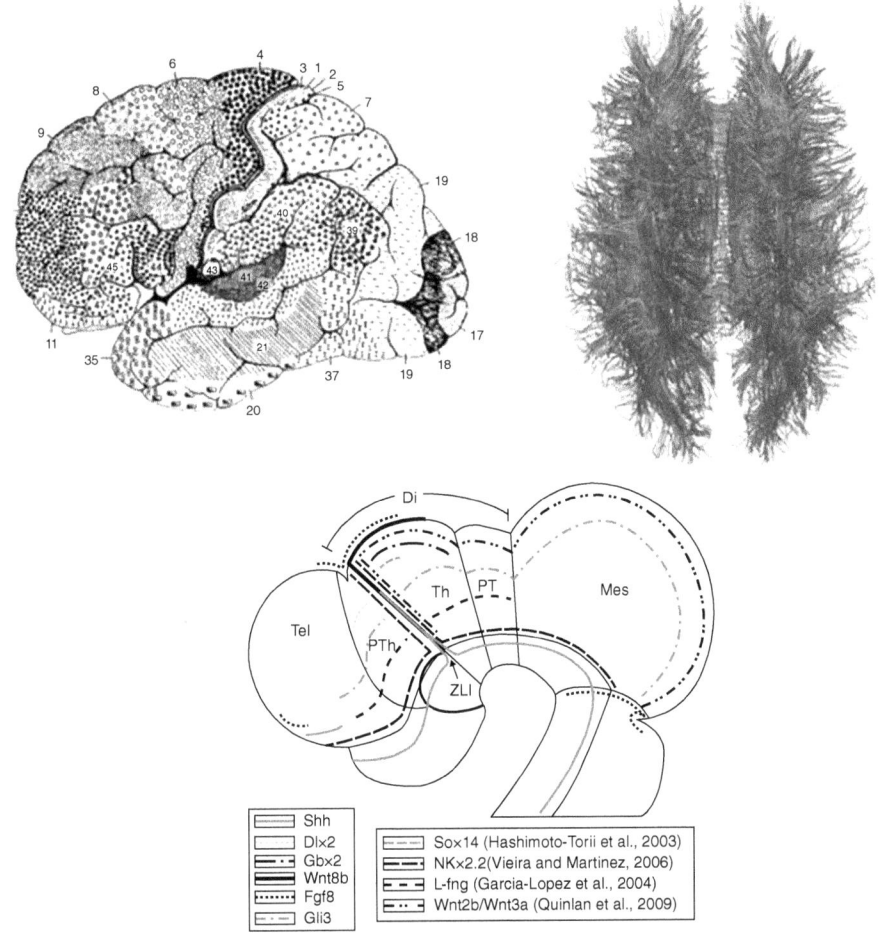

Figure 13.1 **Left**: Cortical areas as originally delineated by Brodmann (1913). **Right**: White matter connections of the cortex. Gigandet et al., 2008. **Bottom**: Divisions of the developing brain as demonstrated by early gene expression. Martínez-Ferre & Martínez, 2012.

misguided, given the information which was at hand. The purpose of our brief, critical retrospective is to disabuse us of any unworthy preconceptions which those early visions of the brain have instilled in our current thinking.

In the present chapter we present an alternative to the traditional, area-based view of brain evolution and development. We will first review in general what is known about the overall scaling of the brain and its parts, at size scales ranging from gross anatomical divisions down to identified cell classes, and the accompanying alterations of development. We will then focus on the evolution of the isocortex as an example, contrasting its emerging combinatorial sources of organization with the area-based view. We will consider various sources of organization in the cortex—columns, layers, areas, and larger-than-area regions— and directionalities in information flow and neuromodulation. We will conclude with a general developmental model of the cortical sheet that can encompass within-cortex, between-individual and cross-species differences in neural architecture.

13.2 Basic Vertebrate Brain Allometry

13.2.1 "Allometry of What?" or "What Should We Measure?"

The first studies of changing brain organization across vertebrates looked at the weights or volumes of brain subdivisions because that was the only comparison possible given the available technology (Jerison, 1973; Northcutt, 1981; Stephan, Frahm, & Baron, 1981). Computer-assisted stereology improved measurements of the total number of neurons and their arrangement in brain structures (Coggeshall, 1992) and, even though this method remains time-consuming, substantial data sets have been generated (e.g., Giannaris & Rosendene, 2012; Pakkenberg, 1993). More recently, neuron and glial number have been counted in relatively large volumes of tissue using flow cytometry techniques (Herculano-Houzel & Lent, 2005). The repeated assertion of flow cytometry studies is that counting neurons (sometimes along with other cell types) provides a more useful unit of measurement of the brain than does comparing volumes (Chapter 2, see also Herculano-Houzel, Collins, & Wong, 2007; Herculano-Houzel, Mota, & Lent, 2006). However, different measures highlight different features and those features could be either the subject of selective pressure or the consequences of developmental mechanisms through which evolution must act.

Single neurons are not "the" definitive measure of information transmission or circuit complexity in the brain. Only in those cases where the transmission of information is solely by action potentials over axons could the number of neurons be a faithful measure of bandwidth or complexity. In structures featuring local, graded computation involving non-spiking communication (such as found in the retina, the cortex and the cerebellum), axodendritic assemblies, rather than whole neurons, often serve as computational units. For example, in the retina, a single A17 amacrine cell provides reciprocal feedback inhibition to hundreds of bipolar cells in isolated parallel circuits (Grimes, Zhang, Graydon, Kachar, & Diamond, 2010). Quite apart from variations in neuronal density, brain structures exhibit very diverse morphologies in their architectures for information processing and assembly. Consider, for example, the three-dimensional stellate arborizations of the neurons of the neocortex, the parallel coursing axons intercepting the close-packed two-dimensional dendrites of the cerebellum's Purkinje cells, or the patch-and-matrix arrangement of the striatum, where patches with particular processing features are interposed in the general matrix of neurons of the caudate and putamen (Brown et al., 2002). All of said structures have different doses of transmission neurons, interneurons and modulating neurons, as well as differences in resting metabolic activity. In addition, the volume of axonal and dendritic connections required to scale up each structure for additional neurons will scale differently for each cytoarchitectural plan. Such diversity makes it plain that using neuron numbers as the single measure of difference between complex structures paints a very incomplete picture.

What if energy consumption, rather than a structure's volume or neuron count, is the implicit variable of concern? Then it is the action potential at the synapse that is by far the greatest consumer of resources (Laughlin, Steveninck, & Anderson, 1998; Logothetis, 2003) and the most relevant piece of information would be the total number of synapses, or perhaps, the volume of neuropil minus the volume of cell bodies. Thus we see that cell number, volume, cellular constituents and energetic requirements all interact to influence both the cost and the functionality of any given structure (cf. Chapter 7 of this volume). Then, it is hardly surprising that no single measure suffices to characterize a region's function or is the obvious subject of selection.

Brain volume, on the whole, does scale as a fairly regular exponent of neuron number (e.g., Zhang & Sejnowski, 2000). However, it will dissociate from neuron number due to a number of factors. The two principal sources of variation are, first, the structure-by-structure differences in neuron assemblies already described above, and second, different adaptations to the "save wire" problem that can become acute at large brain sizes (Cherniak, 1995; Cherniak, Mokhtarzada, Rodriguez-Esteban, & Changizi, 2004). The latter relates to the potential for wild proliferation of the volume of connectivity compared to neuron number if the network architecture of a small brain cannot scale up gracefully. Murre and Sturdy (1995) have calculated that, if a human cortex retained the same per cent connectivity seen in a mouse cortex, that cortex could be approximated as a cube 21 meters on a side, comprised principally of axons with merely a superficial dusting of neurons. Employing "small world" connectivity, whereby a tiny fraction of the network's links being nonlocal ensures a low number of intermediaries between any two neurons, solves this problem (Watts & Strogatz, 1998). However, the parameters generating small world connectivity in differently sized brain networks may well be the subject of selection, e.g. selection on the distribution of axon lengths.

Overall, brain volume, neuron number, and the ratio of neurons versus glia all scale dramatically across vertebrate brains. As we have seen, however, each single feature is a patently inadequate measure of the computational diversity and capacity we are attempting to assess, and it is not clear that any one of these measures ought to be the favored target of selection. The best and only solution is to keep in mind the advantages and drawbacks of each measure, and the computational, energetic, and developmental constraints that bear on changing the size or amount of each kind of neural architecture.

13.2.2 Brain Scaling, Macro Scale

Looking at the overall patterns of brain change, we can infer evolution's trajectory from the variation seen in present-day sharks and rays to current mammals including primates. This corresponds to an evolutionary period of about 450 million years and the pattern of brain change is astonishingly stable: no new brain divisions appear; the same structures always become disproportionately large when the brain gets large (forebrain and cerebellum); and while the brainstem, mesencephalon, forebrain (telencephalon) and cerebellum strongly covary in volume, the olfactory bulb and associated structures (olfactory cortices and hippocampus) vary more independently (Figure 13.2; Yopak et al., 2010).

These general features of brain scaling are now well known and we have given two kinds of accounts of the causes of brain scaling. For the first, given the radical diversity of the functions of homologous parts of the forebrain, for example comparing between chickadees and chimpanzees, a functional reason for the preferential enlargement of these regions must lie at a computational level common to these species and arrived upon early in the vertebrate lineage. We have suggested that the insights into what kinds of computational architectures are both robust and evolvable can be found in the computational and robotics literature (Charvet, Striedter, & Finlay, 2011; Finlay,

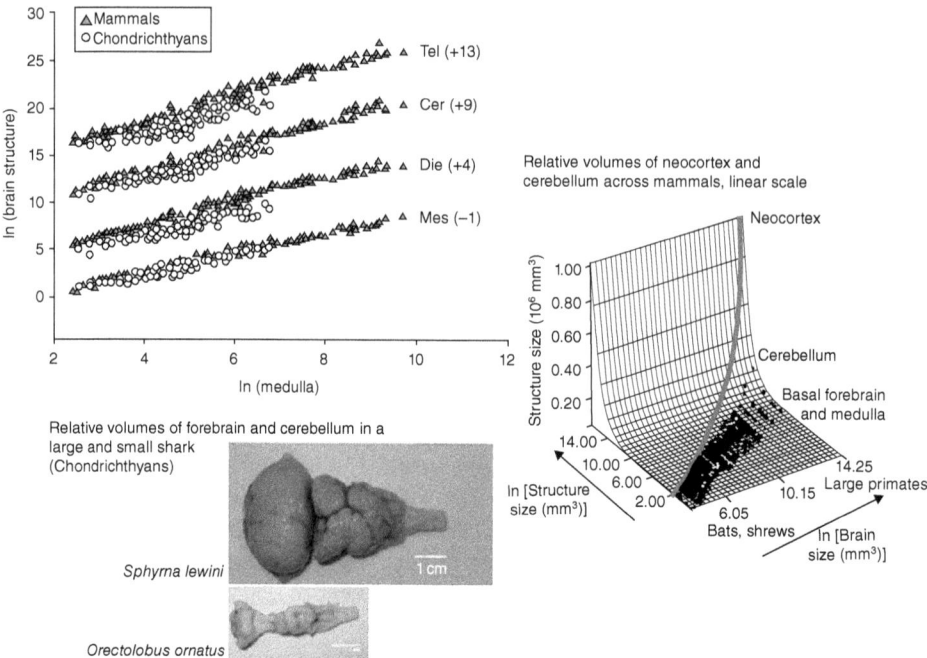

Figure 13.2 Top: Mammals demonstrate a "grade shift" in the relative volume of the forebrain (telencephalon, labelled "Tel") and cerebellum ("Cer") at all brain sizes, but, in both mammals and nonmammals, those two structures enlarge disproportionately in the largest brains, a fact whose magnitude is often hard to appreciate in log scale representations as in the top panel. Note the data series for the respective brain parts have been offset by the quantities in parentheses for easy of viewing. Yopak 2010. Reproduced with permission of National Academy of Sciences. **Bottom left:** Large and small shark brains provide an example of the forebrain's disproportionate enlargement. Digital Fish Library. **Bottom right:** An alternative representation of the relative scaling of brain parts. The vertical axis is in regular (non-log) units to better demonstrate the isocortex's volume scaling. Finlay 1995. Reproduced with permission of American Association for the Advancement of Science.

Hinz, & Darlington, 2011). Regarding the second kind of account, a developmental mechanism capable of producing this repeated change is the conserved segmental organization of the vertebrate brain: the most lateral locations in the developing neural plate, namely the cerebellum and forebrain pallium, produce their neurons for the longest time and so have the potential for disproportionate enlargement (Finlay, Hersman, & Darlington, 1998).

13.2.3 Individual Variability

The developmental account we have given of how extending embryonic development can directly produce the disproportionate enlargment of the cortex and cerebellum should work at large and small scales, that is, over the phylogenetic scale and over the range of variability of individuals. We have investigated this prediction in three separate data sets: the reports of the volumes of multiple brain parts in feral and domestic minks and pigs (Finlay et al., 2011); in laboratory mice of identified genetic strains

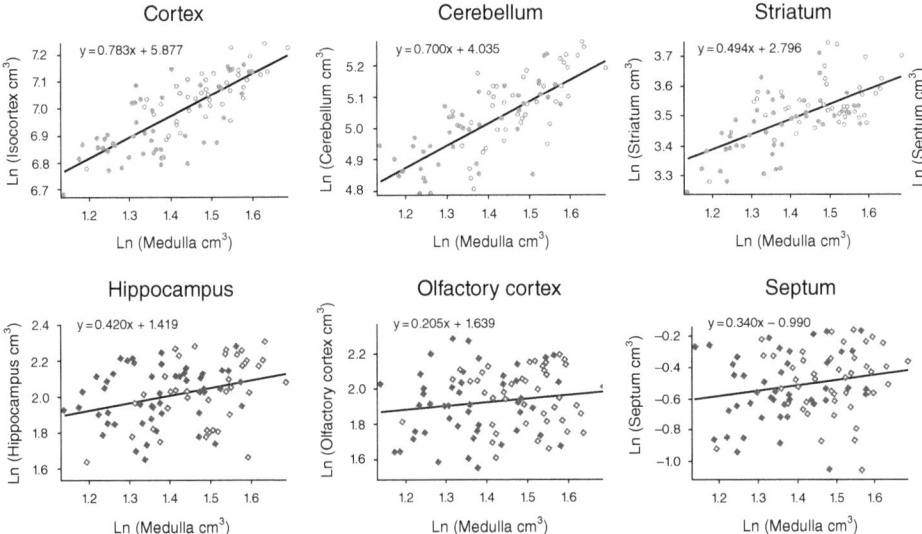

Figure 13.3 Natural-logged values of brain region volumes regressed against the natural-logged values of medulla volumes in N=90 humans. Circles = nonlimbic structure; diamonds = limbic structure; open circles and open diamonds = males; closed circles and closed diamonds = females. Adapted from Charvet 2013, Brain, Behavior and Evolution, Karger AG.

(Finlay et al., 2011); and in several publicly available databases of MRI scans in normal humans (Charvet, Darlington, & Finlay, 2013). In all cases, the patterns of phylogenetic variability were reproduced at the level of individual variation. That is despite the range of variation at the individual level being multiple orders of magnitude reduced from that at the phylogenetic level. In all cases, brain parts scale together with high regularity, with the exception of the olfactory bulb and limbic system (Reep, Finlay, & Darlington, 2007). Here again, the cortex and cerebellum become disproportionately large in the largest brains. In Figure 13.3 we see the scaling and the relative variability of 6 neural structures in N=90 humans plotted against the size of the medulla in each.

13.2.4 Scaling of Cortical Areas

To complete our picture of general trends in phylogenetic variability, individual variability, and developmental mechanisms that link the two, we will consider two aspects of the regularities of scaling of cortex which impact within-cortex structures. First, we will consider the size-scaling of cortical areas. Second, we will examine how the number of cortical areas relates to the total volume of different cortices.

Analyzing cortical areas presents some problems which do not arise in dealing with gross brain divisions. The direct homologizing of a "cortical area" from one species to the next can be impossible. For example, in the cortical areas containing the multiple representations of sensory surfaces, or levels of the motor hierarchy, questions of whether there is a homologous V4 in both the macaque monkey and a human, or a homologue of "Broca's area" in a rat cannot be answered because of the ambiguities in the connectional architecture in smaller and larger brains. There are two kinds of exceptions to this. The first is that primary sensory and motor areas are unambiguously identifiable from one species to the next (Kaas, 2011; Krubitzer &

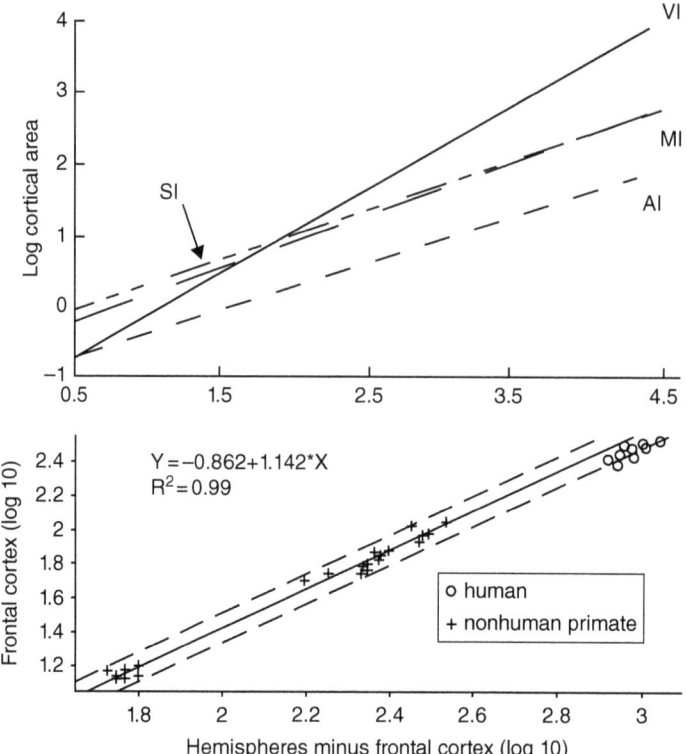

Figure 13.4 Top: Log primary neocortical areas (primary somatosensory, S1; primary visual, V1; primary motor, M1; and primary auditory, A1) are regressed on log(total neocortex size). The intercepts of the lines have been adjusted for easier visualization of the relative slope. The slope for primary visual cortex is greatest. The slope of the regression of log V1 is significantly different from S1 (n = 22, p < 0.021) and from A1 (n = 13, p < 0.011). Kaskan 2005. Reproduced with permission of The Royal Society. **Bottom:** Regressing log(volume of frontal cortex) against log(volume of hemispheres minus frontal cortex) for n = 24 great apes (crosses), Semendeferi et al. arrived at the regression line and intervals of 95% confidence (dashed lines) before overlaying the data points for human subjects (open circles). Semendeferi 2002. Reproduced with permission of Nature Publishing Group.

Seelke, 2012)—they lie in the same relative positions and have the same types of input and output. The second exception is if cortical areas are massed, into large divisions like "frontal" or "parietal," with respect to the primary sensory and motor areas. Then, the large regions massed will be homologous by exclusion. That these two kinds of within-cortex division have regular scaling has been known for some time. Jerison (1973) described the regular, hyperallometric scaling of the frontal cortex with respect to the rest of the brain. The visual cortex also scales somewhat hyperallometrically (Frahm, Stephan, & Baron, 1984). In Figure 13.4, taken from more current work, these two kinds of regularities in cortical scaling can be seen.

In the top panel of the figure, the relative increase in the size of V1 is shown versus primary auditory, somatosensory, and motor cortex, as measured in a sample of diurnal and nocturnal animals (Kaskan et al., 2005). The slope of the allometric equation of V1 (1.086) was significantly higher than the other three (0.697, 0.66 and

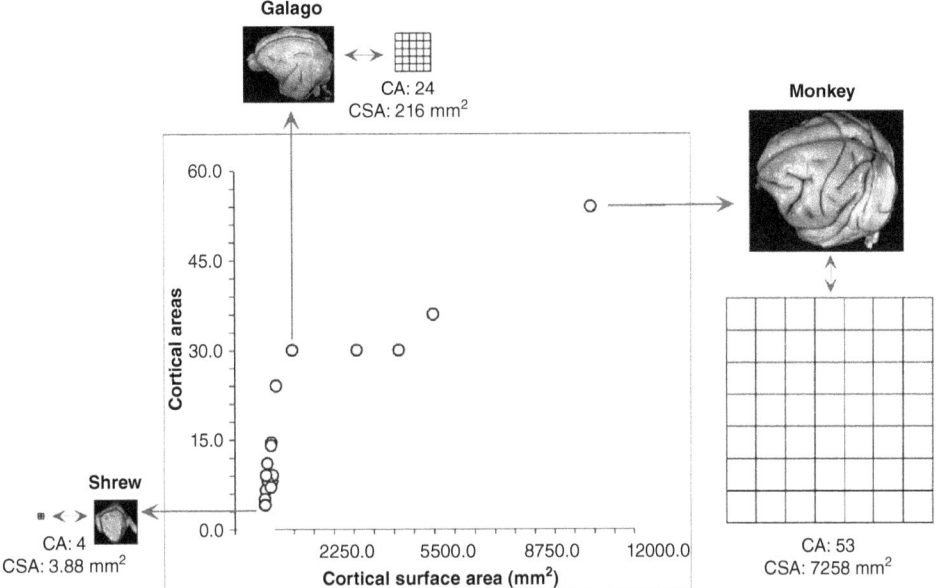

Figure 13.5 The number of cortical areas plotted versus cortex surface area. The approximate total cortical surface area of mapped visual and somatomotor areas (CSA) and number of cortical areas (CA) for the shrew, galago, and macaque are depicted, as an example of small, medium, and large cortex area. Note that the entire cortex of the Galago, comprising 24 areas, could be accommodated within a single cortical area of the rhesus monkey. Reproduced from Finlay et al., 2005. (*See insert for color representation of the figure*).

0.84 respectively). In the bottom panel of Figure 13.4, the predictability of the size of human frontal cortex compared to great apes is plotted (Semendeferi, Lu, Schenker, & Damasio, 2002).

A second feature of cortical scaling is the change in the number of cortical areas in cortices of different volumes. Using the complete maps of cortical areas in 20 mammalian species (where a cortical area is described as a full thalamocortical topographic map in any modality and having distinct patterns of outputs and inputs compared to its neighbors) the relationship in Figure 13.5 can be seen (Finlay, & Brodsky, 2006; Finlay, Cheung, & Darlington, 2005).

Cortical areas increase very rapidly in number up until a cortex size of about 200 mm² in total area; thereafter the rate of increase is slowed (see Figure 13.5). We have suggested that the faster increase in the number of areas in small brains is produced by the scale-dependent segregating mechanisms of Hebbian sorting, very similar to the mechanisms which in large brains produce features like ocular dominance columns. However, quite why there should be a discontinuity in scaling at about the cortical area of a galago as yet demands a precise mechanistic explanation.

13.2.5 Summary of Allometry, Focusing on the Cortex

We have described how, in the vertebrate lineage, the pallium (cortex or its forebrain homologue) is preferentially enlarged whenever brains become large. We have also shown that the same pattern of preferential enlargement is recapitulated in the individual variability of several mammalian species, including humans. The consistent pattern of brain change at these levels suggests a similar developmental mechanism should account for both. Within the cortex, areas that can be identified consistently across species enlarge at predictable allometric rates: the number of overall cortical areas increases with overall cortical volume very rapidly in small to medium-sized brains and then more slowly in the largest. Considering the brain as a whole, and extrapolating from our work with the cortex, the manner and rate in which each brain part scales with overall brain size will depend on the relative size of its progenitor pool and the relative duration over which it proliferates during neurogenesis. The volume produced by neuron cell bodies is further adjusted by the type of interconnectivity each neuronal type maintains.

13.3 Evolutionary Developmental Models for the Cerebral Cortex

Taken together, the regularities in brain scaling discussed above suggest the developmental mechanisms which generate central nervous systems are strongly conserved across species. Firstly, we noted the remarkable regularity of the relationships of brain component sizes to total brain size across species. Secondly, we observed the regular scaling of cortical areas as the cortex gets bigger overall. Thirdly, individual variation in humans was seen to recapitulate the same general pattern of covariation in brain part sizes as is observed across species. What features of brain structure are contributed by the scaling of conserved developmental mechanisms? To address those questions we have created formal, quantitative models of neural development. The models elucidate what avenues are made accessible to selection by the conserved mechanisms shaping the basic landscape of an embryonic nervous system. Given the baselines provided by the model, the extent to which a particular brain part (e.g., cerebral cortex) or cortical area (e.g., a visual or language area) has been a privileged subject of selection can be better evaluated.

For decades, the developmental mechanisms which generate the mammalian cerebral cortex have been the subject of extensive investigations and competing theories, e.g. "protomap" (Rakic, 1988) versus "protocortex" (O'Leary, 1989). More recently, efforts to catalog developmental and adult gene expression across the cortex in multiple species have yielded a burgeoning volume of data, some of which buttress existing theories but many of which demand to be included in more comprehensive mechanistic models (Hawrylycz et al., 2012; Kang et al., 2011). What has become clear is that rather than there being a single, definitive mechanism to coordinate the structure and layout of the cortex, many mechanisms and sources contribute order throughout development (Dehay, & Kennedy, 2007; Sansom & Livesey, 2009; Yamamori, 2011). Early polarization and regionalization of the cortex is directed by morphogens issuing from signaling centers in the cortical primordium (Fukuchi-Shimogori & Grove, 2001). Spatial gradients in the kinetics of neurogenesis change the extent and timing of neuronal production from location to location (Bayer & Altman, 1991; Rakic, 1974, 2002). Molecular signals guide axons of the various sensory modalities to enter the growing cortex at particular locations (Finlay & Pallas, 1989). The axons of those projections are kept in topological register and that orders the various topographic maps in the cortex. The structure of correlations in sensory information flowing through those same projections further refines the adult cortical architecture (Johnson & Vecera, 1996). As we unravel how this complex ontogeny is orchestrated by the genome, formal models for developmental mechanisms will be pivotal in synthesizing data on developmental gene expression patterns (Lewis, 2008).

Considering that developmental mechanisms act in combination to direct the ontogeny of any part of the cortex, it is perhaps not surprising that selection may not be able to address particular circuits or cortical areas without also affecting others. That is to say, rather than being mosaic as implied in the "proper mass" hypothesis, adaptations might necessarily be coordinate in nature. For example, it could be, depending on the developmental mechanisms in play, that the only way to provide an increased number of neurons per cortical column in visual area 1 would be to boost neurogenesis output according to position along the anterior-posterior axis. Such an adaptation would have an impact beyond the borders of that cortical area under direct selection pressure. A reasonable first reaction to the hypothesis that developmental mechanisms admit only such coordinated change is that it sounds highly restrictive, forbidding the selection of a myriad of potentially expedient mosaic adaptations. However, appropriate, "evolvable" developmental mechanisms might provide useful structure. They might leave available to selection the most useful reduced set of parameters from a search space otherwise far too large for genotypes to sample extensively. Assuming the coordinate nature of adaptations raises two questions in particular. Firstly, what changes to the co-operating developmental mechanisms have given rise to the very differently sized cerebral cortices apt to serve the cognitive requirements of mammals in highly diverse niches? Secondly, what are the signatures of these coordinate adaptations—i.e. if features are not independently selected, then what types of co-variation ought we to expect?

To address, in a specific context, the questions of what developmental parameters might change and what resultant coordinate changes arise in the features of the phenotype, we discuss in what follows a quantitative evolutionary-developmental model for neurogenesis in the cerebral cortex (Cahalane, Charvet, & Finlay, 2014). In particular, we are concerned with how the kinetics of cortical neurogenesis are altered both across species, furnishing very differently sized cortices with their requisite numbers of neurons, and across the developing cortex of an individual, giving rise to differing numbers and classes of neurons and hence a cortical architecture which varies across locations.

13.3.1 Changes in Cortical Development across Species

The mechanics of mammalian cortical neurogenesis are patently adaptable: changes in duration and kinetics result in the production of five orders of magnitude more neurons in the largest cortices compared to the smallest. To inform our discussion of what parts or aspects of neurogenesis may be accessible to selection, we will briefly sketch the basic process by which cortical neurons are produced. For our purposes, the salient features of neurogenesis are as follow (for more detailed review, see Rakic, 2002).

A founding population of precursor cells in ventricular zones near the wall of the cerebral vesicles initially undergoes rounds of symmetric division, whereby both daughter cells are precursors, thus swelling the precursor pool. Neurogenesis begins when some divisions in the precursor pool become asymmetric: with some probability a daughter cell is now a differentiated neuron which will not undergo further rounds of cell division and will migrate out of the ventricular zones. We refer to the probability of a daughter cell being a neuron as the "quit fraction." As long as the quit fraction remains close to zero, the precursor population increases approximately exponentially. As neurogenesis proceeds, the quit fraction becomes larger. The precursor population peaks exactly when the quit fraction is one half: Now every precursor that undergoes division is expected to produce one precursor and one neuron, so the growth phase of the precursor pool has ended. Eventually the majority of the cells produced are neurons and the precursor pool becomes further depleted with each round of divisions. Young neurons migrate out of the ventricular zone to populate the developing layers of the cortex in an "inside-out" manner. It has been established that the neurons of the deep cortical layers (VI and V) are produced first. Neurons destined for the progressively more superficial layers (IV through II) subsequently migrate through the already-present layers. These events take place over approximately 8 days in the rat and 60 days in the rhesus macaque (Kornack & Rakic, 1995).

Our model builds upon several existing computational models which aim to reconcile the limited empirical data available on the kinetics of the cortical neurogenesis with data on the mature distribution of cortical neurons (Caviness et al., 2003; Gohlke, Griffith, Bartell, Lewandowski, & Faustman, 2002; Gohlke, Griffith, & Faustman, 2007; Kornack & Rakic, 1998; Nowakowski, Caviness, Takahashi, & Hayes, 2002; Takahashi, Nowakowski, & Caviness, 1996, 1997). Most published models focus on a single species but some comparative studies address a few species. We have developed a model to investigate the key parameters contributing to the variations in neural architecture observed not only across the mammalian order but also between individuals of a given species and across the cortical surface. The core of the model is purposefully simplistic, tracking two populations, namely the precursor pool and the neuronal progeny of those precursors, as the size of each changes during neurogenesis. Given a founder population, the model predicts the time course of neuronal output at a ventricular zone location in terms of amplification of the founder population. Parameters in the model which change over time include the quit fraction (as described above) and the cell-cycle duration. Cell death is also an important phenomenon in the ventricular regions and the developing cortical plate, but, due to limited quantitative data on its magnitude, we have not included it explicitly in the model. Also, studies have suggested that separate compartments within the ventricular region are respectively responsible for producing the deep and superficial layers of the cortex (for a review, see Dehay & Kennedy, 2007). For simplicity, we model the ventricular region as a homogeneous source of neurons, their layer assignment being dependent on their time of origin.

What developmental parameters must change to accommodate the much greater number of neurons required to populate larger cortices? Data collected in this laboratory suggest that, in contrast to a change of roughly five orders of magnitude in adult cortical neuron number, the size of founder populations changes only modestly from small to large cortices, increasing by no more than a factor of two from gerbil to cat (Charvet, Cahalane, & Finlay, 2013). The onus in boosting neuronal output, then, is on the process of neurogenesis which begins after the founder population of precursor cells has been established. We refer to the ratio of the adult neuron number to

- Progenitors undergo cell divisions to produce more progenitors or differentiated cortical neurons.

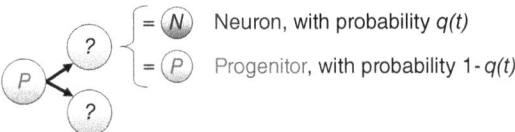

- The relative probability of those outcomes changes during the neurogenetic interval.

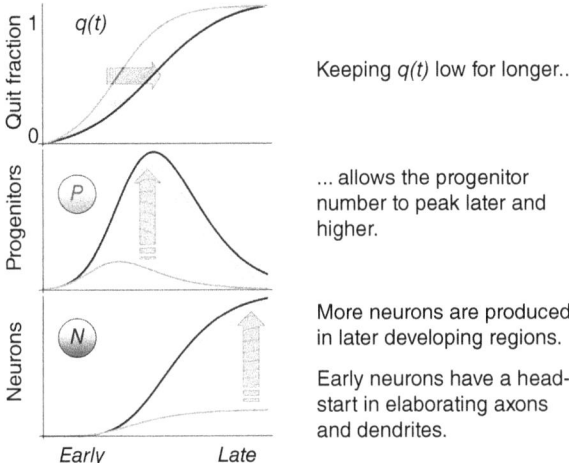

Figure 13.6 Top: Defining the "quit fraction" as the probability that a daughter cell in the ventricular zone is a differentiated neuron. Bottom: Delaying the rise of the quit fraction has a large effect on the peak size of the precursor pool and the total number of neurons produced. (*See insert for color representation of the figure*).

the founder population number as the "amplification" of the founder population. In our model, the amount of amplification is determined by the progression of the cell-cycle duration and the quit fraction. We employ empirical data to fix the duration of neurogenesis (normally longer in larger brains) and to constrain our estimates of the cell-cycle duration (typically increasing during neurogenesis) at each point in time. The model recapitulates the required amounts of amplification by adjusting how the quit fraction changes in time. Suppressing the quit fraction has the effect of prolonging the near exponential growth phase of the precursor pool. Thus, even modest adjustments to the quit fraction can have a dramatic effect on the total amplification, and all the more so in larger brains where the total number of cell cycles is greater (Figure 13.6). The leverage which even slight temporal adjustments to the quit fraction have over total neuron number have been noted in the context of cross-species differences in neuron number (Nowakowski et al., 2002; Takahashi et al., 1997). As to mechanisms governing the quit fraction, under- and over-expression of the cell cycle inhibitor p27 in embryonic mice have been shown to affect the time course of the quit fraction's progression and impact total neuronal output of the proliferative zone (Caviness et al., 2003).

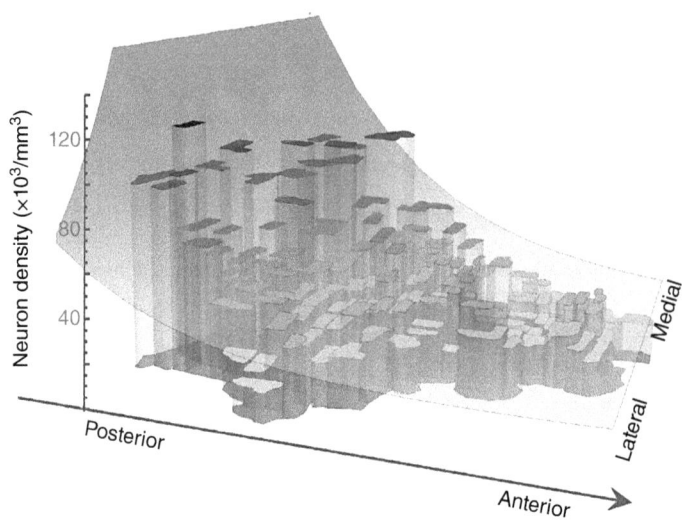

Figure 13.7 The density of neurons was measured in N=141 samples, together comprising the entire (flattened) cortical sheet of a baboon (Collins et al., 2010). Neuron density exhibits marked variation across the cortex of the baboon, the general trend (indicated here by the transparent surface) being to increase along an axis from anterior lateral cortex to posterior medial cortex. Figure redrawn, based on a figure which appeared in Cahalane et al., 2012. (*See insert for color representation of the figure*).

13.3.2 Cross-Cortex Gradients in Neurogenesis

That there is marked, systematic variation in the density of neurons across the cortical sheet (Figure 13.7), and so also in the number of neurons in any column taken orthogonal to the sheet's surface, was not appreciated until recently (Cahalane, Charvet, & Finlay, 2012). A high throughput method (the isotropic fractionator with flow cytometry) enabled the first studies enumerating the cellular contents of very many sites (up to N=141) in the primate cortex (Collins, Airey, Young, Leitch, & Kaas, 2010). These findings are in contrast to the long-standing notion that the cortical architecture is uniform with the exception of higher neuron numbers in primate visual areas (Rockel, Hiorns, & Powell, 1980). The supposedly privileged visual areas were suggested to be a distinguished subject of selection, with tailored neurogenesis to supply their higher neuron number (Dehay, Giroud, Berland, Smart, & Kennedy, 1993). Instead, we posit that the isocortex-wide gradients in neuron density and number are better understood as arising from an isocortex-wide developmental mechanism. Insofar as the visual areas are privileged, it may be largely due to their position on the gradient, typically at the point of highest density.

What developmental mechanisms could account for a gradient in neuron number across the cortex? Observed gradients in the timing of neurogenesis across the rodent, carnivore, and primate cortex (Bayer & Altman, 1991; Jackson, Peduzzi, & Hickey, 1989; Kornack & Rakic, 1998; Luskin & Shatz, 1985; Miyama, Takahashi, Nowakowski, & Caviness, 1997; Rakic, 2002; Smart, Dehay, Giroud, Berland, & Kennedy, 2002) hint that the answer is related to the kinetics of neurogenesis. In rodents and primates alike, the noncingulate isocortex is populated with neurons in an anterior to posterior progression. The longer period of gestation in primates makes the difference in neurogenesis end-dates more notable in those species than in rodents: in the macaque, despite beginning at approximately the same developmental time in all regions, neurogenesis ends as many as three weeks later in posterior cortex. Of what consequence are such timing gradients? We noted, in considering cortical expansion across species, that changes to the time-course of the quit fraction have a potent effect on total neuronal output. Employing the same model, we observe that delaying the rise of the quit fraction increases total neuronal output but those neurons are also born later. Both findings are in agreement with the empirical data: more neurons produced later in posterior regions. This mechanism is not the only determinant of neuron number at a cortical location, and other contributory factors are discussed below, but it is capable of explaining much of the variation observed.

13.3.3 Cross-Species and Cross-Cortex Differences in Neuron Number Arise from the Same Mechanism

How do cross-species differences interact with cross-cortex variations? Modeling the progression of the quit fraction as varying by location across the ventricular zone, consistent with the empirically observed timing gradients in mouse, rat, and macaque (Bayer & Altman, 1991; Miyama et al., 1997; Rakic, 2002), generates in each case a cortex with spatial variations in neuron number consistent with empirical data (Cahalane et al., 2014). Given the lesser variation in neurogenesis duration in small brains, the model predicts little change in neuron number and in the layer distribution of neurons across the rodent cortex. By contrast, the pronounced gradient from anterior to posterior cortex in the primate neurogenesis timing implies a large change in both neuron number and their distribution across layers (Figure 13.8). The model predicts

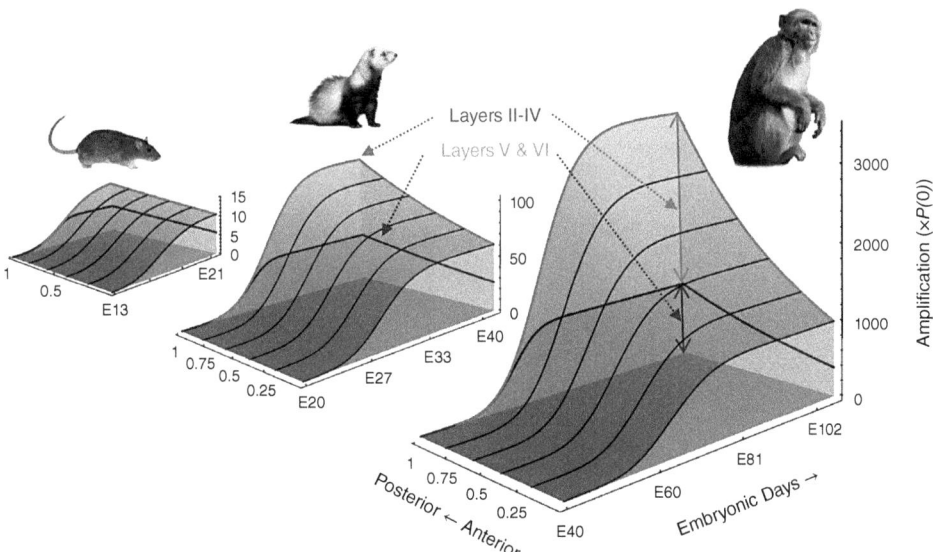

Figure 13.8 Within-cortex gradients in neurogenesis timing lead to pronounced changes in the total number of neurons and in their layer distribution across a large cortex like that of a primate. By contrast, the shorter total duration of neurogenesis and the lesser variation in neurogenesis timing in a small cortex leads to neuronal and layer distributions which are relatively uniform across the cortex. Figure redrawn, based on a figure which appeared in Cahalane et al., 2014. (*See insert for color representation of the figure*).

that the difference of approximately 30% in the duration of neurogenesis leads to a four or five fold change in the number of mature neurons per unit of precursor cells. Moreover, the bulk of neuronal production in posterior regions occurs relatively late in development, whereby, in our model, it contributes predominantly to the superficial cortical layers.

13.3.4 Interaction with Other Mechanisms

We have outlined a model for how the kinetics of neurogenesis could give rise to an embryonic cortex whose architecture varies smoothly and systematically from the anterior to posterior poles (or along other axes). We conjecture that these gradients establish the basic landscape that richer areal and cellular structure is built upon, as prompted by genetic markers, projections from subcortical structures, or other local cues (Kingsbury & Finlay, 2001). We offer the following as an example of how local deviations can be overlaid on the basic landscape set up by the global gradient in neuron number. In their investigation of neuronal densities across the cortex, Collins et al. (2010) noted that areas involved in sensory processing had higher neuron densities than some adjacent areas. Identifying the data points of Collins et al., which related to primary sensory areas in baboon, we used a two-factor model, incorporating each sample's location and whether or not it was from a primary sensory area, to look for significant differences in neuronal density from what a "location-only" model predicted (Cahalane et al., 2012). We found that primary areas have a density of neurons that is 26% higher than that predicted for a non-primary area in the same cortical location (Figure 13.9). So, clearly, a mechanism other than smooth, location dependent changes in neurogenesis is required to explain the variations in neuron density. Lower levels of neuron death during early development have been reported in putative sensory areas (Finlay & Slattery, 1983) relative to other areas. We suggest

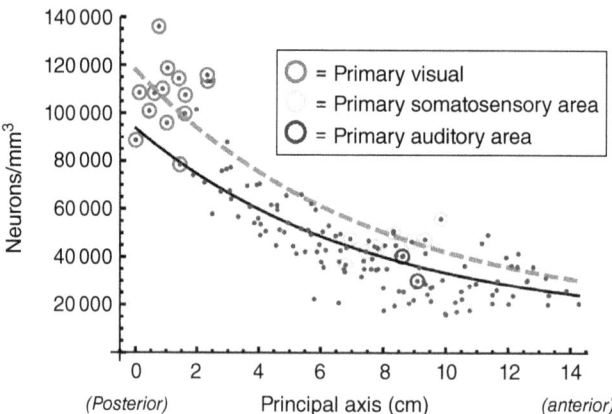

Figure 13.9 Using a two-factor model (location and primary or non-primary area) of neuronal density is better than a location-only model. In the two-factor model, primary sensory areas have a neuronal density 26% higher than would a non-primary sensory area at the same location. The origin of the spatial "principal" axis is at the posterior medial pole of the flattened cortex and it extends towards the anterior lateral pole. Figure redrawn, based on an original which appeared in Cahalane et al., 2014. (*See insert for color representation of the figure*).

that mechanism and possibly others, in combination with the smooth gradients in neurogenesis output described above, may explain the greater number of neurons per unit column in primary sensory regions.

13.4 Structural and Functional Implications of Gradients in Cortical Neurogenesis

To elucidate what variations in neural architecture underlie the observed differences in neuron density, we note that one or both of two factors could contribute to altered density. Firstly, the dosage of densely packed granule cells, particularly in isocortical layer IV, is known to vary across the cortex, being most numerous in posterior regions and being absent in many frontal regions. Secondly, a varying amount of connectivity of pyramidal neurons, with their axonal and dendritic processes occupying relatively more space as connectivity increases, could contribute to reduced neuron density. Evidence of such increased connectivity was found in the increased soma sizes of neurons found in layers II and III in the anterior cortex of four New World monkeys (Cahalane et al., 2012). Alongside the increased soma sizes, there was a trend of decreased neuronal density (similar to the results found by Collins et al. in the study discussed above). Since isocortical layers II and III together are both an important source and target of intra-cortical axonal projections, and because the volume and extent of a neuron's processes can be a factor in determining the volume of its soma (e.g., Elston, Oga, Okamoto, & Fujita, 2009), we conclude that a greater fraction of the cortical volume is devoted to axonal and dendritic processes in anterior regions (see Figure 13.10).

The anterior-to-posterior changes in cortical cellular architecture which we have just described imply corresponding variation in the types of neural processing the respective regions of the cortex are most apt to support. Indeed, the cortical variations we have highlighted are aligned with important functional and processing axes. We note that

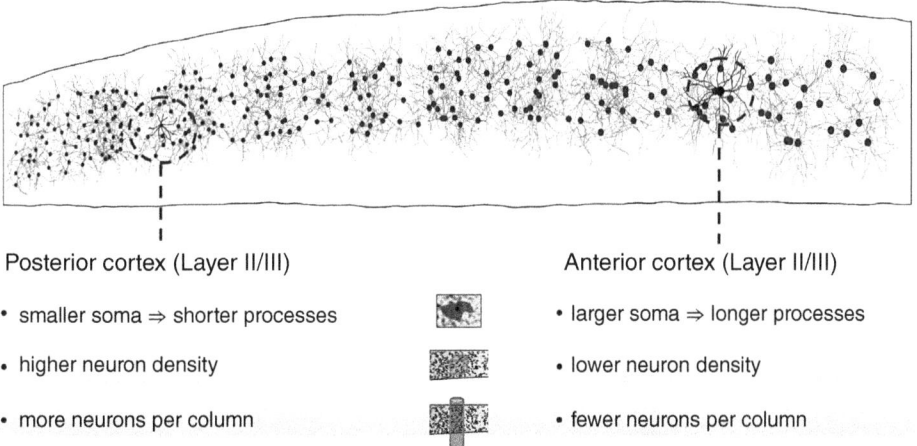

Figure 13.10 Schematic summary of the changes in the cortical architecture of layers II and III as implied by increased neuronal density but decreased neuron size along the anterior-to-posterior axis. Figure redrawn, based on an original which appeared in Cahalane et al., 2012. (*See insert for color representation of the figure*).

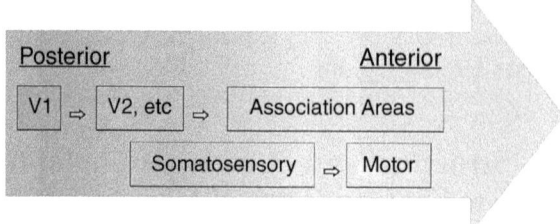

Figure 13.11 Considering the layout of cortical areas, we see that regions with a more integrative role in information processing are typically located anterior to those cortical areas receiving primary sensory input.

higher stages of information processing and integration occur at progressively more anterior locations in the cortex. For example, higher visual areas and association areas integrating visual information are located anterior to the primary visual areas (Van Essen, Anderson, & Felleman, 1992). From somatosensory areas, information flows in the anterior direction to the motor areas where it informs motor control. Information flow in the auditory areas, does not exhibit such a clear alignment with the anterior-posterior axis (Kaas & Hackett, 2004). We note, however, that the auditory areas differ from the other sensory areas by being relatively small and also by not having a spatial topographic map. That frontal regions have more integrative roles in neural processing is also indicated by their structural network connectivity, with a preponderance of the cortex's hubs being located in frontal regions (Modha & Singh, 2010). We conclude that the architectural gradients foster successively higher and more integrative stages of neural processing: as information is represented in successively higher (i.e., progressively anterior) areas, their reduced areal extents and lower numbers of neurons per unit column imply the dimensional reduction or other compression of that information (Figure 13.11). Thus, we see that the developmental mechanisms which lead to within-cortex variations in neural architecture impact cortical function and so are presumably a target of selection.

13.5 In Conclusion

The strikingly regular scaling of brain component sizes within and across species, along with the regular scaling of cortical area sizes across species, are key features of vertebrate brain evolution. As we have argued, such regularities hint at the strong conservation across taxa of the developmental mechanisms which produce the central nervous system. We have given a model, in the specific instance of cortical neurogenesis, for how a conserved mechanism can explain both cross-species scaling and within-cortex variations. We conclude now by addressing the following two questions. Firstly, what further empirical data may soon be available to support or challenge the degree to which coordinate changes, arising from conserved developmental mechanisms, account for the structure of the brain? Secondly, in the midst of so much conservation, what remains variable and available to selection?

Regarding empirical data, and although such a study would present notable challenges in delimiting the shifting borders of cortical areas across individuals, it would be interesting to investigate what regularities exist in the relative scaling of cortical area sizes across individuals of a given species. If conserved mechanisms are at work, we would expect within-species variation in cortical area sizes to obey the same scaling laws as apply across species, an analog to the scaling of brain parts within and across species. Regarding another source of empirical data, it was cross-cortex isotropic fractionator studies in primates which revealed the pronounced gradients in neuron number along the anterior-posterior axis, as discussed above (Collins et al., 2010). In contrast to the primate cortices, stereological measurements of neuron number in this laboratory suggest a much more uniform distribution of neurons across the rodent cortex. The question arises as to whether a cortex whose size is intermediate to that of the rodent and primate would exhibit an intermediate level of gradation in neuron number (Charvet, Cahalane, et al., 2013). Fractionator studies examining multiple sites in the cortices of non-primates would help answer the question of whether gradients in neuron number are an obligatory feature of an enlarged cortex or whether they are a primate innovation.

As to what parameters might be made accessible to selection by conserved developmental mechanisms, by means of which the brains of different groups or species become differentiated from those of common ancestors, we offer the following as possible examples. As alluded to above, even if neuron number must vary smoothly, in a graded manner, across the cortex, then the slope of that gradient might be a subject of selection. The relatively flat neuron distribution seen in rodent cortices could be due to the low number of cell cycles in neurogenesis limiting the slope. However, it is not clear that the ramped distributions seen in primates could not have a slope set independent of cortex size. Another possible target of selection might be those parameters governing the emergence and differentiation of cortical areas during development. Above we mentioned briefly that Hebbian sorting mechanisms likely impact the relationship between the number of cortical areas and the total size of the cortex. Tweaking the parameters in such mechanisms would allow for the selection of different thresholds determining whether areas expand in size or become greater in number as cortical space becomes available. For our final example, we return to the issue of the network architecture of the cortex, and in particular the axonal infrastructure supporting it. We mentioned that a "small-world" architecture would be required to overcome the challenge of maintaining short network path-lengths while minimizing the volume required to accommodate lengthy axons. A small-world architecture implies that there are many more short axons than long axons but the further details of the distribution of axonal lengths remains to be determined. We suggest that various taxa could have settled on different choices for that axon length distribution, i.e. that the parameters determining that distribution may be available as a target of selection. Differences reported between the scaling laws for the cellular content of rodent versus primate cortical neural architecture might in part be the result of different axon length distributions in those taxa.

Acknowledgments

The authors wish to thank their colleagues Christine Charvet and Richard Darlington for helpful conversations and many contributions to the material discussed here.

References

Barton, R. A., & Harvey, P. H. (2000). Mosaic evolution of brain structure in mammals. *Nature*, 405(6790), 1055–1058.

Bayer, S. A., & Altman, J. (1991). *Neocortical development*. New York, NY: Raven Press.

Brodmann, K. (1913). Neue Forschungsergebnisse der Drosshirnrindenanatomie mit besonderer Berücksichtigung anthropologischer Fragen. *Verhandlungen der Gesellschaft Deutscher Naturforscher und Ärzte*, 85, 200–240.

Brown, L. L., Feldman, S. M., Smith, D. M., Cavanaugh, J. R., Ackermann, R. F., & Graybiel, A. M. (2002). Differential metabolic activity in the striosome and matrix compartments of the rat striatum during natural behaviors. *The Journal of Neuroscience*, 22(1), 305–314.

Cahalane, D. J., Charvet, C. J., & Finlay B. L. (2012). Systematic, balancing gradients in neuron density and number across the primate isocortex. *Frontiers in neuroanatomy*, 6, 28. doi: 10.3389/fnana.2012.00028

Cahalane, D. J., Charvet, C. J., & Finlay, B. L. (2014). Modeling local and cross-species neuron number variations in the cerebral cortex as arising from a common mechanism. *Proceedings of the National Academy of Sciences of the USA*, 111(49), 17642–17647.

Caviness, V. S., Goto, T., Tarui, T., Takahashi, T., Bhide, P. G., & Nowakowski, R. S. (2003). Cell output, cell cycle duration and neuronal specification: A model of integrated mechanisms of the neocortical proliferative process. *Cerebral Cortex*, 13(6), 592–598.

Charvet, C. J., Cahalane, D. J., & Finlay B. L. (2013). Systematic, cross-cortex variation in neuron numbers in rodents and primates. *Cerebral Cortex*, 25(1), 147–160.

Charvet, C. J., Darlington, R. B., & Finlay, B. L. (2013). Variation in human brains may facilitate evolutionary change toward a limited range of phenotypes. *Brain Behavior & Evolution*, 81(2), 74–85.

Charvet, C. J., Striedter, G. F., & Finlay, B. L. (2011). Evo-devo and brain scaling: candidate developmental mechanisms for variation and constancy in vertebrate brain evolution. *Brain Behavior & Evolution*, 78(3), 248–257.

Cherniak, C. (1995). Neural component placement. *Trends in Neurosciences*, 18(12), 522–527.

Cherniak. C., Mokhtarzada, Z., Rodriguez-Esteban, R., & Changizi, K. (2004). Global optimization of cerebral cortex layout. *Proceedings of the National Academy of Sciences of the United States of America*, 101(4), 1081–1086.

Coggeshall, R. E. (1992). A consideration of neural counting methods. *Trends in Neurosciences*, 15(1), 9–13.

Collins, C. E., Airey, D. C., Young, N. A., Leitch, D. B., & Kaas, J. H. (2010). Neuron densities vary across and within cortical areas in primates. *Proceedings of the National Academy of Sciences*, 107(36), 15927–15932.

Dehay, C., Giroud, P., Berland, M., Smart, I., & Kennedy, H. (1993). Modulation of the cell cycle contributes to the parcellation of the primate visual cortex. *Nature*, 366(6454), 464–466.

Dehay, C., & Kennedy, H. (2007). Cell-cycle control and cortical development. *National Review of Neuroscience*, 8(6), 438–450.

Elston, G. N., Oga, T., Okamoto, T., & Fujita, I. (2009). Spinogenesis and pruning from early visual onset to adulthood: An intracellular injection study of layer III pyramidal cells in the ventral visual cortical pathway of the macaque monkey. *Cerebral Cortex*, 20(6), 1398–1408. doi: 10.1093/cercor/bhp.203

Finlay, B. L., Cheung, D., & Darlington, R. B. (2005). Developmental constraints on or developmental structure in brain evolution? In Y. Munakatu & M. H. Johnson (Eds.), *Processes of change in brain and cognitive development. Attention and performance XXI* (pp. 131–162). Oxford: Oxford University Press.

Finlay, B. L., & Brodsky, P. B. (2006). Cortical evolution as the expression of a program for disproportionate growth and the proliferation of areas. In J. H. Kaas (Ed.), *Evolution of nervous systems* (pp. 73–96). Oxford: Oxford University Press.

Finlay, B. L., & Darlington, R. B. (1995). Linked regularities in the development and evolution of mammalian brains. *Science*, 268(5217), 1578–1584.

Finlay, B. L., Hersman, M. N., & Darlington, R. B. (1998). Patterns of vertebrate neurogenesis and the paths of vertebrate evolution. *Brain Behavior and Evolution*, 52(4–5), 232–242.

Finlay, B. L., Hinz, F., & Darlington, R. B. (2011). Mapping behavioural evolution onto brain evolution: The strategic roles of conserved organization in individuals and species. *Philosophical Transactions of the Royal Society B, Biological Sciences*, 366(1574), 2111–2123.

Finlay, B. L., & Pallas, S. L. (1989). Control of cell number in the developing mammalian visual system. *Progress in Neurobiology*, 32(3), 207–234.

Finlay, B. L., & Slattery M. (1983). Local differences in the amount of early cell death in neocortex predict adult local specializations. *Science*, 219(4590), 1349–1351.

Frahm, H. D., Stephan, H., & Baron, G. (1984). Comparison of brain structure volumes in insectivora and primates. V. Area striata (AS). *Journal für Hirnforschung*, 25(5), 537–557.

Fukuchi-Shimogori, T., & Grove, E. A. (2001). Neocortex patterning by the secreted signaling molecule FGF8. *Science*, 294(5544), 1071–1074.

Giannaris, E. L., & Rosene, D. L. (2012). A stereological study of the numbers of neurons and glia in the primary visual cortex across the lifespan of male and female rhesus monkeys. *The Journal of Comparative Neurology*, 520(15), 3492–3508.

Gigandet, X., Hagmann, P., Kurant, M., Cammoun, L., Meuli, R., & Thiran, J.-P. (2008). Estimating the confidence level of white matter connections obtained with mri tractography. *PLoS ONE*, 3(12). doi: 10.1371/journal.pone.0004006

Gohlke, J. M., Griffith, W. C., Bartell, S. M., Lewandowski, T. A., & Faustman, E. M. (2002). A computational model for neocortical neurogenesis predicts ethanol-induced neocortical neuron number deficits. *Developmental Neuroscience*, 24(6), 467–477.

Gohlke, J. M., Griffith, W. C., & Faustman, E. M. (2007). Computational models of neocortical neurogenesis and programmed cell death in the developing mouse, monkey, and human. *Cerebral Cortex*, 17(10), 2433–2442.

Grimes, W. N., Zhang, J., Graydon, C. W., Kachar, B., & Diamond, J. S. (2010). Retinal parallel processors: More than 100 independent microcircuits operate within a single interneuron. *Neuron*, 65(6), 873–885.

Hawrylycz, M. J., Lein, E. S, Guillozet-Bongaarts, A. L, Shen, E. H., Ng, L., Miller, J. A., ... Jones, A. R. (2012). An anatomically comprehensive atlas of the adult human brain transcriptome. *Nature*, 489(7416), 391–399.

Herculano-Houzel, S., Collins, C. E., Wong, P., & Wong, K. (2007). Cellular scaling rules for primate brains. *Proceedings of the National Academy of Sciences of the USA*, 104(9), 3562–3567.

Herculano-Houzel, S., & Lent, R. (2005). Isotropic fractionator: A simple, rapid method for the quantification of total cell and neuron numbers in the brain. *The Journal of Neuroscience*, 25(10), 2518–2521.

Herculano-Houzel, S., Mota, B., & Lent, R. (2006). Cellular scaling rules for rodent brains. *Proceedings of the National Academy of Sciences of the USA*, 103(32), 12138–12143.

Jackson C. A., Peduzzi, J. D., & Hickey, T. L. (1989). Visual cortex development in the ferret. I. Genesis and migration of visual cortical neurons. *The Journal of Neuroscience*, 9(4), 1242–1253.

Jerison, H. J. (1973). *Evolution of the brain and intelligence*. New York: Academic Press.

Johnson, M. H., & Vecera, S. P. (1996). Cortical differentiation and neurocognitive development: The parcellation conjecture. *Behavioural Processes*, 36(2), 195–212.

Kaas, J. H. (2011). Reconstructing the areal organization of the neocortex of the first mammals. *Brain Behavior and Evolution*, 78, 7–21.

Kaas, J. H., & Hackett, T. A. (2004). Auditory cortex in primates: Functional subdivisions and processing streams. In M. S. Gazzaniga (Ed.), *The cognitive neurosciences* (3rd ed., pp. 215–232). Cambridge, MA: MIT Press.

Kang, H. J., Kawasawa, Y. I., Cheng, F., Zhu, Y., Xu, X., Li, M., ... Šestan, N. (2011). Spatiotemporal transcriptome of the human brain. *Nature*, 478(7370), 483–489.

Kaskan, P. M., Franco, E. C. S., Yamada, E. S., Silveira, L. C. de L., Darlington, R. B., & Finlay, B. L. (2005). Peripheral variability and central constancy in mammalian visual system evolution. *Proceedings of the Royal Society B, Biological Sciences*, 272(1558), 91–100.

Kingsbury, M. A., & Finlay, B. L. (2001). The cortex in multidimensional space: Where do cortical areas come from? *Developmental Science*, 4(2), 125–142.

Kornack, D. R., & Rakic P. (1995). Radial and horizontal deployment of clonally related cells in the primate neocortex: Relationship to distinct mitotic lineages. *Neuron*, 15(2), 311–321.

Kornack, D. R., & Rakic, P. (1998). Changes in cell-cycle kinetics during the development and evolution of primate neocortex. *Proceedings of the National Academy of Sciences*, 95(3), 1242–1246.

Krubitzer, L. A., & Seelke, A. M. H. (2012). Cortical evolution in mammals: The bane and beauty of phenotypic variability. *Proceedings of the National Academy of Sciences*, 109(Supplement_1), 10647–10654.

Laughlin, S. B, van Stevenink, R. R. de R., & Anderson, J. C. (1998). The metabolic cost of neural information. *Nature Neuroscience*, 1(1), 36–41.

Lewis, J. (2008). From signals to patterns: Space, time, and mathematics in developmental biology. *Science*, 322(5900), 399–403.

Logothetis, N. K. (2003). The underpinnings of the BOLD functional magnetic resonance imaging signal. *The Journal of Neuroscience*, 23(10), 3963–3971.

Luskin, M. B., & Shatz, C. J. (1985). Neurogenesis of the cat's primary visual cortex. *The Journal of Comparative Neurology*, 242(4), 611–631.

Martinez-Ferre, A., & Martínez, S. (2012). Molecular regionalization of the diencephalon. *Frontiers in Neuroscience*, 6, 73.

Miyama, S., Takahashi, T., Nowakowski, R. S., & Caviness, V. S., Jr. (1997). A gradient in the duration of the G1 phase in the murine neocortical proliferative epithelium. *Cerebral Cortex*, 7(7), 678–689.

Modha, D. S., & Singh, R. (2010). Network architecture of the long-distance pathways in the macaque brain. *Proceedings of the National Academy of Sciences*, 107(30), 13485–13490.

Murre, J. M. J., & Sturdy, D. P. F. (1995). The connectivity of the brain: Multi-level quantitative analysis. *Biological Cybernetics*, 73(6), 529–545.

Northcutt R. G. (1981). Evolution of the telencephalon in nonmammals. *Annual Review of Neuroscience*, 4(1), 301–350.

Nowakowski, R. S., Caviness, V. S., Takahashi, T., & Hayes, N. L. (2002). Population dynamics during cell proliferation and neuronogenesis in the developing murine neocortex. In C. Hohmann (Ed.), *Cortical Development* (Vol. 39, pp. 1–25). Berlin, Heidelberg: Springer.

O'Leary, D. D. (1989). Do cortical areas emerge from a protocortex? *Trends in Neurosciences*, 12(10), 400–406.

Pakkenberg, B. (1993). Total nerve cell number in neocortex in chronic schizophrenics and controls estimated using optical disectors. *Biological Psychiatry*, 34(11), 768–772.

Rakic, P. (1974). Neurons in rhesus monkey visual cortex: systematic relation between time of origin and eventual disposition. *Science*, 183(123), 425–427.

Rakic, P. (1988). Specification of cerebral cortical areas. *Science*, 241(4862), 170–176.

Rakic, P. (2002). Neurogenesis in adult primate neocortex: an evaluation of the evidence. *Nature Reviews Neuroscience*, 3(1), 65–71.

Reep, R. L., Finlay, B. L., & Darlington, R. B. (2007). The limbic system in mammalian brain evolution. *Brain, Behavior & Evolution*, 70(1), 57–70.

Rockel, A. J., Hiorns, R. W., & Powell, T. P. S. (1980). the basic uniformity in structure of the neocortex. *Brain*, 103(2), 221–244.

Sansom, S. N., & Livesey, F. J. (2009). Gradients in the brain: The control of the development of form and function in the cerebral cortex. *Cold Spring Harbor Perspectives in Biology*, 1(2). doi: 10.1101/cshperspect.a002519

Semendeferi, K., Lu, A., Schenker, N., & Damasio H. (2002). Humans and great apes share a large frontal cortex. *Nature Neuroscience*, 5(3), 272–276.

Smart, I. H. M., Dehay, C., Giroud, P., Berland, M., & Kennedy, H. (2002). Unique morphological features of the proliferative zones and postmitotic compartments of the neural epithelium giving rise to striate and extrastriate cortex in the monkey. *Cerebral Cortex*, 12(1), 37–53.

Stephan, H., Baron, G., & Frahm, H. D. (1988). Comparative size of brain and brain components. In J. Erwin & H. D. Steklis (Eds.), *Comparative primate biology* (Vol. 4, pp. 1–38). New York: A. R. Liss.

Stephan, H., Frahm, H., & Baron, G. (1981). New and revised data on volumes of brain structures in insectivores and primates. *Folia Primatologica; International Journal of Primatology*, 35(1), 1–29.

Striedter, G. F. (2005). *Principles of brain evolution*. Sunderland, MA: Sinauer Associates.

Takahashi, T., Nowakowski, R. S., & Caviness, V. S., Jr. (1996). The leaving or Q fraction of the murine cerebral proliferative epithelium: a general model of neocortical neurogenesis. *The Journal of Neuroscience*, 16(19), 6183–6196.

Takahashi, T., Nowakowski, R. S., & Caviness, V. S., Jr. (1997). The mathematics of neocortical neuronogenesis. *Developmental Neuroscience*, 19(1), 17–22.

Van Essen, D., Anderson, C., & Felleman, D. (1992). Information processing in the primate visual system: an integrated systems perspective. *Science*, 255(5043), 419–423.

Watts, D. J., & Strogatz, S. H. (1998). Collective dynamics of "small-world" networks. *Nature*, 393(6684), 440–442.

Yamamori, T. (2011). Selective gene expression in regions of primate neocortex: Implications for cortical specialization. *Progress in Neurobiology*, 94(3), 201–222.

Yopak, K. E., Lisney, T. J., Darlington, R. B., Collin, S. P., Montgomery, J. C., & Finlay B. L. (2010). A conserved pattern of brain scaling from sharks to primates. *Proceedings of the National Academy of Sciences*, 107, 12946–12951.

Zhang K., & Sejnowski TJ. (2000). A universal scaling law between gray matter and white matter of cerebral cortex. *Proceedings of the National Academy of Sciences of the United States of America*, 97(10), 5621–5626.

14

Comparative Aspects of Learning and Memory
Michael Koch

14.1 Introduction

Learning and memory refers to the acquisition, consolidation, and retention of information for future use (retrieval). It enables an organism to "predict" future events, on the basis of this information and the current set of external and internal conditions, to adjust its behavior accordingly (Abel & Lattal, 2001). When this information is vital—for example, the recall of the location of a food source or mating partner, or the avoidance of a predator—it is obvious that the biological mechanisms of learning and memory are of considerable adaptive value for all species. Therefore, it is not surprising that several forms of information preservation can be found in animals (memory storage in the nervous and immune systems) and even plants (Rensing, Koch, & Becker, 2009).[1] The question, though, is whether these processes of information storage follow similar general rules, based on similar cellular and physiological processes. Such a general rule was postulated by Donald Hebb (1949, p. 62): "When an axon of cell A is near enough to excite a cell B and repeatedly or persistently takes part in firing it, some growth process or metabolic change takes place in one or both cells such that A´s efficiency, as one of the cells firing B, is increased." It would be interesting to know which of these sorts of mechanisms share common ancestry and which represent parallel paths of evolution.

From a human perspective, it is interesting to ask what differences and similarities exist between learning mechanisms and memory systems of different organisms and how they relate to the human memory system, which is so important for our personality and individuality. This chapter gives a short overview of the neurobiology of learning and memory in different animal species, including both vertebrates and invertebrates. I put a special emphasis on the molecular basis of learning and memory across different time scales. However, memory is a complex phenomenon, involving multiple functions and brain systems, classified according to the way the information is stored and retrieved (see Figure 14.1; Henke, 2010; Kandel, Schwartz, & Jessell, 2000).

As depicted in Figure 14.1, human memory can be divided into short-term (e.g. sensory and working memory) and long-term memory, with long-term subdivided further into explicit (declarative) and implicit (nondeclarative, including procedural) memory. Declarative memory can have episodic or semantic (remembering and knowing) characteristics. The existence of episodic memory in nonhuman

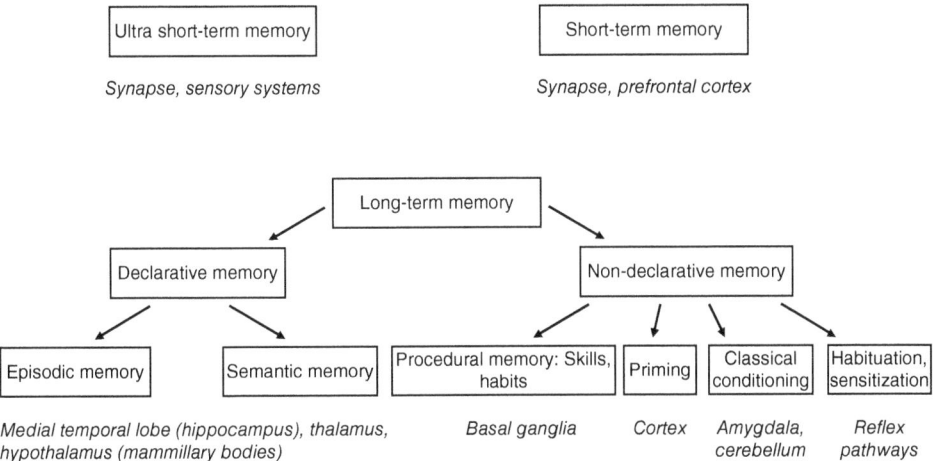

Figure 14.1 Simplified Schematic Diagram of Different Forms of Memories and Their Most Important Neuronal Substrates in Humans. Modified after Henke, 2010; Kandel et al., 2000.

animals has been disputed for a while, but now seems to be proven at least for some vertebrate species (Salwiczek, Watanabe, & Clayton, 2010). To the best of our knowledge—perhaps due to the experimental challenge of testing declarative memory in insects and snails—nothing is known about the existence of these memory forms in invertebrates. Hence, we here focus on nondeclarative memory.

14.2 General Aspects of Learning and Memory

Nondeclarative or implicit learning is broadly categorized into nonassociative and associative forms of information acquisition. In addition, imprinting, priming, and habit formation are observed in several species and are considered special forms of information storage (Horn, 2004; Yin & Knowlton, 2006). Nonassociative learning comprises habituation and sensitization, while associative learning is further divided into classical (Pavlovian) and instrumental conditioning (Kandel et al., 2000).

Habituation is the reduction in response magnitude (most often tested as a simple reflex, such as withdrawal or startle) due to repeated presentation of a stimulus. Sensitization refers to the enhancement or potentiation of a habituated response following a biologically relevant, strong stimulus. Habituation and sensitization are found in both vertebrates and invertebrates (Davis, 1984; Kandel, 2004).

In classical associative delay conditioning, the temporal pairing of a biologically relevant unconditioned stimulus (US) with a neutral conditioned stimulus (CS) leads to prediction that the CS will be followed—with a certain probability—by the US. Control experiments include the nonpaired condition, where the US and the CS are presented in an unrelated way. Intermediate between paired and unpaired training conditions are the so called *trace conditioning* procedures where the US follows the CS after a certain time gap. Here, the predictive value of the CS is determined by the duration of this temporal lag. In general, the predictive value of the CS changes the individual's behavior to approach predicted rewards and avoid predicted aversive situations. Repeated presentation of the CS in the absence of the US after conditioning reduces the ability of the CS to elicit a conditioned response (extinction). On the other hand, repeated presentation of the prospective CS *before* conditioning reduces the degree of CS-US association (latent inhibition). It is not possible in this short chapter to consider in depth all the different temporal aspects of learning (acquisition, consolidation, retrieval) and memory (short-term, intermediate-term, long-term, permanent) when comparing the mechanisms between different species. Therefore, we focus on the mechanisms of acquisition and long-term memory in classical conditioning.

A key cellular mechanism involved in associative learning and memory is long-term potentiation (LTP). First described in 1973 by Bliss and Lomo in the rabbit hippocampus (1973), LTP is now well established as a process by which certain changes in synaptic activity can induce a long lasting and specific increase of the strength of this synapse in vertebrates (Lynch, 2004; Malenka, 2003; Morris, 2003) as well as in invertebrates (Glanzman, 2008; Menzel & Manz, 2005; Roberts & Glanzman, 2003).

14.3 Learning and Memory in Invertebrates

Neurobiological research in invertebrates has the advantage of dealing with relatively "simple" nervous systems, even allowing experimenters to target specific identified neurons. Perhaps the most eminent early studies into the molecular basis of learning and memory were started in the 1960s and 1970s by Eric Kandel and colleagues. He used the marine snails *Aplysia depilans* and *A. californica* for experiments on nonassociative and associative learning (Kandel & Pittenger, 1999; Kupfermann & Kandel, 1969). In this model, the gill-withdrawal reflex is used as a dependent variable which is reduced in magnitude (habituation) after repeated stimulation with a mild water-jet shot onto the skin of the snail. Conversely, the reflex is enhanced after noxious stimulation of the animal's tail (sensitization) or after presentation of an otherwise neutral CS (e.g., mechanical stimulation of the mantle shelf) that has been paired with a noxious tail shock. Habituation and sensitization of the gill-withdrawal reflex in *Aplysia* have been shown to involve, respectively, reduced or enhanced calcium-dependent glutamate release from the presynaptic terminal of sensory neurons, which contact motor neurons that activate the gill muscles and mediate the reflex. Enhanced release is due to heterosynaptic facilitation by serotonergic interneurons. The stimulatory effects of serotonin (5-HT, 5-hydroxytryptamine) on the presynaptic terminal of the sensory neurons are due to activation of different types of calcium channels. This triggers a cascade of intracellular events involving cAMP-dependent Protein Kinase A (PKA) and the stimulation of Protein Kinase C (PKC) by diacylglycerol. These processes are probably convergent to habituation and sensitization in vertebrates (Weber, Schnitzler, & Schmid, 2002).

However, *postsynaptic* signalling events triggered by a rise in intracellular calcium are also important for learning in *Aplysia*. For example, the induction of retrograde messengers such as the gaseous messenger molecule, nitric oxide (NO), (Michel, Green, Eskin, & Lyons, 2011) and the up-regulation of the glutamate receptor, AMPA (α-amino-3-hydroxy-5-methyl-4-isoxazolepropionic acid), in the postsynaptic motor neuron (Glanzman, 2008) are found in *Aplysia*. Structural changes of the synapses, the basis of long-term memory in *Aplysia*, also involve cAMP-response Element Binding Protein (CREB)-mediated gene expression (Bailey & Kandel, 2008; Kandel, 2004).

Others have considered honeybees (*Apis mellifera*) as useful animals for the study of learning and memory (Bitterman, 1976; Lindauer, 1970; Menzel, 1979). Here, the appetitive proboscis-extension reflex to olfactory food stimuli is often taken as the behavioral measure of these cognitive functions. The mushroom bodies, which comprise large parts of the bees' brains, integrate various sensory pathways and are essential for learning and memory. In addition to the mushroom bodies, the antennal lobes have also been found to be involved in memory storage in bees (Giurfa & Sandoz, 2012; Hammer & Menzel, 1998; Menzel, 2012). Learning and memory formation in bees is also divided into different temporal phases and the molecular determinants of these different cellular processes are distinguished (Müller, 2002) resembling those outlined for humans in Figure 14.1. As in other animals, the initial cellular mechanisms involved in learning and memory also include a synapse-specific rise in calcium levels, followed by activation of retrograde messengers such as NO (Müller, 1994, 1996), and finally, activation of several different protein kinases (Giurfa, 2007; Menzel, 2001). For long-term memory it appears that, in honeybees, CREB (or rather its bee orthologue, *Am*CREB) plays the essential role (Eisenhardt et al., 2003; Müller, 2000).

In addition, research on the fruitfly (*Drosophila melanogaster*) has pioneered the neurogenetic basis of learning and memory in invertebrates, especially with respect to synaptic physiology (Davis, 2011; Dubnau & Tully, 1998; Heisenberg, 2003; Heisenberg, Borst, Wagner, & Byers, 1985). As in other memory systems, the initial signal relevant for the selective strengthening of specific synapses is an increased calcium influx into Kenyon cells of the fly's mushroom bodies. The sequel of this event includes the activation of NO-synthase, adenylate cyclase, increased intracellular cAMP levels and subsequent activation of protein kinases (Gerber, Tanimoto, & Heisenberg, 2004; Heisenberg, 2003; Müller, 1997). Here, as in other species, long-term memory formation appears to require CREB-dependent gene transcription and protein synthesis in a subset of neurons (Chen et al., 2012).

14.4 Learning and Memory in Vertebrates

Most vertebrate learning studies are done in mammals, especially rodents (rats and mice) and humans. Therefore, we shall focus on mammalian memory systems here, despite the fact that there are fascinating forms of learning and memory in non-mammalian models, including imprinting and song learning in birds, as well as the formation of declarative memory in various species.

Several brain systems that are relatively conserved across mammals appear to be adapted for learning and memory (probably through anatomical and neurochemical characteristics) including amygdala, hippocampus, cerebellum, basal ganglia, and cerebral cortex (Henke, 2010). This list is by no means complete. For example, consider the formation of long-term pain memories in the mammalian dorsal horn of the spinal cord, associated with hyperalgesia. However, of these regions, the hippocampus is the most famous. Since Scoville and Milner (1957) reported severe anterograde amnesia after bilateral temporal lobe resection in their patient, H.M. (Henry Gustav Molaison, 1926–2008), and especially since the discovery of LTP (Bliss & Lomo, 1973), the hippocampal formation has been the paradigmatic brain structure in which to study learning and memory processes in mammals (van Strien, Cappaert, & Witter, 2009).

In these brain structures, several functional and morphological modifications occur in the course of learning, mainly triggered by an increase in postsynaptic entry of calcium into the cell through NMDA (N-methyl-D-aspartate) glutamate receptors and voltage-gated calcium channels (Wang, Hu, & Tsien, 2006). The NMDA receptor is an ion channel comprised of four subunits that is mainly permeable for sodium and calcium. This channel has the very interesting property of being both ligand- and voltage-gated. At membrane potentials of up to −35 mV the channel pore is blocked by a magnesium ion. However, once the membrane is depolarized > −35 mV, Mg^{2+} is expelled from the channel, so that sodium and calcium can enter the neuron. (Due to the special properties of being ligand- and voltage-gated the NMDA receptor has been termed a "coincidence detector," because the opening properties depend on the temporally coincident input.) This rise in the intracellular calcium concentration triggers short- and long-term changes in second-messenger systems. One might ask how a general increase in the postsynaptic Ca^{2+} concentration can lead to rather specific intracellular effects, ultimately strengthening a particular synaptic input. This is mainly due to the restriction of diffusion of the ion by mechanical barriers in the dendritic spines (nano- and microdomains), by binding to different Ca^{2+}-binding proteins and through differences in the kinetics of Ca^{2+} channels (Burgoyne, 2007).

Short-term plasticity includes retrograde signaling, mostly through NO, which once synthesized from l-arginine, diffuses back to the presynaptic terminal to activate guanylate cyclase, increase the concentration of cGMP, and in turn activate Protein Kinase G. These events enhance the activity of the presynaptic terminal, for example, by phosphorylating calcium channels, and eventually support processes such as LTP (Bon & Garthwaite, 2003; Lange, Doengi, Lesting, Pape, & Jüngling, 2012). Short-term changes also include postsynaptic activation by calcium of adenylate cyclase, leading to an increase in cAMP, activation of several protein kinases, and induction of transcription factors which alter gene activation, finally leading to persistent functional and structural changes of the synaptic strength (Sacktor, 2011; Thomas & Huganir, 2004; Wang et al., 2006). Key molecules in the postsynaptic events that lead to the activity-dependent increase in synaptic strength include the different protein kinases (Mayford, 2007), the extracellular signal-related kinases (ERK 1/2) (Thomas & Huganir, 2004), the calcium/calmodulin-dependent protein kinase II (CaMKII) (Lisman, Yasuda, & Raghavachari, 2012), and several transcription factors, most importantly CREB (Han et al., 2007). On different time scales these events lead to improved information transfer from the pre- to the postsynaptic membrane, for example by the phosphorylation of AMPA and /or NMDA receptors (Salter & Kalia, 2004), the synthesis and postsynaptic insertion of AMPA receptors (Collingridge, Isaac, & Wang, 2004), and by morphological changes of the dendrites (Lamprecht & LeDoux, 2004).

In vertebrates, a distinction is drawn between the memory consolidation on the cellular and the systems level, meaning that the molecular changes described above are just one—perhaps initial—step necessary for the formation of long-term memories. The problem of memory time spans outlasting the biological half-lives and turnover rates of the molecules involved in its formation has long been known and is still not resolved (see, e.g. Crick, 1984; Frankland & Josselyn, 2013). A crucial step is thought to involve dynamic interactions between primary memory sites, including the hippocampus, and the cortex (Chklovskii et al., 2012; Lesburguères et al., 2011; Miyashita, 2004; Redondo & Morris, 2011; Wang et al., 2006), most likely mediated by neuronal synchronisation (Fell & Axmacher, 2011; Seidenbecher, Laxmi, Stork, & Pape, 2003). This shift of long-term memory to new brain sites raises the interesting question of what happens to the initial memory trace, for instance, in the hippocampus, once a particular memory has been successfully shifted to the cortex. Considering that the memory capacity of the hippocampus is limited, it might be that outdated memory representations are somehow erased or overwritten by new ones. Another interesting possibility is that neurogenesis in the adult brain (Doetsch & Hen, 2005; Nilsson et al., 1999; Shors et al., 2001) clears old memories from the hippocampus. This idea has been suggested by Wang and colleagues (2006) and is mainly based on findings indicating that the knockout of the presenilin-1 gene impaired neurogenesis and clearance of outdated memories.

14.5 Differences and Commonalities

It appears that functional and morphological changes of synapses in the course of learning and memory formation are a general feature of all members of the animal kingdom. Of course, there are profound differences in the brain *bauplan* and anatomy between insects, molluscs, and vertebrates since these evolutionary lines and ecological niches diverged millions of years ago.[2] Therefore, there are vast differences in the organization of the brains of different species in general, and in the brain sites most prominently involved in learning and memory in particular. Here, the similarities between invertebrates and vertebrates represent convergent solutions and are more general in the sense that all the structures involved in learning and memory are characterized by a high degree of input convergence, strong intrinsic neuronal communication, and abundance of certain molecular components—for example, voltage- and ligand- gated calcium channels, such as NMDA receptors. Hence, the obvious similarities in pre- and postsynaptic cellular mechanisms of learning and memory across phyla indicate a fairly strong degree of conservation in evolution. One perfect example of a common mechanism, in the context of learning, is the NO signaling pathway, which has been found to be used by all phyla of the animal kingdom (Moroz & Kohn, 2011). Other molecules that are crucial for learning and memory in all known species are calcium (e.g., an increase of the intracellular Ca^{2+} concentration through the release from intracellular stores or by an influx through voltage or ligand gated channels), cAMP, several kinases, and transcription factors (e.g., CREB and immediate-early genes) (see Figure 14.2). In this context, Reissner, Shobe, and Carew (2006) proposed the interesting concept of "molecular nodes" in the communication cascades between the synapse and the cell nucleus, that is, signaling ions or molecules that serve as integrators, or points of intersection for multiple converging inputs and diverging outputs.

Recently, a very interesting review article directly compared the brain systems involved in fear- and relief-learning in Drosophila, rats and humans (Gerber et al., 2014). Relief-learning refers to the observation that the onset and the offset, respectively, of an aversive event (stimulus or context) induce opposite memories: the stimulus onset induces fear, whilst the offset induces relief. This can easily be investigated in classical fear-conditioning paradigms such as fear-potentiated startle in mammals or odor-avoidance in flies. Gerber et al. show a high degree of convergence between fruit

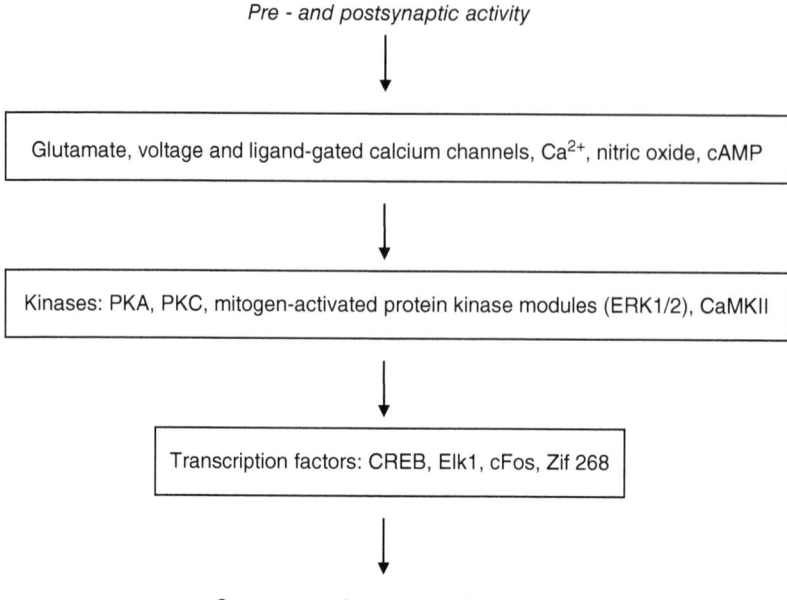

Figure 14.2 Components Involved in General Mechanisms for Communication between the Synapse and the Cell Nucleus Which Mediate Activity-Dependent Changes in both Vertebrates and Invertebrates.

This diagram simply summarizes in a hierarchical way the components that are found across a wide range of animal species. It does not specify the differential involvement of these components in the different temporal characteristics of learning and memory. Short-lasting forms of synaptic plasticity involve a rise in intracellular calcium, triggering the action of retrograde messengers and the modification of existing proteins (such as transmitter receptors or ion channels), for example, by phosphorylation through various kinases. The biological half-life of these modifications is relatively short (minutes to hours). Intermediate-term plasticity is due to longer-lasting effects induced by kinases, for example, the transport of AMPA-receptors into the active zone of a synapse. Long-term plasticity is due to changes in protein synthesis (e.g. new synthesis of AMPA-receptors, growth of dendrites) following the activation of transcription factors. It has to be noted that the "synaptic re-entry reinforcement" hypothesis states that long-term memory formation following short-term plasticity is not simply a linear unidirectional pathway, but that instead synaptic consolidation requires that NMDA receptors (ligand- and voltage-gated calcium channels), have to be repeatedly reactivated after learning so that short-term memory is finally converted into long-lasting memory. (Wang et al., 2006).

flies and mammals, for example, a strong involvement of the dopamine/octopamine reward system in relief-learning.

One of the most important challenges to understanding memory storage and retrieval is the persistent localization of a memory trace ("engram") in the organism's brain. This is still not well understood, given the fact that the proteins modified in the course of learning undergo regular metabolic turnover (Wang et al., 2006), despite the finding of special protein kinases which may have very long biological half-lives (Sacktor, 2011). Here, it is important to distinguish between memory that is merely based on modulatory or structural changes within a fixed sensorimotor pathway, such as found in *Aplysia*, and dynamically distributed memory.

Organisms with more complex brains have the capacity to shift memories from a molecular, fixed-pathway level to a more distributed systems or network level (Wang et al., 2006). Here, a large-scale dynamical cortical activity pattern appears to be the "equivalent" of an engram (Miyashita, 2004). A strict operational approach would argue that it does not make sense to ask for the location or representation of an engram in this network until it is retrieved and can be measured as a pattern of network activity corresponding to this retrieval process. The import point to consider here is that this cortically distributed memory uses the same neuronal computing modules that are involved in sensory perception and motor control (Fuster, 1997), so that memory (irrespective of its temporal characteristics) is "defined by a pattern of connections between neuron populations associated by experience" (Fuster, 2009). This process may have physiological correlates in oscillation patterns reflecting neural synchronization within a neural network (Fell & Axmacher, 2011) that may comprise a huge portion of the brain (Basar, 2006). The direct relationship between neural synchronization and memory retrieval has been shown impressively in a study on fear conditioning in freely moving mice: Here, a tone (CS) previously paired with a foot-shock (US), but not an unpaired tone, synchronized the theta-band oscillations in an amygdala-hippocampal network and induced freezing as a conditioned fear response (Seidenbecher et al., 2003). The question arises whether or not this form of distributed memory representation is also found in invertebrates. To the best of my knowledge, oscillatory brain activity in the context of learning and memory has not been described in *Aplysia*, although molluscs share basic physiological similarities in the generation of cellular potentials and show oscillatory local field potentials (Schütt, Basar, & Bullock, 1992) related to recognition memory (Gelperin, 2006). While the higher invertebrates such as insects and crustaceans show different field potentials patterns than vertebrates, octopus, known for their large brains, appear to be more vertebrate-like (Bullock & Basar, 1988). Interestingly, in the honeybee, olfactory conditioning increased power of the local field potentials induced by the CS in the antennal lobe within the 15–40 Hz frequency band (Denker, Finke, Schaupp, Grün, & Menzel, 2010) and induced learning-related changes in odour representations visible to calcium imaging (Rath, Galizia, & Szyszka, 2011).

Taken together, despite profound diversities in the macroscopic structure and organization of nervous systems of animal species from different phyla, and although these animals may differ in *what* they learn, the general rules for associative and nonassociative learning are fairly similar (e.g., Hebb's rules, see: Bitterman, 1975; Bitterman, Menzel, Fietz, & Schäfer, 1983; Bolles & Beecher, 1988; Menzel, 1983). All animals, thus, appear to exploit the fact that, in the real world, temporally contingent events are often causally related. Likewise, as described in Chapter 4, the molecular modules underlying the different memory stores are also strikingly uniform and likely homologous. Brain structures important for learning and memory naturally differ due to the fundamental differences in the *bauplan* of vertebrates and invertebrates. Species differences naturally occur during adaptive specialization. However, despite these differences, the general organization of memory traces on the systems level bears a remarkable degree of similarity. Hence, the evidence for species-specific learning abilities probably did not arise from profound differences in the cellular mechanisms or biochemical foundations for learning and memory; even memory representation at the systems level appears to share conceptual similarities, provided the brains are sufficiently complex.

Acknowledgments

I wish to thank Drs. B. Mathes (Bremen) and R. Menzel (Berlin) for helpful discussion.

Notes

1. If we consider the experience-dependent remodeling of an organism's response to a stimulus as a simple form of memory (e.g. *Oxford Advanced Learner's Dictionary*: "Memory is the preservation of past experience for future use"), we find plenty of evidence for memory in plants (Thellier, Le Sceller, Norris, Verdus, & Ripoll, 2000; Trewavas, 2003): for example, calcium-dependent storage and recall of meristem production in flax seedlings (Verdus, Ripoll, Norris, & Thellier, 2012).
2. It should be noted here that it is becoming increasingly evident that evolution is not only based on genetic inheritance, but has a strong epigenetic component. Therefore, it appears that empirical knowledge has a strong impact on phylogeny (summarized in several chapters of Bolles & Beecher, 1988 and Heynes & Huber, 2000). This line of thinking goes back to James Baldwin ("Baldwin effect") suggesting that learning and memory affect the rate and direction of evolution (Weber & Depew, 2003). This has to be considered when talking about different or related forms of information storage in different organisms.

References

Abel, T., & Lattal, K. M. (2001). Molecular mechanisms of memory acquisition, consolidation and retrieval. *Current Opinion in Neurobiology*, 11, 180–187.

Bailey, C. H., & Kandel, E. R. (2008). Synaptic remodeling, synaptic growth and the storage of long-term memory in Aplysia. *Progress in Brain Research*, 169, 179–198.

Basar, E. (2006). The theory of the whole-brain-work. *International Journal of Psychophysiology*, 60, 133–138.

Bitterman, M. E. (1975). The comparative analysis of learning. *Science*, 188, 699–709.

Bitterman, M. E. (1976). Incentive contrast in honey bees. *Science*, 192, 380–382.

Bitterman, M. E., Menzel, R., Fietz, A., & Schäfer, S. (1983). Classical conditioning of proboscis extension in honeybees (*Apis mellifera*). *Journal of Comparative Psychology*, 97, 107–119.

Bliss, T. V., & Lomo, T. (1973). Long-lasting potentiation of synaptic transmission in the dentate area of the anaesthetized rabbit following stimulation of the perforant path. *Journal of Physiology*, 232, 331–356.

Bolles, R. C., & Beecher, M. D. (1988). *Evolution and learning*. Hillsdale, NJ: Lawrence Erlbaum Associates.

Bon, C. L., & Garthwaite, J. (2003). On the role of nitric oxide in hippocampal long-term potentiation. *Journal of Neuroscience*, 23, 1941–1948.

Bullock, T. H., & Basar, E. (1988). Comparison of ongoing compound field potentials in the brains of invertebrates and vertebrates. *Brain Research Review*, 13, 57–75.

Burgoyne, R. D. (2007). Neuronal calcium sensor proteins: generating diversity in neuronal Ca^{2+} signalling. *Nature Reviews Neuroscience*, 8, 182–193.

Chen, C.-C., Wu, J.-K., Lin, H.-W., Pai, T.-P., Fu, T.-F., Wu, C.-L., ... Chiang, A.-S. (2012). Visualizing long-term memory formation in two neurons of the Drosophila brain. *Science*, 335, 678–685.

Chklovskii, D. B., Mel, B. W., & Svoboda, K. (2012). Cortical rewiring and information storage. *Nature*, 431, 782–788.

Collingridge, G. L., Isaac, J. T. R., & Wang, Y. T. (2004). Receptor trafficking and synaptic plasticity. *Nature Reviews Neuroscience*, 5, 952–962.

Crick, F. (1984). Memory and molecular turnover. *Nature*, 312, 101. doi: 10.1038/313636a0

Davis, M. (1984). The mammalian startle response. In R. C. Eaton (Ed.) *Neural mechanisms of startle behavior* (pp. 287–351). New York, NY: Plenum.

Davis, R. L. (2011). Traces of Drosophila memory. *Neuron*, 70, 8–19.

Denker, M., Finke, R., Schaupp, F., Grün, S., & Menzel, R. (2010). Neural correlates of odor learning in the honey bee antennal lobe. *European Journal of Neuroscience*, 31, 119–133.

Doetsch, F., & Hen, R. (2005). Young and excitable: the function of new neurons in the adult mammalian brain. *Current Opinion in Neurobiology*, 15, 121–128.

Dubnau, J., & Tully, T. (1998). Gene discovery in Drosophila: New insights for learning and memory. *Annual Review of Neuroscience*, 21, 407–444.

Eisenhardt, D., Friedrich, A., Stollhoff, N., Müller, U., Kress, H., & Menzel, R. (2003). The AmCREB gene is an ortholog of the mammalian CREB/CREM family of transcription factors and encodes several splice variants in the honeybee brain. *Insect Molecular Biology*, 12, 373–382.

Fell, J., & Axmacher, N. (2011). The role of phase synchronization in memory processes. *Nature Reviews Neuroscience*, 12, 105–118.

Frankland, P. W., & Josselyn, S. A. (2013). Memory and the single molecule. *Nature*, 493, 312–313.

Fuster, J. M. (1997). Network memory. *Trends in Neuroscience*, 20, 451–459.

Fuster, J. M. (2009). Cortex and memory, Emergence of a new paradigm. *Journal of Cognitive Neuroscience*, 21, 2047–2072.

Gelperin, A. (2006). Olfactory computations and network oscillations. *Journal of Neuroscience*, 26, 1663–1668.

Gerber, B., Tanimoto, H., & Heisenberg, M. (2004). An engram found? Evaluating the evidence from fruit flies. *Current Opinion in Neurobiology*, 14, 737–744.

Gerber, B., Yarali, A., Diegelmann, S., Wotjak, C. T., Pauli, P., & Fendt, M. (2014). Pain-relief learning in flies, rats, and man: Basic research and applied perspectives. *Learning & Memory*, 21: 232–252.

Giurfa, M. (2007). Behavioral and neural analysis of associative learning in the honeybee: A taste from the magic well. *Journal of Comparative Physiology*, A 193, 801–824.

Giurfa, M., & Sandoz, J.-C. (2012). Invertebrate learning and memory: Fifty years of olfactory conditioning of the proboscis extension response in honeybees. *Learning & Memory*, 19, 54–66.

Glanzman, D. L. (2008). New tricks for an old slug: The critical role of postsynaptic mechanisms in learning and memory in Aplysia. *Progress in Brain Research*, 169, 279–293.

Hammer, M., & Menzel, R. (1998). Multiple sites of associative odor learning as revealed by local brain microinjections of octopamine in honeybees. *Learning & Memory* 5, 146–156.

Han, J.-H., Kushner, S. A., Yiu, A. P., Cole, C. J., Matynia, A., Brown, R. A., ... Josselyn, S. A. (2007). Neuronal competition and selection during memory formation. *Science*, 316, 457–460.

Hebb, D. O. (1949). *The organization of behavior*. New York, NY: Wiley.

Heisenberg, M. (2003). Mushroom body memoir: From maps to models. *Nature Reviews Neuroscience*, 4, 266–275.

Heisenberg, M., Borst, A., Wagner, S., & Byers, D. (1985). Drosophila mushroom body mutants are deficient in olfactory learning. *Journal of Neurogenetics*, 2, 1–30.

Henke, K. (2010). A model for memory systems based on processing modes rather than consciousness. *Nature Reviews Neuroscience*, 11, 523–532.

Heynes, C., & Huber, L. (2000). *The evolution of cognition*. Cambridge, MA: The MIT Press.

Horn, G. (2004). Pathways to the past: The imprint of memory. *Nature Reviews Neuroscience*, 5, 108–120.

Kandel, E. R. (2004). The molecular biology of memory storage: A dialog between genes and synapses. *Bioscience Reports*, 21, 477–522.

Kandel, E. R., & Pittenger, C. (1999). The past, the future and the biology of memory storage. *Philosophical Transactions of the Royal Society of London B*, 354, 2027–2052.

Kandel, E. R., Schwartz, J. H., & Jessell T. M. (2000). *Principles of Neural Science*. New York: McGraw-Hill.

Kupfermann, I., & Kandel, E. R. (1969). Neuronal controls of a behavioral response mediated by the abdominal ganglion of Aplysia. *Science*, 164, 847–850.

Lamprecht, R., & LeDoux, J. (2004). Structural plasticity and memory. *Nature Reviews Neuroscience*, 5, 45–54.

Lange, M. D., Doengi, M., Lesting, J., Pape, H.-C., & Jüngling, K. (2012). Heterosynaptic long-term potentiation at interneuron–principal neuron synapses in the amygdala requires nitric oxide signalling. *Journal of Physiology*, 590, 131–143.

Lesburguères, E., Gobbo, O L., Alaux-Cantin, S., Hambucken, A., Trifilieff, P., & Bontempi, B. (2011). Early tagging of cortical networks is required for the formation of enduring associative memory. *Science*, 331, 924–928.

Lindauer, M. (1970). Lernen und Gedächtnis—Versuche an der Honigbiene. *Naturwissenschaften*, 57, 463–467.

Lisman, J., Yasuda, R., & Raghavachari, S. (2012). Mechanisms of CaMKII action in long-term potentiation. *Nature Reviews Neuroscience*, 13, 169–182.

Lynch, M. A. (2004). Long-term potentiation and memory. *Physiology Review*, 84, 87–136.

Malenka, R. C. (2003). The long-term potential of LTP. *Nature Reviews Neuroscience*, 4, 923–926.

Mayford, M. (2007). Protein kinase signaling in synaptic plasticity and memory. *Current Opinion in Neurobiology*, 17, 313–317.

Menzel, R. (1979). Behavioural access to short-term memory in bees. *Nature*, 281, 368–369.

Menzel, R. (1983). Neurobiology of learning and memory: The honeybee as a model system. *Naturwissenschaften*, 70, 504–511.

Menzel, R. (2001). Searching for the memory trace in a mini-brain, the honeybee. *Learning & Memory*, 8, 53–62.

Menzel, R. (2012). The honeybee as a model for understanding the basis of cognition. *Nature Reviews Neuroscience*, 13, 758–768.

Menzel, R., & Manz, G. (2005). Neural plasticity of mushroom body-extrinsic neurons in the honeybee brain. *Journal of Experimental Biology*, 208, 4317–4332.

Michel, M., Green, C. L., Eskin, A., & Lyons, L. C. (2011). PKG-mediated MAPK signaling is necessary for long-term operant memory in Aplysia. *Learning & Memory*, 18, 108–117.

Miyashita, Y. (2004). Cognitive memory: Cellular and network machineries and their top-down control. *Science*, 306, 435–440.

Moroz, L. L., & Kohn, A. B. (2011). Parallel evolution of nitric oxide signaling: Diversity of synthesis & memory pathways. *Frontiers in Bioscience*, 16, 2008–2051.

Morris, R. G. (2003). Long-term potentiation and memory. *Philosophical Transactions of the Royal Society of London B, Biological Sciences*, 358, 643–647.

Müller, U. (1994). Ca2+/calmodulin-dependent nitric oxide synthase in Apis mellifera and Drosophila melanogaster. *European Journal of Neuroscience*, 6, 1362–1370.

Müller, U. (1996). Inhibition of nitric oxide synthase impairs a distinct form of long-term memory in the honeybee, Apis mellifera. *Neuron*, 16, 541–549.

Müller, U. (1997). The nitric oxide system in insects. *Progress in Neurobiology*, 51, 363–381.

Müller, U. (2000). Prolonged activation of cAMP-dependent protein kinase during conditioning induces long-term memory in honeybees. *Neuron*, 27, 159–168.

Müller, U. (2002). Learning from honeybees: From molecules to behaviour. *Zoology*, 105, 313–320.

Nilsson, M., Perfilieva, E., Johansson, U., Orwar, O., & Eriksson, P. S. (1999). Enriched environment increases neurogenesis in the adult rat dentate gyrus and improves spatial memory. *Journal of Neurobiology*, 39, 569–578.

Rath, L., Galizia, G., & Szyszka, P. (2011). Multiple memory traces after associative learning in the honey bee antennal lobe. *European Journal of Neuroscience*, 34, 352–360.

Redondo, R. L., & Morris, R. G. M. (2011). Making memories last: The synaptic tagging and capture hypothesis. *Nature Reviews Neuroscience*, 12, 17–30.

Reissner, K. J., Shobe, J. L., & Carew, T. J. (2006). Molecular nodes in memory processing: insights from Aplysia. *Cellular & Molecular Life Sciences*, 63, 963–974.

Rensing, L., Koch, M., & Becker, A. (2009). A comparative approach to the principal mechanisms of different memory systems. *Naturwissenschaften*, 96, 1373–1384.

Roberts, A. C., & Glanzman, D. L. (2003). Learning in Aplysia: Looking at synaptic plasticity from both sides. *Trends in Neuroscience*, 26, 662–670.

Sacktor, T. C. (2011). How does PHMz maintain long-term memory? *Nature Reviews Neuroscience*, 12, 9–15.

Salter, M. W., & Kalia, L. V. (2004). SRC kinases: A hub for NMDA receptor regulation. *Nature Reviews Neuroscience*, 5, 317–328.

Salwiczek, L. H., Watanabe, A., & Clayton, N. S. (2010). Ten years of research into avian models of episodic-like memory and its implications for developmental and comparative cognition. *Behavioural Brain Research*, 215, 221–234.

Schütt, A., Basar, E., & Bullock, T. H. (1992). The effects of acetylcholine, dopamine and noradrenaline on the visceral ganglion of Helix pomatia - I. Ongoing compound field potentials of low frequencies. *Comparative Biochemistry & Physiology. Part C: Comparative Pharmacology*, 102C, 159–168.

Scoville, W. B., & Milner, B. (1957). Loss of recent memory after bilateral hippocampal lesions. *Journal of Neurology Neurosurgery & Psychiatry*, 20, 11–21.

Seidenbecher, T., Laxmi, T. R., Stork, O., & Pape, H.-C. (2003). Amygdalar and hippocampal theta rhythm synchronization during fear memory retrieval. *Science*, 301, 846–850.

Shors, T. J., Miesegaes, G., Beylin, A., Zhao, M., Rydel, T., & Gould, E. (2001). Neurogenesis in the adult is involved in the formation of trace memories. *Nature*, 410, 372–375.

Thellier, M., Le Sceller, L., Norris, V., Verdus, M. C., & Ripoll, C. (2000). Long-distance transport, storage and recall of morphogenetic information in plants: The existence of a primitive plant "memory." *Comptes Rendus de l'Académie des Sciences*, 323, 81–91.

Thomas, G. M., & Huganir, R. L. (2004). MAPK cascade signalling and synaptic plasticity. *Nature Reviews Neuroscience*, 5, 173–183.

Trewavas, A. (2003). Aspects of plant intelligence. *Annals of Botany*, 92, 1–20.

van Strien, N. M., Cappaert, N. L. M, & Witter, M. P. (2009). The anatomy of memory: an interactive overview of the parahippocampal-hippocampal network. *Nature Reviews Neuroscience*, 10, 272–282.

Verdus, M. C., Ripoll, C., Norris, V., & Thellier, M. (2012). The role of calcium in the recall of stored morphogenetic information by plants. *Acta Biotheoretica*, 60(1), 83–97. doi: 10.1007/s10441-012-9145-5

Wang, H., Hu, Y., & Tsien, J. Z. (2006). Molecular and systems mechanisms of memory consolidation and storage. *Progress in Neurobiology*, 79, 123–135.

Weber, B. H., & Depew, D. J. (2003). *Evolution and learning. The Baldwin effect reconsidered*. Cambridge, MA: MIT Press.

Weber, M., Schnitzler, H.-U., & Schmid, S. (2002). Synaptic plasticity in the acoustic startle pathway: the neuronal basis for short-term habituation? *European Journal of Neuroscience*, 16, 1325–1332.

Yin, H. H., & Knowlton, B. J. (2006). The role of the basal ganglia in habit formation. *Nature Reviews Neuroscience*, 7, 464–476.

15

Brain Evolution, Development, and Plasticity

Rayna M. Harris, Lauren A. O'Connell, and Hans A. Hofmann

15.1 Brain Evolution and Development

Across animals there is astonishing diversity in the structure and function of nervous systems and the resulting behavior patterns. Not surprisingly, the question of how this diversity has evolved has long fascinated biologists, prompted initially by the observation that allometric relationships exist between the size of the brain—or brain region—and body size across a wide range of vertebrates (Striedter, 2005). Yet it was not until fairly recently that the mechanisms that make such variation possible have become a focus of study. Brain development and plasticity are clearly dynamic processes that change neural structure and function on a variety of time scales, from early patterning of the developing brain and neural changes within an individual's lifetime to changes over evolutionary time. In the present chapter we discuss brain development and plasticity across levels of neural organization in a comparative framework to shed light on brain evolution across and within vertebrates and, to a lesser extent, invertebrates.

With the exception of sponges and placozoans, all animals have a nervous system. During the course of evolution, nervous systems in diverse taxa showed increasing cephalization and regionalization. Cephalization refers to the tendency for nerve cells to concentrate near sensory organs (i.e., mouth, eyes, nose) at the front end of the body. Regionalization refers to the idea that specific brain areas carry out specific functions. These organizational principles are accompanied by an ever increasing complexity in the diversity of neuronal cell types, functions, and connections (Striedter, 2005).

Properties of the environment are often thought to dictate which physical and sensory adaptations will be successful, and much research has focused on how socio-ecological pressures sculpt brains throughout evolution (Pollen & Hofmann, 2008). There is a strong positive correlation between brain size and body size, a phenomenon known as allometry (Snell, 1892; Thompson, 2011). In order to facilitate more robust comparisons across taxa the resulting allometric scaling exponent (usually ranging between 0.67 and 0.75) can be used to calculate an encephalization quotient for each species, which can be highly variable across vertebrate species and has often been associated with cognitive abilities within the context of comparative analyses

(Harvey & Krebs, 1990; Jerison, 1973). Mammals and birds are in the upper portion of this range, and their position is often attributed to the increased size and complexity of the cerebral cortex in terms of the number of layers and neurons. The increase is a continuum, and correlations between large cortex and complex social behavior are very strong. However, is cortical expansion really responsible for increases in cognitive and behavioral complexity? Comparative studies have provided insight into which socioecological variables best explain variation in phenotype (in this case brain size) across populations and species (Pollen et al., 2007; Pollen & Hofmann, 2008).

Two models have been proposed to explain how brains evolve: the adaptationist model (often also referred to as "mosaic evolution") and the developmental constraints model (Pollen & Hofmann, 2008).

The adaptationist model suggests that the brain contains functionally distinct regions (or modules) that mediate particular sets of behaviors (Barton & Harvey, 2000). Selection on a specific set of behaviors should favor a change localized to the brain region mediating that behavior. A few studies have provided support for this model. For example, Barton and Harvey (2000) showed that structure size correlates with functionally related structures in both primates and insectivores. Wang Mitra and Clark (2002) found that the fraction of the adult brain occupied by the telencephalon is significantly larger in socially complex birds, while eating habits, migration patterns, mating type, and vocal learning did not correlate with telencephalic fraction. Reader and Laland (2002) also found that telencephalon size is correlated with innovation frequency and social learning in primates. However, it is important to understand that causal relations are not always clear in these and other studies, and even though these adaptive hypotheses may be plausible, they are difficult to test.

On the other hand, the developmental constraints model recognizes that a common set of genes and developmental processes may regulate the development of a range of functional regions. Finlay and Darlington argue that developmental timing can explain much of the variation in brain structure size. In their model of brain size evolution, they argue that selection for a change in any single brain structure would cause the brain to change as a whole unit (Finlay & Darlington, 1995; Finlay, Darlington, & Nicastro, 2001). They find evidence that brain structure sizes across mammals are strongly correlated with the brain size according to different power relationships, such that the neocortex exponent might explain higher neocortex fraction in primates. The authors posit that shifts in the developmental time of cortical neurogenesis between primates and rodents explain the expansion of the neocortex in primates (Finlay & Darlington, 1995). A synthesis by Striedter (Striedter, 2005) provides support for both models, suggesting that both mosaic evolution and developmental constraints play fundamental roles in driving brain/behavior changes (see also Chapter 13, this volume).

It is important to keep in mind that there are several potential confounds that often make the interpretation of comparative studies susceptible to simplistic adaptationist interpretations (Gould & Lewontin, 1979; Pollen & Hofmann, 2008). First, we usually do not know the selective forces that were at work during a given period of evolution. Second, genetic drift instead of selection can cause changes in neural and behavioral phenotypes. Third, because of their common evolutionary history, traits across species within a hierarchical and branched phylogeny cannot be considered independent, and therefore, in order to draw conclusions from the covariation of traits across taxa, this phylogenetic nonindependence needs to be taken into account (Felsenstein, 1985; Harvey & Pagel, 1991; Pagel, 1999). Since Felsenstein's classic paper, the generally accepted method of overcoming the effect of shared ancestry has been to calculate differences in (extant and ancestral) trait values between sister taxa. Two traits are then considered evolutionarily correlated (i.e., change in one trait has been accompanied by change in the other) if these (standardized) differences—or phylogenetically independent contrasts—in one trait significantly covary with contrasts in the other trait (Garland, Harvey, & Ives, 1992). Even though more sophisticated approaches have since been developed (Freckleton, Harvey, & Pagel, 2002; Pagel & Meade, 2006), the fundamental assumption is that the phylogenetic relationships between the species studied are known. However, even for groups that have been relatively well studied, well-resolved phylogenies often do not exist, and it is of paramount importance to conduct comparative analyses for the different phylogenetic hypotheses if a consensus has not yet been reached.

15.2 Developing Diverse Brains

How can we explain the diversity of the structures that make up vertebrate brains? Beyond the "just so" stories that often characterize the interpretation of the causes and origins of brain diversity (Healy & Rowe, 2007), two problems have vexed this line of research. First, it is not at all obvious how an increase in (relative) size would give rise to functional differences (e.g., increased cognitive abilities, novel sensory specializations, or behavioral complexity). Although a larger number of neurons and/or synapses might well result in greater processing power and/or speed, there is no clear relationship between such measures and behavioral or cognitive outcomes. Second, our understanding of the developmental mechanisms that give rise to the observed variation in brain structure is still very limited. In this context it is also important to keep in mind that differences in brain structure and function can be as much a consequence of environmentally responsive developmental *plasticity* as of genetically driven developmental *control* (Pollen & Hofmann, 2008).

15.2.1 Generating Diversity through Early Patterning

Many studies have suggested that neurogenesis later in development generates diversity, which might result in the differential expansion of various brain areas (see below). Similar to the basic patterning processes that specify the main body axes across all metazoans, the overall spatial and temporal activity patterns of transcription factor networks that establish the main compartments during early brain development are highly conserved (Puelles, Harrison, Paxinos, & Watson, 2013; Puelles & Rubenstein, 2003). This neuromeric model describes the spatiotemporal patterns of highly conserved developmental genes, which divide the developing brain into anteroposterior segments (neuromers) prefiguring adult functional units, and uses this information to identify homologous structures across species (Puelles et al., 2013). Because the genomic control of neural morphogenesis is remarkably conservative, the relationship between embryonic patterns and adult structure is very consistent across vertebrates (see Chapter 12 this volume). This developmental framework has thus been key to resolving putative homology relationships across vertebrates for numerous brain regions (O'Connell & Hofmann, 2011a). However, it should also be noted that many homologies are still considered tentative (Goodson & Kingsbury, 2013) and that comparisons across vertebrates that include teleosts continue to be particularly challenging because actinopterygian (ray-finned fish) forebrains develop via eversion not invagination (Yamamoto et al., 2007).

Given such a conserved theme of brain development, could small variations arising from developmental expression profiles potentially result in substantial, and possibly adaptive, changes in brain structure? This question has received surprisingly little attention. Insights into the developmental processes that give rise to brain diversity can be gained by examining the remarkable phenotypic diversity found in the cichlid fishes from East Africa's Great Lakes, which have undergone the most rapid and extensive adaptive radiations known for vertebrates. They display an astonishing array of phenotypes with little genetic diversification (Renn, Aubin-Horth, & Hofmann, 2004). The extraordinary ecological (e.g., habitat, feeding specialization) and behavioral (e.g., color preferences by females, mating and parental care systems) diversity is correlated with variation in brain structure of a magnitude that exceeds that of all mammals and facilitates comparisons across large social and physical gradients in closely related species of cichlids (Pollen et al., 2007).

In an elegant study in cichlid fishes from Lake Malawi, Sylvester et al. (2010) examined gene expression variation in a regulatory circuit (composed of *six3*, *fezf2*, *shh*, *irx1b*, and *wnt1*) known to specify anterior-posterior brain polarity and to set the boundary limits between the developing fore- and midbrain. There is considerable variation in the expression patterns of these genes between rock-dwelling mbuna (*Labeotropheus fuelleborni*, *Maylandia zebra*, and *Cynotilapia afra*) and sand-dwelling nonmbuna cichlids (*Copadichromis borleyi*, *Mchenga conophorus*, and *Aulonocara jacobfreibergi*), consistent with the differences observed in the relative size of fore- and midbrain structures in adult fish. When the WNT signaling pathway is chemically perturbed in the developing embryo, alterations in this coexpression network are sufficient to give rise to the observed differences in brain development, resulting for instance in a rock-dweller with the forebrain shaped and sized like that of a sand-dweller. These results strongly suggest that evolutionary changes in the patterning of developing brain compartments can establish ecologically and behaviorally relevant differences in the adult brain. Variation in subsequent neurogenesis, which until now has been thought to be the main source of variation in brain structure across species, may then elaborate the construction of diverse brains (Sylvester et al., 2010). Clearly, diversity in early patterning constitutes a potentially important, yet hitherto underappreciated, avenue by which natural selection can act on brain structure and function, possibly releasing the brain to some extent from developmental constraints imposed by cell proliferation mechanisms common across brain regions.

15.2.2 Neuronal Cell Fate and Development

Our understanding of neural development and brain function in part depends on an understanding of the cell fate of a neuron and its location and connectivity in the brain. To illustrate the role of this information in comparative brain development and plasticity we discuss two examples: the specification of dopaminergic neurons in the brain and the caudal migration of gonadotropin-releasing hormone (GnRH) neurons early during development.

Dopamine is an ancient neurochemical that, in diverse species, modulates the selection of behavior patterns such as basic motor programs (Joshua, Adler, & Bergman, 2009; Vidal-Gadea et al., 2011), social behavior (Aragona & Wang, 2009; O'Connell & Hofmann, 2011b), and learning and memory (Hyman, Malenka, & Nestler, 2006; Wise, 2004). In mammals, dopaminergic cell populations are limited to a relatively small number of discrete brain regions, while in teleosts more than 20 groups of dopamine neurons have been described (O'Connell, 2013). How this variation comes about and to which extent it contributes to differences between lineages is not well understood, as these cell populations are not easy to homologize across vertebrates. Nonetheless, gene expression patterns in dopaminergic neurons of the posterior tuberculum are consistent with those of the tetrapod ventral tegmental area (O'Connell, 2013), which releases dopamine into the reward system. Flames and Hobert (2009) proposed a conserved regulatory code that specifies and maintains dopaminergic neurons from *Caenorhabditis elegans* worms to vertebrates, although a detailed evolutionary understanding of these neurons has remained elusive (reviewed in O'Connell, 2013).

GnRH neurons comprise a small population of neuroendocrine cells in the rostral hypothalamus and basal forebrain where they serve as a key regulator of vertebrate reproduction (Gore, 2002a). Like most peptidergic cell groups, they are born in the olfactory placode early in development but migrate caudally as embryogenesis proceeds. They secrete gonadotropin-releasing hormone (GnRH-1), communicate with many areas of the brain, and integrate multiple inputs to control gonad maturation, puberty and sexual behavior. GnRH-1 neurons migrated from olfactory bulb and midbrain. The exact mechanisms of this migration and target finding are under intense study (Sabado, Barraud, Baker, & Streit, 2012), but cell-specific molecular profiling has provided increasing evidence that these neurons are part of an ancient class of neurosecretory cells already present in the last common ancestor of all bilaterian animals (Tessmar-Raible et al., 2007).

15.2.3 Differential Proliferation Dynamics Generate Variation in Cortex Size

The evolutionary expansion of the cerebral cortex in mammals, particularly in primates, has fascinated scientists for some time (Finlay & Darlington, 1995; Reader & Laland, 2002). The increase in cortex size in the lineage leading to humans has been interpreted as the result of variation in neurogenesis later in development, when cells in pre-established compartments proliferate, die, and/or differentiate into mature neurons and glia cells. According to the radial unit hypothesis, simply altering the first of the three phases of cell division that produce cortical excitatory neurons can scale the size of the cortex (Rakic, 1995). In contrast, the intermediate progenitor hypothesis, which seems to have stronger support, suggests that, in the evolutionary expansion of the cortex, proportionately more neurogenesis occurs during the third and final phase of proliferation (Hill & Walsh, 2005; Kriegstein, Noctor, & Martínez-Cerdeño, 2006).

Scientists have begun to unravel the molecular mechanisms regulating the size of the neocortex. Given the importance of differential proliferation dynamics in determining cortex size discussed above, it is no surprise that the mitotic spindle protein, ASPM (abnormal spindle-like microcephaly-associated protein) is a major player in the process (Pulvers et al., 2010). It is known that mutations in ASPM cause microcephaly (decrease in brain size) in some human families (Bond et al., 2003), and that it has undergone positive selection in the primate lineage leading to humans (hominids) (Kouprina et al., 2004). β-catenin is another protein that appears to control cerebral cortex size through its effects on cell proliferation during cortex development via Wnt signaling (Chenn & Walsh, 2002). While these and other studies have identified putative genetic events underlying the evolution of the human brain and its emergent cognitive capacities, allelic variation in *ASPM* or *Microcephalin* does not seem to be associated with IQ in humans (Mekel-Bobrov et al., 2007), which again underscores the previous insight that the functional implications of variation in the size of a brain structure are often unclear. Also, we need to ask what the relative importance of differential proliferation is compared with the initial delineation of the future pallial vs. other areas much earlier during development, as discussed above (§15.2.1). Specifically, variation in early patterning might reduce the developmental constraints that otherwise limit the extent to which natural selection can sculpt neural structure and function in a brain-region-specific manner.

15.2.4 Cortical Development Is Remarkably Plastic

There is an astonishing degree of diversity in cortical organization across vertebrates (Krubitzer & Dooley, 2013). For example, somatosensory cortical maps reflect biological adaptations. In the naked mole rat, the somatosensory cortex is dominated by the representation of teeth (Catania & Remple, 2002), while in the human it is dominated by the mouth, hands, and eyes (Marieb & Hoehn, 2012). However, cortical development is very plastic, and altering the environment can alter the structure of the brain and thereby possibly its function. One study showed that a considerable portion of the developing cortical sheet could be removed and functional regions that would normally appear in the removed area are accommodated elsewhere (Huffman et al., 1999). Studies on humans and other vertebrates that have undergone limb amputations or sensory organ removal show similarly plastic remodeling of the cortex (Farnè et al., 2002; Karlen & Krubitzer, 2009). Clearly, there is a lot of evolutionary and developmental plasticity, but how does it come about mechanistically?

15.3 Neural Circuits, Neurochemicals, and Behavior

To understand how the brain mediates a behavioral output, it is necessary to understand both the changes in gene expression that occur in response to external or internal stimuli and the neural circuitry in which these changes take place. Here we introduce the neural circuits that govern (social) behavior, how neurochemicals and hormones modulate those circuits, and how these hormone-neurotransmitter systems have evolved.

15.3.1 Inferring Homologies for Neural Circuits Underlying Social Behavior

Do "complex" behaviors drive the evolution of complex brains? For example, is the size of the primate neocortex a result of high-quality foraging and Machiavellian social competition, or a simple consequence of body size? Interdisciplinary efforts to combine neuroscience, evolution, and development have given rise to the field of "neuro-evo-devo" and have shed light on the evolutionarily conserved neurochemical circuits that underlie behavior. As described in §15.2.1, comparative work across bilaterians has demonstrated how early developmental patterning partitions functional units of the developing brain. These comparative and integrative approaches have facilitated a mechanistic understanding of the evolution of variation in brain morphology, neural phenotypes, and neural networks that determine brain function and give rise to behavioral diversity across taxa (O'Connell, 2013).

All animals evaluate the salience of external stimuli and integrate them with internal physiological information to produce adaptive behavior. Natural and sexual selection impinges on these processes, yet our understanding of behavioral decision-making mechanisms and their evolution is still very limited. Insights from mammals indicate that two neural circuits are of crucial importance in this context: (1) the social behavior network, consisting of amygdalar and hypothalamic regions that regulate multiple forms of social behavior (sexual behavior, aggression, and parental care), are reciprocally connected, and contain sex steroid hormone receptors (Goodson, 2005; Newman, 1999) and (2) the mesolimbic reward system, which evaluates the salience of an external stimulus and consists mostly of telencephalic brain regions and dopaminergic projections from the midbrain ventral tegmental area (Deco & Rolls, 2005; Wickens, Budd, Hyland, & Arbuthnott, 2007; Wise, 2005). Based on a synthesis of neurochemical, tract-tracing, developmental, and functional lesion/stimulation studies, O'Connell and Hofmann (2011a) delineated homology relationships for most of the nodes of these two circuits across the five major vertebrate lineages (mammals, birds, reptiles, amphibians, and teleost fish; see Figure 15.1B, D). Even though many of these homologies should still be considered tentative (Goodson & Kingsbury, 2013), this comparative analysis of the two neural circuits clearly suggested that these circuits were already present in early vertebrates and that together they form a larger social decision-making network that regulates adaptive behavior. This synthesis provides an strong foundation on which we can build research programs to better understand the evolution of the neural mechanisms underlying reward processing and behavioral regulation (O'Connell & Hofmann, 2011a).

15.3.2 Evolution of Neurochemistry Underlying Behavior

Establishing homology across vertebrate brain regions mediating social behavior has opened exciting new opportunities. In particular, it invites study of how variation across taxa in the neural basis of social decision making might explain observed differences in behavior—as well as how and why these differences evolved. A recent comparison of the social decision-making network across 88 vertebrate species has revealed that, although neurochemical profiles are very much conserved, vertebrate lineages differ more in the spatial distributions of ligands (cell populations that synthesize neuropeptides, neurotransmitters, or steroids) than their receptors (neuropeptide and neurotransmitter receptors, sex steroid hormone receptors)

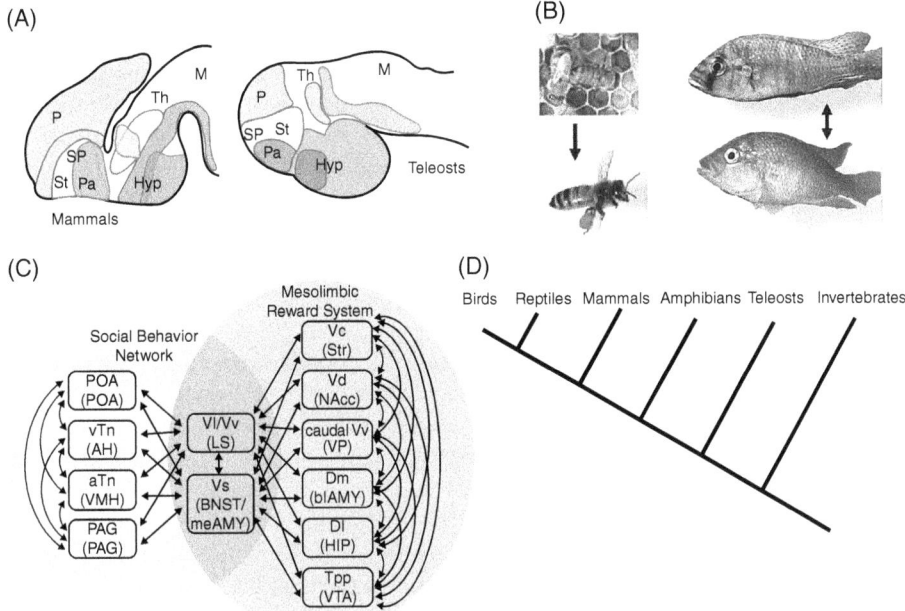

Figure 15.1 Levels and Time Scales of Neural and Behavioral Plasticity.
Understanding how genetic and environmental factors affect neural plasticity requires detailed analyses of development, neural networks, neural connectivity, life history, and evolution. (A) Gene expression patterns of developmental genes (colored regions) are highly conserved and regulate development of homologous brain structures. Adapted from O'Connell 2013. (B) The social decision making network provides a framework for analyzing structurally and functionally homologous brain regions across vertebrates. Adapted from O'Connell & Hofmann 2011b. (C) Life history traits and transitions are highly diverse, but social behavior (e.g., courtship and mating) is present in all kingdoms. Neural activation patterns underlying behavior can be conserved or divergent across species. (D) Phylogenetic relationships between major vertebrate lineages. (*See insert for color representation of the figure*).

(O'Connell & Hofmann, 2012). It is important to note that this large-scale comparative study only noted presence or absence of a neurochemical or gene in a particular brain region; however, quantitative variation in neurochemical gene expression, also seems to be important for variation in behavior within lineages. An extensively studied example is how quantitative variation in the vasopressin receptor expression in different species of *Microtus* voles is linked to differences in mating systems (reviewed in Wang, Young, De Vries, & Insel, 1998).

15.3.3 Neurotransmitter Circuitry Underlying Decision Making

Animals are constantly confronted by challenges and opportunities in their social environment in which they must make adaptive decisions to ultimately maximize their fitness. Before responding to a social stimulus with a behavioral output, animals must first evaluate the salience of a stimulus. The neural circuit in which this evaluation takes place is thought to be the mesolimbic dopamine system (Deco & Rolls, 2005;

Wickens et al., 2007; Wise, 2005), with a key role for dopaminergic projections from the ventral tegmental area of the mammalian midbrain to the forebrain. In mammals, dopaminergic projections from the ventral tegmental area to the nucleus accumbens encode both rewarding and aversive stimuli while projections to the prefrontal cortex encode aversive stimuli selectively (Lammel, Ion, Roeper, & Malenka, 2011). The neuroanatomical components of the dopamine reward circuitry seem to be conserved across vertebrates (O'Connell & Hofmann, 2011a), which is not surprising given its crucial role in evaluating stimuli in many species.

Although dopamine is a key neurotransmitter for encoding the value of stimuli in vertebrates, octopamine (homologous to the vertebrate norepinephrine) plays a prominent role in arthropods (Barron, Sovik, & Cornish, 2010). Studies in honeybees show that an individual's response to a reward (proboscis extension to sucrose) was reduced when injected with dopamine (Mercer & Menzel, 1982), whereas octopamine enhanced the proboscis response and could even substitute for sucrose presentation (Hammer & Menzel, 1998). This study highlights the opposite roles of dopamine and octopamine in arthropod behavior, recently confirmed in pharmacological manipulations of both the cricket *Gryllus bimaculatus* (Mizunami et al., 2009) and honeybee *Apis mellifera* (Farooqui, Robinson, Vaessin, & Smith, 2003; Vergoz, Roussel, Sandoz, & Giurfa, 2007) as well as genetic manipulations of the fruit fly *Drosophila melanogaster* (Schwaerzel et al., 2003), where the octopamine is necessary for reward learning and dopamine is necessary for aversive learning (reviewed in Barron et al., 2010).

15.3.4 Hormonal Modulation of Neural Circuits

Steroid hormones play a pivotal role in brain development and in sex-typical adult behavior. Classically, sex steroid hormones (estrogens, androgens, and progestins) are thought to organize neural circuits of the brain during development and then play an activational role when adult reproductive function is obtained (Arnold & Breedlove, 1985; Phoenix, Goy, Gerall, & Young, 1959). This organizational period refers to a critical time in vertebrate brain development when steroid hormones masculinize/ defeminize or feminize/demasculinize the neural circuits which program behavioral repertoires in adulthood. This valuable framework has also been extended to the organizational and activational effects of the juvenile hormone and other hormones in insects (Elekonich & Robinson, 2000). There are two classes of steroid hormone receptors that mediate these organizational and activational effects in vertebrates. Nuclear sex-steroid hormone receptors are transcription factors that mainly exert their effects through long-term changes in gene transcription (Hall, Couse, & Korach, 2001; Hall & McDonnell, 2005; Nilsson et al., 2001). In additional to the classical role of modulation via gene expression, there are also membrane-bound steroid hormone receptors that transduce fast actions through second messenger cascades (Marino, Galluzzo, & Ascenzi, 2006). In the past decade, work in many social vertebrates has delineated a role of these membrane steroid receptors in mediating behavior and neuronal plasticity (Balthazart, Absil, Gérard, Appeltants, & Ball, 1998; Sisneros, 2009). The social behavior network (Newman, 1999), comprised of several interconnecting brain regions that are responsive to steroid hormones and are involved in aggressive, sexual, and parental behavior in mammals, has been homologized across all other vertebrate classes (Crews, 2003; Goodson, 2005; O'Connell & Hofmann, 2011a) and hence provides an ideal comparative framework.

Peptide hormones also play an important role in modulating behavior (see Chapter 10). Much of this work on social behavior has intensely focused on the nonapeptides vasopressin (vasotocin in most non-mammalian vertebrates) and oxytocin (mesotocin in birds, reptiles and amphibians; isotocin in most fish). Perhaps the best studied example of peptide regulation of social behavior is found in *Microtus* voles, where vasopressin and oxytocin play important roles in pair bonding and parental care (reviewed in Young & Wang, 2004). Moreover, species differences in the vasotocin receptor abundance in several brain regions have been linked to species differences in mating systems (reviewed in Young, Wang, & Insel, 1998). In a broad sense, the actions of nonapeptides in mediating social affiliation transcend vertebrate lineages (Goodson, Kelly, & Kingsbury, 2012; Oldfield & Hofmann, 2011), suggesting their functional role is highly conserved. However, the specific role of these nonapeptides in mediating behavior should not be oversimplified, as their roles can be quite varied across species even within a lineage (Goodson et al., 2012).

Behavioral decision making depends on neural circuits that evaluate the salience of a stimulus and coordinate physiological information into a behavioral output appropriate to the social situation. To achieve this, neural processing impinges on two vertebrate neural circuits that have previously been studied in isolation, the mesolimbic dopamine system and the social behavior network, that can be considered part of a larger social decision-making network (O'Connell & Hofmann, 2011a, 2012). Both sex steroid hormones and neuropeptides orchestrate the functional state of this network to mediate appropriate behavioral outputs. For example, in the monogamous prairie vole (*Microtus ochrogaster*) males have elevated V1a receptor activity in the ventral pallidum and oxytocin receptor in the nucleus accumbens and striatum compared to the polygamous meadow vole (*Microtus pennsylanicus*). The nucleus accumbens, striatum, and ventral pallidum are the main recipients of dopaminergic input from the ventral tegmental area in mammals and thus represent the core of the mesolimbic reward system. Moreover, dopamine release in these brain regions is necessary for pair-bond formation. There are some sex differences in nonapeptide regulation of pair-bonding, however, as females seem to rely more on oxytocin while males seem to rely more on vasopressin (reviewed in Young & Wang, 2004). Steroid hormones may organize these sex differences, as most of the vasopressin and oxytocin cell groups that project to and release neuropeptides in many forebrain regions reside in nodes of the social behavior network (Newman, 1999). In male prairie voles, vasopressin abundance is actively regulated by androgens, as castration severely decreases both the number of vasopressin cell bodies and the density of vasopressin fibers throughout the forebrain (Wang & De Vries, 1993), whereas estrogens seem to regulate oxytocin abundance, as estrogens up-regulate both production of oxytocin and expression of the oxytocin receptor (reviewed in (Cushing & Kramer, 2005). Decades of work on social affiliation in *Microtus* voles highlights the mechanistic approach to studying the evolution of social behavior and teaches us that by studying the interactions of steroid hormones and neuropeptides across the social decision-making network (see Figure 15.1B), we can gain a more detailed view how evolution has sculpted the neuroendocrine mechanism of social decision-making to produce species-specific adaptive behaviors.

Relevant from an evolutionary perspective, invertebrates also produce variants of the highly conserved nonapeptides (annelids: Oumi et al., 1994; cephalopods: Takuwa-Kuroda, Iwakoshi-Ukena, Kanda, & Minakata, 2003; nematodes: Beets et al., 2012; Garrison et al., 2012; beetles: Stafflinger et al., 2008; leeches: Wagenaar, Hamilton,

Huang, Kristan, & French, 2010). Some functional studies suggest that these (likely homologous) nonapeptide cells subserve similar behavioral functions in vertebrates and invertebrates (reviewed in O'Connell, 2013), as experimental manipulation of nonapeptide function affects reproductive behavior in the medicinal leech *Hirudo spp.* (Wagenaar et al., 2010), the nematode worm *Caenorhabditis elegans* (Garrison et al., 2012), and the annelid *Eisenia foetida* (Oumi et al., 1994).

15.4 Timescales of neural plasticity

The brain is incredibly dynamic and varies between individuals of the same species, across an individual's lifetime, and over generations. This variation has profound influences on how an individual responds to a stimulus and explains in part why we see so much diversity in animal behavior. In the following, we discuss how the brain integrates external social and environmental information with internal physiology to produce an appropriate behavioral response. This integration occurs through changes in neural gene expression and organization, altering information processing in animals' brains to promote socially appropriate behavioral responses that ultimately maximize their fitness.

15.4.1 Neural Changes with Social Stimulation

In order to make adaptive decisions about their social environment, animals need to remember social experiences so that they can respond appropriately to the next similar encounter. Many studies present an animal with a behaviorally relevant sensory stimulus and measure electrical activity in neurons of various brain regions. The neural basis of vocal learning in songbirds provides an excellent example, as songs produced by males vary based on the social context. In the zebra finch (*Taeniopygia guttata*), neuronal activity is markedly different in brain regions involved in song learning when the male sings a song directed at a conspecific compared to undirected song (Hessler & Doupe, 1999). Neuronal firing can also encode the salience of a song stimulus. In receptive female canaries (*Serinus canaria*), neurons will respond with increased activity to attractive components of a male courtship song, but not unattractive song components, suggesting that a female's responsiveness to a sexual stimulus can be encoded by neural firing activity (Del Negro, Kreutzer, & Gahr, 2000).

However, electrophysiological recordings are difficult in awake and behaving animals moving in a naturalistic habitat—thus, measuring an animal's neural responses to a social stimulus often requires a different approach. Importantly, populations of neurons can integrate external inputs by other means than via short-term changes in spike frequency. Synaptic inputs, via the activation of 2nd messenger cascades, can result in rapid (taking place within minutes to hours) changes in gene expression, which, in turn, can result in the structural remodeling of synapses and other cellular structures (Loebrich & Nedivi, 2009). The genes that show a change in expression with the shortest latency (within minutes) are termed immediate early genes (IEGs, e.g., *c-fos*, *egr-1*, *c-jun*, and *arc*). IEGs encode transcription factors that are thought to coordinate cellular and ensemble responses to a variety of internal and external stimuli, which eventually result in long-term plastic changes of neuronal function. In the context of functional neuroanatomy, mapping the induction of IEG expression after a neurochemical or behavioral stimulus has become a useful tool for inferring the neural circuitry that governs behavioral responses (Clayton, 2000; Hofmann, 2010). The widespread use of IEGs has accelerated research into the functional neuroanatomy of social behavior and has shed light on how neural responses to social stimuli are conserved even across wide evolutionary distances (Hofmann, 2010). In some monogamous mammalian and cichlid fish species where males care for offspring, paternal behavior is associated with IEG induction in homologous brain regions, specifically the lateral septum and preoptic area (de Jong, Chauke, Harris, & Saltzman, 2009; Kirkpatrick, Kim, & Insel, 1994; O'Connell, Matthews, & Hofmann, 2012). Thus, the neural substrates underlying paternal care appear remarkably conserved, even though paternal care clearly evolved independently in mammals and teleosts (Reynolds, Goodwin, & Freckleton, 2002).

A well-studied example of neural changes with social information is the "winner effect," where physiology and gene expression change, after a social contest, in the brains of both the victor and the vanquished. In many vertebrates, winning an aggressive encounter induces a surge in circulating androgens (Archer, 2006; Goymann, 2009; Hirschenhauser & Oliveira, 2006; Oliveira, 2004; Wingfield, Hegner, Dufty, & Ball et al., 2010) which in turn increases the probability of winning future encounters (Dugatkin, 1997; Hsu, Earley, & Wolf, 2006; Hsu & Wolf, 1999; Rutte, Taborsky, & Brinkhof, 2006). In the male California mouse (*Peromyscus californicus*), winners of a conflict will respond with a rise in circulating testosterone that is accompanied by an increase in androgen receptor expression in brain regions associated with aggression (Fuxjager et al., 2010). Furthermore, the androgen response to victory is more pronounced when the animal wins a fight in its home cage rather than in an unfamiliar environment. This context-dependent social experience is translated in the brain by increasing androgen receptor expression in regions that modulate reward processing when the fight is won in the home cage but not in an unfamiliar location (Fuxjager et al., 2010). This neural plasticity to social interactions may serve to increase future winning ability by preparing the animal for future encounters in a context-dependent manner.

The use of IEGs is even more powerful when placed in the functional context of a particular cellular phenotype. The work by Goodson and colleagues on group size preferences and courtship behavior in Estrildid finches provides an excellent example of dopaminergic neurons involved in encoding the salience of a stimulus. In this family of songbirds, gregarious species have increased IEG induction in dopaminergic cells of the ventral tegmental area when exposed to a same-sex conspecific compared to territorial finches that do not live in social groups (Goodson et al., 2009), suggesting that species differences in sociality are reflected in dopaminergic neurons encoding conspecific presence. In the model cichlid *Astatotilapia burtoni*, males show c-Fos induction in response to a visual challenge stimulus specifically in dopaminergic neurons of area Vc—a putative striatal homologue located ventrally in the central telencephalon—whereas presentation of a chemical challenge stimulus (an androgen metabolite) did not induce c-Fos in this neuron population. These results suggest that different sensory cues are processed in a social-context-specific manner as part of adaptive decision-making processes (O'Connell, Rigney, Dykstra, & Hofmann, 2013). In the monogamous cichlid fish, *Amatitlania nigrofasciata*, males and females provide parental care. To determine what brain regions may contribute to paternal care, O'Connell and colleagues quantified c-Fos induction, and found that single fathers have more c-Fos induction in the forebrain area Vv (putative lateral septum homologue) than did biparental fathers or males that had lost their offspring. While overall preoptic area c-Fos induction was similar between groups, single fathers showed increased c-Fos induction in the parvocellular preoptic isotocin neurons, suggesting that isotocin mediates the increase of paternal care observed after mate removal (O'Connell et al., 2012).

The outcome of social encounters not only alters the brain profiles of the participants, but that of the observing audience as well. In their native Lake Tanganyika, *A. burtoni* females visit leks to watch males fight each other for the chance to mate. After observing the fights, the female chooses a mate (Talling, 1991). To examine the neural effects of observing male interaction, Desjardins and colleagues (Desjardins, Klausner, & Fernald, 2010) set up a laboratory experiment to let a female choose a mate between two attractive males. After displaying her preference for a particular male, the female watched as her male of choice either won or lost a fight. Then the authors measured immediate–early gene induction to assess neural activity in response to observing these male–male interactions. Females who observed their preferred male win a fight experienced IEG induction in brain regions involved in reproduction; however, females that observed their preferred mate lose a fight experienced an increase in neural activity in brain regions related to stress or anxiety (Desjardins et al., 2010). This is one of few studies that examine how the brain responds to observing social information. Clearly, more work needs to be done to determine how the brain response to a social challenges and opportunities over short time scales.

15.4.2 Neural Changes with Reproductive Transition

The brain undergoes remarkable changes as animals undergo reproductive transitions that give rise to the more visible changes in behavior and physiology: puberty, reproductive senescence, social ascension, the ovarian cycle of females, and (in many teleosts) even sex change. One of the main regulators of these processes is the gonadotropic-releasing hormone (GnRH, see §15.2.2). Pulsatile release of GnRH initiates the onset of adult reproductive function (puberty) in most vertebrates (Gore, 2002b). Kisspeptin, a neuropeptide expressed in the hypothalamus, is necessary and sufficient for initiating puberty by increasing GnRH release (Navarro, Castellano, García-Galiano, & Tena-Sempere, 2007). In mammals, the adolescent brain undergoes a major reorganization that coincides with puberty including changes in GnRH and kisspeptin cellular morphology (Ojeda, Lomniczi, Sandau, & Matagne, 2010), substantial changes in brain gene expression (Ojeda et al., 2010; Walker, Kirson, Perez, & Gore, 2012), and an initiation of steroid-hormone-dependent neurogenesis that accentuates sex differences in the relative size of certain reproductive brain regions (Ahmed et al., 2008; Sisk & Foster, 2004).

After the initiation of adult reproductive function in females of many species, cyclical changes in hormones (especially sex steroid hormones) throughout the ovarian cycle coordinate a number of changes in the brain including neurogenesis in the rat hippocampus (Pawluski, Brummelte, Barha, Crozier, & Galea, 2009) and changes in brain size in humans (Hagemann et al., 2011). Surprisingly, little is known about changes in brain gene expression in naturally cycling animals.

As females age into reproductive senescence, ovarian cycles become irregular and eventually cease due to the lack of sex steroids, especially estrogens. Although how these changes alter brain function is not yet well understood, there are profound behavioral implications, including lapses in cognitive abilities and a higher risk for neurodegenerative diseases (Kermath & Gore, 2012). Work by Gore and colleagues suggests a causal role for the hypothalamus in reproductive senescence, as glutamate NMDA receptor regulation of GnRH release changes between young and old female rats (Gore, 2002b).

In many animals, reproductive transitions are associated with radically rapid changes in both the brain and gonads. Especially striking examples are found in teleost fishes that display a wide variety of mating tactics. In the African cichlid fish, *Astatotilapia burtoni*, males are either socially dominant or subordinate. Dominant males have large testes, are brightly colored, and aggressively defend territories where they mate with females. On the other hand, subordinate males have small testes, are dull in coloration, and school in the open water. A single male can alternate between dominant and subordinate status many times throughout its lifespan depending on the immediate social environment (Hofmann, Benson & Fernald, 1999). Such a phenotypic transition is accompanied by drastic changes in sex steroid hormone levels, testes morphology, and brain gene expression (Huffman, Mitchell, O'Connell, & Hofmann, 2012; Maruska, Zhang, Neboori & Fernald, 2013). Huffman and colleagues found that males immediately become aggressive, and testosterone levels increase when they become dominant, whereas reproductive behavior and estradiol levels increase slightly later. Increases in steroid hormone levels are accompanied by increased expression of steroidogenic acute regulatory protein (StAR) in the testis and an increase in testis maturation (Huffman et al., 2012). In a similar paradigm, Maruska and colleagues found that social ascent was accompanied by changes in gene expression of sex-steroid hormone receptors, the enzyme aromatase (which converts testosterone into estradiol), and immediate early genes in specific nodes of social behavior network (Maruska et al., 2013).

Teleost species that change gonadal sex provide perhaps an even more drastic example of brain plasticity in relation to reproduction. The transition from one sex to another is usually socially dependent and based the community sex composition (reviewed in Godwin, 2009). The transition from one sex to another is initiated by the brain (independent of gonads) and is associated with changes in brain expression of neuropeptides, steroid hormone related genes, and neurotransmitter receptors (reviewed in Godwin, 2009, 2010). The molecular mechanisms by which changes in the social environment directly or indirectly alter the sex of the brain are not understood and are currently an area of intense research focus.

15.4.3 Neural Changes over a Lifespan

Some animals undergo fascinating changes in brain and behavior across their lifetime. The nonreproductive females of honeybee (*Apis mellifera*) societies transition through distinct divisions of labor as they age. Workers begin their lives tending to within-hive chores such as nursery/queen care and with age transition to the role of a forager. This age-related transition to foraging is associated with changes in brain morphology and brain gene expression. For example, as workers transition to the role of foragers, the mushroom bodies—a region in the insect brain associated with complex social behavior and memory (Erber, Homberg, & Gronenberg, 1987)—increase in size (Withers et al., 1993). This age-dependent transition of labor roles is also associated with substantial changes in the expression of thousands of genes (Whitfield et al., 2006). The hive-bee to forager transition is accompanied by changes in energy-related genes (Whitfield, Fahrbach, & Robinson, 2006) and genes driven by the actions of juvenile hormone, highlighting the importance of hormones in driving neural plasticity.

15.4.4 Neural Changes across Generations

Evidence of neural plasticity can also be observed across generations. An excellent example of this is the monarch butterfly (*Danaus plexippus*), which exhibits spectacular migratory patterns in the fall and spring that span three to four generations (Brower, 1996). The integration of two sophisticated mechanisms in the brain—a molecular clock and a sun-compass—provides the basis for the navigational feat these animals accomplish during their migration from Canada to Mexico and back (Reppert, Gegear, & Merlin, 2010). As migrating butterflies are always on their maiden voyage, innate genetic programs must govern both the northerly and southerly migration. Fall migrant butterflies are reproductively inactive whereas summer monarchs are reproductively active, a switch triggered by the juvenile hormone and a cascade of hormonally regulated genes involved in longevity, immunity, and metabolism. Additionally, microarray analyses have revealed 40 genes related to migratory behavior (independent of juvenile hormone) that are differentially expressed between summer and fall migrants, each spanning multiple generations (Zhu, Gegear, Casselman, Kanginakudru, & Reppert, 2009).

15.5 Evolution of Mechanisms Underlying Brain Plasticity

Although the striking similarities in neurochemistry and plasticity underlying complex behavior are seen across wide evolutionary distances, differentiating between convergent and conserved traits requires well-established phylogenies in which behavioral mechanisms are well resolved at many branches (Pollen & Hofmann, 2008). However, it has become increasingly clear, in cases of the convergent evolution of behavioral phenotypes, that even across vast evolutionary distances a conserved molecular tool kit can be repeatedly recruited (O'Connell & Hofmann, 2011a; Toth & Robinson, 2007). Signaling molecules such as peptide or steroid hormones and biogenic amines likely acted within a common ancestor to coordinate responses to external (often social) stimuli. Over the course of animal evolution, this simple behavioral framework may have been modified in various ways in order to adapt to new environmental challenges or opportunities that represented rewarding or aversive valence (Barron et al., 2010).

It has become increasingly clear that brain development and plasticity are dynamic processes that occur across diverse time scales with dramatic effects on brain function and behavior. By examining brain development and plasticity in a comparative framework, and across levels of neural organization, the mechanisms by which evolution shapes brain structure and function are beginning to come into focus. While many challenges remain to be overcome when using a comparative framework, recent advances have allowed us to better control for phylogenetic non-independence and thus gain a deeper understanding of the mechanisms that give rise to variation in brain structure and function.

Acknowledgments

We thank the editor, Stephen V. Shepherd, for giving us the opportunity to review the state of our field and for providing helpful guidance during the writing of this chapter, and we thank members of the Hofmann Lab and the 2013 graduate course in Brain, Behavior, and Evolution for discussion. LAO is supported by a Bauer Fellowship from the Faculty of Arts and Sciences at Harvard University, an Adele Lewis Grant Fellowship from the Graduate Women in Science, a L'Oreal For Women in Science Fellowship, a Konishi Research Grant from the International Society for Neuroethology, and NSF. HAH has been supported by NSF, NIH-NIGMS, and the Alfred P. Sloan Foundation.

References

Ahmed, E. I., Zehr, J. L., Schulz, K. M., Lorenz, B. H., DonCarlos, L. L., & Sisk, C. L. (2008). Pubertal hormones modulate the addition of new cells to sexually dimorphic brain regions. *Nature Neuroscience*, 11, 995–997.

Aragona, B. J., & Wang, Z. (2009). Dopamine regulation of social choice in a monogamous rodent species. *Frontiers in Behavioral Neuroscience*, 3, 15. doi: 10.3389/neuro.08.015.2009

Archer, J. (2006). Testosterone and human aggression: An evaluation of the challenge hypothesis. *Neuroscience & Biobehavioral Reviews*, 30, 319–345.

Arnold, A. P., & Breedlove, S. M. (1985). Organizational and activational effects of sex steroids on brain and behavior: a reanalysis. *Hormones & Behavior*, 19, 469–498.

Balthazart, J., Absil, P., Gérard, M., Appeltants, D., & Ball, G. F. (1998). Appetitive and consummatory male sexual behavior in Japanese quail are differentially regulated by subregions of the preoptic medial nucleus. *Journal of Neuroscience*, 18, 6512–6527.

Barron, A. B., Sovik, E., & Cornish, J. L. (2010). The roles of dopamine and related compounds in reward-seeking behavior across animal phyla. *Frontiers in Behavioral Neuroscience*, 4, 163. doi: 10.3389/fnbh.2010.00163

Barton, R. A., & Harvey, P. H., (2000). Mosaic evolution of brain structure in mammals. *Nature*, 405, 1055–1058.

Beets, I., Janssen, T., Meelkop, E., Temmerman, L., Suetens, N., Rademakers, S., ... Schoofs, L. (2012). Vasopressin/oxytocin-related signaling regulates gustatory associative learning in C. elegans. *Science*, 338, 543–545.

Bond, J., Scott, S., Hampshire, D. J., Springell, K., Corry, P., Abramowicz, M. J., ... Woods, C. G. (2003). Protein-truncating mutations in ASPM cause variable reduction in brain size. *American Journal of Human Genetics*, 73, 1170–1177.

Brower, L. (1996). Monarch butterfly orientation: Missing pieces of a magnificent puzzle. *Journal of Experimental Biology*, 199, 93–103.

Catania, K. C., & Remple, M. S. (2002). Somatosensory cortex dominated by the representation of teeth in the naked mole-rat brain. *Proceedings of the National Academy of Sciences of the USA*, 99, 5692–5697.

Chenn, A., & Walsh, C. A. (2002). Regulation of cerebral cortical size by control of cell cycle exit in neural precursors. *Science*, 297, 365–369.

Clayton, D. F. (2000). The genomic action potential. *Neurobiology of Learning & Memory*, 74, 185–216.

Crews, D. (2003). The development of phenotypic plasticity: Where biology and psychology meet. *Developmental Psychobiology*, 43, 1–10.

Cushing, B. S., & Kramer, K. M. (2005). Mechanisms underlying epigenetic effects of early social experience: The role of neuropeptides and steroids. *Neuroscience & Biobehavioral Reviews*, 29, 1089–1105.

Deco, G., & Rolls, E. T. (2005). Attention, short-term memory, and action selection: a unifying theory. *Progress in Neurobiology*, 76, 236–256.

De Jong, T. R., Chauke, M., Harris, B. N., & Saltzman, W. (2009). From here to paternity: Neural correlates of the onset of paternal behavior in California mice (Peromyscus californicus). *Hormones & Behavior*, 56, 220–231.

Del Negro, C., Kreutzer, M., & Gahr, M. (2000). Sexually stimulating signals of canary (Serinus canaria) songs: Evidence for a female-specific auditory representation in the HVc nucleus during the breeding season. *Behavioral Neuroscience*, 114, 526–542.

Desjardins, J. K., Klausner, J. Q., & Fernald, R. D. (2010). Female genomic response to mate information. *Proceedings of the National Academy of Sciences of the USA*, 107, 21176–21180.

Dugatkin, L. A. (1997). Winner and loser effects and the structure of dominance hierarchies. *Behavioral Ecology*, 8, 583–587.

Elekonich, M. M., & Robinson, G. E. (2000). Organizational and activational effects of hormones on insect behavior. *Journal of Insect Physiology*, 46, 1509–1515.

Erber, J., Homberg, U., & Gronenberg, W. (1987). Functional roles of the mushroom bodies in insects. In A. P. Gupta (Ed.), *Arthropod brain: Its evolution, development, structure, and functions*, (pp. 485–511). New York, NY: John Wiley & Sons.

Farnè, A., Roy, A. C., Giraux, P., Dubernard, J. M., & Sirigu, A. (2002). Face or hand, not both. Perceptual correlates of reafferentation in a former amputee. *Current Biology*, 12(25), 1342–1346.

Farooqui, T., Robinson, K., Vaessin, H., & Smith, B. H. (2003). Modulation of early olfactory processing by an octopaminergic reinforcement pathway in the honeybee. *Journal of Neuroscience*, 23, 5370–5380.

Felsenstein, J. (1985). Phylogenies and the comparative method. *The American Naturalist*, 125, 1–15.

Finlay, B. L., & Darlington, R. B. (1995). Linked regularities in the development and evolution of mammalian brains. *Science*, 268, 1578–1584.

Finlay, B. L., Darlington, R. B., & Nicastro, N. (2001). Developmental structure in brain evolution. *Behavioral & Brain Sciences*, 24, 263–278.

Flames, N., & Hobert, O. (2009). Gene regulatory logic of dopamine neuron differentiation. *Nature*, 458, 885–889.

Freckleton, R. P., Harvey, P. H., & Pagel, M. (2002). Phylogenetic analysis and comparative data: a test and review of evidence. *The American Naturalist*, 160, 712–726.

Fuxjager, M. J., Forbes-Lorman, R. M., Coss, D. J., Auger, C. J., Auger, A. P., & Marler C. A. (2010). Winning territorial disputes selectively enhances androgen sensitivity in neural pathways related to motivation and social aggression. *Proceedings of the National Academy of Sciences of the USA*, 107, 12393–12398.

Garland, T., Harvey, P. H., & Ives, A. R. (1992). Procedures for the analysis of comparative data using phylogenetically independent contrasts. *Systematic Biology*, 41, 18–32.

Garrison, J. L., Macosko, E. Z., Bernstein, S., Pokala, N., Albrecht, D. R., & Bargmann, C. I. (2012). Oxytocin/vasopressin-related peptides have an ancient role in reproductive behavior. *Science*, 338, 540–543.

Godwin, J. (2009). Social determination of sex in reef fishes. *Seminars in Cell & Developmental Biology*, 20, 264–270.

Godwin, J. (2010). Neuroendocrinology of sexual plasticity in teleost fishes. *Frontiers in Neuroendocrinology*, 31, 203–216.

Goodson, J. L. (2005). The vertebrate social behavior network: Evolutionary themes and variations. *Hormones & Behavior*, 48, 11–22.

Goodson, J. L., Kabelik, D., Kelly, A. M., Rinaldi, J., & Klatt, J. D. (2009). Midbrain dopamine neurons reflect affiliation phenotypes in finches and are tightly coupled to courtship. *Proceedings of the National Academy of Sciences of the USA*, 106, 8737–8742.

Goodson, J. L., Kelly, A. M., & Kingsbury, M. A. (2012). Evolving nonapeptide mechanisms of gregariousness and social diversity in birds. *Hormones & Behavior*, 61, 239–250.

Goodson, J. L., & Kingsbury MA (2013). What's in a name? Considerations of homologies and nomenclature for vertebrate social behavior networks. *Hormones & Behavior*, 64, 103–112.

Gore, A. C. (2002a). *GnRH: The master molecule of reproduction*. Norwell, MA: Kluwer Academic Publishers. Retrieved from http://books.google.com/books/about/GnRH_The_Master_Molecule_of_Reproduction.html?id=s4wsKR0DS00C&pgis=1

Gore, A. C. (2002b). Gonadotropin-releasing hormone (GnRH) neurons: Gene expression and neuroanatomical studies. *Progress in Brain Research*, 141, 193–208.

Gould, S. J., & Lewontin, R. C. (1979). The spandrels of San Marco and the Panglossian paradigm: A critique of the adaptationist programme. *Proceedings of the Royal Society B, Biological Sciences* 205, 581–598.

Goymann, W. (2009). Social modulation of androgens in male birds. *General & Comparative Endocrinology*, 163, 149–157.

Hagemann, G., Ugur, T., Schleussner, E., Mentzel, H.-J., Fitzek, C., Witte, O. W., & Gaser, C. (2011). Changes in brain size during the menstrual cycle. *PLoS One* 6, e14655. doi: 10.1371/journal.pone.0014655

Hall, J. M., Couse, J. F., & Korach, K. S. (2001). The multifaceted mechanisms of estradiol and estrogen receptor signaling. *Journal of Biological Chemistry*, 276, 36869–36872.

Hall, J. M., & McDonnell, D. P. (2005). Coregulators in nuclear estrogen receptor action: From concept to therapeutic targeting. *Molecular Interventions*, 5, 343–357.

Hammer, M., & Menzel, R. (1998). Multiple sites of associative odor learning as revealed by local brain microinjections of octopamine in honeybees. *Learning & Memory*, 5, 146–156.

Harvey, P., & Krebs, J. R. (1990). Comparing brains. *Science*, 249, 140–146.

Harvey, P. H., & Pagel, M. D. (1991). *The comparative method in evolutionary biology*. Oxford University Press, Oxford.

Healy, S. D., & Rowe, C. (2007). A critique of comparative studies of brain size. *Proceedings of the Royal Society B, Biological Sciences*, 274, 453–464.

Hessler, N. A., & Doupe, A. J. (1999). Social context modulates singing-related neural activity in the songbird forebrain. *Nature Neuroscience*, 2, 209–211.

Hill, R. S., & Walsh, C. A. (2005). Molecular insights into human brain evolution. *Nature*, 437, 64–67.

Hirschenhauser, K., & Oliveira, R. F. (2006). Social modulation of androgens in male vertebrates: Meta-analyses of the challenge hypothesis. *Animal Behavior*, 71, 265–277.

Hofmann, H. A. (2010). The neuroendocrine action potential. Winner of the 2008 Frank Beach Award in Behavioral Neuroendocrinology. *Hormones & Behavior*, 58, 555–562.

Hofmann, H. A., Benson, M. E., & Fernald, R. D. (1999). Social status regulates growth rate: Consequences for life-history strategies. *Proceedings of the National Academy of Sciences of the USA*, 95, 14171–14176.

Hsu, Y., Earley, R. L., & Wolf, L. L. (2006). Modulation of aggressive behaviour by fighting experience: Mechanisms and contest outcomes. *Biological Reviews*, 81, 33–74.

Hsu, Y., & Wolf, L. (1999). The winner and loser effect: Integrating multiple experiences. *Animal Behavior*, 57, 903–910.

Huffman, K. J., Molnár, Z., Van Dellen, A., Kahn, D. M., Blakemore, C., & Krubitzer, L. (1999). Formation of cortical fields on a reduced cortical sheet. *Journal of Neuroscience*, 19, 9939–9952.

Huffman, L. S., Mitchell, M. M., O'Connell, L. A., & Hofmann, H. A. (2012). Rising StARs: Behavioral, hormonal, and molecular responses to social challenge and opportunity. *Hormones & Behavior*, 61, 631–641.

Hyman, S. E., Malenka, R. C., & Nestler, E. J. (2006). Neural mechanisms of addiction: the role of reward-related learning and memory. *Annual Review of Neuroscience* 29, 565–598.

Jerison, H. J. (1973). *Evolution of the brain and intelligence*. Academic Press, New York. Retrieved from http://doi.wiley.com/10.1002/ajpa.1330430123

Joshua, M., Adler, A., & Bergman, H. (2009). The dynamics of dopamine in control of motor behavior. *Current Opinion in Neurobiology*, 19, 615–620.

Karlen, S. J., & Krubitzer, L. (2009). Effects of bilateral enucleation on the size of visual and nonvisual areas of the brain. *Cerebral Cortex*, 19, 1360–1371.

Kirkpatrick, B., Kim, J. W., & Insel, T. R. (1994). Limbic system fos expression associated with paternal behavior. *Brain Research*, 658, 112–118.

Kouprina, N., Pavlicek, A., Mochida, G. H., Solomon, G., Gersch, W., Yoon, Y.-H., ... Larionov V (2004). Accelerated evolution of the ASPM gene controlling brain size begins prior to human brain expansion. *PLoS Biology*, 2, E126.

Kriegstein, A., Noctor, S., & Martínez-Cerdeño, V. (2006). Patterns of neural stem and progenitor cell division may underlie evolutionary cortical expansion. *Nature Reviews Neuroscience*, 7, 883–890.

Krubitzer, L., & Dooley, J. C. (2013). Cortical plasticity within and across lifetimes: How can development inform us about phenotypic transformations? *Frontiers in Human Neuroscience*, 7, 620. doi: 10.3389/fnhum.2013.00620

Lammel, S., Ion, D. I., Roeper, J., & Malenka, R. C. (2011). Projection-specific modulation of dopamine neuron synapses by aversive and rewarding stimuli. *Neuron*, 70, 855–862.

Loebrich, S., & Nedivi, E. (2009). The function of activity-regulated genes in the nervous system. *Physiological Reviews*, 89, 1079–1103.

Marieb, E. N., & Hoehn, K. N. (2012). *Human anatomy & physiology* (9th ed.). San Francisco, CA: Benjamin Cummings. Retrieved from http://www.amazon.com/Human-Anatomy-Physiology-Books-Edition/dp/0321802187

Marino, M., Galluzzo, P., & Ascenzi, P. (2006). Estrogen signaling multiple pathways to impact gene transcription. *Current Genomics*, 7, 497–508.

Maruska, K. P., Zhang, A., Neboori, A., & Fernald, R. D. (2013). Social opportunity causes rapid transcriptional changes in the social behaviour network of the brain in an African cichlid fish. *Journal of Neuroendocrinology*, 25, 145–157.

Mekel-Bobrov, N., Posthuma, D., Gilbert, S. L., Lind, P., Gosso, S. F., Luciano, M., ... Lahn, B. T. (2007). The ongoing adaptive evolution of ASPM and Microcephalin is not explained by increased intelligence. *Human Molecular Genetics*, 16, 600–608.

Mercer, A. R., & Menzel, R. (1982). The effects of biogenic amines on conditioned and unconditioned responses to olfactory stimuli in the honeybee Apis mellifera. *Journal of Comparative Physiology*, 145, 363–368.

Mizunami, M., Unoki, S., Mori, Y., Hirashima, D., Hatano, A., & Matsumoto, Y. (2009). Roles of octopaminergic and dopaminergic neurons in appetitive and aversive memory recall in an insect. *BMC Biology*, 7, 46.

Navarro, V. M., Castellano, J. M., García-Galiano, D., & Tena-Sempere, M. (2007). Neuroendocrine factors in the initiation of puberty: The emergent role of kisspeptin. *Reviews in Endocrine & Metabolic Disorders*, 8, 11–20.

Newman, S. W. (1999). The medial extended amygdala in male reproductive behavior. A node in the mammalian social behavior network. *Annals of the New York Academy of Sciences*, 877, 242–257.

Nilsson, S., Mäkelä, S., Treuter, E., Tujague, M., Thomsen, J., Andersson, G., ... Gustafsson, J. A. (2001). Mechanisms of estrogen action. *Physiological Reviews*, 81, 1535–1565.

O'Connell, L. A. (2013). Evolutionary development of neural systems in vertebrates and beyond. *Journal of Neurogenetics*, 27, 69–85.

O'Connell, L. A., & Hofmann H. A. (2011a). The vertebrate mesolimbic reward system and social behavior network: a comparative synthesis. *Journal of Comparative Neurology*, 519, 3599–3639.

O'Connell, L. A., & Hofmann H. A. (2011b). Genes, hormones, and circuits: an integrative approach to study the evolution of social behavior. *Frontiers in Neuroendocrinology*, 32, 320–335.

O'Connell, L. A., & Hofmann H. A. (2012). Evolution of a vertebrate social decision-making network. *Science*, 336, 1154–1157.

O'Connell, L. A., Matthews, B. J., & Hofmann, H. A. (2012). Isotocin regulates paternal care in a monogamous cichlid fish. *Hormones & Behavior*, 61, 725–733.

O'Connell, L. A., Rigney, M. M., Dykstra, D. W., & Hofmann, H. A. (2013). Neuroendocrine mechanisms underlying sensory integration of social signals. *Journal of Neuroendocrinology*, 25, 644–654.

Ojeda, S. R., Dubay, C., Lomniczi, A., Kaidar, G., Matagne, V., Sandau. U. S., & Dissen, G. A. (2010). Gene networks and the neuroendocrine regulation of puberty. *Molecular & Cellular Endocrinology*, 324, 3–11.

Ojeda, S. R., Lomniczi, A., Sandau, U., & Matagne, V. (2010). New concepts on the control of the onset of puberty. *Endocrine Development*, 17, 44–51.

Oldfield, R. G., & Hofmann, H. A. (2011). Neuropeptide regulation of social behavior in a monogamous cichlid fish. *Physiology & Behavior*, 102, 296–303.

Oliveira, R. F. (2004). Social modulation of androgens in vertebrates: Mechanisms and function. *Advances in the Study of Behavior*, 34, 165–239.

Oumi, T., Ukena, K., Matsushima, O., Ikeda, T., Fujita, T., Minakata, H., & Nomoto, K. (1994). Annetocin: An oxytocin-related peptide isolated from the earthworm, Eisenia foetida. *Biochemical & Biophysica Research Communications*, 198, 393–399.

Pagel, M. (1999). Inferring the historical patterns of biological evolution. *Nature*, 401, 877–884.

Pagel, M., & Meade, A. (2006). Bayesian analysis of correlated evolution of discrete characters by reversible-jump Markov chain Monte Carlo. *American Naturalist*, 167, 808–825.

Pawluski, J. L., Brummelte, S., Barha, C. K., Crozier, T. M., & Galea, L. A. M (2009). Effects of steroid hormones on neurogenesis in the hippocampus of the adult female rodent during the estrous cycle, pregnancy, lactation and aging. *Frontiers in Neuroendocrinology*, 30, 343–357.

Phoenix, C. H., Goy, R. W., Gerall, A. A., & Young, W. C. (1959). Organizing action of prenatally administered testosterone propionate on the tissues mediating mating behavior in the female guinea pig. *Endocrinology*, 65, 369–382.

Pollen A. A., & Hofmann, H. A. (2008). Beyond neuroanatomy: Novel approaches to studying brain evolution. *Brain Behavior & Evolution*, 72, 145–158.

Pollen, A. A., Dobberfuhl, A. P., Scace, J., Igulu, M. M., Renn, S. C. P., Shumway, C. A., & Hofmann, H. A. (2007). Environmental complexity and social organization sculpt the brain in Lake Tanganyikan cichlid fish. *Brain Behavior & Evolution*, 70, 21–39.

Puelles, L., Harrison, M., Paxinos, G., & Watson, C. (2013). A developmental ontology for the mammalian brain based on the prosomeric model. *Trends in Neuroscience*, 1–9.

Puelles, L., & Rubenstein, J. L. R. (2003). Forebrain gene expression domains and the evolving prosomeric model. *Trends in Neuroscience*, 26, 469–476.

Pulvers, J. N., Bryk, J., Fish, J. L., Wilsch-Bräuninger, M., Arai, Y., Schreier, D., ... Huttner, W. B. (2010). Mutations in mouse Aspm (abnormal spindle-like microcephaly associated) cause not only microcephaly but also major defects in the germline. *Proceedings of the National Academy of Sciences of the USA*, 107, 16595–16600.

Rakic, P. (1995). A small step for the cell, a giant leap for mankind: A hypothesis of neocortical expansion during evolution. *Trends in Neuroscience*, 18, 383–388.

Reader, S. M., & Laland, K. N. (2002). Social intelligence, innovation, and enhanced brain size in primates. *Proceedings of the National Academy of Sciences of the USA*, 99, 4436–4441.

Renn, S. C., Aubin-Horth, N., & Hofmann, H. A. (2004). Biologically meaningful expression profiling across species using heterologous hybridization to a cDNA microarray. *BMC Genomics*, 5, 42.

Reppert, S. M., Gegear, R. J., & Merlin, C. (2010). Navigational mechanisms of migrating monarch butterflies. *Trends in Neuroscience*, 33, 399–406.

Reynolds, J. D., Goodwin, N. B., & Freckleton, R. P. (2002). Evolutionary transitions in parental care and live bearing in vertebrates. *Philosophical Transactions of the Royal Society B, Biological Sciences*, 357, 269–281.

Rutte, C., Taborsky, M., & Brinkhof, M. W. G. (2006). What sets the odds of winning and losing? *Trends in Ecology & Evolution*, 21, 16–21.

Sabado, V., Barraud, P., Baker, C. V. H., & Streit, A. (2012). Specification of GnRH-1 neurons by antagonistic FGF and retinoic acid signaling. *Developmental Biology*, 362, 254–262.

Schwaerzel, M., Monastirioti, M., Scholz, H., Friggi-Grelin, F., Birman, S., & Heisenberg, M. (2003). Dopamine and octopamine differentiate between aversive and appetitive olfactory memories in Drosophila. *Journal of Neuroscience*, 23, 10495–10502.

Sisk C. L., & Foster, D. L. (2004). The neural basis of puberty and adolescence. *Nature Neuroscience*, 7, 1040–1047.

Sisneros, J. A. (2009). Steroid-dependent auditory plasticity for the enhancement of acoustic communication: Recent insights from a vocal teleost fish. *Hearing Research*, 252, 9–14.

Snell, O. (1892). Die Abhängigkeit des Hirngewichtes von dem Körpergewicht und den geistigen Fähigkeiten. *Archiv für Psychiatrie und Nervenkrankheiten*, 23, 436–446.

Stafflinger, E., Hansen, K. K., Hauser, F., Schneider, M., Cazzamali, G., Williamson, M., & Grimmelikhuijzen, C. J. P. (2008). Cloning and identification of an oxytocin/vasopressin-like receptor and its ligand from insects. *Proceedings of the National Academy of Sciences of the USA*, 105, 3262–3267.

Striedter, G. F. (2005). *Principles of brain evolution*. Sunderland, MA: Sinauer Associates.

Sylvester, J. B., Rich, C. A., Loh, Y. E., van Staaden, M. J., Fraser, G. J., & Streelman, J. T. (2010). Brain diversity evolves via differences in patterning. *Proceedings of the National Academy of Sciences of the USA*, 107, 9718–9723.

Takuwa-Kuroda, K., Iwakoshi-Ukena, E., Kanda, A., & Minakata, H. (2003). Octopus, which owns the most advanced brain in invertebrates, has two members of vasopressin/oxytocin superfamily as in vertebrates. *Regulatory Peptides*, 115, 139–149.

Talling, J. F. (1991). Lake Tanganyika and its life. In G. W. Coulter (Ed.), *Aquatic conservation: Marine and freshwater ecosystems* (pp. 1–354). British Museum (Natural History) Publications — Oxford University Press. Retrieved from http://doi.wiley.com/10.1002/aqc.3270010210

Tessmar-Raible, K., Raible, F., Christodoulou, F., Guy, K., Rembold, M., Hausen, H., & Arendt, D. (2007). Conserved sensory-neurosecretory cell types in annelid and fish forebrain: Insights into hypothalamus evolution. *Cell*, 129, 1389–1400.

Thompson, D. W. (2011). *On growth and form*. CreateSpace Independent Publishing Platform. Retrieved from http://www.amazon.com/Growth-Form-DArcy-Wentworth-Thompson/dp/146358735X

Toth, A. L., & Robinson, G. E. (2007). Evo-devo and the evolution of social behavior. *Trends in Genetics*, 23, 334–341.

Vergoz, V., Roussel, E., Sandoz, J.-C., & Giurfa, M. (2007). Aversive learning in honeybees revealed by the olfactory conditioning of the sting extension reflex. *PLoS One* 2, e288. doi:10.1371/journal.pone.0000288

Vidal-Gadea, A., Topper, S., Young, L., Crisp, A., Kressin, L., Elbel, E., … Pierce-Shimomura, J. T. (2011). Caenorhabditis elegans selects distinct crawling and swimming gaits via dopamine and serotonin. *Proceedings of the National Academy of Sciences of the USA*, 108, 17504–17509.

Wagenaar, D. A., Hamilton, M. S., Huang, T., Kristan, W. B., & French, K. A. (2010). A hormone-activated central pattern generator for courtship. *Current Biology*, 20, 487–495.

Walker, D. M., Kirson, D., Perez, L. F., & Gore, A. C. (2012). Molecular profiling of postnatal development of the hypothalamus in female and male rats. *Biology of Reproduction*, 87, 19–30.

Wang, S. S.-H., Mitra, P. P., & Clark, D. A. (2002). How did brains evolve? *Nature*, 415, 135–135.

Wang, Z., & De Vries, G. J. (1993). Testosterone effects on paternal behavior and vasopressin immunoreactive projections in prairie voles (Microtus ochrogaster). *Brain Research*, 631, 156–160.

Wang, Z., Young, L. J., De Vries, G. J., & Insel, T. R. (1998). Voles and vasopressin: A review of molecular, cellular, and behavioral studies of pair bonding and paternal behaviors. *Progress in Brain Research*, 119, 483–499.

Whitfield, C. W., Ben-Shahar, Y., Brillet, C., Leoncini, I., Crauser, D., LeConte, Y., ... Robinson, G. E. (2006). Genomic dissection of behavioral maturation in the honey bee. *Proceedings of the National Academy of Sciences of the USA*, 103, 16068–16075.

Wickens, J. R., Budd, C. S., Hyland, B. I., & Arbuthnott, G. W. (2007). Striatal contributions to reward and decision making: Making sense of regional variations in a reiterated processing matrix. *Annals of the New York Academy of Sciences*, 1104, 192–212.

Wingfield, J. C., Hegner, R. E., Dufty, A. M., & Ball, G. F. (2010). The "Challenge hypothesis": Theoretical implications for patterns of testosterone secretion, mating systems and breeding strategies. *American Naturalist*, 136, 829–846.

Wise, P. M., Smith, M. J., Dubal, D. B., Wilson, M. E., Krajnak, K. M., & Rosewell, K. L. (1999). Neuroendocrine influences and repercussions of the menopause. *Endocrine Review*, 20, 243–248.

Wise, R. A. (2004). Dopamine, learning and motivation. *Nature Reviews Neuroscience*, 5, 483–494.

Wise, R. A. (2005). Forebrain substrates of reward and motivation. *Journal of Comparative Neurology*, 493, 115–121.

Withers, G. S., Fahrbach, S. E., & Robinson, G. E. (1993). Selective neuroanatomical plasticity and division-of-labor in the honeybee. *Nature*, 364, 238–240.

Yamamoto, N., Ishikawa, Y., Yoshimoto, M., Xue, H.-G., Bahaxar, N., Sawai, N., ... Ito, H. (2007). A new interpretation on the homology of the teleostean telencephalon based on hodology and a new eversion model. *Brain Behavior & Evolution*, 69, 96–104.

Young, L. J., & Wang, Z. (2004). The neurobiology of pair bonding. *Nature Neuroscience*, 7, 1048–1054.

Young, L. J., Wang, Z., & Insel, T. R. (1998). Neuroendocrine bases of monogamy. *Trends in Neuroscience*, 21, 71–75.

Zhu, H., Gegear, R. J., Casselman, A., Kanginakudru, S., & Reppert, S. M. (2009). Defining behavioral and molecular differences between summer and migratory monarch butterflies. *BMC Biology*, 7, 14.

16

Neural Mechanisms of Communication

Julia Sliwa, Daniel Y. Takahashi, and Stephen V. Shepherd

16.1 We Are Not Alone

Antelopes such as the Thompson's gazelle, observing predators, often engage in 'stotting"—they flee in a series of erratic leaps, legs stiff and straight. This method of escape is inefficient, and is abandoned for the most dangerous predators, but is common in response to slower or less-focused threats. What is happening here? It appears the antelope is sending a signal to its potential predator, who receives it.

All animals live in an environment that includes others. They can be detected by others and can influence the likelihood (and consequences) of this detection by sending signals. Signals are bodily features or behaviors of the signaler that trigger specific behaviors in the receiver. The receiver, signaler, signal, and medium are the four basic building blocks of any communication cycle. Each component can be considered separately, but in the service of communication they are interdependent and defined only in relation to one other. Cycles of reciprocal signal exchange mediate social interactions, but even "asocial" species coordinate reproduction, manage conflict over territory, and may anticipate and influence potential predators and prey.

Communication arose long before the evolution of animals and neurons, yet is a crucial aspect of animal behavior and nervous system evolution. In this chapter, we review general principles of signaling exchange, then detail how these exchanges take place among nonhuman primates and how neural systems act to mediate them. We end by outlining commonalities between primate and nonprimate communication systems, and close by discussing broader implications for the study of social neuroscience.

16.2 The Evolution of Communicative Signals

Signals can be sent through diverse media: Tactile signals are primarily short range, instantaneous, and reciprocal. By contrast, olfactory signals are perceived at spatiotemporal distance (depending on wind/current) and can be displaced from the body. Some media may not be evident to human observers, including infrasonic vibration and ultrasonic hearing, thermal emission of light, and electromagnetic fields generated

Table 16.1 The Relative Spatial Range, Temporal Range, and Reciprocity of Communication Channels by Modality.

	Spatial Range	Temporal Range	Reciprocity
Touch	Local	Instant, transient	Strong
Vision	Easily occluded	Instant, transient	Intermediate
Vibration	Attenuates	Fast, transient	Weak
Chemosensation	Plume (Displaced)	Slow, transient (Enduring)	Weak
Electroception	Attenuates	Instant, transient	Strong

by neuronal activity. Signals can be multimodal. Likewise, receivers can sense signals in diverse modalities and use them all to make inferences about their social environment. Depending on the environments, some signals and not others can be detected. The efficiency, range and directedness of signals vary with their modality and environmental context (see Table 16.1).

Despite signals' diversity, and the diversity of mechanisms required to produce and perceive them, all have common features which determine how they evolve and mediate behavior.

16.2.1 The Building Blocks of a Communication Cycle

Communication requires both a signaler and a receiver. The communicative component that is shared between the signaler and the receivers is the signal: a behavior or bodily transformation of the signaler that influences the receiver's behavior in a given way. For example, consider again the Thompson gazelle's stotting behavior (FitzGibbon & Fanshawe, 1988). Stotting functions as an honest signal: By indicating that the gazelle is aware of the predator and extremely fit, it deters further pursuit. Because the predator prefers not to waste time and energy chasing an alert, athletic animal, both parties benefit—despite the fact one would like the other for lunch.

Not all signals are honest. Among cuttlefish, which can change their shape and their skin color and patterning, males sometimes display typical male courtship patterns to females with one side of the body while simultaneously displaying typical *female* patterns to a rival male with the other—thus fooling the rival into complacence (Brown, Garwood, & Williamson, 2012). In both examples, the signalers influence the receivers into not chasing or attacking them by displaying a signal that triggers a pacific behavior of the receiver.

In many communication situations, a signal triggers a less specific outcome in the receiver: It simply attracts its attention. Often signals feature high contrast from sensory background, behavioral stereotypy, or pattern repetition (Johnstone, 1997), which increase saliency of the performer's state or of its immediate environment to an audience. Sometimes a conspicuous "alerting component" prefixes and attracts receivers' attention to a less-conspicuous signal (Johnstone, 1997). Through different combinations of these components, signals can advertise traits or states of the animal or its environment, and grade from relatively fixed and inflexible to complex and context-dependent.

A *signal* can be defined as a derived feature or behavior (i.e., one that did not exist in ancestors) that adds little of direct adaptive value to the performer, but increases the influence on the receiver. Thus *signaling* can be described as "informing" to exert influence. On the other hand, *receiving* can be described as "interpreting" signals to reduce the receiver's uncertainty about its environment. While signals first arise from incidental information exchanges between signalers and receivers, they become part of the species communicative repertoire either through phylogenetic (evolved) or ontogenetic (learned) processes if they benefit the individuals or their species. In this case, signaling features are made more noticeable through a process of "ritualization" where, the signaling feature is isolated, cued, exaggerated, and/or iterated (Johnstone, 1997).

16.2.2 The Evolution of Communicative Repertoires

Like any organism's features, communicative traits evolve through natural selection (Darwin, 1876). However the evolution of signaling systems takes place simultaneously in both signalers and receivers, and their evolutionary trajectories are incomplete when considered from just one side.

Phylogeny of innate signals has historically been reconstituted by species observation. Darwin observed the expression of emotions across different taxa and found that certain human facial expressions have apparent counterparts in nonhuman primates and other mammals (Darwin, 1872). For example fear would be expressed by wide opening of the mouth and eyes, with upraised eyebrows in many primate species. Some sensory features of innate distress signals, such as aversively loud, irregularlypatterned vocalizations, or overwhelming and persistent scents, also appear to be highly conserved across taxa, both in terms of signaler behavior and receiver response including beating of the heart, trembling of the muscles and cold perspiration (Rendall, Owren, & Ryan, 2009). Studies of behavior in animals raised in isolation, or cross-fostered with other species, directly addressed which signals are determined relatively more by phylogenetic evolution (i.e., are "innate") or relatively more by ontogenetic learning. For example cross-fostered galah cockatoos use their innate alarm calls but use contact calls from their adoptive species of pink cockatoos (Rowley & Chapman, 1986). Most commonly, "innate" signaling behaviors are involved in response to imminent danger, in courtship, and in parental care (Domb & Pagel, 2001; Ghazanfar & Santos, 2004).

The evolution of communication interacts with ontogenetic learning, whereby signal usage is refined during individuals' development. For example young vervet monkeys refine their production of vocalizations, their use in appropriate circumstances, and the response to the vocalizations of others, by observation of adult vervets (Seyfarth & Cheney, 1986). In different species, ontogenetic learning can arise from habituation, Pavlovian association, trial-and-error learning, emulation, imitation, or even active teaching. Ontogenetic learning comprises at least two aspects: (1) developmental learning of species-typical communication patterns and (2) signal production depending on context.

An example of species-typical communication is the use of "semantic" or "referential" signals (Macedonia, 1990; Macedonia & Evans, 1993): Many vertebrates, ranging from monkeys to chickens, produce predator-specific alarm calls (Gyger, Marler, & Pickert, 1987; Pereira & Macedonia, 1991; Seyfarth, Cheney, & Marler, 1980); rhesus monkeys produce specific calls for food discovery/ally recruitment (Gouzoules, Gouzoules, & Marler, 1984; Hauser & Marler, 1993); dolphins and parrots seem to "name" their peers (Wanker, Sugama, & Prinage, 2005; Janik, Sayigh, & Wells, 2006). In some cases animals combine these "semantic" signals in "syntax," in which the meaning of an overall pattern of signals differs from that of their individual parts. Campbell (Ouattara, Lemasson, & Zuberbühler, 2009) or putty-nosed (Arnold & Zuberbühler, 2006) monkeys exchange combinatorial alarm calls. Referential codes have been found to be probabilistic: A "leopard call" is used to signify the presence of a ground predator, but also on occasion in group conflicts—showing that referential calls are not truly indexical but rather are social constructs. However, learning of species-typical signals is most evident on the receivers' side. While alarm calls typically elicit an innate startle from most animals, in some species animals can learn to display specific responses to the type of predator-specific alarm call they hear. For example, Belding's ground squirrels learn to run to the trees at the sound of a "terrestrial predator call" and not to the bushes: They learn to display the same hiding behavior as when they notice the predator themselves (Mateo, 2010). When sight of a specific threat is reliably predicted by a heard and innately arousing signal, learning of an appropriate defensive response is greatly facilitated. In this case initial, innate response to a communication signal is supplemented by further ontogenetically learned association.

The second major type of ontogenetic learning concerns context flexibility in communication. Flexibility in signal production may be evidenced by the generalization of a single signal to novel contexts, by selection among multiple signals within a context, or by the ability to forego signaling in its typical context. Audience effects exemplify how a given signal can be modulated or even inhibited depending on context: Audience effects have been documented in species ranging from chickens (Marler, Dufty, & Pickert, 1986) to bats (Bohn, Smarsh, & Smotherman, 2013), monkeys (Cheney & Seyfarth, 1990; Gouzoules et al., 1984), and chimps (Crockford, Wittig, Roman, Mundry, & Zuberbühler, 2012), all of whom modulate intensity and type of signal production based on the presence of conspecifics. Flexibility is also evident in the signal reception. It includes responding to a same signal differentially depending on who emitted it in what context, as well as the ability to inhibit automatic responses and interpret semantic/syntactic signals. For example, alarm calls emitted by infant baboons will not only be detected and localized by adult baboons, but also interpreted. Adult baboons can infer from a vocalization's acoustical properties the size, the age, and family line of the caller (Cheney & Seyfarth, 1999). With this perceptual ability, baboons respond flexibly to the received call by inhibiting their rescue responses and gazing instead at the juvenile's mother. Here, a motivational cue (the call type) and kinship information (the juvenile's voice) are encoded in the same auditory signal; in other cases, receivers may seek contextual information from other modalities, for example by looking toward the call source (Kirchhof & Hammerschmidt, 2006).

Biological and cultural mechanisms can both propagate communication systems across generations (Levinson, 2006). Until recently, there was little evidence for within-species cultural variation, in the sense that all members of a species were observed to learn and use the same referential calls (Owren, Dieter, Seyfarth, & Cheney, 1993; Wheeler & Fischer, 2012). However, it is only recently that social practices of different groups of individuals within given species have been compared. For example, we now know the mating songs of humpback whales are shared by all males of a given population but differ between groups; moreover, songs can undergo cultural revolution when foreign singers join a local population (Noad, Cato, Bryden,

Jenner, & Jenner, 2000). Cultural transmission can also apply to referential signals, as illustrated when a population of chimpanzees using a particular "food call" abandoned it upon merging with a new group, adopting the call of their hosts (Watson et al., 2015). Cultural transmission feeds back, in turn, upon the transmission of genes. For instance, female killer whales more readily accept sexual advances from males who sing a new dialect rather than males who sing their pod dialect (Baird & Whitehead, 2000; Rendell & Whitehead, 2001). Conversely, Darwin finches mate preferentially with individuals singing their own songs, facilitating speciation along cultural lines within a common environment (Grant & Grant, 1996).

16.3 Primates

Most primates gather to sleep in shared or clustered shelters, and live in groups characterized by complex family and friendship bonds, in which shifting alliances are important in both within- and between-group competition. These conditions are paralleled by routine exchange of signals across multiple sensory channels and by the use of flexible communication, which we will describe below. As such, primates exhibit neural mechanisms for a diverse range of communication systems.

16.3.1 Communicative Repertoire

Like most mammals, primates communicate through olfactory signals found in urine, saliva, scat, vaginal secretions, and sometimes in specialized scent glands. Olfactory signaling is prominent among primates with scent glands, and continues to influence behavior in all species. While odors can travel passively through the air to reach a receiver who is close to the signaler, species with scent glands also engage in "scent marking"— actively applying odorants to environmental substrates to indicate territorial borders (Heymann, 2006). Olfactory signaling also plays an important role in primate courtship and intrasexual competition. Because olfactory signals reliably encode individual phenotype (Smith, 2006) by association with the major histocompatibility system (Knapp, Robson, & Waterhouse, 2006), they advertise signalers' gender, kinship, dominance, and reproductive status. This information is used by receivers, primarily to guide mating behavior (e.g., kin avoidance: Boulet, Charpentier, & Drea, 2009; Olsson, Barnard, & Turri, 2006; Porter, 1998; Weisfeld, Czilli, Phillips, Gall, & Lichtman, 2003).

Touch is the first sense to develop in primates, and touching behavior during nursing and mating is later elaborated to include hugging, grooming, sociosexual behavior, and rough-and-tumble play, each of which have important roles in primate social attachment and normal development (Suomi, Harlow, & Kimball, 1971). Interpersonal touch is crucial for emotional well-being and prosocial behavior in large primate groups, and its prototypical form is social grooming. Social grooming evolved from a cleaning behavior, originally used to remove ticks and parasites, into a social bonding ritual which plays an important role in reducing stress and cementing alliances (Dunbar, 1991, 2010). Conversely, aggressive touch (kicking, biting, slapping, or clasping) not only directly injures or impedes the target but also establishes psychological dominance. Both types of touch pattern relationships within the group (De Waal & Yoshihara, 1983; Judge & De Waal, 1997). In certain human and ape cultures, additional forms of touch have been ontogenetically ritualized (e.g., the handshake used as a greeting). Touch is susceptible to audience effects—sex, rank, kin, and available time influence social grooming (Schino, 2001; Seyfarth, 1977).

Primates also produce postural, orofacial, and manual behaviors which can be seen and/or heard by receivers. Gestural communication includes *expressions* (e.g. Atkinson, Dittrich, Gemmell, & Young, 2004; de Gelder, 2009; de Meijer, 1989; van Hooff, 1967), *attention getters*, and *pointing*.

- Attention-getting and solicitation signals are used flexibly, are sensitive to the audience, and illustrate ontogenetic ritualization (Call & Tomasello, 2007); they are reportedly used deceptively in mangabeys (Coussi-Korbel, 1994). These signals attract attention and may imply communicative intent.
- Expressions comprise facial and bodily postures, primarily expressing behavioral states of *avoidance, affiliation*, or *aggression* (Partan, 2002). In macaques, facial expressions from each category include the silent-bared teeth display or "fear grin," the lipsmack, and the open mouth threat, respectively (Parr & Heintz, 2009).
- Pointing signals defined broadly as ritualized orienting behaviors, include exaggerated stance, reach, or stare that coordinate attention (Meunier, Prieur, & Vauclair, 2013; Shepherd, 2010; Shepherd & Cappuccio, 2011).

Most primates have been shown to not only identify their peers' displays but also the signalers' status including dominance, kinship, group membership, age, gender, and reproductive status through their facial and body features (Deaner, Khera, & Platt, 2005; Gerald, Waitt, & Little, 2009; Mahajan et al., 2011; Pokorny & de Waal, 2009; Schell, Rieck, Schell, Hammerschmidt, & Fischer, 2011; Sliwa, Duhamel, Pascalis, Wirth, 2011; Waitt et al., 2003; Waitt, Gerald, Little, & Kraiselburd, 2006).

Facial and bodily expressions sometimes include vocal displays, and attention getters often include an acoustic component. The three main classes of expression are usually accompanied by different types of vocalizations. Species-specific vocalizations can be produced automatically as a function of an individual's behavioral state; however, call production and response to heard calls can also be modulated by audience, as shown through playback experiments (Cheney & Seyfarth, 1997; Seyfarth et al., 1980). Many primates have been shown to recognize the species, identity, group membership, size, age, and kinship of the signalers (Adachi, Kuwahata, Fujita, Tomonaga, & Matsuzawa, 2006; Bachorowski & Owren, 1999; Ghazanfar et al., 2007; Rendall, Rodman, & Emond, 1996; Sliwa et al., 2011), through the formants present in their calls (Fant, 1960; Fitch & Fritz, 2006).

Humans also communicate through language. Language is amodal: It can be spoken and heard, signed and seen, written and read (by eye or, in Braille, by touch). In language, arbitrary symbols are sequenced to encode conceptual relations, called semantically compositional syntax, which has only been found so far in language (Petkov & Wilson 2012; Wheeler & Fischer, 2012). It has been suggested that language ability was facilitated by increasing use of communicative gesture (Arbib, Liebal, & Pika, 2008; Shepherd & Cappuccio, 2011), and as an adaptive response to (or even a prerequisite for) ever-increasing social complexity (Byrne, 1996; Dunbar, 1992). Language may also directly strengthen social alliances, since human seem largely to have replaced social grooming with small talk (Dunbar, 2010). Attempts to teach language to human-raised great apes showed that, while they fail to learn humanlike grammar (Terrace, Petitto, Sanders, & Bever, 1979; Yang, 2013), they possess the ability to learn to use arbitrary signs and symbols to communicate. Field studies also described both functional reference and proto-syntax in diverse nonhuman species.

Primates display and respond to a wide range of communicative signals, from fast automatic attention-getters to complex, multimodal, and flexible referential signs—many are multisensory. How are these signals produced and perceived by primate brains? To what extent are the neural mechanisms of communication modular, and how are those modules organized? Only a few primate models have received intensive study (e.g., rhesus and long-tailed macaques, marmosets, and humans), yet most primates likely possess specialized neural pathways dedicated to the production and perception of communicative signals.

16.3.2 Neural Mechanisms of Signal Production

In this section, we briefly discuss neural mechanisms by which two specific, relatively well-studied primate signals are produced: facial expressions and vocalizations.

16.3.2.1 Producing facial expressions. One of the most distinctive features of primate communication, compared to that of other mammals, is the central importance of orofacial movements. Primates use a relatively conserved suite of facial muscles (Diogo, Wood, Aziz, & Burrows, 2009) to produce stereotyped species-typical facial postures and movements that communicate behavioral state (Ekman, Sorenson, & Friesen, 1969; van Hooff, 1967). The facial muscles are controlled mainly by two cranial nerves: the trigeminal (V), which primarily innervates jaw muscles related to ingestion, and the facial (VII), which innervates the more superficial "mimetic" muscles. The final common output for facial motor control is the facial nucleus, which (uniquely in anthropoids) receives projections directly from motor cortex (Sherwood et al., 2005). In total, five interconnected facial motor representations exert cortical control over movement; they reside in the primary and supplemental, rostral and caudal cingulate, and ventrolateral prefrontal motor areas (Morecraft, Louie, Herrick, & Stilwell-Morecraft, 2001).

Symptoms of human brain lesions suggest that neural governance of voluntary and automatic facial movements are dissociable (Hopf, Müller-Forell, & Hopf, 1992; Morecraft, Stilwell–Morecraft, & Rossing, 2004). In *volitional* facial paralysis, patients are impaired at voluntarily moving their facial muscles, for example, in order to speak, but are spared when responding reflexively to emotionally provocative stimuli. This condition is associated with lesions in the motor cortical area M1 and its underlying white matter. The opposite condition is *emotional* facial paralysis, in which a patient ceases to produce spontaneous facial expressions on one side of the face, though they can voluntarily approximate them with effort. This condition is usually related to lesions involving the anterior cingulate cortex (ACC), insula, thalamus, striatocapsular region, and pons. These clinical reports suggest the existence of separate emotional and voluntary facial movement centers, in which nonspeech orofacial signaling is governed reflexively by medial motor representations, while speech movements are governed by goal-directed ventrolateral representations.

The neural circuits which govern the production of different communicative facial movements across different social contexts are less well understood. Although limbic system activity is involved in the generation of facial expressions (Weintstein & Bender, 1943), its specific mechanisms remain elusive. Among the limbic and paralimbic areas related to facial movements, the anterior cingulate cortex (ACC) is one of the most studied. Stimulation of ACC can induce both affective facial expressions and vocalization (Smith, 1945), while lesions of ACC are associated with decreased spontaneous vocalization and flattened facial affect (Devinsky et al., 1995); neural activity related to affective vocalization has been recorded both in the anterior ACC gyrus and in the ACC sulcus near facial motor areas (West & Larson, 1995). Because the ACC is also strongly implicated in the processing of reward contingencies, it may transform perceived social risks and rewards into appropriately responsive social behavior (Behrens, Hunt, & Rushworth, 2009; Paus, 2001; Rushworth, Behrens, Rudebeck, & Walton, 2007). In particular, information encoded in the ACC gyrus, where lesions disrupt normal social valuation signals (Rudebeck, Buckley, Walton, & Rushworth, 2006), may influence the adjacent motor representations in the anterior cingulate sulcus.

The ACC is part of a wider network involved in governing adaptive behavior, including orbitofrontal cortex (OFC), premotor cortex, somatosensory cortex and the amygdala. Interestingly, it has been shown that the amygdala also initiates adaptive production of facial signals via its connections to the motor systems (Livneh, Resnik, Shohat, & Paz, 2012).

16.3.2.2 Producing vocalizations. Primate vocalization is the result of coordinated action by several separate effectors, including the diaphragm, chest muscles, ribs, lungs, larynx, and upper vocal cavity. Each of these effectors is responsive to sensory feedback and governed by specific neural nuclei. Respiratory muscles are controlled by motor neurons in the ventral horn of the thoracic and upper lumbar spinal cord, while internal laryngeal muscles are controlled by motor neurons in the nucleus ambiguus. Cell groups below the hypoglossal nucleus control the external laryngeal muscles, and the articulatory orofacial muscles are controlled by motor neurons of the facial nucleus, trigeminal nucleus, ventral horn of the uppermost cervical cord, nucleus ambiguus, and hypoglossal nucleus (see Jürgens (2009) for details). There are very few direct projections between these different motor neuron pools, which suggests that outside brain regions must act to coordinate them. Consistent with this hypothesis, anatomical, microstimulation, recording, and lesioning studies have identified the lateral reticular formation and nucleus retorambiguus as coordinating areas (Hage & Jürgens, 2006; Hannig & Jürgens, 2006; Jürgens, 2000; Lüthe et al., 2000). The reticular formation has direct connections to all phonatory motor neurons, and contains neurons that show activity correlated with the duration and frequency modulation of vocalizations (Kirzinger & Jüergens, 1991). The nucleus retroambiguus is a relay to respiratory centers with direct connections to parabrachial regions; it and the lateral reticular formation are reciprocally connected (Mantyh, 1982; Vanderhorst et al., 2000).

Both of these brainstem coordinating areas receive input from the periaquedutal gray (PAG). Electrical stimulation of PAG produces vocalizations resembling naturally produced species-specific calls, different from the more artificial-sounding vocalizations produced when the lateral reticular formation alone is stimulated (Jürgens & Ploog, 1970). More than half of the neurons in PAG increase their neural activity before vocal onset. Moreover, most of their neural activity is not correlated with the specific acoustic features of vocalizations (Dusterhoft, 2004). These data suggest that PAG works as a vocal initiator rather than a vocal pattern generator

(Hage & Jürgens, 2006). PAG also works as a relay station for sensory-motor interaction, having strong reciprocal connections to superior collicullus and paraleminiscal area. Finally, PAG receives several inputs from limbic structures, including the ACC, amygdala, hypothalamus, hippocampus, midline thalamus, nucleus accumbens, nucleus striae terminalis, preoptic areas, septum, and subcallosal gyrus (Dujardin & Jürgens, 2005, 2006). Lesioning these limbic structures abolishes or significantly reduces their spontaneous vocal production, which can be rescued by direct stimulation of PAG (Jürgens & Pratt, 1979). These structures are therefore likely to initiate and modulate the vocal production and communication, and are in turn likely modulated by activity elsewhere in frontal cortex (see Figure 16.1). This feedforward and hierarchical view is no doubt an oversimplification, not least because sensory feedbacks plays a crucial role at every step of vocal production: For example, if both vagal nerves are cut, disrupting somatosensory feedback from the lungs, then vocalization cannot be elicited by PAG stimulation (Nakazawa et al., 1997).

Although humans and nonhuman primates share these same basic anatomical and neural structures for signaling, some interesting differences are apparent. Humans produce more distinct sound elements than other primates, including distinct consonants and vowels, using specialized movements of lips, tongue, and respiratory muscles (Maclarnon & Hewitt, 2004). One theory holds that that the uniquely-human "descended larynx," which enlarged the human vocal tract relative to other primates, increased the variety of speech formants which could be produced by subtle tongue movement (Fitch, 2000). Moreover, major neural differences between nonhuman primates and humans may explain the latter's relatively greater vocal flexibility. In particular, while extensive lesions to Broca's area homologues do not significantly alter vocal production and communication in macaques (Kirzinger & Jürgens, 1982), they lead in humans to significant speech disorders, including mutism (Trupe et al., 2013) Electrical stimulation of ACC systematically induces vocalization in monkeys (Jürgens & Ploog, 1970; Robinson, 1967; Smith, 1945), but stimulation of ACC in humans has not produced reliable vocalization in humans (Pool, 1954; Pool & Ransohoff, 1949; Talairach et al., 1973), although there are reports of momentary speech arrest (Lewin & Whitty, 1960, and production of mirth and laughter (Caruana et al., 2015). Electrical stimulation of orofacial regions of human primary motor cortex consistently induces vocalization, but repeated attempts in monkeys failed (Jürgens, 1974; Jürgens & Ploog, 1970; Robinson, 1967). These data suggest that the neocortex exerts stronger control over human speech than other communicative signals shared across species, but the exact nature of this difference remains elusive. Even in nonhuman primates, it is likely that cortical centers contribute to the ability to modulate otherwise-reflexive communicative signals based on learned contexts and contingencies.

The primates' ability to produce diverse audiovisual signals necessitates machinery for processing these signals. This suggests the following questions: How do brains read communication signals? How are they processed at the neuronal level in the receiver's brain? What features and feature conjunctions must they encode to do so?

16.3.3 Neural Mechanisms of Signal Perception

Far more is known about mechanisms for signal perception in primates than for signal production. Signal processing in primate brains begins in several pathways representing distinct sensory modalities—though we will see that these modalities are

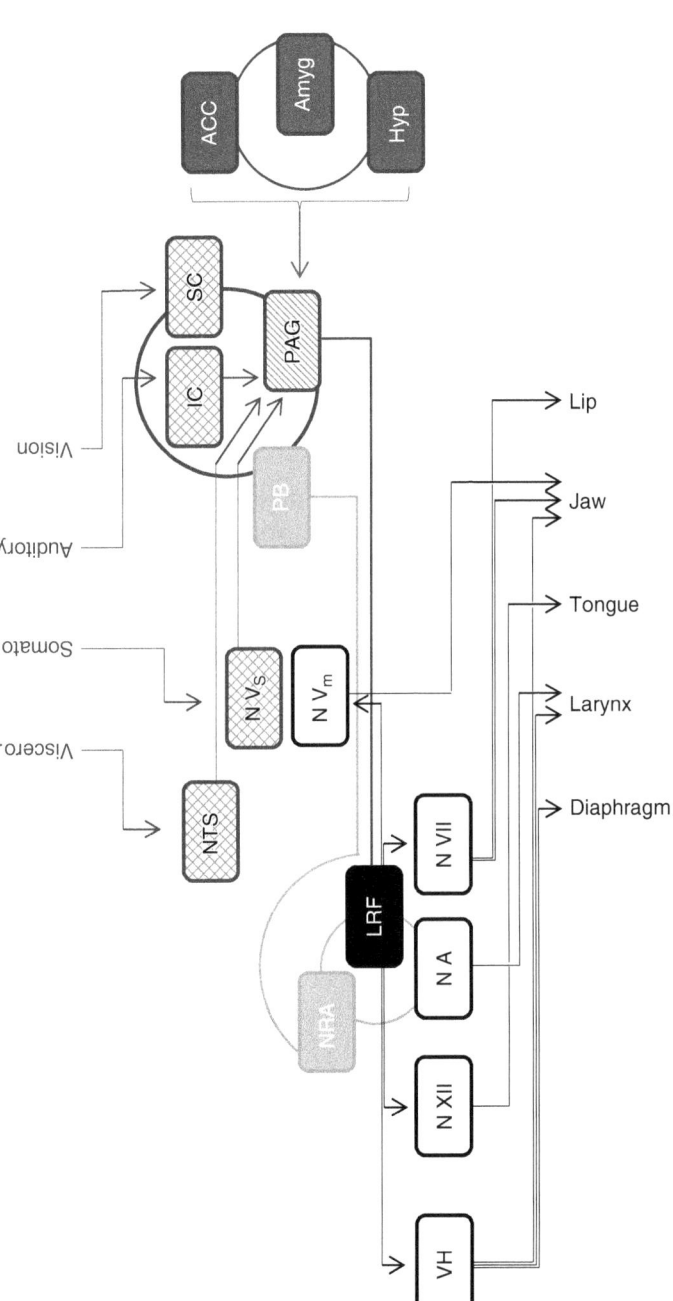

Figure 16.1 Schematic Diagram of the Main Connections between Brain Areas in Pathways Leading to Vocalization. Motor output nuclei are shown in black outline (VH., ventral horn of the medulla; N XII, hypoglossal nucleus; N A, nucleus ambiguus, N VII, facial nucleus; N V$_m$, motor trigeminal nucleus). Respiratory control centers and connections are in ▨ (NRA, nucleus retroambiguus; PB, parabrachial region). The lateral reticular formation (LRF, solid black) plays a key coordinating role among these motor nuclei, while the periaqueductal gray (▨, PAG) plays a crucial role in coordinating behavioral state and initiating affective vocalization. Descending executive control from the limbic system is shown in ▨ (ACC, anterior cingulate cortex; Amyg, amygdala; Hypo, hypothalamus); while incoming sensory input from visceromotor, somatosensory, auditory and visual stimuli (including motor feedback) are shown in ▨ (NTS, solitary nucleus; N V$_s$, sensory trigeminal nucleus; IC, inferior colliculus; SC = superior colliculus). Finally, it must be noted that a crucial projection is omitted: uniquely, anthropoid primates have direct connections from motor cortex to the motor nuclei governing face and jaw movements.

ultimately integrated. While reflexive responses (including orienting toward attention-getters) are processed mainly through subcortical structures, flexible responses rely heavily on cortical processing.

16.3.3.1 Processing signals that influence attention. Attention-getters indicate *where* a signal is sent from. They are processed rapidly by generic pathways, as illustrated by fast orientating to the location of an abrupt sound or movement such as a cleared throat or waved hand. This reflexive orienting behavior is computed in structures that map between sensory and motor reference frames (see Figure 16.2); the earliest include the inferior and superior colliculi (Jay & Sparks, 1984). Similarly, touch gestures, such as the "poke at" or "throw chips" gestures of chimpanzees, can be detected by somatosensory pathways and transformed into orienting coordinates, again by the superior colliculus (Groh & Sparks, 1996). These pathways are used in social signal processing, but are not specific to it. Interestingly, however, socially dedicated subcortical systems contribute to orienting responses (e.g., to faces and eyes), typically through the superior colliculus in concert with the pulvinar nucleus of the thalamus and the amygdala (Johnson, 2005; Sewards & Sewards, 2002; Vuilleumier, Armony, Driver, & Dolan, 2003).

While attention-getters elicit fast reflexive behavior, they and other signals are concomitantly processed through relatively slow cortical pathways. Body, head, eye, and hand direction can trigger sophisticated attention shifts in the receiver. These flexible orienting behaviors require geometric computation of line-of-sight, and involve neuronal computation occurring in cortical networks. For gaze, this network includes the lateral intraparietal cortex (LIP), dorsal posterior inferotemporal corex (PITd) and the posterior superior temporal sulcus (STS) of macaques (Freiwald & Tsao, 2010; Marciniak, Atabaki, Dicke, & Thier, 2014; Roy, Shepherd, & Platt, 2012; Shepherd et al., 2009) and their putative homologues, the intraparietal sulcus and pSTS of humans (Hoffman & Haxby, 2000; Pelphrey, Morris, & McCarthy, 2005). These networks both read the orientation of the signaler and compute the orientation behavior to be undertaken by the receiver, a seeming "mirror mechanism" which will be discussed in more depth in §16.3.3.4.

16.3.3.2 Subcortical processing of communicative signals. Neural processing of phylogenetically conserved signals frequently involves subcortical networks of brain areas (see Figure 16.2) (Sewards & Sewards, 2002). A large body of work has examined the involvement of the amygdala in automatic processing of fear and other emotions in human facial displays (Breiter et al., 1996; Costafreda, Brammer, David, & Fu, 2008; Morris et al., 1996), voices (Dolan, Morris, & de Gelder et al., 2001; Fecteau, Belin, Joanette, & Armony, 2007; Sander & Scheich, 2001), and body postures and movements (De Gelder, Snyder, Greve, Gerard, & Hadjikhani, 2004). Studies in rhesus monkey suggest this pathway is broadly conserved (Gil-da-Costa et al., 2004; Hadj-Bouziane et al., 2012; Hoffman, Gothard, Schmid, & Logothetis et al., 2007; Petkov et al., 2008), and examined the computations performed by individual amygdala neurons (Gothard, Battaglia, Erickson, Spitler, & Amaral, 2007; Kuraoka & Nakamura, 2006, 2007; Leonard, Rolls, Wilson, & Baylis, 1985).

One of the least understood social signaling systems in primate brains is chemosensory. Accessory olfaction is defined as the chemoreceptive system that employs the vomeronasal organ (VNO) and its distinct central projections to the accessory

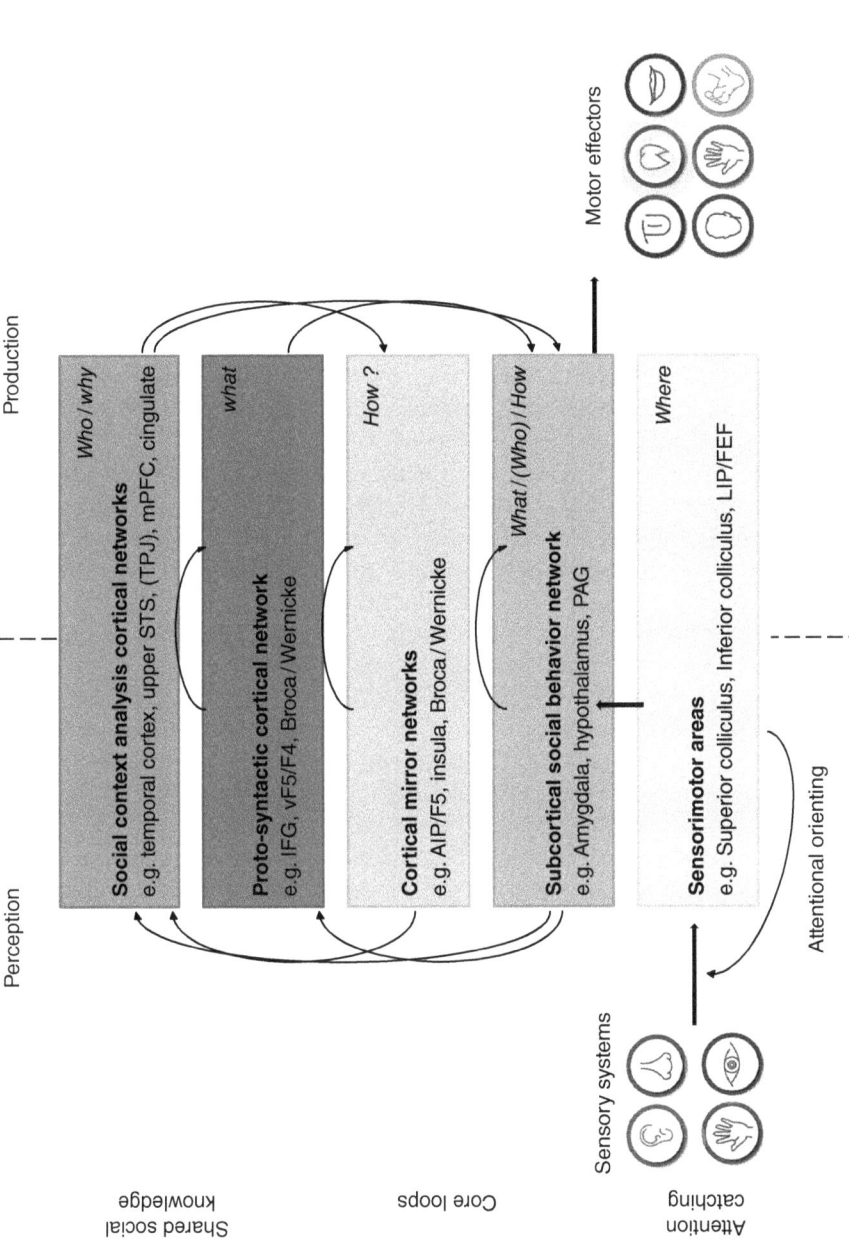

Figure 16.2 Schematic Representation of the Cognitive Processes Involved in Primates' Perceiving and Producing Communicative Signals, and Their Associated Neural Substrate.

The cognitive processes depicted include: *Where the signal comes from*, by sensorimotor areas, *What is being signaled*, by the subcortical social behavior network and the proto-syntactic cortical network, *Who is signaling* and *Why*, by social context analysis cortical networks, and *How the signal is transmitted*, by cortical mirror networks. The figure highlights that the perception and production parts of these different cognitive processes are largely integrated, with the exception of early sensorimotor areas, which are primarily involved in detecting signals and orienting attention. The arrows show streams of signal processing and depict how these different processes interact with and modulate each other.

olfactory bulb (AOB) and limbic/cortical systems. Vomeronasal function is maintained within the Strepsirrhines and tarsiers, reduced in Platyrrhines, and mostly absent in the Catarrhines including humans (Evans, 2006). However, even in humans, in whom the VNO is entirely vestigial (Wyatt, 2014), some hormone-like scents cause sex-differentiated hypothalamic activation (Savic, Berglund, Gulyas, & Roland, 2001). Similarly, in marmosets, sexually arousing odors of ovulating monkeys enhanced activity in the preoptic area and anterior hypothalamus (Ferris et al., 2001), striatum, septum, periaqueductal gray, and cerebellum (Ferris et al., 2004) compared to the odors of ovariectomized monkeys.

16.3.3.3 Cortical networks processing communicative signals. Cortical networks enable extensive computation based on learned patterns and remembered experiences. In primates, this includes flexible analysis of *what is expressed* by a face, body or vocalization and integration with its social context (*who* is expressing it and *why*). The cortical processing pathways operate largely in parallel with subcortical pathways (see Figure 16.2) and advance hierarchically from primary sensory areas toward increasingly complex and integrative neural representations in the anterior temporal lobe and prefrontal cortex (Rosa & Tweedale, 2005).

Studies in the 1970s identified the existence of neurons preferentially active when the macaque subjects were seeing a face (Gross, Bender, & Rocha-Miranda, 1969; Gross, Rocha-Miranda, & Bender, 1972). These neurons were seemingly scattered throughout the temporal lobe, but are now known to cluster into a specialized cortical network for face perception (Bell et al., 2011; Tsao, Freiwald, Tootell., & Livingstone, 2006). This network has been imaged in humans (Haxby et al., 1994; Kanwisher, McDermott, & Chun, 1997 Kriegeskorte, Formisano, Sorger, & Goebel, 2007; McCarthy, Puce, Gore, & Allison, 1997; Puce, Allison, Gore, & McCarthy, 1995; Sergent, Ohta, & MacDonald, 1992; Tsao, Moeller, & Freiwald, 2008), chimpanzees (Parr, Hecht, Barks, Preuss, & Votaw, 2009), macaques (Ku, Logothetis, & Goense, 2011; Logothetis, Guggenberger, Peled, & Pauls, 1999; Moeller, Freiwald, & Tsao, 2008; Tsao, Freiwald, Knutsen, Mandeville, & Tootell, 2003) and marmosets (Hung et al., 2015). In all of these species, it runs through similar regions of the temporal lobe, suggesting cortical specialization for face processing evolved prior to the split between New and Old World monkeys. However, significant differences exist: For instance, face-processing areas discovered in the orbital and ventrolateral frontal cortex of macaques (Tsao, Schweers, Moeller, & Freiwald, 2008) are less evident in humans. The cortical areas devoted to faces are abutted and partially overlapped by patches specialized for bodies, as shown in both humans (Downing et al., 2001) and macaques (imaging: Pinsk, DeSimone, Moore, Gross, & Kastner, 2005; Pinsk et al., 2009; electrophysiology: Bell et al., 2011; Perrett, Smith, Mistlin et al., 1985; Popivanov, Jastorff, Vanduffel, & Vogels, 2014).

Voice processing pathways parallels that for face processing, running dorsally through the temporal lobe in humans (Belin, Zatorre, & Ahad, 2002; Belin, Zatorre, Lafaille, Ahad, & Pike, 2000; Binder et al., 2000; DéMonet, Jiang, Shuman, & Kanwisher, 1992), chimpanzees (Taglialatela, Russell, Schaeffer, & Hopkins, 2009), macaques (Ghazanfar & Rendall, 2008; Gil-da-Costa et al., 2004, 2006; Petkov et al., 2008; Poremba et al., 2004) and marmosets (Sadagopan, Temiz-Karayol, & Voss, 2015). Voice-selective "call detector" neurons were first recorded in squirrel and marmoset monkeys (Wang, 2000; Wang & Kadia, 2001; Wang, Merzenich,

Beitel, & Schreiner, 1995). Voice-selective neurons were next described in the macaque monkey (Kikuchi, Horwitz, & Mishkin, 2010; Rauschecker, Tian, & Hauser, 1995; Recanzone, 2008; Russ, Ackelson, Baker, & Cohen, 2008; Tian, Reser, Durham, Kustov, & Rauschecker, 2001), principally clustered within fMRI-identified voice-selective areas (Perrodin, Kayser, Logothetis, & Petkov, 2011) but also in the insula (Remedios, Logothetis, & Kayser, 2009a) and in prefrontal and orbitofrontal cortex (Cohen, Russ, Gifford, Kiringoda, & MacLean, 2004; Rolls, Critchley, Browning, & Inoue, 2006; Romanski & Goldman-Rakic, 2002). These areas also contribute to the perception of non-vocal auditory social signals: Macaque drumming (an auditory attention-getter) activates both the amygdala and cortical "voice" areas in macaques (Remedios, Logothetis, & Kayser, 2009b).

Chemosensory and tactile cortical pathways have received far less exploration than those for vision and audition, but likely follow similar patterns for flexible odors and touch processing. Cortical processing of chemosensory signals occurs in piriform and insular cortex, extending into the anterior temporal and ventral frontal lobes (Gottfried, 2010), where they may interact with the auditory and visual social processing streams for more nuanced processing of behavioral relevance (Ferris et al., 2004; Pause, 2012). Cortical processing of somatosensory signals likely arises in somatosensory cortex in the parietal lobe, adjacent to and interacting with frontal motor cortex, while affective aspects of somatosensation appear to additionally involve representations in the posterior insula along with the orbitofrontal cortex, anterior insula and anterior cingulate (Gazzola et al., 2012; Gordon et al., 2011; Keysers, Kaas, & Gazzola, 2010).

16.3.3.3.1 Perceiving traits: identity and status. How do neurons from these networks code for essential social information present in faces, bodies and voices about *who* is sending a signal (including identity, gender, age, dominance status, species, kinship, familiarity, sexual state, attractiveness, health state, etc., see Figure 16.2)?

Neuronal activity to faces becomes less dependent on specific viewpoint anteriorly in the temporal lobe (De Souza, Eifuku, Tamura, Nishijo, & Ono, 2005; Freiwald & Tsao, 2010; Perrett, Smith, Potter et al., 1985), suggesting that these anterior areas represent categorical or individual identity (Kriegeskorte et al., 2007; Rotshtein, Henson, Treves, Driver, & Dolan, 2005; Sergent et al., 1992). Within these areas, physical identity appears to be encoded into a "face-map" in a norm-based rather than template-based manner (Freiwald, Tsao, & Livingstone, 2009; Leopold, Bondar, & Giese, 2006), likely built up based on faces perceived through experience (Sugita, 2008). Accordingly, person familiarity is processed by anterior temporal parts of face-selective cortex (Eifuku, De Souza, Nakata, Ono, & Tamura, 2011; Eifuku, Nakata, Sugimori, Ono, & Tamura, 2010; Sliwa, Planté, Duhamel, & Wirth, 2016; Sugiura et al., 2001) and adjacent mnemonic hippocampal complex (Denkova, Botzung, & Manning, 2006; Leveroni et al., 2000; Sliwa et al., 2016). Voice areas, like face areas, appear to process identity in the most anterior regions of the cortical voice-dedicated network in both humans (Andics, McQueen, Petersson, Gál, Rudas, & Vidnyánszky, 2010; Imaizumi et al., 1997; Nakamura et al., 2001) and macaques (Kikuchi et al., 2010; Perrodin et al., 2011; Petkov et al., 2008). Also, similarly, familiar voices activate the para-hippocampal complex (Nakamura et al., 2001; Sliwa et al., 2014; von Kriegstein & Giraud, 2004). Processing identity from body features has been relatively less investigated, but since cortical body areas are located at similar positions in the visual processing hierarchy to the face areas, anterior body regions are expected to extract identity. Accordingly, the human fusiform body area has been found to process static more than dynamic aspects of body appearance (Vangeneugden, Peelen, Tadin, & Battelli, 2014).

Another property of the signaler that is processed by the receiver is the former's attractiveness. Face attractiveness is associated in humans with orbitofrontal cortex, close to regions devoted to processing reward (Ishai, 2007; Kim, Adolphs, O'Doherty, & Shimojo, 2007; Nakamura et al., 1998). Similarly, bodily attractiveness engages orbitofrontal cortex in humans (Sescousse, Redouté, & Dreher, 2010) and may similarly modulate frontal-lobe responses to bodies within other primates.

16.3.3.3.2 Perceiving states: actions and affect. Faces, bodies and voices also signal the *current internal state* of the signaler, perception of which engages not only the amygdala but also the dedicated cortical processing networks (Figure 16.2).

Expressive body postures and movements activate a large network of human cortical areas interconnected with the amygdala, including the body areas of the fusiform gyrus and superior temporal sulcus (STS) (de Gelder, 2006), with the STS especially involved in processing dynamic body features (Vangeneugden et al., 2014). Likewise in macaques an fMRI study found that body areas in the STS are preferentially sensitive to threat signals (de Gelder & Partan, 2009). Facial movements (Furl, Henson, Friston, & Calder, 2014; Pitcher, Dilks, Saxe, Triantafyllou, & Kanwisher, 2011) including gaze (Allison, Puce, & McCarthy, 2000, Hoffman & Haxby, 2000), lip movements (Puce, Allison, Bentin, Gore, & McCarthy, 1998), and facial expressions (Harris, Young, & Andrews, 2012) also selectively activate the STS in humans. In the monkey, homologue regions in the upper STS appear specialized for dynamic facial features (Fisher & Freiwald, 2015; Polosecki et al., 2013). However while neurophysiological recordings in macaques found that neurons in the upper STS are sensitive to and categorize facial expressions (Hasselmo, Rolls, & Baylis, 1989; Sugase, Yamane, Ueno, & Kawano, 1999), fMRI studies of macaques have failed to constrain them to single cortical regions (Furl, Hadj-Bouziane, Liu, Averbeck, & Ungerleider, 2012; Hadj-Bouziane, Bell, Knusten, Ungerleider, & Tootell, 2008; Hadj-Bouziane et al., 2012; Janssens, Zhu, Popivanov, & Vanduffel, 2014; Morin, Hadj-Bouziane, Stokes, Ungerleider, & Bell, 2014; Polosecki et al., 2013; Zhu et al., 2013). Voices also convey the signaler's current state. The ventromedial prefrontal cortex has been found in macaques to host neurons categorizing the different types of vocalizations according to their emotional meaning (Cohen et al., 2004; Cohen, Hauser, & Russ, 2006; Cohen, Theunissen, & Russ, 2007; Gifford, MacLean, Hauser, & Cohen, 2005; Russ & Cohen, 2009; Russ et al., 2008; Russ, Lee, & Cohen, 2007), and similarly prefrontal regions in humans were found to process emotional vocal sounds (Imaizumi et al., 1997).

Importantly, however, this prefrontal processing of emotional utterances in humans is segregated from the pathways processing language and more broadly semantic or indexical non-emotional signals (von Kriegstein, Eger, Kleinschmidt, & Giraud, 2003; von Kriegstein & Giraud, 2004 Vouloumanos, Kiehl, Werker, & Liddle, 2001). Those pathways would have evolved independently (Petkov, Logothetis, & Obleser, 2009), and might be shared across modalities. For instance, nonemotional communicative content in bodily gesture engages the human posterior STS cortical regions but also left inferior frontal regions implicated in semantic processing (Holle, Gunter, Rüschemeyer, Hennenlotter, & Iacoboni, 2008; Straube, Green, Bromberger, & Kircher, 2011). Homologous brain regions in other primates might be involved in proto-syntactic learning (Petkov & Wilson, 2012).

It should be emphasized that multisensory integration is a prominent feature of primate signaling (Ghazanfar & Santos, 2004; Partan, 2002). This is evident from the grouping of similar functional processes in nearby brain regions irrespective of signal modality, as described above. And it is also manifested by the existence of large specialized brain region, such as dorsal regions of the superior temporal sulcus, that are maximally engaged when signals are presented in multiple modalities (Beauchamp, Argall, Bodurka, Duyn, & Martin, 2004; Chandrasekaran & Ghazanfar, 2009; Dahl, Logothetis, & Kayser, 2009).

16.3.4 Integrative Mechanisms of Communication

The framework we have outlined—with separate perception and production of communication signals—only gets us so far. Indeed, we have already noted that some crucial nodes, such as the amygdala, appear to be engaged in both the perception *and* production of communicative signals, enabling rapid responses to urgent signals. Existence of such integrated loops, with nodes having overlapping functions, illustrates Darwin's suggestion of a direct link between the production of facial or body expressions and the emotions felt, for instance, in response to a perceived signal (Darwin, 1872). As easy as it is to link an emotion with an antecedent perception, this view underlines that emotions are just as linked to observable behavioral responses, instead of being private and abstract mental states.

An example illustrating how signal perception, internal feeling, and signal production can be strongly interlinked, both behaviorally and neurally, is the communication of disgust. It has been found that all three components selectively activate a particular cortical region: the insula. Observing a disgusted face reflexively triggers the production of a disgusted face and is accompanied by a feeling of disgust (Lundqvist & Dimberg, 1995)—meaning the three experiences are, to some extent, inseparable. We seem to reflexively "mirror" the affective states of others.

Neuronal signatures of mirroring are widespread within the cortex. The initial case studied gesture understanding within parietal and frontal regions of macaque monkeys. Neurons in these regions were known to be involved in guidance of motor action, but in 1992 Rizzolatti and colleagues announced that some of these neurons are also sensitive to the observed actions of others (di Pellegrino, Fadiga, Fogassi, Gallese, & Rizzolatti, 1992). Later, mirror properties appeared to be common in other primate species as well, and potentially nonprimates (e.g. birds: Prather et al., 2008). "Mirroring" may play a role in representing goals, predicting actions, or motivating imitative, cooperative and competitive behavior (Cook, Bird, Catmur, Press, & Heyes, 2014; Rizzolatti & Craighero, 2004) all of which are present in communication. As noted in §16.3.3.1, mirror neurons are active in gaze following, suggesting a role in communication of attentional states (Shepherd, Klein, Deaner, & Platt, 2009).

All social animals coordinate their movements; some additionally coordinate their actions toward the external world. Mirror activity may thus have general relevance to social coordination, for example by mediating recursive cognitive "contagion" between individuals. During social interactions, humans spontaneously adopt the behavioral states of others, aligning their words and phrases and body postures, synchronizing movements, and reflexively mimicking expressions and gaze directions with parallel changes in emotion and attention (Lakin & Chartrand, 2003). This synchronization is paralleled by a synchronization of *brain states* in successfully communicating individuals (Dikker, Silbert, Hasson, & Zevin, 2014; Hasson, Ghazanfar, Galantucci, Garrod, & Keysers, 2012; Jiang et al., 2012; Stolk et al., 2013; Yun, Watanabe, & Shimojo, 2012). Such contagion of behavioral states is almost certainly not restricted to humans, except perhaps in scope.

By synchronizing concepts between individuals, language may be the king of mirror processes. When we speak, sign or write, we activate similar representations in ourselves and in others (Dikker et al., 2014; Stolk et al., 2013). Humans depend heavily on language, which is classically thought to arise through interactions between Wernicke's area at the left temporo-parietal junction and Broca's area in the left inferior frontal gyrus. The former plays a dominant role in perception (via links to primary auditory cortex), the latter in production (via links to primary motor cortex), and the two are strongly interconnected through the arcuate fascicularis (Geschwind, 1970). A modern understanding stresses dissociable ventral and dorsal components of this network, with the former processing conceptual relations and the later auditory-motor sequences (Hickok & Poeppel, 2007) in part via a prominent thalamo-cortico-striatal loop (Petkov & Jarvis, 2012). Interestingly, the brain areas in which "mirror" neurons have been found include inferior frontal gyrus and inferior parietal cortex—potential homologues of human Broca's and Wernicke's areas. Among the many theories of language origins, several explicitly emphasize a role for action mirroring in language and speech evolution (Arbib, Liebal, & Pika, 2008), for example through the use of pointing to flexibly guide behavior toward the external world (Cappuccio & Shepherd, 2013).

16.4 Comparisons to Other Communication Systems

Most studies of communication have focused on a relatively small number of model species' signaling systems. These include the pheromonal communication of moths and ants, the waggle dance of bees, courtship in laboratory fruit flies, olfactory signaling and social behavior in rodents, vocal signaling amongst Tungara frogs and the impact of eavesdropping by bats, head bobs in lizards, electrical communication amongst electric fish, alarm displays by squirrels (notably including thermal signaling toward pit viper threats), and, of course, birdsong. We here discuss how the mechanisms of primate signaling generalize, first across vertebrates and then across all species.

16.4.1 Mechanisms of Communication in Other Vertebrates

Data from sheep and dogs suggest that primate specializations for audiovisual communication may share deep homology: as social mammals adapted to diurnal conditions, they orient their gaze preferentially toward conspecific faces, which they individuate and remember, and from which they read behavioral states (Leopold & Rhodes, 2010). Moreover, single-cell recordings in sheep temporal cortex found patches that discriminate features of conspecifics faces (as well as those of humans and sheepdogs); as in primate face patches, neurons encode configural cues of identity and facial expression (Tate, Fischer, Leigh, & Kendrick, 2006). Imaging studies conducted in awake dogs show similar temporal-lobe specializations for social processing in the auditory domain (Andics, Gácsi, Faragó, Kis, & Miklósi, 2014). It is unclear to what extent these cortical specializations represent genetic conservation (homology) or evolutionary convergence (analogy), but it appears likely that temporal lobe (and perhaps frontal lobe) cortical processing are involved in audiovisual social signal perception in diverse large-brained mammals. Interestingly, face-to-face interactions in rodents are primarily somatosensory, mediated by vigorous and rhythmic mutual whisking (Brecht & Freiwald, 2012). Whiskers are thought to have evolved early in the mammalian lineage (Grant, Haidarliu, Kennerley, & Prescott, 2013)—this suggests that rhythmic components of orofacial communicative movements such as human speech and monkey lipsmacking (Ghazanfar & Takahashi, 2014) may ultimately derive in part from sensory behavior, instead from ingestive patterns as generally supposed (Andrew, 1963). Production mechanisms involving the frontal cortex and basal ganglia—including those for language—may operate in part by rearranging ancient signaling pathways.

In mammalian taxa, especially those with large brains, subcortical processing appears to subordinate itself to cortical. However, subcortical structures remain involved in social signaling—especially as mechanisms of (relatively innate) signal production, but also of (relatively plastic) signal responsiveness. Subcortical structures spanning the midbrain and ventral prosencephalon play a prominent role in organizing primal drives (aggression, fear, lust, hunger), including species-typical responses to communicative stimuli. A subcortical social brain network was first defined in mammals as a set of interconnected, sex-steroid-sensitive nodes including the medial extended amygdala (amygdala and adjacent bed nucleus of the stria terminalis), lateral septum, medial preoptic area, anterior hypothalamus, ventromedial hypothalamus, periacqueductal gray and associated ventral tegmentum (Newman, 1999). Each of these areas contains small, distributed clusters of neurons encoding components of diverse social behaviors and act combinatorially in communication (Newman, 1999). Despite its evolutionary flexibility, mediating a variety of social organizations, this core architecture appears to be conserved across vertebrates (Goodson, 2005). Notably, it is highly sensitive to social nonapeptides (e.g. oxytocin/vasopresin), which play a pleiotropic role in physiology, stress and social behavior, including pair bonding in voles and finches (Goodson, 2013). These subcortical structures may play a crucial developmental role, even in mammals with large cerebral cortices. By signaling stimulus saliency early in development, while cortex is immature, they may drive appropriate social learning to innately salient social cues (Johnson, 2005).

Birds, like mammals, have complex signaling behaviors to mediate social interactions. While homologies between bird and mammalian brains are still being refined, it appears that they have convergently developed hypertrophied telecephalic structures for perception and learning of flexible communications both with conspecifics and heterospecifics (Cross et al., 2013). As among primates, many birds use short calls for emotional communication, including functionally-referential alarm calls (Kaplan, 2008). These calls rely upon midbrain structures, including the dorsomedial nucleus intercollicularis, a homologue of dorsal PAG (Kingsbury, Kelly, Schrock, & Goodson, 2011). The most studied signaling system in birds, however, is that for bird songs or "long calls," which, in many species, involve both learning and complex generative "grammars." The system responsible for these long calls, like that responsible for human language, coordinates and repurposes more ancient and phylogenetically conserved midbrain structures: in particular, vocal learners have interconnected motor pallial areas, featuring a thalamo-pallial-striatal loop, which overrides midbrain areas to control vocal signaling (Petkov & Jarvis, 2012).

16.4.2 Mechanisms of Communication in Invertebrates

There are relatively few points of contact between systems neuroscience descriptions of primate social interactions and those of invertebrates, despite many interesting examples of sophisticated communicative behavior, such as the dances of bees (Riley, Greggers, Smith, Reynolds, & Menzel, 2005; von Frisch, 1967) and the skin-changing of cephalopods (Brown et al., 2012; Messenger, 2001).

Honey bees, in particular, are fascinating in that they use referential signals (the waggle and round dance) to describe food source directionality and quality. The neural implementation of this signaling pathway is known only in outline, but dances appear to be perceived and to guide behavior via exapted sensory systems (the neck hairs and the Johnstone's organ) which generally play a role in online guidance of flight movements (Ai & Hagio, 2013). These communication systems evolved as honey bees leveraged the information-gathering ability of larger social groups (Donaldson-Matasci, DeGrandi-Hoffman, & Dornhaus, 2013).

Another remarkable invertebrate communication system is that of the cephalopods. Cephalopods including squid, cuttlefish and octopuses manipulate their skin appearance by dilating and contracting skin patches which differentially reflect wavelengths and polarizations of light, and while distinct patches of skin are governed by distinct neurons, there is no obvious central homunculus (Messenger, 2001). Skin-changing can be accompanied by textural and postural changes both to convey social signals and to produce camouflage, including sophisticated mimicry (Brown et al., 2012) as described in §16.2.1. These responses are hierarchically coordinated under the supervision of cephalopods' hypertrophied optic lobes (Messenger, 2001).

The most complete model of signaling in invertebrates may be the fruit fly, *Drosophila melanogaster*. Fruit flies exchange social signals in diverse contexts, and remember the results of interactions they witness or in which they participate (Sokolowski, 2010). Successful courtship between fruit flies involves the exchange of pheromonal, tactile, and vibrational signals (Pavlou & Goodwin, 2013). Recent progress in Drosophila genetics has made it possible to molecularly dissect neural control systems for many of these behaviors, identifying their anatomical pathways.

A potential benefit of broadly comparative approaches to social neuroethology is the potential to unravel deep molecular homologies underlying independently-evolved social behaviors. For example, responses to territorial intrusion in honeybee include molecular motifs that resemble territorial intrusion responses manifested in the tetrapod hypothalamus (Rittschof et al., 2014). Even the most vaunted of social behaviors may build on ancient pathways homologous across many species. For example, the neuropeptide system mediating pair bonds in diverse vertebrates (see above) is also involved in mating drive in nematodes (Garrison et al., 2012).

A final, humbling observation is that these deep molecular homologies do not obey our intuitive sense of communication as an integrated process, consciously and intentionally deployed. Communication evolved before brains. The molecular homologies underlying communication therefore interweave diverse biological functions: intracellular housekeeping, development, arousal, energy and fluid homeostasis, and social drive. Consider that the prototypical insect signaling system—the response to diffusible pheromones—was named for its ambiguous position between a sensory percept and an extrinsic hormone (Karlson & Luscher, 1959). Indeed, some chemosensory signaling pathways appear to involve neurons only incidentally, if at all ("allohomones": Wyatt, 2014). The presence of other life is a fundamental part of an organisms' environment, and the mechanisms by which organisms thrive in social environments have evolutionary roots stretching long before the invention of multicellular organisms, much less brains.

16.5 Conclusions

Primates present a rich range of communication strategies in different modalities that evolved as signaling, perceiving, and signaling back behaviors. Most neural mechanisms subtending their communication strategies have been studied in only three species, macaque monkeys, humans, and marmosets, and the neural mechanisms of some communication strategies are only just starting to be investigated, such as those concerning tactile and chemosensory communication. However, important general principals about audio-visual communication can already be outlined from the existing studies.

The first one concerns the contrasting roles of phylogeny and genetics versus ontogeny and learning: While most signaling behaviors and some production behaviors have been passed through generations and are present at birth in primates, these behaviors subsequently go through an important process of ontogenetic refinement mediated by learning from peers and imitation. This process of ontogenetic ritualization, which will eventually enable fine communication variations between group members, can go as far as creating different cultures in some species. While the role of learning and social motivation for learning in signal production and perception is critical, many more studies are needed to understand the neural mechanisms subtending them in the developing primate brain.

The second important point concerns the creation of a system of coordinated perception and action in the signaler and receiver brains. We saw that both subcortical and cortical loops tightly link perception and production of communicative signals, in reflexive and flexible ways. Additionally studies in humans started unraveling how two brains synchronize each other during face-to-face communication. Similar studies in other primates would be of great importance to understand if this mechanism can serve as a basis for reciprocated tit calls created by marmosets or lipsmacking in macaque monkeys. It would unravel the mechanisms subtending primates' shared intention and interactions in creating common-for-communication specialties. Synchronization might also be the mechanism through which humans have evolved the ability to understand each other minds and mental states. And a greater exploration of these mechanisms in other primates might unravel the phylogenetic process through which this ability emerged.

The third important point is that, when taking a broadly comparative view of social neuroscience, we must consider that social signaling predates the evolution of brains, and that many molecular homologies may thus entangle systems governing metabolic and social aspects of environmental sensation and responsiveness. Moreover, because social signalizing also predates the evolution of multicellularity, these broadly conserved molecular toolkits may be entangled with transcription regulation and development. Rather than denigrating the importance of social signaling for nervous system evolution, it should highlight it: regulation of social interactions, like regulation of energy metabolism and fluid homeostasis, is one of the primal challenges that brains evolved to solve.

Finally, studies of primate communication and its mechanisms are exciting because they carry the possibility of understanding primates' knowledge about their social and outer world. Even though other primates might not have elaborated internal thoughts as humans have, expressed, moreover, through internal language, studies about communication teach us about their perception and categorization of the environment. They provide us with a window to our own perception of the world whether conscious or unconscious, innate or learned.

Acknowledgments

JS was supported by a fellowship from the Bettencourt-Schueller Foundation and from the Human Frontier Science Program Organization (LT000418/2013-L). SVS was supported by a fellowship from the Leon Levy Fellowship in Neuroscience and by NIH K99HD077019. DYT was supported by a Pew Latin American Fellowship and Brazilian Science without Border Fellowship.

References

Adachi, I., Kuwahata, H., Fujita, K., Tomonaga, M., & Matsuzawa, T. (2006). Japanese macaques form a cross-modal representation of their own species in their first year of life. *Primates*, 47, 350–354.

Ai, H., & Hagio, H. (2013). Morphological analysis of the primary center receiving spatial information transferred by the waggle dance of honeybees. *The Journal of Comparative Neurology*, 521(11), 2570–2584.

Allison, T., Puce, A., McCarthy, G. (2000). Social perception from visual cues: Role of the STS region. *Trends in Cognitive Sciences*, 4, 267–278.

Andics. A., McQueen, J. M., Petersson, K. M., Gál, V., Rudas, G., & Vidnyánszky, Z. (2010). Neural mechanisms for voice recognition. *NeuroImage*, 52, 1528–1540.

Andics, A., Gácsi, M., Faragó, T., Kis, A., & Miklósi, Á. (2014). Voice-sensitive regions in the dog and human brain are revealed by comparative fMRI. *Current Biology*, 24(5), 574–578.

Andrew, R. J. (1963). Evolution of facial expression. *Science*, 142(3595), 1034–1041.

Arbib, M. A., Liebal, K., & Pika, S. (2008). Primate vocalization, gesture, and the evolution of human language. *Current Anthropology*, 49(6), 1053–1076.

Arnold, K., & Zuberbühler, K. (2006). The alarm-calling system of adult male putty-nosed monkeys, cercopithecus nictitans martini. *Animal Behaviour*, 72, 643–653.

Atkinson, A. P., Dittrich, W. H., Gemmell A. J., & Young, A. W. (2004). Emotion perception from dynamic and static body expressions in point-light and full-light displays. *Perception*, 33, 717–746.

Bachorowski, J. A., & Owren, M. J. (1999). Acoustic correlates of talker sex and individual talker identity are present in a short vowel segment produced in running speech. *The Journal of the Acoustical Society of America*, 106, 1054–1063.

Baird, R. W., & Whitehead, H. (2000). Social organization of mammal-eating killer whales: Group stability and dispersal patterns. *Canadian Journal of Zoology*, 78, 2096–2105.

Beauchamp, M. S., Argall, B. D., Bodurka, J., Duyn, J. H., & Martin, A. (2004). Unraveling multisensory integration: Patchy organization within human sts multisensory cortex. *Nature Neuroscience*, 7, 1190–1192.

Behrens, T. E. J., Hunt, L. T., & Rushworth, M. F. S (2009). The computation of social behavior. *Science*, 324(5931), 1160–1164.

Belin, P., Zatorre, R. J., & Ahad, P. (2002) Human temporal-lobe response to vocal sounds. *Cognitive Brain Research*, 13, 17–26.

Belin, P., Zatorre, R. J., Lafaille, P., Ahad, P., & Pike, B (2000) Voice-selective areas in human auditory cortex. *Nature*, 403, 309–312.

Bell, A. H., Malecek, N. J., Morin, E. L., Hadj-Bouziane, F., Tootell, R. B. H., & Ungerleider, L. G. (2011). Relationship between functional magnetic resonance imaging-identified regions and neuronal category selectivity. *The Journal of Neuroscience*, 31, 12229–12240.

Binder, J. R., Frost, J. A., Hammeke, T. A., Bellgowan, P. S. F., Springer J. A., Kaufman, J. N., & Possing, E. T. (2000). Human temporal lobe activation by speech and nonspeech sounds. *Cerebral Cortex*, 10, 512–528.

Bohn, K. M., Smarsh, G. C., & Smotherman, M. (2013). Social context evokes rapid changes in bat song syntax. *Animal Behaviour*, 85, 1485–1491.

Boulet, M., Charpentier, M. J., & Drea C. M. (2009). Decoding an olfactory mechanism of kin recognition and inbreeding avoidance in a primate. *BMC Evolutionary Biology*, 9, 281.

Brecht, M., & Freiwald, W. A. (2012). The many facets of facial interactions in mammals. *Current Opinion in Neurobiology*, 22(2), 259–266.

Breiter, H. C., Etcoff, N. L., Whalen, P. J., Kennedy, W. A., Rauch, S. L., Buckner, R. L., ... Rosen, B. R. (1996). Response and habituation of the human amygdala during visual processing of facial expression. *Neuron*, 17, 875–887.

Brown, C., Garwood, M. P., & Williamson, J. E. (2012). It pays to cheat: Tactical deception in a cephalopod social signalling system. *Biology letters*, 8, 729–732.

Byrne, R. W. (1996). Machiavellian intelligence. *Evolutionary Anthropology*, 5, 172–180.

Call, J., & Tomasello, M. (2007). *The gestural communication of apes and monkeys*. Mahwah, NJ: Lawrence Erlbaum.

Cappuccio, M. L., & Shepherd, S. V. (2013). Pointing hand: Joint attention and embodied symbols. In Z. Radman (Ed.), *The hand, an organ of the mind: What the manual tells the mental* (pp. 303–326). Cambridge, MA. MIT Press. p 303.

Caruana, F., Avanzini, P., Gozzo, F., Francione, S., Cardinale, F., & Rizzolatti, G. (2015). Mirth and laughter elicited by electrical stimulation of the human anterior cingulate cortex. *Cortex*, 71, 323–331.

Chandrasekaran, C., & Ghazanfar, A. A. (2009). Different neural frequency bands integrate faces and voices differently in the superior temporal sulcus. *Journal of Neurophysiology*, 101, 773–788.

Cheney, D. L., & Seyfarth, R. M. (1990). Attending to behaviour versus attending to knowledge: Examining monkeys' attribution of mental states. *Animal Behaviour*, 40, 742–753.

Cheney, D. L., & Seyfarth, R. M. (1997). Reconciliatory grunts by dominant female baboons influence victims' behaviour. *Animal Behaviour*, 54, 409–418.

Cheney, D. L., & Seyfarth, R. M. (1999). Recognition of other individuals' social relationships by female baboons. *Animal Behaviour*, 58, 67–75.

Cohen, Y. E., Hauser, M. D., & Russ, B. E. (2006). Spontaneous processing of abstract categorical information in the ventrolateral prefrontal cortex. *Biology Letters*, 2, 261–265.

Cohen, Y. E., Russ, B. E., Gifford, G. W., Kiringoda, R., & MacLean, K. A. (2004). Selectivity for the spatial and nonspatial attributes of auditory stimuli in the ventrolateral prefrontal cortex. *Journal of Neuroscience*, 24, 11307–11316.

Cohen, Y. E., Theunissen F., Russ, B. E., & Gill, P. (2007). Acoustic features of rhesus vocalizations and their representation in the ventrolateral prefrontal cortex. *Journal of Neurophysiology*, 97, 1470–1484.

Cook, R., Bird, G., Catmur, C., Press, C., & Heyes, C. (2014). Mirror neurons: From origin to function. *Behavioral and Brain Sciences*, 37, 177–192.

Costafreda, S. G., Brammer, M. J., David, A. S., & Fu, C. H. Y. (2008). Predictors of amygdala activation during the processing of emotional stimuli: A meta-analysis of 385 pet and fmri studies. *Brain Research Reviews*, 58, 57–70.

Coussi-Korbel, S. (1994). Learning to outwit a competitor in mangabeys (*Cercocebus torquatus torquatus*). *Journal of Comparative Psychology*, 108, 164–171.

Crockford, C., Wittig, R. M., Mundry, R., & Zuberbühler, K. (2012). Wild chimpanzees inform ignorant group members of danger. *Current Biology*, 22, 142–146.

Cross D. J., Marzluff, J. M., Palmquist, I., Minoshima, S., Shimizu, T., & Miyaoka, R. (2013). Distinct neural circuits underlie assessment of a diversity of natural dangers by American crows. *Proceedings of the the Royal Society B, Biological Sciences*, 280(1765). doi: 10.1098/rspb.2013.1046

Dahl, C. D., Logothetis, N. K., & Kayser, C. (2009). Spatial organization of multisensory responses in temporal association cortex. *The Journal of Neuroscience*, 29, 11924–11932.

Darwin, C. R. (1872). *The expression of the emotions in man and animals*. London: John Murray.

Darwin, C. R. (1876). *On the origin of species by means of natural selection, or the preservation of favored races in the struggle for life*. London: Murray.

Deaner, R. O., Khera, A. V., & Platt, M. L. (2005). Monkeys pay per view: Adaptive valuation of social images by rhesus macaques. *Current Biology*, 15, 543–548.

de Gelder, B. (2006). Towards the neurobiology of emotional body language. *Nature Reviews Neuroscience*, 7, 242–249.

de Gelder, B. (2009). Why bodies? Twelve reasons for including bodily expressions in affective neuroscience. *Philosophical Transactions of the Royal Society B, Biological Sciences*, 364, 3475–3484.

de Gelder, B., & Partan, S. (2009). The neural basis of perceiving emotional bodily expressions in monkeys. *Neuroreport*, 20, 642–646.

de Gelder, B., Snyder, J., Greve, D., Gerard, G., & Hadjikhani, N. (2004). Fear fosters flight: A mechanism for fear contagion when perceiving emotion expressed by a whole body. *Proceedings of the National Academy of Sciences of the USA*, 101, 16701–16706.

de Meijer, M. (1989). The contribution of general features of body movement to the attribution of emotions. *Journal of Nonverbal Behavior*, 13, 247–268.

Démonet, J.-F., Chollet, F., Ramsay, S., Cardebat, D., Nespoulous, J.-L., Wise, R., ... Frackowiak, R. (1992). The anatomy of phonological and semantic processing in normal subjects. *Brain*, 115, 1753–1768.

Denkova, E., Botzung, A., & Manning, L. (2006) Neural correlates of remembering/knowing famous people: An event-related fmri study. *Neuropsychologia*, 44, 2783–2791.

de Souza, W. C., Eifuku S., Tamura R., Nishijo H. H., & Ono T (2005) Differential characteristics of face neuron responses within the anterior superior temporal sulcus of macaques. *Journal of Neurophysiology*, 94, 1252–1266.

Devinsky, O., Morrell, M. J., & Vogt, B. A. (1995). Contributions of anterior cingulate cortex to behaviour. *Brain*, 118, 279–306.

de Waal, F. B. M., & Yoshihara, D. (1983). Reconciliation and redirected affection in rhesus monkeys. *Behaviour*, 85, 224–241.

Diogo, R., Wood, B. A., Aziz, M. A., & Burrows, A. M. (2009). On the origin, homologies and evolution of primate facial muscles, with a particular focus on hominoids and a suggested unifying nomenclature for the facial muscles of the Mammalia. *Journal of Anatomy*, 215(3), 300–19.

Dikker, S., Silbert, L. J., Hasson, U., & Zevin, J. D. (2014). On the same wavelength: Predictable language enhances speaker–listener brain-to-brain synchrony in posterior superior temporal gyrus. *Journal of Neuroscience*, 34, 6267–6272.

di Pellegrino, G., Fadiga, L., Fogassi, L., Gallese, V., & Rizzolatti, G. (1992). Understanding motor events: A neurophysiological study. *Experimental Brain Research*, 91, 176–180.

Dolan, R. J., Morris, J. S., & de Gelder, B. (2001). Crossmodal binding of fear in voice and face. *Proceedings of the National Academy of Sciences of the USA*, 98, 10006–10010.

Domb, L. G., & Pagel, M. (2001). Sexual swellings advertise female quality in wild baboons. *Nature*, 410, 204–206.

Donaldson-Matasci, M. C., DeGrandi-Hoffman, G., & Dornhaus, A. (2013). Bigger is better: Honeybee colonies as distributed information-gathering systems. *Animal Behaviour*, 85, 585–592.

Downing, P. E., Jiang, Y., Shuman, M., & Kanwisher, N. (2001). A cortical area selective for visual processing of the human body. *Science*, 293, 2470–2473.

Dujardin, E., & Jürgens, U. (2005). Afferents of vocalization-controlling periaqueductal regions in the squirrel monkey. *Brain Research*, 1034, 114–131.

Dujardin, E., & Jürgens, U. (2006). Call type-specific differences in vocalization-related afferents to the periaqueductal gray of squirrel monkeys (Saimiri sciureus). *Behavioural Brain Research*, 168, 23–36.

Dunbar, R. I. M. (1991). Functional significance of social grooming in primates. *Folia Primatologica*, 57, 121–131.

Dunbar, R. I. M. (1992). Neocortex size as a constraint on group size in primates. *Journal of human evolution*, 22, 469–493.

Dunbar, R. I. M. (2010). The social role of touch in humans and primates: Behavioural function and neurobiological mechanisms. *Neuroscience & Biobehavioral Reviews*, 34, 260–268.

Dusterhoft, F. (2004). Neuronal activity in the periaqueductal gray and bordering structures during vocal communication in the squirrel monkey. *Neuroscience*, 123(1), 53–60.

Eifuku, S., De Souza, W. C., Nakata, R., Ono, T., Tamura, R. (2011). Neural representations of personally familiar and unfamiliar faces in the anterior inferior temporal cortex of monkeys. *PLoS One*, 6, e18913. doi: 10.1371/journal.pone.0018913

Eifuku, S., Nakata, R., Sugimori, M., Ono, T., Tamura, R. (2010). Neural correlates of associative face memory in the anterior inferior temporal cortex of monkeys. *Journal of Neuroscience*, 30, 15085–15096.

Ekman, P., Sorenson, E. R., & Friesen, W. V. (1969). Pan-cultural elements in facial displays of emotion. *Science*, 164(3875), 86–88.

Evans, C. S. (2006). Accessory chemosignaling mechanisms in primates. *American Journal of Primatology*, 68, 525–544.

Fant, G. (1960). *Acoustic theory of speech production*. La Haye, Netherlands: Mouton & Co.

Fecteau, S., Belin, P., Joanette, Y., & Armony, J. L. (2007). Amygdala responses to nonlinguistic emotional vocalizations. *NeuroImage*, 36, 480–487.

Ferris, C. F., Snowdon, C. T., King, J. A., Duong, T. Q., Ziegler, T. E., Ugurbil, K., ... Vaughan, J. T. (2001). Functional imaging of brain activity in conscious monkeys responding to sexually arousing cues. *Neuroreport*, 12, 2231–2236.

Ferris, C. F., Snowdon, C. T., King, J. A., Sullivan, J. M., Ziegler, T. E., Olson, D. P., ... Duong, T. Q. (2004). Activation of neural pathways associated with sexual arousal in non-human primates. *Journal of Magnetic Resonance Imaging*, 19, 168–175.

Fisher, C., & Freiwald, W. A. (2015). Contrasting specializations for facial motion within the macaque face-processing system. *Current biology*, 25, 261–266.

Fitch, W. T. (2000). The evolution of speech: A comparative review. *Trends in Cognitive Science*, 4, 258–267.

Fitch, W. T., & Fritz, J. B. (2006). Rhesus macaques spontaneously perceive formants in conspecific vocalizations. *The Journal of the Acoustical Society of America*, 120, 2132–2141.

FitzGibbon, C. D., & Fanshawe, J. H. (1988). Stotting in Thomson's gazelles: An honest signal of condition. *Behavioral Ecology and Sociobiology*, 23(2), 69–74.

Freiwald, W. A., & Tsao, D. Y. (2010). Functional compartmentalization and viewpoint generalization within the macaque face-processing system. *Science*, 330, 845–851.

Freiwald, W. A., Tsao, D. Y., & Livingstone, M. S. (2009). A face feature space in the macaque temporal lobe. *Nature Neuroscience*, 12, 1187–1196.

Furl, N., Hadj-Bouziane, F., Liu, N., Averbeck, B. B., & Ungerleider, L. G. (2012). Dynamic and static facial expressions decoded from motion-sensitive areas in the macaque monkey. *The Journal of Neuroscience*, 32, 15952–15962.

Furl, N., Henson, R. N., Friston, K. J., & Calder, A. J. (2014). Network interactions explain sensitivity to dynamic faces in the superior temporal sulcus. *Cerebral Cortex*, 25(9), 2876–2882.

Garrison, J. L., Macosko, E. Z, Bernstein, S., Pokala, N., Albrecht, D. R., & Bargmann, C. I. (2012). Oxytocin/vasopressin-related peptides have an ancient role in reproductive behavior. *Science*, 338(6106), 540–3.

Gazzola, V., Spezio, M. L., Etzel, J. A., Castelli, F., Adolphs, R., & Keysers, C. (2012). Primary somatosensory cortex discriminates affective significance in social touch. *Proceedings of the National Academy of Sciences of the USA*, 109(25), E1657–66.

Gerald, M. S., Waitt, C., & Little, A. C. (2009). Pregnancy coloration in macaques may act as a warning signal to reduce antagonism by conspecifics. *Behavioural Processes*, 80, 7–11.

Geschwind, N. (1970). The organization of language and the brain. *Science*, 170, 940–944.

Ghazanfar, A. A., & Rendall, D. (2008). Evolution of human vocal production. *Current Biology*, 18, R457–460.

Ghazanfar, A. A., & Santos, L. R. (2004). Primate brains in the wild: The sensory bases for social interactions. *Nature Reviews Neuroscience*, 5, 603–616.

Ghazanfar, A. A., Turesson, H. K., Maier, J. X., van Dinther, R., Patterson, R. D., & Logothetis, N. K. (2007). Vocal-tract resonances as indexical cues in rhesus monkeys. *Current biology*, 17, 425–430.

Ghazanfar, A. A., & Takahashi, D. Y. (2014). Facial expressions and the evolution of the speech rhythm. *Journal of Cognitive Neuroscience*, 26(6), 1196–207.

Gifford, G. W., MacLean, K. A., Hauser, M. D., & Cohen, Y. E. (2005). The neurophysiology of functionally meaningful categories: Macaque ventrolateral prefrontal cortex plays a critical role in spontaneous categorization of species-specific vocalizations. *Journal of Cognitive Neuroscience*, 17, 1471–1482.

Gil-da-Costa, R., Braun, A., Lopes, M., Hauser, M. D., Carson. R. E., Herscovitch, P., & Martin, A. (2004). Toward an evolutionary perspective on conceptual representation: Species-specific calls activate visual and affective processing systems in the macaque. *Proceedings of the National Academy of Sciences of the USA.*, 101, 17516–17521.

Gil-da-Costa, R., Martin, A., Lopes, M. A., Munoz, M., Fritz, J. B., & Braun, A. R. (2006). Species-specific calls activate homologs of broca's and wernicke's areas in the macaque. *Nature Neuroscience*, 9, 1064–1070.

Goodson. J. L. (2005). The vertebrate social behavior network: evolutionary themes and variations. *Hormones and Behavior*, 48(1), 11–22.

Goodson, J. L. (2013). Deconstructing sociality, social evolution and relevant nonapeptide functions. *Psychoneuroendocrinology*, 38(4), 465–478.

Gordon, I., Voos, A. C., Bennett, R. H., Bolling, D. Z, Pelphrey, K. A., & Kaiser, M. D. (2011). Brain mechanisms for processing affective touch. *Human Brain Mapping*, 922, 914–922.

Gothard, K. M., Battaglia, F. P., Erickson, C. A., Spitler, K. M., & Amaral, D. G. (2007). Neural responses to facial expression and face identity in the monkey amygdala. *Journal of Neurophysiology*, 97, 1671–1683.

Gottfried, J. A. (2010). Central mechanisms of odour object perception. *Nature Reviews Neuroscience*, 11(9), 628–41.

Gouzoules, S., Gouzoules, H., & Marler, P. (1984). Rhesus monkey (*Macaca mulatta*) screams: Representational signalling in the recruitment of agonistic aid. *Animal Behaviour*, 32, 182–193.

Grant, B. R., & Grant, P. R. (1996). Cultural inheritance of song and its role in the evolution of Darwin's finches. *Evolution*, 50, 2471–2487.

Grant, R. A., Haidarliu, S., Kennerley, N. J., & Prescott, T. J. (2013). The evolution of active vibrissal sensing in mammals: evidence from vibrissal musculature and function in the marsupial opossum Monodelphis domestica. *Journal of Experimental Biology*, 216(Pt. 18), 3483–94.

Groh, J. M., & Sparks, D. L. (1996). Saccades to somatosensory targets. II. Motor convergence in primate superior colliculus. *Journal of Neurophysiology*, 75(1), 428–438.

Gross, C. G., Bender, D. B., & Rocha-Miranda, C. E. (1969). Visual receptive fields of neurons in inferotemporal cortex of the monkey. *Science*, 166, 1303–1306.

Gross, C. G., Rocha-Miranda, C. E., & Bender, D. B. (1972). Visual properties of neurons in inferotemporal cortex of the macaque. *Journal of Neurophysiology*, 35, 96–111.

Gyger, M., Marler, P., & Pickert, R. (1987). Semantics of an avian alarm call system: The male domestic fowl, Gallus domesticus. *Behaviour*, 15–40.

Hadj-Bouziane, F., Bell, A. H., Knusten, T. A., Ungerleider, L. G., & Tootell, R. B. H. (2008). Perception of emotional expressions is independent of face selectivity in monkey inferior temporal cortex. *Proceedings of the National Academy of Sciences of the USA*, 105, 5591–5596.

Hadj-Bouziane, F., Liu, N., Bell, A. H., Gothard, K. M., Luh, W.-M., Tootell, R. B. H., ... Ungerleider, L. G. (2012). Amygdala lesions disrupt modulation of functional mri activity evoked by facial expression in the monkey inferior temporal cortex. *Proceedings of the National Academy of Sciences of the USA*, 109, E3640–E3648.

Hage, S. R., & Jürgens, U. (2006). Localization of a vocal pattern generator in the pontine brainstem of the squirrel monkey. *The European Journal of Neuroscience*, 23(3), 840–844.

Hannig, S., & Jürgens, U. (2006). Projections of the ventrolateral pontine vocalization area in the squirrel monkey. *Experimental Brain Research*, 169, 92–105.

Harris, R. J., Young, A. W., & Andrews, T. J. (2012). Morphing between expressions dissociates continuous from categorical representations of facial expression in the human brain. *Proceedings of the National Academy of Sciences of the USA*, 109, 21164–21169.

Hasselmo, M. E., Rolls, E. T., & Baylis, G. C. (1989). The role of expression and identity in the face-selective responses of neurons in the temporal visual cortex of the monkey. *Behavioural Brain Research*, 32, 203–218.

Hasson, U., Ghazanfar, A. A., Galantucci, B., Garrod, S., & Keysers, C. (2012). Brain-to-brain coupling: A mechanism for creating and sharing a social world. *Trends in Cognitive Sciences*, 16, 114–121.

Hauser, M. D., & Marler, P. (1993). Food-associated calls in rhesus macaques (*Macaca mulatta*): II. Costs and benefits of call production and suppression. *Behavioral Ecology*, 4, 206–212.

Haxby, J. V., Horwitz, B., Ungerleider, L. G., Maisog, J. M., Pietrini, P., & Grady, C. L. (1994). The functional organization of human extrastriate cortex: A pet-rcbf study of selective attention to faces and locations. *Journal of Neuroscience*, 14, 6336–6353.

Heymann, E. W. (2006). The neglected sense—Olfaction in primate behavior, ecology, and evolution. *American Journal of Primatology*, 68, 519–524.

Hickok, G., & Poeppel, D. (2007). The cortical organization of speech processing. *Nature Reviews Neuroscience*, 8, 393–402.

Hoffman, E. A., & Haxby, J. V. (2000). Distinct representations of eye gaze and identity in the distributed human neural system for face perception. *Nature Neuroscience*, 3, 80–84.

Hoffman, K. L., Gothard, K. M., Schmid, M. C., & Logothetis, N. K. (2007). Facial-expression and gaze-selective responses in the monkey amygdala. *Current Biology*, 17, 766–772.

Holle, H., Gunter, T. C., Rüschemeyer, S.-A., Hennenlotter, A., & Iacoboni, M. (2008). Neural correlates of the processing of co-speech gestures. *NeuroImage*, 39, 2010–2024.

Hopf, H. C., Müller-Forell, W., & Hopf, N. J. (1992). Localization of emotional and volitional facial paresis. *Neurology*, 42(10), 1918–23.

Hung, C., Yen, C. C., Ciuchta, J. L., Papoti, D., Bock, N. A., Leopold, D. A., & Silva, A. C. (2015). Functional mapping of face-selective regions in the extrastriate visual cortex of the marmoset. *Journal of Neuroscience*, 35(3), 1160–1172.

Imaizumi, S., Mori, K., Kiritani, S., Kawashima, R., Sugiura, M., Fukuda, H., ... Nakamura, K. (1997). Vocal identification of speaker and emotion activates differerent brain regions. *Neuroreport*, 8, 2809–2812.

Ishai, A. (2007). Sex, beauty and the orbitofrontal cortex. *International Journal of Psychophysiology*, 63, 181–185.

Janik, V. M., Sayigh, L. S., & Wells, R. S. (2006). Signature whistle shape conveys identity information to bottlenose dolphins. *Proceedings of the National Academy of Sciences of the USA.*, 103, 8293–8297.

Janssens, T., Zhu, Q., Popivanov, I. D., & Vanduffel, W. (2014). Probabilistic and single-subject retinotopic maps reveal the topographic organization of face patches in the macaque cortex. *Journal of Neuroscience*, 34, 10156–10167.

Jay, M. F., & Sparks, D. L. (1984). Auditory receptive fields in primate superior colliculus shift with changes in eye position. *Nature*, 309, 345–347.

Jiang, J., Dai, B., Peng, D., Zhu, C., Liu, L., & Lu, C. (2012). Neural synchronization during face-to-face communication. *Journal of Neuroscience*, 32, 16064–16069.

Johnson, M. H. (2005). Subcortical face processing. *Nature Reviews Neuroscience*, 6, 766–774.

Johnstone, R. A. (1997). The evolution of animal signals. In J. R. Krebs & N. B. Davies (Eds.), *Behavioral ecology: An evolutionary approach* (4th ed., pp. 155–178). Oxford: Blackwell Science Ltd.

Judge, P. G., & De Waal, F. B. M. (1997). Rhesus monkey behaviour under diverse population densities: Coping with long-term crowding. *Animal Behaviour*, 54, 643–662.

Jürgens, U. (2000). Localization of a pontine vocalization-controlling area. *The Journal of the Acoustical Society of America*, 108, 1393–1396.

Jürgens, U. (2009). The neural control of vocalization in mammals: a review. *Journal of Voice: Official Journal of the Voice Foundation*, 23(1), 1–10.

Jürgens, U., & Ploog, D. (1970). Cerebral representation of vocalization in the squirrel monkey. *Experimental Brain Research*, 10, 532–554.

Jürgens, U., & Pratt, R. (1979). Role of the periaqueductal grey in vocal expression of emotion. *Brain Research*, 167, 367–378.

Kanwisher, N., McDermott, J., & Chun M. M. (1997). The fusiform face area: A module in human extrastriate cortex specialized for face perception. *The Journal of Neuroscience*, 17, 4302–4311.

Kaplan, G. (2008). Alarm calls and referentiality in Australian magpies: Between midbrain and forebrain, can a case be made for complex cognition? *Brain Research Bulletin*, 76(3), 253–63.

Karlson, P., & Luscher, M. (1959). Pheromones: A new term for a class of biologically active substances. *Nature*, 183, 55–56.

Keysers, C., Kaas, J. H., & Gazzola, V. (2010). Somatosensation in social perception. *Nature Reviews Neuroscience*, 11(6), 417–28.

Kingsbury, M. A., Kelly, A. M., Schrock, S. E., & Goodson, J. L. (2011). Mammal-like organization of the avian midbrain central gray and a reappraisal of the intercollicular nucleus. *PloS One*, 6(6), e20720. doi: 10.1371/journal.pone.0020720

Kikuchi, Y., Horwitz, B., & Mishkin, M. (2010). Hierarchical auditory processing directed rostrally along the monkey's supratemporal plane. *The Journal of Neuroscience*, 30, 13021–13030.

Kim, H., Adolphs, R., O'Doherty, J. P., & Shimojo, S. (2007). Temporal isolation of neural processes underlying face preference decisions. *Proceedings of the National Academy of Sciences of the USA*, 104, 18253–18258.

Kirchhof, J., & Hammerschmidt, K. (2006). Functionally referential alarm calls in tamarins (saguinus fuscicollis and saguinus mystax)–evidence from playback experiments. *Ethology*, 112, 346–354.

Kirzinger, A., & Jürgens, U. (1982). Cortical lesion effects and vocalization in the squirrel monkey. *Brain Research*, 233(2), 299–315.

Kirzinger, A., & Jürgens, U. (1991). Vocalization-correlated single-unit activity in the brain stem of the squirrel monkey. *Experimental brain research*, 84(3), 545–560.

Knapp, L. A., Robson, J., & Waterhouse, J. S. (2006). Olfactory signals and the mhc: A review and a case study in lemur catta. *American Journal of Primatology*, 68, 568–584.

Kriegeskorte, N., Formisano, E., Sorger, B., & Goebel, R. (2007). Individual faces elicit distinct response patterns in human anterior temporal cortex. *Proceedings of the National Academy of Sciences of the USA*, 104, 20600–20605.

Ku, S.-P., Tolias, A. S., Logothetis, N. K., & Goense, J. (2011). fMRI of the face-processing network in the ventral temporal lobe of awake and anesthetized macaques. *Neuron*, 70, 352–362.

Kuraoka, K., & Nakamura, K. (2006). Impacts of facial identity and type of emotion on responses of amygdala neurons. *Neuroreport*, 17, 9–12.

Kuraoka, K., & Nakamura, K. (2007). Responses of single neurons in monkey amygdala to facial and vocal emotions. *Journal of Neurophysiology*, 97, 1379–1387.

Lakin, J. L., & Chartrand, T. L. (2003). Using nonconscious behavioral mimicry to create affiliation and rapport. *Psychological Science*, 14, 334–339.

Leonard, C. M., Rolls, E. T., Wilson, F. A., & Baylis, G. C. (1985). Neurons in the amygdala of the monkey with responses selective for faces. *Behavioural Brain Research*, 15, 159–176.

Leopold, D. A., Bondar, I. V., & Giese, M. A. (2006). Norm-based face encoding by single neurons in the monkey inferotemporal cortex. *Nature*, 442, 572–575.

Leopold, D. A., & Rhodes, G. (2010). A comparative view of face perception. *Journal of Comparative Psychology*, 124, 233–251.

Leveroni, C. L., Seidenberg, M., Mayer, A. R., Mead, L. A., Binder, J. R., & Rao, S. M. (2000). Neural systems underlying the recognition of familiar and newly learned faces. *The Journal of Neuroscience*, 20, 878–886.

Levinson, S. C. (2006). Introduction: The evolution of culture in a microcosm. In S. C. Levinson & P. Jaisson (Eds.), *Evolution and culture: A Fyssen Foundation symposium* (pp 1–41). Cambridge, MA: MIT Press.

Lewin, W., & Whitty, C. W. M. (1960). Effects of anterior cingulate stimulation in conscious human subjects. *Journal of Neurophysiology*, 23, 445–447.

Livneh, U., Resnik, J., Shohat, Y., & Paz, R. (2012). Self-monitoring of social facial expressions in the primate amygdala and cingulate cortex. *Proceedings of the National Academy of Sciences of the USA*, 109(46), 18956–61.

Logothetis, N. K., Guggenberger, H., Peled, S., & Pauls, J. (1999). Functional imaging of the monkey brain. *Nature Neuroscience*, 2, 555–562.

Lundqvist, L.-O., & Dimberg, U. (1995). Facial expressions are contagious. *Journal of Psychophysiology*, 9(3), 203–211.

Lüthe, L., Häusler, U., & Jürgens, U. (2000). Neuronal activity in the medulla oblongata during vocalization. A single-unit recording study in the squirrel monkey. *Behavioural Brain Research*, 116, 197–210.

Macedonia, J. M. (1990). What is communicated in the antipredator calls of lemurs: Evidence from playback experiments with ringtailed and ruffed lemurs. *Ethology*, 86, 177–190.

Macedonia, J. M., & Evans, C. S. (1993). Essay on contemporary issues in ethology: Variation among mammalian alarm call systems and the problem of meaning in animal signals. *Ethology*, 93, 177–197.

Maclarnon, A., & Hewitt, G. (2004). Increased breathing control: Another factor in the evolution of human language. *Evolutionary Anthropology: Issues, News, and Reviews*, 13, 181–197.

Mahajan, N., Martinez, M. A., Gutierrez, N. L., Diesendruck, G., Banaji, M. R., & Santos, L. R. (2011). The evolution of intergroup bias: Perceptions and attitudes in rhesus macaques. *Journal of Personality & Social Psychology*, 100, 387–405.

Mantyh, P. W. (1982). The ascending input to the midbrain periaqueductal gray of the primate. *Journal of Comparative Neurology*, 211, 50–64.

Marciniak, K., Atabaki, A., Dicke, P. W., & Thier, P. (2014). Disparate substrates for head gaze following and face perception in the monkey superior temporal sulcus. *eLife*, 3, e03222. Retrieved from https://elife-publishing-cdn.s3.amazonaws.com/03222/elife-03222-v2.pdf

Marler, P., Dufty, A., & Pickert, R. (1986). Vocal communication in the domestic chicken: II. Is a sender sensitive to the presence and nature of a receiver? *Animal Behaviour*, 34, 194–198.

Mateo, J. M. (2010). Alarm calls elicit predator-specific physiological responses. *Biology Letters*, 6, 623–625.

McCarthy, G., Puce, A., Gore, J. C., & Allison, T. (1997). Face-specific processing in the human fusiform gyrus. *Journal of Cognitive Neuroscience*, 9, 605–610.

Messenger, J. B. (2001). Cephalopod chromatophores: Neurobiology and natural history. *Biological Reviews of the Cambridge Philosophical Society*, 76, 473–528.

Meunier H., Prieur J., & Vauclair, J. (2013). Olive baboons communicate intentionally by pointing. *Animal cognition*, 16, 155–163.

Moeller, S., Freiwald, W. A., & Tsao, D. Y. (2008). Patches with links: A unified system for processing faces in the macaque temporal lobe. *Science*, 320, 1355–1359.

Morin, E. L., Hadj-Bouziane, F., Stokes, M., Ungerleider, L. G., & Bell, A. H. (2014). Hierarchical encoding of social cues in primate inferior temporal cortex. *Cerebral Cortex*, 25, 3036–3045.

Morecraft, R. J., Louie, J. L., Herrick, J. L., & Stilwell-Morecraft, K. S. (2001). Cortical innervation of the facial nucleus in the non-human primate: A new interpretation of the effects of stroke and related subtotal brain trauma on the muscles of facial expression. *Brain: A Journal of Neurology*, 124(Pt. 1), 176–208.

Morecraft, R. J., Stilwell–Morecraft, K. S., & Rossing, W. R. (2004). The motor cortex and facial expression: New insights from neuroscience. *The Neurologist*, 10(5), 235–249.

Morris, J. S., Frith, C. D., Perrett, D. I., Rowland, D., Young, A. W., Calder, A. J., & Dolan, R. J. (1996). A differential neural response in the human amygdala to fearful and happy facial expressions. *Nature*, 383, 812–815.

Nakamura, K., Kawashima, R., Nagumo, S., Ito, K., Sugiura, M., Kato, T., ... Kojima, S. (1998). Neuroanatomical correlates of the assessment of facial attractiveness. *Neuroreport*, 9, 753–757.

Nakamura, K., Kawashima, R., Sugiura, M., Kato, T., Nakamura, A., Hatano, K., ... Kojima, S. (2001). Neural substrates for recognition of familiar voices: A pet study. *Neuropsychologia*, 39, 1047–1054.

Nakazawa, K., Shiba, K., Satoh, I., Yoshida, K., Nakajima, Y, & Konno, A. (1997). Role of pulmonary afferent inputs in vocal on-switch in the cat. *Neuroscience Research.*, 29(1), 49–54.

Newman, S. W. (1999). The medial extended amygdala in male reproductive behavior. A node in the mammalian social behavior network. *Annals of the New York Academy of Sciences*, 877, 242–57.

Noad, M. J., Cato, D. H., Bryden, M. M., Jenner, M. N., & Jenner, K. C. S. (2000). Cultural revolution in whale songs. *Nature*, 408, 537–537.

Olsson S. B., Barnard J., & Turri, L. (2006). Olfaction and identification of unrelated individuals: Examination of the mysteries of human odor recognition. *Journal of Chemical Ecology*, 32, 1635–1645.

Ouattara, K., Lemasson, A., & Zuberbühler, K. (2009). Campbell's monkeys concatenate vocalizations into context-specific call sequences. *Proceedings of the National Academy of Sciences of the USA*, 106, 22026–22031.

Owren, M. J., Dieter, J. A., Seyfarth, R. M., & Cheney, D. L. (1993). Vocalizations of rhesus (*Macaca mulatta*) and Japanese (*M. Fuscata*) macaques cross-fostered between species show evidence of only limited modification. *Developmental Psychobiology*, 26, 389–406.

Parr, L. A., Hecht, E., Barks, S. K., Preuss, T. M., & Votaw, J. R. (2009). Face processing in the chimpanzee brain. *Current Biology*, 19, 50–53.

Parr, L. A., & Heintz, M. (2009). Facial expression recognition in rhesus monkeys, macaca mulatta. *Animal Behaviour*, 77, 1507–1513.

Partan, S. R. (2002). Single and multichannel signal composition: Facial expressions and vocalizations of rhesus macaques (macaca mulatta). *Behaviour*, 139, 993–1027.

Paus, T. (2001). Primate anterior cingulate cortex: where motor control, drive and cognition interface. *Nature Reviews Neuroscience*, 2(6), 417–424.

Pause, B. M. (2012). Processing of body odor signals by the human brain. *Chemosensory Perception*, 5(1), 55–63.

Pavlou, H. J., & Goodwin, S. F. (2013). Courtship behavior in Drosophila melanogaster: Towards a "courtship connectome". *Current Opinion in Neurobiology*, 23(1), 76–83.

Pelphrey, K. A., Morris, J. P., & McCarthy, G. (2005). Neural basis of eye gaze processing deficits in autism. *Brain*, 128, 1038–1048.

Pereira, M. E., & Macedonia, J. M. (1991). Ringtailed lemur anti-predator calls denote predator class, not response urgency. *Animal Behaviour*, 41, 543–544.

Perrett, D. I., Smith, P. A., Mistlin, A. J., Chitty, A. J., Head, A. S., Potter, D. D., ... Jeeves, M. A. (1985). Visual analysis of body movements by neurones in the temporal cortex of the macaque monkey: A preliminary report. *Behavioral Brain Research*, 16, 153–170.

Perrett, D. I., Smith, P. A., Potter, D. D., Mistlin, A. J., Head, A. S., Milner, A. D., & Jeeves, M. A. (1985). Visual cells in the temporal cortex sensitive to face view and gaze direction. *Proceedings of the Royal Society of London B, Biological Sciences*, 223, 293–317.

Perrodin, C., Kayser, C., Logothetis, N. K., & Petkov, C. I. (2011). Voice cells in the primate temporal lobe. *Current Biology*, 21, 1408–1415.

Petkov, C. I., & Jarvis, E. D. (2012). Birds, primates, and spoken language origins: Behavioral phenotypes and neurobiological substrates. *Frontiers in Evolutionary Neuroscience*, 4. doi: 10.3389/fnevo.2012.00012

Petkov, C. I., Kayser, C., Steudel, T., Whittingstall, K., Augath, M., & Logothetis, N. K. (2008). A voice region in the monkey brain. *Nature Neuroscience*, 11, 367–374.

Petkov, C. I., Logothetis, N. K., & Obleser, J. (2009). Where are the human speech and voice regions, and do other animals have anything like them? *Neuroscientist*, 15, 419–429.

Petkov, C. I., & Wilson, B. (2012). On the pursuit of the brain network for proto-syntactic learning in non-human primates: Conceptual issues and neurobiological hypotheses. *Philosophical Transactions of the Royal Society B, Biological Sciences*, 367(1598), 2077–2088.

Pinsk, M. A., Arcaro, M., Weiner, K. S., Kalkus, J. F., Inati, S. J., Gross, C. G., & Kastner, S. (2009). Neural representations of faces and body parts in macaque and human cortex: A comparative fMRI study. *Journal of Neurophysiology*, 101, 2581–2600.

Pinsk, M. A., DeSimone, K., Moore, T., Gross, C. G., & Kastner, S. (2005). Representations of faces and body parts in macaque temporal cortex: A functional mri study. *Proceedings of the National Academy of Sciences of the USA*, 102, 6996–7001.

Pitcher, D., Dilks, D. D., Saxe, R. R., Triantafyllou, C., & Kanwisher, N. (2011). Differential selectivity for dynamic versus static information in face-selective cortical regions. *NeuroImage*, 56, 2356–2363.

Pokorny, J. J., & de Waal, F. B. M. (2009). Monkeys recognize the faces of group mates in photographs. *Proceedings of the National Academy of Sciences of the USA*, 106, 21539–21543.

Polosecki, P., Moeller, S., Schweers, N., Romanski, L. M., Tsao, D. Y, & Freiwald, W. A. (2013). Faces in motion: Selectivity of macaque and human face processing areas for dynamic stimuli. *The Journal of Neuroscience*, 33, 11768–11773.

Pool, J. L. (1954). The visceral brain of man. *Journal of Neurosurgery*, 11, 45–63.

Pool, J. L., & Ransohoff, J. (1949). Autonomic effects on stimulating rostral portion of cingulate gyri in man. *Journal of Neurophysiology*, 12, 385–392.

Popivanov, I. D., Jastorff, J., Vanduffel, W., & Vogels, R. (2014). Heterogeneous single-unit selectivity in an fmri-defined body-selective patch. *The Journal of Neuroscience*, 34, 95–111.

Poremba, A., Malloy, M., Saunders, R. C., Carson, R. E., Herscovitch, P., & Mishkin, M. (2004). Species-specific calls evoke asymmetric activity in the monkey's temporal poles. *Nature*, 427, 448–451.

Porter, R. H. (1998). Olfaction and human kin recognition. *Genetica*, 104, 259–263.

Prather, J. F., Peters, S., Nowicki, S., & Mooney, R. (2008). Precise auditory–vocal mirroring in neurons for learned vocal communication. *Nature*, 451, 305–310.

Puce, A., Allison, T., Bentin, S., Gore, J. C., & McCarthy, G. (1998). Temporal cortex activation in humans viewing eye and mouth movements. *Journal of Neuroscience*, 18, 2188–2199.

Puce, A., Allison, T., Gore, J. C., & McCarthy, G. (1995). Face-sensitive regions in human extrastriate cortex studied by functional mri. *Journal of Neurophysiology*, 74, 1192–1199.

Rauschecker, J. P., Tian, B., & Hauser, M. (1995). Processing of complex sounds in the macaque nonprimary auditory cortex. *Science*, 268, 111–114.

Recanzone, G. H. (2008). Representation of con-specific vocalizations in the core and belt areas of the auditory cortex in the alert macaque monkey. *Journal of Neuroscience*, 28, 13184–13193.

Remedios, R., Logothetis, N. K., & Kayser, C. (2009a). An auditory region in the primate insular cortex responding preferentially to vocal communication sounds. *Journal of Neuroscience*, 29, 1034–1045.

Remedios, R., Logothetis, N. K., & Kayser, C. (2009b). Monkey drumming reveals common networks for perceiving vocal and nonvocal communication sounds. *Proceedings of the National Academy of Sciences of the USA*, 106, 18010–18015.

Rendall, D., Owren, M. J., & Ryan, M. J. (2009). What do animal signals mean? *Animal Behaviour*, 78, 233–240.

Rendall, D., Rodman, P. S., & Emond, R. E. (1996). Vocal recognition of individuals and kin in free-ranging rhesus monkeys. *Animal Behaviour*, 51, 1007–1015.

Rendell, L., & Whitehead, H. (2001). Culture in whales and dolphins. *Behavioral and Brain Sciences*, 24, 309–324.

Riley, J. R., Greggers, U., Smith, A. D., Reynolds, D. R., & Menzel, R. (2005). The flight paths of honeybees recruited by the waggle dance. *Nature*, 435(1954), 205–207.

Rittschof, C. C., Bukhari, S. A., Sloofman, L. G., Troy, J. M., Caetano-Anollés, D., Cash-Ahmed, A., Stubbs, L. (2014). Neuromolecular responses to social challenge: Common mechanisms across mouse, stickleback fish, and honey bee. *Proceedings of the National Academy of Sciences of the USA*, 111(50), 17929–17934.

Rizzolatti, G., & Craighero, L. (2004). The mirror-neuron system. *Annual Review of Neuroscience*, 27, 169–192.

Robinson, B. (1967). Vocalization evoked from forebrain in Macaca mulatta. *Physiology & Behavior*, 2, 345–354.

Rolls, E. T., Critchley, H. D., Browning, A. S., & Inoue, K. (2006). Face-selective and auditory neurons in the primate orbitofrontal cortex. *Experimental Brain Research*, 170, 74–87.

Romanski, L. M., & Goldman-Rakic, P. S. (2002). An auditory domain in primate prefrontal cortex. *Nature Neuroscience*, 5, 15–16.

Rosa, M. G., & Tweedale, R. (2005). Brain maps, great and small: Lessons from comparative studies of primate visual cortical organization. *Philosophical Transactions of the Royal Society B, Biological Sciences*, 360, 665–691.

Rotshtein, P., Henson, R. N. A., Treves, A., Driver, J., & Dolan, R. J. (2005). Morphing Marilyn into Maggie dissociates physical and identity face representations in the brain. *Nature Neuroscience*, 8, 107–113.

Rowley, I., & Chapman, G. (1986). Cross-fostering, imprinting, and learning in two sympatric species of cockatoos. *Behaviour*, 96, 1–16.

Roy, A., Shepherd, S. V., & Platt, M. L. (2012). Reversible inactivation of psts suppresses social gaze following in the macaque (*Macaca mulatta*). *Social, Cognitive & Affective Neuroscience*, 19(2), 209–217.

Rudebeck, P. H., Buckley, M. J., Walton, M. E., & Rushworth, M. F. S. (2006). A role for the macaque anterior cingulate gyrus in social valuation. *Science*, 313(5791), 1310–1312.

Rushworth, M. F. S., Behrens, T. E. J., Rudebeck, P. H., & Walton, M. E. (2007). Contrasting roles for cingulate and orbitofrontal cortex in decisions and social behaviour. *Trends in Cognitive Sciences*, 11(4), 168–76.

Russ, B. E., Ackelson AL., Baker AE., Cohen YE (2008). Coding of auditory-stimulus identity in the auditory non-spatial processing stream. *Journal of Neurophysiology*, 99, 87–95.

Russ, B. E., & Cohen, Y. E. (2009). Rhesus monkeys' valuation of vocalizations during a free-choice task. *PLoS One*, 4, e7834. doi: 10.1371/journal.pone.0007834

Russ, B. E., Lee, Y. S., & Cohen, Y. E. (2007). Neural and behavioral correlates of auditory categorization. *Hearing Research*, 229, 204–212.

Sadagopan, S., Temiz-Karayol, N. Z., & Voss, H. U. (2015). High-field functional magnetic resonance imaging of vocalization processing in marmosets. *Scientific Reports*, 5, 10950.

Sander, K., & Scheich, H. (2001). Auditory perception of laughing and crying activates human amygdala regardless of attentional state. *Cognitive Brain Research*, 12, 181–198.

Savic, I., Berglund, H., Gulyas, B., & Roland, P. (2001). Smelling of odorous sex hormone-like compounds causes sex-differentiated hypothalamic activations in humans. *Neuron*, 31, 661–668.

Schell, A., Rieck, K., Schell, K., Hammerschmidt, K., & Fischer, J. (2011). Adult but not juvenile barbary macaques spontaneously recognize group members from pictures. *Animal Cognition*, 14, 503–509.

Schino, G. (2001). Grooming, competition and social rank among female primates: A meta-analysis. *Animal Behaviour*, 62, 265–271.

Sergent, J., Ohta, S., & MacDonald, B. (1992). Functional neuroanatomy of face and object processing. A positron emission tomography study. *Brain*, 115 Pt 1, 15–36.

Sescousse, G., Redouté, J., & Dreher, J.-C. (2010). The architecture of reward value coding in the human orbitofrontal cortex. *The Journal of Neuroscience*, 30, 13095–13104.

Sewards, T. V., & Sewards, M. A. (2002). Innate visual object recognition in vertebrates: Some proposed pathways and mechanisms. *Comparative Biochemistry & Physiology Part A. Molecular & Integrative Physiology*, 132, 861–891.

Seyfarth, R. M. (1977). A model of social grooming among adult female monkeys. *Journal of Theoretical Biology*, 65, 671–698.

Seyfarth, R. M., & Cheney, D. L. (1986). Vocal development in vervet monkeys. *Animal Behaviour*, 34, 1640–1658.

Seyfarth, R. M., Cheney, D. L., & Marler, P. (1980). Vervet monkey alarm calls: Semantic communication in a free-ranging primate. *Animal Behaviour*, 28, 1070–1094.

Shepherd, S. V. (2010). Following gaze: Gaze-following behavior as a window into social cognition. *Frontiers in Integrative Neuroscience*, 4. doi: 10.3389/fnint.2010.00005

Shepherd, S. V., & Cappuccio, M. (2011). Sociality, attention, and the mind's eyes. In A. Seelmann (Ed.), *Joint attention: New developments* (pp. 205–42). Cambridge, MA: MIT Press.

Shepherd, S. V., Klein, J. T., Deaner, R. O., & Platt, M. L. (2009). Mirroring of attention by neurons in macaque parietal cortex. *Proceedings of the National Academy of Sciences of the USA*, 106, 9489–9494.

Sherwood, C. C., Hof, P. R., Holloway, R. L., Gannon, P. J., Semendeferi, K., Frahm, H. D., & Zilles, K. (2005). Evolution of the brainstem orofacial motor system in primates: A comparative study of trigeminal, facial, and hypoglossal nuclei. *Journal of Human Evolution*, 48(1), 45–84.

Sliwa, J., Duhamel, J. R., Pascalis, O., & Wirth, S. (2011). Spontaneous voice-face identity matching by rhesus monkeys for familiar conspecifics and humans. *Proceedings of the National Academy of Sciences of the USA*, 108, 1735–1740.

Sliwa, J., Planté, A., Duhamel, J-R., & Wirth, S. (2016). Independent neuronal representation of facial and vocal identity in the monkey hippocampus and inferotemporal cortex. *Cerebral Cortex*, 26(3), 950–966.

Smith, T. (2006). Individual olfactory signatures in common marmosets (callithrix jacchus). *American Journal of Primatology*, 68, 585–604.

Smith, W. K. (1945). The functional significance of the rostral cingular cortex as revealed by its responses to electrical excitation. *Journal of Neurophysiology*, 8(4), 241.

Sokolowski, M. B. (2010). Social interactions in "simple" model systems. *Neuron*, 65(6), 780–94.

Stolk, A., Verhagen, L., Schoffelen, J.-M., Oostenveld, R., Blokpoel, M., Hagoort, P., ... Toni, I. (2013). Neural mechanisms of communicative innovation. *Proceedings of the National Academy of Sciences of the USA*, 110, 14574–14579.

Straube, B., Green, A., Bromberger, B., & Kircher, T. (2011). The differentiation of iconic and metaphoric gestures: Common and unique integration processes. *Human Brain Mapping*, 32, 520–533.

Sugase, Y., Yamane, S., Ueno, S., & Kawano, K. (1999). Global and fine information coded by single neurons in the temporal visual cortex. *Nature*, 400, 869–873.

Sugita, Y. (2008). Face perception in monkeys reared with no exposure to faces. *Proceedings of the National Academy of Sciences of the USA*, 105(1), 394–398.

Sugiura, M., Kawashima, R., Nakamura, K., Sato, N., Nakamura, A., Kato, T., ... Fukuda, H. (2001). Activation reduction in anterior temporal cortices during repeated recognition of faces of personal acquaintances. *NeuroImage*, 13, 877–890.

Suomi, S. J., Harlow, H. F., & Kimball, S. D. (1971). Behavioral effects of prolonged partial social isolation in the rhesus monkey. *Psychological Reports*, 29, 1171–1177.

Taglialatela, J. P., Russell, J. L., Schaeffer, J. A., & Hopkins, W. D (2009). Visualizing vocal perception in the chimpanzee brain. *Cerebral Cortex*, 19, 1151–1157.

Talairach, J., Bancaud, J., Geier, S., Bordas-Ferrer, M., Bonis, A., Szikla, G., & Rusu, M. (1973). The cingulate gyrus and human behaviour. *Electroencephalography and Clinical Neurophysiology*, 34, 45–52.

Tate, A. J., Fischer, H. H., Leigh, A. E., & Kendrick, K. M. (2006). Behavioural and neurophysiological evidence for face identity and face emotion processing in animals. *Philosophical Transactions of the Royal Society of London B, Biological Sciences*, 361(1476), 2155–72.

Terrace, H. S., Petitto, L.-A., Sanders, R. J., & Bever, T. G. (1979). Can an ape create a sentence? *Science*, 206, 891–902.

Tian, B., Reser, D., Durham, A., Kustov, A., & Rauschecker, J. P. (2001). Functional specialization in rhesus monkey auditory cortex. *Science*, 292, 290–293.

Trupe, L. A., Varma, D. D., Gomez, Y., Race, D., Leigh, R., Hillis, A. E., & Gottesman, R. F. (2013). Chronic apraxia of speech and Broca's area. *Stroke*, 44, 740–744.

Tsao, D. Y., Freiwald, W. A., Knutsen, T. A., Mandeville, J. B., & Tootell, R. B. (2003). Faces and objects in macaque cerebral cortex. *Nature Neuroscience*, 6, 989–995.

Tsao, D. Y., Freiwald, W. A., Tootell, R. B., & Livingstone, M. S. (2006). A cortical region consisting entirely of face-selective cells. *Science*, 311, 670–674.

Tsao, D. Y., Moeller, S., Freiwald, W. A. (2008). Comparing face patch systems in macaques and humans. *Proceedings of the National Academy of Sciences*, 105, 19514–19519.

Tsao, D. Y., Schweers, N., Moeller, S., & Freiwald, W. A. (2008). Patches of face-selective cortex in the macaque frontal lobe. *Nature Neuroscience*, 11, 877–879.

Vanderhorst, V. G., Terasawa, E., Ralston, H. J., & Holstege, G. (2000). Monosynaptic projections from the lateral periaqueductal gray to the nucleus retroambiguus in the rhesus monkey: Implications for vocalization and reproductive behavior. *Journal of Comparative Neurology*, 424, 251–268.

Vangeneugden, J., Peelen, M. V., Tadin, D., & Battelli, L. (2014). Distinct neural mechanisms for body form and body motion discriminations. *The Journal of Neuroscience*, 34, 574–585.

Van Hooff, J. A. R. A. M. (1967). The facial displays of the catarrhine monkeys and apes. In D. Morris (Ed.), *Primate ethology* (new ed. 2006, pp. 7–68). London: Wiedenfeld & Nicolson.

Von Frisch, K. (1967). *The dance language and orientation of bees.* Cambridge, MA: The Belknap Press of Harvard University Press.

von Kriegstein, K., Eger, E., Kleinschmidt, A., & Giraud, A. L. (2003) Modulation of neural responses to speech by directing attention to voices or verbal content. *Cognitive Brain Research*, 17, 48–55.

von Kriegstein, K., & Giraud, A. L. (2004) Distinct functional substrates along the right superior temporal sulcus for the processing of voices. *Neuroimage*, 22, 948–955.

Vouloumanos, A., Kiehl, K. A., Werker, J. F., & Liddle, P. F. (2001). Detection of sounds in the auditory stream: Event-related fmri evidence for differential activation to speech and nonspeech. *Journal of Cognitive Neuroscience*, 13, 994–1005.

Vuilleumier, P., Armony, J. L., Driver, J., & Dolan, R. J. (2003). Distinct spatial frequency sensitivities for processing faces and emotional expressions. *Nature Neuroscience*, 6, 624–631.

Waitt, C., Gerald, M. S., Little, A. C., & Kraiselburd, E. (2006). Selective attention toward female secondary sexual color in male rhesus macaques. *American Journal of Primatology*, 68, 738–744.

Waitt, C., Little, A. C., Wolfensohn, S., Honess, P., Brown, A. P., Buchanan-Smith, H. M., & Perrett, D. I. (2003). Evidence from rhesus macaques suggests that male coloration plays a role in female primate mate choice. *Proceedings of the Royal Society B, Biological sciences*, 270 Suppl 2, S144–146.

Wang, X. (2000). On cortical coding of vocal communication sounds in primates. *Proc Natl Acad Sci U S A.*, 97, 11843–11849.

Wang, X, & Kadia, S. C. (2001). Differential representation of species-specific primate vocalizations in the auditory cortices of marmoset and cat. *Journal of Neurophysiology*, 86, 2616–2620.

Wang, X., Merzenich, M. M., Beitel, R., & Schreiner, C. E. (1995). Representation of a species-specific vocalization in the primary auditory cortex of the common marmoset: Temporal and spectral characteristics. *Journal of Neurophysiology*, 74, 2685–2706.

Wanker, R., Sugama, Y, & Prinage, S. (2005). Vocal labelling of family members in spectacled parrotlets, *forpus conspicillatus*. *Animal Behaviour*, 70, 111–118.

Watson, S. K., Townsend, S. W., Schel, A. M., Wilke, C., Wallace, E. K., Cheng, L., ... Slocombe, K. E. (2015). Vocal learning in the functionally referential food grunts of chimpanzees. *Current Biology*, 25, 495–499.

Weintstein, E., & Bender, M. (1943). Integrated facial patterns elicited by stimulation of the brain stem. *Archives of Neurology and Psychiatry*, 789, 34–42.

Weisfeld, G. E., Czilli, T., Phillips, K. A., Gall, J. A., & Lichtman, C. M. (2003). Possible olfaction-based mechanisms in human kin recognition and inbreeding avoidance. *Journal of Experimental Child Psychology*, 85, 279–295.

West, R. A., & Larson, C. R. (1995). Neurons of the anterior mesial cortex related to faciovocal activity in the awake monkey. *Journal of Neurophysiology*, 74(5), 1856–1869.

Wheeler, B. C., & Fischer, J. (2012). Functionally referential signals: A promising paradigm whose time has passed. *Evolutionary Anthropology: Issues, News, and Reviews*, 21, 195–205.

Wyatt, T. D. (2014). Pheromones and Animal Behavior: Chemical Signals and Signatures (Second). Cambridge, UK: Cambridge University Press.

Yang, C. (2013). Ontogeny and phylogeny of language. *Proceedings of the National Academy of Sciences of the USA*, 110, 6324–6327.

Yun, K., Watanabe, K., & Shimojo, S. (2012). Interpersonal body and neural synchronization as a marker of implicit social interaction. *Scientific Reports*, 2. doi: 10.1038/srep00959

Zhu, Q., Nelissen, K., Van den Stock, J., De Winter, F.-L., Pauwels, K., de Gelder, B., ... Vandenbulcke, M. (2013). Dissimilar processing of emotional facial expressions in human and monkey temporal cortex. *NeuroImage*, 66, 402–411.

17

Social Coordination
From Ants to Apes
Anne Böckler, Anna Wilkinson, Ludwig Huber, and Natalie Sebanz

17.1 Introduction

Coordinating actions with others has significant benefits and allows individuals in groups to achieve outcomes they could not achieve on their own. At the same time, living in groups comes with many challenges that are not encountered by solitary individuals. In the last decade, researchers have become more and more interested in the processes underlying different kinds of coordinated group behavior (e.g., Couzin & Krause, 2003; Sebanz, Bekkering, & Knoblich, 2006; Semin & Smith, 2008; Van Overwalle & Baetens, 2009). This ranges from the entrainment and synchronization of actions (Marsh et al., 2009; Schmidt et al., 2011) to the performance of complementary actions in the context of distributed roles (Knoblich, Butterfill, & Sebanz, 2011; Knoblich & Sebanz, 2008). Coordination can rely on informational couplings that occur spontaneously and increase the probability of coordinated behavior, independently of joint plans. However, individuals sometimes also plan their own part of an action in relation to others' actions, guided by mental representations of desired joint outcomes.

Given the broad scope of coordination phenomena and their underlying mechanisms, understanding coordination clearly asks for an interdisciplinary approach. One ingredient in such an approach is a comparative analysis of coordination in different species. What coordination problems do animals that live in groups confront, what mechanisms underlie the solutions to these coordination problems, and what are the benefits? Coordination problems refer to requirements of social situations in which individuals in a group need to implement or maintain particular relations between their actions. With the present overview we aim to provide a comparative perspective on coordination, exploring whether common solutions to particular coordination problems—such as moving in a group, distributing tasks, coordinating actions in time, and transmitting information—exist across different species. This review thus complements prior work on cooperation (Clutton-Brock, 2009; Dugatkin, 2002; Gigerenzer, 2008) which addressed principles underlying the exchange of resources and services. The study of coordination tries to understand how individuals in a cooperative or neutral (i.e., non-competitive) context adjust their actions to each other in time and space.

We do not aim to provide a comprehensive overview of coordination mechanisms in all existing species and neither can we draw any final conclusions on how such mechanisms evolved. Rather, we will illustrate similarities in the ways some common coordination problems are tackled, drawing on studies that include a range of different, mostly group-living species. As such, this overview provides a basis for speculating on the role of convergent evolution in particular coordination mechanisms and on the cognitive abilities needed to engage in more complex joint actions, such as flexible task distributions or active teaching.

17.2 Moving in Groups

Group motion is a paradigmatic case of coordination. Some group-living animals can cover great distances by marching, flying, or swimming together for up to several months (Alerstam, Hedenstrom, & Akesson, 2003; Ran et al., 2008). On a smaller time scale, individuals within a group manage to get away from predators by moving rapidly, yet unpredictably. To successfully move as one, two main tasks have to be accomplished by crowds, herds, flocks, and schools: Individuals have to achieve and maintain a stable formation and they have to be able to quickly and flexibly respond to sudden changes in the environment by adjusting their movement direction.

17.2.1 Formation

Traveling with a large number of conspecifics requires individuals to move in a manner appropriate to maintaining group cohesion whilst avoiding collision. Moving groups are often too large for any one individual to survey the complete group and in many cases, there is not one individual guiding or leading the others.

Despite the apparent complexity of this task it appears that simple and general principles control individual behavior. The local interaction hypothesis (Couzin & Krause, 2003; Reynolds, 1987) suggests that, rather than keeping track of the entire group, an individual merely aligns its movements with the movements of surrounding individuals. For example, starlings (*Sturnus vulgaris*) adjust their flight pattern to that of six to seven neighbors, irrespective of the distance between them and independent of the size of the flock (Cavagna et al., 2007). In locusts (*Schistocerca gregaria*), alignment of movements increases with increasing density of individuals (Buhl et al., 2006).

Local interaction seems to underlie mammalian herding (Krause & Ruxton, 2002) and much of human crowd behavior as well (Dyer et al., 2008; for a comprehensive review of herding in humans, see Raafat, Chater, & Frith, 2009). Individual wildebeest, for instance, line up direction and speed of movement with that of the herd by simply regulating the distance to neighboring individuals (Chao & Levin, 2007). Typical walking patterns in human pedestrians also emerge from local interactions among individuals of the walking group. When a group of people deliberately walks together (families or friends), group motion can additionally be adjusted to enable communication: Low-density supports walking side by side, while increasing density fosters V-like patterns that facilitate conversation (Moussaïd, Perozo, Garnier, Helbing, & Theraulaz, 2010).

By every individual aligning its movement speed and direction with those of surrounding others, groups can form stable self-organized configurations that do not necessitate signaling (e.g. vocalizations) or a stable group leader. In species characterized by dominance hierarchies, leaders can initiate group movements and modulate their directions or durations while principles of local interaction still hold (Petit, Gautrais, Leca, Theraulaz, & Deneubourg, 2009). For instance, in capuchin monkeys, *Cebus capuchinus*, collective movements follow principles of self-organization despite movements being initiated by single individuals (Petit et al., 2009). It seems that different individuals can take the lead in different instances. Group motion, hence, is based on principles of collective movement and can be influenced by individual leadership.

Moving in formations allows individuals in groups to save energy by making use of aerodynamic principles: When flying in formations, pink-footed geese, (*Anser brachyrhynchus*: Cutts & Speakman, 1994); pelicans (*Pelecanus onocrotalus*: Weimerskirch, 2001); and ducks (*Anus platyrhynchos*: Fish, 1995) appear to save up to 30% of the energy they would need to invest when moving alone. Tuna fish take advantage of induced velocity fields of others' tail strokes by traveling in diamond-shaped formations which allows them to reach a higher speed than an individual could achieve alone (Stöcker, 1999). It has been suggested that similar principles are made use of by professional runners or cyclists who reduce wind resistance and preserve energy by forming groups (Kyle, 1979; Olds, 1998).

17.2.2 Responding to Changes in the Environment

In moving groups, adaptation to sudden changes in the environment requires fast decision making. Few individuals may possess relevant information at a given time (e.g., about the location of a predator), therefore rapid information transmission is crucial. Furthermore, individuals may differ with regard to their preferred movement direction, which necessitates consensus decisions to prevent the group from splitting (Couzin, Krause, Franks, & Levin, 2005).

How does information about movement direction spread across a large group? Evidence from fish (three-spine stickleback, *Gasterosteus aculeatus*) suggests that responses to sudden changes in neighbors' direction are highly nonlinear: If just one individual changes direction the group ignores it. However, the probability of other group members adjusting to sudden changes in direction increases logarithmically as more of their neighbors do (Ward, Sumpter, Couzin, Hart, & Krause, 2008). Thus, the more individuals change direction, the faster information spreads across the whole group. This transmission of information is rapid and efficient and maintains group cohesion without requiring individuals to recognize each other or to employ signals such as vocalizations (Couzin et al., 2005).

In an experiment in which groups of stickleback had to reach consensus over which replica fish to follow, the simple behavioral rule of using the majority's decision proved efficient in reaching near optimal solutions. Accordingly, an increase in group size improved decision quality (Sumpter, Krause, James, Couzin, & Ward, 2008). Experiments on human crowds revealed that a few informed individuals are sufficient to efficiently change the movement behavior of the group (Dyer et al., 2008). Groups of people in these experiments were instructed to walk within a circular arena. People could walk wherever they wanted, but had to stay together as a group. Some individuals received prior information about where to go without leaving the group. Results showed that informed individuals could guide the uninformed group to the designated target without communication or signaling. When conflicting information was provided to different individuals, the group decided in favor of the majority (Dyer et al., 2008). This pattern suggests that group motion in humans is also based on alignment principles and that consensus decision making on movement directions may rely on similar mechanisms of information spreading as found in other group-living animals.

A major advantage of moving in groups is that it helps individuals to escape from predators. The increased size and density of the group are protective factors. The more individuals a group contains, the greater the chance that a predator will be discovered by one of them (the "many eyes" theory, see Krause & Ruxton, 2002). Furthermore, singling out and capturing individuals becomes more difficult for predators the more homogenous the individuals in a group are (the "confusion effect," see Krause & Ruxton, 2002). Animals preying upon fish (Carere et al., 2009), starlings (Krause & Ruxton, 2002), or wildebeest (Chao & Lewin, 2007) make fewer kills when the prey are in larger groups than when they are in smaller ones.

Taken together, simple mechanisms such as local interaction and nonlinear responses to motion changes allow large groups to move as one, rapidly transmit information, save energy, enhance speed, and avoid predation. These processes appear to be common to most group-living animals and play a crucial role in group motion of fish schools, bird flocks, mammal herds, and human crowds.

17.3 Working Together

In group-living species, efficient foraging, hunting, or territory protection is often best achieved when tasks are divided between individuals (Anderson, Theraulaz, & Deneubourg, 2002; Sendova-Franks & Franks, 1999). This part of the chapter will consider the mechanisms underlying the distribution of tasks and the integration of individual contributions.

17.3.1 Distributing Tasks

When transporting large objects, building nests, fighting intruders, or hunting, social animals divide labor, either by partitioning tasks sequentially or by forming teams in which different individuals work simultaneously on different tasks (Anderson & Franks, 2001; Jeanne, 1986; Ratnieks & Anderson, 1999). For example, in termites (*Hodotermes mossambicus*) "workers" cut grass and others subsequently transport it to the nest (Leuthold, Bruinsma, & Huis, 1976).

How are tasks distributed? In large decentralized groups where individual recognition is largely absent, division of labor can be achieved by fixed roles (Anderson & Franks, 2001). For example, when weaver ants (*Oecophylla longinoda*) build nests, some individuals hold the leaves, others produce glue, and yet others glue the leaves together (Hölldobler & Wilson, 1983). Some tasks even demand physiological specialization such as nest defense in *Pheidole pallidula* ants, in which intruders are first pinned down by the "minors" and then decapitated by specially-enlarged mandibles of the "majors" (Detrain & Pasteels, 1992).

Species living in somewhat smaller, stable dominance hierarchies in which individuals recognize each other also divide tasks by allocating roles, but role distribution is somewhat more flexible (Chase, 2002). Cooperative hunting, for instance, can be performed by all group members (e.g. African wild dogs, *Lycaon pictus*, see McFarland, 1985) or by "hunting parties" of variable sizes (e.g. Harris hawks, *Parabuteo unicinaus*, see Bednarz, 1988). The size of the hunting group can be adapted to prey difficulty (Scheel & Packer, 1991) and individuals may take different roles and positions in different attacks, such as lions (*Panthera leo*) circling and killing prey (Stander, 1992).

Group hunting behavior in some species is initiated by dominant individuals. In wolves, *Canis lupus*, the alpha pair initiates hunting trips and performs most of the hunts (White, 2001). Since the dominant pair in free-ranging wolves are usually the ones with most hunting experience (e.g., the parents) this seems a way to enhance success (Mech, 1999). In general, hunting success might be a driving factor in role specialization. Experienced group hunters enhance the likelihood of catching prey by taking stable roles. For instance, evidence suggests that lionesses (*Pathera leo*) circle prey and stalk it towards waiting group members who then catch and kill it. Even though hunting behavior is flexible and depends on the composition of the group, individuals repeatedly occupy similar roles in the hunting formation. In those who circle and stalk, for instance, pairs of experienced hunters tend to take opposite positions in the circle and enhance hunting success by keeping the prey in between them (Stander, 1992).

Roles can also be entirely specialized as a consequence of individuals' expertise in performing particular parts. Bottlenose dolphins, *Tursiops truncates*, for instance, display two separate roles when hunting fish in that there are some individuals forming a barrier and one individual driving prey towards the herding individuals (Gazda, Conner, Edgar, & Cox, 2005). It has been shown that the driving role is always occupied by the same individual. Adopting highly specialized complementary roles enables hunts that are extremely difficult to perform. When chimpanzees, *Pan troglodytes*, chase and circle small monkeys they perform specialized hunting roles that take up to 20 years to learn (Boesch, 2003; Boesch & Boesch, 1989). A similarly slow time course in acquiring particularly sophisticated hunting behaviors is observed in human hunter-gatherer societies (Gurven, Kaplan, & Gutierrez, 2006).

Recent evidence from joint action studies in humans suggests that roles in a joint task are distributed depending on relative task difficulty (Vesper, van der Wel, Knoblich, & Sebanz, 2012), so that the individual taking care of the more difficult part of the joint action makes less effort to coordinate than the individual performing the easier task. It has also been shown that certain joint action tasks favor leader–follower distributions, where complementary roles emerge spontaneously in the service of coordination (Richardson, Harrison, May, Kallen, & Schmidt, 2011).

In humans, where many joint actions involve role specialization, evidence suggests that, when tasks are distributed across two coactors, people keep track of the other's part of the task even when this is not relevant for their own performance (Atmaca, Sebanz, & Knoblich, 2011; Böckler, Knoblich, & Sebanz, 2012; Sebanz, Knoblich, & Prinz, 2003). In these experiments, pairs of participants sit next to each other, each responding to a particular stimulus feature on a computer screen by carrying out a specific action (e.g., one of them pressing an assigned response button when seeing the letter S while the other presses another button when perceiving the letter H). People often show performance patterns similar to when they carry out both parts of the task alone. For instance, when the relevant stimulus letter is surrounded by distracter letters, people perform worse when the distracting letters are linked to the coactor's response (Atmaca et al., 2011). The same pattern is seen when people carry out the same task alone, but respond to both letters by pressing two different response buttons (Eriksen & Eriksen, 1974). Results like these suggest that people, when performing specialized roles, automatically take into account aspects of the task that are relevant for the other. There is some evidence to suggest that the ability to form representations of a coactor's task is linked to the ability to attribute mental states to others (Humphreys & Bedford, 2011; Ruys & Aarts, 2010) but future work is needed to clarify what kind and level of mind-reading is required.

17.3.2 Coordinating Individual Actions

While distributing different parts or roles in joint actions facilitates action planning, many coordinated interactions require fine-grained temporal and spatial adjustments of individuals' actions toward each other. When groups are large or decentralized, an important means of coordination is signaling, for instance by leaving chemical traces in the environment. Using pheromones, ants coordinate foraging (Hölldober & Wilson, 1990; Vittori et al., 2006) by reinforcing shorter paths (Deneubourg, Aron, Goss, & Pasteels, 1990) and marking high-quality food (Jackson & Chaline, 2007). To ensure that this environmental information is available for a long period of time, slow-evaporating chemicals are used, while short-lived chemicals serve to enhance the saliency of particular sites (Jackson & Chaline, 2007). Patterns of contact can also serve as a sign: When a "passive" harvester ant, *Pogonomyrmex barbatus*, at the nest side has had a certain number of interactions with successful foragers returning to the nest, it becomes an "active" forager as well (Greene & Gordon, 2007). This positive feedback loop enables flexible and efficient adaptation of the number of foragers to a given workload.

Signaling also plays an important role in human coordination. When performing tasks that involve a physical coupling between coactors (e.g., pulling on different sides of a pendulum to make it move at certain frequencies) people increase movement forces and produce more overlapping forces, thereby enhancing haptic information flow between them (van der Wel, Knoblich, & Sebanz, 2011). Musicians rely on visual signaling like exaggerating head or arm movements in order to enhance acuity of temporal coordination (Goebl & Palmer, 2009). Recent studies have shown that people modulate their hand movements to indicate which part of an object they are going to grasp (Sacheli, Tidoni, Pavone, Aglioti, & Candidi, 2013). This indicates that instrumental actions can be modified to serve as communicative signals (Csibra & Gergely, 2009; Pezzulo & Dindo, 2011; Pezzulo, Donnarumma, & Dindo, 2013).

Learning about contingencies between one's own and others' actions and following heuristics can also facilitate coordination. For instance, when capuchin monkeys, *Cebus apella*, were able to gain access to food by simultaneously pulling on bars with another agent, they learned the contingency between the presence of another agent and success, and they adjusted their pulling behavior accordingly (Mendres & de Waal, 2000). Similarly, chimpanzees learn which of their previous collaborators was efficient and choose this individual for further collaborations (Melis, Hare, & Tomasello, 2006). Furthermore, coordination can be made more precise, temporally and spatially, when the individuals in a group acquire coordination rules. Group hunts, for instance, require that individuals learn to spatially align with at least one group member and adjust the starting point and/or speed of the attack on the basis of the other hunters' behavior. Although cooperative hunts often appear to involve strategic planning, simple algorithms such as "always move where you can keep prey in between you" may underlie this behavior (Holekamp, Boydston, & Smale, 2000). Thus, learning to quickly respond to others' moves so as to minimize distance to the prey and to keep equidistant from cohunters is at the core of most group hunts, even when they are extremely demanding. Chimpanzees, *Pan troglodytes*, when hunting small monkeys who move fast and unpredictably in all three spatial dimensions, learn how to constantly monitor and anticipate the behavior of both the prey and their cohunters (Boesch & Boesch, 1989; Boesch, 2003). When joint actions require precise temporal coordination, humans have been shown to apply simple heuristics such as speeding up their movements in order to enhance predictability. For instance, pairs of participants were requested to time their responses to stimuli so that their actions were performed simultaneously. By increasing response speed as much as possible, participants reduced the variability of their response times and increased the predictability of their actions (Vesper, van der Wel, Knoblich, & Sebanz, 2011). It has recently been shown that macaque monkeys, *Macaca mulatta*, also reduce the variability of their actions in a joint coordination task (Visco-Comandini et al., 2015).

In addition to signaling and the use of heuristics, research on joint action in humans points towards a crucial role for action prediction based on the observer's own motor repertoire (Sebanz & Knoblich, 2009; Van Overwalle & Baetens, 2009). It is thought that aspects of other coactors' actions can be predicted by running internal forward models that also serve to make predictions for one's own actions (Keller, Knoblich, & Repp, 2007; Wolpert, Doya, & Kawato, 2003). The ability to engage in a "motor simulation" of this kind requires a system that matches perceived actions with corresponding motor representations (Prinz, 1997; Wilson & Knoblich, 2005). There is much evidence for the existence of such a system in macaque monkeys as well as in humans (Rizzolatti & Sinigaglia, 2010). In macaques, cells in the premotor and parietal cortex fire both when the monkey performs an action and when the monkey observes the same actions being performed. Studies in humans have shown that observers selectively engage in motor simulations when they expect that an interaction partner is going to act, in contrast to a person who always acts alone (Kourtis, Sebanz, & Knoblich, 2010). These simulations help predict the timing of a partner's action, as when an individual receiving an object is predicting the onset of the action performed by the person handing over the object to them (Kourtis, Knoblich, Wozniak, & Sebanz, 2014). However, it is unlikely that having a perception–action matching system (e.g., in the form of mirror neurons) is sufficient for performing joint actions, because such a system by itself does not allow one to plan one's actions in relation to others'. It has been argued, for instance, that cognitive processes involved in self–other distinction and in inferring and reasoning about others' beliefs and intentions play important roles in human social interaction (Csibra, 2007; Kanske, Böckler, Trautwein, & Singer, 2016; Eskenazi, Rueschemeyer, de Lange, Knoblich, & Sebanz, 2015).

In sum, by dividing labor based on specialized roles, group-living species enhance reliability (Oster & Wilson, 1978), increase learning and individual efficiency (Stander, 1992), and make outcomes more successful. The use of simple signals allows groups to exhibit a "super-efficiency" (Anderson & Franks, 2001) that could not be attained by single individuals. Applying behavioral heuristics in, for instance, group hunting allows individuals to chase more and larger prey with less effort

(Bednarz, 1988; Creel, 1997; van Horn, Engh, Scribner, Funk, & Holekamp, 2004). In addition, humans corepresent their interaction partners' task when roles are distributed and predict others' actions based on motor simulation processes.

17.3.3 Learning from One Another

Knowledge and skills are vital benefits that can be unequally distributed among group members. How are knowledge and skills transferred? In this part of the chapter, we will explore the possibility that coordination plays a role in the transmission of (cultural) knowledge. Coordination can play a direct role in the transfer of knowledge, for example, when individuals of a group engage in teaching behavior (see below). Somewhat more indirectly, cultural evolution can rely on individuals using information that arises from cues produced by the behavior of others ("public information," see Danchin, Giraldeau, Valone, & Wagner, 2004; Kleinhappel, Burman, John, Wilkinson, & Pike, 2014; in press), for instance, by feeding at sites which are preferred by others.

Many group-living species gain information about the location of food or danger by following the gaze of conspecifics (Itakura, 2004; Frischen, Bayliss, & Tipper, 2007). This behavior has been demonstrated in domestic mammals such as goats (*Capra hircus*: Kaminski, Riedel, Call, & Tomasello, 2005) and dogs (*Canis familiaris*: Miklósi & Soprani, 2006; Miklósi, Topál, & Csányi, 2004), in dolphins (*Tursiops truncates*: Pack & Herman, 2006), in various species of human and nonhuman primates (e.g., Pongo pygmaeus, Gorilla gorilla, Pan paniscus, and Pan troglodytes: Bräuer, Call, & Tommasello, 2005; *Macaca nemestrina*: Ferrari, Fogassi, & Gallese, 2000; Homo sapiens: Friesen & Kingstone, 1998; Eulemur fulvus and Eulemur macaco: Ruiz, Gómez, Roeder, & Byrne 2009; *Pan troglodytes, Cercocebus atys torquatus, Macaca mulatta, Macaca arctoides*, and *Macaca nemestrina*: Tomasello, Call, & Hare, 1998), in birds including ravens (Corvus corax: Bugnyar, Stowe, & Heinrich, 2004) and northern bald ibises (*Geronticus eremite*: Loretto, Schloegl, & Bugnyar, 2009), and even in a species of reptile (*Geochelone carbonaria*: Wilkinson, Mandl, Bugnyar, & Huber, 2010). Thus, by shifting visual attention to where conspecifics are looking, many group-living species gain information from informed individuals (Fitch, Huber, & Huber 2010).

Moreover, it has been shown that mammals, birds, reptiles, fish, and insects can learn skills and acquire knowledge by observing the behavior of conspecifics (Danchin et al., 2004; Galef & Laland, 2005; Huber et al. 2009; Kis, Huber, & Wilkinson, 2015; Laland & Galef, 2009; Leadbeater & Chittka, 2007; Schuster, Wohl, Griebsch, & Klostermeier, 2006; Wilkinson et al., 2010; Zentall, 2004). Recent empirical evidence documents social influences on food choice, tool use, patterns of movement, predator avoidance, mate choice, courtship, and song learning (Galef & Laland, 2005; Hoppitt & Laland, 2008). Some animals, accordingly, are able to copy what they have observed before: that is, they imitate. Marmosets, *Callithrix jacchus*, for instance, copy the novel or otherwise improbable action of a conspecific by using the same body part (Voelkl & Huber, 2000) and by matching the movement trajectory (Voelkl & Huber, 2007). In addition to such process-oriented copying, animals have been found to reproduce the result or effect of a demonstration by applying an action other than that used by the model (product-oriented copying; Tennie, Call, & Tomasello 2009a; Whiten, McGuigan, Marshall-Pescini, & Hopper, 2009).

For example, when observing conspecifics opening a box, keas, *Nestor notabilis*, can learn about the affordances or the operating mechanisms of the objects involved (emulation). This is reflected in keas opening the boxes in their own way after observing conspecifics, instead of copying the demonstrated movements (Huber, Rechberger, & Taborsky, 2001; see also Horner & Whiten, 2005).

Finally, some species engage in teaching (Hoppitt & Laland, 2008). Teaching implies the active transfer of knowledge and skills from an informed to an uninformed individual, a behavior that is costly to the teacher (or at least does not provide any benefits) and is performed only when an uninformed animal is present (Caro & Hauser, 1992). A prerequisite of teaching is attracting the attention of the learner during demonstration by ostensive cuing. This is not only a typical element of human teaching (Böckler, Knoblich, & Sebanz, 2011; Csibra & Gergely, 2009), but can be found in nonhuman animals as well (Range, Heucke, Gruber, Konz, Huber, & Virányi, 2009). A less obvious aspect of teaching is that it often involves a substantial amount of inter-individual coordination. Knowledgeable ants, *Temnothorax albipennis*, for instance, teach uninformed individuals about food sites by running in tandem with them (Franks & Richardson, 2006). During the trip the follower performs orientation loops to memorize landmarks, and the leader slows down to wait (for an alternative explanation of the findings see Leadbeater, Raine, & Chittka, 2006). Similarly, wild meerkats, *Suricata suricatta*, actively teach pups how to handle potentially dangerous prey, such as scorpions, by providing prey to very young pups only after killing or disabling it (removing the sting) and by subsequently monitoring and encouraging the pups' handling of prey (Thornton & McAuliffe, 2006). With older pups, mobile prey is released in front of them, providing them with the opportunity to catch and kill it themselves. This behavior is costly to helpers in terms of loss of prey, but is an efficient and successful teaching strategy that involves a significant amount of coordination (Thornton & McAuliffe, 2006).

Making use of public information, gaze following, social learning (Heyes, 2009; Huber et al., 2009), and teaching allow naïve individuals to gain information and to learn to perform behaviors that would otherwise not have been acquired or not acquired so rapidly. Thus, by means of observing and learning from others, complex behavior is conveyed and behavioral variations are transferred, which has been argued to form the basis of culture (Seyfarth, Cheney, & Bergman, 2005; Whiten & van Schaik, 2007). It remains to be further explored how inter-individual coordination mechanisms facilitate cultural learning.

17.4 Conclusions

In this chapter we have reviewed mechanisms that enable different group-living species to solve coordination problems such as moving together both stably and flexibly, dividing tasks and combining individual actions, and transmitting information and skills. While some of the evolved mechanisms may be unique to a given species, others form the basis of increasingly complex behavior in many different species.

The similarities in how coordination problems are solved raise the question of how mechanisms can be so widespread across species of unrelated ancestry and points toward the role of convergent evolution. Coordination problems constitute an important element of the environments confronting all group-living individuals. Like the development of similar biological features in unrelated species (e.g., the development of wings in birds and bats; Morris, 2003), solutions to coordination problems may have evolved in different species on the basis of similar group-living requirements.

Consider, for instance, the principles underlying stable formation and rapid decision making in group motion. Moving in large flocks, schools, herds, or crowds is enabled by comparable mechanisms across many species, such as by aligning one's own movement with the movement of nearby others and by responding nonlinearly to movements shared by larger groups. During group motion, physical constraints are posed on each individual concerning the way they can move in a given medium (water or air, for instance) and in a group of given size and density. Above all, the amount of information from other group members that an individual can process is limited by the sensory and cognitive system. Aligning movements with immediate neighbors (local interactions) and responding to movement changes in subpopulations in a logarithmic fashion may have proven the most efficient strategies to deal with these internal and external constraints and may therefore have evolved in a range of contexts and species. Interestingly, these simple principles can be elaborated by other factors. Dominance in social hierarchy, for instance, can play a role in the initiation of group motion (Petit et al, 2009) and the shape of a moving group of humans is determined by whether it enhances communication fluency (Moussaïd et al., 2010). Whether these and other additional factors affect group motion in a broader range of species is a question for future research.

A second coordination problem we have addressed concerns the joint performance of tasks. When building or defending their habitat and when hunting together, many group-living individuals divide tasks according to more or less flexible roles and coordinate individual efforts in space and time by signaling (using chemical, auditory, or visual signals), motor simulation, and/or by applying heuristics. To what extent do common mechanisms underlie joint task coordination? How did role-specialization and coordination heuristics evolve in so many different species and across so many different tasks? In order to get at the question of whether convergent evolution is in play, future research needs to address how similar environmental and sensory/cognitive factors contribute to these mechanisms in different species. An obvious factor that may have driven the evolution of some of these mechanisms is the difficulty of catching prey in group hunting (Boesch & Boesch, 1989; Stander, 1992). Having individuals specializing in particular roles and coordinating behaviors in time and space allows groups to perform increasingly complex hunts of prey that is larger, faster, and/or stronger than its predators.

An interesting question, in this context, concerns the role of behavioral flexibility. While some species display strict (or even anatomical) specialization of role, roles in other species are distributed more flexibly. Similarly, some animals coordinate individual actions stringently based on signals while others flexibly adjust their behavior to the actual and anticipated behavior of conspecifics. Behavioral flexibility poses certain cognitive requirements (e.g., to be able to perform different roles, to choose the adequate role at a given time, to switch between roles, to anticipate others' actions) and therefore might be limited to species with particular cognitive capabilities.

It has been suggested that the cognitive requirements for group living—in particular for understanding and keeping track of relationships between individuals in complex social groups—were a driving force in the evolution of larger neocortices (Dunbar, 1998; but see Benson-Amram, Dantzer, Stricker, Swanson, & Holekamp, 2016; Healy & Rowe, 2007; MacLean, Sandel, Bray, Oldenkamp, Reddy, & Hare, 2013, for recent criticisms of this hypothesis) and underlie abilities such as causal reasoning, cognitive control, and imagination (Barrett, Henzi, & Dunbar, 2003; Emery & Clayton, 2004). One can only speculate whether coordination demands also played a direct role in increasing particular cognitive capabilities.

Investigating the extent to which specific mechanisms of coordination are present in different species may provide insight into when and how these mechanisms evolved. In humans, for instance, motor simulation is thought to facilitate predicting and adjusting to others' actions in time and space. Temporal coordination in humans is also enhanced by individuals reducing the variability of their behavior (e.g., by speeding up). If these mechanisms prove to be present in other species, this would shed light on the evolution of, and the cognitive capabilities required for, certain coordination mechanisms.

Finally, the present overview has addressed how information and skills are transferred in group-living animals. In addition to observational learning, some species engage in active teaching. Teaching poses both cognitive (e.g. recognizing an uninformed individual) and energetic (e.g. effort and time) demands on the teacher and often involves a high level of social coordination. At the same time, teaching allows for the cultivation of difficult but beneficial behaviors such as hunting. This, again, raises the question of whether the cognitive demands of teaching drove cognitive evolution in some group-living species, or constitute instead its consequence or byproduct. Though questions like these may not be answered conclusively, they illustrate how formulating and investigating hypotheses concerning the role of particular cognitive requirements in specific forms of coordination might help our understanding of evolution. For instance, do overlapping cognitive abilities underlie both flexible hunting and teaching behavior?

Life in social groups poses behavioral and cognitive challenges—challenges which, in turn, have shaped the evolution of group-living individuals. Because common themes emerge in different species, perhaps suggesting universal constraints on how these problems are solved, it is interesting to analyze coordination problems and their solutions from a comparative perspective. Such studies may pay unexpected dividends: For example, further investigations of the mechanisms that underlie social coordination and joint task performance in a broad range of species may prove fruitful in the design and implementation of robots able to interact flexibly with humans (Hoffman & Ju, 2014). Already, investigating recently discovered mechanisms of human joint action (including motor simulation, task distribution, and task co-representation) in other species is enhancing our understanding of how these mechanisms evolved.

Acknowledgments

Natalie Sebanz would like to acknowledge support from the European Research Council under the European Union's Seventh Framework Program (FP7/2007–2013) - ERC grant agreement 616072, JAXPERTISE.

References

Alerstam, T., Hedenstrom, A., & Akesson, S. (2003). Long-distance migration: Evolution and determinants. *Oikos*, 103, 247–260.

Anderson, C., Boomsma, J. J., & Bartholdi, J. J. (2002). Task partitioning in insect societies: Bucket brigades. *Insectes Sociaux*, 49, 1–10.

Anderson, C., & Franks, N. R. (2001). Teams in animal societies. *Behavioral Ecology*, 12, 534–540.

Anderson, C., Theraulaz, G., & Deneubourg, J. (2002). Self-assemblages in insect societies. *Insectes Sociaux*, 49, 99–110.

Atmaca, S., Sebanz, N., & Knoblich, G. (2011). The joint flanker effect: Sharing tasks with real and imagined co-actors. *Experimental Brain Research*, 211, 371–385.

Baron-Cohen, S. (1991). Precursors to a theory of mind: Understanding attention in others. In A. Whiten (Ed.), *Natural theories of mind: Evolution, development and simulation of everyday mindreading* (pp. 233–251). Oxford: Basil Blackwell.

Barrett, L., Henzi, P., & Dunbar, R. I. M. (2003). Primate cognition: From what now to what if. *Trends in Cognitive Science* 7, 494–497.

Bednarz. J. C. (1988). Cooperative hunting in Harris' Hawks (Parabuteo unicinctus). *Science*, 239, 1525–1527.

Benson-Amram, S., Dantzer, B., Stricker, G., Swanson, E. M., & Holekamp, K. E. (2016). Brain size predicts problem-solving ability in mammalian carnivores. *Proceedings of the National Academy of Sciences of the USA*, 113(9), 2532-2537.

Böckler, A., Knoblich, G., & Sebanz, N. (2011). Observing shared attention modulates gaze following. *Cognition*, 120(2), 292-298.

Böckler, A. Knoblich, G., & Sebanz, N. (2012). Effects of a co-actor's focus of attention on task performance. *Journal of Experimental Psychology: Human Perception & Performance*, 38(6), 1404–1416.

Boesch, C. (2003). Cooperative hunting roles among Taï chimpanzees. *Human Nature*, 13, 27–46.

Boesch, C., & Boesch, H. (1989). Hunting behavior of wild chimpanzees in the Taï national park. *American Journal of Physiological Anthropology*, 78, 547–573.

Bräuer, J., Call, J., & Tomasselo, M. (2005). All great ape species follow gaze to distant locations and around barriers. *Journal of Comparative Psychology*, 119, 145–154.

Bugnyar, T., Stowe, M., & Heinrich, B. (2004). Ravens, Corvus corax, follow gaze direction of humans around obstacles. *Proceedings of the Royal Society B, Biological Sciences*, 271, 1331–1336.

Buhl, J., Sumpter, D. J. T., Couzin, I. D., Hale, J. J., Despland, E., Miller, E. R., & Simpson, S. J. (2006). From disorder to order in marching locusts. *Science*, 312(5778), 1402–1406.

Carere, C., Montanino, S., Moreschini, F., Zoratto, F., Chiarotti, F., Santucci, D., & Alleva, E. (2009). Aerial flocking patterns of wintering starlings, Sturnus vulgaris, under different predation risk. *Animal Behaviour*, 77, 101–107.

Caro, T. M., & Hauser, M. D. (1992). Is there teaching in nonhuman animals? *The Quarterly Journal of Biology*, 67, 151–174.

Cavagna, A., Cimarelli, A., Giardina, I., Orlandi, A., Parisi, G., Procaccini, A., ... Stefanini, F. (2007). New statistical tools for analyzing the structure of animal groups. *Mathematical Biosciences*, 214, 32–37.

Chao, D., & Levin, S. A. (1999). A simulation of herding behavior: the emergence of large-scale phenomena from local interactions. In S. Ruan, G. S. K. Wolkowicz, & J. Wu (Eds.), *Differential equations with applications to biology* (pp 81–95). Providence, RI: AMS.

Chase, I., Tovey, C., Spangler-Martin, D., & Manfredonia, M. (2002). Individual differences versus social dynamics in the formation of animal dominance hierarchies. *Proceedings of the National Academy of Sciences of the USA*, 99, 5744–5749.

Clutton-Brock, T. (2009). Cooperation between non-kin in animal societies. *Nature*, 462(7269), 51–57.

Couzin, I. D., & Krause, J. (2003). Self-organization and collective behavior in vertebrates. In P. Slater, J. S. Rosenblatt, C. T. Snowdon, & T. Roper (Eds.), *Advances in the study of behavior* (pp. 1–75). New York, NY: Academic Press.

Couzin, I. D., Krause, J., Franks, N. R., & Levin, S. A. (2005). Effective leadership and decision-making in animal groups on the move. *Nature*, 433, 513–516.

Creel, S. (1997). Cooperative hunting and group size: Assumptions and currencies. *Animal Behaviour*, 54, 1319–1324.

Csibra, G. (2007). Action mirroring and action interpretation: An alternative account. In P. Haggard, Y. Rosetti, & M. Kawato (Eds.), *Sensorimotor foundations of higher cognition. Attention and performance XXII* (pp. 435–459). Oxford: Oxford University Press.

Csibra, G., & Gergely, G. (2009). Natural pedagogy. *Trends in Cognitive Sciences*, 13, 148–153.

Cutts, C. J., & Speakman, J. R. (1994). Energy savings in formation flight of pink-footed geese. *Journal of Experimental Biology*, 189, 251–261.

Danchin, E., Giraldeau, L., Valone, T. J., & Wagner, R. H. (2004). Public information: From nosy neighbors to cultural evolution. *Science*, 305, 487–491.

Deneubourg, J. L., Aron, S., Goss, S., Pasteels, J. M. (1990). The self-organizing exploratory pattern of the Argentine ant. *Journal of Insect Behavior* 3, 159–168.

Detrain, C., & Pasteels, J. M. (1992). Caste polyethism and collective defense in the ant Pheidole pallidula: The outcome of quantitative differences in recruitment. *Behavioral Ecology & Sociobiology*, 29, 405–412.

Dugatkin, L. A. (2002). Cooperation in animals: An evolutionary overview. *Biology and Philosophy*, 17(4), 459–476.

Dunbar, R. I. M. (1998). The social brain hypothesis. *Evolutionary Anthropology*, 178–190.

Dyer, J. R. G., Ioannou, C. C., Morrell, L. J., Croft, D. P., Couzin, I. D., Waters, D. A., & Krause, J. (2008). Consensus decision making in human crowds. *Animal Behaviour*, 75, 461–470.

Emery, N. J., & Clayton, N. S. (2004). The mentality of crows: Convergent evolution of intelligence in corvids and apes. *Science*, 306, 1903–1907.

Eriksen, B. A., & Eriksen, C. W. (1974). Effects of noise letters upon the identification of a target letter in a nonsearch task. *Perception & Psychophysics*, 16, 143–149.

Eskenazi, T., Rueschemeyer, S. A., de Lange, F. P., Knoblich, G., & Sebanz, N. (2015). Neural correlates of observing joint actions with shared intentions. *Cortex*, 70, 90–100.

Ferrari, P. F., Kohler, E., Fogassi, L., & Gallese, V. (2000). The ability to follow eye gaze and its emergence during development in macaque monkeys. *Proceedings of the National Academy of Sciences of the USA*, 97, 13997–14002.

Fish, F. E. (1995). Kinematics of ducklings swimming in formation: Consequences of position. *The Journal of Experimental Zoology*, 273, 1–11.

Fitch, W. T., Huber, L., & Bugnyar, T. (2010). Social cognition and the evolution of language: Constructing cognitive phylogenies. *Neuron*, 65(6), 795–814.

Franks, N. R., & Richardson, T. (2006). Teaching in tandem-running ants. *Nature*, 439, 153–153.

Friesen, C. K., & Kingstone, A. (1998). The eyes have it! Reflexive orienting is triggered by nonpredictive gaze. *Psychonomic Bulletin & Review*, 5, 490–495.

Frischen, A., Bayliss, A. P., & Tipper, S. P. (2007). Gaze cueing of attention: Visual attention, social cognition, and individual differences. *Psychological Bulletin*, 133, 694–724.

Galef, B. G. J., & Laland, K. N. (2005). Social Learning in animals: Empirical studies and theoretical models. *BioScience*, 55, 489–499.

Gazda, S. K., Conner, R. C., Edgar, R. K., & Cox, F. (2005). A division of labour with role specialisation in group-hunting bottlenose dolphins (Tursiops truncatus) off Cedar Key, Florida. *Proceedings of the Royal Society B, Biological Sciences*, 272, 135–140.

Gigerenzer, G. (2008). Why heuristics work. *Perspectives on Psychological Science*, 3, 20–29.

Goebl, W., & Palmer, C. (2009. Synchronization of timing and motion among performing musicians. *Music Perception*, 26, 427–438.

Greene, M. J., & Gordon, D. M. (2007). Interaction rate informs harvester ant task decisions. *Behavioral Ecology*, 18(2), 451–455.

Gurven, M., Kaplan, H., & Gutierrez, M. (2006). How long does it take to become a proficient hunter? Implications for the evolution of delayed growth. *Journal of Human Evolution*, 51, 454–470.

Healy, S. D., & Rowe, C. (2007). A critique of comparative studies of brain size. *Proceedings of the Royal Society B, Biological Sciences*, 274, 453–464.

Heyes, C. (2009). Evolution, development and intentional control of imitation. *Philosophical Transactions of the Royal Society B: Biological Sciences*, 364(1528), 2293–2298.

Hoffman, G., & Ju, W. (2014). Designing robots with movement in mind. *Journal of Human-Robot Interaction*, 1(1), 78–95.

Holekamp, K. E., Boydston, E. E., & Smale, L. (2000). Group movements in social carnivores. In S. Boinski & P. Garber (Eds.), *On the move: Group travel in primates and other animals* (pp. 587–627). Chicago, IL: University of Chicago Press.

Hölldobler, B., & Wilson, E. O. (1983). Queen control in colonies of weaver ants (Hymenoptera: Formicidae). *Annals of the Entomological Society of America*, 76, 235–238.

Hölldobler, B., & Wilson, E.O. (1990). *The ants.* Berlin: Springer Verlag.

Hoppitt, W., & Laland, K. N. (2008). Social processes influencing learning in animals: A review of the evidence. *Advances in the Study of Behavior*, 38, 105–165.

Horner, V., & Whiten, A. (2005). Causal knowledge and imitation/emulation switching in chimpanzees (Pan troglodytes) and children (Homo sapiens). *Animal Cognition*, 8, 164–181.

Huber, L., Rechberger, S., & Taborsky, M. (2001). Social learning affects object exploration and manipulation in keas, Nestor notabilis. *Animal Behaviour*, 62, 945–954.

Huber, L., Range, F., Voelkl, B., Szucsich, A., Viranyi, Z., & Miklosi, A. (2009). The evolution of imitation: What do the capacities of nonhuman animals tell us about the mechanisms of imitation? *The Philosophical Transactions of the Royal Society B*, 364, 2299–2309.

Humphreys, G. W., & Bedford, J. (2011). The relations between joint action and theory of mind: a neuropsychological analysis. *Experimental Brain Research*, 211(3–4), 357–369.

Itakura, S. (2004). Gaze-following and joint visual attention in nonhuman animals. *Japanese Psychological Research*, 46, 216–226.

Jackson, D. E., & Châline, N. (2007). Modulation of pheromone trail strength with food quality in Pharaoh's ant, Monomorium pharaonis. *Animal Behaviour*, 74, 463–470.

Jeanne, R. L. (1986). The evolution of the organization of work in social insects. *Monitore Zoologico Italiano*, 20, 119–133.

Kaminski, J., Riedel, J., Call, J., & Tomasello, M. (2005). Domestic goats, Capra hircus, follow gaze direction and use social cues in an object choice task. *Animal Behaviour*, 69, 11–18.

Kanske, P., Böckler, A., Trautwein, F. M., & Singer, T. (2015). Dissecting the social brain: Introducing the EmpaToM to reveal distinct neural networks and brain–behavior relations for empathy and Theory of Mind. *Neuroimage*, 122, 6–19.

Keller, P. E., Knoblich, G., & Repp, B. H. (2007). Pianists play better when they play with themselves: On the possible role of action simulation in synchronization. *Consciousness & Cognition*, 16, 102–111.

Kis, A., Huber, L., & Wilkinson, A. (2015). Social learning by imitation in a reptile (Pogona vitticeps). *Animal Cognition*, 18(1), 325–331.

Kleinhappel, T. K., Burman, O. H. P., John, E. A., Wilkinson, A., & Pike, T. W. (2014). Diet-mediated social networks in shoaling fish. *Behavioral Ecology*, 25(2), 374–377.

Kleinhappel, T. K., Burman, O. H. P., John, E., Wilkinson, A., & Pike, T.W. (in press). Social interactions in mixed-species networks. *Behavioural Ecology and Sociobiology*. doi: 10.1007/s00265-016-2099-x.

Knoblich, G., Butterfill, S., & Sebanz, N. (2011). Psychological research on joint action: Theory and data. In B. H. Ross (Ed.) *Psychology of learning and motivation: Advances in research and theory*, 54 (pp. 59–101). London: Academic Press.

Knoblich, G., & Sebanz, N. (2008). Evolving intentions for social interaction: from entrainment to joint action. *Philosophical Transactions of the Royal Society B, Biological Sciences*, 363, 2021–2031.

Kourtis, D., Knoblich, G., Wozniak, M., & Sebanz, N. (2014). Attention allocation and task representation during joint action planning. *Journal of Cognitive Neuroscience*, 26(10), 2275–86.

Kourtis, D., Sebanz, N., & Knoblich, G. (2010). Favouritism in the motor system: Social interaction modulates action simulation. *Biology Letters*, 6, 758–761.

Krause, J., & Ruxton, G. D. (2002). *Living in Groups*. Oxford: Oxford University Press.

Kyle, C. R. (1979). Reduction of wind resistance and power output of racing cyclists and runners travelling in groups. *Ergonomics*, 22, 387–397.

Laland, K. N., & Galef, B. G. (2009). *The question of animal culture*. Cambridge, MA: Harvard University Press.

Leadbeater, E., & Chittka, L. (2007). The dynamics of social learning in an insect model, the bumblebee (*Bombus terrestris*). *Behavioral Ecology and Sociobiology*, 61, 11, 1789–1796.

Leadbeater, E., Raine, N. E., & Chittka, L. (2006). Social learning: Ants and the meaning of teaching. *Current Biology*, 16, R323–R325.

Leuthold, R. H., Bruinsma, O., & Huis, A. V. (1976). Optical and pheromonal orientation and memory for homing distance in the harvester termite Hodotermes mossambicus (Hagen). *Behavioral Ecology and Sociobiology*, 1(2), 127–139.

Loretto, M. C., Schloegl, C., & Bugnyar, T. (2010). Northern bald ibises follow others' gaze into distant space but not behind barriers. *Biology Letters*, 6, 14–17.

Marsh, K. L., Johnston, L., Richardson, M. J., & Schmidt, R. C. (2009). Toward a radically embodied, embedded social psychology. *European Journal of Social Psychology*, 39, 1217–1225.

MacLean, E. L., Sandel, A. A., Bray, J., Oldenkamp, R. E., Reddy, R. B., & Hare, B. A. (2013). Group size predicts social but not nonsocial cognition in lemurs. *PLoS ONE* 8(6), e66359.

Mech, L. D. (1999). Alpha status, dominance, and division of labour in wolf packs. *Canadian Journal of Zoology*. 77, 1196–1203.

McFarland, D. (1985). *Animal behaviour*. Bath: Pitman.

Melis, A., Hare, B., & Tomasello, M. (2007). Chimpanzees recruit the best collaborators. *Science*, 311, 1297–1300.

Mendres, K. A., & de Waal, F. B. M. (2000). Capuchins do cooperate: The advantage of an intuitive task. *Animal Behaviour*, 60, 523–529.

Miklósi, Á., Topál, J., & Csányi, V. (2004). Comparative social cognition: what can dogs teach us? *Animal Behaviour*, 67, 995–1004.

Miklósi, A., & Soprani, K. (2006). A comparative analysis of animals' understanding of the human pointing gesture. *Animal Cognition*, 9, 81–93.

Morris, S. C. (2003). *Life's solution: Inevitable humans in a lonely universe*. Cambridge: Cambridge University Press.

Moussaïd, M., Perozo, N., Garnier, S., Helbing, D., & Theraulaz, G. (2010). The walking behaviour of pedestrian social groups and its impact on crowd dynamics. *PLoS One*, 4, 1–7.

Olds, T. (1998). The mathematics of breaking away and chasing in cycling. *European Journal of Applied Physiology*, 77, 492–497.

Oster, G. F., & Wilson, E. O. (1978). *Caste and ecology in the social insects*. Princeton, NJ: Princeton University Press.

Pack, A. A., & Herman, L. M. (2006). Dolphin social cognition and joint attention: Our current understanding. *Aquatic Mammals*, 32, 443–460.

Petit, O., Gautrais, J., Leca, J. B., Theraulaz, G., & Deneubourg, J. L. (2009). Collective decision-making in white-faced capuchin monkeys. *Proceedings of the Royal Society of London, B*, 276, 3495–3503.

Pezzulo, G., & Dindo, H. (2011). What should I do next? Using shared representations to solve interaction problems. *Experimental Brain Research*, 211, 613–630.

Pezzulo, G., Donnarumma, F., & Dindo, H. (2013). Human sensorimotor communication: A theory of signaling in online social interactions. *PLOS One*, 8. doi: 10.1371/journal.pone.0079876

Prinz, W. (1997). Perception and action planning. *European Journal of Cognitive Psychology*, 9, 129–154.

Raafat, R. M., Chater, N., & Frith, C. (2009). Herding in humans. *Trends in Cognitive Sciences*, 13, 420–428.

Ran, N., Getz, W. M., Revilla, E., Holyoak, M., Kadmon, R. Saltz, D. R., & Smouse, P. E. (2008). A movement ecology paradigm for unifying organismal movement research. *Proceedings of the National Academy of Sciences of the USA*, 105, 19052–19059.

Range, F., Heucke, S. L., Gruber, C., Konz, A., Huber, L., & Virányi, Z. (2009). The effect of ostensive cues on dogs' performance in a manipulative social learning task. *Applied Animal Behaviour Science*, 120(3–4), 170–178.

Ratnieks, F. L., & Anderson, C. (1999). Task partitioning in insect societies. *Insectes Sociaux*, 46, 95–108.

Reynolds, C. W. (1987). Flocks, herds, and schools: A distributed behavioral model. *Computer Graphics* 21, 25–33.

Richardson, M. J., Harrison, S. J., May, R., Kallen, R. W., & Schmidt, R. C. (2011). Self-organized complementary coordination: Dynamics of an interpersonal collision-avoidance task. *BIO Web of Conferences*, 1, 00075. doi: 0.1051/bioconf/20110100075.

Rizzolatti, G., & Sinigaglia, C. (2010). The functional role of the parieto-frontal mirror circuit: Interpretations and misinterpretations. *Nature Reviews Neuroscience*, 11, 264–274.

Ruiz, A., Gómez, J., Roeder, J., & Byrne, R. (2009). Gaze following and gaze priming in lemurs. *Animal Cognition*, 12, 427–434.

Ruys, K. I., & Aarts, H. (2010). When competition merges people's behavior: Interdependency activates shared action representations. *Journal of Experimental Social Psychology*, 46(6), 1130–1133.

Sacheli, L. M., Tidoni, E., Pavone, E. F., Aglioti, S. M., & Candidi, M. (2013). Kinematics fingerprints of leader and follower role-taking during cooperative joint actions. *Experimental Brain Research*, 226, 473–486.

Scheel, D., & Packer, C. (1991). Group hunting behaviour of lions: A search for cooperation. *Animal Behaviour*, 41, 697–709.

Schmidt, R. C., Fitzpatrick, P., Caron, R., & Mergeche, J. (2011). Understanding social motor coordination. *Human Movement Science*, 30, 834–845.

Schuster, S., Wohl, S., Griebsch, M., & Klostermeier, I. (2006). Animal cognition: How archer fish learn to down rapidly moving targets. *Current Biology*, 16, 378–383.

Sebanz, N., Bekkering, H., & Knoblich, G. (2006). Joint action: Bodies and minds moving together. *Trends in Cognitive Sciences*, 10, 70–76.

Sebanz, N., & Knoblich, G. (2009). Prediction in joint action: What, when, and where. *Topics in Cognitive Science*, 1, 353–367.

Sebanz, N., Knoblich, G., & Prinz, W. (2003). Representing others' actions: Just like one's own? *Cognition*, 88, B11–B21.

Semin, G. R., & Smith, E. R. (Eds.). (2008). *Embodied grounding: Social, cognitive, affective, and neuroscientific approaches*. New York: Cambridge University Press.

Sendova-Franks, A. B., & Franks, N. R. (1993). Task allocation in ant colonies within variable environments (a study of temporal polyethism). *Bulletin of Mathematical Biology*, 55, 75–96.

Seyfarth, R. M., Cheney, D., & Bergman, T. (2005). Primate social cognition and the origins of language. *Trends in Cognitive Sciences*, 9, 264–266.

Stander, P. E. (1992). Cooperative hunting in lions: The role of the individual. *Behavioral Ecological Sociobiology*, 29, 445–454.

Stöcker, S. (1999). Models for tuna school formation. *Mathematical Biosciences*, 156, 167–190.

Sumpter, D. J. T., Krause, J., James, R., Couzin, I. D., & Ward, A. J. W. (2008). Consensus decision making by fish. *Current Biology*, 18, 1773–1777.

Tennie, C., Call, J., & Tomasello, M. (2009). Ratcheting up the ratchet: On the evolution of cumulative culture. *Philosophical Transactions of the Royal Society B: Biological Sciences*, 364(1528), 2405–2415.

Thornton, A., & McAuliffe, K. (2006). Teaching in wild meerkats. *Science*, 313, 227–229.

Tomasello, M., Call, J., & Hare, B. (1998). Five primate species follow the visual gaze of conspecifics. *Animal Behaviour*, 55, 1063–1069.

Van der Wel, R. P. D., Knoblich, G., & Sebanz, N. (2011). Let the force be with us: Dyads exploit haptic coupling for coordination. *Journal of Experimental Psychology: Human Perception & Performance*, 37, 1420–1431.

Van Horn, R. C., Engh, A. L., Scribner, K. T., Funk, S. M., & Holekamp, K. E. (2004). Behavioral structuring of relatedness in the spotted hyena (*Crocuta crocuta*) suggests direct fitness benefits of clan-level cooperation. *Molecular Ecology*, 13, 449–458.

Van Overwalle, F., & Baetens, K. (2009). Understanding others' actions and goals by mirror and mentalizing systems: A meta-analysis. *Neuroimage*, 48, 564–584.

Vesper, C., van der Wel, R. P. D., Knoblich, G., & Sebanz, N. (2011). Making oneself predictable: Reduced temporal variability facilitates joint action coordination. *Experimental Brain Research*, 211, 517–530.

Vesper, C., van der Wel, R. P. D., Knoblich, G., & Sebanz, N. (2012). Are you ready to jump? Predictive mechanisms in interpersonal coordination. *Journal of Experimental Psychology: Human Perception & Performance*, 38(1), 48–61.

Visco-Comandini, F., Ferrari-Toniolo, S., Satta, E., Papazachariadis, O., Gupta, R., Nalbant, L. E., & Battaglia-Mayer, A. (2015). Do non-human primates cooperate? Evidences of motor coordination during a joint action task in macaque monkeys. *Cortex*, 70, 115–127.

Vittori, K., Talbot, G., Gautrais, J., Fourcassié, V., Araújo, A. F. R., & Theraulaz, G. (2006). Path efficiency of ant foraging trails in an artificial network. *Journal of Theoretical Biology*, 239, 507–515.

Voelkl, B., & Huber, L. (2000). True imitation in marmosets. *Animal Behaviour*, 60, 195–202.

Voelkl, B., & Huber, L. (2007). Imitation as faithful copying of a novel technique in marmoset monkeys. *PLoS ONE*, 2, e611.

Ward, A. J. W., Sumpter, D. J. T., Couzin, I. D., Hart, P. J. B., & Krause, J. (2008). Quorum decision-making facilitates information transfer in fish shoals. *Proceedings of the National Academy of Sciences of the USA*, 105, 6948–6953.

Weimerskirch, H., Martin, J., Clerquin, Y., Alexandre, P., & Jiraskova, S. (2001). Energy saving in flight formation *Nature*, 413, 697–698.

White, A. B. (2001). Wild and captive wolf (*Canis lupus*) aggression in relation to pack size and territory availability. *Environmental Population and Organismic Biology*, 1–38.

Whiten, A., & van Schaik, C. P. (2007). The evolution of animal "cultures" and social intelligence. *Philosophical Transactions of the Royal Society B: Biological Sciences*, 362, 603–620.

Whiten, A., McGuigan, N., Marshall-Pescini, S., & Hopper, L. M. (2009). Emulation, imitation, over-imitation and the scope of culture for child and chimpanzee. *Philosophical Transactions of the Royal Society B: Biological Sciences*, 364(1528), 2417–2428.

Wilkinson, A., Mandl, I., Bugnyar, T., & Huber, L. (2010). Gaze-following in the red-footed tortoise (*Geochelone carbonaria*). *Animal Cognition*, 13, 765–769.

Wilson, M., & Knoblich, G. (2005). The case for motor involvement in perceiving conspecifics. *Psychological Bulletin*, 131, 460–473.

Wolpert, D. M., Doya, K., & Kawato, M. (2003). A unifying computational framework for motor control and interaction. *Philosophical Transactions of the Royal Society of London B*, 358, 593–602.

Zentall, T. R. (2004). Action imitation in birds. *Learning & Behavior*, 32, 15–23.

18

Social Learning, Intelligence, and Brain Evolution

Sally E. Street and Kevin N. Laland

18.1 Introduction

Social learning—learning influenced by observation of, or interaction with, other animals (Heyes, 1994; Hoppitt & Laland, 2008)—allows individuals to acquire information, concerning, for instance, the location and quality of food, mates, predators, rivals, and pathways, as well as foraging techniques, vocalizations and a variety of social behavior (Byrne, 2009; Heyes & Galef, 1996; Laland, Atton, & Webster, 2011; Laland & Galef, 2009). Social learning is typically adaptive because it can act as a short-cut to acquiring optimal, or high-payoff, behavior, avoiding the relative costs of individual "trial-and-error" learning (Rendell et al., 2010). It is therefore unsurprising that a growing body of experimental and observational studies has reported social learning in a wide range of species, including mammals, reptiles, fish, birds, amphibians and insects (Box & Gibson, 1999; Brown & Laland, 2003; Emery, 2006; Ferrari & Chivers, 2008; Leadbeater & Chittka, 2007; Wilkinson, Kuenstner, Mueller, & Huber, 2010). While animals of many species are capable of social learning, there is variation in the extent to which they are reliant on, and perhaps specially adapted for, the exploitation of socially learned information. Specific social learning processes are rarer than the general capability: for instance, chimpanzees are seemingly capable of imitative learning in which a specific action is replicated with high fidelity (Whiten et al., 2007); other species, like nine-spined sticklebacks, appear to deploy specific social learning strategies, such as payoff-based copying (Coolen, Ward, Hart, & Laland, 2005). It is therefore interesting to consider why species vary in how much they rely on social learning for survival and reproduction, why some species might possess enhanced social learning skills, in terms of mechanisms or strategies, and what role these abilities might have played in brain evolution. It may not be coincidence that taxa with enlarged forebrains and which are commonly regarded as highly intelligent, such as primates, certain radiations of birds and cetaceans, also happen to be heavily reliant on social learning and traditions.

The Wiley Handbook of Evolutionary Neuroscience, First Edition. Edited by Stephen V. Shepherd.
© 2017 John Wiley & Sons, Ltd. Published 2017 by John Wiley & Sons, Ltd.

18.2 Brain Enlargement, Intelligence, and Social Learning

Intelligence can be broadly defined as a cross-domain measure of cognitive ability in learning, problem-solving and abstract reasoning, characterized by behavioral flexibility (Jolly, 1966; Reader, Hager, & Laland, 2011; Reader & Laland, 2002). High intelligence in a species may be therefore be suggested by experimental and observational evidence for the flexible use of such behavior as extractive foraging (in the absence of specialized anatomy), food processing, tactical deception, tool use, causal understanding, problem solving, and complex learning (Byrne, 1995; Emery & Clayton, 2004; Huber & Gajdon, 2006; van Schaik, Deaner, & Merrill, 1999; Visalberghi & Tomasello, 1998; Whiten & Byrne, 1997). While the term "intelligence" is often avoided by contemporary animal behavior researchers, who typically address aspects of cognition in a more domain-specific manner, nonetheless interest in comparative cognition seems to have reached its zenith (Shettleworth, 2010). It seems to be the case that researchers remain interested in intelligent behavior, and its evolutionary origins, but that they are conscious of the difficulties of fair comparison of intelligence across diverse taxa, or suspicious of the notion of general intelligence, leaving them reticent to use the term (Mackintosh, 1988). Although the cognitive capabilities of each species are uniquely adapted to the requirements of their niches, this does not preclude that certain species may justifiably be considered more generally intelligent across domains than others (Reader et al., 2011). For instance, primate genera differ in their performance across diverse laboratory tests of cognition, with great apes consistently outperforming other primates (Deaner, Van Schaik, & Johnson, 2006; Tomasello & Call, 1997). It may be difficult to compare intelligence across species, but the suggestion that all nonhuman vertebrates are equally intelligent (Macphail, 1982) is contradicted by extensive evidence (Byrne, 1995; Deaner et al., 2006; Lefebvre, Reader, & Sol, 2004; Reader & Laland, 2002; Tomasello & Call, 1997). Further, the view that intelligence is a meaningful concept across species does not require the assumption of a *scala naturae*—the idea that intelligence decreases with phylogenetic distance from humans (Jensen, 1980). On the contrary, recent evidence supports the view that there has been convergent evolution of intelligence in distant taxa, especially in primates, corvids, and toothed whales (Emery & Clayton, 2004; Reader et al., 2011; Rendell & Whitehead, 2001).

Taxa considered to possess high intelligence, such as primates, cetaceans and birds; specifically monkeys, apes, toothed whales, songbirds and parrots; have undergone convergent enlargement of the forebrain, relative to sister taxa, in terms of deviations from expected scaling with whole brain and body size (Alonso, Milner, Ketcham, Cookson, & Rowe, 2004; Barton & Harvey, 2000; Cnotka, Gunturkun, Rehkamper, Gray, & Hunt, 2008; Emery, 2006; Jerison, 1973; Marino, McShea, & Uhen, 2004; Rendell & Whitehead, 2001; Zelenitsky, Therrien, & Kobayashi, 2009). Although birds lack the mammalian neocortex, the avian pallium is structurally similar and lesions to the caudolateral nidopallium result in specific cognitive impairment, suggesting that the avian and mammalian forebrains are analogous (Cnotka et al., 2008; Emery, 2006; Huber & Gajdon, 2006). Despite centuries of interest in the evolution of intelligence, the intuitively appealing notion that brain volume and intelligence are linked commands surprisingly little support, although some compelling evidence does exist in birds and primates (Byrne & Corp, 2004; Deaner et al., 2006; Johnson, Deaner, & Van Schaik, 2002; Lefebvre et al., 2004; Lefebvre, Whittle,

Lascaris, & Finkelstein, 1997; Reader et al., 2011; Reader & Laland, 2002; Sol, Duncan, Blackburn, Cassey, & Lefebvre, 2005). However, it remains unclear what cognitive benefits brain size expansion brings. Brain expansion potentially is associated with increases in the number of neurons, local and 'long range' connectivity and the number of cortical areas, thereby in a general sense potentially expanding the amount of 'processing power' available to an animal (Byrne & Bates, 2007; Changizi & Shimojo, 2005; Striedter, 2005). Striedter (2005, p. 11) notes a general rule of brain evolution that "large equals well connected."

Trends in vertebrate brain size evolution commonly follow simple allometric scaling rules, with individual brain regions evolving in concert with each other (Finlay & Darlington, 1995). However, certain regions in some taxa appear to vary in size independently of other brain areas, and appear to be associated with evolutionary changes in behavior (Barton & Harvey, 2000; Striedter, 2005). For example, expansion of the primate neocortex (Barton & Harvey, 2000; Byrne & Corp, 2004) is predicted by various social and ecological variables in primates (Alport, 2004; Amici, Aureli, & Call, 2008; Barton, 1996; Dunbar, 1992). Likewise, comparative studies have demonstrated that hippocampus enlargement is associated with food-caching behavior (Basil, Kamil, Balda, & Fite, 1996; Hampton, Sherry, Shettleworth, Khurgel, & Ivy, 1995; Healy & Krebs, 1992; Krebs, Sherry, & Healy, 1989; Sherry, Vaccarino, Buckenham, & Herz, 1989) and that enlargement of the higher vocal centre is associated with vocal repertoire in birds (Devoogd, Krebs, Healy, & Purvis, 1993; Szekely, Catchpole, Devoogd, Marchl, & Devoogd, 1996). Specific neural expansion in association with adaptive specialization supports the general link between brain size and reliance on specific behavior. It is important to recognize, however, that some variation in brain component size across species may be affected by neural plasticity, reflecting usage within the lifetime; hence within and between species differences in neural structure is not inherently evidence for adaptive specialization (Bolhuis & Macphail, 2001).

Comparative studies show evidence of a relationship between social learning, brain size and intelligence in primates. Across primate species, Reader and Laland (2002) demonstrated that there is a positive correlation between the reported incidence of social learning in a given primate species, corrected for research effort, and both relative and absolute measures of brain volume, in a phylogenetically controlled analysis (Figure 18.1). Biases in research effort were corrected by using, as a corrected frequency, the residuals from a regression of the frequency of social learning against the frequency of published articles on the species in the Zoological Record. Relative brain size was measured as "executive brain ratio": the ratio of the volume of the neocortex and striatum (executive brain) to the volume of the mesencephalon and medulla (brainstem). However, social learning incidence also correlated positively with further measures of brain size, including absolute executive brain volume and residuals of a plot of executive brain against brainstem volume, although the latter was not statistically significant (Reader & Laland, 2002).

In addition to the relationship between social learning and brain size, Reader and Laland (2002) showed that the corrected incidence of behavioral innovation also co-varied with measures of brain size across primates, echoing a similar finding in birds (Lefebvre, Whittle, Lascaris, & Finkelstein, 1997). Moreover, the corrected incidence of social learning is positively correlated with corrected frequencies of behavioral innovation and tool use, suggesting that social learning ability could be a

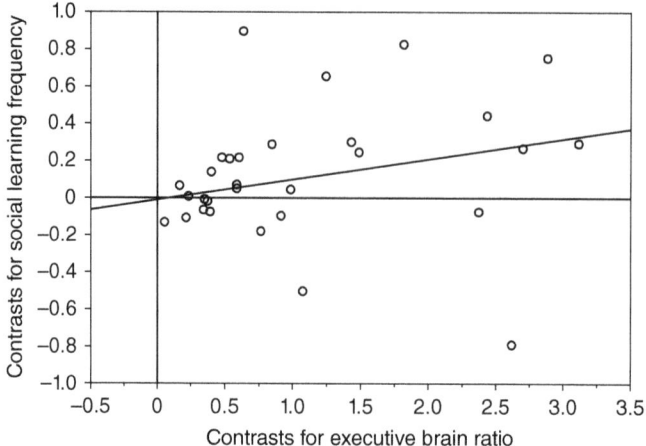

Figure 18.1 Positive Association between Social Learning Frequency (Corrected for Research Effort) and Relative Brain Size (Measured as Executive Brain Ratio) across Primate Species, Using Independent Contrasts to Control for Phylogenetic Non-Independence. Reader 2002. Copyright (2002) National Academy of Sciences, USA.

component of a cross-species general intelligence factor, an idea further explored by Reader et al. (2011). In principal component and factor analyses of various naturalistic indicators of behavioral flexibility (including corrected measures of social learning, tool use, innovation, tactical deception, and extractive foraging, in >60 primate species), Reader et al (2011) extracted a single component/factor, which explained over 65% of the variance. In a further analysis, which included three additional socio-ecological variables (diet breadth, percentage fruit in diet, group size), a major component explained 47% of the variance, on which the five behavioral flexibility measures loaded, plus diet breadth (to a lesser extent); although a second component was also extracted, on which tactical deception, group size, and percentage fruit loaded. These results, which hold when the analyses are conducted at the genus level, when phylogeny is controlled for, and when the apes are removed, strongly imply that aspects of behavioral flexibility co-vary, and are evocative of the notion of a cross-species general intelligence (Reader et al., 2011). This statistical association need not imply that the various measures are reliant on the same brain regions or circuits, although this is a possibility.

Support for the association of high general intelligence and brain size expansion comes from the observation that individual species loadings on the principle component (termed g_s), which can be regarded as measures of the general intelligence of the species, are strongly associated with several measures of brain volume, including neocortex/whole brain ratio, executive brain/brainstem ratio, and absolute neocortex size, although not residuals of neocortex on the rest of the brain (Reader et al., 2011). Recent research suggests that relationships between socio-ecological predictors and brain size vary according to which method of brain measurement (e.g., neocortex/whole brain residuals versus ratios) is employed, although the reasons for this are not well understood (Deaner, Nunn, & Van Schaik, 2000). Further, Reader et al.'s g_s measure was found to correlate positively with measures of performance in laboratory

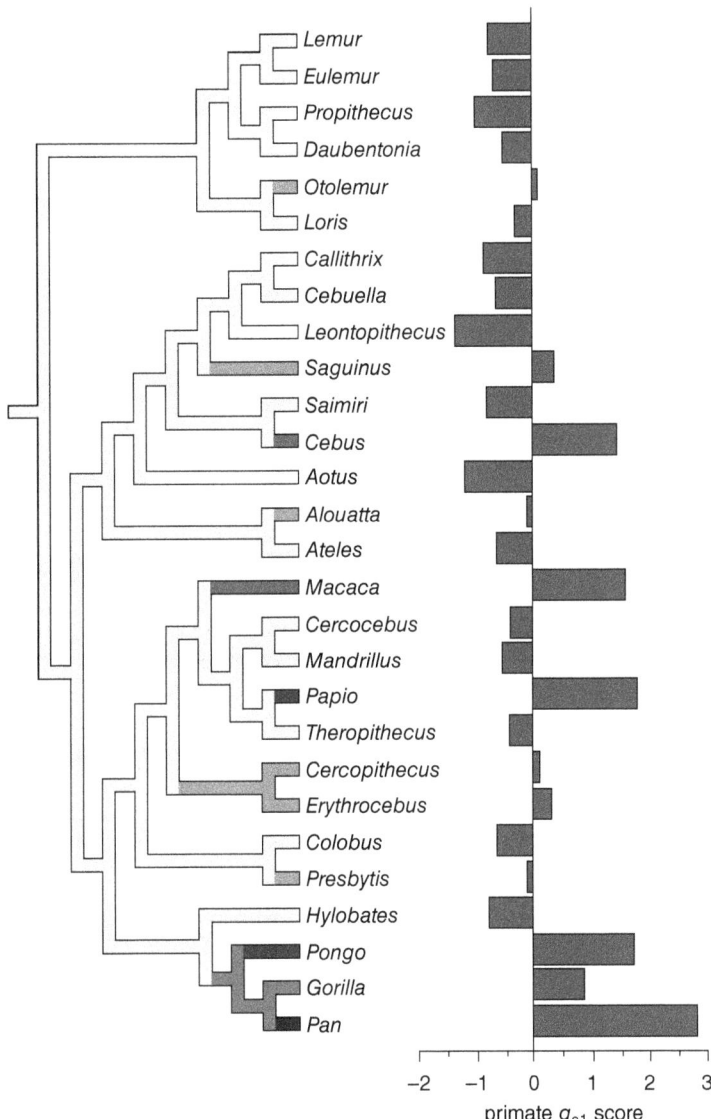

Figure 18.2 Reader et al's (2011) g_s—A Measure of General Intelligence across Primate Species, Mapped on to a Genus-Level Primate Phylogeny. Reader 2011. Reproduced with permission of Kevin Laland. (*See insert for color representation of the figure*).

tests of cognition. Finally, when the g_s measure was mapped onto a primate phylogeny, Reader et al. revealed evidence for convergent evolution of enhanced intelligence in four primates groups—great apes, macaques, baboons, and capuchins—precisely those taxa renowned for reliance on social learning and traditions (Figure 18.2; Cambefort, 1981; Ferrari et al., 2006; Hirata, Watanabe, & Kawai, 2001; Huffman, 1984; Kawai, 1965; Perry, 2011; Petit & Thierry, 1993; van Schaik et al., 2003; Watanabe, 1994; Whiten et al., 1999).

In addition to comparative studies, the taxonomic distribution of certain types of social learning seems to support the idea of a relationship between social learning, intelligence, and large brains. Experimental reports of high-fidelity social learning are concentrated in primates, cetaceans and birds (Hoppitt & Laland, 2008). For example, the ability to learn motor behavior imitatively has been demonstrated in humans, chimpanzees, marmosets and birds, and vocal behavior in cetaceans and birds (Catchpole & Slater, 2008; Hoppitt & Laland, 2008; Janik & Slater, 2000; Rendell & Whitehead, 2001; Voelkl & Huber, 2000). Observational studies and circumstantial accounts of these taxa also attest to their imitative ability, although many such examples have not been systematically or experimentally verified, and it is difficult to extract specific learning processes from observational accounts. However, when specific motor patterns, which are outside of the normal behavioral repertoire or lack apparent utility, are closely reproduced, many researchers infer imitation. In primates, examples include able-bodied chimpanzees reproducing the idiosyncratic actions of disabled individuals (Byrne, 2009; Hobaiter & Byrne, 2010), and the copying of a variety of human behaviors in orang-utans (Russon & Galdikas, 1993). Captive dolphins have been reported anecdotally to replicate spontaneously a variety of actions outside of their behavioral range, including the swimming motion of pinnipeds (Tayler & Saayman, 1973) and tool use of humans (Kuczaj, Gory, & Xitco, 1998). Vocal imitation is a common behavior in cetaceans, demonstrated by examples of captive dolphins and killer whales adopting unfamiliar vocal repertoires, coordinated change over time in humpback whale song and duetting in sperm whales (Rendell & Whitehead, 2001). Songbirds and parrots are highly proficient vocal mimics, many of which are able to closely replicate a wide variety of sounds outside their species' normal vocal range (Emery, 2006; Janik & Slater, 2000). Primates are not known to be especially proficient at vocal imitation, although duetting behavior in gibbons is an exception (Janik & Slater, 2000).

Social learning that results in traditions and "animal culture"—socially transmitted behaviors which endure over many generations and spread throughout a population (Laland & Galef, 2009; Whiten et al., 1999)—is less common than social learning in general, and has been reported in primates, cetaceans, and birds. Studies in which behavioral variation between animal communities is systematically documented have demonstrated considerable ranges of group-specific behaviors, arguably suggestive of socially learned traditions, including in the domains of foraging, social behavior, tool use and/or vocalizations, in chimpanzees (Whiten et al., 1999), orang-utans (Bastian, Zweifel, Vogel, Wich, & Van Schaik, 2010; van Schaik et al., 2003), capuchins (Perry, 2011), spider monkeys (Santorelli et al., 2011), Japanese macaques (Leca, Gunst, & Huffman, 2007), killer whales and sperm whales (Rendell & Whitehead, 2001), bowerbirds (Madden, 2008), passerines (Janik & Slater, 2000) and corvids (Bluff, Kacelnik, & Rutz, 2010). However, traditions are by no means exclusive to larger brained species. Fishes too exhibit traditional use of mating sites, schooling sites, resting sites, and pathways through the reef (Helfman & Schultz, 1984; Laland et al., 2011; Warner, 1988, 1990). There is even preliminary evidence for traditions in insects (Donaldson & Grether, 2007; Leadbeater & Chittka, 2007). Therefore, the current taxonomic spread of animal traditions does not support the idea that traditions necessarily require large brains. There is little reason to assume that socially transmitted traditions necessarily require complex social learning mechanisms (such as high-fidelity copying), since formal theory finds that simpler mechanisms, such as local enhancement, can generate traditions (Franz & Matthews, 2010; Van der Post & Hogeweg, 2008). Furthermore, although group-specific behaviors are suggestive of socially learned traditions, inference of social learning from observational studies and group-specific behaviors is contentious in that ecological, genetic and asocial learning factors can rarely be ruled out as explanations of such behavioral variation. Moreover, behavioral variation is likely to result from multiple causes, which are highly correlated (Laland & Janik, 2006; Langergraber et al., 2011). Therefore, it remains unclear whether the existence of traditions and culture in animals across species is associated with especially high-fidelity copying mechanisms and brain expansion. Some authors have suggested that an association exists between the number of traditions and brain size in animals (Whiten & van Schaik, 2007), but even this is contentious (Laland & Hoppitt, 2003; Laland et al., 2011).

The association of social learning and brain size appears have reached its peak in our own species. Modern humans have both the largest brain size within their order (Striedter, 2005) and a uniquely substantial dependence on culturally acquired behaviors for survival and reproduction (Tomasello, 1999; Whiten, Hinde, Laland, & Stringer, 2011). Humans are especially skilled social learners in that we are capable of copying at exceptionally high fidelity, even as children, when compared to nonhuman apes (Herrmann, Call, Hernandez-Lloreda, Hare, & Tomasello, 2007; Tennie, Greve, Gretscher, & Call, 2010). For example, "two-action" studies (Hopper, Flynn, Wood, & Whiten, 2010) have demonstrated that children learn to solve a task in the specific manner that they saw performed by a demonstrator. In transmission chains of 20 children, each method of solving the task was passed from one child to another by social learning with perfect fidelity. When chimpanzees were tested using the same method, the majority, but not all, of the chimpanzees along the transmission chain conformed to the method shown (Whiten et al., 2005). Human adults too show a high level of precision in copying experimental tasks, down to the specific digits used for each part of a task, whereas macaques (*Macaca nemestrina*) showed little evidence of social transmission in the same experiment (Custance, Prato-Previde, Spiezio, Rigamonti, & Poli, 2006). However, in a relatively simple task involving a door which could either be lifted or swung open, the results of children and chimpanzees were far more similar, with near perfect transmission fidelity in the chimpanzees (Horner, Whiten, Flynn, & De Waal, 2006). Human children, unlike nonhuman primates, have been shown to copy specific details of the sequence of behavior even when these are causally irrelevant to the goal of the task (Lyons, Young, & Keil, 2007; Whiten, McGuigan, Marshall-Pescini, & Hopper, 2009b), an effect which is pronounced in adults (McGuigan, Makinson, & Whiten, 2011). In contrast, chimpanzees seem rather to copy only aspects of the solving the task that are physically necessary (Horner & Whiten, 2005; Whiten et al., 2009a).

In addition to exceptionally high-fidelity copying, humans are seemingly unique in possessing cumulative culture. Cumulative culture is the process by which learned traditions are not only transferred across generations, but increase in complexity, diversity, and efficiency, such that each generation inherits much of the accumulated knowledge of innumerable individuals on which to build further modification and improvement (Sterelny, 2011; Whiten, 2011). Cumulative culture is responsible for the majority of the learned human repertoire, particularly science and technology (Mesoudi, 2011; Tomasello, 1999), and has been instrumental in the evolution of our species' unique reliance on culture for survival and reproduction. Unsurprisingly, cumulative culture in humans has been demonstrated experimentally, in for example, the steady improvement of paper plane flight distances in a laboratory transmission chain study (Caldwell & Millen, 2010). "Ratcheting" effects—in which a learned behavior is modified—have been clearly observed in humans only (Custance et al., 2006; Hopper et al., 2010). In contrast, chimpanzees showed no ability to learn cumulatively in an artificial "honey dipping" task. In the study, chimpanzees learned how to dip a stick to extract honey from the apparatus from a demonstrator. Subsequently, they observed a demonstrator manipulating the device in a previously unseen manner in order to access a higher quality reward. Despite numerous demonstrations, the chimpanzees did not learn the new and improved method (Marshall-Pescini & Whiten, 2008).

Theoretical analyses demonstrate that cumulative culture and high-fidelity copying are mutually reinforcing. Enquist et al. (2010) found a positive acceleratory relationship between the fidelity with which learned information is transmitted between individuals and both the longevity of a cultural trait (how long a transmitted behavior remains in a population) and the number of cultural traits a population can support. In other words, a small increase in transmission fidelity (which we have seen, confers fitness benefits; Rendell et al., 2010) would lead to a big increase in duration of traditions and amount of culture exhibited by a population. Furthermore, a recent theoretical analysis established a link between cumulative culture and teaching (another high-fidelity transmission mechanism), with each promoting the evolution of the other (Fogarty et al., 2011) This supports the argument, originally put forward by Tomasello (1994) that the stable, long-lasting cumulative culture characteristic of humans, but not other animals, requires such high-fidelity social learning mechanisms as teaching, imitation, and language.

In summary, together, comparative studies of social learning, brain size and cognition support the idea that social learning is a composite part of general intelligence, and may be causally related to advanced cognition and expansion of the forebrain in primates. Additionally, experimental and observational studies suggest that high-fidelity, high-efficiency, and the most adaptive forms of copying may be more common in large-brained than small-brained taxa, in particular, in primates, cetaceans and Psittacopasserae birds.

18.3 Why Are Brain Size, Cognition, and Social Learning Related?

Currently, the nature of the relationship between social learning, brain size, and cognition is poorly understood. There have been few investigations of the neural basis of social learning (Hoppitt & Laland, 2013; Rendell, Fogarty et al., 2011). Generally speaking, social learning occurs to a relatively greater extent in species with enlarged brains, but it also occurs in a great many species that have not undergone forebrain expansion and whose cognitive abilities are not thought to be so impressive. For instance, social learning has been documented experimentally in fish, reptiles, amphibians, and insects (Brown & Laland, 2003; Bshary, Wickler, & Fricke, 2002; Ferrari & Chivers, 2008; Wilkinson et al., 2010). Social learning in species with extremely small brains particularly demonstrates that large brains are not a requirement for social learning *per se* (Leadbeater & Chittka, 2007). For example, honeybees are able to acquire flower preferences from other individuals (Leadbeater & Chittka, 2005) and wood crickets can learn to recognize predators from the responses of experienced conspecifics (Coolen, Dangles, & Casas, 2005). Here social learning is often best regarded as merely a special case of associative learning that exploits ancient neural architecture, rather than a distinct process underpinned by specific cognitive and neurological adaptations (Heyes, 1994), although the famous honeybee waggle dance is a striking counter-example. Furthermore, although social learning has tended to be disproportionately reported in species with large brains and behavior suggestive of high intelligence; the effects of anthropogenic factors, such as biased research effort and anthropocentrism, markedly confound this relationship (Laland & Hoppitt, 2003).

However, there are plausible neurocognitive mechanisms that may bridge the gap between social learning ability and brain size that remain to be explored. Such mechanisms need not be restricted to learning processes. Indeed, we envisage that a wide variety of perceptual and cognitive faculties may have been selected or enhanced, in part because they increased the adaptive benefits gained from socially learned behavior, particularly through increased copying fidelity and the strategic use of social learning. Although social learning is a common behavior in a very wide range of animal taxa, varying in brain size and cognitive abilities, animals vary in terms of their reliance on social learning, the specific mechanisms of social learning employed, the social learning strategies deployed (e.g., the contexts in which they copy), and the complexity of behavior transferred. We suspect that large-brained, highly intelligent species tend to exhibit specific processes of social learning that more plausibly require specific cognitive and neurological structures, which are related to forebrain expansion.

The observed relationship between the incidence of social learning and brain size in primates may be an incidental by-product of a more significant relationship between social learning *ability* and brain size. Rather than selecting for more frequent social learning, there may have been selection for better (i.e., more efficient) forms of social learning, with large-brained primates being more effective copiers than small-brained primates. The argument is also supported by the finding from the social learning strategies tournament, as well as analytical theory, that the greater the accuracy of social learning (i.e., the fewer errors in copying), the greater the adaptive advantage of copying relative to trial and error learning (Boyd & Richerson, 1985; Rendell et al., 2010). It would seem that there are likely to be fitness benefits to copying efficiently, generating selection for neural structures and cognitive processes that improve the efficiency with which animals copy. So, although an animal does not need a large brain to socially learn, brain expansion could be advantageous if it helps individuals to best exploit the adaptive benefits of socially acquired behavior. The direction of causality of the relationship between social learning, general intelligence, and brain size is currently unknown. It is difficult to rule out the possibility that brain enlargement has occurred in response to factors unrelated to social learning specialization, and that large-brained species then were able to make facultative usage of their additional processing power through increased social learning. However, Reader et al's findings are consistent with the suggestion that social learning may have driven brain evolution in primates (Reader & Laland, 2002; Whiten & van Schaik, 2007; Wilson, 1985; Wyles, Kunkel, & Wilson, 1983). It is possible that the fitness advantages of social learning, in addition to the ability to invent new behavior (innovation), provided that these abilities have some bases in neural substrate, have generated selection for brain enlargement (Wilson, 1985). This "cultural drive" hypothesis implies that social learning is a driver of brain expansion, culminating in humans, the most innovative, culturally reliant, and largest brained primate (Whiten & van Schaik, 2007).

The cognitive and neurological foundations of social learning, traditions and culture are not well understood. However, forebrain expansion and specific cognitive abilities, if not requisites, may nonetheless offer fitness benefits to animals with large and enduring repertoires of learned behavior. In particular, brain enlargement and high intelligence may support high copying fidelity. Learning behaviors with sufficient copying fidelity that they can be passed along numerous generations may plausibly be enhanced by reliance on specific social learning processes, especially imitation (Hopper et al., 2010). While there is currently little evidence to suggest that, in order to replicate behavior faithfully, cognitive abilities such as perspective-taking, theory of mind, and the ability to extract motor function from visual perception are required, nonetheless it is feasible that these capabilities would promote the ability to imitate and thus the efficiency of social learning. It follows that perceptual and cognitive capabilities may have been favoured by selection because of the fitness benefits they confer through enhancing the fidelity of information transmission (Rendell, Fogarty et al., 2011).

We anticipate that the aforementioned fitness benefits to more efficient copying and high-fidelity transmission might have selected for a number of cognitive and neural structures, including:

- better perceptual systems, to the extent that these allow for more accurate copying of fine-grained motor patterns, or copying from distance;
- more cross-modal mapping and integration across modular structures (or the plasticity to acquire this with experience), to the extent that these help animals address the correspondence problem (the difference in perspective of demonstrator and observer when copying);
- theory of mind, and better comprehension of the intent or goals of others, if this allows for more accurate reproduction of their behavior, and more effective teaching;
- monitoring payoff and frequency dependence, since this allows individuals to implement effective strategies, such as payoff-based and conformist forms of copying;
- sensitivity to social cues, social tolerance and prosociality, to the extent that it enhances transmission fidelity;
- enhanced sequence learning capabilities, to facilitate production imitation, which involves stringing together sequences of action units (Byrne, 1999), and is essential for language learning;
- imitation and teaching capabilities, the latter of which are further enhanced by language, which promote the accuracy with which complex learned knowledge and skills can be transmitted;
- a general enhancement in plasticity, as might be manifest in bootstrapping and Bayesian-style inference, to the extent that this promotes observational causal learning, and the effective processing of socially acquired knowledge (Buchsbaum, Gopnik, Griffiths, & Shafto, 2011).

Theoretical models have shown that social learning is not always beneficial and can be costly, due to the risks of copying errors, receiving bad information, and information becoming outdated in a changing environment (Boyd & Richerson, 1985; Feldman,

Aoki, & Kumm, 1996; Rogers, 1988). Social learning is also parasitic, since its utility depends on the presence of a population of asocial learners from which to sample and learn (Rogers, 1988). This theoretical work establishes that for social learning to be adaptive, in the sense that it increases mean fitness, it cannot be randomly deployed. Rather, using social learning adaptively requires the strategic employment of both social and asocial processes according to evolved rules, known as "social learning strategies" (Laland, 2004). Social learning strategies comprise behavioral biases in learning, regarding under what conditions to copy (Kendal, Coolen, & Laland, 2009; Laland, 2004; Rendell, Fogarty et al., 2011). Theoretical and experimental studies have demonstrated the selective advantages of copying selectively depending on, for example, the costs of individual learning, the behavior of the majority and the number, success and dominance of demonstrators, over copying 'blindly' (Boyd & Richerson, 1985; Kendal et al., 2009; Laland, 2004; Rendell et al., 2010; Rendell, Boyd et al., 2011, Rendell, Fogarty et al., 2011). The results of the social learning strategies tournament showed that, although the amount of information learned strategy correlates with success, the most successful strategies used social learning selectively. In particular, the winning strategy negatively weighted the use of social learning according to the time since information was acquired (Rendell et al., 2010). Empirical evidence shows that many species, including primates, rats, birds, fishes, and insects, use social learning selectively, in terms of choosing when and who to copy from, and choosing between social and individual learning adaptively (Galef, 2009; Horner & Whiten, 2005; Kendal et al., 2009; Whiten, 2011).

The successful use of social learning strategies is likely to exert specific neuro-cognitive demands on individuals. For instance, conforming to the majority behavior requires individuals to estimate the frequency of trait incidence, while payoff-based copying requires an ability to approximate the payoff to others. While such social learning strategies are seen in animals, it is plausible that humans are more effective than other animals in combining strategies. A recent experimental study gives strong support to the suggestion that human social learning is rule-governed, and predicted by evolutionary models. In 4 experiments on human subjects, comprised of 6 separate experimental tasks, Morgan, Rendell, Ehn, Hoppitt, & Laland (2011) found evidence for nine social learning strategies, all of which were anticipated by cultural evolution models, and all of which were adaptive in the sense that they increased the subject's payoff. Moreover, humans may be capable of strategies that other animals are not: for instance, employing time-dependent social learning strategies is likely to be especially computationally challenging (Rendell et al., 2010). The cognitive basis of this ability is largely unknown, but episodic or episodic-like memory, mental time travel, and forward planning are likely candidates. In order to be able to choose when to use socially learned behaviors according to the likely payoffs, it is necessary to be able to recall when information was learned, from whom, and the previous results of behavior, in order to be able to anticipate the likely results of future behavior (de Waal & Ferrari, 2010). Episodic-like memory has been reported in birds (Clayton & Dickinson, 1998; Zinkivskay, Nazir, & Smulders, 2009) and rodents (Babb & Crystal, 2006; Ferkin, Combs, Delbarco-Trillo, Pierce, & Franklin, 2008; Zinkivskay et al., 2009) and true episodic memory in humans (Tulving, 1983), and there are circumstantial accounts of forward planning in chimpanzees (de Waal & Ferrari, 2010). However, it is likely that only humans possess the mental time traveling capability to implement the kind of strategy exhibited by the winner of the social learning strategies tournament

(Rendell et al., 2010), which copied only when its calculations estimated doing so would bring new behavior into its repertoire with a payoff higher than its current behavior.

In addition to cognitive abilities, enhanced perceptual and motor skills are likely to be implicated in the social learning abilities of large-brained species. Forebrain enlargement in primates and birds is associated with visual specialization in response to ecological pressures, corresponding to enlargement of the neocortex and pallium, respectively (Alonso et al., 2004; Barton, 1996; Changizi & Shimojo, 2005; Emery, 2006; Zelenitsky, Therrien, & Kobayashi, 2009). Auditory specialization for echolocation and vocal communication has played a similar role in neocortex expansion in cetaceans (Marino et al., 2004). Visual and auditory specialization expands the potential for social learning in that it helps monitor multiple individuals across large spaces and better perceive social cues. Technical intelligence involves the understanding of the physical properties of objects and causation in addition to fine motor control, and is thus required for the imitation of behaviors involving motor coordination such as bodily movements and tool use (Whiten & Byrne, 1997). Consistently, primates and birds are the most skilled tool users in the animal world. Primates have especially advanced fine motor control over the hands and fingers (Byrne & Bates, 2010). Expansion of the cerebellum—which is involved in motor skills—in mammalian brain evolution is strongly correlated with the neocortex, supporting the seemingly integrated nature of technical and other aspects of cognition such as learning, memory and executive function (Alonso et al., 2004; Barton, 2002; Cantalupo & Hopkins, 2010). Relative cerebellum size increases with the frequency of tool use in primate species (Barton, 2012), and is further suggested by the contrasting cerebellum sizes and tool use reports in chimpanzees and bonobos (Cantalupo & Hopkins, 2010). Further, the association of technical intelligence, social learning and brain size expansion is exemplified by human evolution. The hominid neocortex has not only undergone substantial increases in size, but also in connectivity, and has evolved projections into the medulla and spinal cord (Striedter, 2005). Such connections allow humans to learn intricate routines of movement and complex manual tasks, because the Fodorian executive part of the brain can directly monitor the fingers and feet (Striedter, 2005). The same projections allow exhibit fine control of the tongue, vocal chords, and breathing, without which humans probably could not have learned to speak (Striedter, 2005). These neural connections reinforce our view that social learning has been instrumental in brain evolution, and drove the evolution of various other cognitive capabilities, including tool use and language.

18.4 Conclusions

Evidence from comparative studies of primates and from the general concentration of social learning in large-brained, cognitively advanced species is suggestive of a general link between social learning, brain size, and intelligence. Although social learning itself may not require cognitive and neurological specialization, the extent to which a species is able to utilize and exploit the adaptive benefits of social learning appears to be enhanced by brain expansion and cognitive complexity. Specifically, cognitive and neurological specializations may be required for high-fidelity social learning processes such as imitation and teaching, the use of multiple and complex social learning strategies, especially those which are time-dependent, and in order to support cumulative culture. The cognitive and neurological structures involved in social learning are poorly understood, therefore greater exploration of the neural basis of social learning, for instance through brain-imaging studies, would be of great value.

Acknowledgments

Research supported in part by ERC Advanced Grant (Evoculture) to KNL.

References

Alonso, P. D., Milner, A. C., Ketcham, R. A., Cookson, M. J., & Rowe, T. B. (2004). The avian nature of the brain and inner ear of Archaeopteryx. *Nature*, 430, 666–669.

Alport, L. J. (2004). Comparative analysis of the role of olfaction and the neocortex in primate intrasexual competition. *Anatomical Record Part a—Discoveries in Molecular Cellular and Evolutionary Biology*, 281A, 1182–1189.

Amici, F., Aureli, F., & Call, J. (2008). Fission–fusion dynamics, behavioral flexibility, and inhibitory control in primates. *Current Biology*, 18, 1415–1419.

Babb, S. J., & Crystal, J. D. (2006). Episodic-like memory in the rat. *Current Biology*, 16, 1317–1321.

Barton, R. A. (1996). Neocortex size and behavioral ecology in primates. *Proceedings of the Royal Society B, Biological sciences*, 263, 173–177.

Barton, R. A. (2002). How did brains evolve? *Nature*, 415, 134–135.

Barton, R. A. (2012). Embodied cognitive evolution and the cerebellum. *Philosophical Transactions of the Royal Society B, Biological Sciences* 367, 2097–2107.

Barton, R. A., & Harvey, P. H. (2000). Mosaic evolution of brain structure in mammals. *Nature*, 405, 1055–1058.

Basil, J. A., Kamil, A. C., Balda, R. P., & Fite, K. V. (1996). Differences in hippocampal volume among food storing corvids. *Brain Behavior and Evolution*, 47, 156–164.

Bastian, M. L., Zweifel, N., Vogel, E. R., Wich, S. A., & Van Schaik, C. P. (2010). Diet traditions in wild orangutans. *American Journal of Physical Anthropology*, 143, 175–187.

Bluff, L. A., Kacelnik, A., & Rutz, C. (2010). Vocal culture in New Caledonian crows Corvus moneduloides. *Biological Journal of the Linnean Society*, 101, 767–776.

Bolhuis, J. J., & Macphail, E. M. (2001). A critique of the neuroecology of learning and memory. *Trends in Cognitive Sciences*, 5, 426–433.

Box, H. O., & Gibson, K. R. (eds.) 1999). *Mammalian social learning*. Cambridge: Cambridge University Press.

Boyd, R., & Richerson, P. J. (1985). *Culture and the evolutionary process*. Chicago, IL: University of Chicago Press.

Brown, C., & Laland, K. N. (2003). Social learning in fishes: A review. *Fish and Fisheries*, 4, 280–288.

Bshary, R., Wickler, W., & Fricke, H. (2002). Fish cognition: A primate's eye view. *Animal Cognition*, 5, 1–13.

Buchsbaum, D., Gopnik, A., Griffiths, T. L., & Shafto, P. (2011). Children's imitation of causal action sequences is influenced by statistical and pedagogical evidence. *Cognition*, 120, 331–340.

Byrne, R. (1995). *The thinking ape: The evolutionary origins of intelligence*. Oxford, Oxford University Press.

Byrne, R. W. (1999). Imitation without intentionality. Using string parsing to copy the organization of behavior. *Animal Cognition*, 2, 63–72.

Byrne, R. W. (2009). Animal imitation. *Current Biology*, 19, R111–R114.

Byrne, R. W., & Bates, L. A. (2007). Sociality, evolution and cognition. *Current Biology*, 17, R714–R723.

Byrne, R. W., & Bates, L. A. (2010). Primate social cognition: Uniquely primate, uniquely social, or just unique? *Neuron*, 65, 815–830.

Byrne, R. W., & Corp, N. (2004). Neocortex size predicts deception rate in primates. *Proceedings of the Royal Society of London B, Biological Sciences*, 271, 1693–1699.

Caldwell, C. A., & Millen, A. E. (2010). Human cumulative culture in the laboratory: Effects of (micro) population size. *Learning & Behavior*, 38, 310–318.

Cambefort, J. P. (1981). A comparative study of culturally transmitted patterns of feeding habits in the chacma baboon *Papio ursinus* and the vervet monkey *Cercopithecus aethiops*. *Folia Primatologica*, 36, 243–263.

Cantalupo, C., & Hopkins, W. (2010). The cerebellum and its contribution to complex tasks in higher primates: A comparative perspective. *Cortex*, 46, 821–830.

Catchpole, C., & Slater, P. J. B. (2008). *Bird song: Biological themes and variations*, Cambridge: Cambridge University Press.

Changizi, M. A., & Shimojo, S. (2005). Parcellation and area–area connectivity as a function of neocortex size. *Brain Behavior and Evolution*, 66, 88–98.

Clayton, N. S., & Dickinson, A. (1998). Episodic-like memory during cache recovery by scrub jays. *Nature*, 395, 272–274.

Cnotka, J., Gunturkun, O., Rehkamper, G., Gray, R. D., & Hunt, G. R. (2008). Extraordinary large brains in tool-using New Caledonian crows (Corvus moneduloides). *Neuroscience Letters*, 433, 241–245.

Coolen, I., Dangles, O., & Casas, J. (2005). Social learning in noncolonial insects? *Current Biology*, 15, 1931–1935.

Coolen, I., Ward, A. J. W., Hart, P. J. B., & Laland, K. N. (2005). Foraging nine-spined sticklebacks prefer to rely on public information over simpler social cues. *Behavioral Ecology*, 16, 865–870.

Custance, D., Prato-Previde, E., Spiezio, C., Rigamonti, M. M., & Poli, M. (2006). Social learning in pig-tailed macaques (*Macaca nemestrina*) and adult humans (*Homo sapiens*) on a two-action artificial fruit. *Journal of Comparative Psychology*, 120, 303–313.

De Waal, F. B. M., & Ferrari, P. F. (2010). Towards a bottom-up perspective on animal and human cognition. *Trends in Cognitive Sciences*, 14, 201–207.

Deaner, R. O., Nunn, C. L., & Van Schaik, C. P. (2000). Comparative tests of primate cognition: Different scaling methods produce different results. *Brain Behavior and Evolution*, 55, 44–52.

Deaner, R. O., Van Schaik, C. P., & Johnson, V. E. (2006). Do some taxa have better domain-general cognition than others? A meta-analysis of nonhuman primate studies. *Evolutionary Psychology*, 4, 149–196.

Devoogd, T. J., Krebs, J. R., Healy, S. D., & Purvis, A. (1993). Relations between song repertoire size and the volume of brain nuclei related to song—comparative evolutionary analyses amongst oscine birds. *Proceedings of the Royal Society of London Series B, Biological Sciences*, 254, 75–82.

Donaldson, Z., & Grether, G. (2007). Tradition without social learning: scent-mark-based communal roost formation in a Neotropical harvestman (Prionostemma sp.). *Behavioral Ecology and Sociobiology*, 61, 801–809.

Dunbar, R. I. M. (1992). Neocortex size as a constraint on group size in primates. *Journal of Human Evolution*, 22, 469–493.

Emery, N. J. (2006). Cognitive ornithology: the evolution of avian intelligence. *Philosophical Transactions of the Royal Society B, Biological Sciences*, 361, 23–43.

Emery, N. J., & Clayton, N. S. (2004). The mentality of crows: Convergent evolution of intelligence in corvids and apes. *Science*, 306, 1903–1907.

Enquist, M., Eriksson, K., Laland, K. N., Strimling, P., & Sjostrand, J. (2010). One cultural parent makes no culture. *Animal Behaviour*, 79: 1353–1362.

Feldman, M. W., Aoki, K., & Kumm, J. (1996). Individual versus social learning: Evolutionary analysis in a fluctuating environment. *Anthropological Science*, 104, 209–231.

Ferkin, M. H., Combs, A., Delbarco-Trillo, J., Pierce, A. A., & Franklin, S. (2008). Meadow voles, Microtus pennsylvanicus, have the capacity to recall the "what," "where," and "when" of a single past event. *Animal Cognition*, 11, 147–159.

Ferrari, M. C. O., & Chivers, D. P. (2008). Cultural learning of predator recognition in mixed-species assemblages of frogs: The effect of tutor-to-observer ratio. *Animal Behaviour*, 75, 1921–1925.

Ferrari, P. F., Visalberghi, E., Paukner, A., Fogassi, L., Ruggiero, A., & Suomi, S. J. (2006). Neonatal imitation in rhesus macaques *PLoS Biology*, 4, 1501–1508.

Finlay, B. L., & Darlington, R. B. (1995). Linked regularities in the development and evolution of mammalian brains. *Science*, 268, 1578–1584.

Fogarty, L., Strimling, P., & Laland, K. N. (2011). The evolution of teaching. *Evolution*, 65, 2760–2770.

Franz, M., & Matthews, L. J. (2010). Social enhancement can create adaptive, arbitrary and maladaptive cultural traditions. *Proceedings of the the Royal Society B, Biological sciences*, 277, 3363–3372.

Galef, B. G. (2009). Strategies for Social Learning: Testing predictions from formal theory. In H. J. Brockmann (Ed.), *Advances in the study of behavior* (Vol. 39, pp. 114–151). Amsterdam: Elsevier.

Hampton, R. R., Sherry, D. F., Shettleworth, S. J., Khurgel, M., & Ivy, G. (1995). Hippocampal volume and food-storing behavior are related in parids. *Brain Behavior and Evolution*, 45, 54–61.

Healy, S., & Krebs, J. (1992). Food storing and the Hippocampus in Corvids amount and volume are correlated. *Proceedings of the Royal Society B, Biological Sciences*, 248, 241–245.

Helfman, G. S., & Schultz, E. T. (1984). Social transmission of behavioral traditions in a coral reef fish. *Animal Behaviour*, 32, 379–384.

Herrmann, E., Call, J., Hernandez-Lloreda, M. V., Hare, B., & Tomasello, M. (2007). Humans have evolved specialized skills of social cognition: The cultural intelligence hypothesis. *Science*, 317, 1360–1366.

Heyes, C. M. (1994). Social-learning in animals - categories and mechanisms. *Biological Reviews of the Cambridge Philosophical Society*, 69, 207–231.

Heyes, C. M., & Galef, B. G. J. (eds.) 1996. *Social learning in animals: The roots of culture*, San Diego, C.A.: Academic Press.

Hirata, S., Watanabe, K., & Kawai, M. (Eds.) (2001). *Primate origins of human cognition and behavior*, Heidelberg: Springer-Verlag.

Hobaiter, C., & Byrne, R. W. (2010). Able-bodied wild chimpanzees imitate a motor procedure used by a disabled individual to overcome handicap. *PloS One*, 5. doi: 10.1371/journal.pone.0011959

Hopper, L. M., Flynn, E. G., Wood, L. A. N., & Whiten, A. (2010). Observational learning of tool use in children: Investigating cultural spread through diffusion chains and learning mechanisms through ghost displays. *Journal of Experimental Child Psychology*, 106, 82–97.

Hoppitt, W. J. E., & Laland, K. N. (2008). Social processes influencing learning in animals: A review of the evidence. *Advances in the Study of Behavior*, 38, 105–165.

Horner, V. K., & Whiten, A. (2005). Causal knowledge and imitation/emulation switching in chimpanzees (*Pan troglodytes*) and children. *Animal Cognition*, 8, 164–181.

Horner, V., Whiten, A., Flynn, E., & De Waal, F. B. M. (2006). Faithful replication of foraging techniques along cultural transmission chains by chimpanzees and children. *Proceedings of the National Academy of Sciences of the USA*, 103, 13878–13883.

Huber, L., & Gajdon, G. K. (2006). Technical intelligence in animals: The kea model. *Animal Cognition*, 9, 295–305.

Huffman, M. A. (1984). Stone play of Macaca fuscata. *Journal of Human Evolution*, 13, 725–735.

Janik, V. M., & Slater, P. J. B. (2000). The different roles of social learning in vocal communication. *Animal Behaviour*, 60, 1–11.

Jensen, A. R. (1980). *Bias in mental testing*. London, Methuen.

Jerison, H. J. (1973). *Evolution of the brain and intelligence*. New York, Academic Press.

Johnson, V. E., Deaner, R. O., & Van Schaik, C. P. (2002). Bayesian analysis of rank data with application to primate intelligence experiments. *Journal of the American Statistical Association*, 97, 8–17.

Jolly, A. (1966). Lemur social behavior and primate intelligence. *Science*, 153, 501–506.

Kawai, M. (1965). Newly acquired pre-cultural behavior of the natural troop of Japanese monkeys on Koshima Islet. *Primates*, 6, 1–30.

Kendal, R. L., Coolen, I., & Laland, K. N. (eds.) (2009). *Adaptive trade-offs in the use of social and personal information*. Chicago, IL: University of Chicago Press.

Krebs, D., Sherry, S., & Healy, S. (1989). Hippocampal specialisation of food-storing birds. *Procedings of the National Academy of Sciences of the USA*, 86, 1388–1392.

Kuczaj, S. A., Gory, J. D., & Xitco, M. J. (1998). Using programs to solve problems: Imitation versus insight. *Behavioral and Brain Sciences*, 21, 695–696.

Laland, K. (2004). Social learning strategies. *Learning & Behavior*, 32, 4–14.

Laland, K. N., Atton, N., & Webster, M. M. (2011). From fish to fashion: experimental and theoretical insights into the evolution of culture. *Philosophical Transactions of the Royal Society B, Biological Sciences*, 366, 958–968.

Laland, K. N., & Galef, B. G. (eds.) (2009). *The question of animal culture*, Cambridge, MA: Harvard University Press.

Laland, K. N., & Hoppitt, W. (2003). Do animals have culture? *Evolutionary Anthropology*, 12, 150–159.

Laland, K. N., & Janik, V. M. (2006). The animal cultures debate. *Trends in Ecology & Evolution*, 21, 542–547.

Langergraber, K. E., Boesch, C., Inoue, E., Inoue-Murayama, M., Mitani, J. C., Nishida, T., ... Vigilant, L. (2011). Genetic and "cultural" similarity in wild chimpanzees. *Proceedings of the Royal Society B, Biological Sciences*, 278, 408–416.

Leadbeater, E., & Chittka, L. (2005). A new mode of information transfer in foraging bumblebees? *Current Biology*, 15, R447–R448.

Leadbeater, E., & Chittka, L. (2007). Social learning in insects - From miniature brains to consensus building. *Current Biology*, 17, R703–R713.

Leca, J. B., Gunst, N., & Huffman, M. A. (2007). Japanese macaque cultures: inter- and intra-troop behavioral variability of stone handling patterns across 10 troops. *Behavior* 144, 251–281.

Lefebvre, L., Reader, S. M., & Sol, D. (2004). Brains, innovations and evolution in birds and primates. *Brain Behavior and Evolution*, 63, 233–246.

Lefebvre, L., Whittle, P., Lascaris, E., & Finkelstein, A. (1997). Feeding innovations and forebrain size in birds. *Animal Behavior*, 53, 549–560.

Lyons, D. E., Young, A. G., & Keil, F. C. (2007). The hidden structure of overimitation. *Proceedings of the National Academy of Sciences of the USA*, 104, 19751–19756.

Mackintosh, N. J. (1988). Approaches to the study of animal intelligence. *British Journal of Psychology*, 79, 509–525.

Macphail, E. (1982). *Brain and Intelligence in vertebrates*. Oxford: Clarendon Press.

Madden, J. R. (2008). Do bowerbirds exhibit cultures? *Animal Cognition*, 11, 1–12.

Marino, L., Mcshea, D. W., & Uhen, M. D. (2004). Origin and evolution of large brains in toothed whales. *Anatomical Record Part A—Discoveries in Molecular Cellular and Evolutionary Biology*, 281A, 1247–1255.

Marshall-Pescini, S., & Whiten, A. (2008). Chimpanzees (Pan troglodytes) and the question of cumulative culture: An experimental approach. *Animal Cognition*, 11, 449–456.

Mcguigan, N., Makinson, J., & Whiten, A. (2011). From over-imitation to super-copying: Adults imitate causally irrelevant aspects of tool use with higher fidelity than young children. *British Journal of Psychology*, 102, 1–18.

MesoudI, A. (2011). Variable cultural acquisition costs constrain cumulative cultural evolution. *PloS One*, 6. doi: 10.1371/journal.pone.0018239

Morgan, T. J. H., Rendell, L. E., Ehn, M., Hoppitt, W., & Laland, K. N. (2012). The evolutionary basis of human social learning. *Proceedings of the Royal Society B, Biological Sciences*. 279: 653–662

Perry, S. (2011). Social traditions and social learning in capuchin monkeys (Cebus). *Philosophical Transactions of the Royal Society B, Biological Sciences*, 366, 988–996.

Petit, O., & Thierry, B. (1993). Use of stones in a captive group of Guinea baboons (*Papio papio*). *Folia Primatologica*, 61, 160–164.

Reader, S. M., Hager, Y., & Laland, K. N. (2011). The evolution of primate general and cultural intelligence. *Philosophical Transactions of the Royal Society B, Biological Sciences*, 366, 1017–1027.

Reader, S. M., & Laland, K. N. (2002). Social intelligence, innovation, and enhanced brain size in primates. *Proceedings of the National Academy of Sciences of the USA*, 99, 4436–4441.

Rendell, L., Boyd, R., Cownden, D., Enquist, M., Eriksson, K., Feldman, M. W., … Laland, K. N. (2010). Why copy others? Insights from the social learning strategies tournament. *Science*, 328, 208–213.

Rendell, L., Boyd, R., Enquist, M., Feldman, M. W., Fogarty, L., & Laland, K. N. (2011). How copying affects the amount, evenness and persistence of cultural knowledge: insights from the social learning strategies tournament. *Philosophical Transactions of the Royal Society B, Biological Sciences*, 366, 1118–1128.

Rendell, L., Fogarty, L., Hoppitt, W. J. E., Morgan, T. J. H., Webster, M. M., & Laland, K. N. (2011). Cognitive culture: theoretical and empirical insights into social learning strategies. *Trends in Cognitive Sciences*, 15, 68–76.

Rendell, L., & Whitehead, H. (2001). Culture in whales and dolphins. *Behavioral and Brain Sciences*, 24, 309–382.

Rogers, A. R. (1988). Does biology constrain culture? *American Anthropologist*, 90, 819–831.

Russon, A. E., & Galdikas, B. M. F. (1993). Imitation in free-ranging rehabilitant orangutans. *Journal of Comparative Psychology*, 107, 147–161.

Santorelli, C. J., Schaffner, C. M., Campbell, C. J., Notman, H., Pavelka, M. S., Weghorst, J. A., & AurelI, F. (2011). Traditions in spider monkeys are biased towards the social domain. *PloS One*, 6.

Sherry, D. F., Vaccarino, A. L., Buckenham, K., & Herz, R. S. (1989). The hippocampal complex of food-storing birds. *Brain Behavior and Evolution*, 34, 308–317.

Shettleworth, S. J. (2010). *Cognition, evolution and behavior*. Oxford, Oxford University Press.

Sol, D., Duncan, R. P., Blackburn, T. M., Cassey, P., & Lefebvre, L. (2005). Big brains, enhanced cognition, and response of birds to novel environments. *Proceedings of the National Academy of Sciences of the USA*, 102, 5460–5465.

Sterelny, K. (2011). From hominins to humans: How sapiens became behaviorally modern. *Philosophical Transactions of the Royal Society B, Biological Sciences*, 366, 809–822.

Striedter, G. F. (2005). *Principles of brain evolution*, Sunderland, MA: Sinauer Associates Inc.

Szekely, T., Catchpole, C. K., Devoogd, A., Marchl, Z., & Devoogd, T. J. (1996). Evolutionary changes in a song control area of the brain (HVC) are associated with evolutionary changes in song repertoire among European warblers (Sylviidae). *Proceedings of the Royal Society of London Series B, Biological Sciences*, 263, 607–610.

Tayler, C. K., & Saayman, G. S. (1973). Imitative behavior by Indian Ocean bottlenose dolphins (Tursiops aduncus) in captivity. *Behavior*, 44, 286–298.

Tennie, C., Greve, K., Gretscher, H., & Call, J. (2010). Two-year-old children copy more reliably and more often than nonhuman great apes in multiple observational learning tasks. *Primates*, 51, 337–351.

Tomasello, M. (1994). The question of chimpanzee culture. In R. W. Wrangham, W. C. McGrew, F. B. M. de Waal, & P. G. Heltne (Eds.), *Chimpanzee cultures* (pp. 301–318). Cambridge, MA: Harvard University Press.

Tomasello, M. (1999). *The cultural origins of human cognition.* Cambridge, MA: Harvard University Press.

Tomasello, M., & Call, J. (1997). *Primate cognition.* New York, Oxford University Press.

Tulving, E. (1983). *Elements of episodic memory.* Oxford, Clarendon Press.

Van Der Post, D. J., & Hogeweg, P. (2008). Diet traditions and cumulative cultural processes as side-effects of grouping. *Animal Behavior*, 75, 133–144.

Van Schaik, C. P., Ancrenaz, M., Borgen, G., Galdikas, B., Knott, C. D., Singleton, I., ... Merrill, M. (2003). Orangutan cultures and the evolution of material culture. *Science*, 299, 102–105.

Van Schaik, C. P., Deaner, R. O., & Merrill, M. Y. (1999). The conditions for tool use in primates: implications for the evolution of material culture. *Journal of Human Evolution*, 36, 719–741.

Visalberghi, E., & Tomasello, M. (1998). Primate causal understanding in the physical and psychological domains. *Behavioral Processes*, 42, 189–203.

Voelkl, B., & Huber, L. (2000). True imitation in marmosets. *Animal Behavior*, 60, 195–202.

Warner, R. R. (1988). Traditionality of mating-site preferences in a coral-reef fish. *Nature*, 335, 719–721.

Warner, R. R. (1990). Male versus female influences on mating-site determination in a coral-reef fish. *Animal Behavior*, 39, 540–548.

Watanabe, K. (ed.) 1994). *Precultural behavior of Japanese macques: Longitudinal studies of the Koshima troops*, New York: Kluwer.

Whiten, A. (2011). The scope of culture in chimpanzees, humans and ancestral apes. *Philosophical Transactions of the Royal Society B: Biological Sciences*, 366, 997–1007.

Whiten, A., & Byrne, R. W. (1997). *Machiavellian intelligence II: extensions and evaluations*, Cambridge, Cambridge University Press.

Whiten, A., Goodall, J., Mcgrew, W. C., Nishida, T., Reynolds, V., Sugiyama, Y., ... Boesch, C. (1999). Cultures in chimpanzees. *Nature*, 399, 682–685.

Whiten, A., Hinde, R. A., Laland, K. N., & Stringer, C. B. (2011). Culture evolves Introduction. *Philosophical Transactions of the Royal Society B, Biological Sciences*, 366, 938–948.

Whiten, A., Horner, V., & De Waal, F. B. M. (2005). Conformity to cultural norms of tool use in chimpanzees. *Nature*, 437, 737–740.

Whiten, A., McGuigan, N., Marshall-Pescini, S., & Hopper, L. M. (2009a). Emulation, imitation, over-imitation and the scope of culture for child and chimpanzee. *Philosophical Transactions of the Royal Society B, Biological Sciences*, 364, 2417–2428.

Whiten, A., Mcguigan, N., Marshall-Pescini, S., & Hopper, L. M. (2009b). Emulation, imitation, over-imitation and the scope of culture for child and chimpanzee. *Philosophical Transactions of the Royal Society B, Biological Sciences*, 364, 2417–2428.

Whiten, A., Spiteri, A., Horner, V., Bonnie, K. E., Lambeth, S. P., Schapiro, S. J., & De Waal, F. B. M. (2007). Transmission of multiple traditions within and between chimpanzee groups. *Current Biology*, 17, 1038–1043.

Whiten, A., & Van Schaik, C. P. (2007). The evolution of animal "cultures" and social intelligence. *Philosophical Transactions of the Royal Society B, Biological Sciences*, 362, 603–620.

Wilkinson, A., Kuenstner, K., Mueller, J., & Huber, L. (2010). Social learning in a non-social reptile (*Geochelone carbonaria*). *Biology Letters*, 6, 614–616.

Wilson, A. C. (1985). The molecular basis of evolution. *Scientific American*, 253, 164–173.

Wyles, J. S., Kunkel, J. G., & Wilson, A. C. (1983). Birds, behavior, and anatomical evolution. *Proceedings of the National Academy of Sciences of the USA*, 80, 4394–4397.

Zelenitsky, D. K., Therrien, F., & Kobayashi, Y. (2009). Olfactory acuity in theropods: palaeobiological and evolutionary implications. *Proceedings of the Royal Society B, Biological Sciences*, 276, 667–673.

Zinkivskay, A., Nazir, F., & Smulders, T. V. (2009). What-Where-When memory in magpies (Pica pica). *Animal Cognition*, 12, 119–125.

19

Reading Other Minds
Juliane Kaminski

19.1 Evolving a Theory of Mind

Humans have the ability to attribute mental states to others: that is, to attempt to predict others' knowledge, desires, beliefs and their consequences. To summarize these capacities, Premack and Woodruff (1978) introduced the term "Theory of Mind" (ToM). They called it a "theory" as mental states are not directly observable and therefore need to be inferred. ToM-related skills can be differentiated into three classes: understanding others' perception (e.g., attention, visual or auditory perspective, etc.), understanding others' motivation (e.g., others' goals, intentions, etc.) and understanding others' knowledge (e.g., others' beliefs).

In recent years the question whether nonhuman animals, like humans, have social cognitive capacities became the focus of comparative cognitive research. From an evolutionary perspective, it is most likely that humans share some social cognitive skills, perhaps including mental state attribution, with other species. The so-called "social intelligence hypothesis" formulated by Humphrey (1976) hypothesizes that cognitive capacities are most likely an adaptation to life in complex social groups. In fact, the more complex a social group's structure, the more its constituent individuals can benefit from understanding the other group members' cognitive states. This is because it will allow the individual to make flexible decisions depending on its understanding of the social relationships, and hence to adapt quickly to the constantly-changing social environment. Later, the Machiavellian Intelligence hypothesis, formulated by Whiten and Byrne (1988), added competition as an important driving force for the evolution of social cognitive skills in social species. This hypothesis states that life in groups, and especially competition over resources, puts a constant selection pressure on evolving flexible cognitive skills. As there is a constant struggle to outwit competitors to monopolize resources, Whiten and Byrne hypothesized that social cognitive skills evolved in a kind of arms race between the evolution of measures to manipulate others and the evolution of countermeasures to avoid such manipulation.

If living in complex social groups is seen as the driving force for the evolution of social cognition, then we should expect to find social cognitive capacities, similar to humans', in group-living animals. In recent decades, an important question was therefore to what degree humans share our social cognitive capacities with other animals.

Humans can, in some situations, make predictions and inferences about others' mental states. Humans can predict what others have or have not seen, what others desire, what they believe, and so forth—all often summarized as a "Theory of Mind" (ToM). While some researchers believe that reasoning about mental states is a uniquely human skill, others argue that humans share some social cognitive skills, including mental state attribution, with other species—notably our closest living relatives, the chimpanzees (*Pan troglodytes*) and bonobos (*Pan paniscus*). From an evolutionary perspective, certain social cognitive skills would be beneficial for group-living animals, as they are for humans. Following the Machiavellian Intelligence hypothesis, individuals with some knowledge about others, and the capacity to attribute mental states to others, would be in a better position to outwit their competitors; hence, group living should put a premium on the evolution of social cognitive skills that allow a more flexible understanding of others. However, there are many group-living species, of which few, if any, are thought to have a capacity to attribute mental states to others. This raises the question whether other processes (e.g., associative learning) are sufficient to navigate social groups, even absent a full-fledged ToM.

In order to study the evolutionary history of a certain skill, it is essential to compare the cognitive capacities of different species: For example, to investigate abilities that are particular to humans and our evolutionary history, we need to isolate those that are unique to humans amongst our closest phylogenetic relatives, the other apes. Any cognitive ability that is part of a shared repertoire between related species is likely to be part of their shared evolutionary inheritance from their last common ancestor. When it comes to the evolution of ToM-related skills, the interesting question is whether a complex understanding of others is a widespread phenomenon in the animal kingdom, or whether it is a cognitive capacity unique to humans or shared only with a few other (perhaps closely related) species. While this question remains unresolved, evidence has recently accumulated suggesting that at least one ability—knowing when others can or cannot *see* things—may be a cognitive domain in which the capacities of some animal species are similarly flexible to those of humans.

19.2 Reading Others' Attention

Eye-shaped stimuli are important signals in the animal kingdom. One good example for the importance of eye-shaped signals in the animal kingdom is the Peacock butterfly (*Inachis io*), which has eye-shaped spots on its wings to scare away potential predators. These eyespots are an effective morphological antipredator adaptation that significantly increases individuals' chances of survival (Vallin, Jakobsson, Lind, & Wiklund, 2005), suggesting that attention to eye-like patterns is widespread and can be exploited. However, individuals from this species are most likely not aware that they have this signal. They have very limited control over its presentation to potential predators. They cannot modify the signal based on whether or not the potential predator is in a position to see them. The interaction between both individuals (prey and predator) can therefore be best explained as one example of a sender–receiver relationship in which one individual, the sender, presents a certain signal to which the other individual, the receiver, responds. The sender's signals, as well as the receiver's response, are fixed patterns, shaped by selection processes during evolution. The Peacock butterfly likely has no understanding whatsoever of the predator's mental states.

However, there is evidence that for some species, the eyes signal something about others' attentional states. All great ape species—including chimpanzees, bonobos, gorillas, and orangutans—adjust their gestural communication to the attentional state of a human experimenter. When the human is attentive (e.g., has her head turned towards the subject) they use more visible gestures (such as pointing or reaching) than when the human is not attentive (e.g., has her head turned away). Chimpanzees also use different types of gestures depending on the attentive state of the receiver. They use audible (e.g., hand clapping) instead of visible gestures if others are nearby, but not in a position to see them (Kaminski, Call, & Tomasello, 2004; Liebal, Call, & Tomasello, 2004; Liebal, Pika, Call, & Tomasello, 2004) and use visible gestures (e.g., pointing) when the other is in the position to see them and their eyes are visible (Hostetter, Russell, Freeman, & Hopkins, 2007).

Sensitivity to the eyes as an important signal for others' attention seems to be widespread in the primate family. Rhesus monkeys (*Macaca mulatta*) and also ringtail lemurs (*Lemur catta*), for example, steal less food from a human experimenter whose eyes are open or directed toward them than from one whose eyes are closed or oriented away (Flombaum & Santos, 2005; Sandel, MacLean, & Hare, 2011).

Differentiating others' attentional states is also not restricted to primates and seems to be present in species more distantly related to humans as well. Dolphins (*Tursiops truncatus*) produce more "pointing" (here defined as alignment of the body while remaining stationary for over 2 seconds) if a human is in a position to see them (e.g., oriented toward them) than when he/she is not (Xitco, Gory, & Kuczaj, 2004). Dogs (*Canis familiaris*) also show a high sensitivity to human eyes. When tested in a competitive situation with a human, in which the human forbade them to take a piece of food, dogs took more food when the human was oriented away from the food than when he was oriented toward it, or when the human's eyes were closed as opposed to when they were open, or when the human was distracted as opposed to attentive (Call, Bräuer, Kaminski, & Tomasello, 2003). This was not only true in competitive, but also in more cooperative, contexts in which the dogs had to decide which human to beg from. Here, the dogs directed their begging more toward a human whose eyes were visible than toward a human whose eyes were covered (Gácsi, Miklósi, Varga, Topál, & Csányi, 2004). There is also evidence that different bird species are sensitive to a human's attentional state. Sparrows and jackdaws attend to the presence of the eyes as well as the gaze direction of a human in a competitive situation related to food: When the human's eyes were closed or averted, starlings resumed feeding earlier, at a higher rate, and consuming more, whereas jackdaws were responsive to subtle cues of attention, depending on the social context (i.e., whether the individual was a stranger or familiar to them) (von Bayern & Emery, 2009).

Overall, this shows that a certain level of sensitivity to the status of the eyes is relatively widespread in the animal kingdom among species very distantly related to each other. This could be an indicator that sensitivity to being observed might be an evolutionary ancient and relatively hard-wired behavior with an urgent evolutionary function, but might also suggest that this trait is not homologous in all species and evolved as an analogous trait separately and several times in the animal kingdom.

19.3 Following Others' Gaze

Many species from different taxa not only differentiate whether or not they are being observed, but also attend to where others are looking. For socially living animals, following the gaze of others is beneficial in order to gain information about outside entities. By following another's gaze, the individual can get valuable information about different resources including food, predators, etc. One way to test for this behavior is to see whether an individual follows the gaze direction of another to a specific target outside its own field of view. Various primate species follow the gaze direction of other individuals. For example, all great apes species readily follow the gaze direction of a human experimenter (Bräuer, Call, & Tomasello, 2005). In this study, the human experimenter suddenly shifted her gaze toward the ceiling. Gaze-following behavior in this situation was compared to a control condition during which the experimenter looked straight at the opposite side of the room. Apes looked at the ceiling significantly more often when the human had looked up than when she had not, indicating that they were sensitive to human gaze direction. The ability to follow others' gaze is present not only in apes, but also in various monkey species more distantly related to humans. Emery, Lorincz, Perrett, Oram, and Baker (1997) showed that rhesus macaques were able to locate an object according to the gaze direction of a conspecific depicted on a TV monitor. Tomasello, Call, and Hare (1998) tested several monkey species (including Sooty mangabeys, *Cercocebus atys torquatus*; Rhesus macaques, *Macaca mulatta*; Stumptail macaques, *Macaca arctoides*; and Pigtail macaques, *Macaca nemestrina*) for their ability to follow the gaze of their group members. An experimenter, located in an observation tower, attracted the attention of one individual by presenting food to her. Once this individual had shifted her gaze toward the food, it was recorded whether a nearby subject (that had not seen the food itself) would respond with co-orientation to the conspecific's gaze shift. All monkey species tested in this setting followed the gaze direction of their conspecific. There is also evidence that different New World monkey species, like cotton-top tamarins (*Saguinus Oedipus*), common marmosets (*Callithrix jacchus*) and different lemur species, are responsive to the gaze direction of others (Burkhart & Heschl, 2006; Sandel et al., 2011).

Gaze following is thus widespread among the primates. However, like attention reading, it has also been shown in a wide variety of other mammals including dolphins, seals (*Arctocephalus pusillus*), goats (*Capra hircus*), and dogs and wolves (*Canis lupus*). Dolphins and seals spontaneously attend to the gaze direction of humans (indicated by head direction) in a food search game (Scheumann & Call, 2004; Tschudin, Call, Dunbar, Harris, & van der Elst, 2001). Goats, like primates, follow the gaze of their conspecifics, and dogs seem to be especially sensitive to a human's eyes and gaze direction (Kaminski, Riedel, Call, & Tomasello, 2005). Apart from the mammalian species tested, there also seems to be evidence that species even more distantly related to humans are sensitive to others' gaze direction. Ravens (*Corvus corax*) and rooks (*Corvus frugilegus*) have been shown to follow others' gaze direction. Ravens have been shown to co-orient with the gaze of a human experimenter from an early age. In this test, a human experimenter shifted gaze (head and eye direction) up to a distant location to which the ravens responded with co-orientation (Schlögl, Kotrschal, & Bugnyar, 2007). Recently it was also found that red-footed tortoise (*Geochelone carbonaria*), a solitary living species, follow the gaze of their conspecifics (Wilkinson, Mandl, Bugnyar, & Huber, 2010). This is especially interesting, as another line of research suggests that gaze-following skills may be more sophisticated in species with more complex social structures compared to less socially complex species from the same family. Gibbons, for example, seem to have less sophisticated gaze-following skills than those of great apes, possibly a result of their lack of social complexity as a monogamous species (Liebal & Kaminski, 2012). Conversely, ringtailed lemurs show more gaze-following skills compared to other members of the strepsirrhines, possibly as an adaptation for living in the most complex of strepsirrhine social groups (Sandel et al., 2011).

Taken together, these data show that gaze and gaze direction are important stimuli for a number of species widespread in the animal kingdom. This again suggests a very urgent evolutionary function for gaze following, with a high adaptive value for diverse species. Most likely, the ability to follow gaze helps individuals exploit others for information about important resources like food, mating opportunities, etc. However, the fact that gaze following, like attention reading, has emerged in very distantly related species may suggest that this trait is not homologous in all species, and separately evolved as an analogous trait several times in the animal kingdom.

One important question is to what extent the classical gaze-following behavior—that is, shifting one's gaze direction in response to seeing another individual's gaze shift—a is a more or less learned or inherent automatic response, or truly an indicator of one individual's attention to another individual's "line of sight." If the latter, does this suggest attention to what that individual is seeing, and hence to the other individual's psychological state? One way to test this is to consider the following prediction: If an individual interprets gaze as an indicator of another individual's line of sight, it should, if necessary, relocate to a position from which it can see what the other is looking at. There is evidence that at least some species seem to follow others' gaze not just as an automatic response, but by truly attending to what others are looking at. This is shown by the fact that those species take some effort to track the other's gaze direction to a specific target (by moving towards it) instead of automatically looking in the same direction. Tomasello, Hare, and Agnetta (1999) showed that chimpanzees walk around a barrier in order to track a human's gaze who had just looked behind this barrier. Bräuer et al. (2005) showed that all great apes follow the gaze of a human experimenter behind a barrier by walking around the barrier, presumably to track the human's line of sight. There is also evidence that non-primate species are able to track a human's line of sight. Wolves seem to follow other individuals' gaze around barriers (Range & Viranyi, 2011) and ravens, like apes, will move around a barrier presumably to see what a human is looking at (Bugnyar, Stöwe, & Heinrich, 2004).

However, following gaze around barriers still does not *necessarily* indicate a deeper understanding of seeing in others. Subjects do not have to interpret the other individual's mental states to be successful in this task. Instead of mentally representing that the other individual is seeing something differently, animals may simply have the motivation to look at the same spot the other individual is fixating. Following gaze around barriers may thus indicate representations of spatial relationships, but not necessarily of other minds.

19.4 Perspective Taking

Some mammalian species seem to understand when others' visual access to an object or event is blocked. To test whether chimpanzees have what Flavel, Shipstead, & Croft (1978) define as "Level 1" perspective taking, researchers set up a situation in which two chimpanzees, one dominant over the other, have to compete over two pieces of food. The subordinate chimpanzee, which would normally not have had a chance to gain food with the dominant present, had an advantage: While it had visual access to both pieces of food, the dominant individual could see only one, the other being hidden by a wooden barrier. When given the chance to make a choice, the subordinate chimpanzee preferred to approach the piece of food behind the barrier—the one the dominant could not see—to the piece in the open, visible to the other individual. When the chimpanzees were alone, they chose randomly between the two pieces indicating that their preference for the hidden piece was not merely based on a preference for eating behind an obstacle. In another control condition, the authors showed that chimpanzees would not prefer a piece of food behind a transparent barrier, which potentially protected them from the competitor physically, showing that their preference for the barriers is not due to it being an obstacle that potentially protects them physically (Hare, Call, Agnetta, & Tomasello, 2000).

There is also evidence from other mammalian species that they may have some understanding of when others' line of sight is blocked. Goats and also domestic dogs seem to distinguish between two pieces of reward based on whether another individual has visual access to it or not (Kaminski, Bräuer, Call, & Tomasello, 2009; Kaminski, Call, & Tomasello, 2006). Domestic dogs, for example, distinguish which toy to bring based on the human's visual access to those toys. In this paradigm, the human and the dog sat opposite each other with two toys between them. One toy was placed behind an opaque barrier such that the experimenter had no visual access to it. The other toy was placed behind a transparent barrier such that both the experimenter and the dog had visual access to the toy. Upon the command to fetch, dogs preferred to fetch the toy that was visible to the experimenter. They fetched the visible toy significantly more in this condition than in a control condition where the dog and the experimenter sat on the same side of the barriers and thus had comparable visual access to both toys (Kaminski, et al., 2009). Whether this is based on a true understanding of others' psychological states or based on more simple mechanisms will be the subject of future studies. However, that at least the chimpanzees' behavior cannot be explained by simply perceiving others' eyes as an aversive stimulus is shown by another study. In this study, the chimpanzees are in competition with a human whose eyes they cannot see. The chimpanzees then have to make the decision whether to reach for food through an opaque or a transparent tunnel. As the chimpanzees cannot see the humans' eyes while reaching, their decision has to be based on whether or not the human can potentially see their hand in the tunnel. As chimpanzees preferred to reach through the opaque tunnel, these results suggest that chimpanzees based their behavior on some sensitivity to the visual perspective of the other individual (Melis, Call, & Tomasello, 2006) and did not follow a simple rule, "avoid the piece of food associated with the eyes of the competitor."

From an evolutionary perspective, it is interesting that birds (specifically, corvids), a group of species very distantly related to primates, seem to possess a flexible understanding of others' visual perspective very similar to that of primates. Evidence suggests that these birds seem to have a flexible understanding of others' psychological states, allowing them to form flexible strategies to reduce the probability of others stealing from their hidden caches of food. Scrub jays (*Aphelocoma californica*) and ravens differentiate situations during which they have been observed hiding food from situations where they were able to cache privately (Bugnyar, 2011). When scrub jays have a choice of where to cache while a conspecific is observing, they prefer to hide food in locations which are relatively far from the observer. They also prefer to cache behind an opaque barrier, or in a tray located in the shade, to caching in the open or in a tray located in the light (Dally, Emery, & Clayton, 2004, 2005). The sophisticated cognitive abilities of members of the corvid family, which are very comparable to those of primates, are seen as a good example for analogous evolution as a result of similar selection pressures in the environment. One hypothesis is that it is the complexity of their social environment which put a premium on the evolution of social cognitive skills in corvids, as it has done in primates (Emery, 2004).

19.5 Knowledge Attribution

There is, therefore, plenty of evidence that different animal species understand something about others' current visual perspectives. There is evidence for attention reading, gaze following, and even perspective taking. However, there is also evidence some few species, mainly apes and corvids, not only understand something about others' current visual access, but also about that in their past. One well-known paradigm is the so-called "guesser–knower" paradigm first introduced by Povinelli, Rulf and Bierschwale (1994). The authors conducted a series of experiments which tested whether chimpanzees could take into account what a human had seen in the immediate past. To test this, they confronted chimpanzees with a situation in which they had to distinguish between two human experimenters who informed them about the location of hidden food. One of the experimenters (the knower) witnessed food being placed in one of several containers while the other experimenter (the guesser) waited outside the room. After the guesser reentered the room, the two humans (guesser and knower) pointed to different containers. The chimpanzee was then allowed to choose between the containers, and could potentially base her choice on the information coming from the most reliable source, the knower. In this setting, chimpanzees could only differentiate between humans after several hundred trials, which was most likely the result of discriminating between whether the human was present or absent during baiting. However, one general critique of this paradigm is that it is rather unnatural for chimpanzees: A human indicates the location of food in a very cooperative manner, something that would not occur in a group of chimpanzees. It is highly unlikely that one chimpanzee would indicate the location of food to another chimpanzee with the intention of letting her have it. Kaminski, Call and Tomasello (2008) therefore created a paradigm based on chimpanzees' natural tendency to compete over food. In this paradigm, two individuals, subject and competitor, sat opposite one another, with a sliding board between them that a human could slide back and forth. Each trial began with a hiding event, in which food was hidden under one of three cups while both chimpanzees were watching. Another piece of food was hidden under a second cup, while only the Subject was watching. Hence, while the locations of both pieces of food were known to the Subject, only one of them was known to the Competitor. In some trials, the Competitor was given the first choice with the Subject unable to see this choice being made. After the Competitor had made its choice, it was the Subject's turn. The chimpanzees in this situation preferred the piece of food unknown to the Competitor, presumably because they understood that the other piece, the one the Competitor had information about, was likely to be gone by the time of their own choice. Chimpanzees were similarly successful in this paradigm to six-year old children and adult humans (Kaminski et al., 2008). This finding therefore supports previous studies showing that chimpanzees may take into account what others have seen in their immediate past (Hare, Call, & Tomasello, 2001).

Scrub jays, like chimpanzees, seem to understand others' knowledge states. Dally Emery, and Clayton (2006) presented subjects with a situation in which they had to decide which tray to recover hidden food from. Earlier, the birds were allowed to hide food in one tray in the presence of observer A, with a second tray present but covered. After a delay, the subject was allowed to cache in the other tray with observer B present. After another delay, the subject was given the opportunity to recover their caches from both trays, and had to make the decision which cache to recover based on which observer was present. Interestingly, the birds specifically recovered the caches that observers had seen them make and did not recover any cache if observed by a completely naïve individual, suggesting that it was not simple presence/absence guiding their behavior (Dally et al., 2006). Similar evidence comes from ravens, who seem to be able to predict others' behavior based on what they had observed them observing (Bugnyar, 2011).

19.6 Understanding Others' Beliefs

One ability that is seen as a benchmark for mental state attribution, and therefore theory of mind, is the understanding that others have beliefs and that those beliefs can be true or false. Having an understanding that another individual's belief is false requires an understanding that another person's mental states can be contradictory to one's own and, more importantly, contradictory to reality. So far, there is no evidence that any nonhuman animals can make this distinction.

In one version of a false belief task, chimpanzees were again confronted with a situation in which two individuals had to compete over food. Two chimpanzees sat opposite each other with a sliding board between them, on top of which were three identical cups. The subject, however, had exclusive access to an additional cup to choose from, which was placed sideways to the subject. Two different types of reward were hidden: a preferred high-quality reward was placed in one of the cups on the table between the subjects, and a less-preferred low-quality reward was placed in the additional cup next to the subject. After the initial baiting, which both subjects observed, the high quality reward was manipulated a second time. During this manipulation, the experimenter either lifted the reward and placed it back in its original location or shifted it to a new location. This second manipulation was either witnessed by the competitor or not. Hence, in one of the conditions (the "shift unwitnessed" condition) the competitor has a false belief about the location of the high-quality reward. The competitor was always the first to choose, and the subject did not see the competitor choosing but had to base her decision on what she *guessed* the competitor had done. It turned out that in this setting, subjects did not make the distinction between situations in which the others' belief was true or false (Kaminski, et al., 2008). This is one study of several, all of which indicated that, despite the fact that chimpanzees (and other animals) understand knowledge and ignorance in others, they may not fully appreciate that others have mental representations of the world (Call & Tomasello, 1999, 2008; Kaminski, et al., 2008; Krachun, Carpenter, Call, & Tomasello, 2009).

19.7 Conclusions

Certain social cognitive skills, like reading others' attentional states and following others' gaze directions, seem to be relatively widespread throughout the animal kingdom. This shows that gaze direction and the status of others' attention are a meaningful cue for many socially living animals. However, some of those skills (e.g., gaze following) appear to be automatic reflexive responses which do not necessarily involve any flexible understanding of others' psychological states. The fact that some of those traits are widespread in the animal kingdom suggests that they possess a high survival value, for example, by aiding in the rapid location of predators or avoiding conflict. Other skills, such as the ability to take another's perspective or understand what others have seen in the immediate past, do not seem to be so widespread, and thus may be based on more complex cognitive operations.

However, whether any of these studies can show that animals truly attribute mental states to other individuals is still a highly controversial issue. One criticism of all the studies mentioned above is that the animals in those studies may simply base their strategies on associations formed during the experiment or in earlier life, or have simply read others' behavior and acted based on that information. Instead of having some concept of seeing, animals may simply learn to associate the eyes of their competitor with one piece of food and not the other. The stimulus "eye" may be seen as an aversive stimulus, which the subject then associates with the food, and therefore avoids (the so-called "evil eye hypothesis"). Even though most recent studies try to rule out this associative account, it remains a question whether subjects in those studies need to refer mental states to others in order to solve the problem.

Another nonmentalistic interpretation of the results is that animals do not form concepts of others' *mental states* but rather about others' *behavior*, and that this is sufficient to succeed in all paradigms used with animals so far (Povinelli & Vonk, 2004). This line of thinking suggests that animals follow certain behavioral rules, which they have learned over time: For example, "every time I do x, my conspecific reacts by doing y." The weakness of this approach is that it is not the most plausible explanation across all of the very different studies and paradigms which exist and which provide evidence for animals' understanding of others' psychological states (see Call & Tomasello, 2008 for a discussion of this point). However, all evidence to date also shows that there are strong limits to animals' understanding of others. While some species represent others' *knowledge states*, such as what they may have seen in their immediate past, no nonhuman animal has yet demonstrated the ability to attribute *false beliefs* to others. This suggests that a truly representational theory of mind may be a uniquely human cognitive capacity.

From an evolutionary perspective it is interesting that the most convincing evidence for flexible social cognitive skills comes from two very distantly related groups of species: apes and corvids (Clayton, Dally, & Emery, 2007; Emery & Clayton, 2009). This is interesting, as the morphology of mammalian and bird brains is so substantially different that apes and corvids skills are almost certainly convergent rather than homologous processes (Emery & Clayton, 2009). Similar social cognitive skills therefore may be an adaptation to similar socio-ecological challenges in the social life of these species, for example, in navigating competition over resources and life in a complex social society.

References

Bräuer, J., Call, J., & Tomasello, M. (2005). All great ape species follow gaze to distant locations and around barriers. *Journal of Comparative Psychology*, 119(2), 145–154.

Bugnyar, T., Stöwe, M., & Heinrich, B. (2004). Ravens, *Corvus corax*, follow gaze direction of humans around obstacles. *Proceedings of the Royal Society of London B, Biological Sciences*, 271(1546), 1331–1336.

Bugnyar, T. (2011). Knower–guesser differentiation in ravens: Others' viewpoints matter. *Proceedings of the Royal Society of London B, Biological Sciences*, 278(1705), 634–640.

Burkhart, J., & Heschl, A. (2006). Geometrical gaze following in common marmosets (Callithrix jacchus). *Journal of Comparative Psychology*, 120, 120–130.

Call, J., Bräuer, J., Kaminski, J., & Tomasello, M. (2003). Domestic dogs (*Canis familiaris*) are sensitive to the attentional state of humans. *Journal of Comparative Psychology*, 117(3), 257–263.

Call, J., & Tomasello, M. (1999). A nonverbal false belief task: The performance of children and great apes. *Child Development*, 70(2), 381–395.

Call, J., & Tomasello, M. (2008). Does the chimpanzee have a theory of mind? 30 years later. *Trends in Cognitive Sciences*, 12(5), 187–192.

Clayton, N. S., Dally, J. M., & Emery, N. J. (2007). Social cognition by food-caching corvids. The western scrub-jay as a natural psychologist. *Philosophical Transactions of the Royal Society B, Biological Sciences*, 362, 507–522.

Dally, J. M., Emery, N. J., & Clayton, N. S. (2004). Cache protection strategies by western scrub-jays (*Aphelocoma californica*): Hiding food in the shade. *Proceedings of the Royal Society of London B, Biological Sciences*, 271(Suppl. 6), S387–S390.

Dally, J. M., Emery, N. J., & Clayton, N. S. (2005). Cache protection strategies by western scrub-jays, *Aphelocoma californica*: Implications for social cognition. *Animal Behaviour*, 70(6), 1251–1263.

Dally, J. M., Emery, N. J., & Clayton, N. S. (2006). Food-caching western scrub-jays keep track of who was watching when. *Science*, 310(5780), 1662–1665.

Emery, N. J., Lorincz, E. N., Perrett, D. I., Oram, M. W., & Baker, C. I. (1997). Gaze following and joint attention in rhesus monkeys (*Macaca mulatta*). *Journal of Comparative Psychology*, 111(3), 286–293.

Emery, N. J. (2004). Are corvids "feathered apes"? Cognitive evolution in crows, jays, rooks, and jackdaws. In S. Watanabe (Ed.) *Comparative analysis of minds* (pp. 181–213). Tokyo: Keio University Press.

Emery, N. J., & Clayton, N. J. (2009). Comparative social cognition. *Annual Review of Psychology*, 60, 87–113.

Flavell, J. H., Shipstead, S. G., & Croft, K. (1978). Young children's knowledge about visual perception: Hiding objects from others. *Child Development*, 49(4), 1208–1211.

Flombaum, J. I., & Santos, L. R. (2005). Rhesus monkeys attribute perceptions to others. *Current Biology*, 15(5), 447–452.

Gácsi, M., Miklósi, Á., Varga, O., Topál, J., & Csányi, V. (2004). Are readers of our face readers of our minds? Dogs (*Canis familiaris*) show situation-dependent recognition of humans' attention. *Animal Cognition*, 7(3), 144–153.

Hare, B., Call, J., Agnetta, B., & Tomasello, M. (2000). Chimpanzees know what conspecifics do and do not see. *Animal Behaviour*, 59(4), 771–785.

Hare, B., Call, J., & Tomasello, M. (2001). Do chimpanzees know what conspecifics know? *Animal Behaviour*, 61(1), 139–151.

Hostetter, A. B., Russell, J. L., Freeman, H., & Hopkins, W. D. (2007). Now you see me, now you don't: Evidence that chimpanzees understand the role of eyes in attention. *Animal Cognition*, 10(1), 55–62.

Humphrey, N. K. (1976). The social function of intellect. In P. P. G. Bateson & R. A. Hinde (Eds.), *Growing points in ethology* (pp. 303–317). Cambridge, UK: Cambridge University Press.

Kaminski, J., Bräuer, J., Call, J., & Tomasello, M. (2009). Domestic dogs are sensitive to a human's perspective. *Behaviour*, 146(7), 979–998.

Kaminski, J., Call, J., & Tomasello, M. (2004). Body orientation and face orientation: Two factors controlling apes' begging behavior from humans. *Animal Cognition*, 7(4), 216–223.

Kaminski, J., Call, J., & Tomasello, M. (2006). Goats' behaviour in a competitive food paradigm: Evidence for perspective taking? *Behaviour*, 143(11), 1341–1356.

Kaminski, J., Call, J., & Tomasello, M. (2008). Chimpanzees know what others know, but not what they believe. *Cognition*, 109(2), 224–234.

Kaminski, J., Riedel, J., Call, J., & Tomasello, M. (2005). Domestic goats, *Capra hircus*, follow gaze direction and use social cues in an object choice task. *Animal Behaviour*, 69(1), 11–18.

Krachun, C., Carpenter, M., Call, J., & Tomasello, M. (2009). A competitive nonverbal false belief task for children and apes. *Developmental Science*, 12(4), 521–535.

Liebal, K., Call, J., & Tomasello, M. (2004). Chimpanzee gesture sequences. [abstract from the Spring Meeting of the Primate Society of Great Britain]. *Folia Primatologica*, 75(1), 48.

Liebal, K., Pika, S., Call, J., & Tomasello, M. (2004). To move or not to move: How great apes adjust to the attentional state of others. *Interaction Studies*, 5(2), 199–219.

Liebal, K., & Kaminski, J. (2012). Gibbons (Hylobates pileatus, H. moloch, H. lar, Symphalangus syndactylus) follow human gaze, but do not take the visual perspective of others. *Animal Cognition*, 15(6), 1211–1216.

Melis, A. P., Call, J., & Tomasello, M. (2006). Chimpanzees (*Pan troglodytes*) conceal visual and auditory information from others. *Journal of Comparative Psychology*, 120(2), 154–162.

Povinelli, D. J., Rulf, A. B., & Bierschwale, D. T. (1994). Absence of knowledge attribution and self-recognition in young chimpanzees (*Pan troglodytes*). *Journal of Comparative Psychology*, 108(1), 74–80.

Povinelli, D. J., & Vonk, J. (2004). We don"t need a microscope to explore the chimpanzee's mind. *Mind & Language*, 19(1), 1–28.

Premack, D., & Woodruff, G. (1978). Does the chimpanzee have a theory of mind? *Behavioral and Brain Sciences*, 1(4), 515–526.

Range, F., & Viranyi, Z. (2011). Development of gaze following abilities in wolves (*Canis Lupus*). *PLOS ONE*, 6(2): e16888. doi: 10.1371/journal.pone.0016888

Sandel, A. A., MacLean, E. L., & Hare, B. (2011). Evidence from four lemur species that ring-tailed lemur social cognition converges with that of haplorhine primates. *Animal Behaviour*, 81(5), 925–931.

Scheumann, M., & Call, J. (2004). The use of experimenter-given cues by South African fur seals (*Arctocephalus pusillus*). *Animal Cognition*, 7(4), 224–230.

Schloegl, C., Kotrschal, K., & Bugnyar, T. (2007). Gaze following in common ravens, Corvus corax: Ontogeny and habituation. *Animal Behaviour*, 74(4), 769–778.

Tomasello, M., Call, J., & Hare, B. (1998). Five primate species follow the visual gaze of conspecifics. *Animal Behaviour*, 55(4), 1063–1069.

Tomasello, M., Hare, B., & Agnetta, B. (1999). Chimpanzees, *Pan troglodytes*, follow gaze direction geometrically. *Animal Behaviour*, 58(4), 769–777.

Tschudin, A., Call, J., Dunbar, R. I. M., Harris, G., & van der Elst, C. (2001). Comprehension of signs by dolphins (*Tursiops truncatus*). *Journal of Comparative Psychology*, 115(1), 100–105.

Vallin, A., Jakobsson, S., Lind, J., & Wiklund, C. (2005). Prey survival by predator intimidation: An experimental study of peacock butterfly defence against blue tits. *Proceedings of the Royal Society of London B, Biological Sciences*, 272, 1203–1207.

von Bayern, A. M. P., & Emery, N. J. (2009). Jackdaws respond to human attentional states and communicative cues in different contexts. *Current Biology*, 19(7), 602–606.

Whiten, A., & Byrne, R. W. (1988). The Machiavellian intelligence hypotheses: Editorial. In R. W. Byrne & A. Whiten (Eds.), *Machiavellian intelligence: Social expertise and the evolution of intellect in monkeys, apes, and humans* (pp. 1–9). New York, NY: Clarendon Press/Oxford University Press.

Wilkinson, A., Mandl, I., Bugnyar, T., & Huber, L. (2010). Gaze following in the red–footed tortoise (*Geochelone carbonaria*). *Animal Cognition*, 13(5), 765–769.

Xitco, M. J., Gory, J. D., & Kuczaj, S. A. (2004). Dolphin pointing is linked to the attentional behavior of a receiver. *Animal Cognition*, 7(4), 231–238.

Index

Page numbers in *italics* refer to figures; those in **bold** to tables

accessory olfactory bulb (AOB), primates 456. *see also* olfactory system
acetylcholine (ACh) neurotransmission 281–2, 284–5, 289, 291, **292**, *293*
acorn worms (various species) *184*, 331–332
acoustic communication systems 449
acroterminal domain, forebrain development in vertebrates 360
Actinopterygii. *see* ray-finned fishes
action guidance role of brain x, 1–7, 9, 11, 16. *see also* functionalist embodied perspective
action-oriented representations, embodied cognition 10, 13–16
action potentials, effects of molecular noise 161–166, *162*, *165*, *163*
adaptation/adaptive evolution
  allometric 391, 397
  developmental plasticity 422, *423*, 427
  fossil record 41
  functionalist neuroscience perspective 1, 3, 5, 7
  neurons/nervous system 113, 138, 257, 311, 354, 355
  signaling pathways 54
  social coordination 480, 483
  social learning 503, 505, 506
  theory of mind 514, 515, 518, 523
adaptationist model 423. *see also* functional specialization theory
adenosine triphosphate (ATP), neurotransmission systems 289

Aequorea victoria (crystal jelly) 97, 99, 100, 107, 114
affordances /affordance competition hypothesis 10, 14, 15–16
Aglantha digitale (hydrozoan jellyfish) 94, *95*, *98*, 97–107, *104*, 112–116
agnathae 248, 268–269, *269*, 377
alarm calls 446–447, 461
allometry xi, 388–390, 395–396, 404–405
  cortex *389*, 393–395, *394*, *395*
  brain size 28–29
  brain volume 390–391
  comparative phylogenetics 397–402, *401*
  developmental parameters 396–403, *403*, *404*, 422, 424
  macro scale 391–392, *392*
  neurogenesis gradients *400*, 400–404, *401*
  neuron number/density 391, 397, 398–402, *400*, *403*
  quit fraction 398, *399*
  social learning 497
  variability of individuals 392–393, *393*
  see also scala naturae
all-or-nothing sensory messages 112–113
amines, neurotransmission systems 282, 284, 287, 288, *293*, 298
amino acids, neurotransmission 288–289, 298. *see also* GABA; glutamate
AMPA receptors 59, 72, 412, 414, *416*
amphibians *178*, 263–264
Amphimedon queenslandica (sponges) 283–285

Amphioxus (lancelets) 332–334, 376
amygdala
  learning and memory *411*, 417
  nervous system architecture 247, 259–265, *261*, 267, 268
  primate communication 451–4, *453*, *455*, 457, 459, 461
  social decision making 428
  vertebrate development 369, 378
anamniotes 263–269, *267*, *269*. *see also* amphibians; fish
Anemonia viridis (snakelocks anemone) 97. *see also* sea anemones
annelid worms
  central nervous system 180, 189–192, *190*
  developmental perspectives 317–319, 432
  neurotransmission systems **294**
anterior cingulate cortex (ACC) 450–451, 452, *453*
anterior neural ridge (ANR) 368–369, 376, 378
anterior visceral endoderm (AVE) 376, 377
anteroposterior patterning, bilateral nervous system 130–133, *131*
anthozoans 96–97. *see also* sea anemones
anthropocentrism x, 32, 503
ants, social coordination 481, 483, 486
Aphelocoma californica (scrub jay) 520, 521
Apis mellifera. *see* honeybee
Aplysia spp. (marine snail) 412–413, 416, 417
apoptosis, signaling pathways 57–59, 61, 64, **65–69**, 70, 77, 79, 80
arealization, cortical 388, 402, 404
Aristotle x, 21. *see also scala naturae*
arthropods
  central nervous system 193–197, *194*
  locomotor neuronal circuitry 216–219, *218*
  neural development 325–328, *327*
  neurotransmission systems 291
  visual systems 219–222, *221*
  see also insects
artificial neural network modeling 8
ascidians (sea squirts) 334–335
ASPM (abnormal spindle-like microcephaly-associated protein) 427
association neurons 5
associative learning 411
attention, awareness of 515–517. *see also* gaze direction
attention getters, communication systems 449, 450, 454

attractiveness, perception in primates 458
autophagy signaling pathways 61, 64, **68**, 79–80
axon density, evolutionary constraints 156
axon diameter, evolutionary constraints 155, 158, *159*, 167

bacteria, signaling pathways 50, 50–53, 53–56, *54*, 57
basal ganglia, vertebrate 254–257, *256*, 257–259
basiepithelial nervous system, invertebrate 177–180, *178*
behavioral flexibility 487, 496, 498
beliefs, understanding others 521–522. *see also* theory of mind
Beroe ovata (comb jellies) 93
beta alanine, neurotransmission 288–289
bilateral nervous systems
  basic architectural types *178*
  common patterning mechanisms 136–139, *137*
  complexity 126, *139*
  conservation of developmental mechanisms 129, 130–136, *131*, *134*
  diversity 128–129
  neurotransmission systems *283*, 291–297
  nervous system evolution 23, *24*, 24
  olfactory circuits 136–138, *137*
biogenic amines, neurotransmission systems 282, 284, 287, 288, *293*, 298
bioinformatic analysis, signaling pathways 53
bioluminescence 288
biomediators, neurotransmission systems 281–282, *283*
birds
  communication systems 446–447, 448, 461, 463
  development of evolutionary neuroscience 21, *22*, 26–28
  learning and memory 413
  nervous system architecture 249–263, *252*, *253*, *255*, *256*, *258*, *261*
  neural basis of song 432
  neuronal circuitry *252*, *253*, *255*, *256*, *258*
  social coordination 480, 482
  social learning 497, 506
  theory of mind 516–520
  transmission of cultural knowledge 485, 486
BMP signalling pathway 133–136, *134*

body form transformations, invertebrates 307. *see also* larval stages
body/mind dichotomy 4. *see also* functionalist embodied perspective
body size, and brain size. *see* allometry
body temperature, evolutionary constraints 156–157
Bothrioplana semperi (flatworm) *186*
bottlenose dolphin (*Tursiops truncates*) 482
Bougainvillia superciliaris (hydrozoan jellyfish) 103–105
brachiopods, neural development 314–315
brain fissures, fossil record 42
BrainMap database 12
brain mapping, development of evolutionary neuroscience 30–31
brain regionalization, vertebrates 361, 364. *see also* prosomeric model
brain scaling. *see* allometry; *scala naturae*
brain size
   and body size. *see* allometry
   development of evolutionary neuroscience 28–30, 32
   scaling rules 44–45
   social learning 496–502, *498*, 502–506
   vertebrates 249
brain structure/function
   development of evolutionary neuroscience 26–28
   embodied cognition 11–16
   see also nervous system
Broca's area 11, 12
Brooks, R. 8–9
bryozoa, neural development 315–316
B system, evolution of neurons 102, 105

cadherin, signaling pathways 25, 64, **66**, 71, 72, 77, 78, 80
Caenorhabditis elegans (nematodes)
   central nervous system *178*
   developmental perspectives 426, 432
   neural development 324–325, 336
   neurotransmission systems 291, 297
Calliactis parasitica (sea anemone) 94, 96–97, 115
Callithrix jacchus (marmoset) 485
canary (*Serinus canaria*), neural basis of song 432
Canis lupus (wolf) 482, 517, 518
Canis familiaris. see dog
canonical signaling pathways 61, 62, **65–69**
Capitella spp. (annelid worm) 180, 318

capuchin monkey (*Cebus* spp.), social coordination 480, 483
Carassius auratus (goldfish) 265
cartilaginous fish, nervous system architecture 243, 267–268
Carybdea rastonii (jellyfish), evolution of neurons 111
cell-cycle control, signaling pathways 51, 54, 57–59, *58*
cell-cycle duration, allometry 398
cell proliferation, and diversity generation 426–427
cellular communication 279. *see also* neurotransmission systems
centralization of nervous system, functionalist neuroscience 4, 5
central pattern generators (CPGs), invertebrates *207*, 207–208, 211, *212*, 213, 214, 216–219, *218*
cepahalochrodates 332–334, 376
cephalization, developmental perspectives 422
cephalopods 24
   communication systems 445, 462
   functionalist neuroscience perspective 6–7
   neural development 129, 191, 323, 337, 431
cerebellum 26
   evolutionary constraints *159*
   individual variability *393*
   nervous system architecture 237, *243*, 245
cerebral cortex 397, 402
   allometry *389*, *393*, 393–397, *394*, *395*, *400*, 400–404, *401*
   anatomy *389*
   comparative phylogenetics 397–400, 423
   development of evolutionary neuroscience 32
   developmental perspectives 426–427
   fossil record 41
   functionalist neuroscience perspective 3–4, 5, 6
   neuron number 391, 397, 398, *400*
   primate communication 456–457
   quit fraction *399*
cerebral ganglion, functionalist neuroscience perspective 5, 6
cerebralism, action guidance role of brain 2, 3, 7, 8
cetaceans. *see* dolphins/whales
channel noise. *see* noise
channelopathies 157

Chelophyes appendiculata (siphonophore) 96, 103, *105*, 106–108, 113, 116
chemical neurotransmission 279, 280. *see also* neurotransmission systems
chemosensory communication systems **445**, 454–456, 457
chimpanzee (*Pan troglodytes*)
  social coordination 482, 484
  social learning 495, 500, 501, 502, 505
  theory of mind 515, 516, 518, 519, 520–522
choline acetyltransferase (ChAT) neurotransmission system 281–282
Chomsky, N. 8
chordates 332–335, 375–376. *see also* lancelets; vertebrates
cichlid fish 425, 433–434, 435
circuits, neuronal. *see* neuronal circuitry
Cisek, P. 14
cladistic approach
  development of evolutionary neuroscience 24, 25, 30, 32
  nervous system origins 125–126, 129
  study methodology *39*, 40, 42–43
  vertebrates *241*
classical conditioning 411, *411*, 412, 415–417
Clava multicornis (sessile hydrozoan) 140
cnidaria
  evolution of neurons 88, 89, 90, 94–95, *95*, *98*, *101*, *104*, *105*, 108, *109*
  nerve nets *175*
  nervous system evolution 23, 24, *24*, *127*, 127–128, 139–141
  neural development 308–310, *309*
  neurotransmission systems 283, 285–290, **292**
  swimming *210*
  see also jellyfish; sea anemones
co-evolution, neurotransmission systems 296
cognition, and brain size 502–506
cognitive function of nervous system 29, 30, 32, 74. *see also* functionalist embodied perspective
cognitive neuroscience 8
cognitive revolution 3, 7
Columba spp. (pigeons), neuronal circuitry *252*, *253*, *255*, *256*, *258*
columnar model, forebrain development in vertebrates 356
comb jellies. *see* ctenophora
common descent. *see* homology

communication 90
communication systems xii, 444–446, **445**, 463–464
  comparative phylogenetics 460–463
  evolution 446–448
  primates 448–460, *453*, *455*
  social coordination 481, 483, 484
  see also neurotransmission systems
comparative brain mapping, development of evolutionary neuroscience 30–31
comparative neurobiology 23, 280–281
comparative phylogenetics 25
  allometry 397–400, *401*, 401–402
  communication systems 447, 460–463
  craniates 249–263, *250*, *252*, *253*, *255*, *256*, *258*, *260*, *261*
  development of evolutionary neuroscience 21–24, *22*, 27, 30
  developmental perspectives 422–4234, 428, 428–429, *429*, 430
  invertebrates *179*, 181–197, *184*, *186*, *190*, *194*, 308, 336
  nervous system origins 125–128, *127*
  neurotransmission systems **292**, *293*, **294**, *295*
  social learning 496, *499*, 499–500, 502
  study methodology 38–40, *39*
  theory of mind 515, 523
  vertebrates 240, *241*
  see also molecular phylogenetics; *scala naturae*
complexity
  developmental perspectives 423, 428
  nervous system origins 126, *139*
  neurotransmission systems 279
  social coordination 488
computational model, allometry 391, 398
computer metaphors of the brain x–xi, 153–154
  and embodied cognition 6, 7–8, 11, 13
  see also constraints (neuronal evolution); signaling pathways
conditioning, classical 411, *411*, 412, 415–417
confusion effect, predators 481
connectivity 279
  allometry 391
  fossil record 42
  scaling rules 45
  see also neurotransmission systems; signaling pathways
consciousness 2, 60, 75

conservation, developmental mechanisms 130–136, *131*, *134*, 396, 424, 425
constraints, neuronal evolution xi, 153–154, *154*, 167–169
  action potential properties 161–163, *162*, *163*
  axon density 156
  axon diameter 155, 158, *159*, 167
  effects of body temperature 156–157
  neuron anatomy 166–167, *167*
  noise vs. metabolic cost trade-offs 164–166, *165*, *168*
  summary of trade-offs *168*
  system biology approach 155, 160, 163
  see also noise
continuity, principle of
  functionalist neuroscience perspective 3–4, 11
  signaling pathways 50
convergence. *see* homoplastic (convergent) evolution
coordination phenomena. *see* social coordination
cortical areas, allometry *389*, 393–397, *394*, *395*, *404*, 404–405. *see also* cerebral cortex; prefrontal cortex
corvid family 496, 500, 519, 520, 521, 523
co-transmission, neurotransmission systems 280
cranial nerves, vertebrates 240–245, *241*, *243*, **244**
craniates
  amniotes 249–263, *250*, *252*, *253*, *255*, *256*, *258*, *260*, *261*
  anamniotes 263–269, *267*, *269*
  forebrain development 376–379
  see also lampreys; vertebrates
CREB transcription factors, learning and memory 414, 415, *416*
cricket (*Gryllus bimaculatus*) 430
criticality, signaling pathways 64, **68–69**, 79–80
cross-talk 62, 70, 298. *see also* signaling pathways
crumpling behaviour, medusae 99, 107, 108
crystal jelly (*Aequorea victoria*) 97, 99, 100, 107, 114
ctenophora (comb jellies)
  evolution of neurons 90, 93
  nervous system evolution 23, *127*, 128, 139–141
  neural development 310–311
  neurotransmission systems *283*, 285–286, **292**
cubozoa (box jellyfish)
  evolution of neurons 110–111
  neural development 309
  swimming 208–209
culture, animal
  communication systems 447–448
  evolution xii, 485, 505
  social coordination 485–486, 488
  social learning 500–501, 503–504
cumulative culture, humans 501–502
cuttlefish, communication systems 445.
  see also cephalopods
Cyanea capillata (lion's mane jellyfish) 110, 286
cyborgs, humans as 10
cytoplasmic processors, signaling pathways 62–64
cytoskeleton, signaling pathways 51, 57, 59–61

Danaus plexippus (monarch butterfly) 436
Danio rerio (zebrafish) 243, 251, *258*, 265, 266, *267*
Darwin, Charles 1, 2, 21, 22, 88
decision making
  action-oriented representations 13–16
  developmental perspectives xii, 428, 429–430, 431
  neurotransmission systems 429–430
  social coordination 480–481, 487
declarative memory 410, *411*
definitions
  cephalization 422
  functionalist neuroscience 11
  intelligence 496
  nervous system 125
  neurons 90, 117
  regionalization 422
deuterostomia
  invertebrates *179*, 183–185, *184*
  nervous system origins 125–126, *127*, 129, 139–141
  neural development 329–332, *330*
  neurotransmission systems **292**
developmental constraints model 423.
  see also constraints (neuronal evolution); timescales
developmental mechanisms, allometry 396–403, *403*, *404*

developmental perspectives
  functionalist neuroscience 3, 4
  signaling pathways 64, **65–66**, 70, 74, 75–78
  study methodology 46
developmental plasticity xii, 422–424, *429*
  comparative phylogenetics 422–423, 423–424, 430
  diversity 424–427
  and diversity generation 424–427
  evolution of mechanisms 436
  learning and memory 414
  neuronal circuitry 427–432
  signaling pathways 72–73
  social learning 504
  timescales 423, 432–436
diencephalon 26, 237, *238*, *243*, 246, 364
differentiation of neurons, functionalist perspective 4, 11–12, 14
distributed brain function 13–15. *see also* functional specialization theory
diversity
  bilateral nervous system 128–129
  developmental perspectives 424–427
  invertebrates 336
  patterning processes 424–425, 427, 428
  see also comparative phylogenetics
dog (*Canis familiaris*)
  communication systems 460–461
  theory of mind 516, 517, 519
dolphins/whales *39*, 44
  communication mechanisms 446, 447, 448
  social coordination 482, 485
  social learning 495, 496, 500, 502, 506
  theory of mind 516, 517
domain recombination, signaling pathways 52–53
Donnan-Gibbs effect 56–57
dopaminergic neurons 426, 428–430, 431
dorsal ventricular ridge 26–27, 28, 379
dorsoventral patterning 133–136, *134*, 358
The Dragons of Eden (Sagan) 26
Drosophila. *see* fruit fly
dynamical systems theory, embodied cognition 9, 10

ecdysozoa
  central nervous system 192–197, *194*
  neural development 323–328
  nervous system origins *127*, 139–141
  neurotransmission systems **292**
  see also arthropods; nematodes

echinoderms (sea urchins)
  neural development 329–332, *330*
  neurotransmission systems 291, **292**, *293*, **294**
echuria (spoon worms), neural development 319–320
The Ecological Approach to Visual Perception (Gibson) 8
Edinger, Ludwig 21, 22, 26
Eisenia foetida (earthworm) 432
electrical synapses. *see* gap junctions
electrical transmission, signaling pathways 67, 79
electroception, communication systems **445**
electron microscopy, neurotransmission systems 286
embodied cognition. *see* functionalist embodied perspective
embryogenesis
  law of recapitulation 21–23, 240
  signaling pathways 62, **65**, **66**, **68**, 70
encephalization quotient (EQ) 29–30, 32, 422
endocytic matrix, signaling pathways 51, 59–61
endoprocta 313–316
energy needs. *see* metabolic costs
entoprocta, neural development 316
environmental factors/interactions
  action-oriented representations 14
  cognitive systems 10, 11, 12–13, 15
  developmental perspectives 422
  embodied cognition 9
  social coordination 480–481
Eperetmus typus (marine worm) 103
Ephydatia muelleri (sponge) 91, 284
epigenetics. *see* gene expression
episodic memory 410–411, *411*, 505
epithelial cells, evolution of neurons 88–89, *89*, 108–110
epithelial-mesenchymal transition (EMT), signaling pathways 70
estrogens, neurotransmission systems 296–297. *see also* steroid hormones
Eudorcas thomsonii (Thompson gazelle) stotting behavior 444, **445**
eukaryotic signaling pathways. *see* signaling pathways, eukaryotic
Euphysa (jellyfish), evolution of neurons 108
evolution, concept of 21
evolutionary constraints. *see* constraints, neuronal evolution

evolutionary neuroscience, development of
    the discipline 21–23, 32–33
  brain size 28–30, 32
  brain structure 26–28
  comparative brain mapping 30–31
  humans 31–32
  nervous system 23–25, *24*
excitatory post-synaptic potentials (EPSPs)
    71–72, 114
exocytosis, chemical synapses 280, 281,
    284, 287
expressions, communication systems 449
extracellular signaling 23, 55, 58, 59, 60,
    61. *see also* communication systems
eye-shaped spots, Peacock butterfly 515.
    see also gaze following
eyes. *see* visual systems

facial-expressions/facial-processing
  communication systems 449, 456, 457
  functionalist neuroscience perspective 13
  neural mechanisms of signal production
      450–451
fish
  developmental plasticity 433–434, 435
  diversity 425
  learning/transmission of cultural
      knowledge 485
  nervous system architecture *243*, 251,
      258, 264–267, *267*
  neurotransmission systems 280
  social coordination 480, 481
flatworms (platyhelminths)
  central nervous system *186*, 187–18
  nervous system origins 126, *127*, 129
  neural development 181, *309*, 311–313
flight 480
floor plate secondary organisers, vertebrates
    351, 365–367
flow cytometry 390
Fodor, J. 8
force fields, signaling pathways 60
forebrain, vertebrate xi
  allometry *392*
  early evolution 375–379
  nervous system architecture 246–247
  prosomeric model 350–361, *351*
  secondary organisers 350–354, 361–375,
      362, 363
forebrain–midbrain boundary (FMB),
    nervous system architecture 237
form mapped to function. *see* functional
    specialization theory

fossil record
  nervous system origins 125–126
  study methodology 41–42
frog, central nervous system *178*
fruitfly (*Drosophila melanogaster*)
  central nervous system 177, *178*
  communication systems 462
  conservation of developmental mechanisms
      130, *131*, 132, *134*, 134–135
  decision making 430
  learning and memory 413, 415
  neural development 336
  nervous system evolution 23, 24
  neurotransmission systems **292**, *293*
functionalism, action guidance role of brain
    1, 1–7, 9
functionalist embodied perspective 1, 2,
    7–9, 16
  action-oriented representations 10, 13–16
  brain structure/function 11–16
  definition 11
  differentiation of neurons 4, 11–12, 14
  dynamical systems theory 9, 10
  environmental role in cognitive systems
      10, 11, 12–13, 15
  intelligent bodies 10
functional magnetic resonance imaging
    (fMRI) 8
functional specialization theory
  allometry 388, *389*
  developmental plasticity 422, 423
  embodied cognition 4–6, 12–15
fusiform face area, functionalist neuroscience
    perspective 11–12

GABA (gamma-amino-butyric acid)
  cnidaria 310
  neurotransmission systems 281, 282,
      284–285, 289, 291, **292**, *293*, 298
gap junctions 60, *67*, 100, 199, 209, 280,
    283, 286. *see also* synapses
Garstang, Walter 22
Gasterosteus aculeatus (stickleback) 480
gaze following 458, 460
  learning/transmission of cultural
      knowledge 485, 486
  theory of mind 516, 517–518, 520, 522
gene expression
  developmental plasticity 427, 429, *429*,
      430–436
  forebrain development in vertebrates 363
  neurotransmission systems 281
  signaling pathways 64

genetics
  developmental perspectives 423
  neural morphogenesis 424
  signaling pathways 52–53, 64
  see also molecular phylogenetics
Geodia cydonium (sponges) 284
gestural communication, primates 449.
  see also pointing
Gibson, J. J. 8
gill-withdrawal reflex, marine snails 412
glass sponge (*Rhabdocalyptus dawsoni*) 91, *92*, 92
glia
  invertebrates 205, *206*
  neurotransmission systems 297–298
glucocorticoid (GR) receptors 296
glutamate receptors 91, 284, *293*, 412, 414
  development of evolutionary neuroscience 25
  learning and memory 412, *416*
  neurotransmission systems 281, 282, 284, 285, 289, **292**, *293*, 298
  signaling pathways 59, 72
  see also NMDA (N-methyl-d-aspartate) receptors
glycine, neurotransmission systems 289, **292**
GnRH hormone 290, 293, **294**, 426, 435
goldfish (*Carassius auratus*) 265
G protein-coupled receptors (GPCRs)
  neurotransmission systems 288, **294**
  signaling pathways **67**, 78
graded responses, evolution of neurons 112, 113
gradients, developmental timing/neuron number *400*, 400–404, *401*
group motion/movement 479–481. see also social coordination
growth, functionalist neuroscience perspective 3, 4
growth, theory of 28–29
Gryllus bimaculatus (cricket) 430

habituation responses 411, *411*, 412
Haeckel, Ernst 21, 22
Haller's rule, brain size 28
Hebb, Donald 410
hemichordata (acorn worms) *184*, 331–332
herding behaviour, local interaction hypothesis 479
Hertwig, O. and R. 89, 90

high-fidelity copying, social learning 500, 501, 502
hindbrain, cranial nerves in vertebrates 240–245, *241*, *243*, **244**
hippocampus
  individual variability *393*
  learning and memory 413–414, 415
  vertebrates *255*, *260*
Hippopodius (siphonophore) 108
histamine, neurotransmission systems **292**
histogenetic areas, vertebrates 352, 353, 355
Hodgkin, A. L. 157, 168
Hodotermes mossambicus (termite) 481
homology (common descent)
  allometry 391
  communication systems 461, 462–463
  developmental perspectives 423–425, 428, 431–432
  invertebrates 308
  learning and memory 410, 415–417, *416*
  neurotransmission systems 296
  vertebrates 257–263
homoplastic (convergent) evolution
  developmental plasticity 436
  mammals 39
  social coordination 487
  social learning 496
  vertebrates 236, 257–263
honeybee (*Apis mellifera*)
  communication systems 462
  decision making 430
  developmental plasticity 435–436
  learning and memory 413, 417
  social learning 502–503
hormones. see neurohormones
Hox gene *131*, 132
humans
  communication systems 449, 452, 457–458, 461
  development of evolutionary neuroscience 31–32
  developmental perspectives 427
  learning and memory 410, 413, 415
  neurotransmission systems **292**, *293*, 293, **294**
  social coordination 479, 480, 482–483
  social learning 501–502, 505
  theory of mind 514–515
hunting, social coordination 481–482, 483–484
Huxley, A. F. 28, 157, 168
Hydra, neurotransmission systems 288

hydrozoa
  evolution of neurons 94, *95*, *98*, 97–107, *104*, 112–116, *109*
  neural development 309
  see also jellyfish
hypothalamus 238, 358–360, 372–373, 428

imaging techniques, embodied cognition 8, 12, 13
immediate early genes (IEGs) 432–433
immune system signaling pathways 50
implicit learning (nondeclarative memory) xii, 410, 411, *411*
imprinting 411
Inachis io (peacock butterfly), 515
individual variability, allometry 392–393, *393*
information processing, insects 219–222, *221. see also* computer metaphors
information revolution, signaling pathways 50
information transmission 485–486, 488. *see also* culture; teaching
insects
  butterfly eyespots 515
  developmental plasticity 436
  neural development 325–328, *327*
  neuronal circuitry 219–222, *221*
  neurotransmission systems 280, **294**
  see also fruit fly
integrative nervous system. *see* neural integration
intelligence, social learning 496–502, *499*
'Intelligence without representation' (Brooks) 8
intelligent bodies 10. *see also* functionalist embodied perspective
intracellular signaling. *see* signaling pathways
invaginated invertebrates 177–180, *178*
invertebrates
  basic architectural types 177–180, *178*, *179*
  central nervous system xi, 174–177, *176*
  chordate 332–335
  communication systems 462–463
  comparative phylogenetics *179*, 181–197, *184*, *186*, *190*, *194*
  development of evolutionary neuroscience 33
  learning and memory 412–413
  morphology 197–205, *198*, *201*, *203*, *206*
  nervous system evolution 23–25, *24*

  nervous system origins 125, 173–174
  neural development xi, 307–308, 336–338
  neuronal circuitry *207*, 207–222, *210*, *212*, *215*, *218*, *221*
  neuropeptides 431–432
  neuropil 177, *178*, 180–181, *181*
  see also arthropods; cnidaria; ctenophora; ecdysozoa; echinoderms; lophotrochozoa; molluscs
ion channels
  constraints on evolution 153, 155–158, 160, *161*, *162*, 163–164, 166–167
  evolution of neurons 93, 101–102, 115–116
  learning and memory 414, *416*
  neurotransmission systems 288
  noise vs. metabolic cost trade-offs 164–166, *165*
  signaling pathways 56–57, **67**
isthmic secondary organisers, vertebrates 374–375

James, W. 11–3, 9
jay (*Aphelocoma californica*) 520, 521
jellyfish (various species)
  CNS complexity *139*
  evolution of neurons 97–112, *98*, *101*, *104*, *105*, 114, 116
  neural development 309
  neuronal circuitry 207
  neurotransmission systems 286
  swimming 208–209
  see also hydrozoa

kea (*Nestor notabilis*) 486
kinases 79, *416. see also* MAPK signaling pathways
knowledge states, theory of mind 521, 522

lamprey 248, 268–269, *269*, 377
lancelet 332–334, 376
language, human 8, 9, 10, 11, 12, 449, 452
larval stages, invertebrates 307–308, *309*, *309*, *314*, *317*, *330*
last eukaryotic common ancestor (LECA), signaling pathways 51, 52, 53, 59, 71
law of recapitulation 21–23, 240
layers, cortex 389, 398, 401–403, *401*, *403*
learning and memory xii, 410–412, *411*
  common descent 415–417, *416*
  communication systems 446, 447

evolution of neurons 112, 115
invertebrates 412–413
signaling pathways 50
social coordination 485–486, 488
vertebrates 413–415
see also social learning
leeches, locomotor neuronal circuitry 209–214, *212*
Lemur catta (ringtail lemur) 516
Leuckartiara octona (hydroid jellyfish) 103
life cycles
developmental plasticity 435–436
invertebrates 307–308, 309. *see also* larval stages
lion (*Panthera leo*) 482
lion's mane jellyfish (*Cyanea capillata*) 110, 286
local interaction hypothesis, motion/movement 479
local interneurons, invertebrates 204
localization of function. *see* functional specialization theory
longitudinal zones, vertebrates 239, 352, 353, 356–358, 365, 366, 376
long-term depression (LTD), signaling pathways 72
long-term potentiation (LTP)
learning and memory 412, 413–414
signaling pathways 72
lophotrochozoa
central nervous system 185–192, *186*, *190*
nervous system origins *127*, 139–141
neural development 311–323, *314*, *317*, 338
neurotransmission systems **292**
Lottia gigantea (limpet) 293, 295

macaque (rhesus) monkey (*Macaca mulatta*) 484, 516, 517
Machiavellian Intelligence hypothesis 514, 515
Mackie, G. O. 108, 109
Macrostomum lignano (flatworm) 181, *186*
mammals
brain size 29, 30, 32
brain structure 26, 27, 28
communication systems 444–448, 460–461
comparative brain mapping 30–31
cranial nerves **244**
deducing evolution from extant species 42–44

learning/transmission of cultural knowledge 485
nervous system architecture 249–263, *252*, *253*, *255*, *256*, *261*
nervous system evolution 24, 25
neurotransmission systems 280
phylogenetics 21, 22, *22*, 38–40, *39*
social coordination 482
social learning 497
theory of mind 517, 519
see also specific species by name
'many eyes' theory, predation 481
MAPK signaling pathways 57, *58*, **58**, **65**, 67, **68**, 72
Margulis, Lynn 50
marmoset (*Callithrix jacchus*) 485
medial ganglionic eminence (MGE), vertebrates 371, 377
medusae. *see* jellyfish
membrane lipids, signaling pathways 57
membrane potential, evolutionary constraints 153
membrane receptors, signaling pathways 56, 62, *63*, 75, 76
memory. *see* learning and memory
mental representations, embodied cognition 10, 13–16
mental state attribution, theory of mind 514, 515, 521
mesencephalon, vertebrate 26, 237, *238*, 239, 240, 242, *243*, 245–246
mesenchymal-epithelial transition (MET), signaling pathways 70
mesolimbic reward system, social decision making 428, 429–430
metabolic costs
allometry 390–391
evolutionary constraints 153–154, *154*, 155, *167*, 166–167
vs. noise trade-offs 164–166, *165*, *168*
signaling pathways 50–51
metazoans. *see* cnidaria; ctenophora; deuterostomia; ecdysozoa; lophotrochozoa
metencephalon, vertebrate 237, *238*
Metridium senile (sea anemone) 97
mice (*Mus* spp.)
learning and memory 413
nervous system evolution 23
nervous system origins 130, *131*, 132, *134*, 135, 136
microtubules (MT), signaling pathways 60

Microtus spp. (voles) 431
midbrain–hindbrain boundary (MHB), vertebrate 237
midbrain, vertebrate 26, 237, *238*, 239, 240, 242, *243*, 245–246
migration areas, vertebrate forebrain 352
mind/body dichotomy 4. *see also* functionalist embodied perspective
mineralocorticoid (MR) receptors, neurotransmission systems 296
miniaturization of neural circuits 154, *159*, 167
mirroring 459–460, 484
mitochondria, signaling pathways 50–51, 53, 57, 58
Mnemiopsis leidyi (comb jelly) 90
mode selection, evolution of neurons 112, 114
molecular phylogenetics
　deducing evolution from extant species 43
　development of evolutionary neuroscience 22–25, 27
　nervous system origins 126, 128, 130, *131*, 130–135
　neurotransmission systems 286
　study methodology *39*, 39–40
　see also comparative phylogenetics
molluscs
　central nervous system 189–192, *190*
　neural development *317*, 321–323
　neurotransmission systems 293, **294**, 295
　swimming 214–216, *215*
monarch butterfly (*Danaus plexippus*) 436
morphological computation, functionalist neuroscience perspective 6
mosaic evolution 388, 397, 423. *see also* functional specialization theory
motion/movement, social coordination 479–481, 487
motor system. *see* sensorymotor system
Muggiaea atlantica (hydrozoan jellyfish) 103, 105–106, 116
multipolar neurons, invertebrates 200, *201*
myelencephalon, vertebrates 237, *238*, *243*

Nanomia spp. (siphonophore) 103, 108–110, *109*, 116
natural selection
　action guidance role of brain 1, 8
　communication systems 446
　developmental perspectives 423
　functionalist neuroscience perspective 11

nematodes (round worms)
　central nervous system 194
　developmental perspectives 432
　neural development 324–325
　neurotransmission systems 291, **292**, 293, *293*, **294**
　see also Caenorhabditis elegans
Nematostella vectensis. see starlet sea anemone
neocortex 26–29, 41
Neoturris breviconis (marine worm) *98*, 99
nerve nets *24*, 24, *175*
　basic architectural types 173, 174, *175*, *176*, 177, 178
　evolution of neurons 89, 93, 94, 96–97, 100, 110
　nervous system origins 128
　neurotransmission systems 289
nervous system xi, 125–128
　comparative phylogenetics 422
　complexity 126, *139*
　conservation of developmental mechanisms 130–136, *131*, *134*
　definition 125
　development of evolutionary neuroscience 23–25, *24*
　diversity 128–129
　evolution of neurons 88, 94–95
　invertebrates 23–25, *24*, *127*, 139–141, 125, 173–174
　olfactory circuits 136–138, *137*
　patterning mechanisms 136–139, *137*
　phylogenetics 125–128, *127*
　signaling pathways 50
　vertebrate 125, 246–247
　see also brain structure/function
Nestor notabilis (kea) 486
neural integration
　evolution of neurons 112–117, *404*
　medusae 97–108, *98*, *101*, *104*, *105*
　primate communication 459–460
neural mechanisms
　signal perception in primates 452–459, *453*, *455*
　signal production in primates 450–452
neurites 166–167, 173, 174, 180, 181, *181*, *194*, 202
neuroblasts 326
neurochemistry 428–429. *see also* neurohormones; neuropeptides
neurogenesis, allometry 396–404, *400*, *401*
neurohormones

developmental plasticity 428, 430, 431,
    433–435
  modulation of neural circuits 427, 428,
    430–432
  neurotransmission systems 280, 282, 293,
    294, 296–298
  signaling pathways 64, **66**, 77–78
neuromeres, vertebrates 237, *238*,
    239–240, 352
neuromeric paradigm, forebrain development
    in vertebrates 354
neuronal circuitry
  developmental plasticity 427–433
  primates, communication systems *453*,
    454–457, *455*
  vertebrates 251–257, *252, 253, 255, 256*
neurons xi, 88–90
  allometry 391, 397–399, *400, 401*,
    401–402, *403*
  anthozoa 96–97
  cnidaria 88, 89, 90, 94–95, *95, 98*, 108
  ctenophora 90, 93
  cubozoa and schyphozoa 110–111
  definitions 90, 117
  developmental perspectives 425–426
  epithelial conduction 88–89, *89*,
    108–110
  functionalist neuroscience perspective
    4, 5
  invertebrates 200–204, *201, 203*
  porifera 91–92, *92*
  signaling pathways 61–70, *63*, **65–69**
  social learning 497
  see also constraints (neuronal evolution);
    neuronal circuitry
neuropeptides
  invertebrates 431–432
  modulation of neural circuits 431
  neurotransmission systems 280, 287–288,
    290, 293, **294**, 295–296
neuropil 125, 140, 177, *178*, 180–181, *181*
neurotransmission systems xi, 279–281,
    298, 336
  bilateria *283*, 291–297
  biomediators 281–282, *283*
  cnidaria 285–290, 310
  decision making 429–430
  glia 297–298
  phylogenetics **292**, *293*, **294**, *295*
  porifera 283–285
neurotrophins, signaling
    pathways 72, 76–77

nitric oxide (NO)
  learning and memory 412, 415, *416*
  neurotransmission systems 284, 289,
    291, **292**
NMDA (N-methyl-d-aspartate) receptors
  developmental plasticity 435
  learning and memory 414, 415, *416*, 435
  neurotransmission systems 282
  signaling pathways 59, **66**, 72
noise (random variability) 153–154,
    154, 161
  action potential properties 161–163,
    162, 163
  axon density 156
  axon diameter 155, 158, *159*
  channelopathies 157
  effect on other neurons 163–164
  effects of body temperature 156–157
  vs. metabolic cost trade-offs 164–166,
    165, 168
  system biology approach 155, 160, 163
nonassociative learning 411. *see also* learning
    and memory
non-canonical signaling pathways 61,
    62, **65–69**
nondeclarative memory xii, 410, 411, *411*.
    see also learning and memory
non-spiking neurons, invertebrates 204
norepinephrine (noradrenaline) 280, 288,
    298, 430
nuclear-to-layered hypothesis 27
nucleus, signaling pathways 64, **66**, 77–78

O system, neuronal integration in
    medusae 102
ocular systems 219–222, *221*, **445**
olfactory systems
  common patterning mechanisms
    136–138, *137*
  communication systems 444, **445**, 448
  development of evolutionary
    neuroscience 29
  fossil record 41
  individual variability *393*
  primate communication 456
one-component systems (1CS), prokaryote
    solute sensing *54*, 54–55
ontogenetic learning, communication
    systems 446, *447*
ontogeny recapitulates
    phylogeny 21–23, 240
onychophora (velvet worms) 328

oscillators. *see* pacemakers
osmotic function of nervous system 51, 52, 56, 57, 60, 64, **67**, 74. *see also* solute sensing; solvent sensing
oxytocin, neurotransmission systems 290, 293, *295*

pacemakers
    evolution of neurons 89, 94, 100, 102–103, 106, 110–112, 114, 116
    invertebrate neuronal circuitry *207*, 207–208, 209, *210*, 211
    neurotransmission 289
    origin of nervous system 140
Pantin, Carl 89, 90, 96
parallel evolution, mammals 39
Pan troglodytes. see chimpanzee
Panthera leo (lion) 482
Paramecium (unicellular organism), neurotransmission systems 281, 282
Parker, G. H. 88
partial eversion hypothesis, vertebrates *250*, 266
Passano, L. M. 89
patterning processes
    bilateral nervous system 130–133, *131*, *137*
    and diversity generation 424–425, 427, 428
peacock butterfly (*Inachis io*), 515
peduncular hypothalamus, vertebrates 359
peptide hormones. *see* neuropeptides
perception-action paths, action-oriented representations 14
periaquedutal gray (PAG), primate communication 451–452
peripheral nervous system (PNS), invertebrates 174–177, *176*
perspective-taking 504, 519–520. *see also* theory of mind
phoronid (horseshoe worms), neural development 313–314
phylogenetics. *see* comparative phylogenetics; molecular phylogenetics
pigeon (*Columba* spp.), neuronal circuitry *252*, *253*, *255*, *256*, *258*
piggyback effect, evolution of neurons 108–110
plasticity. *see* developmental plasticity
platyhelminths. *see* flatworms
Playnereis (bristle worm) *134*, 135, *136*, 318
Pleurobrachia bachei (sea gooseberry) 90, 286, 311

pointing
    communication systems 449
    theory of mind 516
Polyorchis spp. (bell jellies) 97, *98*, 99–102, *101*, 105, 112, 114, 116
polypide replacement 316
porifera (sponges)
    evolution of neurons 90–92, *92*
    nervous system origins 128
    neurotransmission systems 283–285, *283*, *293*
positional selector mechanisms, vertebrates 364
positron emission tomography (PET)
    comparative brain mapping 31
    embodied cognition 8
postsynaptic density (PSD) 71–72
postsynaptic signalling, marine snails 412
prechordal plate secondary organisers, vertebrates 369–370
predation
    social coordination 481
    theory of mind 515–516
prefrontal cortex
    action guidance role of brain 13
    development of evolutionary neuroscience 31
    developmental plasticity 430
    learning and memory *411*
    nervous system architecture 259, 262
    neural communication 456, 458
preisthmus 361
premotor systems, vertebrates 247–248
primary organizers, vertebrates 352, 361, 376
primates 448, *453*, *455*, 463–464
    communicative repertoire 448–450
    integrative mechanisms 459–460
    neural mechanisms of signal perception/production 450–459, *453*, *455*
    social learning 496, 497, 506
    theory of mind 515, 516, 517, 518, 520, 523
priming 411, *411*
Principles of Psychology (James) 1, 2
Proboscidactyla (hydroids), evolution of neurons 105
procedural memory 410, *411*
progression indices, brain size 29
progressive phylogenesis 23
prokaryotes, signaling pathways 50–56, *54*, *57*

proper mass hypothesis 388, 397. *see also* allometry
prosencephalon 238, *238*, 461
prosomeric model, vertebrate forebrain 350–361, *351*
  alternative models 354–358
  detailed morphology 358–361
  secondary organisers 350–354, *362, 363*
protein-domain recombination, signaling pathways 52–53
proteomic studies, signaling pathways 71
proto-nervous system, evolution 88, 108
protostome supergroup, nervous system origins 125–126, 129
protosynapses 25, 71, 74
psychomotor neurons, functionalist neuroscience perspective 5
Ptychodera flava (acorn worm), central nervous system *184*
purinergic receptors, neurotransmission systems 289, **292**

quit fraction, cerebral cortex 398, *399*

Ramon y Cajal, S. 1, 3–7, 10, 279
random action potentials (RAP) 156
random variability. *see* noise
rat (*Rattus* spp.)
  developmental plasticity 434
  learning and memory 413, 415
  nervous system architecture *243*
  neuronal circuitry 252, *253, 255, 256*
ray-finned fishes (Actinopterygii)
  diversity 425
  vertebrates 264–267, *267*
recapitulation theory 21–23, 240
referential communication systems 446, 447, 448, 450, 461
reflexes xi
  evolution of neurons 88–89, *89*, 91–92, *92*
  functionalist neuroscience perspective 4, 7
regionalization of brain. *see* functional specialization theory
Renilla koellikeri (sea pansy) 91–92, 287–288
representations, mental 10, 13–16
reproductive transitions, developmental plasticity 434–435
reptiles
  development of evolutionary neuroscience 21, *22*, 22, 26, 27, 28

learning/transmission of cultural knowledge 485
nervous system architecture 257–263, *260*
reward systems, social decision making 428, 429–430
RFamide-related peptides (RFaPs) 288, 290, 293, 310
Rhabdocalyptus dawsoni (glass sponge) 91, *92*, 92
rhesus monkeys (*Macaca mulatta*) 484, 516, 517
rhombencephalon, vertebrate 237, *238*, 239, 240–245, *243*
rhopalia (nerve centres) 110, 111, 114, 139, *140*, 174, 208, *210*
rhythmic activity, evolution of neurons 112, 114
ringtail lemur (*Lemur catta*) 516
rodents. *see* mice; rats; voles
Romanes, George 88, 90
roof plate secondary organisers, vertebrates *351*, 367–368

Saccoglossus cambrensis (acorn worm) *184*
Sagan, Carl 26
Sarsia (jellyfish) 102, 107
scaffold proteins 52, 284
scala naturae x, 21, 22, 26, 27, 33, 236, 496. *see also* comparative phylogenetics
scaling, brain 389, 391–395, *392*, 396, 404, 405. *see also* allometry
scaling rules, study methodology 44–45
scent marking 448. *see also* olfactory system
Scharrer, E. and B. 280
schyphozoa/sciphoza. *see* jellyfish
scrub jay (*Aphelocoma californica*) 520, 521
scyphomedusae. *see* jellyfish
sea anemones (various species)
  evolution of neurons 94, 96–97, 115
  nerve nets 96–97
  neurotransmission systems 285, 287–290, **292**, *293*, **294**, 295–296
sea gooseberry (*Pleurobrachia bachei*) 90, 286, 311
sea pansy (various species) 91–92, 287–288
sea slugs (nudibranchs) **292**, *293*
sea squirts (ascidians) 334–335
sea urchins. *see* echinoderms
secondary organisers, vertebrates 350–354, 361–375, *362, 363*
segmental paradigm, forebrain development in vertebrates 354

semantic communication systems 446–447
semantic memory 410, *411*
sense organs
  cranial nerves in vertebrates **244**
  embodied cognition 6, 10
  fossil record 41
  insect visual systems 219–222, *221*
  nervous system evolution 25,
    136–139, *137*
  see also sensorimotor systems
sensitization 411, *411*, 412
sensorimotor systems
  action guidance role of brain 7, 12, 14, 15
  development of evolutionary
    neuroscience 27
  functionalist neuroscience perspective 4, 15
  signaling pathways 50, 74
septum, individual variability *393*
Serinus canaria (canary) 432
serotonin-based neurotransmission systems
  285, 288, 291–293, **292**, 310
sex steroid hormones. *see* steroid hormones
sheep (*Ovis aries*) 460
signaling pathways x, 49–50, 74–75
  catalog of 75–80
  cell-cycle control 51, *54*, 57–59, *58*
  evolutionary constraints 155
  cytoskeleton/endocytic matrix 51,
    57, 59–61
  fundamental pathways 61–70, *63*, **65–69**
  genetic recombination 52–53
  prokaryotes 50–56, *54*, 57
  prokaryotic-eukaryotic evolutionary
    transition 50–53
  solute sensing 51–56, *54*
  solvent sensing 51, 56–57
  synaptic 70–73
  see also neurotransmission systems
siphonophores, evolution of neurons
  103–106, *105*
sipuncula (peanut worms), neural
  development 320–321
small world connectivity, allometry 391
smell, sense of. *see* olfactory system
snakelocks anemone (*Anemonia viridis*) 97
social behaviour
  developmental perspectives 428
  developmental plasticity 432–434, 435
  social coordination 488
social brain, signaling pathways 74–75
social coordination xii, 478–479, 486–488
  individual actions 483–485

learning/transmission of cultural
  knowledge 485–486, 488
motion/movement 479–481, 487
task distribution 481–483, 487
social decision making. *see* decision making
social learning xii, 495, 506–507
  brain size and cognition 496–506, *498*
  comparative phylogenetics 496, *499*,
    499–500, 502
  intelligence 496–502, *498*, *499*
socioecological variables
  developmental perspectives 422, 423
  social learning 498
  theory of mind 514–515, 523
solute sensing, signaling pathways 51–56, *54*
solvent sensing signaling pathways
  51, 56–57
somatosensory cortex 31, 41, 42, 251, *253*,
  427, 451, 457
sonic hedgehog protein/gene *58*, 75, 76,
  239, 363, 366
speed of response, evolutionary constraints
  153, *154*, 156–157, *168*
Spencer, H. 1, 2, 3, 4
spiking neurons, invertebrates 204
sponges. *see* porifera
spoon worms (echuria), neural development
  319–320
SPREs (signaling pathway response
  elements) 64
starlet sea anemone (*Nematostella vectensis*)
  96–97, 285, 287–288, 289, 290
starling (*Sturnus vulgaris*) 479
states of mind
  perception in primates 458–459
  theory of mind 514, 515, 521
status, primate communication 457–458
steroid hormones
  developmental plasticity 428, 430, 431,
    433–435
  modulation of neural circuits 428, 430
  neurotransmission systems 296–297
stickleback (*Gasterosteus aculeatus*) 480
stochastic modelling, evolutionary
  constraints 160, 163
Stomatoca atra (hydrozoans), evolution of
  neurons 99
stotting behavior, Thompson gazelle
  444, 445
stress/criticality, signaling pathways 64,
  68–69, 79–80
striatum, individual variability *393*

Strongylocentrotus purpuratus 291
study methodology x, 38
   deducing evolution from extant species 42–44
   developmental perspectives 46
   fossil record 41–42
   phylogenetics 38–40, *39*
   scaling rules 44–45
Sturnus vulgaris (starling) 479
subcortical structures, communication systems 454–456, *455*, *456*, 461
subepithelial structures, invertebrates 177–180, *178*
subpallial secondary organisers, vertebrates 370–372, 377–378
swimming
   cnidaria 100–107, *104*, *105*, 208–209, *210*
   molluscs 214–216, *215*
synapses 70–73, 279
   allometry 390
   evolutionary constraints 164
   invertebrates 197–199, *198*, 202–204, *203*
   nervous system evolution 25
   see also neurotransmission systems; signaling pathways
systems biology approach, evolutionary constraints 155, 160, 163

tactile/touch communication systems 444, **445**, 448–449, 457
Taeniopygia guttata (zebra finch) 432
task distribution, social coordination 481–483, 487
taurine 288–289, 310
teaching by animals 502, 504, 506. *see also* transmission of cultural knowledge
telencephalon 26, 237, *238*, *243*, *258*, *261*, *263*, 268, 360, 423
temperature, evolutionary constraints 156–157
tensegrity 60
tentacle length, hydrozoa 103–106
termites (*Hodotermes mossambicus*) 481
Tetrahymena (protozoan), neurotransmission systems 281, 282
theory of growth, brain size 28–29
theory of mind x, xii, 514–515, 522–523
   attention/eye direction 515–517
   comparative phylogenetics 515, 523
   gaze following 517–518
   perspective-taking 519–520
   social learning 504
   understanding others' beliefs 521–522
theory of the new head 240
Thompson gazelle (*Eudorcas thomsonii*), stotting behavior 444, 445
three-component systems (3CS), prokaryote solute sensing *54*, 54–55
timescales
   action guidance role of brain 2, 8–9
   developmental perspectives xii, 423, 432–436
   social learning 505, 507
tissue physiology, signaling pathways 64, **66**, 78–79
tool use by animals 485, 496, 497, 498, 500, 506
touch/tactile communication systems 444, **445**, 448–449, 457
trace conditioning 411
traditions. *see* culture
traits
   deducing evolution from extant species 43
   perception in primates 457–458
transcriptional switches, signaling pathways 64
transforming growth factor beta 133
transmission of cultural knowledge 485–486, 488. *see also* culture
tripartite brain 24, 25, 33
Tripedalia cystophora (jellyfish) *139*
trochozoa, neural development 316–323, *317*. *see also* annelids; molluscs
trophic function of nervous system 53, 74
tunicates, neural development 334–335
Tursiops truncates (bottlenose dolphin) 482
two-component systems (2CS), prokaryote solute sensing 54–55

understanding others 521–522. *see also* theory of mind
unicellular neurotransmission systems 281–282
unicellular to multicellular transition, nervous system origins 125
unipolar neurons, invertebrates 200, *201*
urbilateria, nervous system evolution 23, *24*, 24

variability of individuals, allometry 392–393, *393*
variability, random. *see* noise

vasopressin 290, 293, *295*, 368, 429, 431
velvet worm (onychophora), neural
    development 328
ventricular hypothalamic organ (VHO) 372
vertebrates xi, 236, 269–270
    allometry *392*
    amniotes 249–263, *250, 252, 253, 255,
        256, 258, 260, 261*
    anamniotes 263–269, *267, 269*
    ancestral origins 240–248, *241, 243*, **244**
    communication systems 444–447,
        461–462
    development of evolutionary neuroscience
        23, 25, 26, 27, 33
    early evolution 375–379
    learning and memory 413–415
    morphological/functional units of nervous
        system 236–240, *238*
    nervous system architecture 246–247
    nervous system origins 125
    prosomeric model 350–361, *351*
    secondary organisers 350–354, 361–375,
        362, 363
    social learning 497
    see also amphibians; birds; fish; mammals;
        reptiles
vesicles, vertebrate 236, *238*
vibration, communication systems **445**

visual systems 219–222, *221*, **445**
vocalization 480, 500
    primates 446, 447, 449–452, *453*,
        456–458
voles (*Microtus* spp.) 431
voltage-gated ion channels. *see* ion channels
vomeronasal organ (VNO) 454–456

Weber–Fechner law 62
Welker, Wally 30–31
whales. *see* dolphins/whales
whiskers, communication systems 461
winner effect 433
WNT signaling pathway **65, 66**, 70, 75
    developmental plasticity 425, 427
    vertebrate forebrain development 362,
        367, 368, 372–5, 378
wolf (*Canis lupus*) 482, 517, 518
wound healing, signaling pathways 70
Wozniak, Robert 3

Xenopus laevis (frogs) *178*

zebra finch (*Taeniopygia guttata*) 432
zebrafish (*Danio rerio*) *243*, 251, *258*, 265,
    266, *267*
zona limitans intrathalamica (ZLI) 133,
    239, *351*, 361, 373–374

www.ingramcontent.com/pod-product-compliance
Ingram Content Group UK Ltd.
Pitfield, Milton Keynes, MK11 3LW, UK
UKHW052118190426
11946UKWH00024B/106